Fundamentals of Materials Science

Zhengming Sun · Peigen Zhang · Wei Liu ·
Wei He

Fundamentals of Materials Science

From Basic Theories to Critical Properties

Chemical Industry Press

Springer

Zhengming Sun
School of Materials Science
and Engineering
Southeast University
Nanjing, China

Peigen Zhang
School of Materials Science
and Engineering
Southeast University
Nanjing, China

Wei Liu
Changchun Institute of Applied Chemistry
Chinese Academy of Sciences
Changchun, China

Wei He
School of Materials Science
and Engineering
Southeast University
Nanjing, China

ISBN 978-981-96-3903-8 ISBN 978-981-96-3904-5 (eBook)
https://doi.org/10.1007/978-981-96-3904-5

Jointly published with Chemical Industry Press
The print edition is not for sale in China (Mainland). Customers from China (Mainland) please order the print book from: Chemical Industry Press.

This Springer imprint is published by the registered company Springer Nature Singapore Pte Ltd.
The registered company address is: 152 Beach Road, #21-01/04 Gateway East, Singapore 189721, Singapore

Foreword by Bingbo Wei

This universe consists of space and substance. Materials are those substances with which humans create objects and components for devices, machines, and even infrastructures. Integrating chemistry, physics, mathematics, biology, and other disciplines, materials science is the basis of a series of technological domains, including renewable and sustainable energy, communication and information, aerospace, and transportation. The role of materials science will be more significant in the future, as humans face tough challenges from climate change, environmental concerns, resource depletion, and so forth. The brighter future of this planet relies on the boom of more advanced materials.

As a fundamental component of modern technologies, materials science can be a booster to numerous innovations. For example, a dramatic reduction in aircraft weight was made possible with the invention of advanced composite materials, by which 25% less fuel consumption per passenger was achieved for the Boeing Dreamliner compared with a similar-sized aircraft. White light-emitting diode (LED) lights were made possible with the recent invention of blue light LED materials (Nobel Prize 2014). The popular use of electric vehicles was made possible by the discovery of electrode materials for lithium-ion batteries (Nobel Prize 2019). Materials could also be a bottleneck in some technologies. For example, Moore's law (the number of transistors per chip doubles every 18 months) has seen the steady progress of integrated circuits (ICs) since its invention in 1958; this was powered by the advance of materials science. However, this law is now breaking due to the limitation of silicon, a typical semiconductor material.

Materials science is arguably the most interdisciplinary field. For example, the rigorous development of computational materials science, which uses modeling, simulation, theory, and informatics to understand and predict the structure-property relationship of materials, has been shaping the whole discipline with primary goals of discovering new materials, determining material behavior and mechanisms, explaining experimental data, and developing materials theories.

Materials science is also a rapidly developing discipline that requires updated textbooks to reflect its state-of-the-art development. Southeast University (SEU), where I obtained my master's degree in materials science, always embraces challenges as opportunities. The School of Materials Science and Engineering at SEU has advantages in metallic materials, construction materials, and functional materials.

The school offers the course of Fundamentals of Materials Science in English to sophomores since 2016. To simultaneously accommodate the teaching needs of undergraduates in universities and reflect the latest advances of the subject, a new textbook is necessary. The writing of this textbook was proposed by a group of authors who have taught this course for many years in English. This group is headed by Prof. Zhengming Sun, who had spent over 20 years in Europe, Japan, and the USA on research and teaching in materials science.

That the textbook is published in English is out of the following considerations. Accompanying the rising economy of China, an increasing number of international communication and cooperation with the world are becoming routine, which requires a large amount of talent with a global vision. Higher education, as an indispensable part of the talent training system, comes to the forefront of globalization, which calls for courses offered in English. Materials science was shaped up in the 1950s, and the formation was initiated in Europe and America. Therefore, many basic concepts and terminologies were originally expressed in English. To have effective communication with international peers, one must correctly understand and use the terms and terminologies in English. In addition, most research papers are currently published in English. Therefore, learning this course in English is a smart move and probably a more reasonable way for students to better understand its basics and to communicate with their peers worldwide.

I am happy to see the birth of this long-waited textbook. In addition to traditional contents such as crystal structures, defects, diffusion, phase diagrams, and phase transformation, chapters of electronic structures and computational materials science which are seldom seen in other textbooks on the market, are included in this book, offering readers fundamental knowledge on new trends in materials science. At the end of the book, selected Nobel Prizes of representative innovations on materials are introduced, which is a good attempt to inspire the young generation. This textbook was born by the need of the School of Materials Science and Engineering at SEU, but it has general applicability to undergraduates and postgraduates majoring in materials science, materials engineering, and related disciplines, I would like to see more students find this book helpful.

魏炳波
2022.9.24

Bingbo Wei (魏炳波)
Professor
Northwestern Polytechnical University
Academician
Chinese Academy of Sciences
Chairman
Chinese Materials Research Society
(C-MRS)
Xi'an, P. R. China

Foreword

Materials science is a discipline that shapes the foundation of modern technology. Currently, in the context of global challenges such as sustainable energy, high-tech development, and environmental protection, the significance of advanced materials is increasingly evident. This represents both opportunities and challenges for students and professionals in the field of materials science.

This textbook extends from the basic concepts of materials science to the application scenarios of materials, covering various types of materials including traditional metals, ceramics, polymers, composites, and emerging materials. It delves deep into core issues such as atomic structure, bonding methods, and the electronic structures that determine material properties. Additionally, it actively explores cutting-edge topics like nanotechnology and quantum materials, aiming to showcase the forefront of developments in the field of materials science. Through groundbreaking cases of Nobel Prizes in the materials domain, the textbook allows readers to gain a clear insight into both the theoretical aspects of materials and disruptive technological innovations, whether it's the development of superconductors or the revolution of lithium-ion batteries. These examples not only enrich the learning experience but also highlight the significant impact of materials science on our daily lives.

The initial impetus for writing this book was Southeast University's need to cultivate leaders with a global perspective. Recognizing the globalization of modern education and technological development, especially the history of the development and formation of materials science, this textbook is written in English to accommodate a broader teaching environment and student needs. This not only helps non-native English speakers master professional English but also makes the book accessible to students and educational institutions worldwide. We hope this endeavor will help students pave successful career paths in a globally integrated world. During the writing process, we engaged in in-depth discussions with several experts in the field to ensure the scientific accuracy and foresight of the content. Each chapter is written by teachers with extensive teaching experience.

We sincerely hope this textbook becomes a valuable guide and mentor for students and professionals in the field of materials science as they explore the mysteries of the subject, serving those who aim to use today's materials to solve tomorrow's questions. May it inspire more students to delve into the world of materials science and explore the limitless possibilities of future material technologies. We wish readers a fruitful journey through the world of materials science, where each reading deepens their understanding.

Finally, heartfelt thanks are due to colleagues and experts who cared for and assisted in the writing and publishing of this book. Without their selfless contributions, this book would not have been possible.

Nanjing, Jiangsu, P. R. China Zhengming Sun

Introduction

This textbook provides readers with fundamental knowledge from atomic structures to material applications, organized into 14 chapters that cover various aspects of materials science and engineering. Chapter 1, "Introduction to Materials Science," highlights the importance and applications of materials science, discusses material classifications including metals, ceramics, polymers, and composites, and analyzes the role of materials science in human civilization and its development. Chapter 2, "Electronic Structure," explains the evolution of atomic structure models, atoms and electrons, interatomic bonding, and energy band theory. Chapter 3, "Introduction to Crystallography," offers a detailed look at the long-range order in crystals, classification of crystal systems, Bravais lattices, and stereographic projection methods. Chapter 4, "The Structure of Solid Materials," discusses the crystal structures of metals, ceramics, and polymers, providing methods for calculating theoretical densities and analyzing polymer structures. Chapter 5, "Crystal Defects," explores types of defects in crystalline materials such as vacancies, dislocations, grain boundaries, and twin boundaries, and their impact on material properties. Chapter 6, "Diffusion," elaborates on diffusion laws, mechanisms, and factors affecting diffusion. Chapter 7, "Material Surfaces and Interfaces," addresses the definitions, classifications, surface tension, wetting phenomena, and the structure and characteristics of interfaces in composite materials. Chapter 8, "Phase Diagrams," thoroughly discusses the construction and application of unary, binary, and ternary phase diagrams, including critical concepts like the lever rule and phase transformations. Chapter 9, "Solidification and Crystallization," analyzes the solidification and crystallization processes of metals, ceramics, and polymers, including nucleation and growth mechanisms. Chapter 10, "Solid-State Phase Transformations," covers the thermodynamics and kinetics of solid-state phase transformations, introducing diffusional and diffusionless transformation mechanisms, and the microstructure and properties of steel alloys phases such as pearlite, bainite, and martensite. Chapter 11, "Functional Character of Materials," discusses the thermal, electrical, magnetic, and optical properties of materials, such as heat capacity, thermal expansion, conductivity, and magnetism. Chapter 12, "Deformation and Stress–Strain Behavior of Solid Materials," introduces methods for measuring mechanical properties and typical deformation and strengthening behaviors of different material types. Chapter 13, "Computational Materials Science," details the applications of density functional theory, molecular dynamics, and the Monte Carlo method in materials science. Finally, Chap. 14, "Nobel Prizes and Materials Science," reviews Nobel Prize-winning projects that have significantly contributed to materials science, such as semiconductors, superconductors, optical fibers, LEDs, and lithium-ion batteries, showcasing the scientific background and future developments of these technologies.

Contents

Introduction to Materials Science

1

Learning Objectives

1. Identify the three main categories of solid materials and explain the unique chemical characteristics of each.
2. Explain the role of materials in human civilization and its evolution.
3. Define materials science and engineering, enumerate the four to five components that play a role in the design, production, and use of materials, and briefly explain the interdependence of these components.
4. List five categories of material properties that determine their suitability for various applications.
5. Explain the three critical factors that are taken into consideration during the material selection process.
6. Mention a few examples of advanced materials and the corresponding technologies.
7. Appreciate the value of traditional materials and the direction of their future development.
8. Be informed of the key areas of focus in the advancement of materials science and engineering.

Materials science investigates the structure-property-processing-performance relationships of materials. The importance of materials science cannot be overstated as it lays the foundation for countless technological innovations and advancements that have improved our daily lives. From the creation of advanced composite materials that have reduced the weight of aircraft and thus improved fuel efficiency, to the discovery of electrode materials for lithium-ion batteries that have made electric vehicles a reality, the impact of materials science on our world is far-reaching. In the 1970s, people recognized information, materials and energy as the driving

Z. Sun et al., *Fundamentals of Materials Science*,
https://doi.org/10.1007/978-981-96-3904-5_1

forces behind modern civilization. Decades later, advancements in new materials, information technology, and biotechnology have continued to shape and define our world. Today, the focus is on developing advanced materials, such as quantum materials, metamaterials, and materials for energy and environmental innovations which hold the key to a better future for all. It is clear that materials science is an essential discipline that will play a critical role in shaping our future. Its study is crucial for the development of new and innovative technologies that will improve our quality of life and address some of the greatest challenges of our time.

1.1 Understanding Materials: The Importance of Materials Science and Engineering

Materials are the substances utilized by humans to construct objects, devices, components, machines, and other products. Not all substances can be classified as materials, such as fuels, raw chemicals, industrial chemicals, food, and medicine. The definition of materials is not absolute and can be subject to interpretation. For example, explosive and solid rocket propellants are referred to as "energetic materials" due to their role in artillery and rockets.

We should study materials because they play a crucial role in various fields and industries. As applied scientists or engineers, we encounter design challenges related to materials in our daily work. For example, mechanical engineers might need to design a transmission gear, civil engineers might work on designing a concrete building, chemical engineers may be involved in designing components for an oil refinery, electrical engineers might work on designing a lithium-ion battery, and integrated circuit chip engineers face challenges related to integrated circuit design. A comprehensive understanding of materials is essential for solving these design problems effectively and efficiently.

Choosing the right material from the numerous options available can be a challenging task. The ultimate decision is usually based on several key factors. Firstly, the operating conditions must be taken into account, as they determine the necessary properties of the material. Most materials do not possess the perfect combination of properties, so compromises must often be made between different characteristics. For example, high strength often comes at the cost of reduced ductility. Secondly, the degradation of material properties over time should be considered. For instance, exposure to high temperatures, radiation, or corrosive environments can lead to a significant decrease in mechanical strength. Lastly, cost is a major consideration and plays a crucial role in the material selection process. The ideal combination of properties may be attainable only at an excessively high cost, making it difficult to find practical applications. In such cases, compromises must be made between properties and cost.

Materials science plays a crucial role in shaping the world we live in and determining the future of technology. Its impact can be seen in various fields, where advancements in materials have allowed for new innovations and breakthroughs.

For instance, the invention of advanced composite materials has made it possible to dramatically reduce the weight of aircraft, resulting in a 20% reduction in fuel consumption per passenger on the Boeing Dreamliner. The development of blue light LED materials earned the 2014 Nobel Prize and paved the way for the widespread use of white LED lights. The discovery of electrode materials for lithium-ion batteries was recognized with the Nobel Prize in 2019 and made the widespread use of electric vehicles a reality. On the other hand, the limitations of silicon, a typical semiconductor material, are challenging the progress of Moore's law, which states that the number of transistors per chip doubles every 24 months.

1.2 Classification of Materials

Materials are not only important and ubiquitous, but also diverse, making it challenging to establish a unified classification standard.

Materials can be broadly classified into two categories: structural and functional. Structural materials are primarily used in load-bearing applications and are valued for their mechanical properties, although they must also meet certain requirements for physical or chemical properties such as luster, thermal conductivity, radiation resistance, corrosion resistance, and oxidation resistance. Conversely, functional materials are known for their unique physical, chemical, or biological properties. It is not uncommon for a material to possess both structural and functional properties, such as iron, copper, and aluminum.

Materials can also be classified into two further categories: traditional and advanced. Traditional materials are mature materials that have been widely used and mass-produced in industry, such as steels, cements, and plastics. These materials are commonly referred to as "fundamental materials" because of their widespread use, high output value, and their fundamental role in many key industries. On the other hand, advanced materials are still in development and are often used in high-tech applications. There is no clear boundary between advanced and traditional materials, as traditional materials can become advanced with the adoption of new technologies and improvement in performance, while advanced materials can become traditional with long-term production and use.

Based on their applications, materials can also be categorized into electronics materials, aerospace materials, nuclear materials, construction materials, energy materials, and biomaterials, among others.

Materials, based on their physical and chemical properties, are conveniently classified into three main categories: metals, ceramics, and polymers. This classification scheme is primarily based on their interatomic bonding. Most materials can be classified into one of these categories. For example, beverage cans are made of aluminum alloy (metal), glass (ceramic), and plastic (polymer) as shown in Fig. 1.1. Composite materials, which are engineered combinations of two or more different materials, are also widely used. The following provides a brief overview of these material classifications and their representative characteristics.

(a) (b) (c)

Fig. 1.1 Beverages cans marketed in aluminum cans (**a**), glass bottles (**b**), and plastic bottles (**c**)

1.2.1 Metals

Metals, or metallic materials, are made up of one or more metallic elements such as gold, silver, copper, iron, or tin, and sometimes a small amount of nonmetallic elements such as carbon, nitrogen, or oxygen. Their atoms are usually arranged in an orderly fashion and generally have higher densities than ceramics and polymers. These materials are generally stiff and strong, yet ductile and resistant to breakage, making them ideal for use in structural applications. They have many delocalized electrons, which are not attached to specific atoms, giving metals good electrical and thermal conductivity, a lustrous appearance, and making them non-transparent to visible light. Some metals, like iron, cobalt, and nickel, also have desirable magnetic properties. Figure 1.2 showcases a few items made from metallic materials.

(a) (b) (c)

Fig. 1.2 A few metallic items. **a** A coin (nickel); **b** a spoon (stainless steel); **c** a mask unearthed at Sanxingdui (gold)

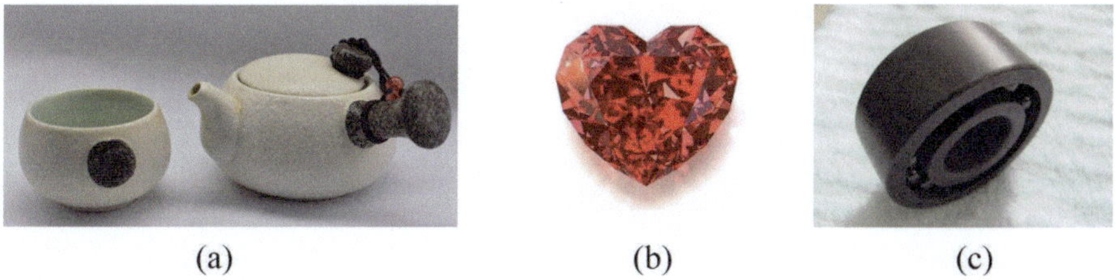

Fig. 1.3 Some products made of ceramics. **a** A set of tea ware (porcelain); **b** ruby jewelry (aluminum oxide); **c** a silicon nitride bearing

1.2.2 Ceramics

Ceramic materials are typically inorganic, non-metallic solids that often consist of compounds such as oxides, carbides, and nitrides. Some common examples of ceramic materials include alumina (Al_2O_3), silica (SiO_2), silicon carbide (SiC), silicon nitride (Si_3N_4), and aluminum nitride (AlN). Ceramics can be further divided into traditional ceramics, such as porcelain and cement, and advanced or fine ceramics, which are used in high-technology applications like microelectronics. When it comes to mechanical behavior, ceramics tend to be stiff and strong, with high Young's moduli compared to metals and polymers. They are also typically very hard. However, ceramics are also brittle and prone to fracture, though advanced ceramics are being developed with improved fracture resistance for use in cookware, cutlery, and even engine parts. Ceramics are typically thermal and electrical insulators, and they are more resistant to high temperatures and harsh environments than metals and polymers, making them ideal for use in engine parts and spacecraft. In terms of optical characteristics, ceramics can be transparent, translucent, or opaque, depending on their microstructure, and some oxide ceramics (e.g., Fe_3O_4) exhibit magnetic behavior. Figure 1.3 showcases a few common ceramic products.

1.2.3 Polymers

Plastics and rubber are well-known polymeric materials. Polymers are typically composed of organic compounds based on carbon, hydrogen, and non-metallic elements such as oxygen, nitrogen, fluorine, and silicon. These compounds have large molecular structures, often a backbone of carbon atoms, and are chain-like in nature. Common polymers include polyethylene (PE), nylon, polyvinyl chloride (PVC), polycarbonate (PC), polystyrene (PS), and silicone rubber. Polymers have the lowest density among all materials, including metals and ceramics. While they are not as stiff or strong as metals or ceramics, their low density means that their stiffness and strength per unit mass are comparable to those of metals and ceramics. Additionally, many polymers are highly ductile and pliable, allowing them to be formed into complex shapes. They are also chemically inert and unreactive in

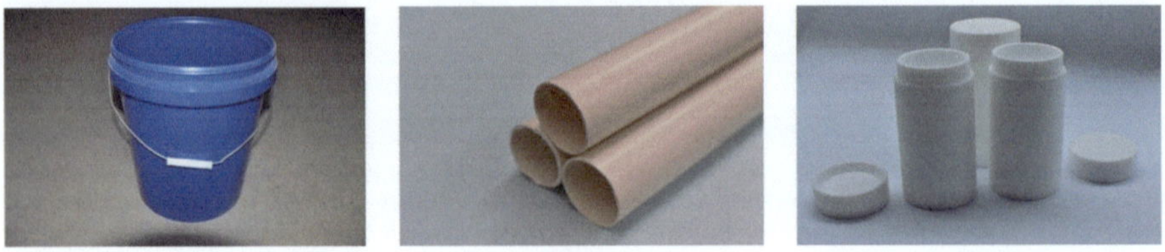

(a) (b) (c)

Fig. 1.4 Some polymer articles. **a** A plastic (polyethylene, PE) bucket; **b** polyvinyl chloride (PVC) pipes; **c** Teflon pressure vessels (polytetrafluoroethylene, PTFE)

many environments. However, a drawback of polymers is their tendency to soften and/or degrade at moderate temperatures. Figure 1.4 illustrates a few items made from polymers.

1.2.4 Composites

A composite material is made up of two or more components that come from the three main categories of metals, ceramics, and polymers. The purpose of a composite material is to combine properties from each component to create a new material that has characteristics that are not found in any single component. There are many different types of composite materials made from combinations of metals, ceramics, and polymers. Many natural materials, such as wood, bamboo, bone, shell, and ancient building materials like mud-and-straw, are naturally occurring composites. However, most of the composites we discuss are man-made or synthetic. On the other hand, many natural materials serve as inspiration for the development of advanced composites, also known as biomimetic materials.

One well-known example of a composite is fiberglass, in which tiny glass fibers are embedded in a polymeric material such as epoxy or polyester. The glass fibers provide strength and stiffness, while the polymer is more flexible. This combination of materials makes fiberglass stiff, strong, and sometimes referred to as "glass steel" in Chinese due to its high strength, similar to steel.

Another important type of composite is carbon fiber reinforced polymer (CFRP), which is made up of carbon fibers embedded in a polymer. These materials are stiffer and stronger than fiberglass composites, but also more expensive. CFRPs are used widely including aircraft, aerospace systems, and high-tech sporting equipment such as bicycles, golf clubs, badminton rackets, and fishing rods. They are also being used in the body and bumpers of cars. In the Boeing 787 Dreamliner, half of the materials used are CFRP composites, and in the Chinese-made passenger jet C919 (Fig. 1.5), 12% of the structural components are CFRP composites.

Fig. 1.5 China's first domestically manufactured single-aisle passenger jet, the C919, incorporates a large quantity of CFRPs

1.3 Materials Driving Human Progress: A Historical Perspective

The evolution of human civilization has demonstrated that materials are the backbone of social progress and a key hallmark of human advancement. They are intertwined with civilization and culture and have played a critical role in shaping history. The renowned leader, Mao Zedong, captured this sentiment in a poem, as depicted in Fig. 1.6, which reflects on the different stages of human history by referring to three key materials: stone, bronze, and iron.

贺新郎·读史

人猿相揖别。只几个石头磨过，小儿时节。铜铁炉中翻火焰，为问何时猜得？不过几千寒热。人世难逢开口笑，上疆场彼此弯弓月。流遍了，郊原血。

Fig. 1.6 Part of a poem "reading the history" by late Chairman Mao, wherein the human history is summarized by referencing to the Stone Age, Bronze Age, and Iron Age

(Translation: Separating early humans from apes, only stone tools were used did humanity turly begin. The flames of bronze and iron soared in the furnace, making the passage of a few thousand years. Human history, akin to daily life, bears few smiles but more sorrow, often witnessing conflicts with bows drawn like a full moon, saturating the fields with blood.)

Materials play a critical role in the advancement of human society. The history of material discovery and utilization has proven that each time a significant material is discovered and utilized, humanity's ability to harness and transform nature is enhanced, resulting in significant increases in social productivity, living standards, and advancements in both material and spiritual civilization.

1.3.1 The Stone Age

Humanity's use of stones and natural ceramics as tools dates back over 3 million years, marking the start of the Old Stone Age (Paleolithic Period), followed by the Middle Stone Age (Mesolithic Period) (Fig. 1.7). Approximately 10,000 years ago, humans began to craft stones into intricate vessels and tools, signaling the beginning of the New Stone Age (Neolithic Period). During this time, humans also started to wear fur, a natural polymer, for clothing. Additionally, the Chinese utilized silk, another natural polymer, for clothing 8,000 years ago, while Indians began growing cotton 4,500 years ago. Although copper and gold were known to humans during this time, their limited availability made them have little impact on daily life.

This Neolithic Age also saw the advent of the first Agricultural Revolution, around 10,000 BC. This historic transition involved moving from a hunter-gatherer society to a settled agricultural lifestyle, commonly referred to as the Neolithic Revolution.

Fig. 1.7 Stone tools humans made during the Paleolithic Period

Fig. 1.8 A 43.7 cm high pot crafted approximately 5000 years ago, unearthed in 1974, Lanzhou, Gansu

Approximately 8,000 to 9,000 years ago, humanity was still in the New Stone Age, but they had invented pottery, a man-made ceramic made by molding and firing clay (Fig. 1.8). Pottery served not only as vessels but also as decorative pieces. Although there is no historical concept of a "pottery age", pottery's impact on civilization was immense. In the process of firing pottery, copper and tin were accidentally discovered and processed, although people were unaware that these metals were formed from copper and tin oxides reduced by carbon at high temperatures. This led to the creation of bright, castable bronze, marking the start of heavy metal use and becoming a significant milestone in human civilization.

1.3.2 The Bronze Age

The Bronze Age, marked by widespread use of bronze, started around 3000 BC in Greece and India, 2500 BC in Egypt, and in the mid-nineteenth century BC in Ancient Babylon. In China, the use of bronze wares dates back to 2700 BC, with a rich history spanning over 5000 years, reaching its peak in the Shang and Zhou dynasties (seventeenth century BC to third century BC). Bronze artifacts such as the 833-kg tetrapod (Fig. 1.9) found in Anyang, Henan, the bronze chime bells discovered in Sui County, Hubei, and the bronze chariot and horse excavated in

Fig. 1.9 Simuwu (also called Houmuwu) tetrapod cast during Shang Dynasty in memory of the queen, unearthed in 1939 and reunearthed in 1946, from Anyang, Henan, with a mass of 832.84 kg, height of 133 cm, length of 110 cm, width of 79 cm; chemical composition: 84.77% Cu, 11.64% Sn, 2.79% Pb

Xi'an, all bear witness to the remarkable level of metallurgical and manufacturing skills of the ancient Chinese.

However, our ancestors soon realized that bronze was not as strong or durable as they wanted when used as tools, and so they started to combine bronze with meteoric iron. A bronze axe with a meteoric iron edge (Fig. 1.10) serves as one of the earliest examples of combining different metals to create tools through brazing.

Fig. 1.10 An 8.7 cm long bronze axe with a blade of meteoric iron, manufactured at the beginning of Zhou Dynasty, unearthed in 1931 in Henan, PR China

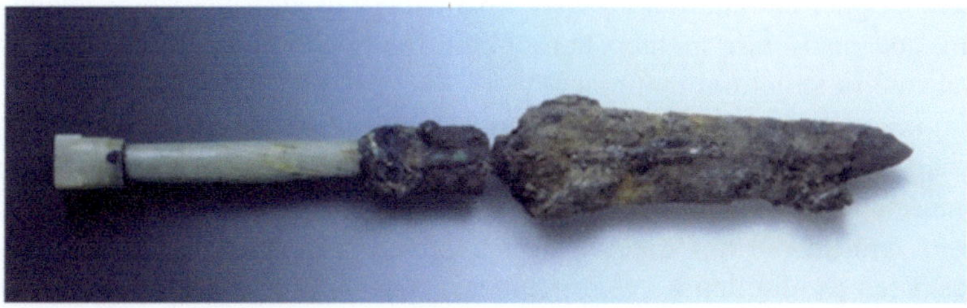

Fig. 1.11 An iron sword with a jade handle made in the Western Zhou Dynasty (1046—771 BC), unearthed in the tomb of Guo Ji, Sanmenxia, Henan Province, PR China

1.3.3 The Iron Age

The beginning of the Iron Age varies globally, with estimates ranging from the 13th to fourteenth centuries BC. However, widespread use of iron tools became more common than bronze tools by the end of the third millennium BC. The earliest evidence of artificial iron production in China dates back to the ninth century BC.

By the end of the Spring and Autumn Period (770—476 BC), significant advancements were made in the production of pig iron in China. The development of ductile cast iron through annealing pig iron and the invention of steel-making techniques greatly boosted China's productivity, playing a crucial role in the development of agriculture, water conservation, and military affairs during the Warring States, Qin, and Han Dynasties.

As early as the second century AD, China's steels and silks were well-known in the Roman Empire. The pig iron technology spread from the Yellow River valley and the Yangtze River basin to other regions, such as the Korean Peninsula, Japan, and Northern Europe, from the sixth century AD to the seventh century AD, promoting the advancement of world civilization.

However, there are fewer iron tools and wares on display in museums compared to bronze tools, primarily due to iron's susceptibility to corrosion. Figure 1.11 depicts an iron sword from the Western Zhou Dynasty (1046—771 BC).

1.3.4 Steels Change the World

In the eighteenth and nineteenth centuries, steam engines and electric motors revolutionized industry and led to increased demand for metals, particularly iron and steel. In the mid-nineteenth century, the invention of converter steelmaking and open-hearth refining significantly increased world steel output, from 60,000 tons in 1850 to 28 million tons in 1890. This paved the way for advancements in machinery manufacturing, railway transportation, and the textile industry. With the advent of electric furnace steelmaking, special steels, including high manganese steel, high-speed steel, silicon steel, and austenitic nickel–chromium stainless steel, were produced. Copper and aluminum were also widely used, followed by magnesium, titanium, and many rare metals. Throughout the twentieth century, metals played a crucial role in structural materials.

It is noteworthy that some metal elements have identical names in Chinese and Japanese, such as gold (金 in Chinese, 金 in Japanese), silver (银 in Chinese, 銀 in Japanese), copper (铜 in Chinese, 銅 in Japanese), iron (铁 in Chinese, 鉄 in Japanese), and tin (锡 in Chinese, 錫 in Japanese). However, metals with shorter histories, such as aluminum (铝 in Chinese, アルミニウム in Japanese), magnesium (镁 in Chinese, マグネシウム in Japanese), and titanium (钛 in Chinese, チタン in Japanese), have different names in the two languages, reflecting the historical relationship between the two neighboring countries.

1.3.5 Polymers Make Life Comfortable

The development of organic chemistry in the late nineteenth century led to the invention of rayon, which was created by imitating Chinese silk. In the early twentieth century, synthetic organic polymers emerged one after the other, such as phenolic resin (bakelite) in 1909, polystyrene in 1920, polyvinyl chloride in 1931, and nylon in 1941. The rapid growth of polymers can be attributed to their excellent performance, high availability, low construction costs, and fast returns. In 2019, the annual output of the world's three main organic synthetic materials (resins, fibers, and rubber) surpassed 370 million tons. The performance of organic materials has continuously improved and their added value has significantly increased, particularly for special polymers with vast potential, which are making their way into various fields of functional materials.

However, the excessive consumption of plastics has had a negative impact on the environment, particularly on marine life. Many countries are proposing bans on single-use plastics, targeting commonplace items such as straws, cotton buds, cutlery, balloon sticks, and drink stirrers. On January 8, 2008, the Chinese government introduced regulations on the manufacture and use of plastic bags. "Single-use plastic bags, which pose a threat to marine life, should be banned or phased out globally. There is no excuse for manufacturing them anymore, anywhere," said Achim Steiner, Executive Director of the UN Environment Programme, which provides environmental policy advice to UN member states.

1.3.6 Ceramics Are Ancient, Yet Future Materials

Ceramics have been utilized for various purposes since ancient times, including building materials, containers, and decorations. They are characterized by advantages such as high abundance, low density, high hardness, good corrosion resistance, low thermal expansion, high-temperature resistance, and good wear resistance. By mid-twentieth century, advanced ceramics such as Si_3N_4, SiC, ZrO_2, etc. were introduced to the market. In recent years, research and development of advanced ceramics has gained significant momentum, leading to the prediction of a "New Ceramic Era." Despite these advantages, ceramics can be brittle and difficult to handle, and the cost of some advanced ceramics can be prohibitively high. Nevertheless, ceramics have been identified as having great potential in the field of functional materials.

1.3.7 Composites Prevail

Composites have been a part of human life for a long time, with natural materials such as wood, leather, and bamboo being composite materials themselves. Humans have been creating composites for thousands of years, using mud mixed with crushed straw to construct houses, for example. Reinforced cement, a composite of brittle and ductile materials, is another example. In recent decades, advances in resin-based and metal-based composites have been made, with resin-based composites being widely used and metal-based composites being more limited due to production difficulties and high cost. Ceramic-based composites have also been developed to enhance the performance of ceramics, particularly in terms of toughness. Carbon has the highest service temperature, reaching up to 2500 °C, but its poor thermal shock resistance restricts its applications. Carbon–carbon composite materials were developed to overcome these limitations and improve its mechanical properties. As a result, they are widely used in the military industry and consumer sectors.

1.3.8 Functional Materials Pave the Way to the Silicon Age

Functional materials have been of great value to human civilization since ancient times. As early as the Warring States Period (third century BC), natural magnetite was utilized to make a type of compass called Sinan. In the Song Dynasty, the invention of the compass through magnetization of a steel needle was a key technology that led to the development of navigation. Additionally, functional materials are the cornerstone of information and automation technologies. The advent of semiconductor materials in particular has had a profound impact on modern civilization. In 1947, the first transistor with amplifying effects was invented, and more than a decade later, integrated circuits were successfully developed. Computers based primarily on silicon materials offered better performance, reduced size, and lower cost, leading to the rise of the Information Age. Given that silicon is a critical material in microelectronics, this era is also referred to as the Age of Electronic Materials or the Silicon Age, highlighting the pivotal role that materials play in human civilization.

1.4 The Emergence and Meaning of Materials Science

The term "material" has been in existence for a long time, but the field of "materials science" only emerged in the 1950s. This was due to the successful launch of a satellite by the former Soviet Union in 1957, which shocked the US government and public. They believed that the lack of advanced materials may have contributed to their technological lag, leading to the establishment of over ten materials research centers in universities. This resulted in an increased focus on materials research, which was based on advanced scientific theories and experimental techniques. The formation of "materials science" can be attributed to the advancement in science and technology.

Materials science is a multi-disciplinary field that combines knowledge from various areas such as solid-state physics, inorganic chemistry, organic chemistry, physical chemistry, and others. This has led to a deeper understanding of the structure and properties of substances. The development of fields like metallurgy, metallography, ceramics, polymer science, and others has also contributed to the growth of materials science. As a result, a strong foundation has been established for the study of materials, their processing, structure, performance, and correlations.

Additionally, before the term "materials science" was coined, metals, ceramics, polymers, and composite materials were already well-established fields. However, there are many similarities between these materials, and advancements in one area can often shed light on others. For example, the martensite phase transformation, which is a key theory in the heat treatment of steels, was first proposed by metal scientists but was later discovered in zirconium oxide ceramics, leading to more effective toughening of the ceramic. Another example is the sol–gel method, which involves the decomposition of metal organic compounds to produce high-purity oxide particles at the nanoscale, and has become a useful method for improving ceramic performance. These materials also have similar testing equipment and production devices, such as optical microscopes, electron microscopes, surface testers, physical property testing equipment, and mechanical property testers.

By utilizing the strengths of various materials, they can complement and substitute each other to maximize their properties and performance. However, in the past, metals, ceramics, and polymers were distinct and separate fields with limited interaction, obstructing the growth and creativity in the development of composite materials. The absence of mutual understanding was a major hindrance that delayed progress and innovation.

Given the need for material development and commonality, the discipline of materials science has emerged. However, due to differences in disciplinary foundations, there are still significant differences between the subjects of different materials, especially between inorganic and organic materials. The rise of soft matter science has helped bridge the gap between soft and hard materials, creating a balance.

Materials Science is a branch of science that explores the relationship between the structure and properties of materials, as well as the investigation of natural laws. It is considered both a basic and an applied science, as its goal is to serve the economy and improve human lives. The practical application of materials requires rational technological processes and mass production. Hence, the term "Materials Science and Engineering" was introduced, emphasizing the focus on the processing and engineering problems that arise during manufacturing and usage.

As a result, many universities have renamed their Department of Metallurgy or Department of Materials to the "Department of Materials Science and Engineering". Some universities have chosen to emphasize either "Materials Science" or "Materials Engineering" in their names, while others, like the University of Cambridge, have retained the term "Metallurgy" and renamed their department as "The Department of Metallurgy and Materials Science." Additionally, journals and books related to Materials Science and Engineering have emerged. The "Encyclopedia of Materials Science and Engineering" edited by scientists from the Massachusetts Institute of Technology in 1986, defines Materials Science and Engineering as the study of the relationship between composition, structure, processing, properties, and application. Therefore, the four basic components of Materials Science and Engineering are composition-structure, synthesis-processing, properties, and performance. These components are often depicted as a tetrahedron, known as the "Materials Science Tetrahedron," as shown in Fig. 1.12a.

In this four-component model, composition is evidently not equal to structure, since materials with the same composition may possess different structures, properties and performance when prepared with different synthesis or processing. To account for this, a five-component model was introduced. This model is depicted as a hexahedron in Fig. 1.12b, and includes composition, synthesis/processing, structure, properties, and performance.

The five-component model of Materials Science and Engineering has two main characteristics. Firstly, it highlights the relationship between properties and performance. Performance is a direct function of properties, and environmental factors such as loading, atmosphere, medium, and temperature can significantly

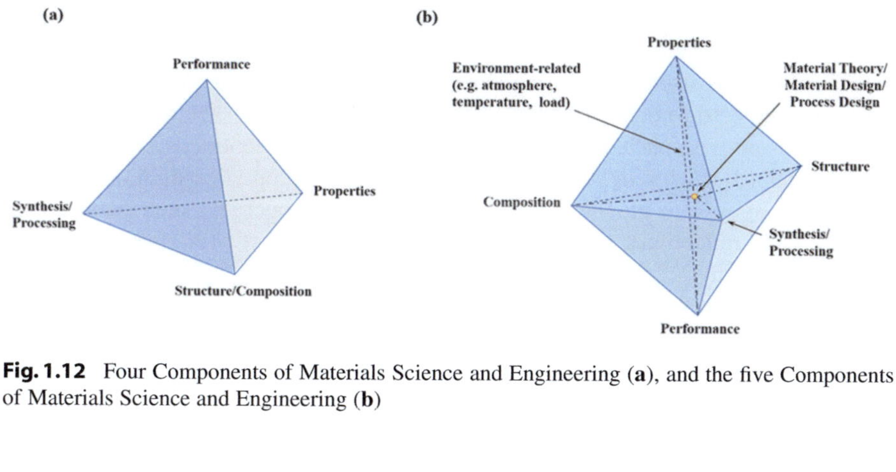

Fig. 1.12 Four Components of Materials Science and Engineering (**a**), and the five Components of Materials Science and Engineering (**b**)

impact performance. For example, some materials may perform well in general environments but suffer from decreased performance in corrosive media or may perform well when smooth but have poor performance with notches, particularly for high-strength materials. Therefore, considering environmental factors is crucial in materials engineering. Secondly, theories and materials design occupy a central position in the hexahedron model. Each component in the model has its own theories, which enable the modeling of materials design or processing design to improve performance, conserve resources, reduce pollution, and lower costs. This is the ultimate goal of the development of Materials Science and Engineering. It's worth mentioning that the four-component model is still widely used globally, while the five-component model is used occasionally.

The term "**composition**" in materials science refers to the chemical makeup of a material, including the types and proportions of elements it contains. The composition of a material can have a significant impact on its physical and chemical properties, such as its density, melting point, thermal conductivity, and reactivity. In materials science and engineering, the composition of a material is one of the important factors that are considered in the design and development of new materials for specific applications.

The term "**structure**" in materials science refers to the way the internal components of a material are arranged, including its particles, atoms, phases, defects, and interfaces. The subatomic structure involves the arrangement of electrons within individual atoms and their interactions with nuclei. The atomic structure pertains to the organization of atoms or molecules in a material. The microscopic structure, also known as microstructure, can be directly observed through a microscope. The macroscopic structure, or macrostructure, refers to the structural element that can be seen with the naked eye.

The term "**property**" in materials science refers to the behavior of a material in response to external stimuli. This can manifest in various ways, such as deformation when subjected to forces, or changes in conductivity when subjected to an electric field. Properties are characteristic traits of a material, reflecting the type and magnitude of its response to specific stimuli, and are typically independent of the material's shape or size. Materials can be classified into five categories based on their properties: mechanical, electrical, thermal, magnetic, and optical. Mechanical properties involve the material's response to applied forces or loads, such as elasticity, strength, and toughness. Electrical properties, such as conductivity and dielectric constant, reflect the material's behavior in an electric field. Thermal properties, such as heat capacity and thermal conductivity, pertain to the material's behavior in relation to heat. Magnetic properties demonstrate the material's response to magnetic fields, while optical properties, such as index of refraction and reflectivity, describe the material's interaction with electromagnetic or light radiation.

The term "**processing**" refers to the series of operations that convert raw materials into an intermediate or a finished product. In materials science and engineering, processing involves transforming a material from its initial state into a desired final form. This may include operations such as shaping, forming, heat treatment, joining, and surface finishing. Processing is an important aspect of materials science and engineering as it affects the material's properties and performance.

The term "**synthesis**" refers to the creation of a new material from its constituent elements. In materials science and engineering, synthesis involves combining two or more elements to produce a new material with desired properties. This may involve methods such as solid-state synthesis, liquid-phase synthesis, vapor-phase synthesis, and others. Synthesis is a crucial aspect of materials science and engineering as it allows for the creation of new materials with unique properties and functionalities.

The term "**performance**" in materials science refers to the capability of a material to carry out its intended function under specified conditions. It encompasses the overall effectiveness and efficiency of the material in its operating environment. In materials science and engineering, performance is often linked to the properties of the material and can be affected by various environmental factors such as temperature, humidity, corrosive conditions, and mechanical stress. The performance of a material can also be influenced by the way it is processed, synthesized, and fabricated, making it an important aspect of material design and development. Examples of performance-related properties include strength, durability, wear resistance, and electrical conductivity.

The principles of processing-structure-properties-performance can be illustrated by the three thin-disk specimens of alumina ceramic material shown in Fig. 1.13. The optical properties of the specimens vary significantly, with the left one being transparent, the center one being translucent, and the right one being opaque. Although the three samples are made of the same composition, the differences in their optical properties are a result of variations in their processing and thus their structure. The left sample was sintered at a low heating rate, resulting in fewer pores and a higher degree of perfection, which led to its transparency. The right sample, on the other hand, was sintered at a high heating rate and had a large number of pores that scattered the reflected light, making it opaque. The center specimen has properties that fall between the other two.

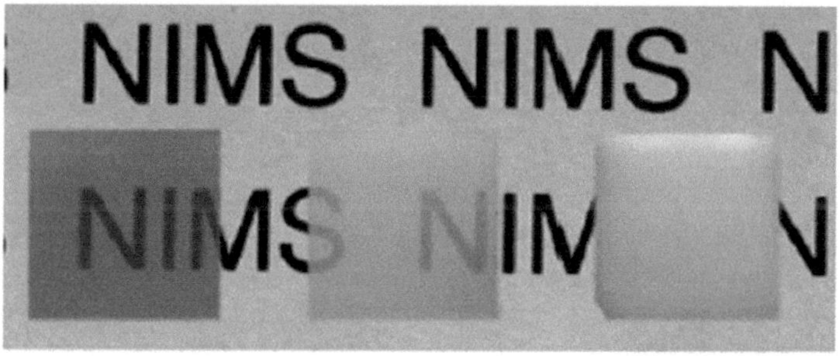

Fig. 1.13 Three thin-disk specimens of alumina sintered at different heating rates

The left disk with the fewest pores is transparent, the right disk with more pores is opaque, and the center disk is in between.

It's clear that the structures of these three specimens are distinct in terms of microstructure and pores, which influence their optical transmittance properties. The variation in structure corresponds to the different processing techniques used during preparation. Because optical transmittance is a critical factor for the intended application, it's crucial to carefully select the appropriate processing method to achieve the desired microstructure, properties, and performance.

In conclusion, materials science encompasses three key attributes. Firstly, it's a multidisciplinary field, arising from the integration of physics, chemistry, metallurgy, metallography, ceramics, polymer chemistry, and computational science. Secondly, it's closely linked to application, with the goal of discovering new materials, improving material performance and quality, and reducing costs and pollution through efficient material usage. Lastly, it's a continuously evolving field, unlike some established disciplines.

1.5 The Importance of Traditional Materials in Economic Growth and Sustainability

Materials have been a driving force in the progress of human society and serve as the physical foundation and precursor to social modernization. Despite the common perception of traditional materials industries as a "sunset industry," they are essential to the economy. Steel, for example, has long been regarded as a key indicator of a country's strength, and even today, some developed countries still consider it a pillar industry due to its unmatched properties and relatively low cost. As seen in Table 1.1, the price-to-specific strength ratio of steel is lower than that of other materials, except for wood. Despite rising prices of many mechanical products and materials, the cost of steel remains unchanged due to advancements in fabrication technology, reduced energy consumption, and increased labor productivity.

Table 1.1 The relationship between prices and specific strength of some structural materials

Material	Steel	Al	Ti	Cement	Al_2O_3	Carbon fiber	Wood	Polypropylene
Specific strength	5.2	11.1	13.3	0.8	97.4	160.9	12.5	3
Price/specific strength (relative value)	1	3	16.7	6.4	4	5.2	0.8	3.8

The construction industry, which is crucial in every country, relies on traditional building materials such as cement, glass, wood, brick, tile, and stone. The automobile industry is also gaining significance in many countries, with materials like steel, iron, aluminum, plastics, and glass making up 90%–95% of automotive structural materials. Traditional industries such as synthetic fibers, resins, plastics, and rubber production are becoming increasingly important to the economy. Additionally, industries such as machinery manufacturing, shipbuilding, and locomotives are all based on steel and other traditional materials.

1.5.1 Traditional Materials Are Generally Used Widely and in Large Quantities

The widespread use and high demand for traditional materials make their significance undeniable. Even minor improvements can yield considerable economic benefits. For instance, it is estimated that extending the service life of roads and bridges in the United States by just 1% could generate 30 billion dollars in revenue. Similarly, reducing the cost of each beverage can by one cent could result in a profit of 600 million yuan, based on an annual consumption of 60 billion cans. This underscores the critical importance of enhancing the performance and reducing the costs of traditional materials.

1.5.2 Traditional Materials Are a Major Consumer of Mineral Resources

The estimated reserves of iron may last only 200–300 years, aluminum for nearly a hundred years, and copper, lead, and zinc for only several decades. In addition to the exploration of new mineral deposits, it is essential to find substitutes for key materials. For instance, magnesium, which is abundant and has properties similar to aluminum, can be considered as a substitute for aluminum. The magnesium reserved in seawater alone is estimated to be 2.1×10^{15} tons, which is considered to be inexhaustible. This makes it even more crucial to focus on research that aims to replace aluminum with magnesium.

1.5.3 The Processing of Traditional Materials Often Leads to Severe Pollution

Statistics show that nearly 80% of toxic wastewater in the United States comes from the chemical industry, steel smelting, and nonferrous metal smelting. The energy-intensive cement production is another major source of pollution. In 1997, China produced 500 million tons of cement, accounting for 40% of the world output, which led to the emission of nearly 400 million tons of CO_2, 49 million tons of SO_2, 1 million tons of NO_x, and more than 11 million tons of dust.

These emissions contribute to the greenhouse effect and cause acid rain. Moreover, the CO_2 emissions during steel smelting are at least twice those of cement. With the growing demand for carbon neutrality, the traditional materials industries face numerous challenges, and novel technologies in materials science and engineering are expected to play a crucial role in addressing these challenges.

Traditional materials play a crucial role in the economy and sustainable development, and it is important to address the challenges associated with their processing. To achieve this, the following approaches must be considered:

(1) Adopting resource-saving and environmentally friendly production processes, with a goal towards "zero emission" production.
(2) Developing eco-friendly materials that are recyclable, degradable, and save resources while reducing pollution.
(3) Improving the performance and durability of traditional materials to save resources and reduce pollution, such as increasing the strength of steels and developing corrosion-resistant and high-strength cements.
(4) Investing in new technologies to enhance the production process, leading to higher labor productivity, better product quality, and lower costs, while also upgrading the competitiveness of traditional materials.

In conclusion, there are ample opportunities for further scientific and technological advancements in traditional materials that can contribute to a sustainable future.

1.6 Advanced Materials: Driving the Wheel of Social Progress

The development and implementation of advanced materials serve as indicators of advancements in science, technology, and industry. A few examples are provided to demonstrate the progression of modern science and technology and the crucial role advanced materials play in these fields.

1.6.1 The Development of Electronic Technology

The impact of materials on modern technology is exemplified by the evolution of electronic devices. From the invention of the electron tube in 1906, which gave rise to the development of radio, television, and computers, to the invention of the semiconductor transistor in 1948, which brought about miniaturization, energy savings, and increased reliability, materials have played a crucial role in shaping the evolution of electronic technology. The advent of integrated circuits in 1958 was a game-changer, allowing for an exponential reduction in the size and cost of computers, making them accessible to a wider audience. Today, microcomputers are 650,000 times smaller, 120,000 times more energy-efficient, and 580,000

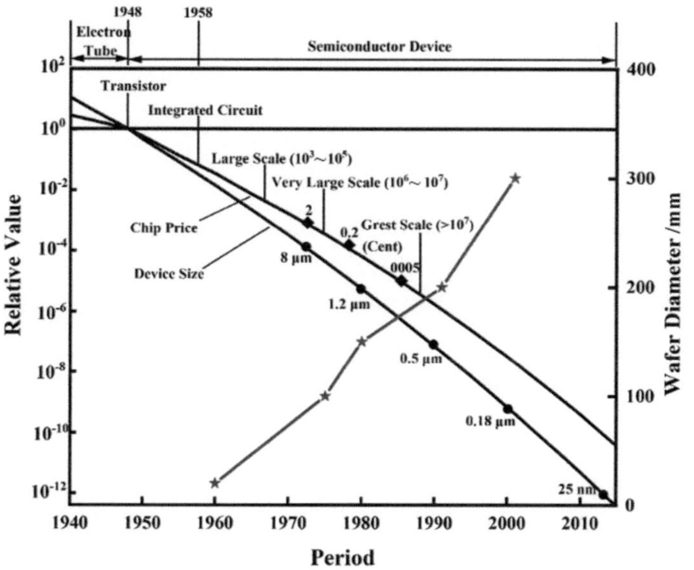

Fig. 1.14 Development of the integrated circuit

times lighter than computers based on electronic tubes from 1946, thanks to the improvement in materials science and technology.

The progress of chip integration has played a crucial role in the miniaturization of electronic devices. As seen in Fig. 1.14, the size of silicon wafers has decreased as chip integration has improved, resulting in a lower unit price. This is due to the reduction of feature length and improvement in manufacturing technologies and materials. To maintain this progress, the feature length must continue to be reduced, and the wafer size must increase, putting higher demands on the quality of silicon single crystals. This has led to an increase in chip density and improved performance of computers, including faster computing speeds of up to 4.42×10^{17} per second (as of 2021).

The advancement of information storage technology has also been greatly influenced by materials. The requirement for high-capacity, high-density, fast, and erasable storage can only be met through materials improvement. Additionally, the accuracy of computer control in industrial automation relies on the sensitivity of sensors, making high-precision and stable-performing sensing materials key components.

1.6.2 The Birth of Optical Fiber Communication

In 1966, C.K. Kao foresaw the potential of optical communication by predicting that it could become a reality once the loss in optical fiber transmission was

Table 1.2 Development of optical fiber communication

Technical indicators	First generation (Short wavelength)	Second generation (Long wavelength)	Third generation (Long wavelength)	Fourth generation (Ultralong wavelength)	Fifth generation (Ultralong wavelength)
Wavelength/μm	0.83	1.3	1.55	2 ~ 5	1 ~ 12
Optical fiber	Fused quartz	Fused quartz	Fused quartz (+ Er)	Fluoride glass	Fluoride glass/ Chalcogenide glass
Laser	GaAlAs	InGaAsP	InGaAsP	GaInAsSb	InAs-AlGaInAsSb/ InP-AlGaInAsP
Detector	Si	Ge, InGaAsP	InGaAsP	InAsGaSb	InAs-AlGaInAsSb/ InP-AlGaInAsP
Loss/ $dB \cdot km^{-1}$ Bit rate/ $Gb \cdot s^{-1}$	2 ~ 3 4.5×10^{-2}	0.5 ~ 1 2 $Gb \cdot s^{-1}$	<0.1 4 $Gb \cdot s^{-1}$	0.1 ~ 0.001 1×10^4	0.1 ~ 0.001 $>1 \times 10^5$
Transmission distance/km	~10	~30	~400	>500	>10,000

reduced to less than 20 dB/km. This prediction came to fruition in 1970 with the invention of quartz doped with alumina, which paved the way for the rapid development of optical fiber communication. Since then, it has advanced at a faster pace compared to integrated circuits and has largely replaced coaxial cables due to its numerous advantages, such as its lightweight, low material consumption, high security, and immunity to electromagnetic interference. As a result, optical cables have become ubiquitous and are now an integral part of modern information highways and networks, found both underwater and in our homes. The optical fiber communication system requires not only optical fibers but also laser devices and detectors, all made of compound semiconductors. Table 1.2 chronicles the development of the optical fiber communication system, with the fifth generation focusing on increasing its capacity through the use of space-division multiplexing (SDM) and the advancement of optical soliton communication.

1.6.3 Advances in Aerospace and Deep Ocean Technology

Economic globalization in the twenty-first century has been greatly facilitated by the growth of e-commerce, enabled by the expansion of information networks and the efficient and high-capacity transportation systems. The former is a direct outcome of advancements in electronics and optoelectronic technology, while the latter is driven by the continuous improvement of aviation, ground, and water transportation technology. Materials play a crucial role in these advancements. For instance, the advent of jet technology in the 1940s was made possible by the development of high-temperature, high-performance structural materials for military use. The improvement of superalloys has increased the temperature tolerance of turbines, and advancements in materials with high specific strength and stiffness have enhanced the safety, payload, and endurance of airliners. Furthermore, the life of aircraft engines has been prolonged and fuel consumption reduced, with an estimated two-thirds of the progress in airplanes being attributed to materials and the performance of aeroengines being largely dependent on materials.

Aerospace and satellites are key indicators of the progress of science and technology, and materials used in these fields must possess not only high specific strength and stiffness but also withstand extreme conditions such as ultrahigh temperatures, radiation, particle clouds, and atomic oxygen erosion. Reducing the weight of spacecraft by just one kilogram can save hundreds of kilograms in launch costs. For instance, the recent successful landing and return of the Chang'e 5 spacecraft from the moon is a testament to the use of advanced materials (as illustrated in Fig. 1.15).

As scientists and engineers continue to explore the possibilities of expediting outer space, they are also making efforts to reach the depths of the ocean. On November 10, 2020, three Chinese divers set a national record by diving to the bottom of the Mariana Trench in the western Pacific Ocean, a feat made possible through the use of high-strength titanium alloys for the dive chamber and high-performance floating materials, among many of the advanced materials employed (as shown in Fig. 1.16).

Fig. 1.15 The landing of Chang'e 5 on the moon on December 1, 2020, collecting lunar soil and returning to the earth, after unmanned docking on the lunar orbit 380,000 km above, bringing back the soil near the second cosmic velocity

Fig. 1.16 China's deep-sea manned submersible Fendouzhe, or Striver, set a national diving record of 10,909 m on November 10, 2020. In the Challenger Deep, an 11,000-m chasm was located at the bottom of the Mariana Trench in the western Pacific Ocean

1.6.4 Quantum Materials and Metamaterials

The twentieth century saw a major transformation in the computation and information fields with the advent of semiconductors, and the twenty-first century is set to be shaped by quantum materials. Quantum materials, a term used in condensed matter physics, are materials whose properties cannot be described by classical and semiclassical approaches. With a relatively short history of just 40 years since the 1980s (as depicted in Fig. 1.17), these materials include Mott insulators, topological materials, multiferroics, superconductors, and quantum spin liquids.

At the atomic level, the interactions between the five fundamental degrees of freedom: spin, charge, topology, orbital, and lattice, in quantum materials create complex electronic states (as illustrated in Fig. 1.18). These interactions result in non-classical and exotic properties that challenge existing physics theories and open up new opportunities for applications. Quantum materials show great potential in revolutionizing energy, information, medical, sensing, and related technologies, and are at the forefront of the second quantum revolution.

These strong couplings lead to emergent functions with cross correlations among different physical observables and to developments toward the applications of emergent functions such as mottronics, magnetoelectrics, topological electronics, and quantum computing.

Metamaterials are a rapidly developing class of materials that are artificially engineered to possess properties not found in nature. These materials consist of

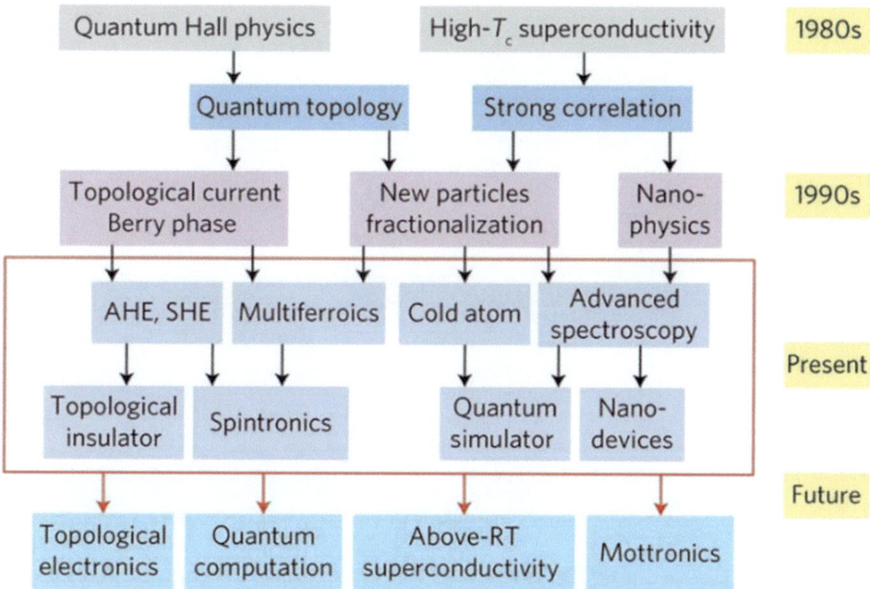

Fig. 1.17 Brief history of the research on the physics of quantum materials and functions

meta-atoms or metamolecules, which can be arranged either periodically or randomly and are often embedded in or layered with dielectric materials (Fig. 1.19). As the meta-atoms are structured on a scale smaller than the wavelength of external stimuli, metamaterials can be considered as effective media.

So far, metamaterials have shown great versatility and functionality in various fields such as electromagnetic, optical, acoustic, thermal, and mechanical engineering. Electromagnetic metamaterials, for instance, have been extensively studied and have exhibited unique behaviors such as negative refraction, superlensing, reverse Doppler shifting, and backward wave propagation. These properties make electromagnetic metamaterials useful in applications such as energy harvesting, object cloaking, high-speed communication, imaging, and antenna design.

In recent years, new types of metamaterials have emerged, including metasurfaces, coding and digital metamaterials, programmable metamaterials, and planar plasmonic metamaterials. For example, metasurfaces with ultrathin subwavelength structures have been precisely fabricated using the latest nanofabrication technology and have enabled the realization of applications like skin cloaking. In the future, metamaterials may evolve into a multidisciplinary field encompassing materials science, artificial intelligence, machine learning, and micro- and nano-technologies, leading to even more potential technological applications.

It is evident that materials, particularly advanced materials, play a critical role in advancing technology. Without semiconductors and other functional materials, the information society as we know it would not exist. Similarly, without materials with high temperature resistance, high specific strength, and high specific

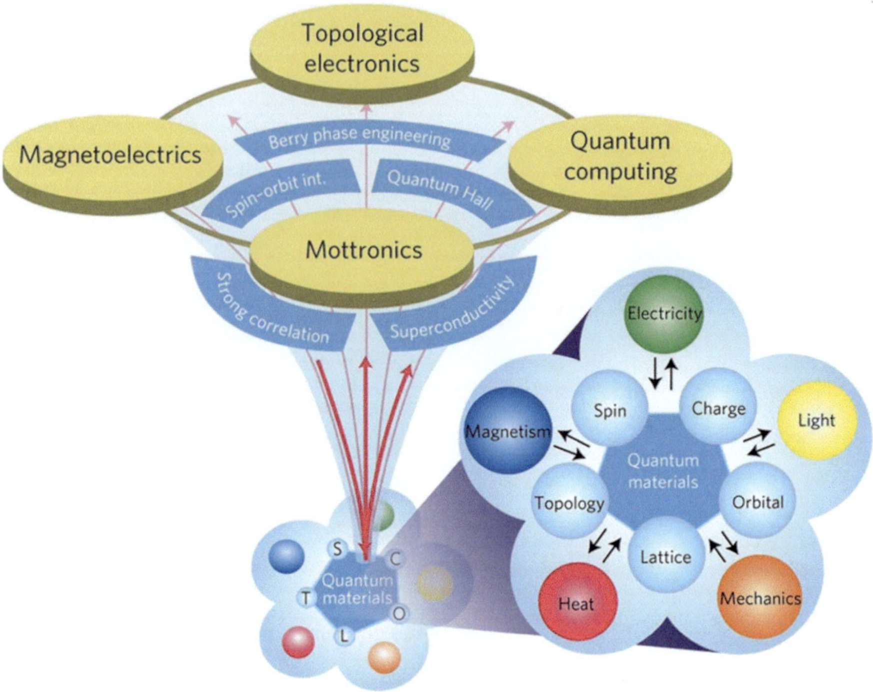

Fig. 1.18 The bottom pentagon shows five degrees of freedom of strongly correlated electrons in solids

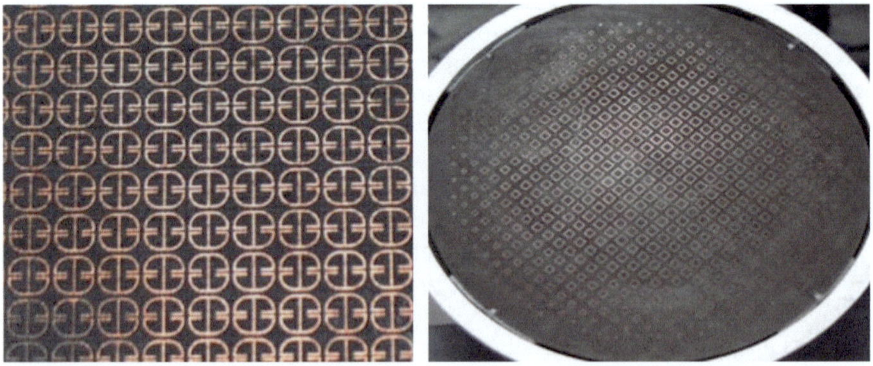

Fig. 1.19 Two typical metamaterial structures in the microwave regime. The left one is a periodic structure, which is equivalent to a homogeneous medium. The right one is a non-periodic structure, which is equivalent to an inhomogeneous medium

stiffness, the aerospace industry would not have made the strides it has. Additionally, without high-performance metallic materials and polymer materials, deep ocean technology would not have advanced to its current state.

1.7 Focus Areas in Materials Science and Engineering Development

The field of materials science and engineering is dedicated to the advancement and improvement of materials, with a focus on the following areas.

1.7.1 Materials Design

The objective of a materials engineer is to achieve the desired performance of a product through thoughtful materials design. The era of "materials by design" is emerging, with the focus shifting from empirical design to scientific design. The concept of materials design gained popularity in the 1980s and has since gained more attention as related fields have developed. Materials design can be broken down into several levels, starting with quantum design. This is the fundamental level as it deals with phenomena caused by the movement of electrons, such as luminescence, which are the basis of functional materials. The second level, atomic design, is the foundation of nanotechnology and deals with the atomic arrangement that determines a material's physical and chemical properties. Atomic manipulation may also be used in the next generation of chips. The third level, microdesign, involves the design of structures at the micron scale, including phase transformations of metals and grain boundary control. Finally, macrodesign is the fourth level, focusing on objects that are millimeter-sized or larger.

The ultimate aim of materials design is to produce high-quality materials through the rational selection of composition and optimization of production processes to meet specific requirements. However, materials design is a complex process, often involving challenges such as the difficulty in accurately modeling materials that are in non-equilibrium states, such as segregation and phase transitions during solidification. Properties can be highly sensitive to changes in structure, and a minor change in structure can have a significant impact on performance. Additionally, information about some properties, such as brittleness and crack initiation and propagation, is limited.

The advancement of science and technology, coupled with a deeper understanding of materials and the increased power of intelligent and supercomputers, is set to bring about a new era of scientifically developed materials. This will require the establishment of comprehensive databases and knowledge bases, as well as the development of practical physical models, through close collaboration between applied mathematicians, physicists, chemists, materials scientists, and engineering technicians.

1.7.2 Development of Materials Processing Technology

The advancement of suitable processing technology is crucial for the successful application of new materials. Despite the discovery of high-temperature superconductors more than 35 years ago, their widespread use in the electric power industry remains limited due to the absence of a low-cost and stable wire production process. Similar challenges exist for other advanced materials, like C_{60}, which despite its potential, has not yet made it out of the laboratory. Optimizing the production processes for traditional materials can also improve their quality, reduce costs and minimize pollution, thereby increasing their competitiveness in the market.

Molecular beam epitaxy technology enables precise control of the thickness of thin films, down to a few atomic layers, making "atomic engineering" and "energy band engineering" a possibility. The adoption of rapid cooling technology ($10^4 \sim 10^8$ K/s) has opened up new avenues for the development of metal materials, including the production of metallic glasses, which exhibit improved strength, anti-abrasion properties, corrosion resistance, and magnetic properties. The discovery of quasicrystals, which challenged classical concepts in crystallography, is also a result of rapid cooling.

Innovative material processing technologies often involve extreme conditions, such as zero gravity, high magnetic fields, strong shock waves, ultrahigh pressure, ultrahigh vacuum, and forced rapid cooling, which can enhance material performance.

1.7.3 Application of Materials

The advancement of materials science and engineering is largely driven by the expansion of materials' application fields. While materials with exceptional performance in laboratory research may not necessarily have practical value, it is crucial to conduct extensive application research to assess their suitability.

In order to fully understand materials, it's important to consider the following factors when evaluating their potential applications:

(1) Performance: The properties and capabilities of the material must be evaluated to determine if it meets the requirements of the intended application.
(2) Durability and reliability: The material must be able to withstand normal use and operating conditions over time without failing.
(3) Environmental compliance: The material's production and use must be in line with environmental regulations and standards.
(4) Cost: The cost of the material and its production must be taken into account to determine if it is economically viable for the intended application.

It's important to take a holistic approach when evaluating the suitability of materials for a particular application, considering all these factors and weighing their relative importance.

The priorities for materials and their applications vary based on several factors. For instance, when mass producing certain materials, such as building or packaging materials, cost becomes the primary concern. In this case, a cost-effective production process and straightforward inspection procedures are essential. On the other hand, for materials used in crucial technologies like aerospace and biomedicine, quality, safety, and reliability take precedence, and the inspection cost may even surpass the material cost. For example, superalloys are used in turbine blades and disks in aeroplanes. In case they break while flying, they can lead to aircraft loss and loss of life. Hence, these materials must have a long lifespan and meet strict reliability requirements.

In addition to quality and stability, production cost and price also determine a material's competitiveness. There are many factors that can impact price, but yield is a significant determinant. A high yield can not only increase production automation but also stabilize product quality, further improving yield rate and reducing costs. The only way to boost production is by expanding the material's scope of application.

1.7.4 Development of Advanced Materials and High-Tech Industries

High technology drives the advancement of society, strengthens national defense, and transforms traditional industries and develops key industries. Advanced materials play a crucial role as the driving force and foundation of high technology. Hence, it is vital to prioritize their development.

(1) Increased Focus on Information Materials

In the age of information, functional materials are gaining more attention to enable the creation of miniature, low-power consumption, multifunctional, and intelligent devices. Information functional materials are involved in the generation, acquisition, storage, transmission, conversion, processing, and display of information. They are primarily used in the fields of computers, communication, and control (3C) and are characterized by high demands, rapid development, and diverse varieties.

A computer is the central component of control equipment, and its critical component is the sensor. Therefore, the sensitivity and stability of sensing materials play a crucial role in determining the computer's accuracy. Sensing materials are developed based on the material's response to physical, chemical, and biological changes, making them diverse and rapidly developing. Table 1.3 lists various inorganic, non-metallic sensing materials and the associated response mechanisms.

(2) Advanced Structural Materials: An Ongoing Focus

Structural materials typically require large quantities of resources and energy, leading to significant pollution during processing, creating a major challenge for sustainable development. Improving the performance of structural materials can

Table 1.3 Inorganic nonmetallic sensitive materials

Properties	Principle	Material	Properties	Principle	Material
Oxygen content	Ionic conduction	$Zr_{1-x}CaO_{2-y}$	Voltage	Grain boundary tunnel	$ZnO \cdot Bi_2O_3$
Humidity	Surface ionic conduction	$MgCr_2O_4 \cdot TiO_2$	PTC thermosensitive resistance	Grain boundary phase transformation	$Ba_{1-x}Ce_xTiO_3$
Pressure	Piezoelectricity	$PbZr_{1-x}TiO_3$	Chemical	Surface electronic conduction	$ZrO \cdot CuO$
Acidity	Surface chemical reaction	IrO_{2-x}	Optical	Light resistance	CdS

help mitigate these problems by extending their service life and reducing material consumption. For transportation machinery, a higher operating temperature results in greater engine efficiency. Thus, the research and development of high-temperature materials is a top priority for engine materials. High specific strength and stiffness are crucial for enhancing mechanical properties. Wear resistance, corrosion resistance, fatigue resistance, and aging resistance are essential for prolonging service life. In terms of material types, metallic materials remain the dominant choice in the early twenty-first century. For advanced ceramic materials, toughness needs to be further improved and costs reduced. Organic materials have been developed further, thanks to their renewability, superior performance, and the ability to realize molecular design. There are various types of advanced composite materials, some of which are widely used, such as carbon fiber reinforced plastics (CFRP), while others are still limited by their high cost and difficulties in manufacturing, such as metal matrix composites. Some ultrahigh-temperature materials, such as carbon–carbon composites, require the resolution of oxidation issues before they can be more widely applied.

(3) Emphasis on Energy Materials

The use of energy is a hallmark of human progress and there is an urgent need to exploit green, renewable energy sources to address the fossil fuel crisis and environmental pollution. With its low density and susceptibility to climate change, solar energy received by the Earth's surface is 10,000 times more than the total energy generated by humanity. Hence, priority should be given to the development of materials with high photoelectric conversion efficiency, low cost, and long lifespan. Deuterium, an isotope of hydrogen, is virtually limitless in seawater and is seen as the ultimate energy source for humanity. Several methods, including plasma ignition and laser ignition, have been used to achieve controlled thermonuclear fusion and electricity generation with fusion energy (Fig. 1.20). In this

Fig. 1.20 Schematic illustration of the magnetic confined thermal nuclear fusion reactor (**a**) and the Tokamak reactor at the Institute of Plasma Physics, Chinese Academy of Sciences, Hefei, Anhui, PR China

application, materials with high radiation resistance, high temperature resistance, and resistance to helium embrittlement are crucial to success.

In addition to the development of new energy sources, energy conservation is also crucial, such as through the use of superconductive technology. Metals such as NbTi alloy and Nb_3Sn are the main superconducting materials, but they require cooling by liquid helium (4 K), making the cost too high for widespread adoption. In the 1980s, oxide superconductors were discovered to exhibit superconductivity at liquid nitrogen temperatures and are referred to high-temperature superconductors. However, the wires and blocks produced are still far from practical industrial use. Superconducting power transmission can reduce line losses and superconducting power storage can improve efficiency.

(4) Polymers—A Growing Field

Polymer materials are experiencing significant growth due to their versatility, abundance, and high-performance capabilities. These materials are not just used as structural components but also play an important role as functional materials, spanning a range of applications from semiconductors to superconductors. Additionally, there is ongoing research into various optical polymer materials. Moreover, ferromagnetic polymers are being studied as well. However, it's still necessary to address some of the shortcomings of polymers, such as stability, aging resistance, and flame retardancy.

(5) The Emergence of Biomaterials

Biomedical materials are receiving more and more attention as biotechnology advances and life expectancy increases along with a desire for improved quality of life. Progress is being made in areas such as organ replacement, sustained drug release, and tissue engineering. Additionally, bionics is a growing field and biomimetic materials have the potential to be self-healing and intelligent. The principles of biotechnology used in biomaterials are expected to be applied to industrial production in the future, with the goal of reducing the use of high-temperature, high-pressure, and energy-intensive methods. Catalysis has already made some progress in this regard, and in 2021, Chinese scientists successfully synthesized starch from carbon dioxide. There is also a plan to use biotechnology in the production of food on a factory scale, which could help solve the world food problem and address excess carbon dioxide.

(6) Nanomaterials Processing: An Urgent Issue

As materials reach the nanometer scale, they exhibit unique physical, chemical, and biological properties due to size effect, grain boundary effect, and quantum effect. This has led to the synthesis of nanostructured structural materials with greatly improved strength and reduced brittleness, and nanostructured functional materials that show efficient catalytic effects and changeable optical, electrical, magnetic, and thermal properties for better performance and wider application. As nanotechnology is expected to shape every aspect of industry, economy, and defense in the same way information technology and biotechnology have, it is poised to lead the next industrial revolution. However, despite its advantages in modifying structural and functional materials, nanotechnology, particularly in the areas of nanoelectronics and nanomedicine, is still in its early stages and much progress needs to be made to fully realize its potential.

1.7.5 Scientific Instruments and Testing Devices

The progress of science and technology is greatly dependent on the constant invention and improvement of scientific instruments. As demonstrated by numerous Nobel Prize winners in physics, many have benefited from the creation of new instruments or the utilization of novel testing methods. The quality of industrial products also often depends on the precision of detection devices. For this reason, industrialized nations place great emphasis on the development and production of scientific instruments and testing equipment.

In the field of material analysis, the use of microscopy has greatly improved over the years. Optical microscopes were initially used in metal material research in 1863 and have since been replaced by more advanced techniques such as scanning electron microscopy (SEM), high-resolution electron microscopy (HREM), and scanning transmission electron microscopy (STEM). More recently, scanning probe microscopy, including atomic force microscopy (AFM), has emerged as a powerful tool for observing the behavior of single atoms and molecules in various structures. These advanced analysis devices, in combination with high-performance computers, have allowed for a more direct and rapid characterization of material structures.

Similarly, the field of crystal structure analysis has advanced significantly since the discovery of X-ray diffraction patterns by Laue et al. in 1912. The technique has since been expanded to include electron and neutron diffraction, becoming a crucial tool for understanding the atomic structure of materials. Spectrometers, based on various principles, have also emerged as effective devices for characterizing material structures and properties.

Materials scientists must not only make use of existing equipment but also strive to invent new inspection and testing devices. This is because commercialized equipment is often at least 5 years behind and breakthroughs are difficult to achieve with existing tools alone.

In conclusion, the significance of materials will never diminish. As science and technology continue to develop, the relationship between materials and devices will become increasingly close. This will lead to the emergence of new, creative materials and artificial synthetic materials. However, predicting the emergence of new materials can often be difficult, as evidenced by the discovery of high-temperature superconductors and C_{60}.

1.8 The Structure of This Textbook

The first chapter of this book, "Introduction to Materials Science" by ZM Sun, stands out compared to other introductory materials science textbooks. While other texts tend to provide a brief overview, and many Chinese textbooks offer little to no introduction at all, this chapter provides extensive coverage. The aim of this extensive coverage is to give students a comprehensive understanding of what materials are, the importance of studying materials, the different classifications of materials, the concept and basic components of materials science, and the future of materials science and engineering.

Building upon this foundation, the following three chapters delve into the structures of materials. Chapter 2, written by W Liu, focuses on electron configurations in atoms, interatomic interactions, and energy bands in solid materials. Chapter 3, written by PG Zhang and ZM Sun, introduces crystallography, with a focus on the spatial arrangement of atoms in crystals. And Chap. 4, written by PG Zhang, L Pan and ZM Sun, examines the structure of solid materials at the atomic scale, highlighting the typical crystal structures of metals, ceramics, and polymers.

The pursuit of perfection is a common goal among many; however, it is a concept that is often unattainable. The same is true for materials, which inevitably coexist with defects. Chapter 5, written by WW Sun, explores the concept of crystal defects - imperfections in the regular geometric arrangement of atoms in a crystalline solid. This includes point defects, line defects, and planar defects. Understanding these defects is crucial to learning many aspects of materials science, such as diffusion, deformation, strengthening, surfaces and interfaces, and phase transformations.

The movement of atoms or molecules within a material is a crucial aspect of their physical and/or chemical behavior, with diffusion being the fundamental method of movement. Chapter 6, written by QS Ruan, explores the concept of diffusion, which is the directional migration of substances on a macroscopic scale. At temperatures above 0 Kelvin (absolute zero), particles in any system are in thermally induced motion, with diffusion being the most important method of substance transfer in solids. The essence of diffusion is the random movement of particles.

An interface, whether it be a grain boundary or phase boundary, plays a significant role in the performance of materials and is discussed in Chap. 7, written by W Zhang. The chapter covers the basics of interface thermodynamics, bonding and adhesion, functionalization, and analysis.

In response to changes in the environment, primarily temperature, a material may change from one phase to another. A phase diagram is a critical tool for materials scientists and engineers, and is covered in detail in Chap. 8, written by WB Tian and ZM Wang. This chapter is taught for 18 h, the longest of all chapters in the book, as it is deemed essential for understanding materials science. The process of phase transformations is further explored in Chaps. 9 and 10, written by W He, covering solidification and crystallization as well as solid-state phase transformations.

The properties of materials are divided into two chapters. Chapter 11, written by W Liu, covers the physical properties of materials, including electrical, optical, thermal, and magnetic properties. The mechanical properties, including the deformation and stress–strain behavior of solid materials, are described in Chap. 12, written by WB Tian, LP Guo, and L Pan.

Computational materials science, which involves multiple disciplines such as materials, chemistry, physics, and computing, is explored in Chap. 13, written by W Liu. This field acts as a bridge between theory and experiment in materials science, potentially transforming traditional, time-consuming, labor-intensive materials research into a more efficient and demand-driven process. It enables the translation of experimental results into quantitative theory, the prediction of material properties directly from theoretical models, and an improvement in research efficiency.

The final chapter, Chap. 14 written by ZM Sun, QS Ruan, W He, L Pan, W Liu, WB Tian, and W Zhang, focuses on some recent Nobel Prizes related to materials science. By exploring the stories behind these groundbreaking discoveries, students will gain an understanding of how advancements in materials science have transformed industries such as electronics, energy storage, and communication. The chapter highlights the significance of semiconductors in the electronics industry, superconductors as a potential game-changer for human beings, the potential applications of fullerene, the correlation between computational methods and quantum mechanics, the impact of optical fiber on communication, the importance of graphene as a star material, the challenges and triumphs of blue light-emitting diodes, the changing landscape of the automobile and energy storage industries due to lithium-ion batteries, and the remarkable properties of conductive polymers. These stories and the underlying science are intended to inspire and motivate students to continue exploring the uncharted territories in the field of materials science.

1.9 Questions

Q1.1 What can you learn from the Simuwu tetrapod (Fig. 1.9). (Aspects that may be considered, but not limited to, include history, culture, art, prosperity of a dynasty, power, patriotism.)

Q1.2 Select one or more of the following modern items or devices and conduct an Internet search to determine what specific material(s) is (are) used and what specific properties this (these) material(s) possess(es) in order for the device or item to function properly. Finally, write a short essay in which you report your findings.

(1) Cell phone/digital camera batteries
(2) Cell phone displays
(3) Solar cells
(4) Wind turbine blades
(5) Fuel cells
(6) Automobile engine blocks (other than cast iron)
(7) Automobile bodies (other than steel alloys)
(8) Space telescope mirrors
(9) Military body armor
(10) Sports equipment
(11) Soccer balls
(12) Basketballs
(13) Lightweight bicycle frames

Q1.3 Classify each of the following materials as to whether it is a metal, ceramic, or polymer. Justify each choice: (a) brass; (b) magnesium oxide (MgO); (c) Plexiglass; (d) polychloroprene; (e) boron carbide (B_4C); and (f) cast iron.

Q1.4 What is your understanding of materials science and engineering and why do you think we should study fundamentals of materials science.

Q1.5 Name a materials technology, in our present society or in the future (fictious), which you think is the most important to the development of human society.

Bibliography

Callister WD, Rethwisch DG (2015) Fundamentals of materials science and engineering: an integrated approach. John Wiley & Sons, New York

Feng D, Shi CX, Liu ZG (2004) Introduction to materials science. Chemical Industry Press, Beijing

Wang DM, Zhao R, Xiao B et al (2020) Review and outlook of the world and China's petrochemical industry in 2019. Int Petrol Econ 28(5):41–47

Electronic Structure 2

Learning Objectives

1. Describe the evolution of the atomic structure models and note the differences.
2. Write the electron configuration of a given atom.
3. Draw the schematic diagram of all subshells.
4. Plot the attractive, repulsive, and net energy versus interatomic distance between two atoms.
5. Describe the primary interatomic bonding and secondary bonding.
6. Describe the definition of energy band, Brillouin Zone, carrier, Fermi level, and mobility.

The properties of a material depend primarily on its electronic structure. The electron states in a solid material depend heavily on its electron configurations and inter-atomic interactions. This chapter, as the basic preparation of academic discussion on materials, first focuses on the electron configurations in atoms, including some functional concepts, atomic structure models, and periodic functions with the increasing atomic number; then, analyze the inter-atomic interactions, including the bonds between atoms, the characteristics of the electron cloud after bonding, classification of bonding, and the energy band in solid materials.

2.1 Atomic Structure

The discovery of subatomic particles is the prerequisite to discover and understand the properties of materials. In this section the electronic structure of isolated atoms, that is, atoms in the gas phase, will be discussed.

Z. Sun et al., *Fundamentals of Materials Science*,
https://doi.org/10.1007/978-981-96-3904-5_2

2.1.1 Atomic Composition

An atom is composed of the nucleus and the outer region of the atom. The nucleus consists of protons and neutrons. The outer region of the atom holds its electrons in certain orbitals around the nucleus. A proton and an electron have equivalent charge of 1.60×10^{-19} C, but in opposite sign, that is, they are respectively charged positively and negatively. The volume of nucleus in an atom is negligibly small, yet it contains most of the mass. The mass of proton and neutron is almost identical, 1.67×10^{-24} kg, while the mass of the electrons is much smaller than that of the protons and neutrons, being 9.11×10^{-31} kg. In addition, the mass of a proton or a neutron is approximately equal to one atomic mass unit (amu), defined as 1/12 of the mass of carbon (^{12}C).

The number of protons (Z) characterizes the elements. For a free atom in gas or an intact atom, the number of the protons is equal to that of electrons, so that the net charge is zero. When the net charge is non-zero, the atoms that gain (lose) electrons are called anion (cation). Although the protons are fixed for a given element, the neutrons number (N) could vary, such as hydrogen (0 neutrons), deuterium (1 neutron), tritium (2 neutrons). Since a proton and a neutron have the same mass, whereas an electron has a mass of 6 to 7 orders of magnitude smaller than a proton, the relative atomic mass A of an element:

$$A \cong Z + N$$

Therefore, the mass of an atom or a molecule of a substance can be expressed in terms of the relative atomic mass or the molar mass. A mole of matter contains 6.0221×10^{23} (Avogadro constant) atoms or molecules. These two atomic scales have the following relationship:

$$1\,\text{amu/atom(or molecule)} = 1\,\text{g/mol}$$

For example, carbon has an atomic mass of 12 amu/atom, or 12 g/mol.

Isotopes are atoms that have the same atomic number but have different neutron numbers. The aforementioned hydrogen, deuterium and tritium are typical isotopes of the No.1 atom in the periodic table. The application of isotopes is widespread in isotope geochemistry, cosmochemistry, paleoclimatology, and so on. For example, isotope analysis is the determination of isotope signature, the relative abundances of isotopes of a given element (such as ^{14}C) in a particular sample. Another example is isotopic substitution, which can be used to track the chemical reaction via observing the movement of isotopic atoms in the kinetic process.

2.1.2 Atomic Structure Model

Here, we review five benchmarks of atomic model evolution (see Fig. 2.1). In 1803, Dalton, a British scholar, proposed the solid sphere model, depicting an atom

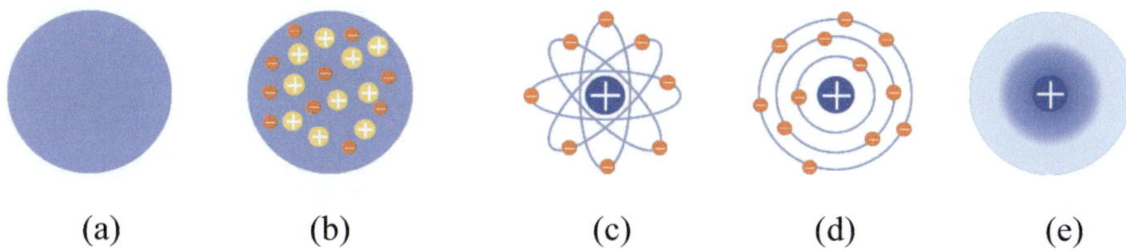

Fig. 2.1 Benchmarks of atomic structure model evolution. **a** Solid sphere model proposed by Dalton; **b** Plum-pudding model by Thomson; **c** Nuclear model by Rutherford; **d** Planetary model by Bohr; **e** Wave-Mechanical model by Schrödinger

with a solid sphere where every part of the atom is equal. In 1904, Thomson came up with the "plum-pudding" model, in which atoms were shaped like spheres filled with positive charges, with electrons embedded in them like raisins. The concept of the nucleus was first proposed by Rutherford in 1911. Followed the nuclear model, Bohr proposed that atoms were like planetary systems (also called Bohr atomic model), with each electron having its own fixed orbit. In the quantum age, Schrödinger proposed the wave-mechanical model and induced the electron cloud, which holds that the electrons are dispersed around the nucleus.

2.1.3 Energy Level and Electron Configuration

① Energy Level

With the development of quantum mechanics, the path of the electron around the nucleus is found to be uncertain, but the orbital of the electron is not arbitrary. Schrödinger equation successfully solved the problem of describing electron movement by introducing the concept of wave function to replace the fixed orbit in classical physics. The wave function can effectively describe the probability of an electron's presence at various locations around the nucleus, from which the orbital of the electrons around the nucleus could be obtained. The orbital is determined by four quantum numbers: principal quantum number (n), secondary quantum number (l), magnetic quantum number (m), and spin quantum moment (m_s).

Among the four quantum numbers, the most important is the principal quantum number n ($n = 1, 2, 3, \cdots$), which is the main parameter for determining the distance and energy of an electron from the nucleus. In the first shell, the principal quantum number of the electrons is $n = 1$, while the principal quantum number of the electrons in the second and third shells is $n = 2$ and 3, respectively. As n increases, the energy of the electron increases successively (see Fig. 2.2).

The electrons in the same shell can be divided into several subshells with different energy levels according to the secondary quantum number l: $l = 1, 2, 3, \cdots$. These subshells are also commonly denoted by lowercase letters: s, p, d, and f. Quantum orbitals are not necessarily spherical, and secondary quantum numbers

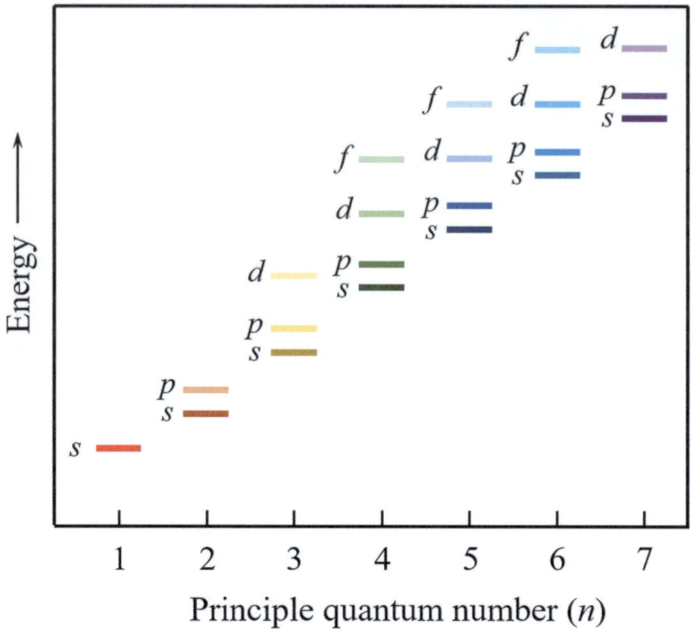

Fig. 2.2 Schematic diagram of the relative energies of the electrons for the shell and subshell

reflect the shape of orbitals. Each orbital, represented by the four secondary quantum numbers *s*, *p*, *d* and *f*, has a different angle distribution around the nucleus, so it is also called angular quantum number or orbital quantum number (see Fig. 2.3). Energy levels of orbitals vary with the secondary quantum numbers (see Fig. 2.2). Orbitals with the same n but different l have different energy levels, and their energy levels rise in the order *s*, *p*, *d*, and *f*. The number of subshells on each shell increases with principal quantum number. The first shell has only one subshell *s*; the second shell has two subshells *s* and *p*; the third shell consists of *s*, *p* and *d* subshells; and the fourth shell consists of *s*, *p*, *d* and *f* subshells.

To analyze the shape and distribution of electron cloud in the various states, the wave function of the state is required, which can be expressed by:

$$u_{nlm} = R_{\boldsymbol{nl}}(r) Y_{\boldsymbol{lm}}(\varphi, \theta) \tag{2.1}$$

where $R_{nl}(r)$ and $Y_{lm}(\varphi, \theta)$ are the radial distribution function and angular distribution function, respectively. $|R_{nl}(r)|^2$ represents the probability of finding an electron with (n, l, m) state at r, called radial probability factor. $|Y_{lm}(\varphi, \theta)|^2$ represents the probability of finding an electron with (n, l, m) state at (φ, θ).

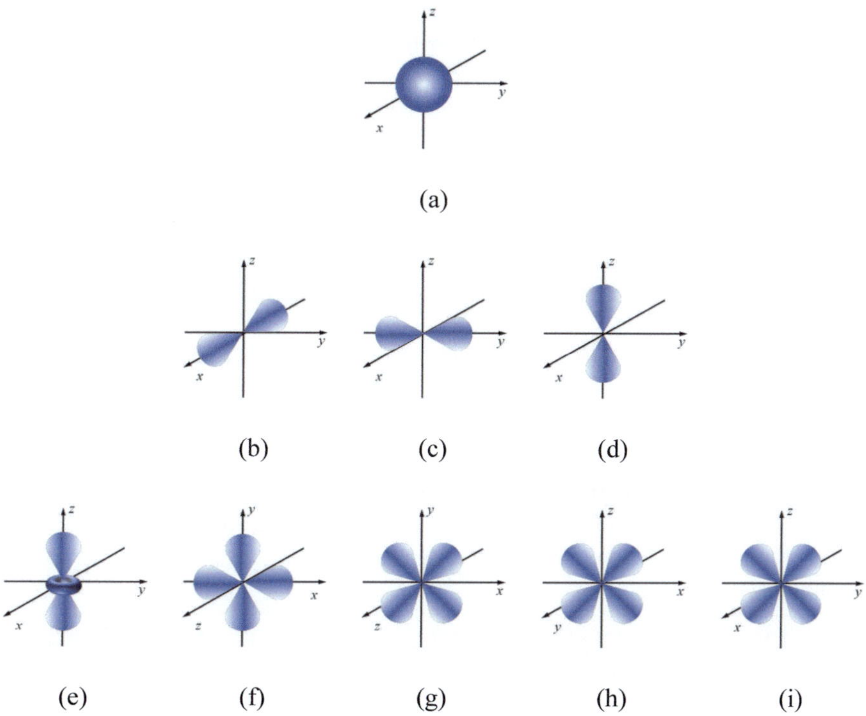

Fig. 2.3 Schematic diagram of the subshells including **a** s state, **b–d** three p states, and **e–i** five d states

The magnetic quantum number is expressed by m_l, $m_l = 0, \pm 1, \pm 2, \pm 3, \cdots$. An s subshell corresponds to a single energy state, whereas p, d, f subshells have 3, 5, 7 energy states, respectively. The magnetic quantum numbers in s, p, and d subshells and their corresponding shapes are shown in Fig. 2.3. Electrons in the same subshell with two spin orientations (up or down) can be classified by spin quantum moment ($m_s = +\frac{1}{2}, -\frac{1}{2}$). Two spin orientations have the same energy in the absence of an external magnetic field. But under an external magnetic field, the energy of these orbitals with different spatial orientations will be different.

② Electron Configurations

The electron states, i.e., their energy levels, in an atom have been described with four quantum numbers, in the above section. However, how these states are filled with electrons needs to be further determined by various rules, as follows.

Pauli exclusion principle. Each quantum state of an electron in an atom can contain at most two electrons, and the two electrons should be in opposite spin orientations. So, the s, p, d, and f orbitals can hold 2, 6, 10, and 14 electrons, respectively (see Table 2.1).

Table 2.1 Maximum number of electrons held in some of the electron shells and subshells

Principal quantum number	Subshell	Number of states	Per subshell	Per shell
1	*s*	1	2	2
2	*s*	1	2	8
	p	3	6	
3	*s*	1	2	18
	p	3	6	
	d	5	10	
4	*s*	1	2	32
	p	3	6	
	d	5	10	
	f	7	14	

Lowest energy states principle. Not all possible orbitals in an atom are occupied by electrons. For most atoms, electrons tend to fill the lowest energy orbitals, leaving the system in the lowest energy state, which is called lowest energy principle.

Hund's rules. Before any orbital is doubly occupied, every orbital in a subshell is singly occupied, and all the electrons in singly occupied orbitals have the same spin.

Following these three rules, an atom is said to be in the ground state when all the electrons are in the lowest energy orbital. However, it is also possible for the electrons to be in a high energy state, which is called the excited state. The configuration or structure of an electron in an atom is a representation of the occupied states of the atomic orbitals. In general, the electron configuration of an atom is expressed by the number of electrons in each subshell. For example, the electron configuration of H atom is $1s^1$, and that of C atom is $1s^22s^22p^2$. Table 2.2 lists the electron configurations of the atoms of elements 1–18.

2.1.4 Periodic Table of Elements

A periodic table of elements classifies all the elements according to their electron configurations (see Fig. 2.4). The elements are situated with increasing atomic number, in seven horizontal rows called periods. The elements with similar valent electrons are arranged in a group of column. In this way, the electron configurations between two adjacent atoms, either in same row or in the same column, bear some similarities. As a result, the chemical and physical properties between these two atoms also show some similarities. It is thus also implied that the properties of the elements in the same row or in the same column vary gradually from one element to the next.

Table 2.2 A list of electron configurations for the first 18 elements

Elements and atomic number	Principal quantum number and secondary quantum number					
$n =$	1	2		3		
$l =$	0	0	1	0	1	2
1 H	1					
2 He	2					
3 Li	2	1				
4 Be	2	2				
5 B	2	2	1			
6 C	2	2	2			
7 N	2	2	3			
8 O	2	2	4			
9 F	2	2	5			
10 Ne	2	2	6			
11 Na	2	2	6	1		
12 Mg	2	2	6	2		
13 Al	2	2	6	2	1	
14 Si	2	2	6	2	2	
15 P	2	2	6	2	3	
16 S	2	2	6	2	4	
17 Cl	2	2	6	2	5	
18 Ar	2	2	6	2	6	

The elements positioned in first and second column, group IA and IIA, are alkali metal, and alkaline earth metals, which have one or two more electrons in the outer subshell than the corresponding stable electron configurations. The group IIIB through IIB are called transition metals, which have partially filled *nd* electrons and in some cases one or two $(n + 1)s$ electrons. Groups IIIA through VIA are intermediate between the metal and nonmetals, which are the important component of semiconductors. Group VIIA, the last group, called halogens, have one less electron than in the stable electron configurations.

In ancient times, both Easterners and Westerners believed that the matter consisted of a few basic "elements". In ancient China, these "elements" refer to gold (金), wood (木), water (水), fire (火), and earth (土), while these "elements" are earth, gas, water, fire in ancient Greek. In the eighteenth century, Antoine Lavoisier discovered that water can be made by hydrogen (H) and oxygen (O). In the following years, more than 60 elements were discovered. But many questions remain unanswered thereafter: How many elements are there in the world? What is the relationship between the elements? How do we find new elements? In 1869, Mendeleev had a "magic" dream: when the 63 discovered elements are

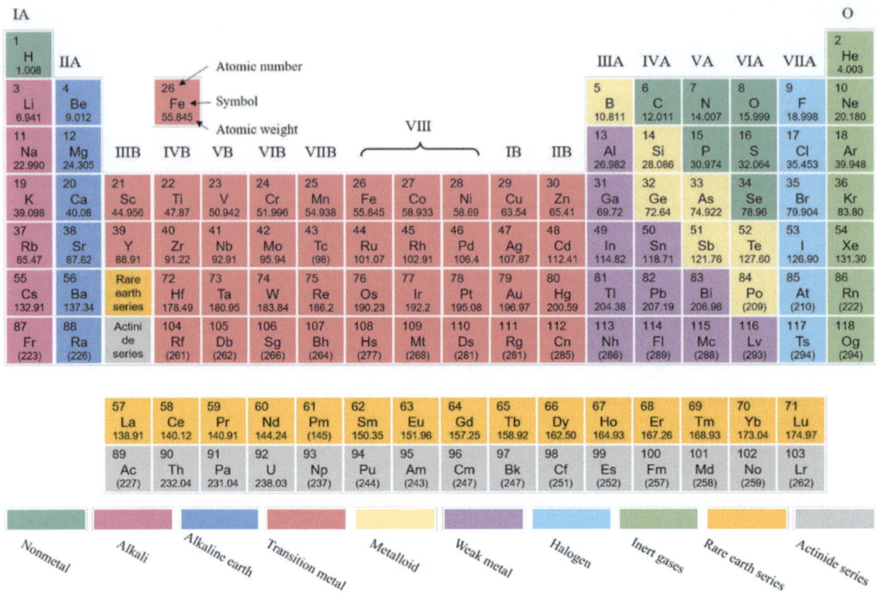

Fig. 2.4 Periodic table of the elements. The numbers in parentheses are the atomic weights of the most stable or common isotopes

placed in a corresponding grid to form a table, they are arranged as the increasing atomic mass and appeared to change regularly as the mass increased. Following the periodic table, more and more elements were discovered.

2.2 Interatomic Bonding

Understanding interatomic bonding is the basis of the analysis and use of materials, and the bonding determines the shape of materials as well as their physical and chemical properties. This section starts with the understanding of the interaction between two isolated atoms as they get closer, and then deals with the bonding strength and angle.

2.2.1 Bonding Orbitals and Antibonding Orbitals

When two hydrogen atoms move close enough, their electron clouds overlap to form molecular orbitals. How do electron clouds form? Do they enhance each other or weaken each other? It depends on the state of the two valence electrons. Figure 2.5 shows that if two valence electrons are in a symmetric state, then the two atoms will attract each other, and their electron clouds will strengthen each other. The resultant electron cloud is called a bonding orbital because it attracts two nuclei to form a stable hydrogen molecule. If two valence electrons are in an antisymmetric state, they repel each other, and their electron clouds weaken each other. The resultant electron cloud is called an antibonding orbital. According to this principle of enhancement and weakening, it is not difficult to obtain the synthetic results of various electron clouds. As an example, we will discuss the synthesis results of two *s*-state electron clouds and two *p*-state electron clouds as shown in Figs. 2.6, 2.7, and 2.8, respectively. Each atomic orbital of an isolated atom is split into two molecular orbitals, one lower in energy than the original energy level and one higher in energy. The high energy level is called antibonding orbitals and the lower energy level is called bonding orbitals. As a result of the spherical nature of *s* state, there is only one case where two *s*-states approach (see Fig. 2.6). Figure 2.6b show the bonding orbitals formed when two *s*-state electron clouds partially overlap and strengthen each other, while Fig. 2.6c show the antibonding orbitals formed when two *s*-state electron clouds partially overlap and weaken each other. For the *p*-states, there are two approaching ways, along the molecular axis (p_z) and perpendicular to the molecular axis (p_x and p_y). For the p_z-state, the bonding and antibonding orbitals are shown in Fig. 2.7b and c. Conversely, the p_x and p_y-states form molecular orbitals with their shoulders, and the formed bonding orbitals and antibonding orbitals are shown in Fig. 2.8.

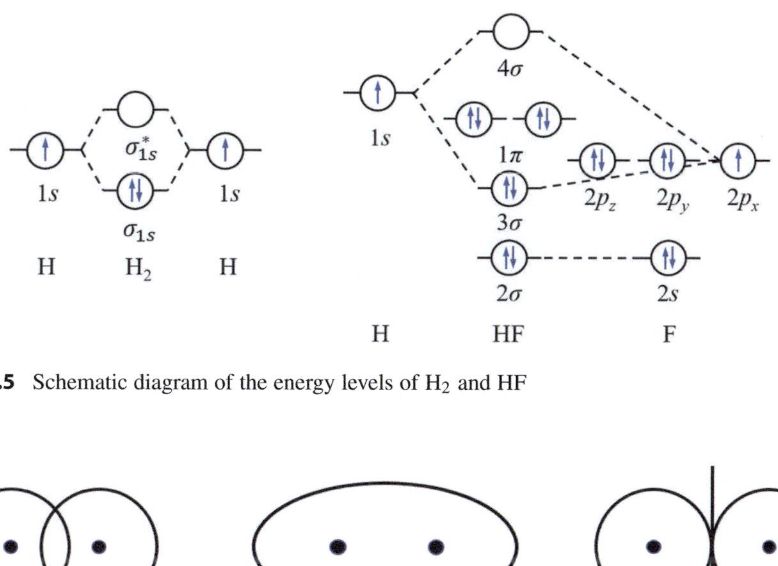

Fig. 2.5 Schematic diagram of the energy levels of H_2 and HF

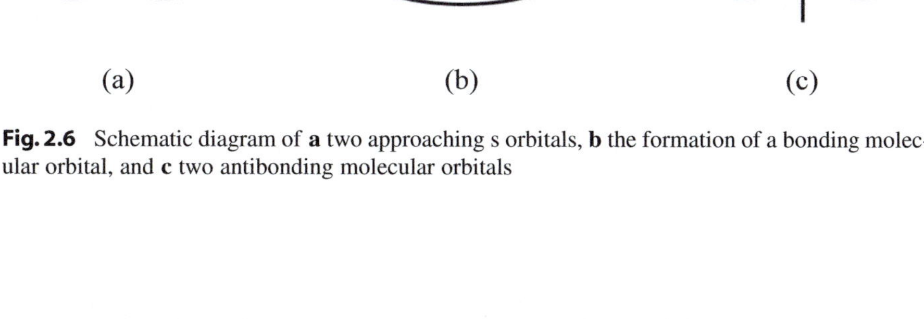

Fig. 2.6 Schematic diagram of **a** two approaching s orbitals, **b** the formation of a bonding molecular orbital, and **c** two antibonding molecular orbitals

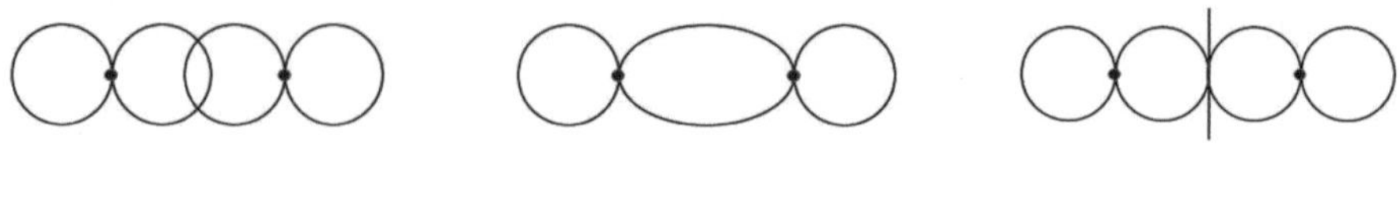

Fig. 2.7 Schematic diagram of **a** two approaching p orbitals along with molecular axis, **b** the formation of bonding molecular orbitals, and **c** antibonding molecular orbitals

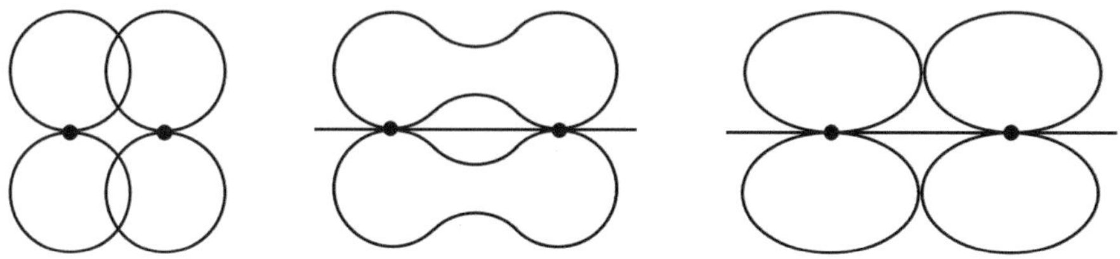

Fig. 2.8 Schematic diagram of **a** two approaching p orbitals perpendicular to the molecular axis, **b** the formation of bonding molecular orbitals, and **c** antibonding molecular orbital

These diagrams indicate that some molecular orbitals have the region of maximum electron cloud density located on the molecular axis, and such orbitals are called σ orbitals (where the region of maximum density is between the two nuclei) or σ^* orbitals (where the region of maximum density is on the outside of the two nuclei). There are also molecular orbitals where the maximum electron cloud density region is not on the molecular axis but is distributed as in Fig. 2.8b. Such atomic orbitals are called π orbitals (where the maximum density region is on the inner side of the two nuclei) or π^* orbitals (where the maximum density region is on the outer side of the two nuclei). Normally, the π bond between two atoms is called small π bond, while the large π bond contains multiple atoms. In the large π bond, there are multiple parallel p orbitals, which can overlap each other to form a polycentric π bond. In this situation, the p electrons are no longer confined to two atoms, but are moving among multiple atoms. Both σ and π orbitals are bonding orbitals, and both σ^* and π^* orbitals are antibonding orbitals. They are characterized by a plane perpendicular to the molecular axis between the two nuclei, and the density of the electron cloud is zero at all points on this plane.

The molecular orbital is also an important concept to understand the state of an atom. HOMO (Highest Occupied Molecular Orbital) and LUMO (Lowest Unoccupied Molecular Orbital) are the highest energy of the bonding orbitals and lowest energy of antibonding orbitals, respectively. The energy difference between the HOMO and LUMO is termed the HOMO–LUMO gap, which can be used to predict the strength and stability of transition metal complexes.

2.2.2 Orbital Hybridization

Bonding electrons are formed by pairing two electrons with opposite spins. If the valence electrons in the outer layers of the atom are already paired, they can no longer participate in bonding. For example, the electronic structure of an isolated Be atom in the ground state is $1s^22s^2$. Since the spins of all the s-electrons are already paired, it seem difficult form a typical covalent bond in a normal way. However, it is a fact that Be in the vapor state can form a Be_2 molecule, although it is not as stable as a Li (much less than H_2) molecule. The explanation for this phenomenon is that when two Be atoms combine to form a molecule, first one of the $2s$ electrons of each atom is excited to the $2p$ state (p_x, p_y or p_z), forming a $1s^22s^12p^1$ structure. The unpaired $2s^1$ and $2p^1$ orbitals are then hybridized to form a new *s-p* hybrid atomic orbital. Finally, two *s-p* hybrid atomic orbitals superimpose to form a hybrid molecular orbital.

Such energy change in this process shows that it takes energy to excite the paired electrons from the 2 *s* state to the $2p$ state, but energy is released when the hybrid molecular orbital is formed from the hybrid atomic orbital. For the formation of the hybrid orbital, the energy released must be greater than the energy consumed. This requires the excitation energy not to be too large, so that the common hybrid orbitals are often hybrids of *s* and *p* states of the same shell, or hybrids of *ns*, *np* and $(n-1)d$ states.

There are three common ways in which *s*-state electrons hybridize with *p*-state electrons, namely *sp*, sp^2 and sp^3 hybridization. The structures of these hybridization orbitals are shown in Fig. 2.9. Here we take carbon element as an example. The first example is diamond, in which each carbon atom is connected to four surrounding carbon atoms and its molecular orbital is formed by the sp^3 hybridized atomic orbital of the central carbon atom and the sp^3 hybridized orbital of the four surrounding carbon atoms, as shown in Fig. 2.10a. The second example is graphene, which begins with the $2s$ electrons of each carbon atom being excited to form a $1s^22s^12p_x^12p_y^12p_z^1$ structure, where the $2s^1$, $2p_x^1$, $2p_y^1$ atomic orbitals hybridize to form sp^2 hybridized atomic orbitals, as shown in Fig. 2.10b. When adjacent carbons are approaching, the hybridized orbital of each carbon atom overlaps with that of the surrounding carbon atoms to form a C–C σ bond. However, each carbon atom also has a $2p_z$ orbital that is not involved in hybridization and the $2p_z$ orbital is perpendicular to the atomic plane of the graphene, so the p_z orbitals of all the carbon atoms cross over laterally to form large π bonds, like benzene. The third example is graphyne, a carbon isotope containing both sp^2 hybridization and *sp* hybridization discovered by the team of Academician Li Yuliang of the Chinese Academy of Sciences in 2010, as shown in Fig. 2.10c. Graphyne is produced by a coupling reaction catalyzed by hexynyl-benzene on Cu metal surfaces where the benzene ring carbon is sp^2 hybridized. The alkynyl carbons are *sp* hybridized. First the $2s$ electrons of the alkyne-based carbon are excited to form $1s^22s^12p_x^12p_y^12p_z^1$, and the *s*- and p_x-orbitals along the molecular axis overlap with the *s*- or p_x-orbitals of the immediately adjacent carbon atom, respectively, to form

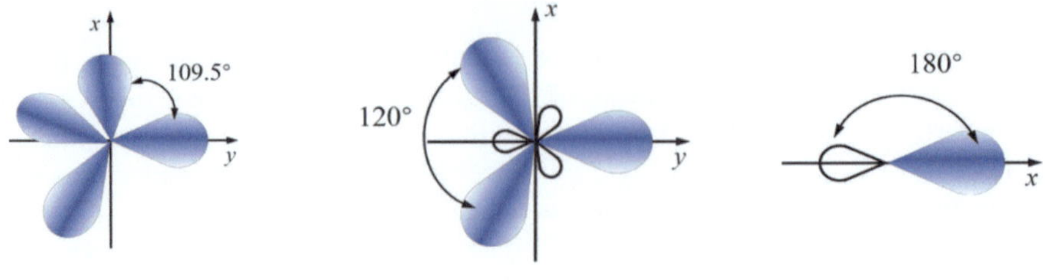

Fig. 2.9 Structure of electron cloud of hybridization orbitals with **a** sp^3, **b** sp^2, and **c** sp configurations

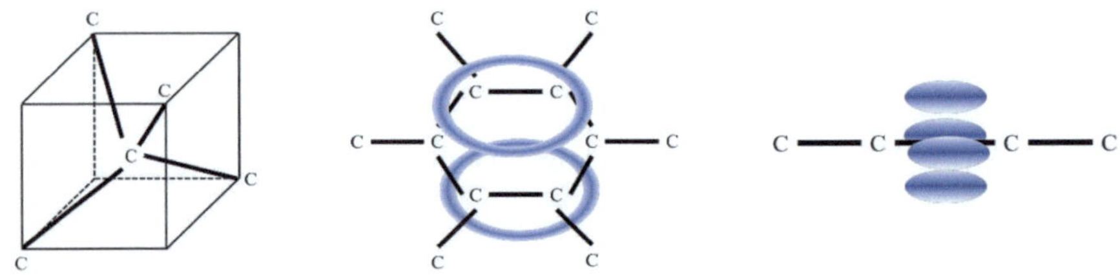

Fig. 2.10 Bonding and structure of electron cloud of **a** diamond, **b** part of graphene; **c** C≡C in graphyne

a C–C σ bond. The p_y and p_z orbitals that are not involved in the hybridization overlap, forming two π orbitals.

2.2.3 Bond Strength and Bond Angle

We understand that covalent bonding has a distinctive feature, being directional. For example, the four covalent bonds (molecular orbitals) of diamond or methane are 109.5° to each other in regular tetrahedron form, as shown in Fig. 2.10. The σ bonds of graphene are located within the graphene face at 120° to each other, forming the planar triangle where three atoms covalently bonded to each carbon atom. In contrast, the *sp*-hybridized C atom in graphyne is attached to two other atoms and is 180° to each other. The directionality of the covalent bonds arises from the asymmetric distribution of the electron clouds of the *p*, *d* and *f* states. For example, since the three *p*-state electron clouds (p_x, p_y, p_z) are orthogonal to each other, the molecular orbitals formed by the superposition of the *p*-state atomic orbitals should also be oriented in three mutually orthogonal directions. However, if the *p* and *s* states hybridize, the four hybridized orbitals cannot be orthogonal

Table 2.3 Hybrid bonds and their directionality with various number of bonds

Number of bonds	Involved orbitals	Directionality of bonds	Angle
2	*s*, *p*	Line	180°
3	*s*, *p*, *p*	Plane triangle	120°
	p, *p*, *p*	Triangular pyramid	109.5°
4	*s*, *p*, *p*, *p*	Tetrahedron	109.5°
	s, *p*, *p*, *d*	Square planar	90°, 180°
5	*s*, *d*, *d*, *d*, *d*	Rectangular pyramid	90°
6	*s*, *p*, *p*, *p*, *d*, *d*	Octahedron	90°, 180°
8	*s*, *p*, *p*, *p*, *d*, *d*, *d*, *f*	Square antiprismatic	/

to each other, yet they are still directional bonds and not spherically symmetric bonds. Usually, we refer to the angle between the bonds as the bond angle and the distance between the atoms connected by a covalent bond as the bond length. The common hybrid orbitals and their orientations are listed in Table 2.3.

By definition, the bond energy for a diatomic molecule, D_{X-Y}, is determined by:

$$XY_{(g)} \rightarrow X_{(g)} + Y_{(g)} \quad D_{X-Y} = \Delta H^{\circ} \tag{2.2}$$

For the same two atoms, the bond energy has the order: triple bonds > double bonds > single bonds. The bond length decreases with increasing bond energy.

2.3 Classification of Bonding

Particles (atoms, ions, or molecules) in space are arranged according to certain rules and combined with each other by bonding. The essence of bonding is the binding force between particles, which is the resultant force of attraction and repulsion between particles. When the particle interaction is in equilibrium position, the resultant force is zero, and the corresponding energy is at the lowest point of the potential curve, as shown in Fig. 2.11.

Attraction is derived from the Coulombic attraction between opposite charges, which is a long-range force. The repulsive force consists of two parts, the Coulombic repulsive force between the like charges and the Pauli repulsive force between the electrons. According to the Pauli exclusion principle, when two atoms come close to each other, their electron clouds overlap, which increases the kinetic energy of some electrons, leading to a rise in total energy. To reduce the total energy of the system, the electrons should occupy more space, resulting in a repulsive force between the electrons, which is a short-range force.

The binding energy of atoms is usually defined as the interaction energy of atoms at equilibrium positions. The magnitude of the binding energy corresponds to the work required to separate the two atoms completely. The greater the binding

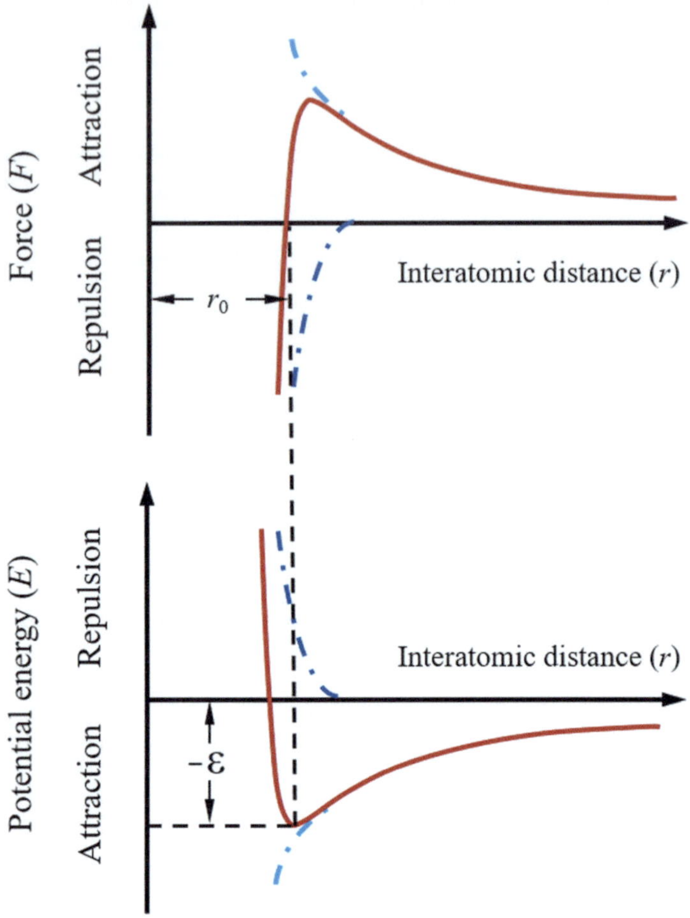

Fig. 2.11 Schematic diagrams of (top) attractive, repulsive, and net forces and (down) attractive, repulsive, and net potential energies as a function of interatomic distance. The light blue, blue, and red curves indicate the attractive, repulsive, and net forces or net potential energies, respectively

energy, the more stable the binding bond between the atoms. According to the different bonding forms between the atoms, the bonds are divided into covalent, ionic, metallic bonds, and weak bonds including van der Waals force and hydrogen bond.

2.3.1 Primary Interatomic Bonds

Chemical bonds are the forces that hold atoms together in molecules or crystals. The three primary types of chemical bonds are covalent bonds, ionic bonds, and metallic bonds. Covalent bonds involve the sharing of electrons between atoms, while ionic bonds arise from the transfer of electrons between atoms with opposite charges. Metallic bonds occur in metals and involve a delocalized sharing of electrons among all the atoms in the material.

① Covalent Bond

When two atoms have the similar ability to gain or lose electrons, the atoms might gain electrons to become negative ions or lose electrons to become positive ions. These atoms are bonded with each other by sharing electron pairs to form molecules or crystals, and the shared valence electrons belong to both two adjacent atoms, allowing each atom to achieve a stable, filled outmost shell. This type of bonding in the form of shared electrons pairs is called covalent bonding, as shown in Fig. 2.12. The Lewis structure is proposed to show the bonding between atoms of a molecule and the lone pairs of the electrons in the molecules, as shown in Fig. 2.13.

Covalent bonds are divided into polar covalent bonds and non-polar covalent bonds. A polar covalent bond is a covalent bond in which electrons are unevenly distributed between two atoms with different electronegativities, resulting in one

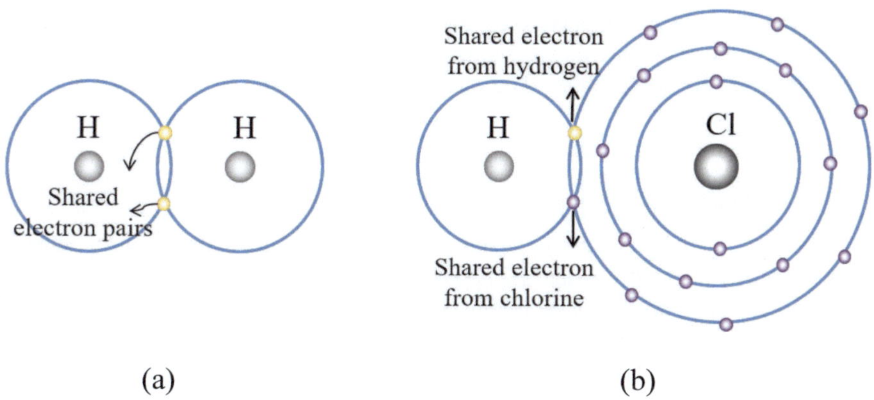

Fig. 2.12 Schematic representation of **a** non-polar covalent bonding in H_2 and **b** polar covalent bonding in HCl

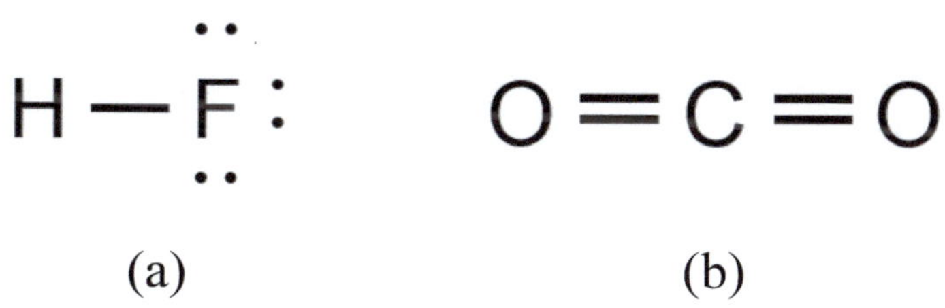

Fig. 2.13 Lewis structure of **a** HF, and **b** CO_2

end of the bond having a partial positive charge and the other end having a partial negative charge, creating a charge separation, and a non-polar covalent bond is a covalent bond between two atoms with the same or very similar electronegativities. In this type of bond, electrons are evenly distributed between the two atoms, resulting in no charge separation, meaning that no partial positive or partial negative charges are generated between the two atoms. The bond between two hydrogen atoms in a hydrogen molecule is the most typical non-polar covalent bond, and the crystals formed by elements of group IV such as carbon, silicon and germanium are all non-polar covalent crystals. Polar covalent bonds can exist between H atom and the atoms with high electronegativity such as F, Cl, N, and O.

According to the theory of covalent bond, the more electron clouds overlap, the stronger the bond energy and the more stable the covalent bond will be. Therefore, when atoms form covalent bonds, they would form in the direction where the electron clouds overlap the most, which means that covalent bonds are directional. For a particular atom, the number of its covalent bonds is certain, which is determined by the number of valence electrons, thus covalent bonds are saturated.

Due to the directivity of covalent bond, it is difficult to change the relative positions of atoms, which leads to a poor plasticity for covalent bond materials. The strong covalent bond (second to ionic bond) makes the covalent bond crystals possess a stable structure and the properties of high melting and boiling point, high hardness, and brittleness. The shared electron pairs are restrained between the adjacent atoms, and cannot move freely in the crystals, thus these materials usually possess poor conductivity.

② Ionic Bond

Ionic bonding usually occurs between active metallic elements (such as metallic elements of main group I, II and III and transition metal elements with low valence state) and active nonmetallic elements (such as group VI and VII) The essence of ionic bonding is the Coulombic attractive force between positive and negative ions. Metal atoms are easy to lose their outer electrons, while nonmetal atoms are easy to gain electrons. Thus, the metal atom would easily give up their outer electrons to the non-metallic atoms, forming a stable filled shells on their outmost shells. In this situation, the metal and non-metal atoms become positive ions and negative ions, respectively. Consequently, the positive and negative ions are bonded to each other through the electrostatic attraction between them. Most of salts, bases, and metal oxides are ionically bonded crystals.

For ionic crystals, the electron cloud is symmetrically and uniformly distributed around the core, forming a spherical shape. For the ions in ionic crystals, there exists the same force of interaction with the opposite ions in all directions, thus the ionic bond is nondirectional. As shown in Fig. 2.14, positive and negative ions are alternately arranged in the crystal of ionic bond, indicating that the ionic bond is unsaturated.

Due to the strong electrostatic attraction between positive and negative ions in ionic crystals, the bonding of ionic crystals is strong, thus the materials bonded by

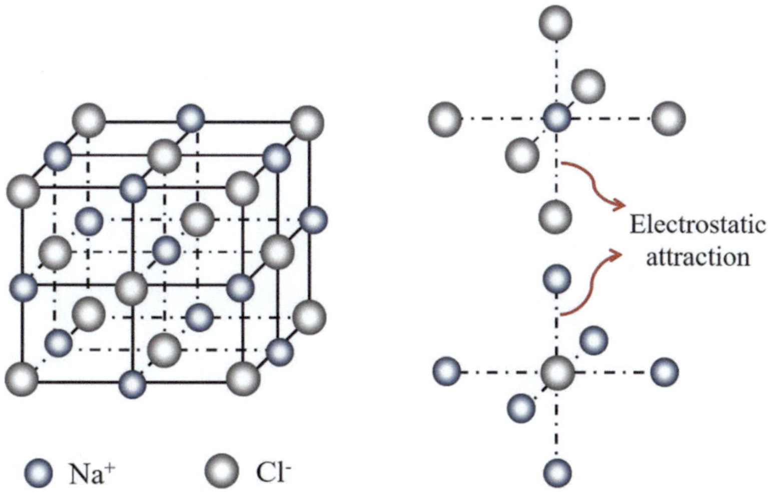

Fig. 2.14 Schematic representation of ionic bonding in sodium chloride (NaCl)

ionic bond possess high melting point and high hardness. As the outer electrons are firmly restrained around the ion, the conductivity of ionic bonded materials at room temperature is very poor and ionic materials usually exhibit insulating property. In the molten state, all ions can move freely, thus ionic materials can conduct electricity. The energy of visible light is not sufficient to excite the outer electrons, thus ionic crystals are usually colorless and transparent. When an external force is applied, it can disrupt the electric balance between ions, and the ionic bond will be destroyed, which leads to the fracture of the materials at the macro level. In other words, the ionic bonded materials are typically brittle.

③ Metallic Bond

The metal atoms typically have a few electrons in their outermost shell, usually no more than three, which are easy to escape from the binding of the nucleus and become free electrons. The valence electrons are shared by all atoms and can move freely throughout the entire crystal, forming a so-called "electron cloud", while all metal atoms become metal ions losing their outermost valence electrons. The positive metal ions are bonded to these free electrons through Coulombic attraction, which is called metallic bond, as shown in Fig. 2.15.

Metallic bonds have no saturation or directionality, and there is no specific requirement for the arrangement of atoms in the metal crystal and the metal atoms tend to form the closed compact structure with lowest energy.

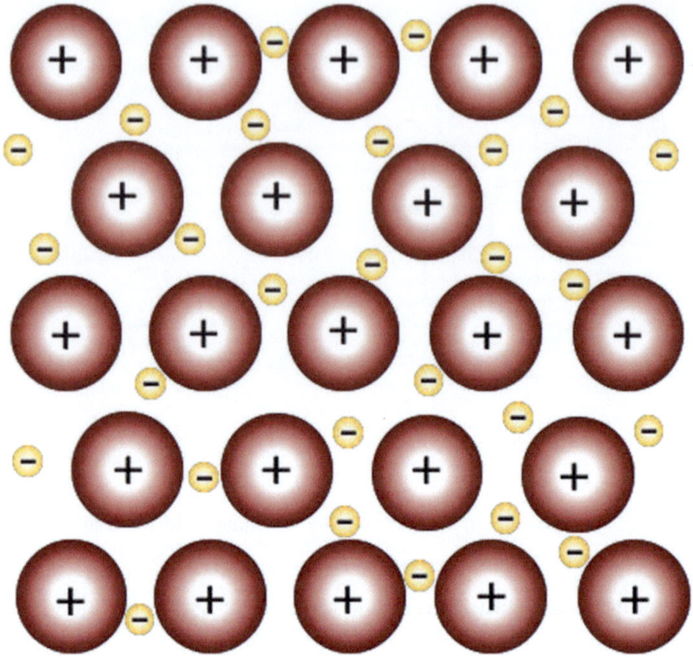

Fig. 2.15 Schematic illustration of metallic bonding

When metal deforms under external force, the position of metal atoms or positive ions changes. Since free electrons can move freely in the crystal, the metallic bonds are difficult to break under such force, thus metal typically has good ductility. The vibration of free electrons and positive metal ions in the crystal makes the metal have good thermal conductivity. When an external electric field is applied, free electrons in the metal will move along the opposite direction of the electric field, generating an electric current and giving metal excellent electrical conductivity. With the increase of temperature, the vibration amplitude of metal positive ions increases, which can hinder the directional motion of free electrons. Therefore, the electrical resistance of metal increases with temperature. Free electrons can absorb the energy of light waves to induce transitions, and thus the metals usually exhibit a metallic luster and opaqueness.

2.3.2 Secondary Bonding

The aforementioned primary bonds, i.e., the covalent bonds, ionic bonds and metallic bonds, are strong bonding forces. In addition, there are weaker types of bonding, including van der Waals forces and hydrogen bonds, which are physical bonds, and called secondary bonds.

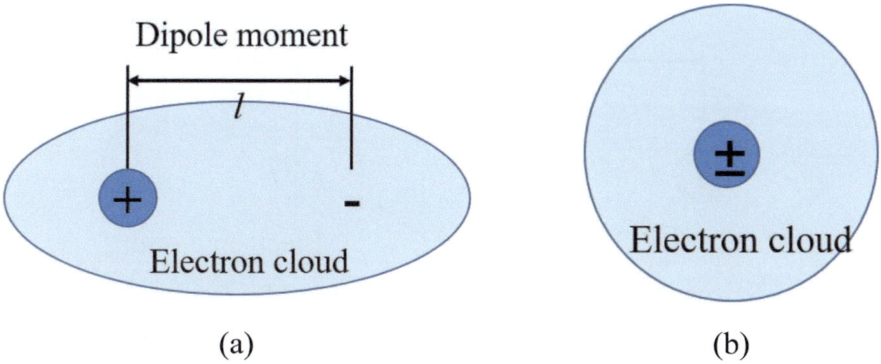

Fig. 2.16 Schematic representations of **a** polar molecule with dipole moment and **b** nonpolar molecule without dipole moment

① van der Waals Force

van der Waals forces are ubiquitous in nature, and their bond energies are one or two orders of magnitude lower than those of other chemical bonds, with a range of about 0.3–0.5 nm. van der Waals force are non-direactional and unsaturable, and it contains three components: orientation forces, induced forces, and dispersion forces.

Orientation force

Many polar molecules possess a permanent dipole moment because their positive and negative charge centers do not coincide, as shown in Fig. 2.16a. When two polar molecules come close to each other, the positive end of dipole moment in one molecule is electrically attracted to the negative end of dipole moment in its neighboring molecule. The essence of orientation force is the electrostatic attraction between the permanent dipole moments of polar molecules.

Induced force

It is generally accepted that nonpolar molecules do not have a dipole moment as their centers of positive and negative charges coincide. However, when a polar molecule approaches a non-polar molecule, the positive and negative charges of the non-polar molecule are shifted under the dipole electric field of the polar molecule. As a result, the positive and negative charge centers of the non-polar molecule no longer coincide, that is, the non-polar molecule generates an induced dipole moment, as shown in Fig. 2.17. The electrostatic attraction between the induced dipole moment and the permanent dipole moment of the polar molecule is called the induced force.

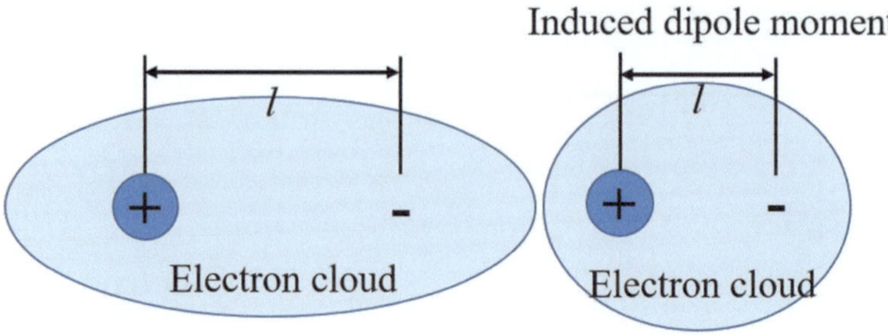

Fig. 2.17 Schematic illustration of induced dipole moment

Dispersion force

Nonpolar molecules are generally considered to have no dipole moment because the electron cloud surrounding the nucleus is regarded as spherically symmetric and uniformly distributed, resulting in no permanent dipole moment at a macro level. However, in fact, at a certain moment during the movement of electrons around the nucleus, the electrons are not uniformly distributed in space, and the instantaneous centers for positive and negative charges do not coincide. In this case, the instantaneous dipole moment is generated, and the electrostatic attraction between the dipole moments caused by the instantaneous dipole moment is called the dispersion force.

Orientation and induced forces exist only in polar molecules, and they are weak or negligible in non-polar molecules. The attractive energy from van der Waals forces is inversely proportional to the sixth power of distance between the molecules r, so the van der Waals forces only work when the distance between the molecules is small. The repulsive energy between molecules is inversely proportional to the 12th power of r, so the repulsive energy decreases rapidly with increasing distance. The binding energy of van der Waals force is much weaker than that of covalent bond and ionic bond, so the molecular crystal usually possesses low melting point, low boiling point and high compressibility. Graphite is a crystal composed of various binding forms.

At room temperature, graphite has a layered structure. Carbon atoms in each layer are bound by C–C covalent bond, and the layers are bound together by van der Waals force, as shown in Fig. 2.18. In 2004, Noveselov and Geim successfully obtained a freestanding graphite monolayer by mechanical exfoliation, named as graphene, and won the 2010 Nobel Prize in Physics for this work. Subsequently other two-dimensional materials were prepared through the similar method, such as transition metal dichalcogenides (TMDCs), black phosphorus. Recently, the vertical van der Waals heterostructures stacked by different 2D materials, have attracted a wide interest, and been widely applied in the nano-electron, photoelectron, sensors, photocatalysis, and energy conversion.

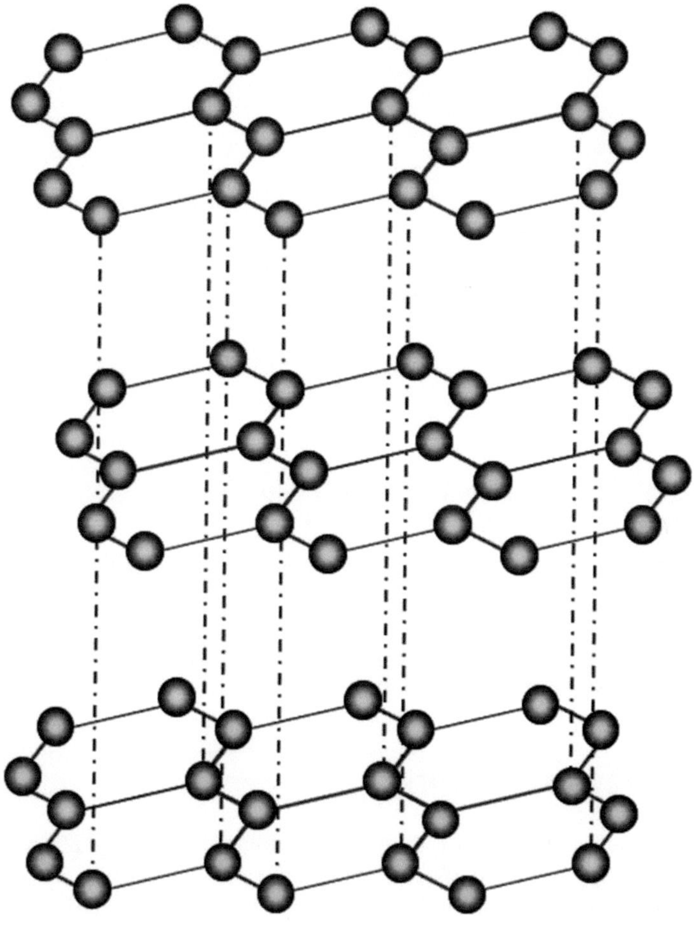

Fig. 2.18 Structure of graphite

② Hydrogen Bonding

Hydrogen bonds are polar molecular bonds, which usually exist in HF, H_2O, NH_3 and other molecules. Since there is only one electron surrounding the nucleus of hydrogen atom, its shared valence electron is closer to the highly electronegative atom X (such as F, Cl, O, N, etc.), when the hydrogen atom forms a molecule with atom X bound by a polar covalent bond. In this case, hydrogen atom tends to be positively charged, then the hydrogen atom can attract the highly electronegative atom Y in another molecule, forming a hydrogen bond with the hydrogen atom as the center, which can be expressed as X–H···Y, as shown in Fig. 2.19.

Hydrogen bond is essentially the electrostatic attraction between the hydrogen atom in one molecule and the highly electronegative atom in another molecule. Hydrogen atom is a bridge between two highly electronegative atoms. Unlike van der Waals forces, hydrogen bonds are directional and saturated. X–H can only

Fig. 2.19 Hydrogen bonds of water (H_2O) molecules in solid ice

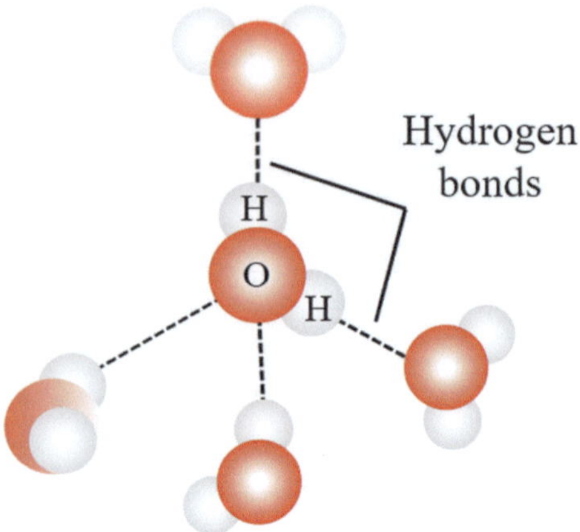

bond to one Y atom, and only if X–H···Y is in a straight line, the hydrogen bond possesses the strongest bonding force.

Hydrogen bonds exist widely in inorganic and organic substances, and lead to the association between molecules in molecular liquids. Some of the abnormal properties of water (such as boiling point up to 373 K, density anomalies above melting point) are related to hydrogen bonding. Ice is a crystalline molecule formed by hydrogen bonds between water molecules.

2.4 Energy Band Theory

Energy band theory is a major theoretical basis for studying electron motion in solids, which qualitatively illustrates the universal characteristics of electron motion. For example, this determines whether a solid is an insulator, a conductor, or a semiconductor.

2.4.1 Energy Band

In the potential field of the nucleus and other electrons, electrons of an atom are separated into different energy levels, forming so-called electron shells, and each shell corresponds to a certain energy. When atoms are close to each other to form a crystal, there is an overlap of the electron shells of different atoms. In this situation, the electron is no longer confined to one atom, but can transfer from one atom to another adjacent atom, so that the electron can move throughout the crystal. This motion is called the common motion of the electron.

For a crystal, each atom is perturbed by the potential field of the surrounding atoms. The original electron energy level of each atom is split into N new energy levels close to the original energy level, and these N new energy levels form an energy band, as shown in Fig. 2.20. Each split band is called the allowed band,

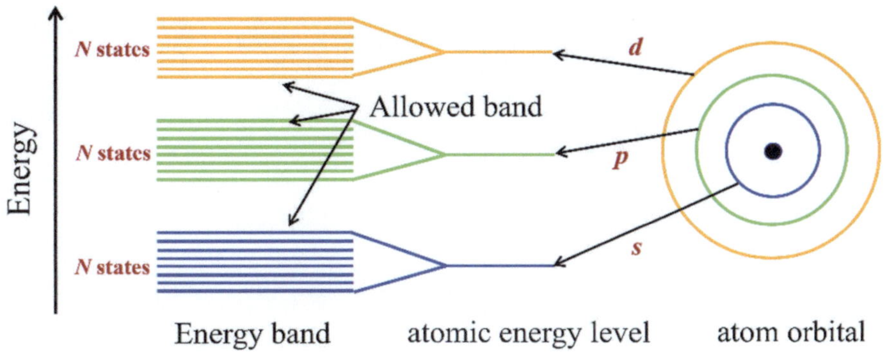

Fig. 2.20 Atomic energy level splits to form electron energy band consisting of N states

and the energy levels in the allowed band allow electrons to occupy. There are no energy levels between the allowed bands, which is called the band gap without the occupation for electrons.

The band filling with electrons in the crystal also obeys the lowest energy principle and Pauli exclusion principle. The energy levels with the lower energy band are firstly occupied by electrons, which are called filled band. Higher energy bands that are partially or completely unoccupied by electrons are called empty band. The inner energy levels of atoms are filled with electrons, so its corresponding energy band in a crystal is filled band; while the outer energy levels of atoms are unoccupied by electrons, the corresponding energy band in the crystal is empty band.

The conductance of a solid is, in essence, the energy exchange between the electrons of neighboring atoms in a solid under external fields. The variation of energy for the electrons means the transition of state for electrons in the band. The energy band can clearly illustrate the difference in conductive properties of conductors, semiconductors, and insulators. The filled band is occupied by valence electrons, also known as the valence band. As there is no empty state in the valence band, the electrons in the valence band do not conduct electricity. An empty band has many empty electron states, so electrons can move freely in the empty band, which is also called conduction band.

As shown in Fig. 2.21, the first two graphs show the energy bands of two typical metal conductors. For metal conductors, there exist many empty electron states near the highest occupied state, thus the electrons in the valence band can be easily excited to the empty electron states and become conductive electrons that can move freely. Therefore, the metal exhibits good electrical conductivity. The energy bands of semiconductors and insulators are similar: there is a band gap between the filled band and empty band. The band gap of semiconductor is narrow, thus the electron state in valence band does not change under the general external field. Therefore, semiconductor materials do not conduct electricity under normal conditions. The band gap of semiconductor materials is usually less than

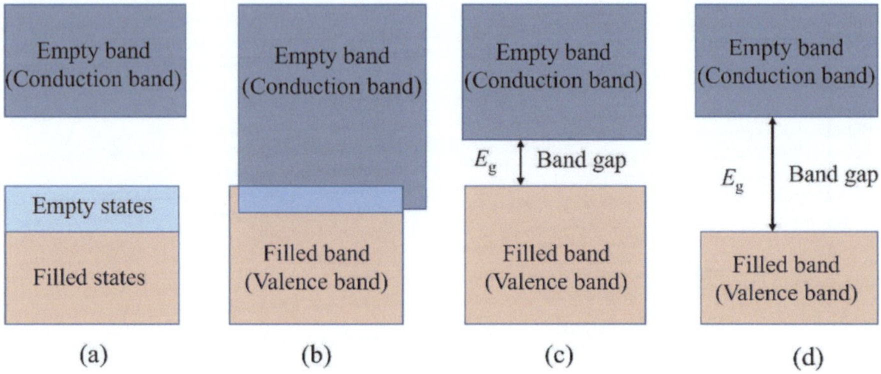

Fig. 2.21 Band structure of **a** metals such as copper, **b** metals such as magnesium, **c** semiconductors, and **d** insulators

2 eV. Upon the excitation of light, electric field, temperature and other external fields, the electrons in the valence band obtain enough energy to jump to the conduction band and become the electrons that can move freely. In this case, the semiconductor conducts electricity. The band gap of the insulator is too large to excite the electrons in the valence band to jump to the conduction band at room temperature, so the conductive performance of the insulator material is very poor. Generally, the band gap of insulators is several electron volts or more, for example, the band gap of diamond is 6–7 eV, and the band gap of silicon dioxide is about 8 eV.

2.4.2 Brillouin Zone

① Reciprocal Lattice

The concept of reciprocal lattice is very useful for the description and analysis of X-ray diffraction and lattice vibration or electron motion in crystals. The reciprocal lattice is a lattice consisting of periodic reciprocal lattice points, which correspond to family of crystal planes in the lattice: (1) the direction of the vector of the reciprocal lattice is the normal direction of a crystal plane family; (2) The modulus of the vector is the reciprocal (the integer of reciprocal) of the spacing d between crystal planes of one family of crystal planes.

Assume that the basis vectors of the lattice are $\boldsymbol{a}_1$, $\boldsymbol{a}_2$ and $\boldsymbol{a}_3$, and the basis vectors of the reciprocal lattice are $\boldsymbol{b}_1$, $\boldsymbol{b}_2$ and $\boldsymbol{b}_3$, which satisfy the following relationship:

$$\begin{cases} \boldsymbol{a}_1 \cdot \boldsymbol{b}_1 = \boldsymbol{a}_2 \cdot \boldsymbol{b}_2 = \boldsymbol{a}_3 \cdot \boldsymbol{b}_3 = 2\pi \\ \boldsymbol{a}_1 \cdot \boldsymbol{b}_2 = \boldsymbol{a}_1 \cdot \boldsymbol{b}_3 = \boldsymbol{a}_2 \cdot \boldsymbol{b}_1 = \boldsymbol{a}_2 \cdot \boldsymbol{b}_3 = \boldsymbol{a}_3 \cdot \boldsymbol{b}_1 = \boldsymbol{a}_3 \cdot \boldsymbol{b}_2 = 0 \end{cases} \tag{2.3}$$

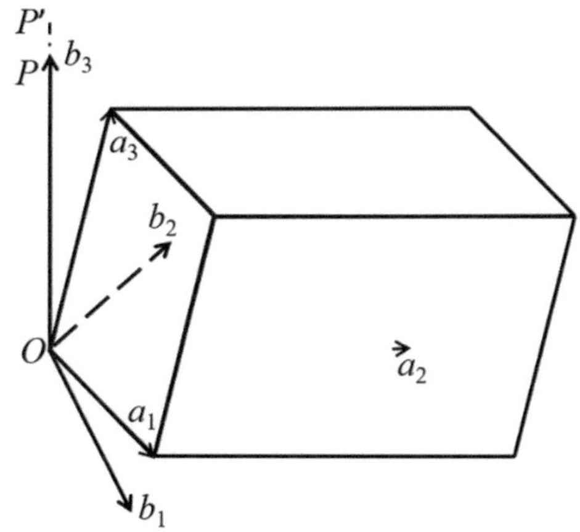

Fig. 2.22 Basis vectors of the reciprocal lattice

And then the basis vectors of the reciprocal lattice $\boldsymbol{b}_1$, $\boldsymbol{b}_2$ and $\boldsymbol{b}_3$ and can be defined as:

$$\begin{cases} \boldsymbol{b}_1 = 2\pi \frac{\boldsymbol{a}_2 \times \boldsymbol{a}_3}{\boldsymbol{a}_1 \cdot \boldsymbol{a}_2 \times \boldsymbol{a}_3} \\ \boldsymbol{b}_2 = 2\pi \frac{\boldsymbol{a}_1 \times \boldsymbol{a}_3}{\boldsymbol{a}_1 \cdot \boldsymbol{a}_2 \times \boldsymbol{a}_3} \\ \boldsymbol{b}_3 = 2\pi \frac{\boldsymbol{a}_1 \times \boldsymbol{a}_2}{\boldsymbol{a}_1 \cdot \boldsymbol{a}_2 \times \boldsymbol{a}_3} \end{cases} \tag{2.4}$$

where $\boldsymbol{a}_1 \cdot \boldsymbol{a}_2 \times \boldsymbol{a}_3$ is the volume of regular lattice Ω. Obviously we can derive the reciprocal lattice from the regular lattice and vice versa, as shown in Fig. 2.22.

② Brillouin Zone

Brillouin zone is an important physical concept which is commonly used in lattice vibration and energy band theory. There is a periodic relationship between the energy of the electron and the wave number in the crystal. Because the length of the reciprocal lattice has the same dimension as the wave number, it can be regarded as that the energy of the electron in the crystal possesses the periodicity of the reciprocal lattice. The Brillouin zone is the periodic repeating unit divided from the space of reciprocal lattice. The region closest to the origin is the first Brillouin zone. The second Brillouin zone is the region enclosed by the interface of the first Brillouin zone and the second closest perpendicular bisection plane to the origin, and so on.

Figure 2.23 shows the first Brillouin zone of some common structures. The first Brillouin zone of the simple cube is still a simple cube, and the reciprocal lattice of the simple hexagonal structure is still a simple hexagonal lattice. The first Brillouin zone of a body-centered cubic structure is a rhombic dodecahedron. The first Brillouin zone of a face-centered cubic structure is a truncated octahedron, a dodecahedron with 8 regular hexagons and 6 squares. The volume of these first Brillouin zones is exactly the size of the primitive cell of reciprocal lattice.

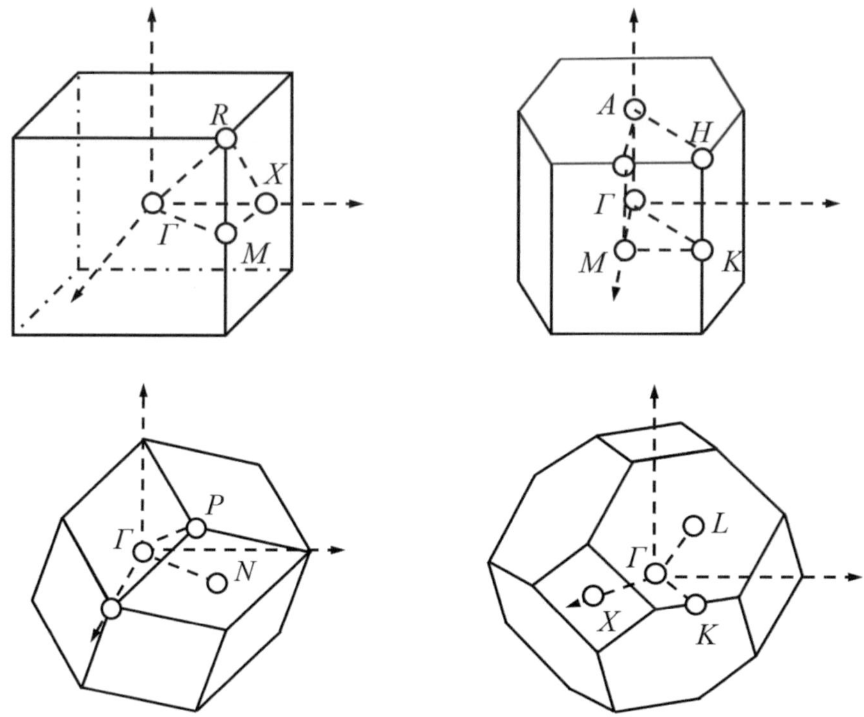

Fig. 2.23 The first Brillouin zone of common structures

2.4.3 Carriers and Femi Level

① Carriers

Carriers are particles in solids that can move freely and conduct electricity. The carriers in metal are free electrons. In semiconductors, there are two types of carrier charges: conduction band electrons and valence band holes, and both contribute to the current. In the fully occupied bands, the applied electric field won't change the state of the electrons, so the valence band electrons do not conduct electricity. When some electrons in the valence band are excited to the conduction band, both the valence band and the conduction band are partially occupied. Upon the external electric field, the electrons in both the conduction band and the valence band can conduct electricity. However, it is inconvenient to calculate the current generated by the electrons in the valence band, so the concept of "hole" is introduced.

Take silicon, a typical semiconductor material, as an example. There are four valence electrons in the outer shell of a silicon atom, which form covalent bonds with four adjacent silicon atoms to form a crystal. When some electrons in the maximum of valence band are excited to conduction band, there leave some empty states in the valence bands, which is equivalent to the electron of covalent bond gains enough energy to escape from the covalent bond and become a conduction electron, while there appears an electron vacancy in the covalent bond. The minimum energy required for an electron to escape from the covalent bond is the band gap E_g.

When valence band lose an electron of k_e state and other energy levels in the valence band are occupied by electrons, we use J to represent the actual current density caused by the electrons in the valence band. Since the current density is zero when the valence band is in a filled state, there is:

$$J = 0 - (-q\boldsymbol{v}(k_e)) = q\boldsymbol{v}(k_e) \tag{2.5}$$

This equation shows that the actual current caused by the loss of an electron $-q$ with velocity $\boldsymbol{v}$ (k_e) in the valence band is equivalent to the current generated by a positively charged particle $-q$ moving at velocity $\boldsymbol{v}$ (k_e). This positively charged particle can be used to describe the conduction of electrons in the valence band, and this hypothetical positively charged particle is called "hole".

(2) Fermi level

The distribution of free electrons in the crystal satisfies the Fermi–Dirac distribution:

$$f(E) = \frac{1}{e^{E-E_F/kT} + 1} \tag{2.6}$$

In this equation, $f(E)$ is the Fermi–Dirac distribution function, which is used to represent the probability of free electrons occurring at energy E in thermal equilibrium, E_F is the Fermi energy level at temperature T, and k is the Boltzmann constant.

Figure 2.24 shows the Fermi–Dirac distribution function at different temperatures. When T = 0 K, there is: if $E < E_F$, $f(E) = 1$; if $E > E_F$, $f(E) = 0$. In other words, at absolute zero, all energy states below the Fermi level are occupied by electrons, and energy states above the Fermi level are empty, without electrons occupied. Therefore, the Fermi energy level is the highest energy level that the free electrons can occupy at absolute temperature. When T > 0 K, there is: if $E = E_F$, $f(E) = 1/2$; If $E < E_F$, $1/2 < f(E) < 1$; If $E > E_F$, $0 < f(E) < 1/2$. This means that when T > 0 K, the energy levels below E_F are mostly occupied with electrons, and the energy levels above E_F are mostly empty. For an unfilled energy level, we can be deduced that it is near E_F. This is because that when the temperature rises, electrons near the Fermi level absorb heat and jump to higher energy states, while electrons farther away from the Fermi level are difficult to be excited.

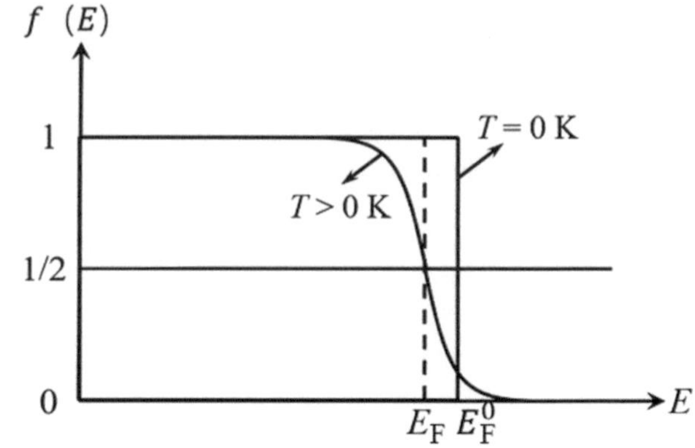

Fig. 2.24 The Fermi–Dirac distribution function at different temperatures

2.4.4 Mobility and Scattering

① Mobility

When applying external electric field, the electrons in the crystal move along the opposite direction of the electric field to form a current. The motion of electrons upon the electric field force is called drift motion, and the speed of this oriented motion is called drift velocity. When the electric field in the crystal is constant, the electron has a constant average drift velocity ($\overline{v}_d$). The average drift velocity is proportional to the intensity of electric field, and its relationship can be written as:

$$\overline{v}_d = \mu\xi \tag{2.7}$$

where μ is called mobility, which is used to describe the average drift velocity of the electron per unit field intensity, in units of $m^2/(V{\cdot}s)$ or $cm^2/(V{\cdot}s)$. Since electrons are negatively charged, it is generally opposite to the field ξ, while it is customary to set the mobility to a positive value:

$$\mu = \left|\frac{\overline{v}_d}{\xi}\right| \tag{2.8}$$

Table 2.4 shows the electrical mobility for several bulk and 2D semiconductor materials, whose magnitudes are ranging from 100 to 2,000,000 $cm^2/(V{\cdot}s)$.

② Scattering

The scattering that the carriers suffer during the motion in the crystal originates from the destruction of periodic potential field of the crystal. In actual crystals, there are various impurity atoms and defects, and the lattice atoms constantly

Table 2.4 Room-temperature electrical mobility for common bulk and 2D semiconductor materials

Materials	Electrical mobility/$cm^2 \cdot (V \cdot s)^{-1}$
Si	1,350
Ge	3,800
GaAs	8,000
Graphene	2,000,000
TMDCs	100–1,000
Black phosphorus	~1,000

vibrate at their lattice points, which would cause the scattering of carriers. Carrier scattering is manifested as the variation of the magnitude and direction of their velocity. There are two main scattering mechanisms of carriers: lattice scattering (phonon scattering) and ionized impurity scattering.

When the temperature is higher than zero at the thermodynamic temperature, the atoms in the crystal have a certain thermal energy, and vibrate irregularly at their lattice positions. The vibration of the lattice destroys the ideal periodic potential field. An ideal periodic potential field in a solid allows electrons to move freely throughout the crystal without being scattered. However, the thermal vibration of the atoms destroys the potential function, and the carrier electrons and holes would interact with the vibrating lattice atoms. This lattice scattering is also known as phonon scattering.

Another scattering mechanism that affects carrier mobility is ionization impurity scattering. In the actual crystal, it is inevitable that there are various impurity atoms. In practical applications, doping with impurity atoms is often used to tune or change the properties of semiconductors. At room temperature, these impurity atoms are ionized, so there is a Coulombic interaction between the ionized impurities and the electrons or holes. The Coulombic interaction causes collision and scattering of carriers, which can also change the carrier velocity characteristics.

2.5 Questions

Q2.1 Please provide a concise summary of the significance of the four quantum numbers.

Q2.2 Please elucidate the electron configuration within an atom.

Q2.3 Could you explain the distinction between a molecular orbital and an atomic orbital?

Q2.4 Can you elucidate why hydrogen fluoride (HF) exhibits a higher boiling point compared to hydrogen chloride (HCl)?

Q2.5 Could you elaborate on the distinctions in crystalline structure, bonding strength, and bonding angle between graphite and diamond?

Q2.6 What constitutes the fundamental nature or core aspect of bonding?

Q2.7 Could you enumerate all the types of binding present in solids and outline their respective differences?

Q2.8 Can you sketch the first Brillouin zone for the simple cubic structure and determine its volume, given the lattice's basis vectors: $a_1 = a_i$, $a_2 = a_j$, $a_3 = a_k$?

Q2.9 In the context of energy band theory, could you elucidate the distinctions among metals, semiconductors, and insulators?

Q2.10 Calculate the drift velocity of electrons at room temperature when the electric field strength is 250 V/m. Consider that the electron mobility of graphene at room temperature is approximately 2,000 $m^2/(V{\cdot}s)$.

Bibliography

Bransden BH, Joachain CJ (2003) Physics of atoms and molecules (Second edition), 2nd ed. Prentice Hall, Hatliw

Huang K (2009) Solid state physics. Peking University Press, Beijing

Kittel C, Hellwarth RW (1957) Introduction to solid state physics. Phys Today 10(6):43–44

Kragh H, Brown LM (2000) Quantum generations: a history of physics in the twentieth century. Phys Today 53(5):56–57

Li G, Li Y, Liu H et al (2010) Architecture of graphdiyne nanoscale films. Chem Commun 46(19):3256–3258

Pauling L (1931) The nature of the chemical bond. Application of results obtained from the quantum mechanics and from a theory of paramagnetic susceptibility to the structure of molecules. J Am Chem Soc 53(4):1367–1400

Pauling L (1960) The nature of the chemical bond. North-Holland, New York

Introduction to Crystallography 3

Learning Objectives

1. Recognize the importance of long-range order in crystalline materials, which distinguishes them from amorphous materials.
2. Identify the seven crystal systems based on the symmetry of their unit cells.
3. Recall the names of common crystal systems.
4. Comprehend the concept of Bravais lattice.
5. Determine the position of a given point within a crystal.
6. Specify the Miller indices for crystalline directions and planes located within a unit cell.
7. Construct crystallographic directions and planes using Miller indices.
8. Understand the stereographic projection.

There are three primary states of matter: solids, liquids, and gases. Solids can be classified into two distinct types - amorphous and crystalline. The majority of solid materials are crystalline, and modern technologies are, in fact, underpinned by the crystalline materials. Consequently, it is crucial for scientists and engineers in materials science to have a comprehensive understanding of crystals.

Crystallography is a natural science discipline that primarily focuses on the study of crystals, including theoretical and experimental analyses of atomic arrangements within crystalline materials. The discipline originated in the seventeenth century, driven by the symmetry and aesthetics of crystals. Early developments relied on conjecture, argument, and simple experiments, with a few endeavors of simple experiments, and were closely linked to mineralogy. Mineralogists provided the original experimental evidence of crystal periodicity. Today, modern crystallography is primarily based on X-ray diffraction analyses of crystals that act as optical gratings. From crystal symmetry and shape, it was deduced

Z. Sun et al., *Fundamentals of Materials Science*,
https://doi.org/10.1007/978-981-96-3904-5_3

that crystals are made up of ordered arrangements of atoms and molecules. X-ray crystallography has allowed chemists to determine atomic-level internal structures of minerals, large complex molecules, and biological macromolecules such as proteins and DNA.

Crystallography has had a significant impact on related disciplines, revolutionizing our knowledge of matters, and pushing forward the development of solid-state physics. It is now considered a central part of modern applied science, and the field of materials science and engineering benefits greatly from crystallography. The applications of crystallography are numerous, and it underpins a diverse range of industries, including aeronautics, automobiles, computers, cosmetics, healthcare, and mining, among others. Despite its contributions to modern science and technology, crystallography remains relatively unknown to the general public, which was one of the reasons behind the proclamation of 2014 as the International Year of Crystallography by the United Nations General Assembly (UNGA). United Nations Educational Scientific and Cultural Organization (UNESCO) led and coordinated educational and capacity-building activities during the Year, in collaboration with the International Union of Crystallography (IUCr). In 2014, X-ray crystallography celebrated its centennial, and the International Year of Crystallography highlighted the continued importance of the field in addressing post-2015 development issues, such as food security, safe drinking water, healthcare, sustainable energy, and environmental remediation.

The story of modern crystallography dates back to 1912, when the X-rays, discovered in 1895, changed all that about the science about crystals. Not long after this discovery, it was found that these rays could "see" the structure of crystals in a non-destructive way. In 1912 in Germany, Max Laue, Walter Friedrich, and Paul Knipping showed that an X-ray beam incident on a crystal was scattered to form a regular pattern of spots on a film (a phenomenon called diffraction). Thus, it was proved that X-rays are a kind of wave and furthermore, this was accepted as solid and direct evidence of the underlying order of atoms in the crystal. The breakthrough gave birth to the modern crystallography, and also made the first Nobel Prize on this topic to Laue in 1914. Another pioneer in this domain was William Lawrence Bragg (WLB) who in 1912 showed how the observed diffraction pattern could be used to determine the positions of atoms in the crystal, thus setting a landmark on the course of crystallography, and effectively launching a completely new scientific discipline, X-ray crystallography. Working with his father, William Henry Bragg (WHB), he quickly determined the crystal structures of several materials. Both father and son shared Nobel prize in 1915, the second one about the crystallography investigation by X-rays.

The inner order of crystals is reflected by their common fundamental properties, namely, homogeneity, anisotropy, symmetry, self-confinement, stability, etc. The homogeneity of crystals means that each part of a crystal has the same set of properties regardless of the parts' size, and this homogeneity is not the same as that of amorphous solids, where it is the result of statistical effects. The anisotropy refers to the dependence of the properties of crystals on the directions along which it is measured. The symmetry of crystals has two types, microscopic and macroscopic, and the former means the periodicity of their building blocks and the latter relates to their shape (same parts of a crystal). The self-confinement of crystals states that crystals always take particular forms as they grow under a given set of conditions. The stability of crystals means that a crystal does not simultaneously transit to another state. It is a necessary result from thermodynamics and also means that the crystals have the minimum internal energy at a particular set of conditions compared with other states of the matter with the same composition.

3.1 Crystalline Materials and Their Periodicity

Shown in Fig. 3.1 is quartz, a typical example of crystalline materials that belong to the trigonal crystal system and have a six-sided prism shape with pyramids at each end. Quartz is the second most abundant mineral in the earth's continental crust after feldspar. Crystalline materials have a regular internal arrangement of atoms or atom groups, which is the signature of all crystals, and they exhibit definite melting points, regular shapes that often possess symmetry elements, and are capable of diffracting X-rays, as demonstrated in Fig. 3.2.

Periodicity refers to the regular arrangement of atoms or atom groups in crystals, where the same atoms or atom groups appear regularly in every certain distance of atomic scale, regardless of the direction along which a crystal is observed. The distance is called the period associated with that observation direction, which may vary along various directions. The periodicity is demonstrated in Fig. 3.2b and c, and it is more evident in 2D cases as shown in Fig. 3.2e and f.

Fig. 3.1 Samples of natural quartz

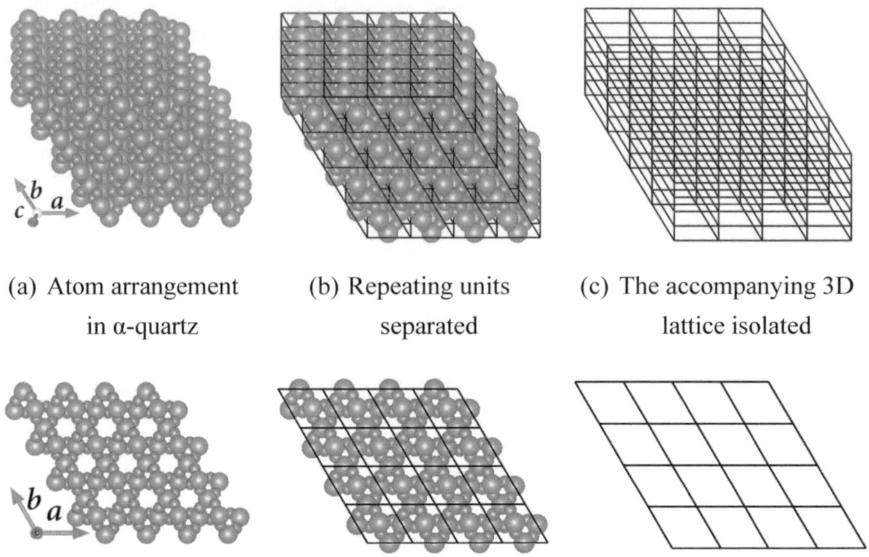

(a) Atom arrangement in α-quartz (b) Repeating units separated (c) The accompanying 3D lattice isolated (d) Projected along ***c*** axis (e) the repeating unit in a plane perpendicular to ***c*** (f) The 2D lattice in the plane perpendicular to ***c***

Fig. 3.2 The hard-sphere model of quartz, its repeating unit and the accompanying lattice for 3D and 2D projections

3.2 Crystal Structures and Space Lattices

As shown in Fig. 3.2c, the regularly arranged atoms found in crystalline materials may be treated geometrically as spatially periodically distributed points. The points construct a space lattice, each of the points is called a lattice point, and the distance between two adjacent points is the period of the lattice along the direction. By applying the abstract space lattice back to a given crystal, each lattice point may be associated with an atom or a group of atoms. Regardless of what atom or atom groups are situated at the lattice points, they have exactly the same environmental conditions, **meaning that all the lattice points are equal**. In an infinite lattice, from whichever lattice point, no difference can be found towards any direction with respect to any other lattice point. This core characteristic of a space lattice can be intuitively understood by the illustrations shown in Fig. 3.3.

In a 3-dimensional (3D) lattice, as shown in Fig. 3.3c, connecting adjacent points can form a 3D network, which is also known as a crystal lattice. For a given crystalline material, a group of points can be abstracted out of its crystal structure regardless of its size, shape, or properties. In reality, a crystal structure

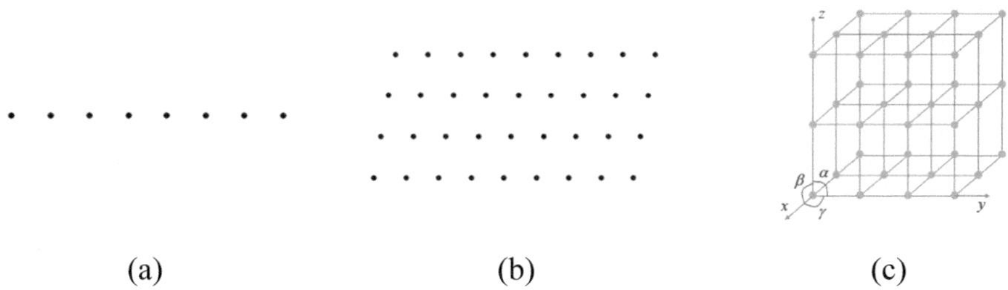

Fig. 3.3 Examples of space lattices. Infinite lattices expanding in space of different dimensionalities: **a** 1D, **b** 2D, and **c** 3D

can be obtained by attaching a specific kind of structural motif or basis to each lattice point in a space lattice. To put it simply, this can be formulated as follows.

$$\text{Space Lattice} + \text{Motifs} \rightarrow \text{Crystal Structure}$$
$$\text{or, Crystal Structure} = \text{Motifs} + \text{Lattice Points}$$

Hence, it is apparent that diverse crystal structures can possess the same space lattice. This is exemplified by sodium chloride and copper, which share a face-centered cubic lattice, while having different motifs positioned at the lattice points, resulting in distinct crystals. The space lattice emphasizes the regular and periodic nature of crystals, which constitutes their fundamental essence. As the space lattice plays a crucial role in determining the properties of crystalline materials, comprehending it is crucial and beneficial.

For a given 3D lattice, as shown in Fig. 3.3c, it can be built up by consecutively and closely stacking its basic unit, a parallelepiped (Fig. 3.4), along the directions $\boldsymbol{a}$, $\boldsymbol{b}$ and $\boldsymbol{c}$. The parallelepiped is called a unit cell. Then, what factors determine the size and shape of a given lattice? Apparently, as long as the size and shape of the unit cell of a lattice are defined, the lattice can be well constructed by simple translation of the unit cell. First, as shown in Fig. 3.4, the lengths of $\boldsymbol{a}$, $\boldsymbol{b}$ and $\boldsymbol{c}$ (basis vectors) determine the size of the parallelepiped, and then the three angles α, β and γ, associated with the relative orientation of the three basis vectors, determine the shape of the parallelepiped. Therefore, all the six parameters ($\boldsymbol{a}$, $\boldsymbol{b}$, $\mathbf{c}$, and α, β, γ) are called lattice constants, or lattice parameters. In the context of crystallography, $\boldsymbol{a}$, $\boldsymbol{b}$ and $\boldsymbol{c}$ are called crystal axes.

There are a number of ways a unit cell can be selected for a given lattice. For simplicity, taking a 2D lattice for example, multiple ways in which a parallelepiped can be selected such as, a, b, c, d, e, etc. in Fig. 3.5. This situation also happens to any 3D lattice. However, to better represent the symmetry and periodicity of the lattice, there are several rules to comply with as its unit cell is selected.

- The unit cell with the highest level of geometrical symmetry should be chosen, allowing for the translation of the unit cell along the three crystal axes to produce the entire lattice without any redundant or missing points.

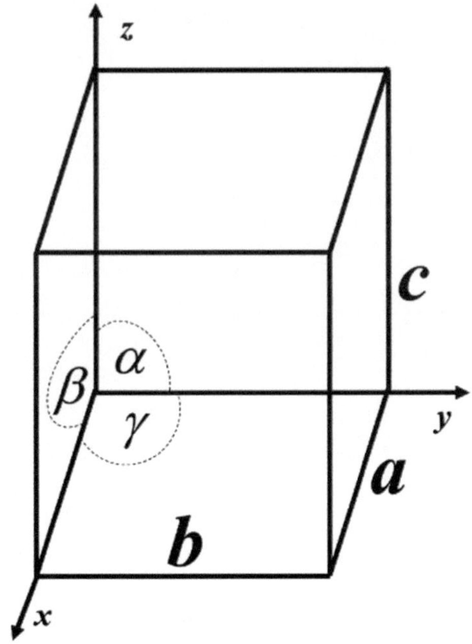

Fig. 3.4 The unit cell of a lattice and the lattice constants: axial lengths and in-teraxial angles

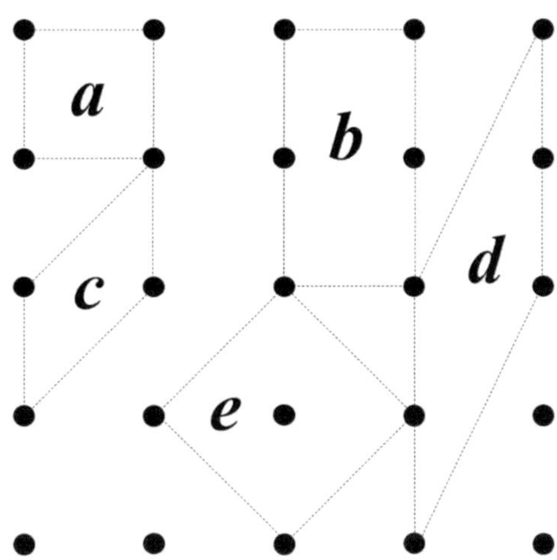

Fig. 3.5 Multiple ways a parallelepiped can be selected in a 2D lattice

- Whenever possible, the three angle parameters should include as many right angles as possible.
- In addition to the above conditions, the unit cell should have the smallest volume possible. In solid state physics, a primitive cell is often preferred because it has the smallest volume, although it may not reflect all the symmetry elements of the lattice.

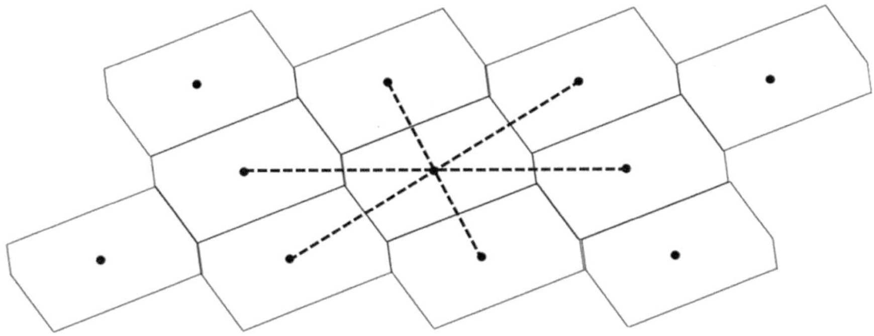

Fig. 3.6 Construction of Wigner–Seitz cells for a 2D lattice

Although there are alternative choices of unit cell for a given lattice, all the alternatives can generate the same lattice using only translation. And since the unit cell reflect the symmetry of the lattice, it brings convenience for the investigation of the lattice structure. It is worth noting that there are primitive cell and conventional cell. The primitive cell corresponds exactly to one lattice point, containing only one copy of motif in a real crystal assuming this lattice, and therefore it is the smallest possible unit cell. The primitive cell is usually not sufficient to reflect the full symmetry of the crystal structure, so a conventional cell is often used. A conventional cell may include more than one lattice points (thus accommodating more than one motif), demonstrating the full symmetry of the lattice. In addition to the parallelepiped primitive cells, there is a special type of primitive cell known as Wigner–Seitz cell. In a Wigner–Seitz cell, a single lattice point is at the center of the cell, and the shape is often not a parallelogram or parallelepiped. It can be constructed, taking a 2D lattice for example, as depicted in Fig. 3.6. Firstly any lattice point is selected; secondly, the closest neighbors of the lattice point are connected with the point by linear segments; finally, all the perpendicular bisectors of the linear segments are constructed. The resultant enclosed area is the Wigner–Seitz cell. This cell is also able to build the whole lattice by pure translation operation. The primitive cell constructed this way for a reciprocal lattice in momentum space is called the Brillouin zone.

3.3 Bravais Lattice

The first question that comes out about space lattices may be how many distinct space lattices exist in total. This is equivalent to asking how many ways atoms or groups of atoms can be arranged in space.

Beginning with a simple scenario, the one-dimensional (1D) space, which is linear in nature, can be beneficial. On a line, infinite points are equally spaced, as shown in Fig. 3.3a. This case can be called a linear lattice. Mathematically, the linear lattice can be treated by Eq. 3.1, where m can be any integer, and $\boldsymbol{a}$ is a translation vector (basis vector), as follows.

$$R = m\boldsymbol{a} \tag{3.1}$$

And therefore, $\boldsymbol{R}$, a vector, is actually a set of all the lattice points for this linear lattice. And each of the lattice points can be found by translation operations. Obviously, there is only one type of 1D lattice (linear lattice). An example can be found in a row of ions arranged in NaCl, an ionic compound consisting of two types of ions, as shown in Fig. 3.7.

The case becomes slightly complex when discussing the 2-dimensional (2D) space, where, for the plane lattices, the mathematical expression of Eq. 3.1 shall be modified to Eq. 3.2, where n and $\boldsymbol{b}$ are the counterparts of m and $\boldsymbol{a}$, respectively, to expand the lattice to 2D space. Note that $\boldsymbol{a}$ is not parallel with $\boldsymbol{b}$, but they also do not need to be perpendicular to each other. Therefore, a plane lattice has two basis vectors in the 2D case.

$$R = m\boldsymbol{a} + n\boldsymbol{b} \tag{3.2}$$

Lattices with isomorphic symmetry groups are considered equivalent. There are five possible 2D lattices with isomorphic symmetry groups: square, hexagonal, rectangular, centered rectangular, and oblique lattices, as shown in Fig. 3.8. The five 2D Bravais lattices can be grouped in four families, as the rectangular family is accepted in two different types of lattices, as shown in Fig. 3.8c and d. Any

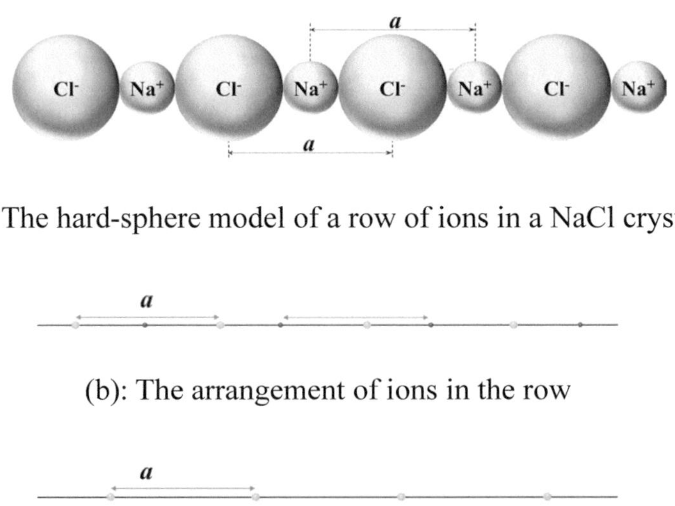

(a): The hard-sphere model of a row of ions in a NaCl crystal

(b): The arrangement of ions in the row

(c): The linear lattice attached to the row of ions

Fig. 3.7 A row of ions (cations and anions) in a NaCl crystal, and the corresponding linear space

attempt to create a new type of plane lattice would coincide with one of the five types shown in Fig. 3.8.

One can obtain an example of a plane lattice by examining an ion layer in a NaCl crystal, as illustrated in Fig. 3.9. An isolated ion plane of a NaCl crystal is shown in Fig. 3.9a, and its accompanying plane lattice can be found by connecting the centers of the ions, as done in Fig. 3.9b and c. In particular, the ions of the same type are connected, and each of the two sets of ions (Na^+ and Cl^-) forms a square lattice, Fig. 3.9c. If a pair of cation and anion is treated as a whole thing (a motif), then one can abstract out a single square lattice.

In the real world, crystalline materials usually exist in 3D space. Correspondingly, they relate to 3D lattices, for example, lattices shown in Fig. 3.2c. The situation becomes more complex in contrast to the previous ones. However, the mathematical treatment is still applicable, as long as a third basis vector is added, leading to an updated expression of Eq. 3.3 as follows,

$$R = h\boldsymbol{a} + k\boldsymbol{b} + l\boldsymbol{c} \tag{3.3}$$

where the h, k and l are any integers and $\boldsymbol{a}$, $\boldsymbol{b}$ and $\boldsymbol{c}$ are primitive translation vectors or basis vectors which lie in different directions (not necessarily mutually perpendicular).

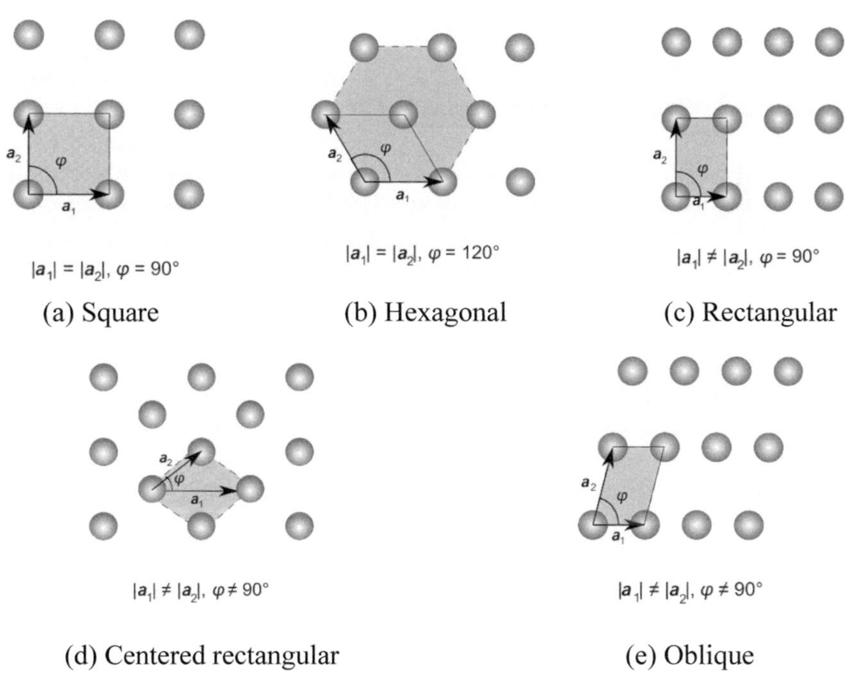

Fig. 3.8 The five different types of 2-dimensional Bravais lattices

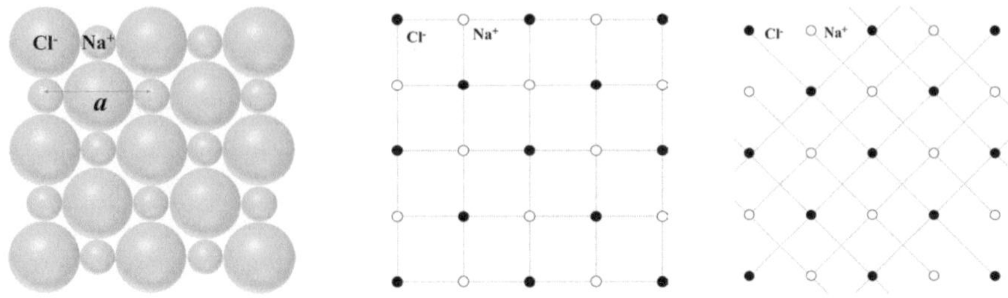

(a) An ion layer in a NaCl crystal (b) The accompanying plane lattice (c) Directly showing the square lattice

Fig. 3.9 An ion layer in a NaCl crystal and its accompanying plane lattice

For 3D space lattices, it becomes very hard to find out the number of the different types of lattices by exhaustive algorithm. But, according to the relationship characteristics of the six lattice parameters, all the possible lattices can be categorized into seven systems, as listed in Table 3.1 (strictly speaking, this is resulted from the consideration of symmetry of the lattices).

Although there are totally six lattice parameters ($\boldsymbol{a}$, $\boldsymbol{b}$, $\boldsymbol{c}$, and α, β, γ, as shown in Fig. 3.4) for a given lattice, it often needs fewer parameters to determine the lattice. The **cubic lattice**, for example, only has one variable (the length of $\boldsymbol{a}$) since its unit cell is a cube. The **hexagonal lattice** has two variables, the lengths of $\boldsymbol{a}$ and $\boldsymbol{c}$, which are conveniently investigated using four axes ($\boldsymbol{a_1}, \boldsymbol{a_2}, \boldsymbol{a_3}, \boldsymbol{c}$) that will be discussed in detail in subsequent sections. The **trigonal lattice** has three axes of equal length, with none perpendicular to another, and all six faces having the same size and shape. The **trigonal lattice** has two variables, a and α. The **tetragonal lattice** has two variables, the lengths of a and c. The **orthorhombic lattice** has three variables, a, b, and c. The monoclinic lattice has four variables, three variable edges and one variable angle. The **triclinic lattice** has six variables, all six lattice parameters need to be determined.

How many space lattices are in each of the seven systems? The number of lattices depends on the arrangement of lattice points in a given lattice. Starting from the constraint that it looks all the same from each of the lattice points, and the fact that the corner points, face centers, and body center of any given lattice out of the seven systems contained in Table 3.1 are identical points, there are 28 (4 times 7) lattices, namely, primitive (P, 7), base-centered (C, 7), face-centered (F, 7) and body-centered (I, 7).

The 28 lattices are not all distinct and can be reduced to 14 based on their symmetry, and their conventional unit cells are listed in Table 3.2. Auguste Bravais, a French physicist, in 1848 first answered this question. And the term "Bravais lattice" in crystallography is named after him. In geometry, a Bravais lattice can be generated by a set of discrete translation operations in Euclidean space described

Table 3.1 Classification of lattices by symmetry

Lattice system	Axial lengths and angles	Unit cell geometry	Examples of crystal
Cubic (isometric)	$a = b = c$; $\alpha = \beta = \gamma = 90°$	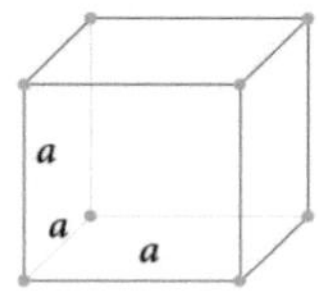	Au, Cu, Al, NaCl, α-Fe
Hexagonal	$a = b \neq c$; $\alpha = \beta = 90°$, $\gamma = 120°$	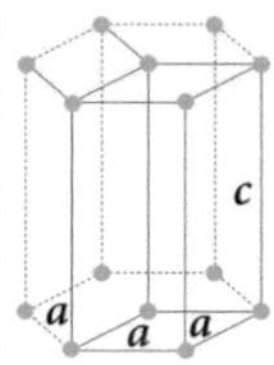	Zn, Cd
Trigonal (Rhombohedral)	$a = b = c$; $\alpha = \beta = \gamma \neq 90°$	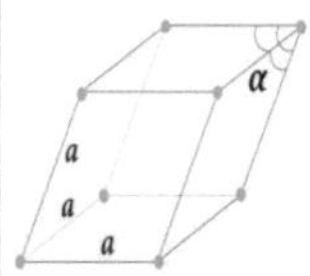	As, Sb, Bi, Calcite
Tetragonal (Rhombic)	$a = b \neq c$; $\alpha = \beta = \gamma = 90°$	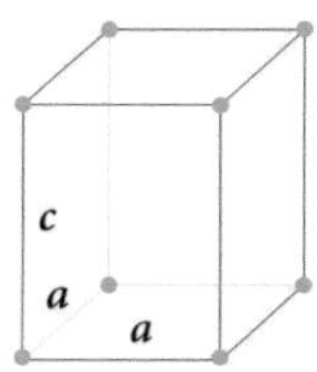	β-Sn, TiO_2 (Rutile and Anatase)
Orthorhombic	$a \neq b \neq c$; $\alpha = \beta = \gamma = 90°$	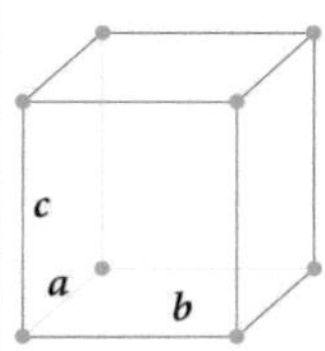	α-S, Ga, Fe_3C

(continued)

Table 3.1 (continued)

Lattice system	Axial lengths and angles	Unit cell geometry	Examples of crystal
Monoclinic	$a \neq b \neq c$; $\alpha = \gamma = 90° \neq \beta$	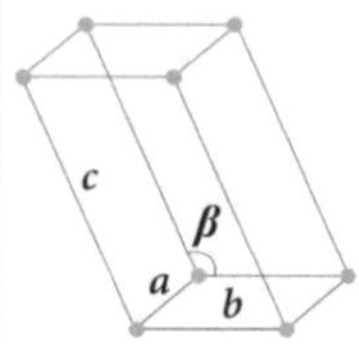	β-S
Triclinic	$a \neq b \neq c$; $\alpha \neq \beta \neq \gamma \neq 90°$	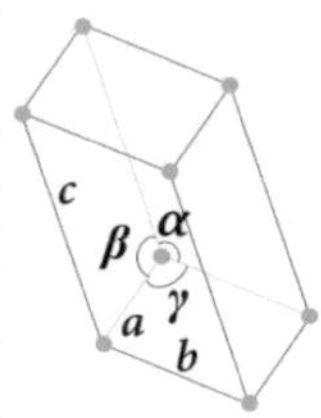	K_2CrO_7

by Eq. 3.3. The lattices defined by the set of lattice points generated by the basis vectors are called translation lattices.

There are three types of unit cells belonging to the cubic system, namely, simple cubic (SC), body-centered cubic (BCC), and face-centered cubic (FCC). The monoclinic system has two types of unit cells, simple and base centered unit cells. The rhombohedral, hexagonal, and triclinic systems all have only one simple type of unit cell. In the orthorhombic system, all four types are drawn in Table 3.2. In the tetragonal system there are only simple and body-centered unit cells; it seems that the face-centered tetragonal unit cell is missing but it can be constructed from four body-centered tetragonal unit cells.

3.4 Crystallographic Directions and Planes

A plane that passes through an atomic plane in a crystal is referred to as a crystallographic plane, while any vector linking two lattice points in a crystal is called a crystallographic direction. Since the atomic arrangement along various crystallographic planes and directions may differ, several properties of crystalline materials depend on them. Therefore, it is frequently crucial to specify a particular point, crystallographic direction, or plane of atoms within a crystal during an investigation.

A unit cell, shown in Fig. 3.4, is used to determine the values of the index. A right-handed coordinate system consisting of three (x, y, and z) axes is established, with its origin point located at one of the corners and coinciding with the unit cell edges. For the systems (hexagonal, trigonal, monoclinic, triclinic) whose axes are not mutually perpendicular, the corresponding x-y-z coordinate systems

Table 3.2 The 14 Bravais lattices

Lattice system	Primitive lattice (P)	Base-centered lattice (C)	Body-centered lattice (I)	Face-centered lattice (F)
Cubic	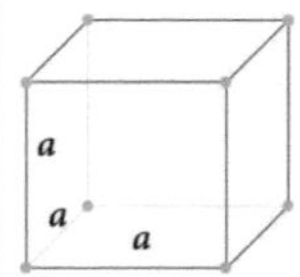	N/A	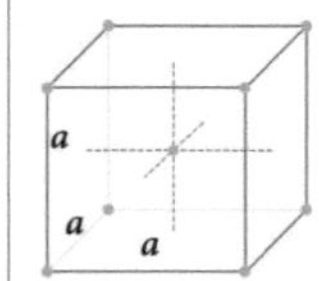	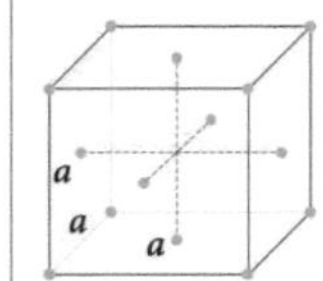
Hexagonal	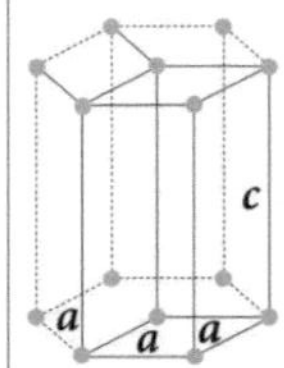	N/A	I = P	F = P
Trigonal (Rhombohedral*)	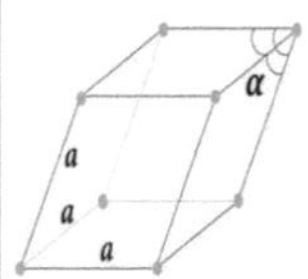	N/A	I = P	F = P
Tetragonal	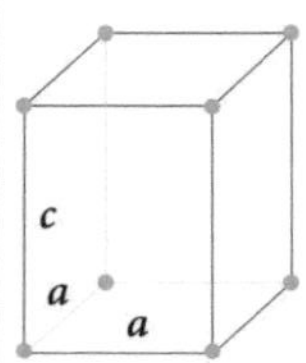	C = P	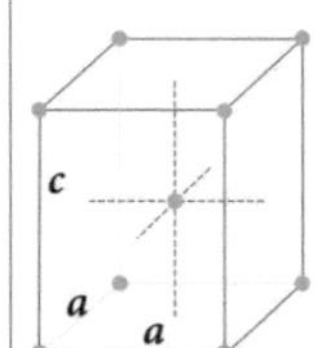	F = I

(continued)

Table 3.2 (continued)

Lattice system	Primitive lattice (P)	Base-centered lattice (C)	Body-centered lattice (I)	Face-centered lattice (F)
Orthorhombic	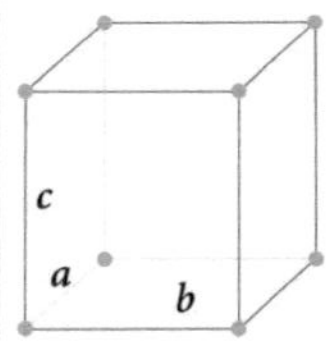	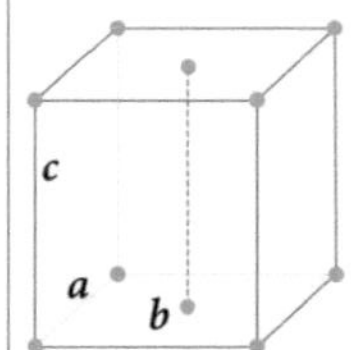	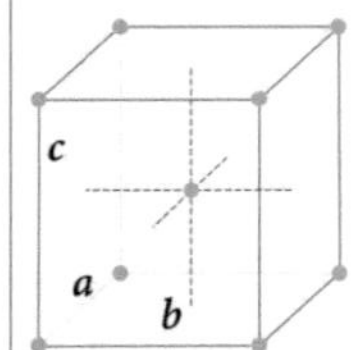	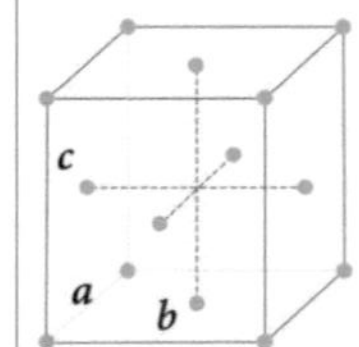
Monoclinic	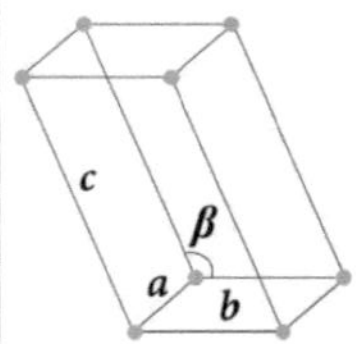	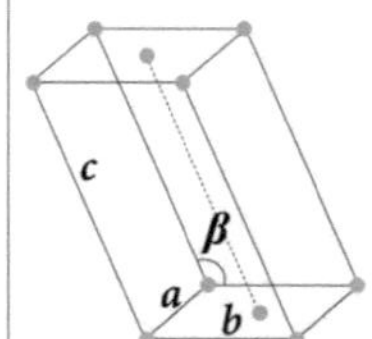	I = C	F = C
Triclinic	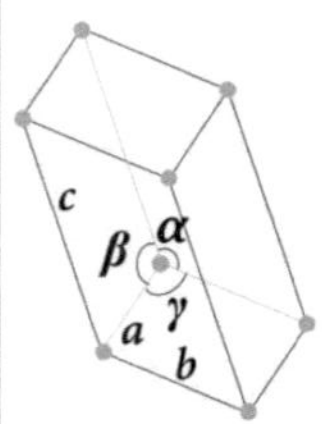	C = P	I = P	F = P

* *Note* Rhombohedral system is a sub-category of hexagonal system

are also established as in the familiar Cartesian coordinate scheme. Points, directions, and planes in any 3D lattice can be determined by a set of three numbers. The numbers for crystallographic directions and planes are called Miller indices, providing a notation system in crystallography for directions and planes in Bravais lattices. The following subsections will provide a detailed explanation of determining crystallographic direction and plane.

3.4.1 Point Coordinates

Every lattice point is uniquely fixed with respect to the origin of the lattice by the vector given by Eq. 2.1, in which only the three numbers, h, k, and l, are needed since the basis vectors of the lattice are given. And, for any given point located within a unit cell, its position can be specified in terms of three coordinates as fractional multiples of the unit cell edge lengths (a, b and c).

This simple process is diagrammatically illustrated in Fig. 3.10. In the same way that the projection of a vector along axes is determined when solving a geometry problem, the coordinates of the point P within the unit cell can be found out. Since its projection along the axes are, ma, nb, and lc, the coordinates of point P can be designated as $[m, n, l]$. Locating a point in a unit cell is the basis for indexing crystallographic directions and planes.

In Table 3.3, the right-handed coordinate system for each of the 14 Bravais lattices is established, and the locations of the lattice point(s) in each of the lattices are listed. In the simple configuration of the seven lattice systems, there is only one lattice point, [0, 0, 0], in their unit cell, and all of their lattice points can be found by translation according to Eq. 3.3. For the complex lattices with base

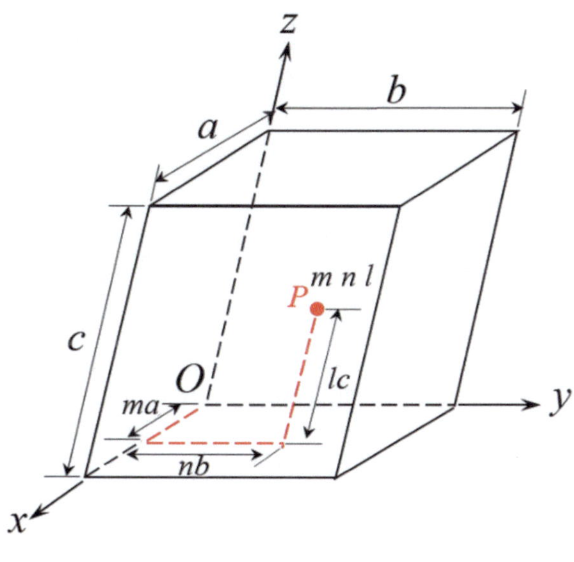

Fig. 3.10 Determination of the coordinates of a point located within a unit cell in terms of the edge lengths of the cell

Table 3.3 Establishment of the right-handed coordinate systems for the 14 Bravais lattices and the locations of their lattice point(s) within a unit cell

Types of structure	Locations of lattice points
Simple cubic (SC)	[0, 0, 0]
Body-centered cubic (BCC)	[0, 0, 0] [1/2, 1/2, 1/2]
Face-centered cubic (FCC)	[0, 0, 0] [1/2, 1/2, 0] [1/2, 0, 1/2] [0, 1/2, 1/2]
Simple hexagonal	[0, 0, 0]
Simple trigonal	[0, 0, 0]
Simple tetragonal	[0, 0, 0]
Body-centered tetragonal	[0, 0, 0] [1/2, 1/2, 1/2]
Simple orthorhombic	[0, 0, 0]
Body-centered orthorhombic	[0, 0, 0] [1/2, 1/2, 1/2]
Base-centered orthorhombic	[0, 0, 0] [1/2, 1/2, 0]
Face-centered orthorhombic	[0, 0, 0] [1/2, 1/2, 0] [1/2, 0, 1/2] [0, 1/2, 1/2]
Simple monoclinic	[0, 0, 0]
Base-centered monoclinic	[0, 0, 0] [1/2, 1/2, 0]
Simple triclinic	[0, 0, 0]

centers, body center, or face centers occupied, their unit cells contain more than one lattice point, such as four lattice points within a face-centered cubic unit cell.

3.4.2 Indices of Crystallographic Directions [uvw]

To determine a crystallographic direction in a 3D lattice, one can express it as a linear combination of components along the ***a***, ***b*** and ***c*** axes using Eq. 3.3. This involves assigning values to the three indices of *u, v,* and *w*.

(1) Establish a coordinate system with the origin on the direction to be determined.
(2) Select a point (other than the origin) on the direction.
(3) Find the coordinates (*x, y, z*) of the selected point.
(4) Reduce the three numbers to the smallest integer values, *u, v,* and *w*, by multiplying or dividing by a common factor. The *u*, *v*, and *w* correspond to the *x, y,* and *z* axes, respectively.

(5) Enclose the three integers in square brackets, [*uvw*], without any separation marks.
(6) Use a bar over the index number to represent a negative Miller index value, for example, [$\bar{u}vw$] means the first index is a negative number.

By the procedure stated above, the common directions notated by Miller indices for a cubic unit cell are shown in Fig. 3.11. However, when finding the Miller indices of a crystallographic direction, the direction may not pass through the origin of the coordinate system. In such cases, a translation along the vector or the coordinate is necessary, but no rotation is permitted.

All parallel direction vectors in a lattice have the same direction indices. Directions are considered to be crystallographically equivalent if the atom spacing along each of them is the same. Therefore, all parallel directions in a crystal are crystallographically equivalent. In addition to the sets of parallel directions, some sets of nonparallel directions with different indices are also possible to be crystallographically equivalent if they have the same atomic spacing. For example, in a cubic unit cell, shown in Fig. 3.11, the three directions along the face diagonals, [$\bar{1}\bar{1}0$], [$\bar{1}10$], and [$\bar{1}01$], meet the necessary condition, namely having the same atomic spacing, and they are crystallographically equivalent. A set of equivalent lattice directions can be denoted by a special code, *<uvw>*. It denotes all lattice directions equivalent to the lattice direction, including [*uvw*] itself. Such a set is called an equivalent family of lattice directions.

It is not difficult to find the six equivalent directions belonging to the <100> family of lattice directions in a crystal having cubic unit cell, actually the three edges of the unit cell. Likewise, the notations<110> and <111> are used to

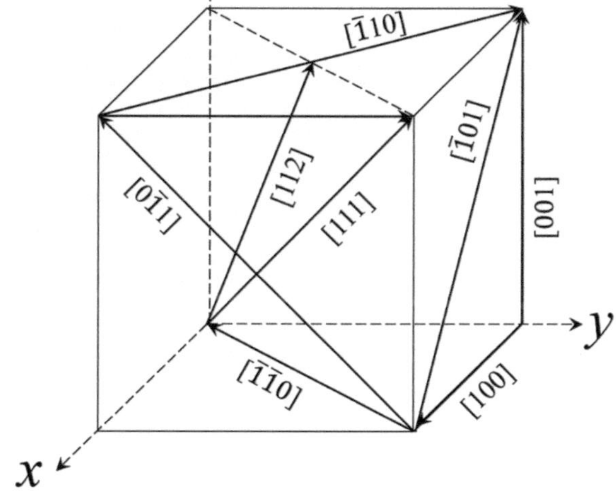

Fig. 3.11 The common directions with a lattice. E.g., [100], the direction along *x* axis; [$\bar{1}\bar{1}0$], the *x*–*y* face diagonal of the unit cell; [111], the body diagonal

represent the cubic face diagonals and body diagonals, respectively.

$$
\begin{aligned}
<100>:&\ [100], [010], [001], [\bar{1}00], [0\bar{1}0], \text{ and } [00\bar{1}] \\
<110>:&\ [110], [101], [011], [\bar{1}10], [1\bar{1}0], [\bar{1}01], [10\bar{1}], \\
&\ [\bar{1}\bar{1}0], [\bar{1}0\bar{1}], [0\bar{1}\bar{1}], [01\bar{1}], [0\bar{1}1] \\
<111>:&\ [111], [\bar{1}\bar{1}\bar{1}], [\bar{1}11], [1\bar{1}1], [11\bar{1}], [1\bar{1}\bar{1}], [\bar{1}1\bar{1}], [\bar{1}\bar{1}1],
\end{aligned}
$$

To clarify, directions with the same indices in a cubic crystal, regardless of order and sign, such as the <110> listed above, and [123] and $[3\bar{1}2]$ are equivalent. Thus, for the general case (u, v and w are not equal to each other, and they are all non-zero), a set of 48 equivalent directions exists. However, this is not the case for crystal systems that are not cubic. For example, in crystals of tetragonal symmetry ($a = b \neq c$), [100] and [010] directions are equivalent, while [100] and [001] are not.

For crystals belonging to the hexagonal crystal system, another common structure for crystalline materials, their crystallographic equivalent directions do not have the same set of indices, as they do in the cubic systems. To simplify dealing with these crystals, a Miller-Bravais coordinate system with four axes is adopted, as shown in Fig. 3.12. In the Miller–Bravais scheme, there are three axes, $\boldsymbol{a_1}$, $\boldsymbol{a_2}$, and $\boldsymbol{a_3}$, contained within the same plane (the basal plane) at 120° to each other, and the z axis is perpendicular to this basal plane, thus perpendicular to all the three axes in that plane.

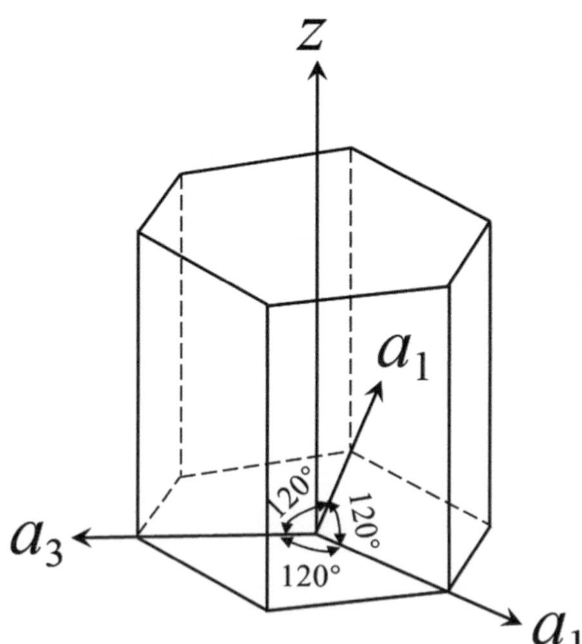

Fig. 3.12 Coordinate system construction of Miller-Bravais scheme for a hexagonal unit cell

Directional indices, steps stated previously still applicable for the Miller–Bravais scheme, shall consist of four numbers, [*uvtw*]. By convention, the first three numbers, *u*, *v*, and *t*, pertain to projections along the $\boldsymbol{a}_1$, $\boldsymbol{a}_2$, and $\boldsymbol{a}_3$ axes, respectively, and *w* to *z* axis. Since three parameters are sufficient to determine a vector in 3D space, there must be a redundant number among the four. To convert from the three-axis coordinate system, [*u'v'w'*], to the Miller-Bravais system, [*uvtw*], the following procedure is used.

$$\begin{aligned} u &= \frac{1}{3}(2u' - v') \\ v &= \frac{1}{3}(2v' - u') \\ t &= -(u + v) \\ w &= w' \end{aligned} \tag{3.4}$$

The auxiliary index in the Miller-Bravais system is t, as shown in the conversion formulae. By the set of formula, [110] corresponds to [$11\bar{2}0$], and [210] corresponds to [$10\bar{1}0$]. Then, using the [*uvtw*] indices, equivalent lattice directions in hexagonal crystals have the same set of numbers without regard to their order and sign.

3.4.3 Indices of Crystallographic Planes (Hkl)

When a plane intersects the three axes (***a***, ***b***, and ***c***) of a given 3D lattice, it has intercepts of *m* (***x***-axis), *n* (***y***-axis), and *p* (*z*-axis) in terms of the edge lengths of its unit cell. The intercept numbers do not have to be integral, and they determine the position of the lattice plane. For example, shown in Fig. 3.13, a plane has three intersects ($m = 1/3$, $n = 2/3$, and $p = 1$), and its orientation and position are completely specified. By convention, the Miller indices of a crystal plane are normalized as follows.

(1) Determine the intercepts of the plane in terms of the edge lengths of the unit cell of the lattice, *m*, *n*, and *p*. The three intercepts may be fractions. E.g., the 1/3, 2/3, and 1 for the plane shown in Fig. 3.13;
(2) Form the reciprocals of the three intercepts, resulting in three numbers, namely, 1/*m*, 1/*n*, and 1/*p*. E.g., 3, 3/2, and 1, corresponding the example stated above;
(3) Find the least common denominator (LCD) and multiply each reciprocal by this LCD, resulting three integers (usually denoted by *hkl*) without common divisor;
(4) Enclose the three integers within parentheses to obtain the Miller indices of the plane, such as (632) for the example given, putting nothing in between them.

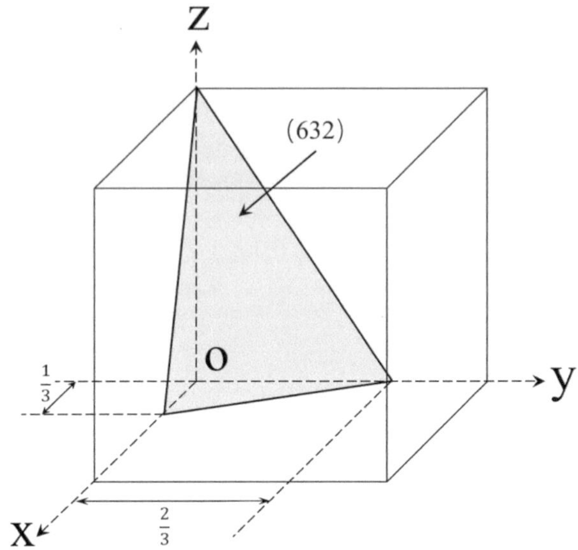

Fig. 3.13 A unit cell of a cubic lattice. A plane having intersects of *a*/3, 2*a*/3, and *a* on *x*, *y*, and *z* axes, respectively

The three integers written complying with the convention as stated above are called Miller indices of the plane. In cases where a negative intercept occurs, the same notation as used for crystal direction is employed, with a bar placed over the corresponding Miller index. For example, ($3\bar{1}2$) plane has a negative intercept on *y*-axis.

The geometrical meaning of the Miller indices is very clear, as shown in Fig. 3.14. When a plane is parallel to any axis, the corresponding intercept of the plane is infinity, resulting to a zero reciprocal, and thus a zero Miller index for the axis, e.g., the planes (001), (010), and (100). When a plane passes through the origin, it shall have zero intercepts, e.g., the plane ABFE in Fig. 3.14. This poses a challenge since the steps listed above cannot be applied directly. The plane must be translated away from the origin by translating the plane or the coordinate axes. A new coordinate system (*x'*-*y'*-*z'*) is used to circumvent this inconvenience for these planes.

Similar to the crystal direction, the notation (*hkl*) represents not only a particular lattice plane, but an infinite set of parallel planes with a constant interplanar spacing. It should be noted that (*hkl*) and ($\overline{\text{hkl}}$) indices actually define the same parallel set of planes. Only when the orientation and distance between the planes are of interest, there is no need to separately consider (*hkl*) and ($\overline{\text{hkl}}$).

In cubic lattices, planes and directions with equal Miller indices are mutually perpendicular, and this important relationship is true only for cubic lattices. For example, [111] direction (body diagonal of unit cells) is perpendicular to the (111) plane. The interplanar spacing (often referred by *d*-spacing) between two closest parallel planes with the same Miller indices is designated as d_{hkl}, where *h, k,* and *l* are the Miller indices of the parallel planes of the set. The spacing d_{hkl} represents the distance from a plane to its nearest neighbor in the same set. As a general rule,

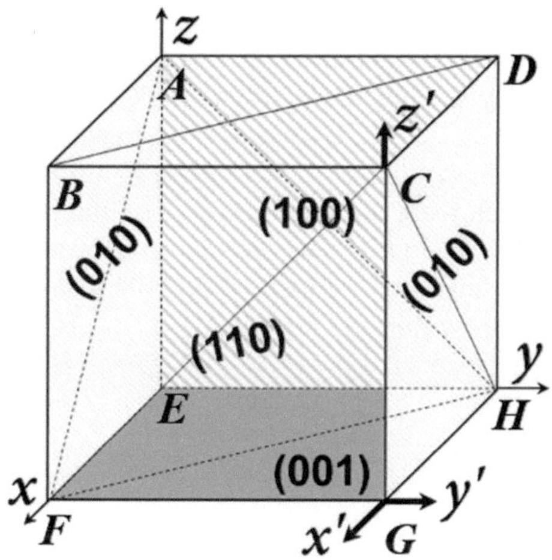

Fig. 3.14 Indexing of common planes in a unit cell of a cubic lattice

it can be said the higher the density of lattice points in each member of the (*hkl*) family, the larger the corresponding d_{hkl}. Lattice planes with low values of *h*, *k* and *l* have high lattice point density and larger d_{hkl}.

Similar to the equivalent lattice directions ([*uvw*] and<*uvw*>), a family of planes, such as {*hkl*}, contains all planes that have the same atomic arrangement and are considered crystallographically equivalent. These families are designated by indices enclosed in braces, and include not only the plane with the specified indices (*hkl*), but also its equivalent ones. For instance, the plane family of {111} contains (111), $(11\bar{1})$, and so on. Sets of symmetry-equivalent plane families, like (*hkl*), (*h'k'l'*), and (*h"k"l"*), are equivalent if they have the same lattice plane spacing, which is $d_{hkl} = d_{h'k'l'} = d_{h''k''l''}$. In cubic crystals, arbitrary permutation and sign change of *h*, *k*, and *l* indices result in a maximum of 48 equivalent plane families within the {*hkl*} set. If certain *h*, *k*, and *l* have the same value or one or two are zero, some of the listed families become "degenerated." For example, the set {111} ($h = k = l = 1$) has only eight families of lattice planes: (111), $(\bar{1}11)$, $(\bar{1}11)$, $(1\bar{1}1)$, $(\overline{111})$, $(\bar{1}1\bar{1})$, $(\bar{1}1\bar{1})$, $(\overline{11}1)$ and the set {100} comprises only six families: (100), (010), (001), $(\bar{1}00)$, (010), $(00\bar{1})$.

For hexagonal crystals, to express their equivalent families of lattice planes from the notation {*hkl*} is less straightforward than for cubic crystals. To express their equivalent families of lattice planes, an additional index *i* is inserted in the notation {*hkl*} before the "*l*", resulting in a quaternary notation (*hkil*), whereas *h*, *k* and *l* refer to the basis vectors $\boldsymbol{a_1}$, $\boldsymbol{a_2}$ and $\boldsymbol{c}$ in the same way as described in the previous section for lattice directions, and the additional index *i* refers to an additional auxiliary basis vector $\boldsymbol{a_3} = -(\boldsymbol{a_1} + \boldsymbol{a_2})$. This is known as Miller-Bravais indices and provides an intuitive derivation of the equivalent families of lattice planes. The additional index *i* is linearly dependent on $\boldsymbol{a_1}$ and $\boldsymbol{a_2}$ and has the numerical relationship $|\boldsymbol{a_1}| = |\boldsymbol{a_2}| = |\boldsymbol{a_3}| = \boldsymbol{a}$, where $\boldsymbol{a}$ is the basis vector.

Therefore, for each (hkl) plane of the hexagonal crystals, one always can add i to (hkl) according to $i = \bar{\mathrm{h}} + \bar{\mathrm{k}}$, deriving $(hkil)$.

By adding the auxiliary index i, the $(hkil)$ index can be incorporated into a permutation scheme similar to that used for cubic crystals. The set of equivalent families of lattice planes $\{hkil\}$ in hexagonal crystals can be determined as follows: (1) allowing for an arbitrary permutation of h, k, and i (but not l), (2) also allowing for h, k, and i to be simultaneously replaced by $\bar{\mathrm{h}}$, $\bar{\mathrm{k}}$ and $\bar{\mathrm{i}}$, and (3) additionally allowing l to be replaced by $\bar{\mathrm{l}}$. This leads to a maximum of 24 families of lattice planes within the set $\{hkil\}$:

$$(hkil), (ihkl), (kihl), (khil), (hikl), (ikhl),$$

$$(\bar{h}\bar{k}\bar{i}l), (\bar{i}\bar{h}\bar{k}l), (\bar{k}\bar{i}\bar{h}k), (\bar{k}\bar{h}\bar{i}l), (\bar{i}\bar{h}\bar{k}l), (\bar{i}\bar{k}\bar{h}l)$$

$$(hk\overline{il}), (ihk\bar{l}), (kih\bar{l}), (khi\bar{l}), (hik\bar{l}), (ikh\bar{l})$$

$$(\bar{h}\bar{k}\bar{i}\bar{l}), (\bar{i}\bar{h}\bar{k}\bar{l}), (\bar{k}\bar{i}\bar{h}\bar{l}), (\bar{k}\bar{h}\bar{i}\bar{l}), (\bar{h}\bar{i}\bar{k}\bar{l}), (\bar{i}\bar{k}\bar{h}\bar{l})$$

Using the formula for d_{hkl} in hexagonal crystals results in the same value of d_{hkl} for all 24 families of lattice planes discussed above. As with cubic crystals, degeneracy can occur in hexagonal crystals, meaning the number of equivalent families of lattice planes is often less than 24, if certain values of h, k, i or l are equal to each other or 0.

3.5 Linear and Planar Densities

As previously explained, the equivalence of crystallographic directions and planes arises from their sharing of the same atomic packing, which results in equal linear and planar densities. In other words, the linear and planar densities serve as the one- and two-dimensional counterparts, respectively, of the atomic packing factor (APF).

Linear density (LD) is defined as the number of atoms contained in a unit length along a specific crystal direction, with the centers of the atoms located on the crystal direction. It can be determined counting the number of atoms along a given direction and dividing by the length (Eq. 3.5). And similarly, the planar density (PD) is determined by Eq. 3.6.

$$\mathrm{LD} = \frac{\text{numer of atoms centered on direction vector}}{\text{length of the direction vector}} \tag{3.5}$$

$$\mathrm{PD} = \frac{\text{numer of atoms centered on a plane}}{\text{area of the plane}} \tag{3.6}$$

Taking the face-centered cubic structure as an example, the bottom surface of its unit cell is shown in Fig. 3.15. The crystal direction [110] passes through atoms x, y, and z. Among them, the atom y is completely located on the direction, while atoms x and z are shared with adjacent cells along the [110] direction, so only half of them are located on the crystal direction. As a result, according to Eq. 3.5, in the FCC cells, the linear density along [110] direction is calculated in terms of the unit cell parameter *a* as shown in Eq. 3.7. In the same Figure, it is easy to count the number of atoms centered on the bottom face of the FCC unit cell, i.e., the (001) plane. There are 5 atoms on this crystal plane, with the center atom completely located on the plane and the other 4 atoms sharing with the neighboring crystal cells (only 1/4 of them on the plane). Therefore, according to Eq. 3.6, the PD of (001) plane in an FCC cell is calculated as shown in Eq. 3.8.

$$-\mathrm{LD}_{[110]} = \frac{1+\frac{1}{2}\times 2}{\sqrt{2}a} = \frac{\sqrt{2}}{a} \tag{3.7}$$

$$-\mathrm{PD}_{(001)} = \frac{1+4\times\frac{1}{4}}{a^2} = \frac{2}{a^2} \tag{3.8}$$

Tables 3.4 and 3.5 list the LD and PD for common directions and planes of FCC and BCC lattices, respectively. The information about LD and PD is crucial in determining the crystal structure, which in turn affects its physical and chemical properties. For example, high LD and PD indicate closely packed atoms, resulting in stronger interatomic bonds and higher mechanical strength. Conversely, low LD and PD indicate looser packing and weaker bonds, resulting in lower

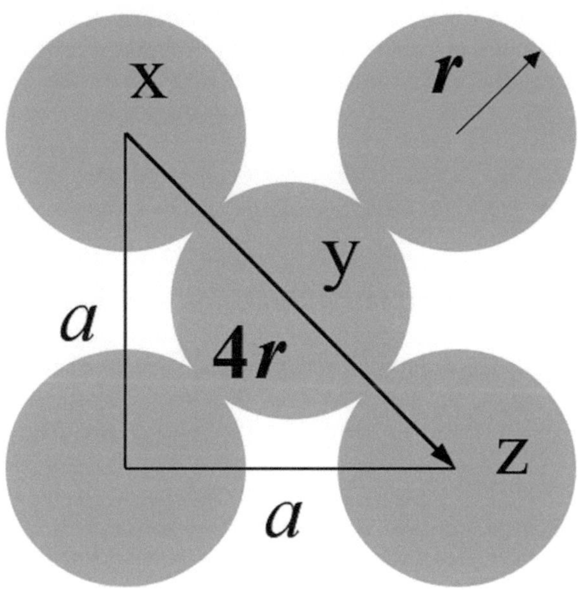

Fig. 3.15 Geometric relationship between atomic radius and lattice parameter in a cubic unit cell

Table 3.4 Some common crystal directions: their atomic arrangements and linear densities

Directions	FCC lattice		BCC lattice	
	Arrangements of atoms	LD (atoms/ unit length)	Arrangements of atoms	LD (atoms/ unit length)
<100>	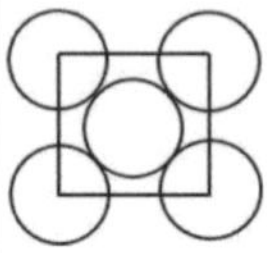	$\frac{2\times\frac{1}{2}}{a}=\frac{1}{a}$	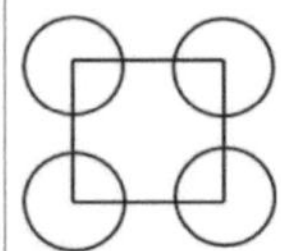	$\frac{2\times\frac{1}{2}}{a}=\frac{1}{a}$
<110>		$\frac{1+2\times\frac{1}{2}}{\sqrt{2}a}=\frac{\sqrt{2}}{a}$		$\frac{2\times\frac{1}{2}}{\sqrt{2}a}=\frac{\sqrt{2}}{2a}$
<111>		$\frac{2\times\frac{1}{2}}{\sqrt{3}a}=\frac{\sqrt{3}}{a}$		$\frac{2\times\frac{1}{2}+1}{\sqrt{3}a}=\frac{2\sqrt{3}}{a}$

mechanical strength. These concepts have broad applications in various fields, and comprehending their importance is critical for developing new materials and technologies.

LD and PD are important factors in determining the anisotropy of single crystal materials. Different atomic densities and binding forces are observed on planes and directions with different LD and PD, leading to different physical, chemical, and mechanical properties in different directions. This explains why crystals generally exhibit anisotropy. Cleavage properties are a common example of this phenomenon, where crystals have the tendency to cleave along specific orientations, often related to high planar density surfaces exposed on the crystal surface. Cleavage fractures often occur on these specific surfaces and in these specific directions, and the crystal surface tends to be composed of these surfaces.

3.6 Interplanar Spacings and Angles

The distance between adjacent (*hkl*) lattice planes in a set of parallel planes is defined as the interplanar spacing d_{hkl}. Contained in Table 3.6 are the formulae giving the d_{hkl} in terms of Miller indices and their lattice parameters for all the seven crystal systems.

The *d*-spacing is a fundamental aspect of X-ray diffraction technology and serves as a unique identifier, like a fingerprint, for a specific crystal.

Sometimes, the angles between two crystal planes are of interest. It can be calculated using the normal of the planes. With the help of scientific calculators,

Table 3.5 Some common crystal planes: their atomic arrangements and planar densities

Planes	FCC lattice		BCC lattice	
	Arrangements of atoms	LD (atoms/unit length)	Arrangements of atoms	LD (atoms/unit length)
{100}	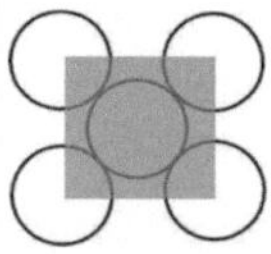	$\frac{4\times\frac{1}{4}+1}{a^2}=\frac{2}{a^2}$	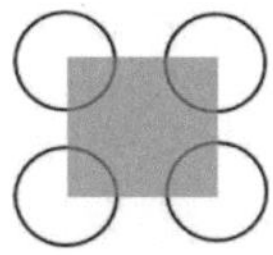	$\frac{4\times\frac{1}{4}}{a^2}=\frac{1}{a^2}$
{110}	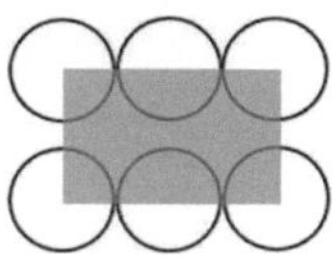	$\frac{4\times\frac{1}{4}+2\times\frac{1}{2}}{\sqrt{2}a^2}=\frac{\sqrt{2}}{a^2}$	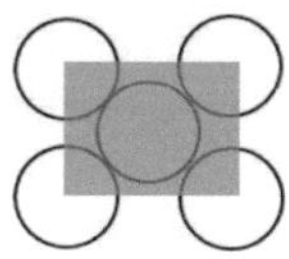	$\frac{4\times\frac{1}{4}+1}{\sqrt{2}a^2}=\frac{\sqrt{2}}{a^2}$
{111}		$\frac{3\times\frac{1}{6}+3\times\frac{1}{2}}{\frac{\sqrt{3}}{2}a^2}=\frac{4\sqrt{3}}{3a^2}$	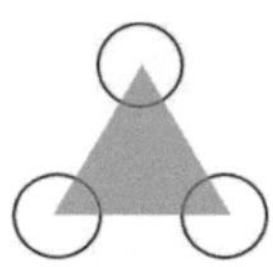	$\frac{3\times\frac{1}{6}}{\frac{\sqrt{3}}{2}a^2}=\frac{\sqrt{3}}{3a^2}$

Table 3.6 The interplanar spacings (d short for d_{hkl}) in seven crystal systems

Crystal system	$1/d^2$
Cubic	$\frac{h^2+k^2+l^2}{a^2}$
Hexagonal	$\frac{l^2}{c^2}+\frac{4(h^2+hk+k^2)}{3a^2}$
Trigonal	$\frac{(h^2+k^2+l^2)\sin^2\alpha+2(hk+kl+hl)(\cos^2\alpha-\cos\alpha)}{a^2(1-3\cos^2\alpha+2\cos^3\alpha)}$
Tetragonal	$\frac{h^2+k^2}{a^2}+\frac{l^2}{c^2}$
Orthorhombic	$\frac{h^2}{a^2}+\frac{k^2}{b^2}+\frac{l^2}{c^2}$
Monoclinic	$(\frac{h^2}{a^2}+\frac{k^2\sin^2\beta}{b^2}+\frac{l^2}{c^2}-\frac{2hl\cos\beta}{ac})\csc^2\beta$
Triclinic	$\frac{\frac{h^2}{a^2}\sin^2\alpha+\frac{k^2}{b^2}\sin^2\beta+\frac{l^2}{c^2}\sin^2\gamma+\frac{2kl}{bc}(\cos\beta\cos\gamma-\cos\alpha)+\frac{2hl}{ac}(\cos\gamma\cos\alpha-\cos\beta)+\frac{2hk}{ab}(\cos\alpha\cos\beta-\cos\gamma)}{1-\cos^2\alpha-\cos^2\beta-\cos^2\gamma+2\cos\alpha\cos\beta\cos\gamma}$

spreadsheets and computer programs, the tabulations of interplanar angles are no longer necessary. The expression for determining interplanar angles can be used to easily find the values of interest. And the expressions can be found in a handbook on crystallography.

3.7 The Weiss Zone Law

A group of crystallographic planes that intersect in parallel edges is known as a zone. The edges are parallel to a line, referred to as the zone axis, which is represented by a zone symbol notation similar to Miller indices (Fig. 3.16).

The mathematical condition for a vector [*uvw*] to lie in a crystallographic plane (*hkl*) is expressed by Zone law, also known as Weiss zone law. This condition can be determined entirely through elementary vector operations. Figure 3.17 represents the plane (*hkl*) with the normal to the plane ***OP***.

In Fig. 3.17, a general vector, ***r***, lying in the plane (*hkl*) can be expressed as a linear combination of any two unparallel vectors lying in the same plane, such as ***AB*** and ***AC*** according to Eq. 3.9 as follows.

$$\boldsymbol{r} = \lambda \boldsymbol{AB} + \mu \boldsymbol{AC} \tag{3.9}$$

In the equation, expressing ***AB*** and ***AC*** in terms of ***a***, ***b*** and ***c***, resulting in Eq. 3.10.

$$\boldsymbol{r} = \lambda\left(\frac{\boldsymbol{b}}{k} - \frac{\boldsymbol{a}}{h}\right) + \mu\left(\frac{\boldsymbol{c}}{l} - \frac{\boldsymbol{a}}{h}\right) = -\frac{(\lambda+\mu)}{h}\boldsymbol{a} + \frac{\lambda}{k}\boldsymbol{b} + \frac{\mu}{l}\boldsymbol{c} \tag{3.10}$$

The derived equation can be re-expressed as Eq. 3.11.

$$\boldsymbol{r} = u\boldsymbol{a} + v\boldsymbol{b} + w\boldsymbol{c} \tag{3.11}$$

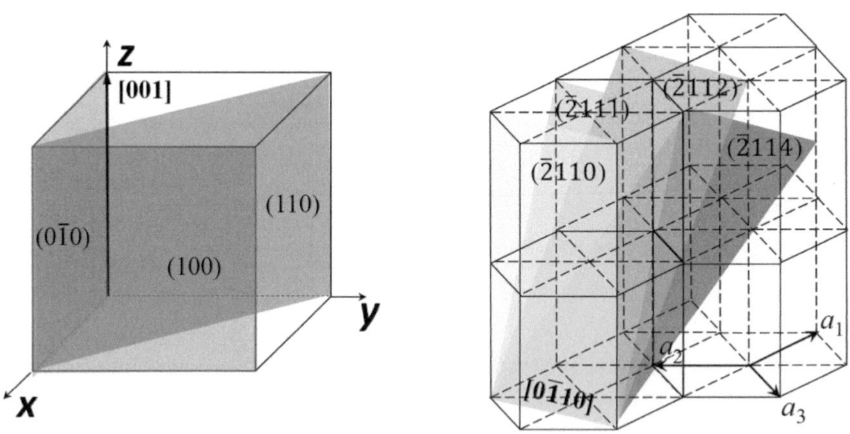

Fig. 3.16 Zones in cubic and hexagonal unit cells

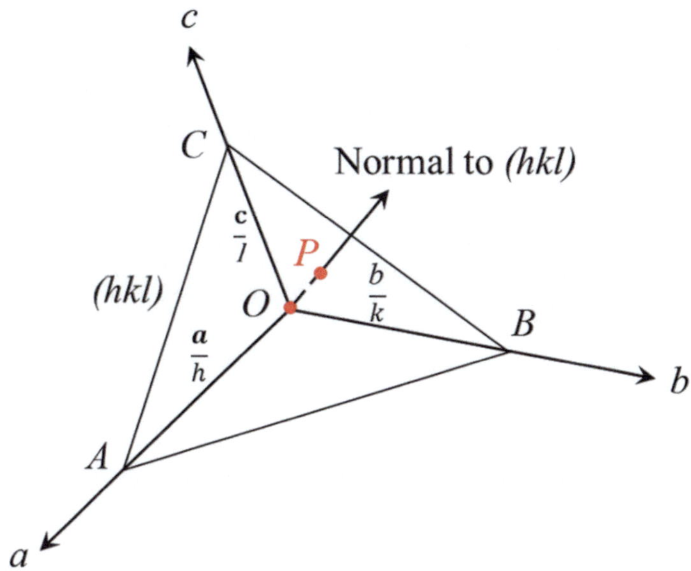

Fig. 3.17 The figure assisting the derivation of the Weiss zone law

And, $\boldsymbol{r}$ is a general vector $[uvw]$ lying in (hkl). Thus, it gives an equation set as follows.

$$\begin{aligned} u &= -\frac{(\lambda + \mu)}{h} \\ v &= \frac{\lambda}{k} \\ w &= \frac{\mu}{l} \end{aligned} \tag{3.12}$$

Out of the set of Eq. 3.12, it obviously gives a new expression after cancelling out the two parameters, λ and μ, as follows.

$$hu + kv + lw = 0 \tag{3.13}$$

A vector $[uvw]$ lies in the plane (hkl) as long as the Eq. 3.13 holds. This is called the **Weiss zone law**. This derivation shows that the Weiss zone law applies to any orientation of the x-, y-, and z-axes relative to each other (the angles between the axes are irrelevant).

A number of important crystal lattice planes usually lie in the same zone, meaning that they intersect one another all in parallel lines. For instance, in cubic lattices, the crystal planes (100), $(0\bar{1}0)$ and (110) are all parallel to the direction [001]. Therefore, they are in the zone [001], and [001] is a common direction lying in all the planes. The orientation of these planes is perpendicular to the direction [001], and this is not a coincidence but rather determined by the Weiss zone law.

To determine the indices of the zone $[uvw]$ in which two crystal planes with known Miller indices, such as $(h_1k_1l_1)$ and $(h_2k_2l_2)$, lie, one needs to solve the simultaneous equation set described in Eq. 3.14.

$$\begin{aligned} -h_1u + k_1v + l_1w &= 0 \\ -h_2u + k_2v + l_2w &= 0 \end{aligned} \tag{3.14}$$

Since it is only the ratios u:v:w which are of interest, these equations can be solved, giving Eq. 3.15.

$$\begin{aligned} u &= k_1l_2 - k_2l_1 \\ v &= l_1h_2 - l_2h_1 \\ w &= h_1k_2 - h_2k_1 \end{aligned} \tag{3.15}$$

It is also evident that for two planes $(h_1k_1l_1)$ and $(h_2k_2l_2)$ lying in the same zone $[uvw]$, their linear combination, given by Eq. 3.16 also lies in the same zone $[uvw]$.

$$(mh_1 + nh_2)u + (mk_1 + nk_2)v + (ml_1 + nl_2)w = 0 \tag{3.16}$$

Taking linear combinations of the indices of two planes belonging to a specific zone generates the indices of another plane in that same zone. Thus, the indices formed by these linear combinations provide the indices of the other plane. It's worth noting that m and n can be either positive or negative in general.

3.8 The Symmetry of Crystals

The first question regarding symmetry is its definition. Symmetry is a concept that we can perceive in our daily lives, whether we are aware of it or not. It is an almost universal feature in nature and human civilization and is considered one of the most basic principles governing the natural world. Most organisms exhibit some form of symmetry in their appearance, while buildings are often designed with symmetry, not only for aesthetic reasons. Figure 3.18 displays a butterfly with a symmetrical shape, while Fig. 3.19 shows the well-known Forbidden City, which features symmetrically arranged buildings. In general, symmetry gives people a sense of harmony and beautiful proportion and balance.

To utilize symmetry, a precise definition is necessary. In mathematics, symmetry is defined more rigorously and generally refers to an object that remains unchanged under certain transformations, such as reflection, rotation, and so on. In this context, we will first examine symmetry in everyday objects before discussing symmetry in crystals.

A symmetry element is a geometrical entity about which a symmetry operation is carried out. These elements can be in the form of a point, axis, or plane (mirror). Figure 3.20 illustrates that a benzene molecule possesses an axis (c), which is

Fig. 3.18 A butterfly having a perfect symmetrical shape

Fig. 3.19 The forbidden city

perpendicular to the molecule plane and passes through its center. If the molecule is rotated by $\pi/6$ around this axis, it will appear the same as it did before the rotation. A symmetry operation is the movement of an object (such as a molecule or crystal) that results in the object appearing identical to its original state. The existence of a symmetry operation implies that a corresponding symmetry element

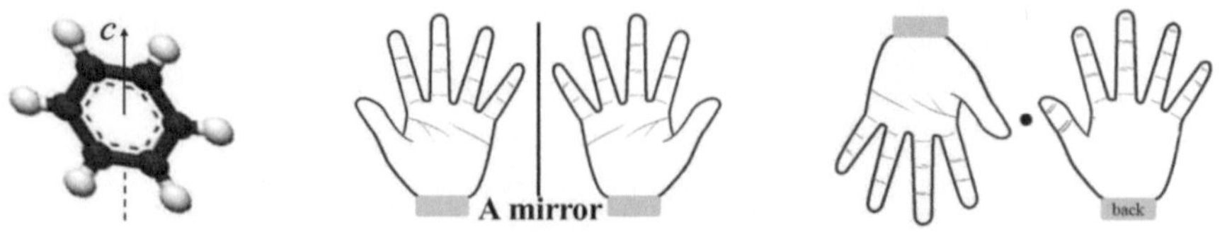

Fig. 3.20 The symmetry elements of axis, plane and point (the two hands should be considered as a whole)

is present, and conversely, the presence of a symmetry element suggests that a certain symmetry operation or set of operations is possible.

3.8.1 Point Symmetry Operations

In general, symmetry operations at a macroscopic level are not translational and can include rotation, inversion, and a combination of the two. These operations are known as point symmetry operations.

Mathematically, symmetry operation is a geometric transformation exerted on a crystal, transforming a point r to r' according to Eq. 3.17 by a transformation matrix $\boldsymbol{A}$ stated in Eq. 3.18.

$$r'(x_1', x_2', x_3') = \mathbf{A}r(x_1, x_2, x_3) \tag{3.17}$$

$$\boldsymbol{A} = \left(a_{ij}\right) = \begin{pmatrix} a_{11} & a_{12} & a_{13} \\ a_{21} & a_{22} & a_{23} \\ a_{31} & a_{32} & a_{33} \end{pmatrix} \tag{3.18}$$

About the transformation described by Eq. 3.17, some points can be summarized as follows.

It is an orthogonal transformation, meaning $\boldsymbol{A}^{\mathrm{T}}\boldsymbol{A}$ equals unit matrix ($\boldsymbol{I}$). This kind of symmetry operations is rigid, and the distance between any two points in the crystals retains the same before and after the transformation. If an orthogonal transformation cannot make difference to a crystal, this transformation is a symmetry operation to the crystal. The symmetry operations of a crystal reflect its symmetry, and the higher the number of symmetry operations, the greater the symmetry level of the crystal.

The symmetry operations of a 3D object are typically expressed as rotation about an axis, inversion around a center of symmetry, or a combination of both operations. The transformation matrices for the rotation and inversion are listed in Eqs. 3.19 and 3.20, respectively, assuming the rotation axis and angle as z and θ, respectively.

$$\boldsymbol{A} = \begin{pmatrix} \cos\theta & -\sin\theta & 0 \\ \sin\theta & \cos\theta & 0 \\ 0 & 0 & 1 \end{pmatrix}, |\boldsymbol{A}| = 1 \tag{3.19}$$

$$A = \begin{pmatrix} -1 & 0 & 0 \\ 0 & -1 & 0 \\ 0 & 0 & -1 \end{pmatrix}, |A| = 1 \tag{3.20}$$

3.8.2 Symmetry Elements

To describe the symmetry of an object, it is more convenient to find all its symmetry elements rather than listing all its symmetry operations. Two notations are commonly used for this purpose: the Schönflies Notation and the so-called International Notation. The former is sufficient for specifying point groups in three dimensions and is commonly used in spectroscopy to describe the symmetry of a molecule. The latter, developed by Charles-Victor Mauguin and Carl Hermann, is the optimal choice for describing the symmetry of crystals. The different types of point operations include identity, rotation, inversion, and reflection.

The ***Identity***, the symmetry operation that can describe all objects, is a trivial symmetry operation that leaves an object unchanged. This symmetry operation is denoted by the symbol 1 in the International Notation or E in the Schönflies Notation. Its transformation matrix $\boldsymbol{A}$ is in Eq. 3.21, containing 1's along the diagonal and 0's off the diagonal, and the matrix is called unit or identity matrix.

$$A = \begin{pmatrix} 1 & 0 & 0 \\ 0 & 1 & 0 \\ 0 & 0 & 1 \end{pmatrix} \tag{3.21}$$

The ***Rotation*** is a kind of operation about an axis. If an object can be rotated by an angle of 180° (π in radians) around an axis and return to its original state, it is considered to have twofold rotational symmetry, denoted as $2(C_2)$ in both the International (Schönflies) notations. Generally, if an object can be rotated by an angle of $2\pi/n$ (or its multiples) around an axis and return to its original state, then this axis is the n-fold axis of the object, and the rotation is referred to as a pure or proper rotation, denoted as $n(C_n)$.

The ***Inversion,*** sometimes called centrosymmetry, is a more complex operation than rotation because it transforms right hands to left hands, and vice visa, as shown in Fig. 3.20 (the rightmost sketch). Every position in space having the coordinates (x, y, z) is given the new coordinates $(-x, -y, -z)$ by the inversion operator $\bar{1}$ (i). The inversion operation can be written as Eq. 3.17 with its transformation matrix $\boldsymbol{A}$ shown in Eq. 3.20.

The ***Reflection***, sometimes called a mirror reflection, is denoted by the symbol m in the international notation and σ in the Schönflies notation. This type of symmetry operation is shown in Fig. 3.20 (the middle sketch). The symmetry element for reflection is called a mirror plane. For a given point in space, by reflection operation, it finds its counterpart at the line going through itself and perpendicular onto the mirror plane, and both the point and its mirror counterpart have the same distance to the mirror plane along the perpendicular line. Thus, if the left hand is drawn palm up on the left side of the mirror, as shown in Fig. 3.20 (the middle one), the mirror image is a right hand, also palm up, found on the other side of the mirror plane. For the coordinate system shown, its transformation matrix ***A*** associated with Eq. 3.17 is as shown in Eq. 3.22, which means the mirror plane (*x–z* plane) is perpendicular to the *y* axis. From the perspective of symmetry relation, one's right hand is the enantiomorphic (mirror) image of the left hand.

$$\boldsymbol{A} = \begin{pmatrix} 1 & 0 & 0 \\ 0 & -1 & 0 \\ 0 & 0 & 1 \end{pmatrix} \quad (3.22)$$

A subscript is commonly added to the Schönflies symbol for a mirror operation (σ) to indicate the orientation of the mirror plane. For instance, if the principal axis is defined to be the *c*-axis, σ_h refers to a mirror reflection through a plane perpendicular to this principal axis, where the subscript *h* stands for horizontal, assuming that the principal axis is vertical.

The ***Rotation-inversion***, the last type of symmetry operation, is the most complex one. This type of symmetry is known as improper rotations. A rotation-inversion operation is in fact a compound operation, which is the product of two separate operations. It is important to understand that generally, each of the two individual operations is not by itself a symmetry operation for a particular crystal or molecule, but the product (the compound one) of the two is.

Another reason making improper rotations complex is their notations and treatment approaches. In the International system, they are called rotation-inversions, while in the Schönflies system, they are referred to as rotation-reflections. In the Schönflies system, an *n*-fold improper rotation axis, denoted as S_n, is an axis of rotatory reflection, and the operation is a combination of rotation by 360°/*n* followed by reflection in a plane normal to the S_n axis; in the international notation system, an *n*-fold improper rotation axis, denoted by $\bar{n}$ is an axis of rotatory inversion, and the operation is a combination of rotation by 360°/*n* followed by inversion through a point. It is important to note that the effect of operations S_n and $\bar{n}$ can be different because they are defined differently.

To carry out the symmetry operation of rotation-inversion complying with the international notation, we must first perform a rotation operation $n(C_n)$, followed immediately by an inversion operation $\bar{1}$ (*i*), without any operation in between. The whole operation (two steps) can be written as a product, $\bar{1}n$ (iC_n).

Let us analyze the operation $\bar{4}(S_4^3)$ through visualization. We begin with a left hand, palm up, pointing to the top, which is labeled as (a) in Fig. 3.21. We rotate it counterclockwise through 2π/4 to position (b) and then immediately invert it through the origin to position (c). The resulting object from the entire operation is a right hand, palm down, pointing to the right, indicating the creation of a new object. To further clarify, we perform the operation again, starting with the result shown in (c) in Fig. 3.21. The second round of the operation results in a left hand, palm up, pointing to the bottom, labeled as (d) in the figure. It turns out that $\bar{4}^2$

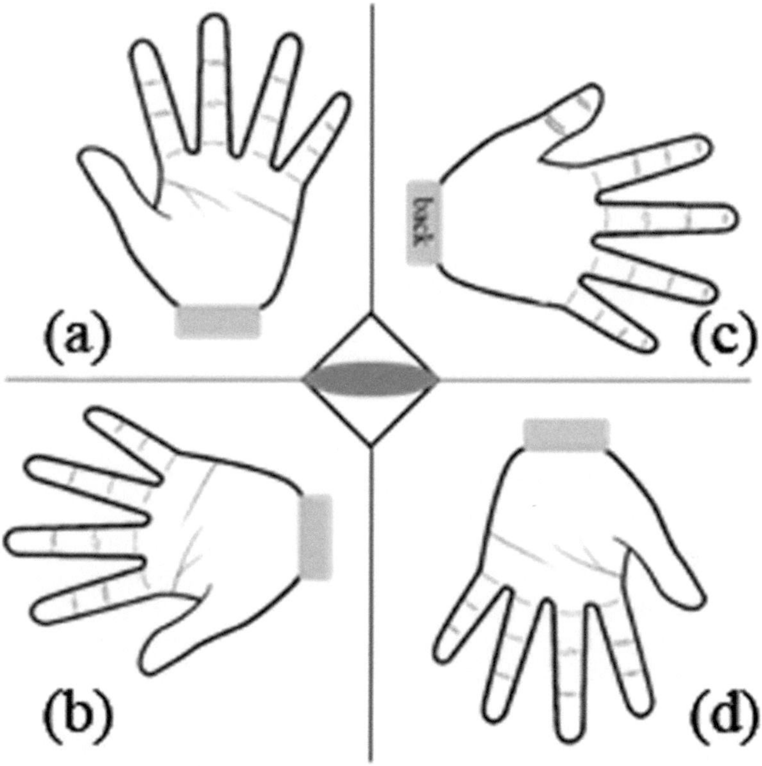

Fig. 3.21 Visualizing the operation of $\overline{4}$'. **Note* Differentiating between the operation and the point symmetry of a group of objects is important. Herein, we find all objects (hands with different orientations) by repeating a symmetry operation $\overline{4}$, but the complete assembly of objects shown in the figure has the point symmetry of $\overline{4}$

$= 2$ (or $S_4^2 = C_2$), meaning that successively applying the operation of $\overline{4}$ twice equals the operation of $2(C_2)$. We continue to reapply the operation of $\overline{4}$, and it turns out that $\overline{4}^4 = 1$.

To summarize, the various types of symmetry elements as well as their operators are tabulated in Table 3.7.

3.8.3 Symmetry Elements in Crystals

A crystal is characterized by its periodic arrangement of atoms, and the defining symmetry of a crystal is the translational symmetry. Rotational symmetry, on the other hand, may or may not be present in a crystal, but if it is, it must be consistent with the translational symmetry. Interestingly, in the case of quasi-crystals, which

Table 3.7 The macroscopic symmetry elements and their operation

Elements	Descriptions	Operations
$1(E)$	Identity	No change (rotation by 2π about any axis)
$n(C_n)$	n-fold axis of rotation	Rotation by $2\pi/n$ leaves the object unchanged
$m(\sigma)$	Plane of symmetry	Reflection in the plane leaves the object unchanged
$\overline{1}(i)$	Center of symmetry	Inversion through the center of symmetry leaves the object unchanged
$\overline{n}$	n-fold axis of improper rotation	The rotary inversion operation consists of rotating by $2\pi/n$ about the axis, followed by inversion through a point on the axis, leaving the object unchanged

lack translational periodicity, they are not bound by this restriction and can have any rotational symmetry, including fivefold.

Before we delve into the symmetry of crystals, it is helpful to look at tiling floors. While we are accustomed to seeing floors tiled by equilateral triangles, squares, or regular hexagons, as shown in Fig. 3.22a–c, a tiling of regular pentagons is less common. In each of these cases, the tiles are regular polygons, meaning they have sides of the same length and angles between them. Adjacent tiles share whole edges, resulting in no overlap between tiles and no gaps in the floor.

Regular pentagons are not commonly used in tiling floors of buildings. In contrast, triangles, squares, and hexagons are the only regular shapes that can tessellate by themselves, meaning they can be used to tile a floor without leaving any voids or overlapping parts. Other tessellations using regular shapes are possible but require the use of more than one type of shape. It is also possible to tessellate with pentagons, but not all of the pentagons used are regular.

To achieve a vertex-to-vertex tessellation with regular polygons, the interior angle of the polygon must be a divisor of 360°. Since 108° is not a divisor of

(a) Regular triangle tiles (b) Square tiles (c) Regular hexagon tiles (d) Regular pentagons (e) Regular octagons

Fig. 3.22 **a**–**c**: The three regular tilings (and only three); **d** and **e**: packing of objects displaying 5- and eightfold symmetry, which do not fill the plane completely

360°, it is impossible to tile the Euclidean plane with regular pentagons, as shown in Fig. 3.22d. Moreover, it can be observed that equilateral polygons with more than 6 edges cannot tessellate the plane in this manner, as shown in Fig. 3.22e. The underlying mathematical principles governing the existence or non-existence of such tilings can provide insight into the corresponding scenarios in crystals.

Objects like spheres where there is no translation symmetry as in a crystal can have higher-fold rotation axes, such as fivefold or sevenfold. In fact, there is only one object in the geometrical universe with perfect symmetry, and it is a sphere. It has infinite planes and rotational axes of symmetry passing through its center, and any rotation (no matter how little/much) on these axes leaves the sphere unchanged. But, restricted by its translation symmetry, only particular symmetry elements of macroscopic symmetry are found in crystals. That means symmetry breaking happens as materials crystallize, during which, some elements of the symmetry have been lost. The possible rotation axes in crystals are restricted to onefold, twofold, threefold, fourfold, and sixfold, which can be proven mathematically. For a given crystal, we choose an axis by which the crystal is rotated by θ. Then we find a lattice plane perpendicular to the rotation axis in the crystal. In the lattice plane, two basis vectors $\boldsymbol{a}_1$ and $\boldsymbol{a}_2$ are selected, and therefore, any lattice point ($\boldsymbol{R}$) lying in the plane can be represented by Eq. 3.23.

$$\boldsymbol{R} = m\boldsymbol{a}_1 + n\boldsymbol{a}_2 \tag{3.23}$$

Put the origin at lattice point A, as shown in Fig. 3.23, and from origin A, by translation of a basis vector $\boldsymbol{a}_1$, another lattice point B is found. If the rotation by θ is a symmetry operation for the crystal, point B' is also a lattice point for the crystal. Since all lattice points are equivalent in a crystal, same operation $(-\theta)$ applied to point B can also map point A to another lattice point A'. It is self-evident from its production steps stated above that $AB//A'B'$ in Fig. 3.23.

In the crystal, same direction (parallel directions) has the same periodicity, and therefore, Eq. 3.24 is obtained.

$$\overline{B'A'} = k\overline{AB}, \text{ where } k \text{ is an integer.} \tag{3.24}$$

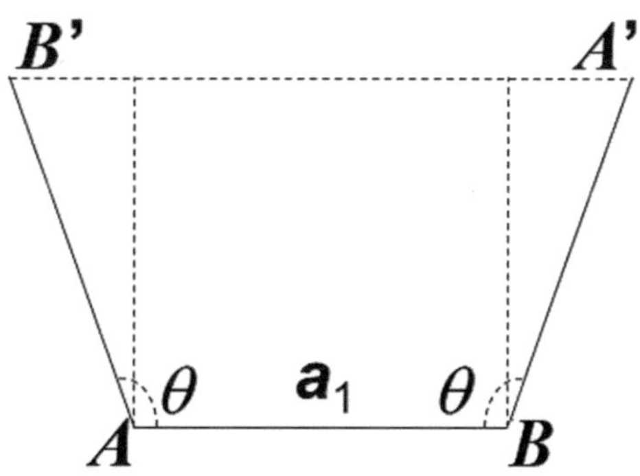

Fig. 3.23 Illustration of rotating a crystal by θ

Furthermore, the geometrical relation between the two segments ***AB*** and ***A'B'*** can be calculated by Eq. 3.25.

$$\overline{B'A'} = \overline{AB} + 2\overline{AB}\cos(\pi - \theta) = \overline{AB}(1 - 2\cos\theta) \tag{3.25}$$

Compare the two expressions of the length of segment *A'B'*, the following Eq. 3.26 is derived.

$$k = 1 - 2\cos\theta \tag{3.26}$$

Since $|\cos\theta| \le 1$, the trace of $(1-2\cos\theta)$ is bounded between -1 and 3. And within this range, it is easy to find all possible values for the integer *k*, as contained in Table 3.8.

In a similar way, it proves that $\overline{5}$ and any $\overline{n}$ higher than $\overline{6}$ cannot be compatible with the translation symmetry existing in crystals. For other possible $\overline{n}$ for crystals, we examine $\overline{1}$, $\overline{2}$, $\overline{3}$, $\overline{4}$ and $\overline{6}$, as shown in Fig. 3.24. It is seen that $\overline{1} = i$, $\overline{2} = m(\sigma)$, $\overline{3} = 3 + i$, and $\overline{6} = 3 + m$. This means that, for example the $\overline{3} = 3 + i$, a system having $\overline{3}$ has both the symmetry elements of 3 and *i*, and vice versa. But this case is not true for a system having $\overline{4}$, meaning $\overline{4}$ is an independent symmetry element.

In summary, only eight basic macro symmetry elements may independently exist in crystals as follows:

$$1, 2, 3, 4, 6, \overline{1}, \overline{2} \text{ and } \overline{4}$$

Table 3.8 All possible integer *k* and the corresponding rotation angle in Eq. 3.26

Integer *k*	θ (in rad)	*n* (in $\theta = 2\pi/n$)
-1	0, or 2π	1
0	$2\pi/6$	6
1	$2\pi/4$	4
2	$2\pi/3$	3
3	$2\pi/2$	2

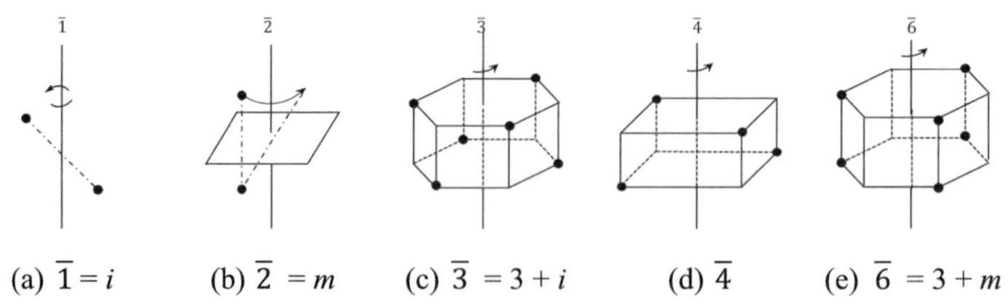

Fig. 3.24 Systems with different $\overline{n}$ symmetry elements

3.8.4 The 32 Crystallographic Point Groups

The combinations of symmetry operations enable us to describe the symmetry of 3D objects, such as molecules, crystals. Mathematically, a set of macro symmetry op-erations satisfying certain conditions is called a point group. The term "point" implies that all symmetry operations must act about one point in common, which is stationary throughout the application of all symmetry operations. This point is known as the origin.

Before we discuss the point groups for crystals, some concepts about group are helpful. In mathematics, a group is formed by a set of operations (Eq. 3.27) only if the four operational rules, namely closure, associativity, identity, and inverse axioms are satisfied.

$$G = \{E, g_1, g_2, \ldots\} \tag{3.27}$$

Closure axiom. The product of any two operations is also a member of the set. That is, if g_i and g_j belong to the set G, their product $g_k = g_i g_j$ also belongs to the set.

Associativity axiom. The multiplication of operations is associative. That is, for any g_i,g_j, g_k belonging to the set G, $g_i(g_j g_k) = (g_i g_j) g_k$.

Identity axiom. The set G includes an identity $1(E)$, which, for any g_i belonging to the set G, has $E g_i = g_i E = g_i$.

Inverse axiom. For each operation g_i in the set G, there is an inverse g_i^{-1}, such that $g_i g_i^{-1} = 1(E)$.

Group theory is a systematic course, and fortunately, to discuss the 32 crystallographic point groups, we do not need to cover the entire theory. The word "crystallographic" here implies that the analysis pertains to crystals, and only rotations (1, 2, 3, 4, and 6) that are consistent with the translational symmetry of crystals are allowed. This restriction on rotation symmetry leads to 32 point groups, collections or sets of symmetry operations, as listed in Table 3.9.

The crystals will be categorized based on their point groups. Crystals that share the same set of symmetry elements belong to the same symmetry system. To simplify things, we will only focus on the symmetry of crystal lattices and ignore the symmetry of their basis for now. By group theory, it can be strictly proven that there are only 7 classes of crystal lattices based on their macro symmetry. It is known that the unit cells of crystal lattices can obviously reflect their macro symmetry, and the symmetry of a crystal lattice must contain the relation between its lattice parameters, called the a, b, c and α, β and γ.

3.8.5 Microscopic Symmetry Elements

In addition to the operations of the point group (describing macroscopic symmetry elements), there are microscopic symmetry elements in crystals caused by

Table 3.9 Crystal classes, Point groups and their Notations

Crystal systems	Point group type	
	International notation	Schönflies notatoin
Triclinic	1	C_1
	$\bar{1}$	C_i
Monoclinic	2	C_2
	m	C_s
	2/m	C_{2h}
Orthorhombic	222	D_2
	mm2	C_{2v}
	mmm	D_{2v}
Trigonal	3	C_3
	$\bar{3}$	C_{3i}
	32	D_3
	3m	C_{3v}
	$\bar{3}$m	D_{3d}
Tetragonal	4	C_4
	$\bar{4}$	S_4
	4/m	C_{4h}
	422	D_4
	4mm	C_{4v}
	$\bar{4}$2m	D_{2d}
	4/mmm	D_{4h}
Hexagonal	6	C_6
	$\bar{6}$	C_{3h}
	6/m	C_{6h}
	622	D_6
	6mm	C_{6v}
	$\bar{6}$m 2	D_{3h}
	6/mmm	D_{6h}
Cubic	23	T
	m$\bar{3}$	T_h
	432	O
	$\bar{4}$3m	T_d
	m$\bar{3}$m	O_h

the introduction of translational operation. The microscopic symmetry elements include pure translations (moving a point along a vector), screw axes (rotating a point around an axis while translating parallel to the axis), glide planes (reflecting a point through a plane while translating it parallel to the plane).

Translation Axis. The fundamental characteristic of a crystal is that its internal structure possesses the characteristics of a spatial lattice, which is the periodicity of the internal structure of the crystal. If the lattice is translated along any two points' vectors in the lattice, the lattice will be restored. The symmetric group formed by combining these translation operations is called a translation group and can be expressed by the following Eq. 3.28.

$$T_{mnp} = m\boldsymbol{a} + n\boldsymbol{b} + p\boldsymbol{c}, \text{ where } m, n, p = 0, \ \pm 1, \ \pm 2, \ldots \tag{3.28}$$

Therefore, the spatial lattice reflects the geometric form of the periodicity of the crystal's internal structure, and the expression of the translation group reflects the algebraic form of the periodicity of the crystal's internal structure. The correspondence between the lattice and the corresponding translation group is also reflected in the following two important properties. The vector connecting any two points in the lattice must belong to a translation vector in the translation group $\boldsymbol{T}$. Any vector belonging to the translation group $\boldsymbol{T}$ must pass through two lattice nodes. Therefore, it can be concluded that the combination of translation group vectors still belongs to the translation group.

The difference between the macroscopic symmetry and microscopic symmetry of a crystal lies in the fact that macroscopic symmetry operations require at least one point to remain fixed, while microscopic symmetry operations require all points to move. Therefore, macroscopic symmetry cannot reflect the translational part of microscopic symmetry. However, once the macroscopic symmetry elements are combined with translation, new microscopic symmetry elements can be formed.

Screw Axis. A screw axis is an imagined line around which identical parts inside the crystal rotate periodically, accompanied by translational motion along the axis. A screw axis is a composite symmetry element consisting of an imaginary line and a direction parallel to it. The corresponding symmetry transformation is a combination of rotation around this line by a certain angle and translation along this line direction. The screw axis has a rotation order n that can only be 1, 2, 3, 4, 6, and the translation transformations it contains have a translation distance equal to s/n, where s is a natural number less than n, along the direction of the screw axis. The international notation for a screw axis is generally represented as n_s. Based on the order of the rotation axis and the translation distance, screw axes can be divided into 2_1, 3_1, 3_2, 4_1, 4_2, 4_3, 6_1, 6_2, 6_3, 6_4, 6_5, a total of 11 types of screw axes.

Glide Plane. The slip plane is an imaginary plane. The identical portions within the crystal are repeated by translating along a straight line parallel to this plane and then reflecting. A glide plane is also a composite symmetry element of an imaginary plane and a direction parallel to this plane. The corresponding symmetry transformation involves reflection across this plane and a combined translation along this line, where the distance of translation is equal to half of the distance between lattice points in that direction.

These microscopic symmetry elements, along with rotation axes, mirror planes, and inversion centers, contribute to the overall symmetry classification of a crystal structure. They help describe the arrangement and repetitive patterns of atoms within the crystal lattice.

3.8.6 Space Groups

To consider the whole symmetry characteristics of a crystal, space groups are powerful. The symmetry of a crystal is completely described by a space group. There are 230 distinct space groups.

$$\text{Space group} = \text{translation} + \text{point group}$$

In crystallography, space groups are also called the crystallographic or Fedorov groups, and describe the symmetry of the crystal. A definitive source regarding 3-dimensional space groups is the International Tables for Crystallography.

A space group is a set of symmetry operations that define how a crystal can be transformed or translated, while maintaining its repeating pattern. The symmetry operations include rotations, reflections, and translations. There are a total of 230 unique space groups that can describe the structure and symmetry of all known crystals.

The physical and optical properties, reactivity, and stability of a crystal are determined by its space group. A crystal with a higher degree of symmetry due to its space group can exhibit specific physical properties, such as improved optical properties or higher mechanical strength.

When analyzing a crystal structure, the first step is to determine its space group. This is usually done using X-ray diffraction, electron diffraction, or neutron diffraction. These techniques produce a diffraction pattern, which can be analyzed to determine the space group of the crystal. Once the space group has been determined, it can be used to predict the arrangement of atoms in the crystal, and this information can be used to design new materials or understand the properties of existing ones.

To sum up, space groups are a key concept in crystallography that are vital for analyzing and comprehending crystal structures. They furnish a mathematical explanation of the symmetry, orientation, and periodicity of crystals, and can be employed to deduce various properties of crystals, such as their physical and optical characteristics, reactivity, and stability.

3.9 Stereographic Projection

It is important to identify the orientation of crystals, as well as the angles between directions and planes, when conducting crystal investigations. To make this more convenient, 3D crystals are often projected onto a 2D plane using various methods.

To develop a method for projecting crystals and determining their orientation, some definitions are necessary. Firstly, the interfacial angle between two crystal planes, as shown in Fig. 3.25, is defined as the angle between lines perpendicular to the planes, which are called the poles to the crystal planes. Secondly, imagine a crystal inside a sphere, from which we draw lines perpendicular to each face (poles to the face). The (010) face is defined as having a Φ angle of 0° and the Φ angle for any other face is measured clockwise from the b axis in the plane of the equator. The ρ angle is defined as the angle between the $\boldsymbol{c}$ axis and the pole to the crystal face, measured downward from the North pole of the sphere. In the diagram shown in Fig. 3.26, a crystal face has a ρ angle measured in the vertical plane containing the axis of the sphere and the face pole, and a Φ angle measured in the horizontal equatorial plane. The (010) face has a ρ angle of 90°.

We need a systematic way to define crystallographic angles. For this we use a spherical projection. As an example, the ρ and Φ angles for the (111) crystal face in a crystal model is shown in Fig. 3.27. The ρ angle is measured in the vertical plane that includes the c axis and the face's pole, whereas the Φ angle is measured in the horizontal plane in a clockwise direction from the b axis.

In general, the angles ρ and Φ of the spherical projection are provided for each crystal face. With this information, the angles between any two faces can be easily determined using trigonometry or the stereonet method, which is discussed below.

It is the angle between crystal faces that reveals the orderly arrangement of atoms within a crystal and its internal symmetry, which then determines its external symmetry. This can be visualized through the use of stereographic projection for projecting crystallographic angles.

Fig. 3.25 Interfacial angle between two crystal faces

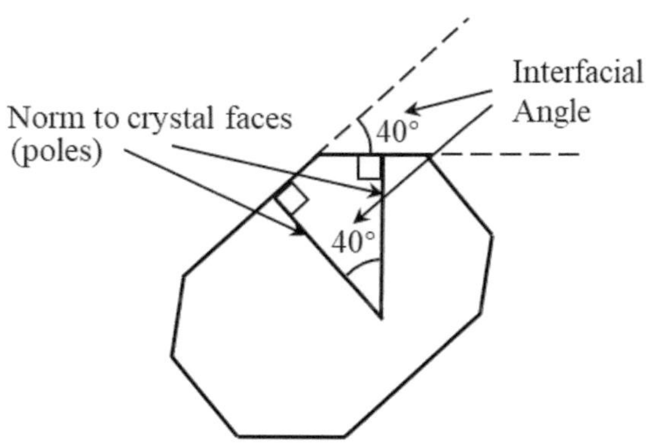

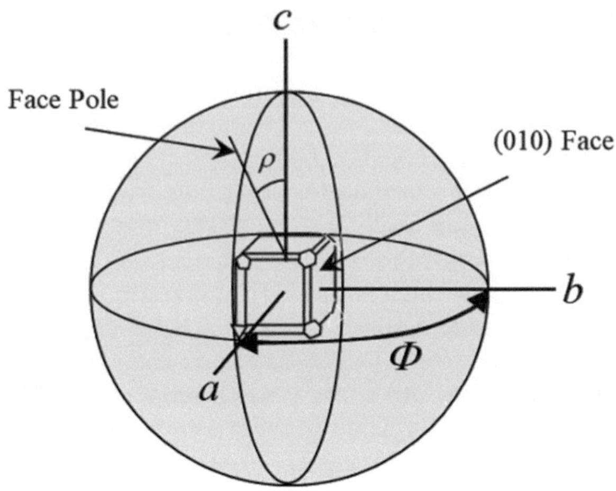

Fig. 3.26 Spherical projection

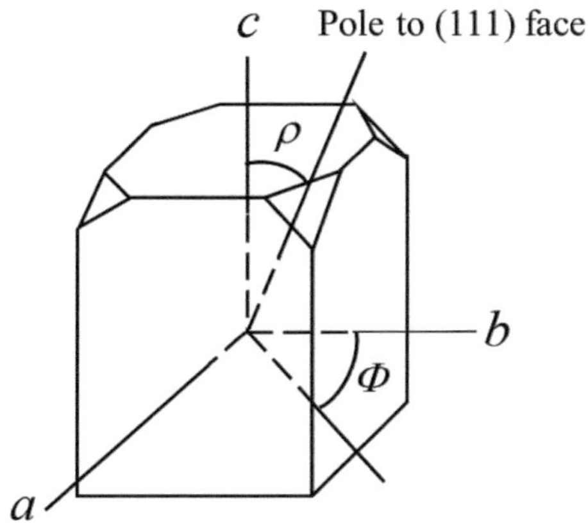

Fig. 3.27 The ρ and Φ angles for the (111) crystal face in a crystal

Stereographic projection is a technique employed in both crystallography and structural geology to illustrate the angular relationships among crystal faces and geological structures, respectively. While the present discussion focuses on the crystallographic use of the technique, it is similar to the method employed in structural geology.

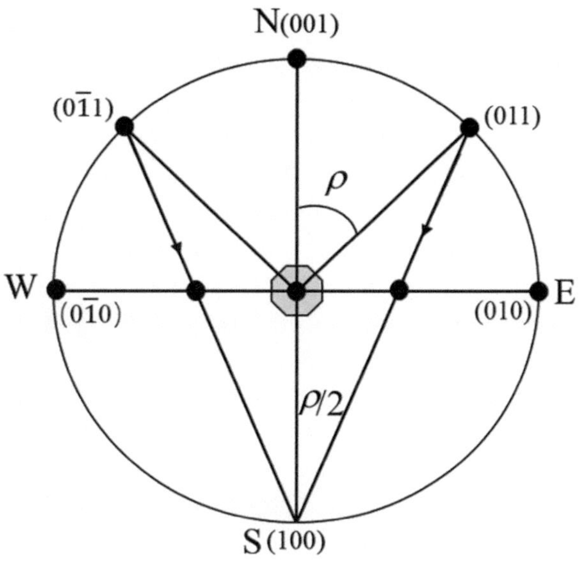

Fig. 3.28 A cross-section of the sphere

Consider a crystal inside a sphere, and a cross-section of the sphere shown in Fig. 3.28. The crystal is oriented so that the pole to the (001) face (*c* axis) is vertical and points to the North pole of the sphere. The pole to the (011) face intersects the outside of the sphere, and we draw a line from that point on the sphere to the South Pole. The point where this line intersects the equatorial plane is where we plot the point, and the stereographic projection is then seen on the equatorial plane.

Figure 3.29 illustrates the stereographic projection of an isometric crystal's faces. Note how the ρ angle is measured as the distance from the center of the projection to the position where the crystal face plots. The Φ angle is measured around the circumference of the circle, in a clockwise direction away from the *b* crystallographic axis or the plotting position of the (010) crystal face.

A stereographic net or stereonet, as shown in Fig. 3.30, is used to simplify the plotting of the stereographic projection. It is important to note that the North (N) and South (S) poles on the stereonet do not correspond to the North and South Poles as defined in the projection discussed earlier.

As per our projection definition, the N and S poles would locate directly above and below the stereonet's center. The stereonet has several components, which we define below.

(1) The Primitive Circle is the one surrounding the stereonet.
(2) Great Circles are the curved lines linking the points marked N and S on the stereonet. The E-W and N-S axes, along with the Primitive Circle, are also

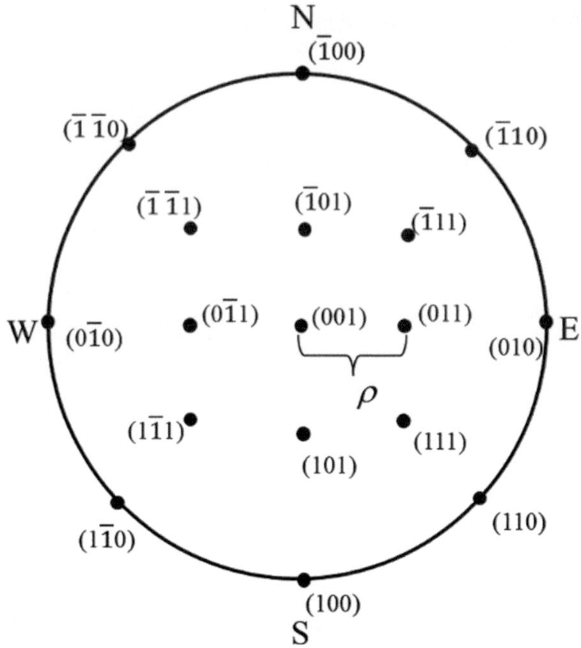

Fig. 3.29 The stereographic projection for faces of an isometric crystal

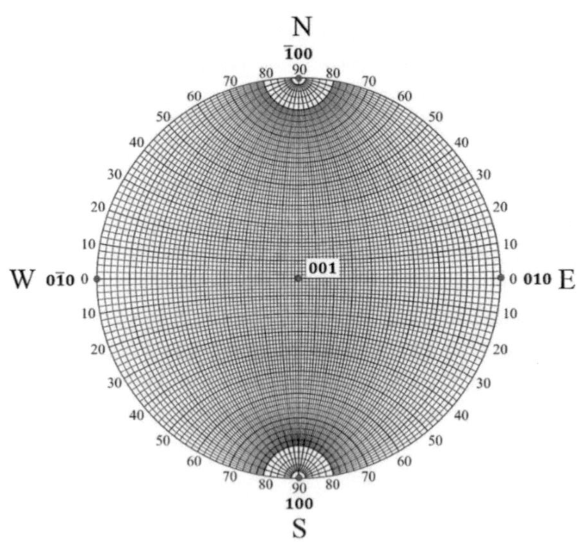

Fig. 3.30 The stereonet

great circles. Measurements of angular relationships between points can be taken only on Great Circles.

(3) Small Circles are the highly curved lines that bend upward and downward on the stereonet.

3.10 Questions

Q3.1 What fundamental distinction exists between crystalline and amorphous materials?

Q3.2 Define the concept of a Bravais lattice.

Q3.3 How does a unit cell differ from a crystal structure?

Q3.4 Enumerate and classify the seven crystal systems.

Q3.5 In the provided figure displaying a unit cell for a hypothetical metal:

(1) What crystal system does this metal crystal belong to?
(2) Which Bravais lattice is associated with this metal crystal?

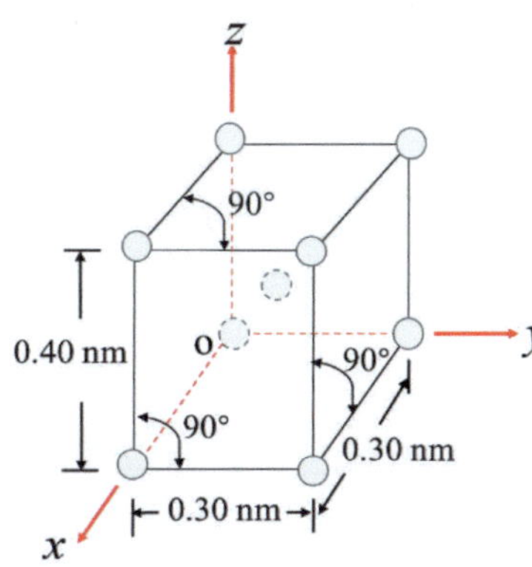

Q3.6 Understanding directions and planes within crystals holds significance in materials investigation.

(1) Sketch a triclinic unit cell and identify the direction [211] within that unit cell.
(2) Find the Miller indices for the crystallographic directions depicted in the provided figure of a cubic unit cell.
(3) Determine the Miller indices for the crystallographic directions displayed in the given figure of a hexagonal unit cell, and subsequently convert the Miller indices to the Miller-Bravais scheme.

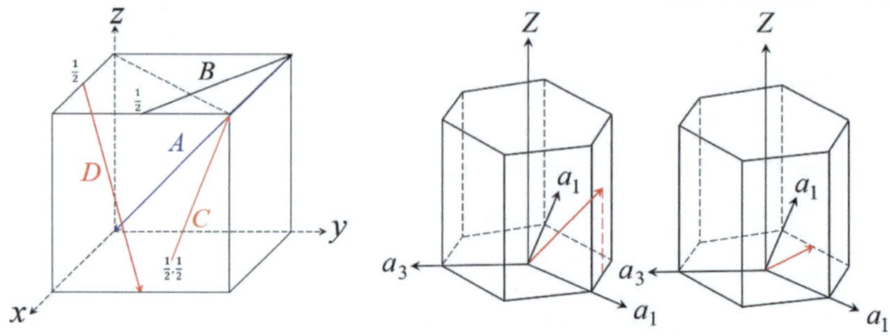

(4) In a hexagonal unit cell, find the $[11\bar{2}4]$, $[10\bar{1}2]$ and $[10\bar{1}0]$ directions.
(5) Identify the Miller indices for the planes depicted in the provided figure of a unit cell.

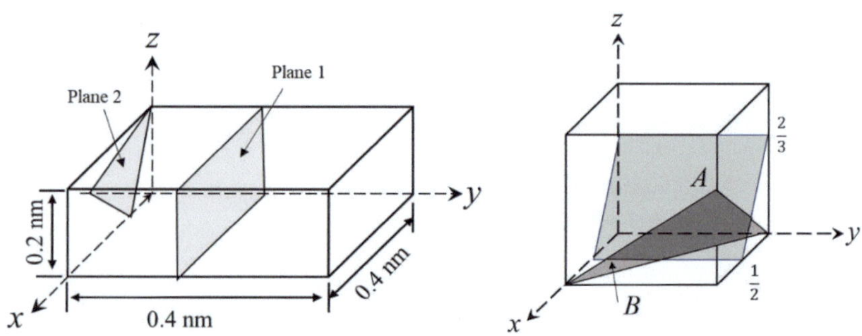

(6) Draw all the members in the direction family of <100> and all the members in the family of crystal plane of {110} in FCC structure.

Q3.7 Demonstrate the Weiss zone law using your own methodology.

Q3.8 Develop a spreadsheet enabling users to input lattice parameters for automatic determination of d-spacing.

Q3.9 Outline the formulas employed for computing linear and planar densities in a crystal.

Q3.10 Explain the relationship between the arrangement of atoms in a crystal and its linear and planar densities.

Q3.11 Discuss how the linear and planar densities of a crystal impact its mechanical properties.

Q3.12 Determine if the (110), (112), and (312) planes belong to the same zone. If affirmative, identify the direction indices of the zone axis.

Q3.13 Enumerate all symmetrical elements of each shape in the following manner.

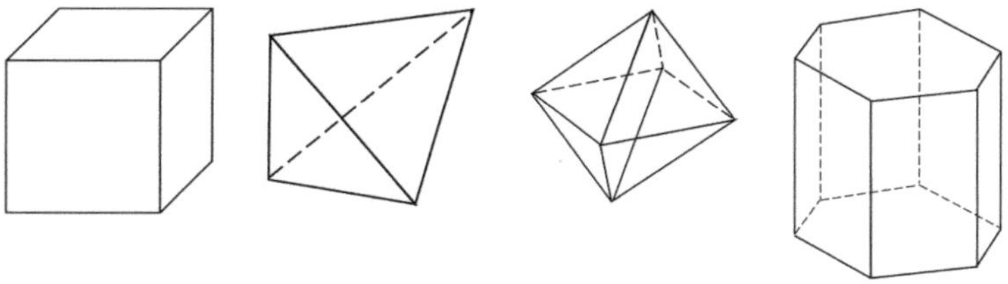

(a) Cubic (b) Tetrahedron (c) Octahedron (d) Hexagonal prism

Q3.14 In a mineral sample exhibiting a cubic crystal system, a stereographic projection of the crystal structure is captured from a direction perpendicular to the (111) plane. The plotted points on the projection are (100), (110), and (111). Utilizing the stereographic projection method, ascertain the crystal axes and determine the indices of the crystal structure. Note: Solving this problem requires proficiency in crystal systems, crystal axes, and stereographic projections. Additionally, the student should be adept in performing vector calculations, including vector addition and cross products, within a three-dimensional space.

Bibliography

Callister WD Jr, Rethwisch DG (2010) Materials science and engineering: an introduction, 8th edn. John Wiley & Sons, New York

Callister WD Jr, Rethwisch DG (2018) Fundamentals of materials science and engineering: an integrated approach, 5th edn. John Wiley & Sons, New York

Hu GX, Cai X, Rong YH (2010) Fundamentals of materials science, 3rd edn. Shanghai Jiao Tong University Press, Shanghai

Pan JS, Tong JM, Tian MB (2011) Fundamentals of materials science. Tsinghua University Press, Re.ed. Beijing

Tao J, Yao ZJ, Xue F (2018) Fundamentals of materials science, 2nd edn. Chemical Industry Press, Beijing

Yu YN (2012) Fundamentals of materials science, 2nd edn. Higher Education Press, Beijing

The Structure of Solid Materials

4

Learning Objectives

1. Describe the three most common crystal structures for metallic materials.
2. Understand the concept of alloy.
3. Describe the typical crystal structures found in ceramic materials, and acquire knowledge of the formation conditions using the radius ratio rule.
4. Calculate the theoretical density for both metallic and ceramic materials.
5. Depict the structural units, molecular shape, and copolymer structures in polymers.
6. Illustrate head-to-head configurations, three stereoisomerisms, and geometrical isomerism of polymers.
7. Differentiate between configurations and conformations.
8. Calculate the molecular weight and its distribution of polymers.
9. Briefly describe the amorphous and crystalline structure of polymeric materials.
10. Comprehend the interstitial sites in the common crystal structures.
11. Understand the concept of polymorphism.
12. Understand the difference between single crystals and polycrystalline materials.
13. Explain nanocrystals and quasicrystals.
14. Understand the structure of polymeric materials.
15. Explain the X-ray diffraction phenomenon of crystals and utilize it to analyze the structure of crystalline materials.

Z. Sun et al., *Fundamentals of Materials Science*,
https://doi.org/10.1007/978-981-96-3904-5_4

In the Materials Science Tetrahedron, the fundamental elements of a material are interdependently related. Among these elements, structure means how some internal components of the material are arranged, including subatomic, atomic, microscopic, and macroscopic components, in terms of dimensionality. The structure of a material determines its properties, potential applications, processability, performance within applications, environmental impact, and other aspects. Hence, an understanding of material structure is crucial for materials scientists and engineers.

4.1 Introduction

Crystalline and amorphous are two main classifications for solid materials. The key distinction between them is the periodic arrangement of motifs (atoms, molecules, ions or their groups) found in crystalline materials, which is absent in amorphous materials. Crystalline solids have a highly ordered, three-dimensional arrangement of atoms or groups, resulting in anisotropic properties and a well-defined melting point. Examples include diamond, silicon ingot, and gold. In contrast, amorphous solids have a disordered arrangement of particles with varying intermolecular forces and distances, and lack a defined geometric shape. They are isotropic and often referred to as supercooled liquids. Glass is a common example of an amorphous solid in everyday life. Understanding the structural differences between these two types of solids is crucial for materials scientists and engineers.

Metallic materials and their alloys, along with ceramics and polymeric materials, form the basis of many modern engineering projects and advanced technologies. This chapter provides an overview of the structures of solid materials, which are essential for understanding their properties.

4.2 Metallic Crystal Structures

Metals exhibit a shiny and lustrous appearance when freshly prepared or polished, and are ubiquitous in our daily lives due to their numerous desirable properties. The majority of elements in the periodic table are metals, as illustrated in the periodic table (Fig. 2.4). A stair-stepped amphoteric line begins between Boron (B, $Z = 5$) and Aluminum (Al, $Z = 13$), dividing the elements into two groups. Elements to the left of this line are classified as metals, with the exception of hydrogen, which, despite being the lightest known gas, exhibits certain properties that suggest it may also be a metal.

The atomic bonding in metals is nondirectional and referred to as metallic bonding. This allows for minimal restrictions in the number and position of nearest-neighbor atoms during crystallization. Because the atoms in an elemental metal are identical and the sea of electrons shields the cation cores, elemental metals tend to be densely packed to lower bond energy. As a result, most elemental metals crystallize in one of the three relatively simple crystal structures upon cooling: body-centered cubic (BCC), face-centered cubic (FCC), and hexagonal close-packed (HCP).

4.2.1 The Body-Centered Cubic Crystal Structure

The body-centered cubic (BCC) crystal structure belongs to the cubic system (also known as isometric system) and has cubic symmetry. Its unit cell is usually a cube, as shown in Fig. 4.1a, with identical motifs occupying all eight corners and the center of the cube, as shown in Fig. 4.1b. In the hard-sphere model, cube center and corner atoms touch one another along cube diagonals. On its (110) plane, as shown in Fig. 4.1c, the unit cell length (represented by a) has a clear relationship with the radius of atoms (represented by R). This relationship can be considered as a pure solid geometry problem:

$$a = \frac{4R}{\sqrt{3}} \tag{4.1}$$

Table 4.1 contains information on the lattice constants and atomic radii of several metals, including chromium, iron (α-Fe), tungsten (W), and others, which exhibit the BCC crystal structure at room temperature.

To gain a better understanding of metallic crystals, it is necessary to determine the number of atoms present within each unit cell. This can be accomplished easily for a given crystal structure by considering how the atoms are shared with neighboring unit cells. In the case of BCC, there are two types of atoms in the unit cell: corner atoms and cube center atoms. The interior atom is entirely contained within the unit cell, while each corner atom is shared with eight adjacent unit cells, with only one-eighth of it belonging to the unit cell under consideration. Based on this analysis, the number (N) of atoms in a BCC unit cell is the sum of the number of interior atoms (N_i) and one-eighth of the number of corner atoms (N_c): $N = N_\mathrm{i} + N_\mathrm{c}/8 = 2$.

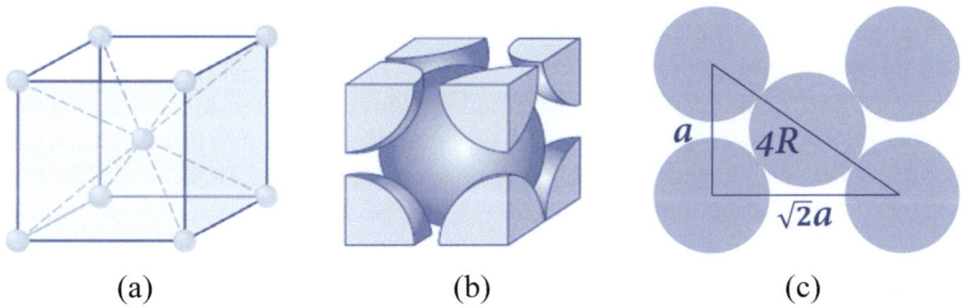

Fig. 4.1 **a**: the unit cell of BCC structure with atoms shown in reduced size, **b**: atoms completely within a unit cell, **c** the (110) crystalline plane of a BCC crystal

Table 4.1 Selected metals having the BCC crystal structure at 300 K and their lattice constants and atomic radii

Metal	Lattice constant a (nm)*	Atomic radius R (nm)**
Chromium (Cr)	0.289	0.125
Iron*** (Fe)	0.287	0.124
Molybdenum (Mo)	0.315	0.136
Potassium (K)	0.533	0.231
Sodium (Na)	0.429	0.186
Tantalum (Ta)	0.330	0.143
Tungsten (W)	0.316	0.137
Vanadium (V)	0.304	0.132

* Lattice constants usually depend on temperature, and can be determined by diffraction techniques
** Atomic radius (R) is calculated from lattice constants by using Eq. 4.1
*** Alpha form of iron (α-Fe)
(*Source* William F. Smith, Javad Hashemi, and Francisco Presuel-Moreno, Foundations of materials science and engineering, Mcgraw-Hill Publishing, 2006, p. 99)

Having established the number of atoms within a unit cell and the relationship (Eq. 4.1) between the lattice parameter a and atomic radius R, we can now determine the atomic packing factor (APF). The APF is defined as the sum of the volumes of all the atoms associated with a unit cell, divided by the volume of the unit cell. When treated using the hard-sphere model, this becomes a relatively straightforward geometry problem.

$$APF = \frac{N_{\text{atoms}} \cdot V_{\text{atom}}}{V_{\text{unitcell}}} \tag{4.2}$$

Therefore, the APF for BCC structure is calculated as follows:

$$APF_{\text{BCC}} = \frac{2 \cdot \frac{4\pi}{3} \cdot R^3}{a^3} = \frac{\sqrt{3}\pi}{8}$$

The atomic packing factor (APF) of BCC is approximately 68%, indicating a 32% empty space in the hard-sphere model. APF is a measure of packing efficiency for a particular crystal structure, and it is directly related to the coordination number (CN) of the crystal. In the hard-sphere model, an atom's CN is determined by the number of its nearest-neighbor or touching atoms. For BCC, its CN is 4, as evidenced by isolating a unit cell, as shown in Fig. 4.1b.

4.2.2 The Face-Centered Cubic Crystal Structure

The face-centered cubic (FCC) crystal structure, like the BCC structure, also belongs to the cubic system. However, the way atoms are positioned within their

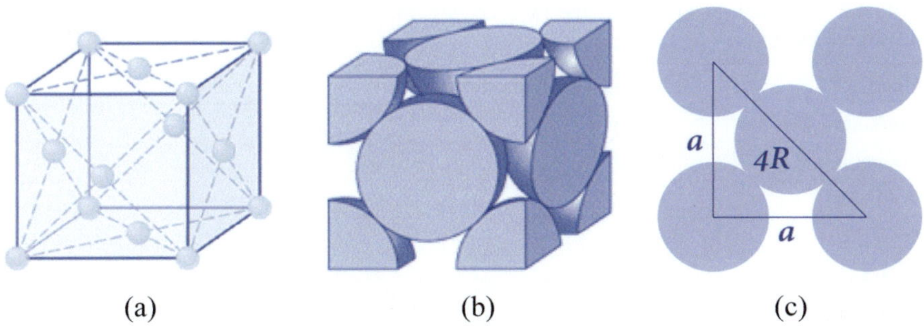

Fig. 4.2 **a**: the unit cell of FCC structure with atoms shown in reduced size, **b**: atoms completely within a unit cell, **c** the (100) crystalline plane of an FCC crystal

unit cells is different between the two structures. In the FCC structure, all atoms are situated in a cube where both the corners and the face centers are occupied. The atoms are in contact along the diagonal of the cube faces and from its (100) plane, as depicted in Fig. 4.2. The relationship between atomic radius (R) and cube length (a) can be easily established from this.

$$a = 2\sqrt{2}R \tag{4.3}$$

The same analysis used for the BCC structure can also be applied to determine the number of atoms in a single unit cell of FCC, but with the difference that FCC unit cells have face atoms instead of cube center atoms. There are 6 face atoms (N_f) since each face of the parallelepiped is associated with one atom, and each face atom is shared with two adjacent unit cells. Thus, the number of atoms within a single FCC unit cell is calculated as $N = N_f/2 + N_c/8 = 4$. The *APF* of the FCC structure can be determined by the definition of APF (Eq. 4.2) in a similar way as that for BCC.

$$APF_{\text{FCC}} = \frac{4 \cdot \frac{4\pi}{3} \cdot R^3}{a^3} = \frac{\sqrt{2}\pi}{6}$$

The *APF* for FCC is about 74%, indicating that 26% of the FCC unit cell is empty based on the hard-sphere model's geometry. This value is greater than that of BCC, indicating that the FCC has a higher packing efficiency. The coordination number for FCC structure is also higher than that of BCC and is 12, as shown in Fig. 4.2. It is noteworthy that 74% is the closest packing factor possible for identical hard spheres. The HCP structure, to be discussed, shares the highest *APF* with the FCC structure, indicating that both structures have some common characteristics concerning the arrangement of the atoms inside.

Several metals such as copper (Cu), silver (Ag), and gold (Au) have the FCC crystal structure, as shown in Table 4.2. At elevated temperatures (912 °C to 1394

Table 4.2 Selected metals having the FCC crystal structure at 300 K and their lattice constants and atomic radii

Metal	Lattice constant a (nm)*	Atomic radius R (nm)**
Aluminum (Al)	0.405	0.143
Copper (Cu)	0.362	0.128
Gold (Au)	0.408	0.144
Lead (Pb)	0.495	0.175
Nickel (Ni)	0.352	0.125
Platinum (Pt)	0.393	0.139
Silver (Ag)	0.409	0.144

* Lattice constants usually depend on temperature, and can be determined by diffraction techniques
** Atomic radius (R) is calculated from lattice constants by using Eq. 4.3
(*Source* William F. Smith, Javad Hashemi, and Francisco Presuel-Moreno, Foundations of materials science and engineering, Mcgraw-Hill Publishing, 2006, p. 101)

°C), some metals that otherwise have different crystal structures may crystallize with the FCC structure, including iron (Fe).

4.2.3 The Hexagonal Close-Packed Crystal Structure

In addition to the two structures belonging to isometric system as they are discussed above, the third common metallic crystal structure is the hexagonal close-packed (HCP) crystal structure shown in Fig. 4.3. As discussed in the previous chapter, there is only one Bravais lattice belonging to the hexagonal crystal system, and it is the simple hexagonal structure. However, there are no examples of elements with this crystal structure because the *APF* is too low. Atoms can have a more stable condition by lowering their energy through packing in the HCP structure.

The HCP crystal structure possesses hexagonal symmetry, and its unit cell depicted in Fig. 4.3 displays six atoms that create regular hexagons on the top and bottom faces surrounding a central atom. An additional plane, situated between the top and bottom planes, contains three more atoms.

The unit cell of the HCP crystal structure contains two face atoms ($N_f = 2$), twelve corner atoms ($N_c = 12$) and three interior atoms ($N_i = 3$). Due to its hexagonal symmetry, the corner atoms are shared with only six unit cells instead of eight as in cubic crystal structure. Therefore, the number of atoms in a single unit cell of the HCP crystal structure is calculated as $N_f/2 + N_c/6 + N_i = 1 + 2 + 3 = 6$.

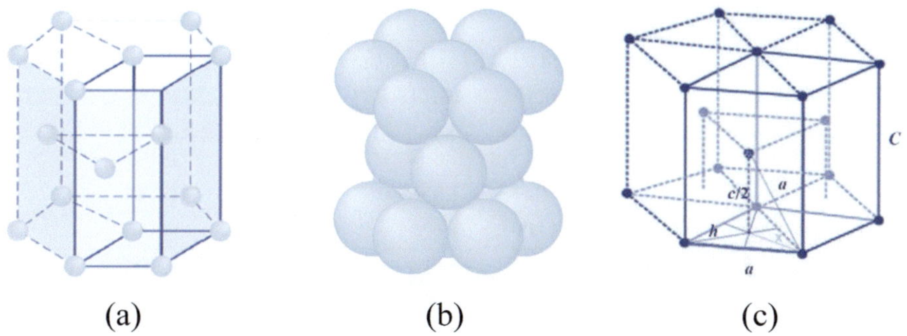

Fig. 4.3 **a**: the unit cell of HCP structure with atoms shown in reduced size, **b**: atoms within a unit cell, **c** the geometrical relation between its two lattice parameters **a** and **c**

The ratio of the height c of the HCP unit cell to the edge length a of its bottom plane is commonly referred to as the c/a ratio. By considering the solid geometry of its unit cell depicted in Fig. 4.3c, the relationship between its lattice parameters a and c can be determined. Since the atoms touch each other along its basal side, the value of a can be related to the atomic radius (R) as follows:

$$a = 2R \tag{4.4}$$

In a HCP unit cell, the two nearest corner atoms touch the atoms in the mid-plane, forming a right triangle ABO with lengths of $2R$, $\frac{2\sqrt{3}R}{3}$, and $c/2$ for AB, AO, and BO, respectively. Solving the right triangle gives the relationship between $\boldsymbol{c}$ and atomic radius (R) as:

$$c = \frac{4\sqrt{6}}{3}R \tag{4.5}$$

Consequently, for an ideal HCP structure, the c/a ratio is calculated as $\frac{2\sqrt{6}}{3}$, which is typically rounded to 1.633. Using the definition of APF (Eq. 4.2), the value for the APF of an ideal HCP crystal structure can be calculated as follows:

$$APF_{\text{FCC}} = \frac{6 \cdot \frac{4\pi}{3} \cdot R^3}{24\sqrt{2}R^3} = \frac{\sqrt{2}\pi}{6}$$

The atomic packing factor (APF) for an ideal HCP crystal structure is the same as that of FCC, both having a value of 0.74, and each atom in the HCP crystal structure is surrounded by twelve other atoms, also same as the case for FCC crystal structure. Thus, the coordination number of both structures is twelve, and the atoms are packed as closely as possible. The stacking sequence of their crystallographic planes reveals the intimate connection between the two structures as stated in the previous section.

Table 4.3 Selected metals having the HCP crystal structure at 300 K and their lattice constants (*a* and *c*), *c*/*a* ratios, and atomic radii

Metal	Lattice constants			% Deviation from the ideal value (1.633)	Atomic Radius *R* (nm)
	a (nm)	*c* (nm)	*c*/*a*		
Cadmium (Cd)	0.2973	0.5618	1.890	15.7	0.149
Zinc (Zn)	0.2665	0.4947	1.856	13.6	0.133
Magnesium (Mg)	0.3209	0.5209	1.623	−0.66	0.160
Cobalt (Co)	0.2507	0.4069	1.623	−0.66	0.125
Zirconium (Zr)	0.3231	0.5148	1.593	−2.45	0.160
Titanium (Ti)	0.2950	0.4683	1.587	−2.81	0.147
Beryllium (Be)	0.2286	0.3584	1.568	−3.98	0.113

(*Source* William F. Smith, Javad Hashemi, and Francisco Presuel-Moreno, Foundations of materials science and engineering, Mcgraw-Hill Publishing, 2006, p. 103)

Metals with the HCP crystal structure are not uncommon, and several well-known metals, such as zinc (Zn) and magnesium (Mg), assume this structure. Table 4.3 shows selected metals with the HCP crystal structure and their lattice constants at room temperature.

4.2.4 Crystal Structures of Alloys

All crystals discussed thus far consist of identical atoms throughout the crystal. However, such pure crystals are virtually nonexistent, and nearly all metallic materials used are alloys. An alloy is a combination of metals, or a metal combined with one or more other elements (metals and/or nonmetals), forming a material in which the constituent elements are bonded together by chemical bonds. The components of an alloy include all the elements, whether they are metals or not, contained within it. Alloys can be classified as binary, ternary, and so on, depending on the number of components they contain. Alloy series, on the other hand, are alloys that have the same set of components but with varying content. Alloys are widely used in practical applications, as there are few pure metals used. Commonly used metals, including aluminum, iron, copper, and magnesium, have various alloys with different alloying elements and concentrations. Metal/nonmetal alloying elements are added to a metal to form an alloy in materials engineering, providing increased strength, higher corrosion resistance, and/or other desired properties. Alloys are used extensively, such as iron alloys used in virtually everything, including buildings, automobiles, and surgical tools, and titanium alloys used in the aerospace industry and as artificial bone. Understanding the structure of alloys can help control and design them to meet specific applications, as alloys are designed to reduce overall material cost while preserving important properties or to impart synergistic properties to the constituent metal elements.

① Alloy Phase

Different from pure metals, at a given condition (temperature and pressure), inside an alloy of a certain composition (the overall concentration) there may be zones having various local composition, structure and thus properties. Alloy phase, or phase for short (a region of a material throughout which all physical and chemical properties are essentially uniform), is used to refer to a part in an alloy where the composition, structure and properties are homogeneous. In some cases, regions with the same structure but continuously varying composition are also considered to be the same phase. The concept of alloy phases is fundamental in materials science and will be discussed in detail in later chapters of this book. A homogeneous alloy consists of a single phase, while a heterogeneous alloy has two or more phases.

Alloy phases can be classified into two categories based on their crystal structures, namely intermetallic compounds and solid solutions. An alloy can either be a solid solution of the host metal or a mixture of several metallic phases, and the overall mixture is usually referred to as the microstructure of the alloy.

② Solid Solutions

Solid solutions are the simplest type of alloy and refer to a family of materials that have a range of composition. The term "solution" is used to describe the intimate mixing of components at the atomic level through chemical bonds, which distinguishes solid solutions from physically mixed homogeneous materials. As crystalline materials, solid solutions possess a crystal structure, concentration, solubility, and order. These materials can be grouped based on these characteristics, and on the basis of structure, solid solutions can be broadly classified into two types: substitutional and interstitial solid solutions, which have different formation mechanisms. The types of solid solutions, substitutional and interstitial alloys, depend on the types of solutes involved. In substitutional alloys, solute atoms substitute for host atoms at lattice points, as shown in Fig. 4.4a and b. In interstitial solid solutions, solute atoms fill the interstitial voids of the host crystal, as illustrated in Fig. 4.4a and c. It is also common for solid solutions to have both types of solute atoms, as shown in Fig. 4.4d, in practical applications.

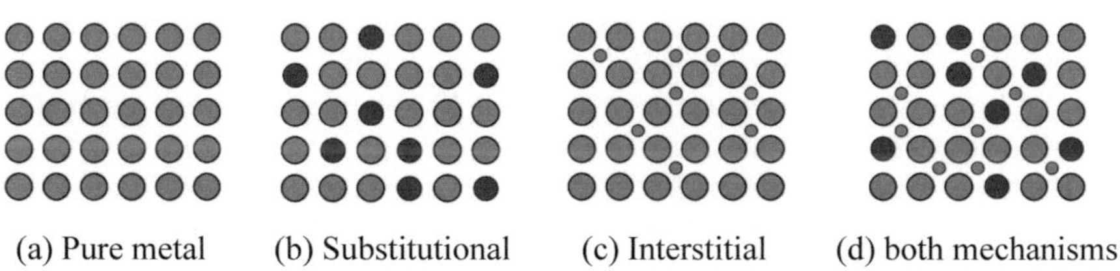

Fig. 4.4 Illustrations of a pure metal phase (**a**), a substitutional solid solution phase (**b**), an interstitial solid solution phase (**c**), and a solid solution having two mechanisms (**d**)

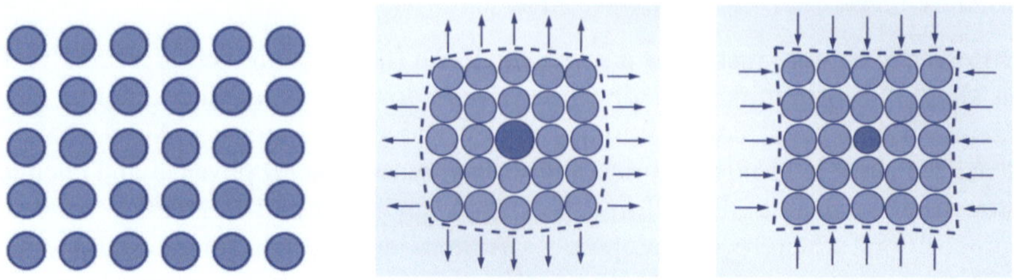

(a) (b) (c)

Fig. 4.5 Local stress in substitutional solid solutions. **a** Pure metal, having identical atoms. Some host atoms substituted by larger (**b**) or smaller (**c**) alloying element atoms

The solubility (composition) of an alloy can be restricted to a specific range, or continuously change throughout the entire range, exemplified by the CuNi alloy, depending on the physicochemical properties of the parent elemental crystals (solute/solvent). Solute atoms may incorporate into the solvent crystal lattice by substituting solvent atoms in the lattice or squeezing into the voids between solvent atoms. Both types of solid solution affect the properties of the resultant alloy by distorting the crystal lattice and interrupting the physical and chemical homogeneity of the host crystal. Figure 4.5 illustrates how the unit cells (crystal structure) of the host crystal often expand or contract to accommodate solute atoms with different sizes from the host atoms.

The ability of an alloy system to form solid solutions depends on various characteristics of the parent crystals, such as their chemical, crystallographic, and other properties. While some systems can readily form solid solutions, others cannot. Complete substitutional solubility can occur on the same crystal structure in certain binary systems, such as the Cu-Ni system. The Hume-Rothery rules, which take into account the crystal structure, atom size, electronegativity, and relative valency of the parent pure metals, can predict and explain the formation and solubility of solid solutions. These rules are still widely used today. The following are the four factors considered by the empirical Hume-Rothery rules for determining the solubility of solid solutions:

(1) Crystal structure: The crystal structures of the solute and solvent elements should be similar to form substitutional solid solutions. To have an appreciable solid solubility, the elemental crystals of both solute and solvent much share the same crystalline structure.
(2) Atom size: Solute atoms with a size difference greater than 15% from solvent atoms will likely have low solubility. Most systems with complete solid solubility have a size difference of less than 8%. The size difference is determined by Equation 4.6.

$$\left|\frac{R_{\text{solvent}} - R_{\text{solute}}}{R_{\text{solute}}}\right| \leq 15\% \tag{4.6}$$

(3) Electronegativity: Similar electronegativities of solute and solvent atoms favor greater solubility. Large differences in electronegativity between the solute and solvent atoms can cause the formation of compounds instead of a solid solution.
(4) Valency: This is the final one in the set the Hume-Rothery rules, and this one sees most exceptions. Elements with the same valency are more likely to form a solid solution than those with different valency.

The copper-nickel alloy is a well-known example of substitutional solid solutions. Both copper and nickel can form solid solutions across all concentrations, which can be explained by the Hume-Rothery rules. Both components have the same FCC crystal structure in the pure metal phase, atomic radii of 0.128 and 0.125 nm, respectively, electronegativities of 1.9 and 1.8, respectively, and common valences of +1 (sometimes +2) for copper and +2 for nickel. However, for systems that deviate from the most favorable Hume-Rothery rules, it is unlikely to form solid solutions throughout the entire composition range, exemplified by the lead–tin system.

In interstitial solid solutions, interstitial atoms are accommodated in the interstices of host element crystal structure, which is usually dense leaving small voids. The size of the interstitial atom must be significantly smaller than that of host atoms, with a radius ratio less than 0.41. Small atoms like H and elements from the upper right of the periodic table (B, C, N, etc.) are commonly found in this type of alloys. The solid solubility is limited due to lattice distortions and stresses, with a low maximum allowable concentration of alloying element. The iron-carbon system is a significant example of interstitial solid solutions, with solid solubility of 0.10%, 2.06%, or 0.02% depending on the temperature, indicating a maximum of 0.02% carbon can dissolve in an α-Fe solid solution (at 723 °C).

③ Structure of Compounds

Intermetallic compounds are metallic alloys consisting of ordered compounds between two or more metallic elements. They are classified as intermediate phases because they always lie between two solid solutions on binary phase diagrams. Intermetallics generally exhibit high-temperature mechanical properties and are hard and brittle compared to their constituent metals. They have well-defined stoichiometry and crystal structure, and are sometimes referred to as intermetallic alloys, ordered intermetallic alloys, or long-range-ordered alloys. Intermetallics can be either stoichiometric or nonstoichiometric, and may contain non-metallic elements with different crystal structures. However, carbides and nitrides are not considered intermetallics. In some treatments, cementite (Fe_3C), also known as an interstitial compound, is included in the classification of intermetallics.

Intermetallic compounds are utilized in various applications, owing to their exceptional blend of physical and chemical properties. Their features, such as high melting points, high electrical and thermal conductivity, and low thermal expansion coefficients, make them ideal for utilization in high-temperature and high-stress environments. Moreover, intermetallic compounds are highly resistant to oxidation and corrosion, making them suitable for applications that demand durability, such as aerospace and automotive components.

Intermetallic compounds are characterized by their complex crystal structures resulting from various arrangements of metal atoms. The range of compositions over which the free energy curve has a meaningful existence depends on the structure of the phase and the nature of interatomic bonding, which can be metallic, covalent, or ionic. Three primary factors determine the structure of intermediate phases, namely, the relative atomic size, valency, and electronegativity. Therefore, intermetallic compounds can be classified into three types, namely valence compounds, electron phases, and size-factor compounds.

Valence compounds are a type of intermetallic compound that exhibit a stable octet configuration, whereby the atoms bond in a manner that allows each atom to have 8 electrons in its outermost shell. This bonding is dictated by the valence of the atoms, which represents the number of electrons in their outermost shell.

Normal valence compounds are intermetallic compounds that are formed by certain metals and electronegative IVA, VA, VIA families based on the chemical atomic valence rule. They can be expressed by molecular formula, usually AB, A_2B (or AB_2), or A_3B_2 type, such as MgSe, Mg_2Sn, and Mg_2Si, formed by divalent Mg and tetravalent Sn, Ge, Si. The crystal structure of normal valence compounds often corresponds to the ionic compound structure of similar molecular formula, such as NaCl type, ZnS type, and CaF type. The stability of normal valence compounds is related to the difference in electronegativity between the components. The smaller the difference in electronegativity, the less stable the compound and the more it tends to be bonded by metallic bonds; the greater the difference in electronegativity, the more stable the compound and the more it tends to be bonded by ionic bonds. For instance, the most stable compound among the above four normal valence compounds is Mg_2Si, which has a melting point of 1102 °C and is a typical ionic compound. In contrast, the melting point of Mg_2Pb is only 550 °C, and it exhibits typical metallic properties with increasing resistance value at higher temperature. Hume-Rothery et al. found that the valence electron concentration (*VEC*) of many compounds controls their structures. The *VEC* is defined as the ratio of the total number of valence electrons to the total number of atoms in the crystal. For example, in the A_mB_n compound, the atomic valences of A and B are p and q, respectively, and then the *VEC* is calculated as shown in Eq. 4.7.

Hume-Rothery and colleagues discovered that the *VEC* is a key factor that influences the structures of many compounds. *VEC* is determined by dividing the total number of valence electrons by the total number of atoms in the crystal. For an A_mB_n compound with A and B having atomic valences p and q, respectively, *VEC* can be calculated using Eq. 4.7.

$$VEC = \frac{mp + nq}{m + n} \tag{4.7}$$

The compounds whose *VEC* are the same share the same crystal structure, and these compounds are called electron compounds or Hume-Rothery phase. Initially, the electron compounds were discovered when investigating alloys (e.g., Cu–Zn, Cu–Al, Ag–Sn) consisting of elements belonging to IB group (Cu, Ag, Au) and IIA, IIIA and IVA groups, and afterwards in the alloys between transition metals (e.g., Fe–Al, Ni–Al and Co–Zn) this kind of compound was also found.

Electron phases, despite having chemical formulas, do not adhere to valence rules and have variable composition within a certain range. They can be considered as a solid solution centered on a compound, and their electron concentration also varies within a certain range. Electronic compounds primarily exhibit metallic properties due to the bonding between atoms being predominantly metallic.

Size-factor compounds. The type of compounds is related to the difference in atomic size of the constituent elements. When two elements with a large difference in atomic radii form a compound, they tend to form interstitial compounds or interstitial phases, while when the difference is moderate, they tend to form Topological Close-Packed Phase.

Non-metallic elements with small atomic radii such as C, H, N, B, etc. can form interstitial phases or interstitial compounds with metal elements (mainly transition metal elements). This mainly depends on the ratio of the non-metal (X) and metal (M) atomic radii R_X/R_M; when $R_X/R_M < 0.59$, it forms a phase with a simple crystal structure called interstitial phase; when $R_X/R_M > 0.59$, it forms a phase with a complex crystal structure, usually called an interstitial compound.

The atomic radii of H and N are only 0.046 nm and 0.0071 nm, respectively, so they meet the condition of $R_X/R_M < 0.59$ with all transition metal elements, so the hydrides and nitrides of transition metal elements are interstitial phases. As for B, its atomic radius is 0.097 nm, which is relatively large, so the borides of transition metal elements are interstitial compounds. As for C, it is in an intermediate state, some carbides such as TiC, VC, NbC, WC, etc. are interstitial phases with a simple structure, while Fe_3C, Cr_7C_3, $C_{23}C_6$, Fe_3W_3C, etc. are interstitial compounds with a complex structure.

Interstitial phases have a relatively simple crystal structure, such as face-centered cubic (FCC), close-packed hexagonal (HCP), a few are body-centered cubic (BCC) or simple hexagonal structures, and they are different from the structure of the constituents. In the crystal, metal atoms occupy normal positions, while non-metallic atoms regularly distribute in the interstices of the lattice, forming a new crystal structure. What interstitial position the non-metallic atoms occupy in the interstitial phase mainly depends on the size factor of the atoms. When $R_X/R_M < 0.414$, it can enter the tetrahedral interstice; if $R_X/R_M > 0.414$, it enters the octahedral interstice. The formula of interstitial phases is usually M_4X, M_2X, MX and MX_2. In dense packing structures (FCC and HCP), the ratio of octahedral and tetrahedral interstitials to the number of atoms in the cell is 1 and 2, respectively. When the octahedral interstitials are filled with non-metallic atoms, the composition of the interstitial phase is exactly MX, with a NaCl-type structure (the MX compound can also have a zinc blende structure, with non-metallic atoms occupying half of the tetrahedral interstitials); when the tetrahedral interstitials are filled with non-metallic atoms (occurring only in hydrides), the MX_2 interstitial phase is formed, such as TiH_2 (in MX_2, hydrogen atoms can also be paired and filled in octahedral interstitials, such as ZrH_2); in M_4X, the metal atoms form a FCC structure, and the non-metallic atoms occupy one octahedral interstitial in each cell; in M_2X, the metal atoms are arranged in dense hexagonal packing (some also have FCC, such as W_2N, MoN, etc.), and non-metallic atoms occupy half of the octahedral interstitial positions or one quarter of the tetrahedral interstitial positions. M_4X and M_2X can be considered as structures with unfilled interstitials.

Interstitial phases can be viewed as solid solutions, with compositions that can vary within a certain range. They can also dissolve each other and their constituent elements. Interstitial phases with the same crystal structure and a difference in metal atom radii of less than 15% can form infinite solid solutions, such as TiC–ZrC, TiC–VC, ZrC–NbC, VC–NbC, etc. The bonding between atoms in interstitial phases is covalent and metallic, even when the fraction of non-metallic component atoms is greater than 50%. They exhibit metallic characteristics and have high melting points and high hardness, making them important components in alloy tool steels and hard alloys.

When the ratio of non-metallic atom radius to transition metal atom radius R_X/R_M is greater than 0.59, the resulting phase often has a complex crystal structure, which is referred to as an interstitial compound. Typically, carbides formed by transition metals such as Cr, Mn, Fe, Co, Ni and carbon elements are interstitial compounds. Common interstitial compounds include M_3C (such as Fe_3C, Mn_3C), M_7C_3 (such as Cr_7C_3), $M_{23}C_6$ (such as $Cr_{23}C_6$), and M_6C (such as Fe_3W_3C, Fe_4W_2C). The metal elements in interstitial compounds are often substituted by other metal elements to form solid solutions based on compounds. For example, $(Fe, Mn)_3C$, $(Cr, Fe)_7C_3$, $(Fe, Ni)_3(W, Mo)_3C$, etc.

Fe_3C is a basic phase in iron-carbon alloys, known as cementite. The ratio of the atomic radii of C and Fe is 0.63, and its crystal structure is shown in Fig. 4.6, which is orthorhombic and the three lattice constants are not equal. The unit cell contains 16 atoms, 12 of which are Fe atoms and 4 are C atoms, in accordance with the Fe:C = 3:1 relationship. The Fe atoms in Fe_3C can be substituted by metal atoms such as Mn, Cr, Mo, W, and V to form an alloy cementite, while the C in Fe_3C can be substituted by B but not by N.

Interstitial compounds exhibit both covalent and metallic bonding between atoms, and possess high melting points and hardness (though lower than interstitial phases). They are mainly used as strengthening phases in steel. Furthermore, it is important to note that only transition metals to the left of Fe in the periodic table can form carbides (including interstitial phases and interstitial compounds) in steel. Elements with fewer d-electrons have a greater affinity for carbon, resulting in more stable carbides.

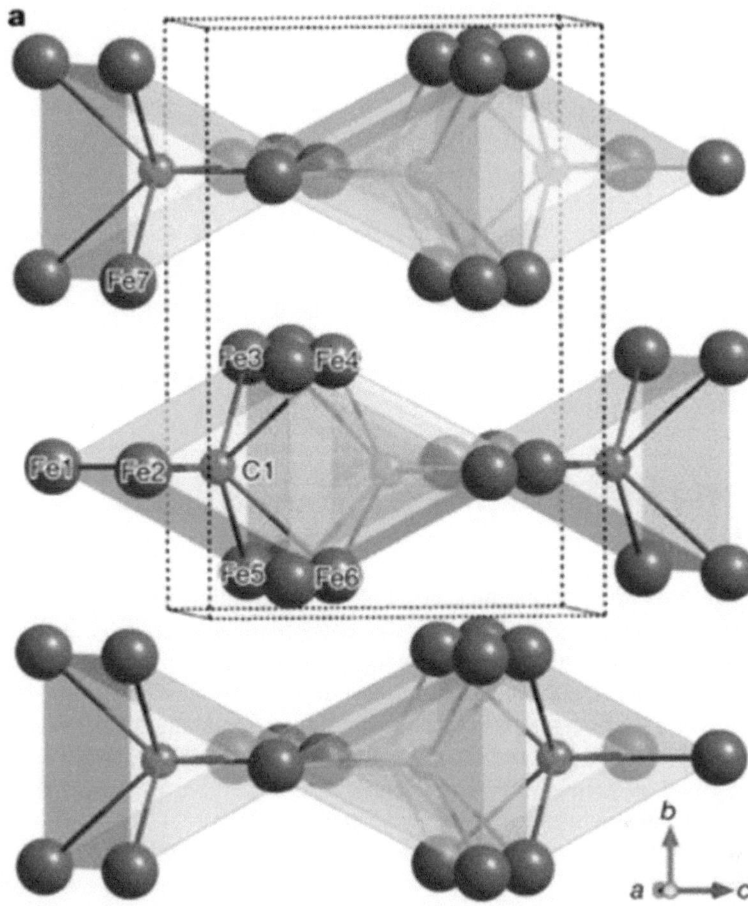

Fig. 4.6 The Fe_3C structure represented as polyhedra of atoms

4.3 Crystal Structure of Ceramics

Ceramic materials have more complex crystal structures compared to metallic materials due to several reasons. Firstly, ceramics are composed of at least two elements, often with ions of varying sizes. Secondly, while the metallic materials have nondirectional, unsaturated metallic bonds, the atomic bonding in ceramics can range from purely ionic to completely covalent, as displayed in Table 4.4. Thirdly, the arrangement of ions in ionic crystals needs to maintain electroneutrality. As a result, the crystal structures of ceramic materials are considerably complex and diverse, even though they still belong to the 14 Bravais lattices.

The properties of ceramics are highly dependent on their composition and bonding. Understanding the crystal structure of ceramics can offer valuable insights into

Table 4.4 Percent ionic character of the interatomic bonds for several common compounds

Compound	Percent ionic character	Compound	Percent ionic character
CaF_2	89	MgO	73
NaCl	67	Al_2O_3	63
SiO_2	51	Si_3N_4	30
ZnS	18	SiC	12

(*Source* William D. Callister, Jr. and David G. Rethwisch, Materials science and engineering, and introduction, 8th ed., Foundations of materials science and engineering, Wiley, 2009, p. 453.)

their mechanical and physical properties. This section will examine the influence of ion size on the crystal structure of ionic compounds, as well as several common and complex crystal structures found in ceramic materials.

4.3.1 The Radius Ratio Rule for the Crystal Structures of Ceramics

The crystal structures of ceramics in which ionic bonding predominates may be treated as being composed of electrically charged ions instead of atom as it is in metallic materials. The minimum energy principle governs the way ions pack together as tightly as possible, resulting in a crystal structure that depends on the relative sizes of the constituent ions and the need to maintain charge neutrality.

In typical ionic solids, cations are normally smaller than anions since the former are formed by giving up electrons and the latter receive the electrons. Anions occupy lattice sites, while cations occupy interstitials between anions. Cations are surrounded by anions to maintain charge neutrality and to shield Coulombic repulsion between anions. The coordination number (CN) is defined as the number of nearest neighbors surrounding a central cation, and it is also equal to the number of anions surrounding the cation. The CN for ionic solids is similar to that for metallic solids, where all atoms are identical. The charge balance is necessary to form a stable structure, and the central cation tends to be surrounded by the largest possible number of anions that can contact it.

Four diagrams illustrated in Fig. 4.7 are used to explain the effect of radius ratio of ions on the packing methods of ionic compounds. In an octahedral configuration, surrounding the central cation, there are four anions in the plane of the diagram, and other two anions (out of the plane, not shown). The first configurations shown in Fig. 4.7a and b have the central cation surrounded by the six anions, leading to a stable structure. Whereas, the cases shown in Fig. 4.7c are not stable structures, because of the inappropriate ratio between cation and anion radii, denoted by R_C and R_A, respectively. And the configuration represented in Fig. 4.7a has the minimum radius ratio.

Intuitionally, greater cation radius generally leads to a greater coordination number. However, when the cation and anion radii are too close, the anions pack

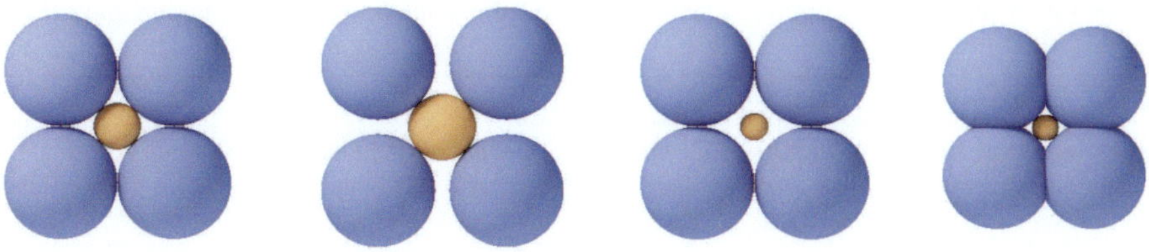

(a) Stable (b) Stable (c) Unstable

Fig. 4.7 The three diagrams illustrate the influence of different radius ratios of ions on the packing and stability of ionic crystal structures. In each of the three cases, the darker, smaller circles inside represent cations, surrounded by anions. **a**: the radius ratio is just right, and the cations and anions in this configuration fit well; **b**: the cation in this case is too large to form a perfect contact for the ions in this layer, and thus its packing would be different; **c** the cation in this case is too small, and it cannot well fit the interstitials in between anions (left). If the cation moves closer to the anions, the crowding of anions in turn would result in strong repulsion among them (right)

closely together, leading to Coulombic repulsion that can destabilize the structure. Consequently, the system changes to some other favorable arrangement. And if the ions in ionic compounds are assumed to be hard spheres, the radii ratio (R_C/R_A) at which anions just touch each other, as well as central cation, is named the critical or minimum R_C/R_A. The assumption that atoms/ions can be treated as hard spheres, and neighbor atoms/ions are tangent is very useful to investigate the crystal structures. By the hard-sphere model, for each stable configuration, the critical R_C/R_A for a given structure can be determined geometrically.

Linus Pauling derived the minimum radius ratio for which a cation contacts a given number of anions using geometrical consideration. If the ratio is too small, it cannot form a stable three-dimensional structure, and only linear or planar configurations with coordination numbers of 2 and 3, respectively, can be obtained, as shown in Fig. 4.8. And obviously, for a periodic structure, the minimum coordination number is 2, and it has linear structure, Fig. 4.8a.

Ionic crystals have various voids which are occupied by cations of different sizes. The larger voids, such as cubic sites, are occupied by larger cations while smaller voids, such as tetrahedral sites, are occupied by smaller cations. Different coordination numbers result in different coordination geometries. Typical configurations have the CN of 4, 6 and 8, and shown in Table 4.5 are the critical radius

(a) CN=2, No critical radius ratio (b) CN=3, and critical radius ratio: 0.155

Fig. 4.8 The low coordination numbers and their corresponding configurations. Anions presented by larger spheres, and cations by the smaller ones in between

Table 4.5 The common coordination numbers and the related configurations and the critical radius ratios

CN	4	6	8
Distribution of anions surrounding central cation	Corners of tetrahedron	Corners of octahedron	Corners of cube
Geometric representation			
Critical radius ratio (R_C/R_A)	0.225	0.414	0.732
Representatives	Zinc Sulfide	Sodium Chloride	Cesium Chloride

(*Source* William F. Smith, Javad Hashemi, and Francisco Presuel-Moreno, Foundations of materials science and engineering, Mcgraw-Hill Publishing, 2006, p. 589)

ratio associated with each CN and the typical ionic compounds that exhibit the corresponding configuration.

The radius ratio rule is the first of the five rules created by Linus Pauling in 1929 to predict and explain crystal structures of ionic compounds. Other rules include the electrostatic valence rule (2nd), sharing of polyhedron centers, edges, and faces (3rd). Using the radius ratio rule to predict crystal structures requires values for ion radii, which should be obtained from the same reference ion to ensure correct relative size. The oxide ion is often used as the standard ion due to its combination with many elements and comparably low polarizability. Additionally, it is important to point out that ionic radii vary depending on the coordination number.

4.3.2 Typical Ceramic Structures

Some ceramic materials, often referred to as AX compounds, just consist of two kinds of ions in equal numbers, and despite having a similar formula, they can have various crystal structures. These structures are typically named after a well-known ceramic material that has that structure, such as the cesium chloride (CsCl), sodium chloride (NaCl), and zinc blende (ZnS).

① Rock Salt Crystal Structure

Sodium chloride, or salt, is a typical ionic compound with the chemical formula NaCl, and has a crystal structure known as the rock salt crystal structure. It represents a large family of compounds. The cations and anions with equal numbers in sodium chloride crystals are highly ionically bonded. As illustrated in Fig. 4.9a, in the rock salt crystal structure, negative Cl^- ions occupy the regular sites in a FCC crystal structure, and positive Na^+ ions situate the interstitial sites constructed by

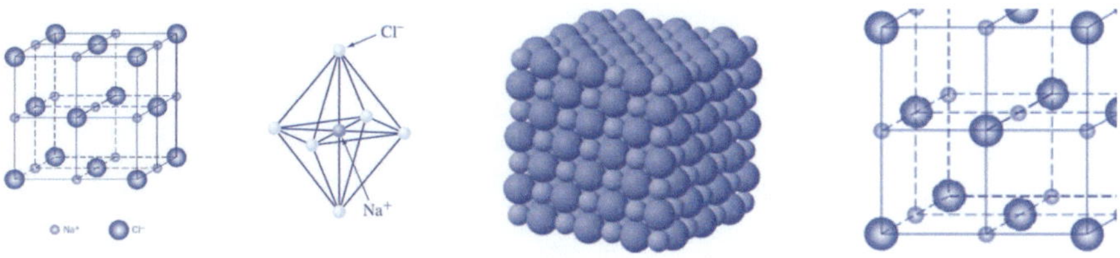

(a) (b) (c)

Fig. 4.9 **a** The unit cell of rock salt crystal structure with atoms shown in reduced size; **b** cations and ions packed within a unit cell; **c** the geometrical relation between its lattice parameter *a* and the radii of ions *r* and *R*, displayed in it (100) plane

the Cl^- ions. The rock salt crystal structure is constructed by nesting two FCC lattices for both cations and anions. An isolated octahedron, shown in Fig. 4.9b, demonstrates that each central Na^+ cation is surrounded by six Cl^- anions, and the octahedron coordination, that is CN is 6, can be easily identified.

The formation of the NaCl crystal structure follows Pauling's radius ratio rule, as evidenced by the data in Table 4.6, with an ionic radius ratio of 0.541, falling between 0.414 and 0.732. The rock salt crystal structure is found in many alkaline halides (such as KCl, LiCl, LiF, NaBr), some oxides (such as MgO, FeO), and other binary compounds (such as ScN).

② Cesium Chloride Crystal Structure

The ionic compound cesium chloride has an equal number of Cs^+ and Cl^- ions, as indicated by its chemical formula. The unit cell of the compound is illustrated in Fig. 4.10. Although the unit cell seems to have a BCC crystal structure at first glance, this is not the case for CsCl because the atom at the body center is not the same kind of atom as those at the corners of the cells. Instead, the crystal structure of CsCl can be described as two interpenetrating simple cubic (SC) structures,

Table 4.6 Ionic radii for several cations and anions

Ion	Radius/nm	Ion	Radius/nm	Ion	Radius/nm
Al^{3+}	0.057	Na^+	0.098	V^{4+}	0.061
Ba^{2+}	0.034	Ni^{2+}	0.078	V^{5+}	~0.04
Ca^{2+}	0.106	Pb^{4+}	0.084	Zn^{2+}	0.083
Cs^+	0.165	Si^{4+}	0.039	Br^-	0.196
Fe^{2+}	0.087	Sn^{4+}	0.074	Cl^-	0.181
Fe^{3+}	0.067	Ti^{2+}	0.076	F^-	0.133
K^+	0.133	Ti^{3+}	0.069	I^-	0.220
Mg^{2+}	0.078	Ti^{4+}	0.064	O^{2-}	0.132
Mn^{2+}	0.091	V^{3+}	0.065	S^{2-}	0.174

(*Source* C. J. Smithells (ed.), Metals Reference Book, 5th, ed., Butterworth, 1976.)

(a) (b) (c)

Fig. 4.10 **a**: the unit cell of CsCl structure with atoms shown in reduced size, **b**: cations and ions packed within a CsCl crystal, **c** the geometrical relation between its lattice parameter *a* and the radii of ions *r* and *R*, displayed in it (110) plane

with one SC structure having its corners located at the body centers of the cells associated with the other SC structure. The CsCl crystal has a primitive cubic lattice with a two-ion (Cs^+/Cl^-) basis and a ratio of one between the numbers of cations and anions, which means that the cell looks the same regardless of whether we start with Cs^+ or Cl^-.

The coordination number of cesium chloride can be easily identified as 8 from its unit cell, Fig. 4.10b, unlike the rock salt structure. The geometrical relation between the radii of ions and the structural constants is clearly shown in Fig. 4.10c, which displays its crystalline plane of (110).

According to Pauling's radius ratio rule, when the ratio of the radii of both ions is close to one ($R_C/R_A > 0.732$), the CsCl crystal structure is formed. Therefore, the CsCl crystal structure is also observed in compounds such as CsBr, CsI, RbCl (at high temperatures), and various binary metallic alloys (e.g., AgZn, BeCu, and MgCe).

When heated to elevated temperature, CsCl may undergo a phase transition, which is a common phenomenon in solid materials known as polymorphism (discussed in detail in a later section). Normally, CsCl has a simple cubic crystal structure (α-CsCl), but as the temperature increases above 445 °C, it begins to convert to a rock salt structure (β-CsCl).

③ Zinc Blende (ZnS) Crystal Structure

The zinc blende structure is named after the zinc ore mineral sphalerite, and is a polymorph of zinc sulfide (ZnS), represented by β-ZnS. Its unit cell is presented in Fig. 4.11a. Ignoring Zn atoms inside, an FCC structure is easily identified. In fact, similar in the rock salt structure, there are two interpenetrating FCC lattices formed by the two types of atoms in ZnS crystal. However, unlike the rock salt structure, the zinc blende structure comprises of two interpenetrating FCC lattices that are oriented differently with respect to one another. This orientation results in a tetrahedral coordination of the atoms, where each atom has four nearest neighbors of opposite charge forming a regular tetrahedron. Half of the tetrahedral sites in the FCC structure are occupied due to stoichiometry, and the filled tetrahedra share

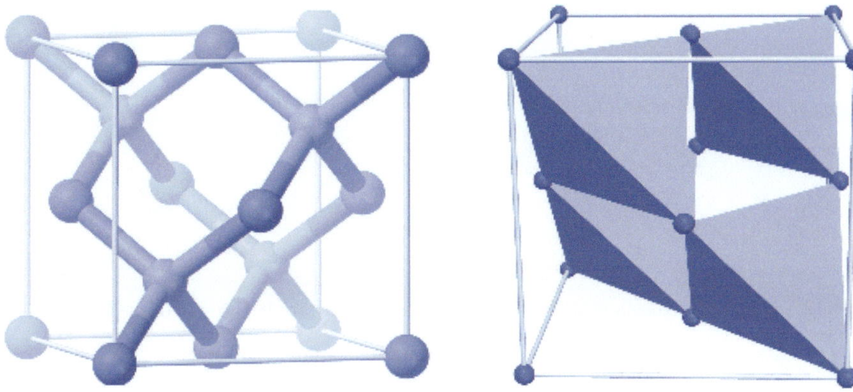

Fig. 4.11 **a** the unit cell of ZnS, with Zn atoms inside the cube; **b** cations and ions packed within a unit cell

corners. Overall, ignoring the difference between Zn and S atoms, the zinc blende structure is the same as the diamond cubic structure, which is a carbon polymorph.

Checking with the Pauling's radius ratio rule, the R_C/R_A equals 0.477 for ZnS, which shall correspond to a coordination number of 6. Thus, the coordination number (4 for Zn and 4 for S) in the zinc blende structure is inconsistent with the Pauling's radius ratio rule. The discrepancy can be explained by the method via which the radii are determined and the deviation of the atomic bonding of ZnS from ionic nature, as contained in Table 4.4.

The zinc blende crystal structure is a widely observed structure that can be found in numerous compounds. Many oxides and sulfides adopt this structure, including compound semiconductors like GaAs, CdTe, ZnTe, and SiC, as well as some oxides and nitrides such as ZnO, BeO, and BN.

4.3.3 Silicate Ceramics

Because of their extremely rich existence in the shallow surface of earth, and their properties are suitable to make pottery, silicates were used by human beings at the dawn of civilization. The vast majority of minerals in the earth's crust are silicates, owing to the high abundance of oxygen and silicon, which make up over 73 wt% of the earth's mass. Silicates include a vast array of minerals, such as quartz, mica, olivine, and clay minerals, and find applications in electrical insulation, refractory linings, and structural components due to their high temperature resistance, toughness, and hardness. Other desirable properties include chemical stability, resistance to wear, and biocompatibility. Therefore, it is important to discuss the structure of this class of minerals.

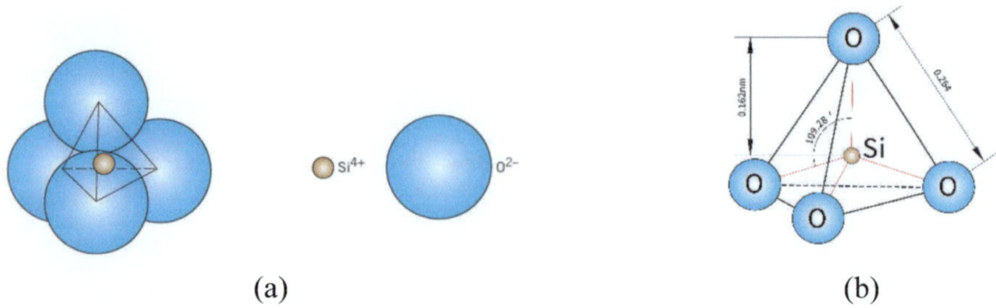

Fig. 4.12 **a** The configuration of the silicon-oxygen tetrahedron; **b** The size of the tetrahedron

In discussing the structure of silicates, it is often more convenient to use arrangements of tetrahedra (Fig. 4.12), where each silicon atom is bonded to four oxygen atoms at the corners of a tetrahedron, with the silicon atom positioned in the center. This tetrahedron is considered the fundamental unit of silicates and is often treated as a negatively charged entity. Silicates are not considered purely ionic because the Si–O bonds have significant covalent character, which makes them directional and relatively strong. Despite this covalent character, each tetrahedron has a formal charge of −4 because each oxygen atom requires an additional electron to achieve stability. By combining these units in one, two, or three dimensions, different silicate structures can be formed.

① Silica and Silica Glasses

The arrangement of these silicon-oxygen tetrahedra in a uniform and orderly pattern results in a crystalline structure. The three main crystalline forms of silica are quartz, cristobalite (Fig. 4.13a), and tridymite. Silica is highly unreactive, making it useful for a variety of applications, including the production of glass, ceramics, and refractories. The strong covalent bonds between silicon and oxygen give silica its high melting and boiling points, strength, and hardness, making it ideal for use in construction materials. Silica is also a key component in many industrial products, such as silicon wafers used in the electronics industry and filler materials in rubber and plastic products. Additionally, it is an important ingredient in the production of everyday items such as toothpaste, soap, and cosmetics.

Inorganic glasses used for common objects, such as windows and containers, are typically made by combining silica glass with other oxides, such as CaO and Na_2O. Unlike polyhedral networks, the cations from these added oxides are incorporated into the network and modify it, as shown in Fig. 4.13b.

Silica has various unique properties that have important applications in scientific, medical, and environmental fields. Silica is utilized in chromatography to separate complex mixtures, and it is used to create highly porous silica gel, which is valuable in applications like desiccation. Additionally, silica has environmental uses in water treatment and in mineral fertilizers to improve soil fertility.

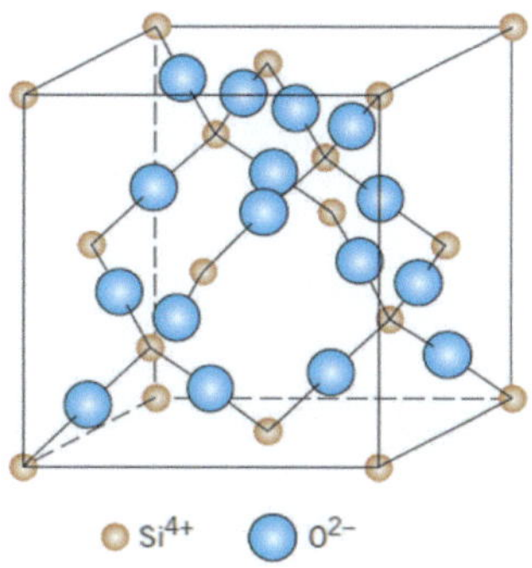

(a)

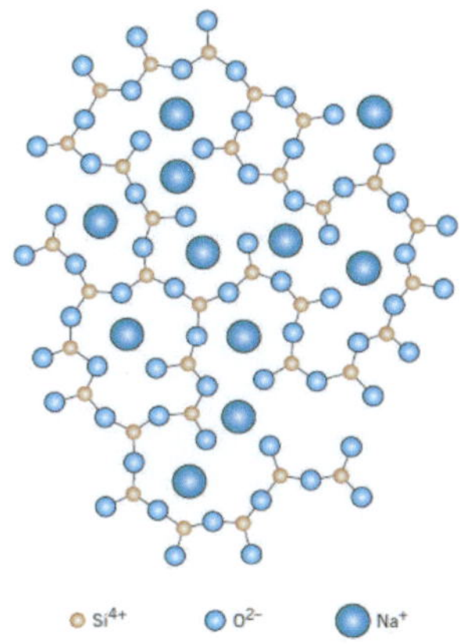

(b)

Fig. 4.13 **a** The arrangement of silicon and oxygen atoms in a unit cell of cristobalite, a polymorph of SiO_2; **b** Schematic representation of ion positions in a sodium–silicate glass

In conclusion, silica's versatility and unique properties make it an ideal material for a wide range of applications across various industries, from construction to biomedical applications. Its abundance and value make it a material of great importance in our modern world.

② Silicates

The basic building blocks of all these silicate minerals is the silica tetrahedron as shown in Fig. 4.12. The silicate minerals have complex structures formed by connecting the building blocks in various ways, sharing one, two or three corner oxygen atoms of the tetrahedra among other tetrahedra. Some of these structures, depicted in Fig. 4.14, have formulas $(SiO_4)^{4-}$, $(Si_2O_7)^{6-}$, $(Si_3O_9)^{6-}$, and so forth, while others exhibit single-chain structures as shown in Fig. 4.14e. Positive cations like Ca^{2+}, Mg^{2+}, and Al^{3+} serve two functions. They first balance the negative charges from the units to attain charge neutrality, and second, they bond the tetrahedra together through ionic bonding.

Simple silicates are minerals composed of isolated silica tetrahedra, which are the most basic in terms of structure and characterized by high thermal and chemical stability. They can be classified further based on their chemical composition, such as silicates containing magnesium (forsterite), iron (fayalite), or aluminum (diaspore).

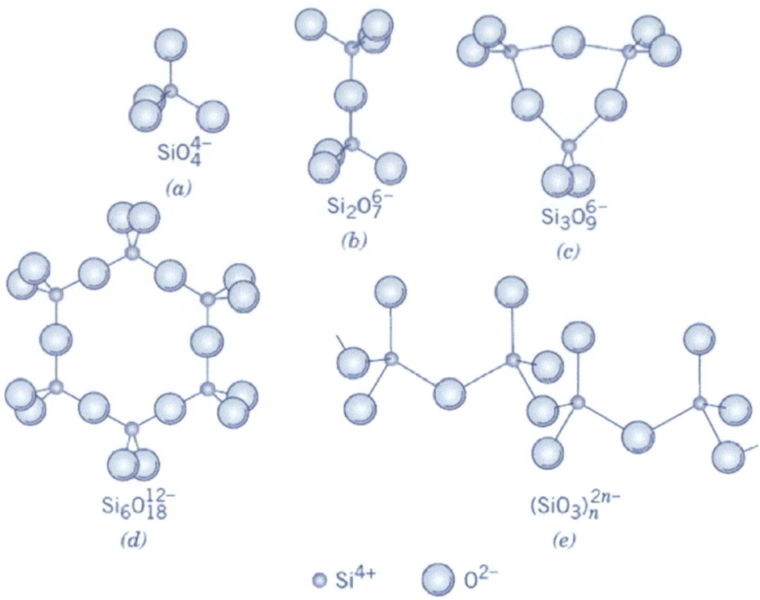

Fig. 4.14 **a** Different silicate structures formed through the sharing of vertices of silicate tetrahedron

Layered or two-dimensional silicates with a sheet structure can be formed by sharing three oxygen ions among the silica tetrahedra, as depicted in Fig. 4.15. The repeating unit formula for the layered structure is $(Si_2O_5)^{2-}$. The negative charge is linked to the oxygen atoms that are not bonded and extend out of the plane of the page. To maintain electrical neutrality, a second planar sheet structure with an excess of positive ions is usually required, which bonds to the unbonded oxygen atoms from the (Si_2O_5) sheet. These materials are known as sheet or layered silicates, and their fundamental structure is characteristic of clays and other minerals.

There are several different types of layered silicates. Kaolinite, one of the most widespread clay minerals, has a simple two-layer silicate sheet structure, as seen in Fig. 4.16, with the silica tetrahedral layer $(Si_2O_5)^{2-}$ electrically balanced by an adjacent $Al_2(OH)_4^{2+}$ layer, as indicated in its formula $Al_2(Si_2O_5)(OH)_4$.

Kaolinite clay becomes hydrated kaolin by adding water, which is more plastic and easier to work with because the water molecules inserted between layers of kaolinite can eliminate partial hydrogen bonds, as shown in Fig. 4.17a and b. Hydrated kaolin is commonly used in the production of ceramics, paper, and cosmetics, as well as in the formulation of catalysts and as a filler in various industrial applications. The similar layered structure can be found in many other minerals, such as pyrophyllite and white mica as shown in Fig. 4.17.

When the four oxygen atoms in silica tetrahedra are all shared, a network silicate (or framework silicate) is formed. The most common and well-known network

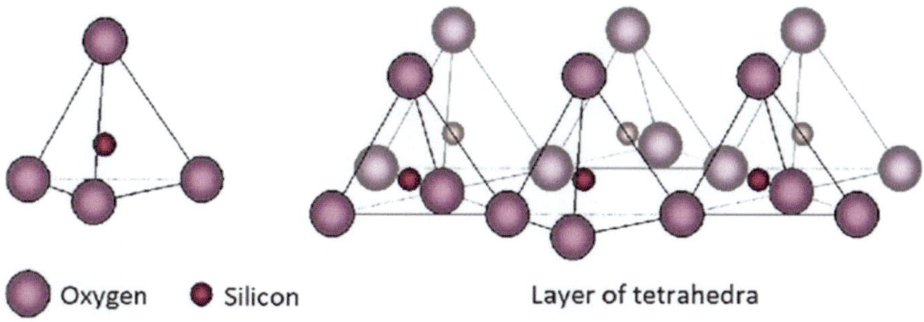

Fig. 4.15 **a** Schematic illustration of the two-dimensional silicate sheet structure, which has a repeating unit of $(Si_2O_5)^{2-}$

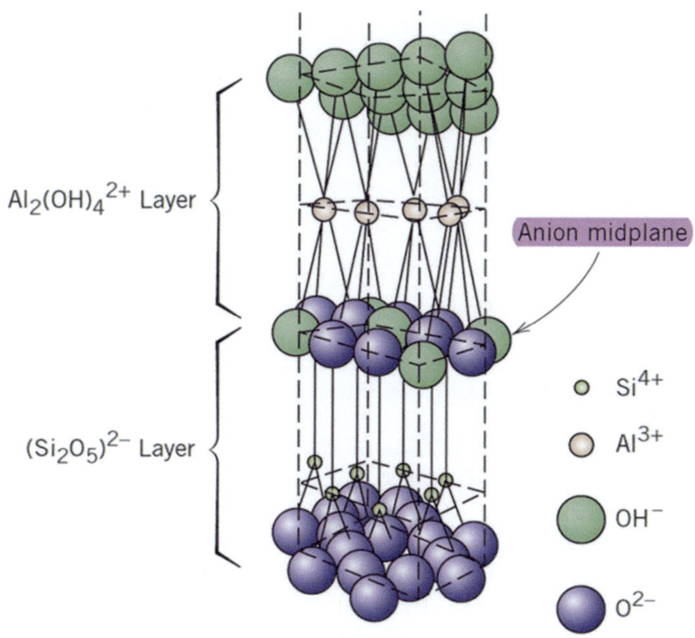

Fig. 4.16 **a** The structure of kaolinite clay (Adapted from W. E. Hauth, "Crystal Chemistry of Ceramics," American Ceramic Society Bulletin, Vol. 30, No. 4, 1951, p. 140.)

silicate is quartz, which is a transparent, colorless mineral that is used in many industrial and decorative applications. Other examples of network silicates include feldspar, mica, and clay minerals. These minerals have different physical properties, such as color, transparency, and cleavage, which are useful for identifying them and classifying them into different groups.

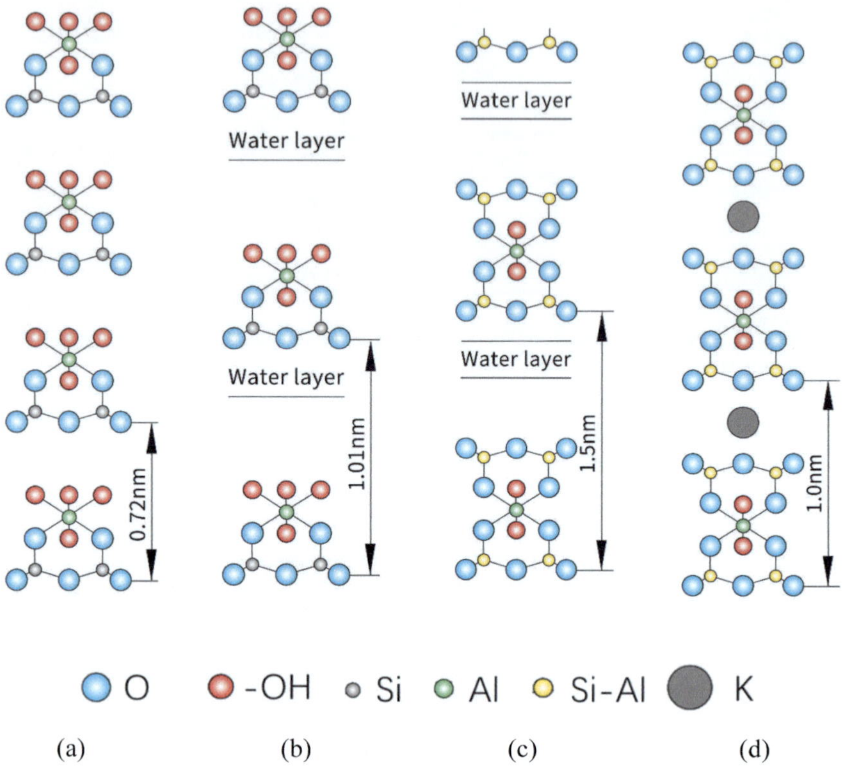

Fig. 4.17 **a** and **b**: projection of layers of kaolinite and hydrated kaolin, respectively; **c** and **d**: the structure of pyrophyllite and white mica, respectively

4.4 Structure of Polymers

Polymers are macromolecules composed of a large number (about 10^3 to 10^5) of structural units derived from monomers and connected through chemical bonds. Strong interactions such as van der Waals forces, hydrogen bonds, electrostatic forces, and hydrophobic interactions exist between polymer chains. Polymer chains can bend, twist, and kink, leading to complex and diverse polymer structures. Unlike solids composed of small molecules, polymer crystal structures are less ordered due to the giant and flexible nature of polymer molecules. However, compared to liquids made of small molecules, amorphous polymer structures are more ordered. Polymer structures can be classified into chain and condensed state structures, as shown in Fig. 4.18.

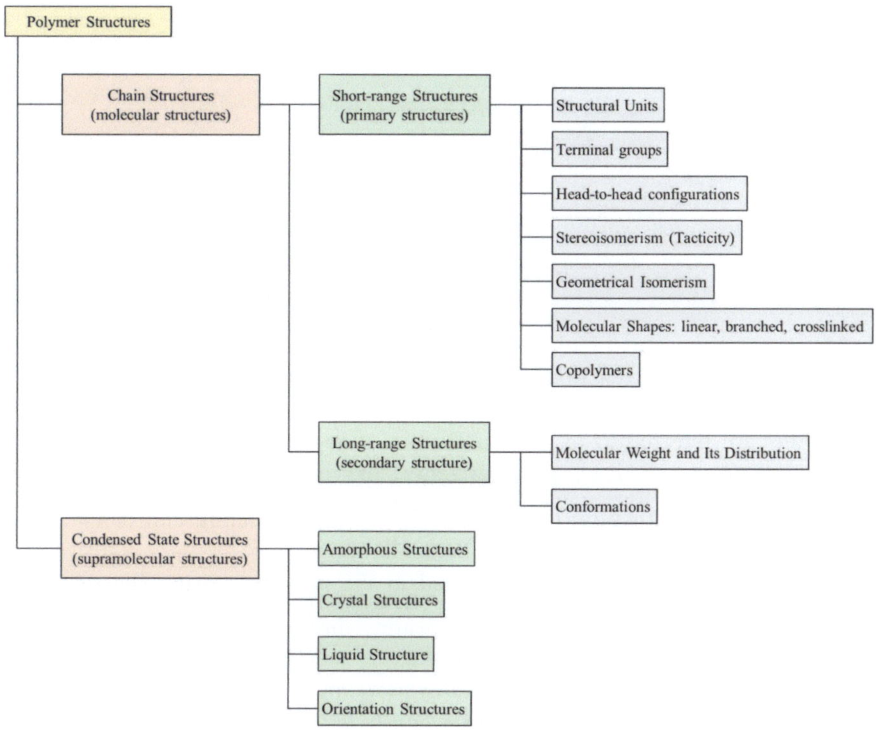

Fig. 4.18 Classification scheme for polymer structures

4.4.1 Chain Structures

The chemical structure of a single polymer molecule is referred to as a chain structure, which can be further divided into short-range and long-range structures. Short-range structures are also known as primary structures, which include constitution and configuration. Constitution refers to the sequence of atoms and bonds in a molecule, while configuration refers to the spatial arrangements of chemically bonded atoms. Changes to the constitution and configuration require the breaking and reorganization of chemical bonds.

In polymer science, the long-range structure, also known as the secondary structure, pertains to the molecular weight and morphology. It is important to note that the terms "short-range" and "long-range" do not refer to distance in space, but rather distance along the polymer chain, as depicted in Fig. 4.19.

① Short-range Structures

In this section, constitutions and configurations of structural units of polymer chains will be introduced.

Fig. 4.19 Illustration of "short-range" and "long-range" in polymer science

(1) Monomer, Structural Unit, and Repeating Unit

During polymerization reactions, monomers are linked together to form polymers with long chains, and the chemical parts of monomers that remain in the polymer chains are referred to as structural units. For example, when the monomer propylene undergoes addition polymerization, the resulting polymer is polypropylene (PP), with the structural unit $-CH_2-\underset{\underset{CH_3}{|}}{CH}-$. The repeating unit of PP is also $-CH_2-\underset{\underset{CH_3}{|}}{CH}-$, as these units are repeated in the polymer chain. The degree of polymerization (DP) is the number of repeating units in a polymer, which in the case of PP is represented by n. It is important to note that condensation polymers are synthesized from two or more types of monomers, resulting in multiple types of structural units but only one repeating unit (e.g., PET in Table 4.7).

(2) Terminal Groups

Polymers are composed of long chains of chemically bonded monomers. The chemical groups at both ends of the polymer chain are called terminal groups, which are not part of the repeating unit. Although terminal groups are present in small quantities, they have a significant impact on the polymer's chemical properties, particularly its thermal stability. Polymer chain scission often begins at terminal groups. For example, the terminal carboxyl and acyl chloride groups of polycarbonate (PC) can cause degradation of PC at high temperatures, resulting in poor thermal stability. In addition to chemical properties, terminal groups also affect molecular motion. (Definition of molecular motion can be found in Sect. 12.5).

(3) Head-to-Head and Head-to-Tail Configurations

In condensation and ring-opening polymerization, the bonding of structural units generally follows a head-to-tail configuration. However, in addition polymerization, there are two possible bonding configurations between structural units, namely "head-to-head" and "head-to-tail", as shown in Fig. 4.20. It should be noted that there can also be random bonding configurations that have both head-to-head and head-to-tail connections. The isomers resulting from different bonding configurations between the structural units are referred to as sequence isomers.

Table 4.7 Summary of some common polymers and their chemical structures

Polymers		Repeating unit	Structural unit	Monomer
Polyethylene (PE)	$[-CH_2-CH_2-]_n$	$-CH_2-CH_2-$	$-CH_2-CH_2-$	$CH_2=CH_2$
Polypropylene (PP)	$[-CH_2-CH(CH_3)-]_n$	$-CH_2-CH(CH_3)-$	$-CH_2-CH(CH_3)-$	$CH_2=CH(CH_3)$
Polytetrafluoroethylene (PTFE)	$[-CF_2-CF_2-]_n$	$-CF_2-CF_2-$	$-CF_2-CF_2-$	$CF_2=CF_2$
Polybutadiene (PB)	$[-CH_2-CH=CH-CH_2-]_n$	$-CH_2-CH=CH-CH_2-$	$-CH_2-CH=CH-CH_2-$	$CH_2=CH-CH=CH_2$
Polyisoprene (PIP)	$[-CH_2-C(CH_3)=CH-CH_2-]_n$	$-CH_2-C(CH_3)=CH-CH_2-$	$-CH_2-C(CH_3)=CH-CH_2-$	$CH_2=C(CH_3)-CH=CH_2$
Polyethylene Terephthalate (PET)	$[-O-C(=O)-C_6H_4-C(=O)-O-CH_2-CH_2-]_n$	$-O-C(=O)-C_6H_4-C(=O)-O-CH_2-CH_2-$	$-C(=O)-C_6H_4-C(=O)-$ $-O-CH_2-CH_2-O-$	$HO-C(=O)-C_6H_4-C(=O)-OH$ $HO-CH_2-CH_2-OH$
Polyamide 6 (PA6)	$[-NH-(CH_2)_5-C(=O)-]_n$	$-NH-(CH_2)_5-C(=O)-$	$-NH-(CH_2)_5-C(=O)-$	(ring: $-NH-(CH_2)_5-C(=O)-$, =O)
Polyamide 6 (PA66)	$[-NH-(CH_2)_6-NH-C(=O)-(CH_2)_4-C(=O)-]_n$	$-NH-(CH_2)_6-NH-C(=O)-(CH_2)_4-C(=O)-$	$-C(=O)-(CH_2)_6-C(=O)-$ $-NH-(CH_2)_6-NH-$	$HO-C(=O)-(CH_2)_6-C(=O)-OH$ $H_2N-(CH_2)_6-NH_2$
Polydimethylsiloxane (PDMS)	$[-Si(CH_3)_2-O-]_n$	$-Si(CH_3)_2-O-$	$-Si(CH_3)_2-O-$	$Cl-Si(CH_3)_2-Cl$

Fig. 4.20 **a** Head-to-tail and **b** head-to-head (or tail-to-tail) configurations of polyolefin. R represents a side group, such as $-CH_3$, $-Cl$, and $-$phenyl

Bonding configurations have a significant impact on polymer properties. Polyvinyl formal, also known as Vinylon, is used as fibers and requires head-to-tail arrangements between its structural units to achieve good crystallization performance and high strength.

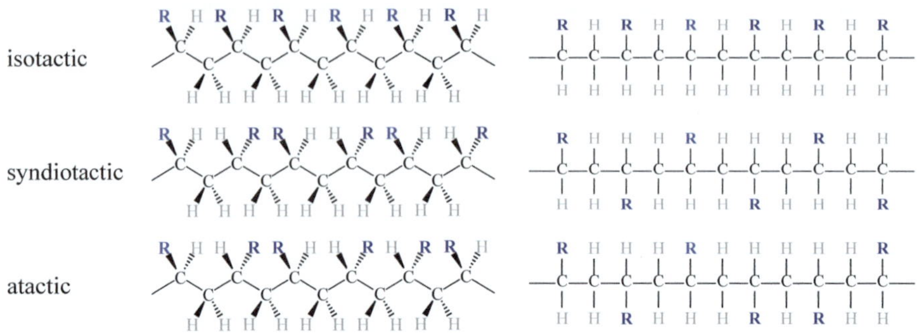

Fig. 4.21 Stereoisomerism of polymers: **a** isotactic, **b** syndiotactic, and **c** atactic configurations

(4) Stereoisomerism (Tacticity)

Polymers containing a chiral carbon atom (C^*) in each structural unit of $-CH_2-C^*HR-$ exhibit two kinds of optical isomers, which can be bonded in three ways. When all structural units have the same optical isomer configuration, the corresponding polymer is called ***isotactic***. In this configuration, all of the R side groups are located on the same side of the polymer chain (Fig. 4.21a). If the two kinds of optical isomers are alternately bonded, then the corresponding polymer is ***syndiotactic***, in which the R groups are alternatively situated on both sides of the polymer chain (Fig. 4.21b). A random bonding of optical isomers produces an ***atactic*** polymer with R groups arranged randomly on both sides of the chain (Fig. 4.21c). These bonding configurations have a significant influence on polymer properties.

Stereoisomerism, which refers to differences in spatial configuration, can have a significant impact on the properties of polymers. For example, isotactic polystyrene (PS) can crystallize and has a high melting point of 240 °C due to its regular structure, while atactic PS cannot crystallize and has a lower softening temperature of only 80 °C due to its irregular structure. Similarly, isotactic or syndiotactic polypropylene (PP) has a regular structure, which results in fast crystallization and the ability to form fibers, while atactic PP cannot crystallize and behaves more like rubbery elastomers.

(5) Geometrical Isomerism

Geometrical configurations in polymer chains are influenced by double bonds. In the case of polyisoprene, as shown in Fig. 4.22, which is produced by 1,4-addition polymerization, carbon atoms connected with double bonds are unable to rotate around the main chain, resulting in the formation of *cis* and *trans* geometric isomers. *Cis*-polyisoprene, also known as natural rubber, possesses excellent rubber properties, whereas *trans*-polyisoprene behaves like a semi-crystalline plastic. A similar phenomenon occurs in *cis*- and *trans*-polybutadiene provides a detailed comparison of *cis*- and *trans*- polyisoprene and polybutadiene properties (Table 4.8).

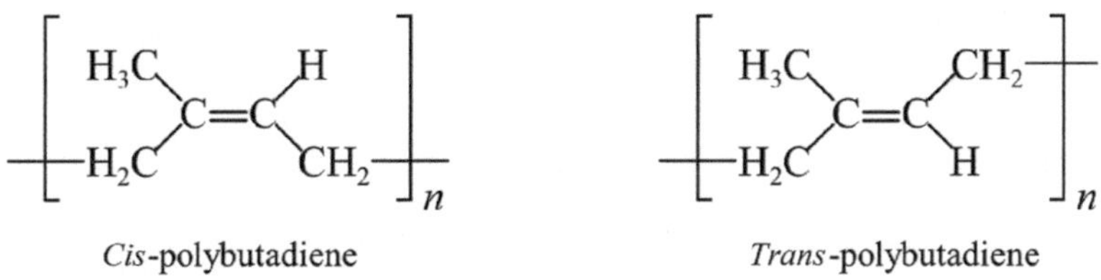

Fig. 4.22 Geometrical configurations of polyisoprene

Table 4.8 The effect of geometrical configuration on melting point and glass transition temperature

Polymer	Melting point T_m/°C		Glass transition temperature T_g/°C	
	cis 1,4	*trans* 1,4	*cis* 1,4	*trans* 1,4
Polyisoprene	30	70	−70	−60
Polybutadiene	2	149	−108	−80

(6) Molecular Shapes

Polymer chains can be classified into three main types based on their shape: linear, branched, and cross-linked.

(1) ***Linear polymers***. Ethylene-based polymers, unvulcanized rubbers, and polycondensates obtained from bifunctional monomers typically have linear molecular structures, as shown in Fig. 4.23a. Linear polymers have relatively free movement between their molecular chains. In addition, most linear polymers are soluble in organic solvents and are thermoplastics, meaning they can be melted by heating.
(2) ***Branched polymers*** have side-branch chains attached to their main chains, such as star and comb polymer chains (Fig. 4.23b). Unlike linear polymers, branched polymers have adverse effects on physical and mechanical properties. However, the synthesis of dendrimers, also known as hyperbranched polymers, has changed the negative perception of branching. Dendrimer chains have a highly regular structure, which eliminates the negative effects of branching and improves processing due to their smaller hydrodynamic volume, closer molecule packing, and lower viscosity at the same relative molecular weight. It is worth noting that side groups connected to linear polymers are not considered branches.
(3) ***Cross-linked polymers***. In polymer science, there are three main types of molecular structures: linear, branched, and cross-linked. Linear polymers have movable chains and are generally soluble in organic solvents, while branched polymers have side chains of the same or similar length as the main chain. Star and comb polymers are examples of branched polymers. On the other

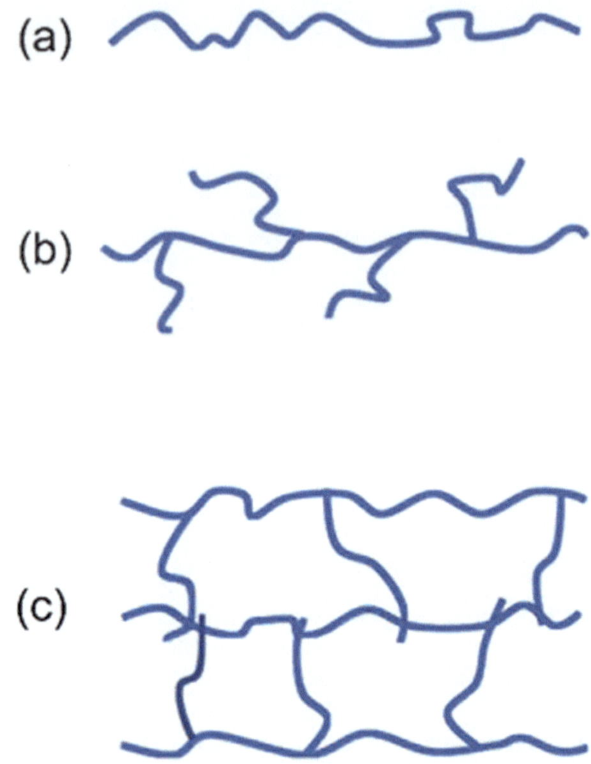

Fig. 4.23 Schematic illustrations of **a** linear, **b** branched, and **c** cross-linked micromolecular structures

hand, cross-linked polymers have a 3D network structure formed by chemical bonds or short branched chains (Fig. 4.23c). Cross-linking can greatly improve the mechanical and thermal properties, as well as solvent resistance and dimensional stability of polymers. The transformation of linear chains to cross-linked structures leads to a leap in polymer properties.

(7) Copolymers

In polymer chemistry, when a single monomer is involved in the polymerization process, it is referred to as homo-polymerization, and the resulting polymer is a **homopolymer** (Fig. 4.24a). On the other hand, if two or more types of monomers are involved, the process is known as copolymerization, and the resulting product is a copolymer.

Copolymerization is a significant technique for synthesizing new polymers and modifying existing ones. There are various types of copolymers. ***Random copolymers*** have structural units that are randomly bonded together (Fig. 4.24b). In ***alternating copolymers***, the two structural units are bonded alternately (Fig. 4.24c).

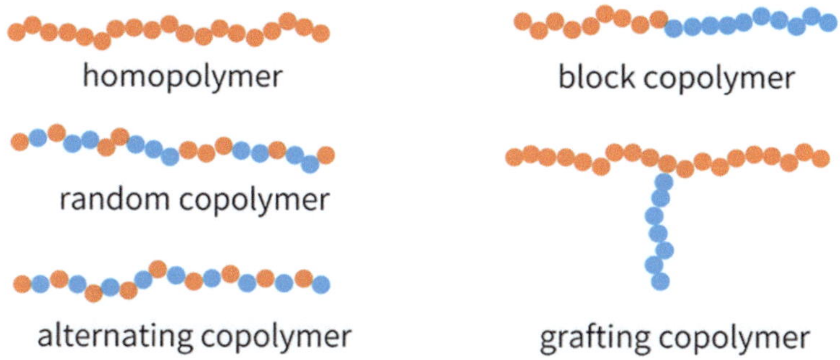

Fig. 4.24 Schematic illustration of homopolymers, random copolymers, alternating copolymers, block copolymers, and grafting copolymers

Block copolymers have identical structural units clustered in blocks along the polymer chains [Fig. 4.24d]. In ***graft copolymers***, the side chains made up of identical units are grafted onto a homopolymer chain made up of different units (Fig. 4.24e).

Copolymers exhibit distinct physical and chemical properties in comparison to homopolymers due to the variation in arrangement of structural units, leading to altered interactions between structural units and molecules.

② Long-range Structures

(1) Molecular Weight and Its Distribution

When used as materials, polymers are typically required to exhibit a certain level of mechanical strength. The molecular weight (also referred to as molecular mass, molar mass, or relative molar mass) of a polymer is a crucial factor that impacts its mechanical strength. Therefore, during the synthesis and processing of polymers, molecular weight is a key parameter for evaluating their quality. In general, the mechanical strength of polymers tends to increase as molecular weight increases, as illustrated in Fig. 4.25.

Polymers are typically produced as a mixture of homologues with varying molecular weights, unlike small molecules such as ethanol and benzene that have fixed molecular weights. Average molecular weight is used to represent the molecular weight of polymers, and the most commonly used measures are number-average molecular weight and weight-average molecular weight.

Number-average molecular weight ($\overline{M}_{\mathrm{n}}$) is defined as that the total mass of polymer m divided by the total molecule numbers n. It can be calculated according to

$$\overline{M}_{\mathrm{n}} \equiv \frac{m}{\sum n_i} = \frac{\sum n_i M_i}{\sum n_i} = \frac{\sum m_i}{\sum (m_i/M_i)} = \sum x_i M_i$$

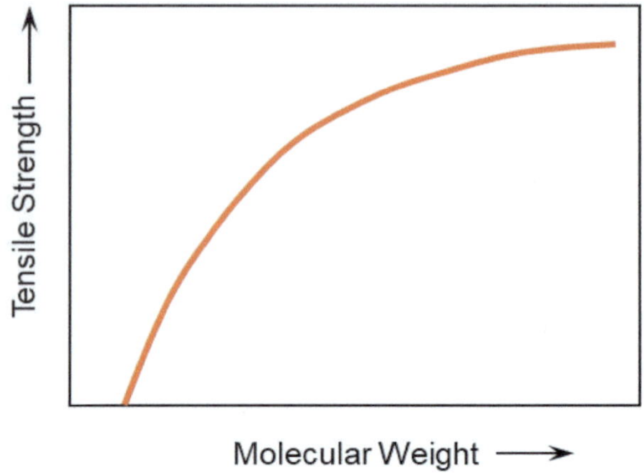

Fig. 4.25 The relationship between mechanical strength and molecular weight of polymers

where n_i, m_i, and M_i are the molecular number, molecular mass and mean molecular weight of size range *i*, and x_i is the number fraction of size range *i* to the total molecular numbers.

Weight-average molecular weight ($\overline{M}_w$) is also called mass-average molecular weight. It is determined by light scattering method and can be calculated as:

$$\overline{M}_w = \frac{\sum m_i M_i}{\sum m_i} = \frac{\sum n_i M_i^2}{\sum n_i M_i} = \sum w_i M_i$$

where w_i is the weight fraction of size range *i* to the total molecular numbers.

Polymers typically have a distribution of molecular weights, which is referred to as polydispersity. There are two common methods for defining polydispersity: the molecular weight distribution index and the molecular weight distribution curve.

The ***molecular weight distribution index*** reflects the width of the molecular weight distribution and is defined as the ratio of weight-average molecular weight ($\overline{M}_w$) to number-average molecular weight ($\overline{M}_n$). When polymers have identical molecules, $\overline{M}_w$ is equal to $\overline{M}_n$, so $\overline{M}_w/\overline{M}_n = 1$. The index ranges from 1.5 to 2.0 for some polymers, but for others, it can range from 20 to 50, depending on the synthesis method. The higher the index, the wider the distribution, and the more uneven the molecular weight.

The ***molecular weight distribution curve*** is illustrated in Fig. 4.26, with $\overline{M}_n$ and $\overline{M}_\omega$ marked on the abscissa in ascending order. The number-average molecular weight is close to the most probable molecular weight. Even when the average molecular weight is the same, the distribution may not be identical because the percentage of the same molecular weight part is not necessarily the same.

The molecular weight distribution is a crucial factor that impacts polymer properties. A low molecular weight fraction can lower the solidification temperature and weaken the polymer, whereas an excessively high molecular weight makes plasticization difficult. The molecular weight distribution must be suitable

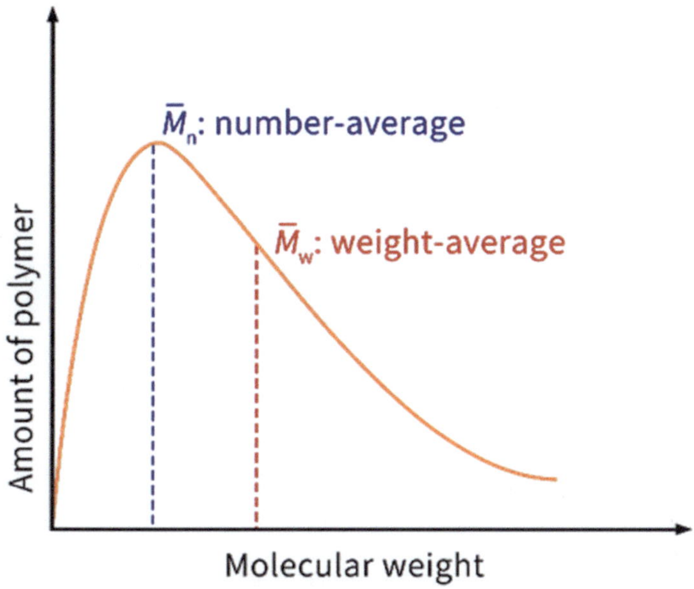

Fig. 4.26 Molecular weight distribution curve of polymers

for different polymer materials. For example, synthetic fiber requires a narrow distribution, while synthetic rubber can have a wide distribution.

(2) Rotational Isomeric States and Chain Flexibility

Polymers such as polyethylene (PE), PP, and PS have main chains that consist of many single bonds. These single bonds, such as C–C, can rotate around their axis, a process known as internal rotation. However, due to the fixed bond angle between adjacent bonds (e.g., 109°28' for C–C bond), free rotation of the single bond is not possible. Figure 4.27 shows that fixing the first C–C bond (σ_1) in the z-axis means that the second C–C bond (σ_2) still has many rotational options due to the internal rotation of σ_1. Similarly, the rotation of σ_2 leads to the revolution of σ_3. As there are many σ bonds in a single polymer chain capable of internal rotation, there can be countless forms of polymer molecules, which are referred to as conformations. ***Conformation*** arises from the internal rotation of single bonds, and the thermal motion of molecules leads to their statistical nature (Table 4.9). According to statistical laws, the probability of straight molecular chains is extremely low, while the probability of curled up chains is high.

The ***random coil*** is the irregularly curled up conformation of polymer chains caused by the internal rotation of single bonds, where the freer the internal rotation, the greater the tendency for curling up.

In reality, C–C single bonds with completely free internal rotation do not exist because carbon chains always carry other atoms or groups. When these atoms or groups are in close proximity, the repulsive forces generated between the outer

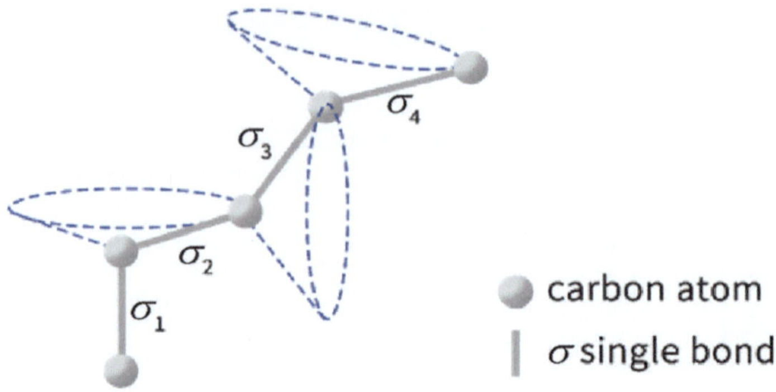

Fig. 4.27 Internal rotation conformation of polymer chain

Table 4.9 Comparison of conformation and configuration

Conformation	Configuration
Spatial arrangement formed by internal rotation	Geometric arrangement fixed by chemical bond
Conversion of conformation is resulted from internal rotation of single bond, and thermal motion is enough to make it happen	Configuration change involves the broken of chemical bond
Unstable	Stable

electron clouds of the atoms prevent them from getting any closer. This impedes the internal rotation of single bonds, which will require some energy to overcome the resistance. Thus, the internal rotation of single bonds is never completely free.

The flexibility of a polymer chain, which is the primary cause of the distinctive properties of high polymers compared to small molecules, can be understood from both static and dynamic perspectives.

The flexibility of polymer chains is affected by several factors, such as the main chain structure, substituent structure and arrangement, and remote interactions.

4.4.2 Condensed Structures

In the context of polymers, the term "condensed state" refers to the stable form of a substance that is composed of a large number of atoms or molecules held together by binding forces. The structure of a single polymer molecule is referred to as the polymer chain structure, while the structure of the entire polymer, which describes how a large number of polymer chains are arranged, is referred to as the condensed state structure.

The investigation of condensed state structures in polymers is of both theoretical and practical importance. An accurate understanding of condensed state structures provides the foundation for developing theories on bulk properties of polymers. Additionally, exploring the characteristics and formation conditions of condensed state structures and their correlation with material properties can facilitate the acquisition of materials with desired condensed state structures and properties through controlling processing and molding conditions.

① Contents of Condensed State Structures

The condensed state structures of polymers comprise amorphous, crystalline, liquid crystal, and oriented structures.

(1) Amorphous Structures

Arranging many thin and long polymer chains, which are different in size, variable in shape, and extremely flexible, into a three-dimensional ordered structure is a challenging task. Therefore, the amorphous state is the first choice for the condensed state of molecular chains. Glassy polymers, rubber, and the amorphous regions in crystalline polymers are all in the amorphous state, indicating that achieving an ordered structure is not always necessary for the bulk properties of polymers.

The comprehension of amorphous structures is continually evolving, and several models have been proposed to date. Among the current models, two are representative and are briefly explained below.

Random coil model. Flory proposed that in amorphous polymers, molecular chains have the same conformation as in solution and appear as random coils, as shown in Fig. 4.28. The size of the coils is equivalent to the size of the polymer molecule in the θ state, and the coil molecules are randomly interpenetrated and entangled. Since there is no ordered structure in the stacking of chain segments, the condensed state structure of the amorphous polymer is homogeneous.

Folded chain fringed micelle particle model. Yeh et al. proposed a model that suggests the existence of a certain degree of local order in amorphous polymers, which consists of two phases: the particle phase and the intergranular phase (Fig. 4.29). The particle phase can be divided into two parts: the ordered area, where molecular chains are arranged in parallel with each other, and the grain boundary area, which surrounds the ordered area and is composed of the bent part, the chain end, the tangled points of the folded chain, and the connection chain. The intergranular phase includes random coils, low-molecular substances, molecular chain ends, and connection chain. The size of the ordered area ranges from 2 to 4 nm, while the grain boundary area is about 1 to 2 nm, and the intergranular phase is in the range of 1 to 5 nm. According to this model, a single molecular chain can pass through several particles and intergranular phases.

Research on the amorphous structure of polymers remains inconclusive due to limited study means and insufficient understanding. It is a subject of ongoing exploration and debate in the polymer science community.

Fig. 4.28 Random coil model. To distinguish each polymer chains, they are illustrated as solid line, dash lines, short dash lines, etc.

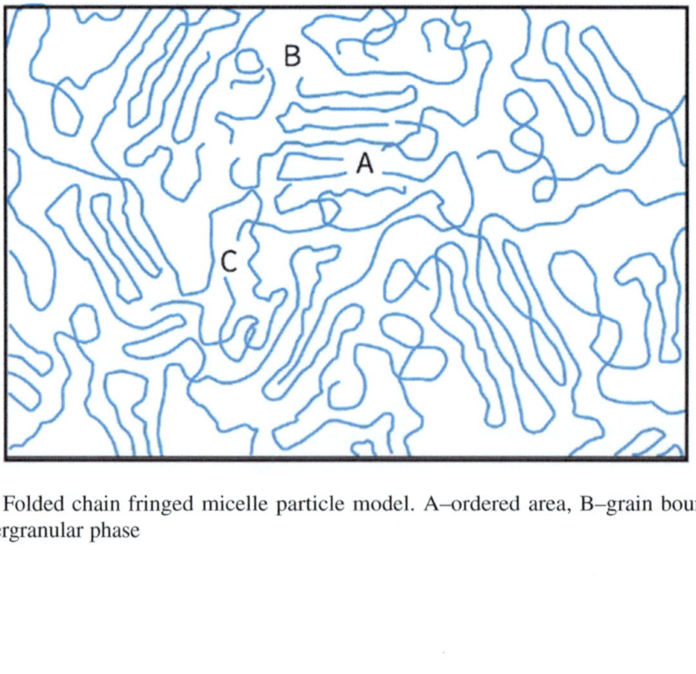

Fig. 4.29 Folded chain fringed micelle particle model. A–ordered area, B–grain boundary area and C–intergranular phase

(2) Crystal Structures

Remarkably, despite the challenges posed by their variable shapes, polymers can exhibit three-dimensional ordered structures. Although most polymers lack a regular shape, they contain small crystal grains with a well-defined structure as evidenced by X-ray diffraction. In particular, Keller was the first to produce polyethylene flake crystals in micrometer dimensions in 1957, followed by macroscopic single crystals of polydiacetylenes with sizes on the order of centimeters, which marked a significant breakthrough in the study of polymer crystalline structure.

The crystalline structure of polymers is relatively well-understood due to its three-dimensional ordering and the powerful research tool of X-ray diffraction. Research in this area includes the determination of crystal structures, crystal morphology, crystallization processes, kinetics, and more.

(3) Liquid Structures

The formation of liquid crystals requires an asymmetric molecular shape of the polymer chain and suitable intermolecular interactions to maintain a parallel and ordered arrangement. The solution of liquid crystals contains ordered structures in certain regions, which can form an ordered micro-domain structure during spinning. This leads to the production of highly-oriented and high-strength synthetic fibers, such as the Kevlar fiber from DuPont that is prepared by liquid crystal spinning.

(4) Single Chain Condensed Matter

The field of single-chain condensed matter is a new area of research that has been mainly developed by polymer physicists in China. It is focused on the study of the crystallization behavior of polymer chains in their single-chain form, as well as the morphology and mechanical properties of the single-chain glassy and highly elastic states, and the transition from single-chain condensed matter to multi-chain condensed matter.

Single-chain condensed matter is a unique phenomenon of macromolecules. The high relative molecular mass of typical macromolecular chains, reaching 10^5-10^6, allows them to occupy a volume of 10^3-10^4 nm^3. Moreover, the critical volume of a primary crystal nucleus is about 10 nm^3, indicating that a crystallizable polymer chain can fold and arrange to form a small nano-sized crystal through nucleation and growth under favorable conditions.

The condensed particles of a single molecular chain have no intermolecular entanglement and are the smallest unit for related research. Studying the characteristics of the single-chain condensed state can provide us with a deeper understanding of the structure, morphology, and various physical and mechanical properties of polymer materials from a new perspective. For instance, entanglement is a natural structural factor and the most important feature of polymers, which affects the crystallization process and should be considered. A single molecular chain is the simplest system of polymer materials, and by studying single-stranded single crystals, oligo-chain, and multi-chain crystallization, it may offer information about the impact of entanglement on the crystallization process.

(5) Orientation Structures

Due to its geometric asymmetry, the polymer chain can be easily oriented by an external force field, resulting in a distinct oriented structure that influences its performance. The orientation structure can be characterized by factors such as the type and degree of orientation, as well as the distribution of orientation units. Techniques for measuring the degree of orientation are also important in characterizing oriented polymers.

The state of high global chain orientation but nearly random segmental orientation of amorphous polymer is referred to as the GOLR state. Although this concept was originally proposed by Soviet scholars, experimental verification of the GOLR state was conducted by Chinese scholars.

At a temperature slightly above the glass transition temperature (T_g) and under a low rate of uniaxial stretching, the amorphous polymer chains can be oriented along the stretching direction, resulting in a high global chain orientation. However, due to the disparate relaxation times of chain segments, local orientation may be completely relaxed during the stretching process, while the overall molecular chain orientation persists. Upon rapid quenching of the polymer to T_g or below, the molecular chain motion is frozen, leading to a state of high global chain orientation but nearly random segmental orientation, which is known as the GOLR state. This concept was originally proposed by Soviet scholars, but its experimental proof was completed by Chinese researchers.

The unique molecular chain orientation in the GOLR state results in a variety of distinct polymer properties. The properties that reflect the orientation of local chain segments, including optical birefringence, infrared dichroism, and sound velocity transmission, are isotropic. On the other hand, properties related to the orientation of large molecular chains, such as thermal expansion coefficient, heat conduction, and stress–strain behavior, are anisotropic.

② Crystal Morphologies

The crystalline states of polymers are diverse and mainly include single crystals, lamellae, spherulites, dendrites, twin crystals, fibrous crystals, string crystals, etc.

(1) Single Crystals

In general, obtaining single crystals of polymers requires specific conditions. Flake-shaped crystals with uniform shapes can be obtained through slow crystallization from highly dilute solutions with concentrations of 0.01% or lower. These crystals typically have a thickness of approximately 10 nm and a size ranging from a few microns to tens of microns, or even larger, as observed under an electron microscope.

(2) Multilayered Dendritic Structures

When the temperature of crystallization is low, the concentration of solution is high, or the relative molecular weight of the polymer is too large, single crystals can no longer be formed during the polymer's crystallization from the solution. Instead, excessive growth of crystals results in more complex crystal forms. Under these conditions, polymer diffusion becomes the controlling factor for crystal growth. The prominent edges and corners of crystals are geometrically advantageous as they can accept crystalline molecules from a larger solid angle. Therefore, they tend to grow forward, becoming thinner and sharper than the rest of the grains, resulting in dendritic growth.

(3) Spherulitic Morphology (spherocrystal)

Spherulites are among the most common forms of polymer polycrystals, typically obtained from concentrated solutions or during cooling and crystallization of melts. When unhindered in growth, spherulites exhibit a spherical shape. Their diameter usually falls between 0.5 and 100 μm, although they can occasionally reach centimeter-level sizes. Spherulites larger than 5 μm can be easily observed under an optical microscope.

The optical characteristic of spherulites is that a black cross extinction pattern (i.e., Maltese Cross) can be observed under a polarizing microscope, and sometimes a series of concentric circular extinction patterns are superimposed on the black cross extinction pattern. Many studies have been conducted on the growth process of spherulites. Figure 4.30 illustrates various stages of spherulite growth. It is just a multilayer lamella (Fig. 4.30a) in the initial stage of nucleation, then gradually expands and grows outwards (Fig. 4.30b, c), and continuously bifurcates and grows into bundles (Fig. 4.30d), forming spherical crystals at last (Fig. 4.30e). In fact, this is still an early stage, and the final spherulites are usually much larger. The detailed diagram of the spherulite structure is shown in Fig. 4.31.

(4) Shish-kebab Structure (involving lamellae)

The string crystal, also known as the Shish-Kabab structure, is formed by crystallizing a polymer solution while stirring. The resulting structure is composed of a centerline with a straight chain structure, and crystal pieces with folded chains growing at intervals around the centerline, as shown in Fig. 4.32. The higher the stirring speed, the greater the proportion of straight chain crystals in the formed string crystals, resulting in higher melting points of the crystals. String crystals provide excellent properties such as high strength, solvent resistance, and corrosion resistance due to their centerline structure. For instance, polyethylene string crystals have a breaking strength of 374.4 MPa, an elongation rate of 22%, and an elastic modulus six times that of ordinary polyethylene fiber, reaching 1999.2 MPa. Moreover, the string crystal structure is found in polymer films obtained through high-speed extrusion and quenching, which greatly improves the modulus and transparency of the film. The discovery of polymer string crystals has significant practical implications for understanding the relationship between structure and performance in fiber spinning and film forming processes.

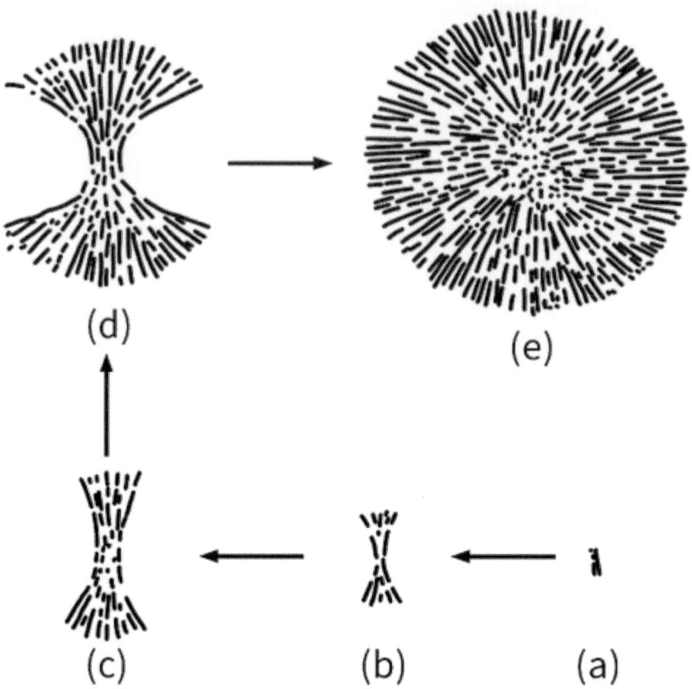

Fig. 4.30 Schematic diagram of spherulite growth process

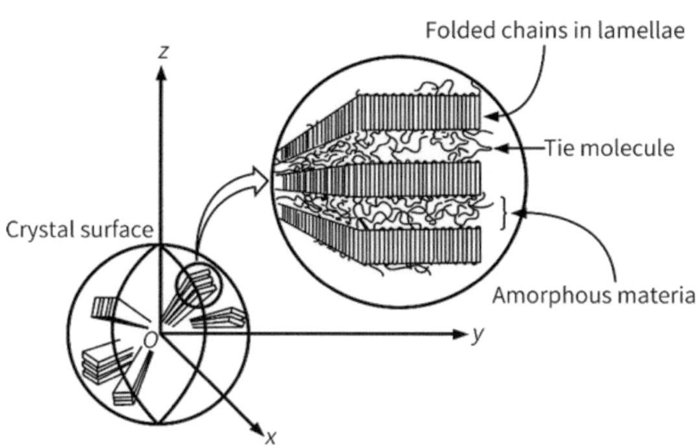

Fig. 4.31 Detailed schematic diagram of spherulitic structure

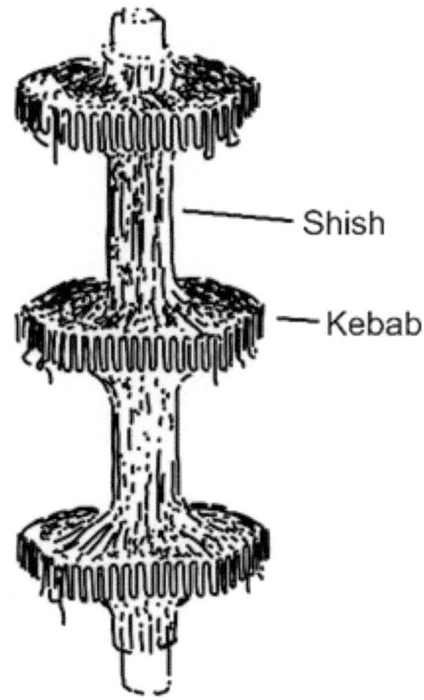

Fig. 4.32 Schematic diagram of Kebab crystal structure

(5) Extended-Chain Crystals

Under high temperature and pressure, polymers can crystallize into stretched chain lamellae consisting of fully extended polymer chains arranged in parallel, with thickness equal to the length of the molecular chain. For example, polyethylene crystallized above 200 °C and 405.3 MPa pressure yields straight chain lamellae with 97% crystallinity, a 140.1 °C melting point, and density over 0.99 g/cm^3. The lamellae are around 103 –104 nm thick, equivalent to the length of a straightened molecular chain, and remain constant regardless of heat treatment conditions. This indicates the stretched chain structure is the most thermodynamically stable aggregate structure in polymers.

③ Structure Models of Crystalline Polymers

Over time, as our understanding of polymer crystals has increased, various models have been developed based on experimental evidence to explain different phenomena and investigate the connection between crystal structure and polymer properties.

(1) The Fringed Micelle Model

Bryant et al. proposed the Fringed Micelle Model in 1947 based on their X-ray study of many crystalline polymers. The model suggests that imperfect crystalline structures exist in crystalline polymers where crystalline regions and amorphous regions coexist and intersect with each other, as depicted in Fig. 4.33. The model

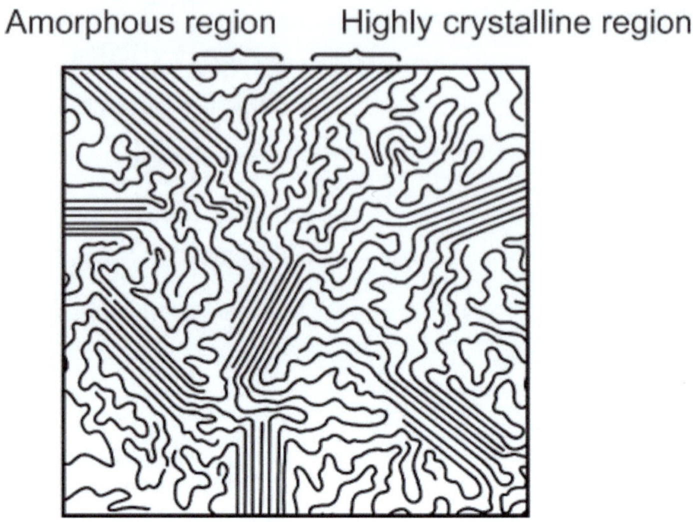

Fig. 4.33 Schematic diagram of fringed micelle model of semi-crystalline polymer

explains various experimental observations, including the X-ray diffraction pattern and the difference in density between the two phases. The stretched polymer exhibits optical birefringence due to the orientation of molecular chains in the amorphous region. Unevenness in chemical reactions and physical effects is produced due to the permeability difference between the two regions, and the melting limit of crystalline polymers is determined by the crystallite size.

(2) The Folded-Chain Model

In 1957, Keller from Britain prepared a polyethylene single crystal (shown in Fig. 4.34), which led to the development of the folded chain model for the crystalline state of polymers.

The folded chain model is a comprehensive single crystal model, but most of the time, semi-crystalline polymers are obtained due to regular folding of polymer chains, forming perfectly arranged lamellae, and irregular segments between

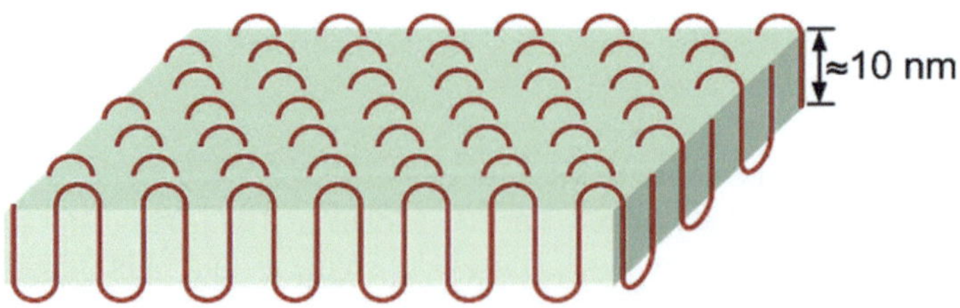

Fig. 4.34 Schematic diagram of crystal piece of high polymer. (The thickness of the lamella is about 10 nm)

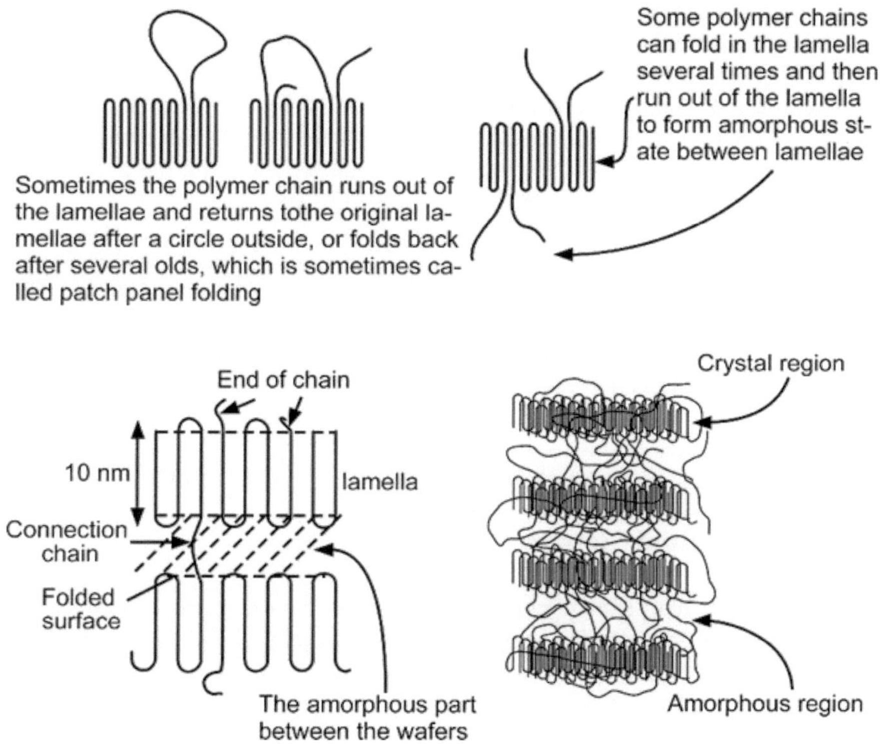

Fig. 4.35 Several situations that may occur during polymer chain folding: Regular folding forms lamellae, irregular folding (patch panel folding) and the connection chains between the folds of polymer chains and the crystal pieces form amorphous regions of crystalline polymers

the lamellae that create amorphous regions. Polymer chains can also pierce out to participate in folding of another lamella, forming non-neighbor or patch panel folding, creating more amorphous regions (Fig. 4.35). Therefore, irregularly arranged amorphous regions exist in crystalline polymers.

4.5 Theoretical Density Calculation

The previous sections have discussed various crystal structures, and understanding their structural information can aid in determining the properties of crystalline materials. Theoretical density is one such property. Theoretical density refers to the maximum density of a material or element that could be achieved if there were no voids or contaminants present in the material. For metallic materials, theoretical density can be calculated using Eq. 4.8.

$$\rho = \frac{nA}{VN} \tag{4.8}$$

where ρ is the theoretical density (g·cm^{-3}), n is the number of atoms per unit cell, A is the atomic weight (g·mol^{-1}), V is the volume per unit cell (cm^3), and N is Avogadro's number (6.022×10^{23} mol^{-1}).

In addition to metallic materials, it is possible to calculate the theoretical density of crystalline ceramic materials, even though their crystal structure is more complex. A modified form of Eq. 4.8 can be used to determine the density of a ceramic (compound) material, which is given by Eq. 4.9.

$$\rho = \frac{n'(\sum A_{\mathrm{C}} + \sum A_{\mathrm{A}})}{VN} \tag{4.9}$$

where n' is the number of formula units per unit cell, $\sum A_{\mathrm{C}}$ is the sum of the atomic weights of all cations in the formula unit, $\sum A_{\mathrm{A}}$ is the sum of the atomic weights of all anions in the formula unit, V is the unit cell volume, and N is Avogadro's number.

4.6 Interstitial Sites in Crystal Structures

According to the hard-sphere model of crystal structures, atoms cannot occupy all the volume of crystals. The maximum atomic packing factor is 74% for FCC and HCP structures, resulting in interstitial voids in all crystals. The shape of these voids (triangular, tetrahedral, octahedral, or cubic) depends on the number of atoms surrounding them. The size, number, and location of interstitial voids are essential features of a crystal and can be used to analyze complex alloy and ceramic structures. In the common metal structures, such as BCC, FCC, and HCP, tetrahedral and octahedral interstices are the two most significant.

4.6.1 Tetrahedral Interstitial Site

A tetrahedral interstital site is located at the center of a tetrahedron that is formed by four lattice atoms. Three atoms, touching each other, are in plane; the fourth atom sits in the symmetrical position on top, as shown in Fig. 4.36.

Tetrahedral interstitial sites in BCC crystal structure is formed by two adjacent atoms located at the body center of unit cells and two atoms on the shared edge, as shown in Fig. 4.36a. In the unit cell, the coordinates of its center are (1/2, 1/4, 1), locating on the (001) plane. Its 2 edges along [100] and [001], respectively, have the same lengths of a, but the other 4 edges have a different length of $\frac{a\sqrt{3}}{2}$, which therefore means that it is not a regular tetrahedron. Based on the cubic symetrey, there are 12 tetrahedral interstitial sites in a BCC unit cell.

In the FCC crystal structure, each of the 8 identical subcubes obtained by dividing the unit cell contains a tetrahedral interstitial at its geometric center, located at coordinates (1/4, 1/4, 1/4), as shown Fig. 4.36b. The edge length of the regular tetrahedral is $\frac{a\sqrt{2}}{2}$. Due to the cubic symmetry, there are 8 identical tetrahedral

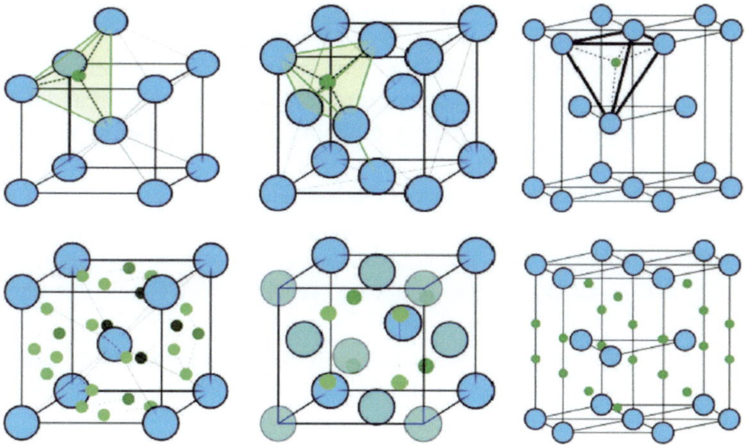

(a) A tetrahedral interstitial site among neighboring atoms and all tetrahedral interstitial site positions within BCC unit cell.

(b) A tetrahedral interstitial site among neighboring atoms and all tetrahedral interstitial site positions within FCC unit cell.

(c) A tetrahedral interstitial site among neighboring atoms and all tetrahedral interstitial site positions within HCP unit cell.

Fig. 4.36 Tetrahedral interstitial sites in BCC, FCC and HCP structures

interstitial sites in a FCC unit cell, with the center of each site located on the body diagonal and a quarter of the way along the diagonal from the corner atoms.

In a HCP unit cell, there are two tetrahedral interstitials on the central axis and six on the three lines parallel to the axis that pass through the middle three atoms. Additionally, 12 tetrahedral interstitials are located on six vertical edges, which are equally shared by three unit cells, as shown in Fig. 4.36c. Therefore, each HCP unit cell contains a total of 12 tetrahedral interstitial sites, which is exactly 50% more than the number of FCC sites. This is not surprising as both FCC and HCP structures are close-packed, and the HCP structure is exactly 50% larger than the FCC structure.

4.6.2 Octahedral Interstitial Site

According to the hard-sphere model of crystal structures, an octahedral interstitial site is located between six atoms. Four of these atoms are situated in a plane, while the remaining two occupy symmetrical positions above and below the plane, as depicted in Fig. 4.37.

In the BCC crystal structure, an octahedral interstitial site is formed by two adjacent atoms located at the body center of unit cells and four vertex atoms of a face, as shown in Fig. 4.37a. The coordinates of an octahedral interstial site are (1/2, 1/2, 1), and according to the symetrical feature of FCC cell, the centers of the cube faces and the midpoints of its edges are equivalent points, and therefore, there are 6 octahedral interstitial sites in a unit cell. Its center is $\frac{a\sqrt{2}}{2}$ away from the four vertex atoms on (001) plane, but $\frac{a}{2}$ away from the two body-center atoms. Therefore, the interstitial site is not a regular octahedron.

In FCC crystal structure, an octahedral interstitial is formed by 6 atoms located at the center of cube faces, as shown in Fig. 4.37b. Due to the symmetric equivalence of the body center of a FCC cell with the midpoints of the 12 edges, there

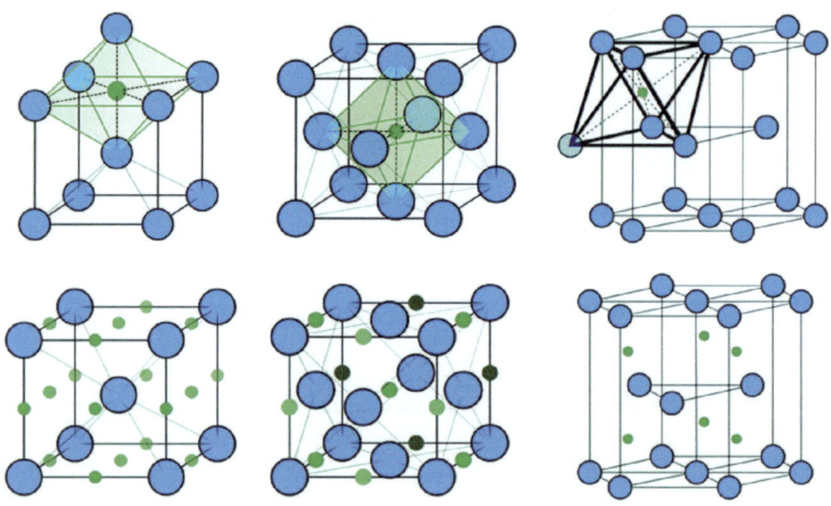

(a) A tetrahedral interstitial site among neighboring atoms and all tetrahedral interstitial site positions within BCC unit cell.

(b) A tetrahedral interstitial site among neighboring atoms and all tetrahedral interstitial site positions within FCC unit cell.

(c) A tetrahedral interstitial site among neighboring atoms and all tetrahedral interstitial site positions within HCP unit cell.

Fig. 4.37 Octahedral interstitial sites in BCC, FCC and HCP structures

Table 4.10 Tetrahedral and octahedral sites in typical crystal structures

Crystal structure	Octahedral sites		Tetrahedral sites	
	No. of sites/no. of atoms	*r*/*R*	No. of sites/no. of atoms	*r*/*R*
BCC	3	0.155	6	0.291
FCC	1	0.414	2	0.225
HCP	1	0.414	2	0.225

are 4 octahedral interstitial sites present in a FCC cell. The length of all edges of the octahedron is $\frac{a\sqrt{2}}{2}$.

As for the HCP structure, the geometrical configuration of the octahedral intersitial site is the same as that of FCC structure. As shown in Fig. 4.37c, the coordinates of the center of the octahedron is (1/3, −1/3, 1/4), and there are 6 symetrical positions of the center in a HCP cell. Consequently, there are 6 octahedral interstitial sites in an HCP cell.

Let R represent the atomic radius of a crystal, and r represent the radius of the inscribed sphere, then the relationship between R and r in different crystal structures is as follows in the Table 4.10.

4.7 Polymorphism

In materials science and related disciplines, polymorphism refers to the phenomenon in which a chemical substance can exist in more than one crystalline structure or form, such as diamond, graphite, carbon nanotubes, and fullerenes for carbon. Given a set of building blocks, such as carbon atoms, various structures, such as diamond and graphite, can be constructed. Polymorphism should be distinguished from isomerism, which refers to two or more compounds with the same chemical formula but different structure and properties due to different spatial arrangements in their molecular structure. Allotropy is another related concept, which describes the polymorphism of elemental solids. As a special case of polymorphism, allotropy is a structural variation of the same element due to its valency.

Polymorphism, especially allotropy, is commonly exemplified using carbon. Two well-known carbon allotropes, diamond and graphite, are often used to explain the concept since it is easy to tell the difference just from their appearance, as shown in Fig. 4.38.

Polymorphism is a common phenomenon in crystalline materials, and nearly every compound has different polymorphic forms. The different structures are determined by the packing of the building blocks, which affects the interatomic forces within the crystal. As a result, even though they share the same building blocks, the various polymorphs of a compound are distinct materials with different physical and chemical properties, such as melting points, color, hardness, density, and electrical conductivity, among others.

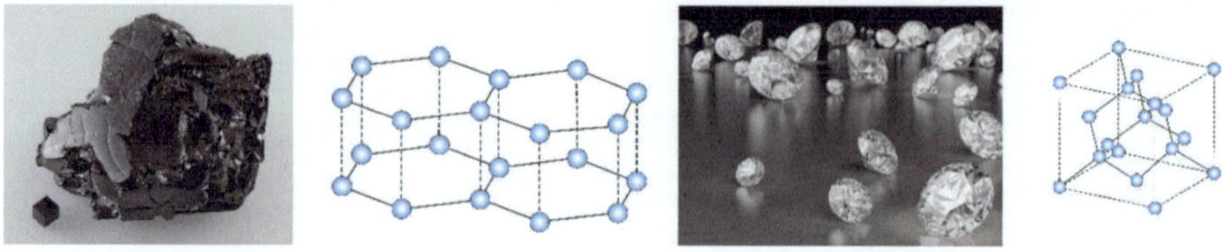

(a) graphite (b) diamond

Fig. 4.38 Photos and crystal structures of the two polymorphs of carbon. **a** graphite, and **b** diamond. (*Source* https://phys.org/news/2011-09-diamonds-emerge-graphite.html)

Polymorphism is often influenced by changes in temperature and pressure, which can cause the rearrangement of building blocks in crystalline materials. Each polymorph of a given composition is stable under a specific temperature range. Although the polymorphs of iron cannot be easily distinguished by their appearance, they do have multiple crystallographic forms that make iron a useful material for forming alloys such as steel. Polymorphs can greatly expand the range of applications for a given material.

Iron exhibits different crystal structures and properties depending on temperature. At room temperature, iron is in the BCC structure, called ferrite (α-Fe), which is relatively soft and can dissolve only a small amount of carbon (less than 0.021 wt% at 910 °C). Ferrite (α-Fe) undergoes a phase transition with temperature exceeding 910 °C from BCC structure to FCC structure, which is called austenite (γ-Fe). Austenite is also relatively soft, but can dissolve considerably more carbon (as much as 2.03 wt% at 1154 °C) than ferrite. Before reaching its melting point (1535 °C), iron undergoes change of structure again above 1400 °C, at which iron starts to go back to BCC structure. Along with the phase transition, different structures of iron appear and therefore display quite different properties which can be taken advantages of for various application purposes.

The above examples discuss polymorphism in elemental crystals, also known as allotropy. A unique example of polymorphism is found in silicon carbide (SiC), which has over 250 identified polymorphs, including some with lattice constants larger than 300 nm, over a thousand times greater than typical SiC polymorphs.

4.8 Polycrystals

In materials science, for a solid, when the periodic arrangement of its components (building blocks) is perfect (i.e., the long-range order extends throughout the whole body of the solid without any interruption), as schematically shown in Fig. 4.39a, the solid is called a single crystal (monocrystalline material). In a single crystal, all unit cells interlock in the same way and thus have the same orientation. Since its lattice is continuous and unbroken to the edges of the crystal, there are no boundaries inside. Conversely, a polycrystalline material, as depicted

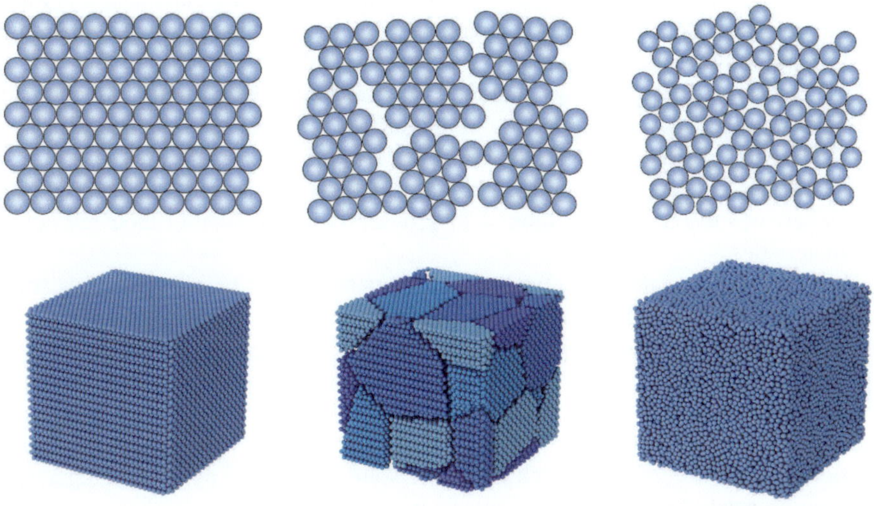

Fig. 4.39 Schematic diagrams for a single crystal material (**a**), a polycrystal material (**b**), and an amorphous material (**c**). The upper row showing the atomic arrangement in a plane, and the row below showing a 3D representation of these states of materials

in Fig. 4.39b, contains multiple single crystals with different orientations, resulting in grain boundaries.

Monocrystalline materials offer unique advantages due to the absence of grain boundaries, resulting in anisotropic mechanical, optical, and electrical properties, which can be harnessed in various technological areas. The semiconductor industry commonly uses silicon single crystals, and advanced information technologies are emerging with the use of single crystals such as GaN and SiC. Single crystal turbine blades exhibit excellent mechanical properties and reduced creep, making them suitable for close tolerance part applications under high temperatures. The jewel-like internal perfection of single crystals has made them useful in areas like photovoltaics, smartphones, and other emerging technologies.

4.9 Amorphous Materials

Solids are characterized by a 3-dimensional arrangement of atoms, ions, or molecules, which can form a repeating crystal lattice and produce a crystalline solid, or a more random arrangement resulting in an amorphous solid. The essential difference between crystalline and amorphous materials lies in the long-range order which exists in crystals, and is absent in amorphous solids. Crystalline materials exhibit long-range atomic order, meaning that their atomic structure is periodic and repetitive over an extended range. In contrast, amorphous materials lack this long-range order and have a disordered atomic arrangement, although some short-range order may be present, as shown in Fig. 4.39c. Crystalline solids have well-defined edges and faces, diffract X-rays, and tend to have sharp melting points. In contrast, amorphous solids have irregular or curved surfaces, do not give well-resolved X-ray diffraction patterns, and melt over a wide range of temperatures. There is a vast difference between crystalline and amorphous solids. To highlight the differences between the two types, Table 4.11 provides a comparison of their respective characteristics.

It is not unreasonable sometimes amorphous solids can be considered as some kind of liquids. A hundred years ago, scientists began to speculate on whether or not solids could flow, but they were unable to design an experiment due to a lack of appropriate materials. In 1927, pitch was discovered to be the perfect material to carry out such an experiment. Pitch is a type of substance that is so viscous it appears to be solid, even though it is actually a fluid that flows extremely slowly at room temperature. While pitch may feel solid and brittle at room temperature and can be easily broken with a hammer, it is actually a liquid.

The pitch drop experiment, as shown in Fig. 4.40, initiated by Professor Thomas Parnell at the University of Queensland, Australia, in 1927, measures the flow of pitch over many years at room temperature. Pitch was poured into a glass funnel

Table 4.11 Difference between crystalline solids and amorphous solids

Attribute	Crystalline solids	Amorphous solids
Long-range order	They have a definite and regular geometry and consist of both long-range as well as short-range order of its constituent particles	The particles of the constituents in the amorphous solids are arranged irregularly. They do not possess any kind of definite geometry and have a short-range order
Melting point	They tend to have high and distinct melting points	Amorphous solids do not have sharp melting points
External form	The external forms of the crystals tend to have regularity when they are formed	Amorphous solids do not have an external regularity when they are formed
Surface	Crystals tend to give a clean surface when they are cut with a knife	The amorphous solids usually exhibit irregular cut-edges
Thermal property	They possess a definite heat of fusion	Amorphous solids do not have any specific heat of fusion
Mechanical performance	The mild distorting forces cannot deform the crystalline solids since they are very rigid in nature	Amorphous solids, on the other hand, do not possess any rigidity. The deformation could be achieved even by bending or compressing them
State	Crystalline solids are known to be the true solids	Amorphous solids are in fact supercooled liquids or pseudo solids
Isotropism/ anisotropism	Crystalline solids tend to show anisotropism	Amorphous solids tend to show isotropism

Fig. 4.40 The pitch drop experiment carried out in 1927 by Professor Thomas Parnell, demonstrating that some substances which appear solid are actually highly viscous fluids. (Pitch drop experiment - Wikipedia)

and left to settle for three years before the stem was unsealed in 1930. The pitch has since been flowing out of the funnel very slowly. It took eight years for the first drop to fall, and more than 40 years for another five to follow. Videos of the experiment can be watched on the homepage hosted by the university.

Amorphous materials encompass glass, gel, amorphous metals and alloys, amorphous semiconductors, amorphous carbon, and certain polymers. They can be broadly classified into two categories: glass and other amorphous materials. Glass is unique due to its glass transition temperature (T_g), which sets it apart from other amorphous materials. Metallic glass, also known as amorphous metals and alloys, is a prime example of glass. It forms from a supercooled liquid state of possible crystal materials by controlling the cooling rate. The viscosity of the material increases rapidly as the temperature drops below the critical temperature, T_g, which is the temperature for solidification into glass. T_g is a temperature range that changes with the cooling rate and typically falls between one-third and one-half of the melting point.

Amorphous alloy, also known as metallic glass, is a new member of amorphous materials and is also the simplest amorphous material. It can be regarded as material formed by disordered accumulation of metal atoms using modern rapid solidification metallurgy technology. They possess excellent mechanical, physical, and chemical properties that combine the best features of both metals and glasses. Unlike traditional metallic alloys that crystallize during cooling, the atoms in amorphous alloys are arranged in a disordered, glassy state, giving them their name. This disordered state is achieved by preventing the nucleation and growth of crystal phases during the solidification process, resulting in a microstructure that is more like a very viscous liquid. As a result, amorphous alloys are sometimes referred to as frozen melts or liquid metals.

4.10 Nanocrystals

Nanocrystalline materials, or nanocrystals, have become increasingly important in various technological areas with the development of nanotechnology. They are polycrystalline materials with sizes ranging from 1 to 100 nm, occupying the transition region between atomic clusters and macroscopic objects. While their size is larger than atomic clusters, they are smaller than ordinary micro powders. Nanoparticles, which are microscopic particles smaller than the wavelength of visible light and similar in size to bacteria, can only be observed with a high-powered microscope.

Nanocrystals have minimal crystal grains that contain a large number of atoms on the interface. When the grain size is smaller than 5 nm, more than 50% of the atoms are located at the grain boundary. The nanocrystalline interface differs from the ordinary polycrystalline interface, and is closely related to the interfacial and thermodynamic properties of nanocrystals. In particular, zero-dimensional crystals are nanoscale particles known as quantum dots. Roughly speaking, the size of the three dimensions of the quantum dot is below 100 nm, and it looks like a tiny dot. The movement of the internal electron in all directions is limited, so the quantum confined effect is particularly significant. There are various methods to synthesize quantum dots, with chemical methods being the primary approach. They exhibit a wide excitation spectrum and narrow emission spectrum, allowing for adjustable color and superior photochemical stability. Quantum dots are a promising fluorescent probe and have garnered attention for their use in semiconductor electronic devices and optoelectronic devices beyond edge emission.

4.11 Quasicrystals

Quasicrystals represent a novel solid-state structure that resides between crystalline and amorphous states. As is widely recognized, crystalline structures, where atoms are arranged in three-dimensional periodicity, inherently lack fivefold symmetry axis or axis with higher order than 6 due to the violation of translational symmetry.

In 1974, British mathematician Roger Penrose developed a quasi-periodic tiling using two types of rhomboid units with interior angles of 36 and 72 degrees, respectively, to create quasi-periodic patterns based on strict assembly rules, as depicted in Fig. 4.41. In the early 1980s, the Penrose tiling was applied in crystallography to generate the Fourier transform map of quintic symmetry and propose the notion of quasi-lattice.

In 1982, experimental evidence of quasicrystals was observed in the quenched $Al_{86}Mn_{14}$ alloy by Shechtman et al., as shown in Fig. 4.42. They observed a spot

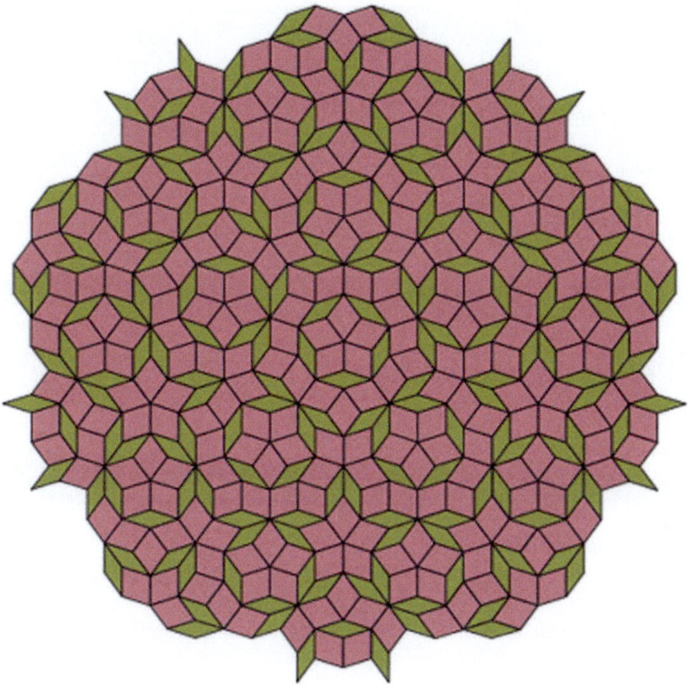

Fig. 4.41 Penrose puzzle

distribution with icosahedral symmetry in the electron diffraction pattern of an $Al_{86}Mn_{14}$ alloy. The brightness of the spots was comparable to that of a crystalline structure, indicating the presence of a fivefold symmetric axis within the alloy. Currently, quasicrystals have been discovered in dozens of alloy systems, exhibiting not only fivefold symmetry but also eightfold, tenfold, and 12-fold symmetric structures. These findings demonstrated that quasicrystals constitute a novel type of solid structure. Consequently, quasicrystals can be defined as intermediate structures between crystals and amorphous solids that exhibit the same long-range order as crystals, but lack the translational symmetry of crystals. In 1992, the definition of crystal was revised by the International Society of Crystallography to include any solid with a substantially discrete diffraction pattern.

4.12 Questions

Q4.1 What disparities exist in the structures and bonding conditions between diamond and graphite? How do these differences correlate with their chemical properties and performance?

Q4.2 Provide an explanation of the Hume-Rothery rules and their practical utility.

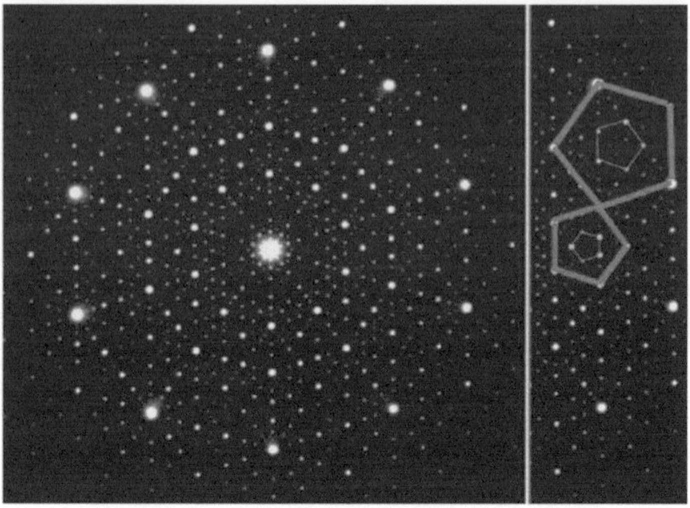

Fig. 4.42 Electron diffraction pattern of quenched $Al_{86}Mn_{14}$ alloy

Q4.3 Discuss the classifications and fundamental characteristics associated with silicate structures.

Q4.4 Based on their respective atomic mass, density, and atomic radius as indicated below, determine the crystal structures (simple cubic, body-centered cubic, face-centered cubic) for metals A, B, and C.

Metals	Atomic mass/(g/mol)	Density/(g/cm^3)	Atomic radius/nm
A	77.4	8.22	0.125
B	107.6	13.42	0.133
C	127.3	9.23	0.142

Cementite (Fe_3C) is an intermetallic compound of iron and carbon, with lattice constant of a = 0.4514 nm, b = 0.5080 nm and c = 0.6734 nm. It is known that the density of cementite is 7.66 g/cm^3, calculate the number of Fe and C atoms in the unit cell.

Q4.5 Compute the theoretical density of CsCl, considering the measured density is 3.99 g/cm^3.

Q4.6 Determine the packing density of the Al_2O_3 crystal structure, considering it as an HCP (Hexagonal Close-Packed) structure with given lattice parameters (a = 0.4759 nm, c = 1.2989 nm) and a density of 3.99 g/cm^3.

Q4.7 Calculate the lattice constant for cristobalite given its density of 2.32 g/cm^3. Additionally, determine the number of ions contained within one unit cell.

Q4.8 Illustrate the monomer, repeating unit, and structural unit for the following polymers: polyethylene, polyisoprene, and polyamide 6.

Q4.9 Sketch all possible isomers of polymers synthesized through the addition polymerization of isoprene.

Q4.10 Using the provided molecular weight data for a series of polystyrene (PS) materials in the table, compute (a) the number-average molecular weight, (b) the weight-average molecular weight, and (c) the molecular weight distribution of the blend.

Composition i	Average molecular weight (M_i)	W_i
A	12,000	0.10
B	21,000	0.19
C	35,000	0.24
D	49,000	0.18
E	75,000	0.11
F	102,000	0.08
G	122,000	0.06
H	146,000	0.04

Bibliography

Callister WD Jr, Rethwisch DG (2010) Materials science and engineering: an introduction, 8th edn. John Wiley & Sons, New York

He MJ (2007) Polymer physics, 3rd edn. Fudan University Press, Beijing

He PS (2012) Polymer structure and properties. Science Press, Beijing

Hu GX, Cai X, Rong YH (2010) Fundamentals of materials science, 3rd edn. Shanghai Jiao Tong University Press, Shanghai

Pan ZR (2011) Polymer chemistry, 5th edn. Chemical Industry Press, Beijing

Pan JS, Tong JM, Tian MB (2011) Fundamentals of materials science. Tsinghua University Press, Re.ed. Beijing

Tao J, Yao ZJ, Xue F (2018) Fundamentals of materials science, 2nd edn. Chemical Industry Press, Beijing

Yu YN (2012) Fundamentals of materials science, 2nd edn. Higher Education Press, Beijing

5 Crystal Defects

Learning Objective

1. Understand the importance of defects in crystalline materials.
2. Learn how to classify the defects.
3. Describe the vacancies, dislocations, grain boundaries, twin boundaries and stacking faults.
4. List the different dislocation motions and know how to draw the schematic diagram of the motions.
5. Calculate the hardening effect by varying the dislocation density.
6. Describe how the dislocations interact with other defects, like voids, grain boundaries and precipitates.
7. Know the main methods for defects observations.
8. Follow the state of art research trends on dislocations study.

5.1 Introduction

It is impossible to find a substance in the world that is completely free of defects. In fact, our understanding of materials begins with the knowledge of their defects. It is the defects that create the variety of materials, and the exceptional properties of materials are largely related to these defects. Therefore, understanding the structures of defects in materials is crucial for developing new materials or for selecting appropriate materials for applications. Crystal defects can be defined as deviations from perfect atomic arrangements, and the greater the deviation, the higher the energy of the defect. There are several types of crystal defects, including point defects, line defects (i.e., dislocations), and planar defects. Point defects include the vacancy, interstitial atoms, and substitutional atoms. Line defects, or

Z. Sun et al., *Fundamentals of Materials Science*,
https://doi.org/10.1007/978-981-96-3904-5_5

dislocations, refer to lines where entire rows of atoms in a crystal are arranged abnormally. Planar defects are defects that form at the boundaries between two adjacent grains or stacking faults.

5.2 Major Defects in the Crystal

Solid materials are made up of a large number of atoms (molecules, ions). These atoms are arranged in a specific manner in space. Ideally, a crystal has a three-dimensional periodic structure with a strict spatial lattice. However, the atoms in real materials cannot be precisely located at the lattice sites due to interference from various factors. For example, thermal vibrations or external forces can cause an atom to move away from its original crystal site, resulting in local lattice incompleteness. The impurities in a crystal inevitably affect its mechanical properties (strength and ductility), chemistry (corrosion), and physical properties (light and electricity).

Crystal defects are often classified into the following three major categories based on the shape and size of the incomplete region of the crystal structure: point defects, line defects, and planar defects (Fig. 5.1); sometimes, volume defects are also mentioned. The simplest type of point defect is a missing atom, which is called a vacancy. Another important type of point defect is an interstitial atom, which is an extra atom that cannot be located at the lattice sites of other atoms. Vacancies and interstitial atoms are types of defects found in a pure crystal. Another type of imperfection is a dislocation, a line defect, which is the most important defect that may run through the entire length of the crystal. Dislocations can be further divided into three styles: edge dislocation, screw dislocation, and mixed dislocation. There is a wide variety of planar defects, and the three main ones are grain boundary, twin boundary, and stacking fault. The movement of dislocations can be affected by all the defects mentioned above, making it a very complex process.

5.2.1 Point Defects

In a perfect crystal, the atoms should occupy a lattice point (with thermal vibration), and a point defect refers to a lattice point where there is no atom (which is a vacancy defect, see Fig. 5.2a) or an interstitial position where there is an additional atom (which is an interstitial defect, see Fig. 5.2b). Sometimes, a lattice point is occupied by a substitutional atom that is different from the atoms in other lattice points, and this is called a substitutional defect, which also belongs to the category of point defects (the α atom in Fig. 5.2b).

In terms of thermodynamics, vacancy as a point defect is stable, whereas a perfect crystal is not thermally stable. The number of vacancies can significantly

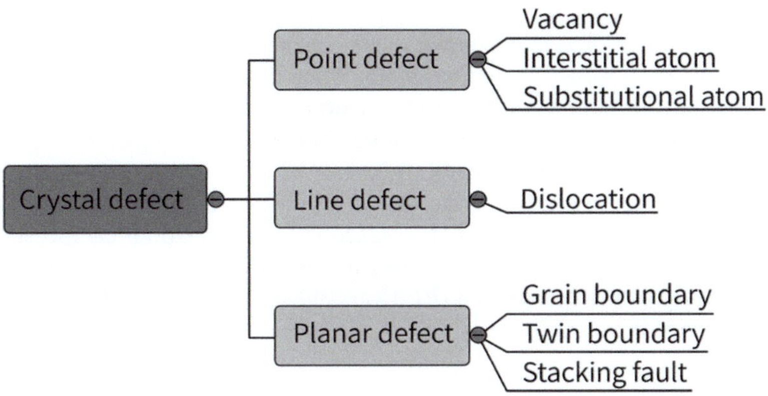

Fig. 5.1 The classification of crystal defects

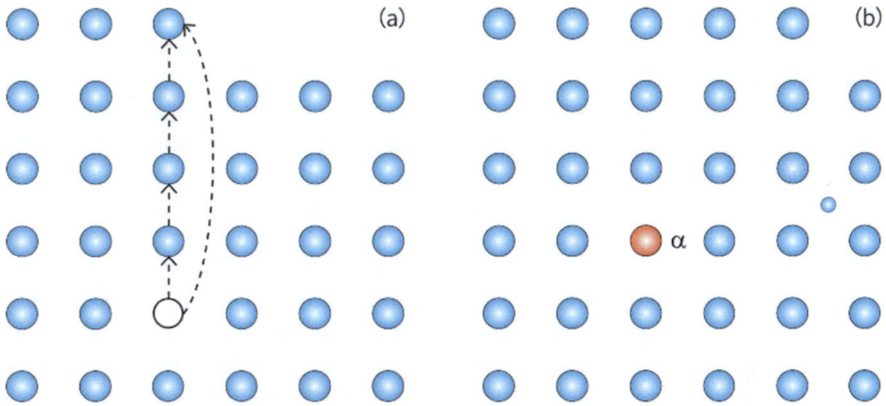

Fig. 5.2 **a** Vacancies, **b** Interstitial atoms (blue) and substitutional atoms (red)

increase with rising temperature, making it important for heat treatments. In metals, double vacancies and multiple vacancies can also form, and vacancies play an essential role in many metal properties and diffusions.

An interstitial atom is formed when an atom departs from its equilibrium position and enters the gap between two lattice positions. Since this gap is small, such an insertion can cause surrounding atoms to deviate from their equilibrium positions, leading to lattice distortion and point defect formation. Interstitial atoms can be either the same type (self-interstitial atom) or a different type (foreign interstitial atom) from the surrounding atoms. The formation energy of self-interstitial atoms is usually larger than that of vacancies, making the formation of self-interstitial atoms less common than vacancies. However, foreign interstitial atoms, such as small carbon atoms in steels, are more likely to form due to space limitations in the solvent lattice.

Foreign interstitial atoms can affect nearly all material properties, including the electrical properties of semiconductors and mechanical properties of metals and alloys. Intentionally added interstitial atoms often control the type and concentration of charge carriers in semiconductors, with conductivity increasing if a solute with excess electrons is doped with saturated bonds. For example, adding a small amount of B atoms to Si significantly improves its conductivity at room temperature. In metals, foreign interstitial atoms can also increase mechanical strength, an important hardening mechanism. For instance, adding interstitial carbon atoms to steels can significantly enhance their strength, resulting in the formation of martensite with a body-centred tetragonal (BCT) crystal structure, which is a very strong material employed in lots of important applications.

5.2.2 Line Defects

A line defect is a type of one-dimensional defect characterized by a small size in two directions and a long extension in the other direction. The most common type of line defect is a dislocation, which is a local, one-dimensional irregularity in the arrangement of atoms.

The formation of dislocations is the main reason for the significant difference between the theoretical strength and the actual strength of crystals. The theoretical strength is the stress required to deform the entire crystal to move all the atoms at once, but if the atoms can be moved step by step, the required stress is much lower. Dislocations allow crystals to undergo plastic deformation under low external stresses.

Although line defects are different from point defects (such as vacancies) and planar defects (such as grain boundaries), they are still closely related to them, and dislocations greatly affect the yield strength of real crystals. In a defect-free, especially dislocation-free crystal, the yield strength should be equal to the theoretical strength. Figure 5.3a represents a perfect crystal, while Fig. 5.3b is a defected crystal with a dislocation at the boundary *CD*, where the number of atoms above and below the boundary is different. Without external stress, this imperfect crystal remains unchanged. However, when a small shear stress is applied to the upper and lower parts of the crystal (Fig. 5.3c), the atoms on the dislocation move to equilibrium positions (Fig. 5.3d). Only the atoms on and near the dislocation need to move from their original positions to achieve this deformation, while the positions of other atoms remain unaffected. As a result, the required stress is low. The dislocation continues to move until it reaches the surface of the crystal and disappears. The amount of deformation in the crystal is equal to the distance between the two atoms (Fig. 5.3d).

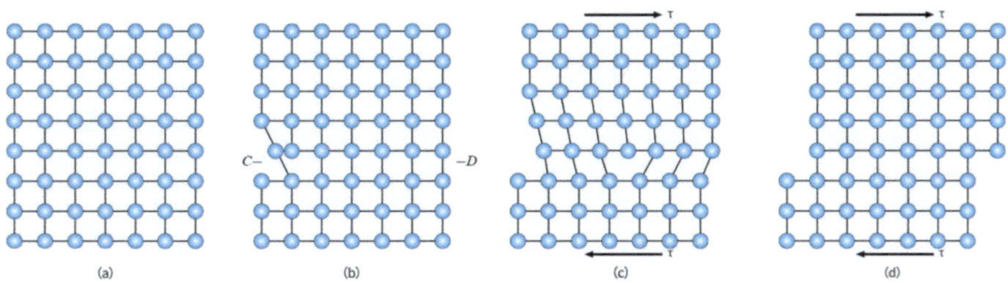

Fig. 5.3 The existence of dislocation makes the crystal easy to be deformed

5.2.3 Planar Defects

Planar defects are significant defects that can have a considerable impact on the properties of metals and other crystalline materials. Crystalline materials are typically composed of multiple domains, each having a relatively perfect atomic arrangement. However, the atomic arrangements are significantly disrupted at the interfaces between domains. Planar defects are the two-dimensional interphase planes that separate these crystalline domains. For instance, commonly used metals are often polycrystalline, and each grain is not perfect since it comprises subgrains. Grain boundaries and subgrain boundaries can be regarded as defects due to the anomalous arrangement of atoms on them.

Planar defects can be divided into several types, including low-angle grain boundaries, high-angle grain boundaries, twin boundaries, and stacking faults. Dislocations can describe low-angle grain boundaries, while twin boundaries are a special orientation of the boundary between two adjacent grains, where the two adjacent grains have a mirror relation separated by the twin boundary. Additionally, a stacking fault can be described as two twin boundaries in close-packed face-centred cubic crystals, where the close-packed arrangement is ABCABCABC.... If the arrangement is ABCACABC..., it indicates the formation of a stacking fault.

5.3 Vacancies

One important type of point defect is vacancy. A vacancy occurs when an atom is missing from its original lattice position, leaving an empty lattice site, as present in Fig. 5.4a illustrates a perfect crystal, while Fig. 5.4b shows a crystal with a vacancy defect, where one atom is absent from its lattice site.

Vacancies are a common occurrence in crystalline solids, and are present in any material at temperatures above absolute zero temperature (0 K). The number of vacancies in a material increases exponentially with temperature. However, there is an equilibrium vacancy concentration, which represents the ratio of the number of vacancies to the total number of atoms in the crystal. At the melting point of certain metals, this ratio can be approximately 1:1000.

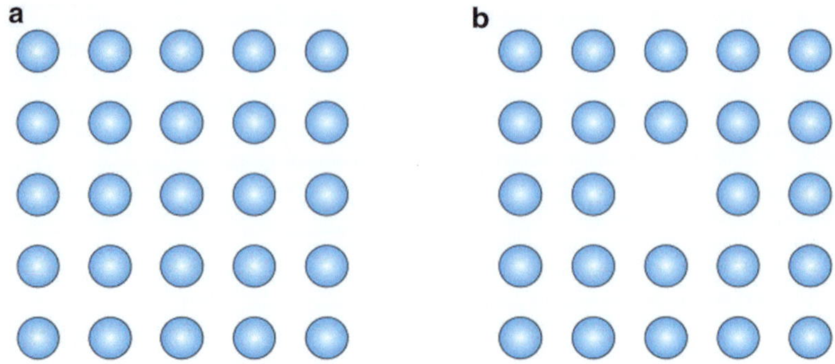

Fig. 5.4 **a** Perfect crystal, and **b** crystal with a vacancy defect

The temperature dependence of the number of vacancies can be described by

$$N_{\text{V}} = N\exp(-Q_{\text{V}}/k_{\text{B}}T) \tag{5.1}$$

where N_{V} is the vacancy concentration, Q_{V} is the energy required for vacancy formation, k_{B} is the Boltzmann constant, T is the temperature in Kelvin, and N is the concentration of atomic sites, which can be calculated by

$$N = mN_{\text{A}}/M \tag{5.2}$$

where m is mass, N_{A} is Avogadro constant, and M the molar mass.

In addition to temperature, vacancies can also arise from the following causes:

1. Non-directional solidification during casting leads to the formation of vacancies.
2. Irradiation or sputtering of crystalline materials results in the formation of vacancies.
3. Plastic deformation is a cause of vacancies.

Vacancies can significantly impact the properties of crystalline materials. As they result from missing atoms within the crystal structure, vacancies can decrease material density, thereby reducing the bulk modulus and the Young's modulus. The formation of vacancies can also increase the hardness of the crystalline solid by putting neighboring atoms under tension, but high vacancy concentrations can lead to reduced ductility. Furthermore, vacancies can alter the thermal and electrical resistivity of the solid and affect other physical properties, such as melting point and color.

5.4 Dislocations

5.4.1 History of Dislocations

In the early 1900s, the concept of dislocation was introduced to explain phenomena that were previously unexplainable, capturing the interest of numerous scientists. As a result, the study of dislocations became a crucial field, with subsequent research providing deeper insights into their nature. Figure 5.5 presents a brief history outlining the evolution of dislocation theory from 1907 to 1998.

In 1907, Volterra solved an elastic problem of discontinuous internal stress in a type of elastomer and introduced the concept of dislocations.

In 1926, Frankel found that there is a significant difference between the rigid shear strength of the theoretical crystal model and the measured critical shear stress.

In 1934, Michael Polanyi, Geoffrey Taylor, and Egon Orowan, separately achieved the same result at approximately the same time, and published papers on dislocation model. In particular, Taylor introduced Volterra dislocations into crystals.

In 1939, J.M. Burgers proposed to characterize dislocations by Burgers vector, meanwhile, he also introduced the concept of screw dislocations into crystals.

In 1940, Peierls proposed the dislocation lattice model, which was then further corrected by Nabarro in 1947. In this model, they broke through the scope of general elasticity, put forward the concept of dislocation width, and estimated the stress required for dislocation initiation.

In 1947, Cottrel raised the interaction between solute atoms and dislocations, which was then used to explain the yield behavior of low carbon steels. This is the first time that problem of the mechanical properties has been solved with the help of dislocation theory. In the same year, Shockley described the process of the formation of extended dislocations in FCC crystals, which opens the door for

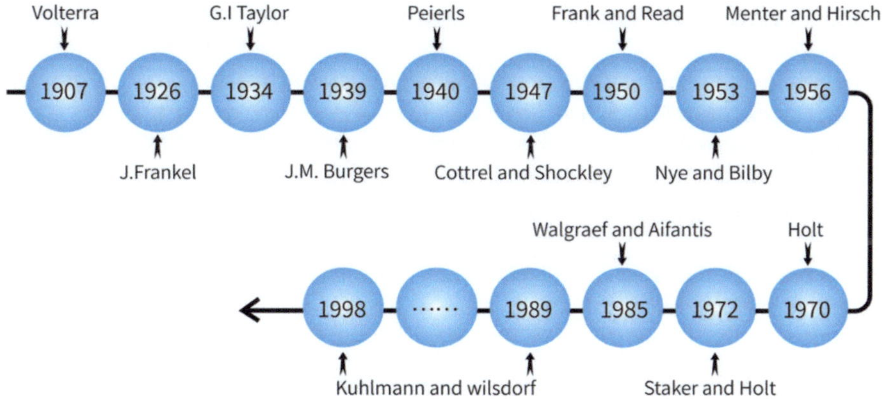

Fig. 5.5 The development of dislocation theory

studying the effect of dislocations on the mechanical properties of close packed crystals.

In 1950, Frank and Read raised the mechanisms of dislocation multiplication.

Nye in 1953, Bilby in 1954 and Kroener later proposed the infinitely small dislocation model.

In 1956, Menter directly observed the existence of dislocations using electron microscopy in platinum titanium cyan crystal. In the same year, Hirsch also observed dislocations in transmission electron microscropy (TEM) by phase contrast method.

Various analytical models have been proposed since the 1970s to explain the mechanisms of the formation and evolutions of complex patterns of dislocations that appear when crystalline materials deform. Holt proposed an irreversible kinetic model in the early 1970s, while Kulhmann-Wilsdorf came up with a low energy dislocation model in 1982. Walgraef and Aifantis proposed the reaction–diffusion model in 1985. Kuhlmann-Wilsdorf later studied the Bipolar cell wall and found that it was an unstable structure that could be destroyed by plastic deformation.

Electron microscopy technology has played an important role in the field of materials science. With the fast development of experimental techniques, researchers have gained a clearer understanding of dislocations. The properties of dislocations, such as dislocation distributions, arrangements, densities, and movements, have been extensively studied. TEM has dominated the experimental techniques in the past 50 years, with Fig. 5.6 showing some TEM image examples of dislocations in different alloys. Other experimental techniques, such as surface methods, decoration methods, X-ray diffraction, field ion microscopy, and atom probe tomography, have also provided useful information about dislocation. However, unlike field ion microscopy and atom probe tomography, other methods do not resolve the arrangement of atoms at the dislocation, instead relying on features such as the strain field in the crystal around the dislocation to make it visible.

5.4.2 Two Basic Types of Dislocations

There are two fundamental types of dislocations: edge dislocations (Fig. 5.7b) and screw dislocations (Fig. 5.7c). In this section, we will focus on dislocations in simple cubic crystals.

An edge dislocation can be characterized by the appearance of an additional half atomic plane perpendicular to a specific crystal plane. To create an edge dislocation, a particular lattice plane (e.g., the *xoz* plane in Fig. 5.7b) is cut to a certain position, resulting in a displacement between the upper and lower parts of the lattice that is equivalent to one atomic distance. The displacement direction is parallel to the *x*-axis, forming the edge dislocation. The interatomic force is then restored in the crystal to close the upper and lower parts that are cut by the above-mentioned plane.

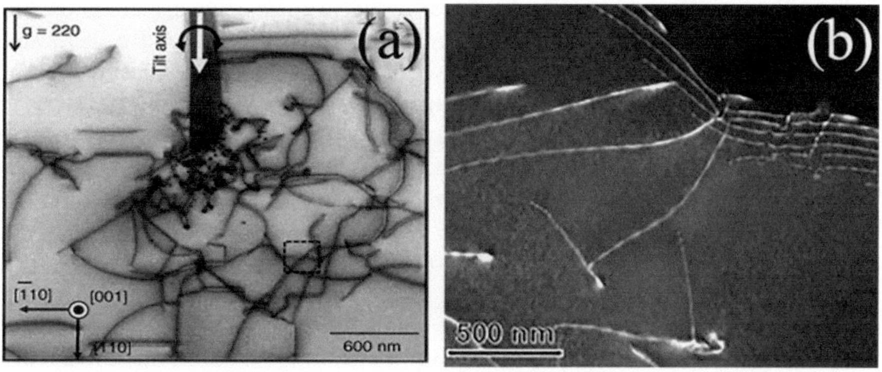

Fig. 5.6 Examples of dislocations in various alloys characterized by TEM. **a** A scanning TEM (STEM) image of dislocations around a crack tip in Si single crystal; **b** a weak-beam TEM dark field image showing a dislocation boundary in a quenched Al-Cu-Mg alloy

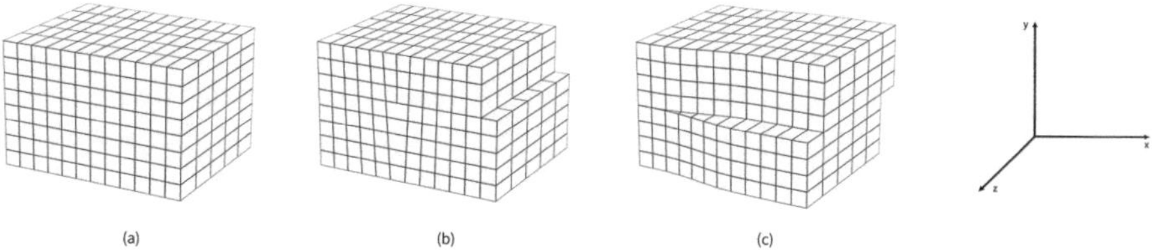

Fig. 5.7 **a** a simple cubic crystal, **b** formation of an edge dislocation, and **c** formation of a screw dislocation in the crystal

In contrast to edge dislocations, the screw dislocation differs in the relative displacement of the upper and lower parts of the crystal when cut by a particular lattice plane. Specifically, the screw dislocation results in a displacement along the z-axis that is equal to one atomic distance.

5.4.3 Burgers Vector

The Burgers vector, $\boldsymbol{b}$, is a crucial concept in understanding dislocations. The physical meaning of $\boldsymbol{b}$ is the vector sum of dislocated arrangement of atoms caused by dislocations, and it represents the magnitude and direction of the deformation that a dislocation could cause.

To determine the Burgers vector of an edge dislocation, we can follow the method illustrated in Fig. 5.8. In this figure, a perfect crystal and a dislocated crystal with an edge dislocation are shown in Fig. 5.8a and b, respectively. The dislocation line is perpendicular to the paper. To create a Burgers circuit, we start from a certain point *M* at a distance from the dislocation in Fig. 5.8b, move counter-clockwise around the dislocation, and return to *M* (*MNOPQ*). In a perfect crystal, if starting from the same point *M* in Fig. 5.8b, and also walking in the same manner from *M* with the same number of steps with that in Fig. 5.8a, we will stop at point *Q*, which is an atomic distance away from *M*. Therefore, the vector from *Q* to *M*, ***QM***, is defined as the Burgers vector of the dislocation and is written as ***b***. The Burgers vector is unique to each dislocation and remains constant regardless of the size and shape of the circuit. However, the direction of the Burgers vector changes with the direction of the circuit. For example, if we start from *M* in Fig. 5.8 and walk clockwise around the dislocation, the resulting Burgers vector ***b*** will have the opposite direction but the same magnitude as the previous one.

The Burgers vector of a screw dislocation can also be determined using a Burgers circuit, as illustrated in Fig. 5.9. By following the same steps as in Fig. 5.8, a closed circuit *MNOPQ* is formed in the dislocated crystal (Fig. 5.9b), while the circuit in the perfect crystal (Fig. 5.9a) remains open. This closed circuit allows us to determine the Burgers vector of the screw dislocation. And the ***QM*** is the Burgers vector of the screw dislocation. It should be noted that the Burgers vector of the screw dislocation is parallel to the dislocation line.

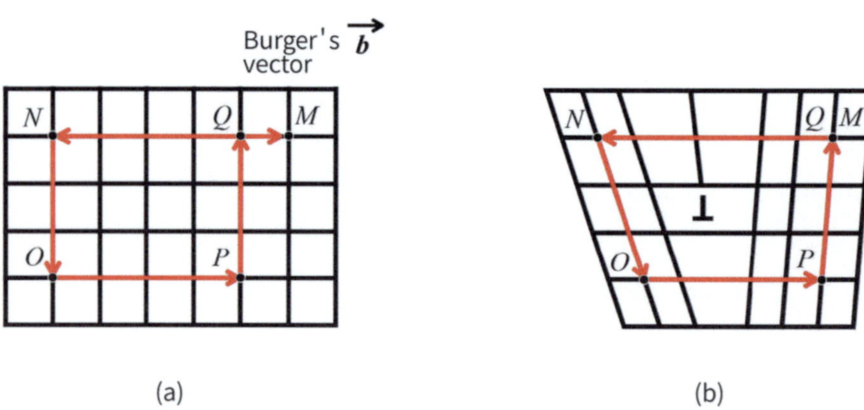

Fig. 5.8 The way to define the Burgers vector of an edge dislocation. **a** A circuit in a perfect crystal; **b** a circuit in a dislocated crystal with an edge dislocation

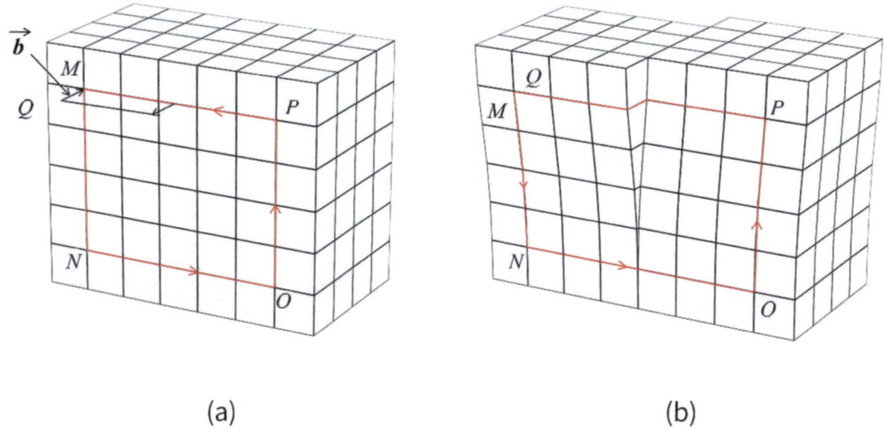

Fig. 5.9 The way to define the Burgers vector of a screw dislocation. **a** A circuit in a perfect crystal; **b** a circuit in a dislocated crystal with a screw dislocation.

5.4.4 Dislocation Motion

① Motion of an Edge Dislocation

Before delving into the motion of dislocations, it is important to briefly touch upon the external forces and stresses that induce crystal slip. Plastic deformation is common in many material processing techniques such as rolling, extrusion, forging, and sheet metal forming, all of which involve slip and twinning. Although the external forces and stresses imposed during material processing can be highly complex, the existence of shear stress is the fundamental factor underlying plastic deformation. Even a simple uniaxial tensile test can cause shear stress in the tensile specimen.

Consider the cylindrical specimen in Fig. 5.10 with a cross-sectional area of A. After being stretched, slip occurs along the shadowed surface with an area of $A/\cos\varphi$, where φ is the angle between the normal of the slip plane and the external force, $\boldsymbol{F}$. The external force $\boldsymbol{F}$ can be decomposed into two components acting on the slip plane: $\boldsymbol{F}\cos\varphi$ is the normal force, and $\boldsymbol{P}\sin\varphi$ is the tangential force. Therefore, the stress induced by the external force consists of the normal stress $(\boldsymbol{F}/A)\cos^2\varphi$, which tends to break atomic bonds on both sides of the slip plane, and the shear stress $(\boldsymbol{F}/A)\cos\varphi\sin\varphi$, which tends to cause atoms on both sides to slip. As the shear stress induced by the external force is often not aligned with the inherent slip direction of the crystal, it is crucial to determine the partial shear stress in the slip direction. The partial shear stress is

$$\tau = (\boldsymbol{F}/A)\cos\varphi\sin\varphi \tag{5.3}$$

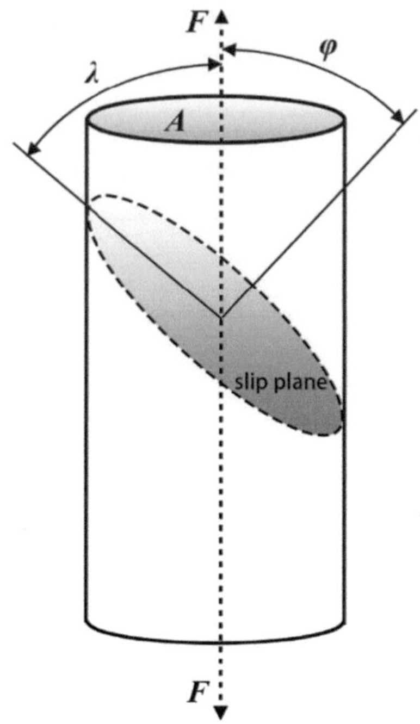

Fig. 5.10 Geometry of slip in a cylindrical crystal

Normally, this formula can be abbreviated as

$$\boldsymbol{\tau} = (\boldsymbol{F}/A)\cos\varphi\cos\lambda \tag{5.4}$$

where λ is the angle between the tensile direction and sliding direction. To induce movement in an edge dislocation, a shear stress must be present, but it can arise from various sources. Tangential force is not a necessary requirement, as long as the applied forces or stresses can generate sufficient shear stress to facilitate edge dislocation slip. What kind of shear stress drives the sliding of an edge dislocation? As shown in Fig. 5.11, if a pair of shear stresses, perpendicular to the dislocation line (parallel to the Burgers vector) and on the plane parallel to the dislocation's slip plane, are applied, the dislocation can move. The experiments show such shear stress is low. Therefore, if the external force-induced shear stress attains this low shear stress, the dislocation can slide. In Fig. 5.11, under 0.5~5 MPa shear stress, the dislocation line progressively moves towards the right. As it exits the right surface, a half-atomic plane protrudes on the right, indicating the occurrence of Burgers vector slip. Conversely, if the stress is applied in the opposite direction, the dislocation line moves to the left. In the absence of the above-described shear stress, the edge dislocation will remain stationary.

The motion mode of edge dislocations known as climbing is another way in which they can move. Experimental evidence has shown that the slip plane of a crystal is typically the plane with the highest packing density of atoms, and the

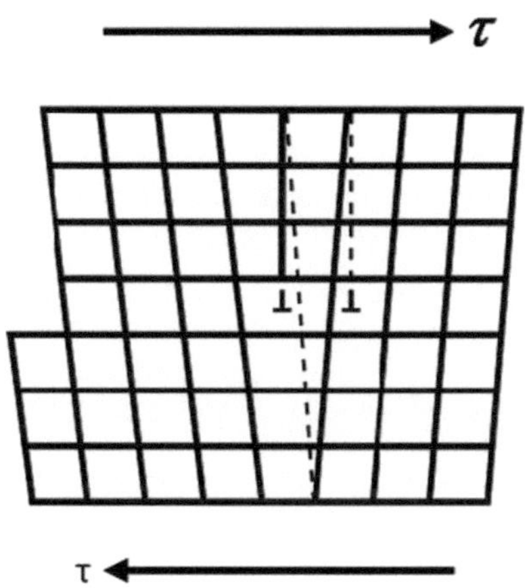

Fig. 5.11 Motion of an edge dislocation under shear stress

slip direction is along the direction of highest packing density. When an edge dislocation is located on the slip plane with its Burger's vector in line with the highest packing density direction, it can easily slip provided there is sufficient applied shear stress, as described earlier. However, if the edge dislocation line and Burger's vector do not lie on a slip plane, the dislocation will not be able to slip due to a barrier. Nevertheless, whether an edge dislocation is able to slip or not, it can still move by the migration of matter, that is, the diffusion of atoms or vacancies. This mode of motion is depicted in Fig. 5.12, where atoms on the periphery of the extra half atomic plane (i.e., the dislocation line) either leave by diffusion, or atoms outside the edge dislocation attach to the dislocation line through diffusion. These processes result in the rise or fall of dislocations on the extra half atomic plane. Climbing can also occur in dislocations that are already slipping or even dislocations located on the slip plane. However, climbing is generally more challenging and slower than slip, and it is mainly observed at higher temperatures. Moreover, the tensile and compressive stresses acting on the extra half atomic plane also affect the climbing process, with tensile stress aiding the migration of atoms on the dislocation line and the removal of vacancies, while compressive stress works in the opposite way. The motion of an edge dislocation through climbing is sometimes referred to as non-conservative motion, while slip is known as conservative motion, depending on whether the motion takes place on the plane formed by the dislocation line and Burger's vector.

② Motion of a Screw Dislocation

In terms of screw dislocation motion, there are two modes, slip and cross slip.

Slip of a screw dislocation is illustrated in Fig. 5.13, where the white circle represents the atom beneath the slip plane and the black dot represents the atom above it. To facilitate visualization, each column of atoms is numbered. Under a small applied shear stress above and below the paper, most atoms remain in their

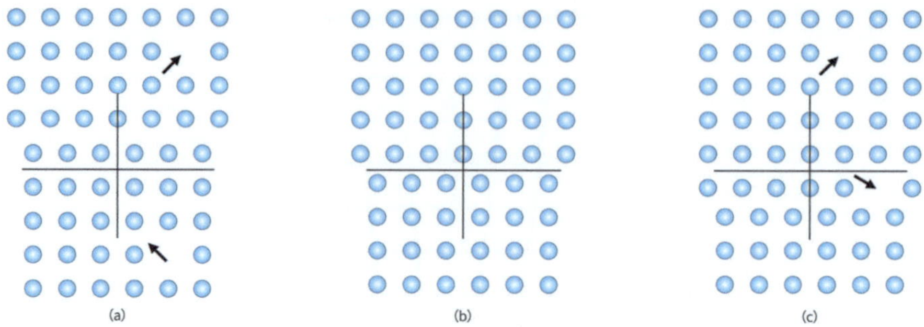

Fig. 5.12 Climb of an edge dislocation, **a** positive climb, **b** no climb occurred, and **c** negative climb

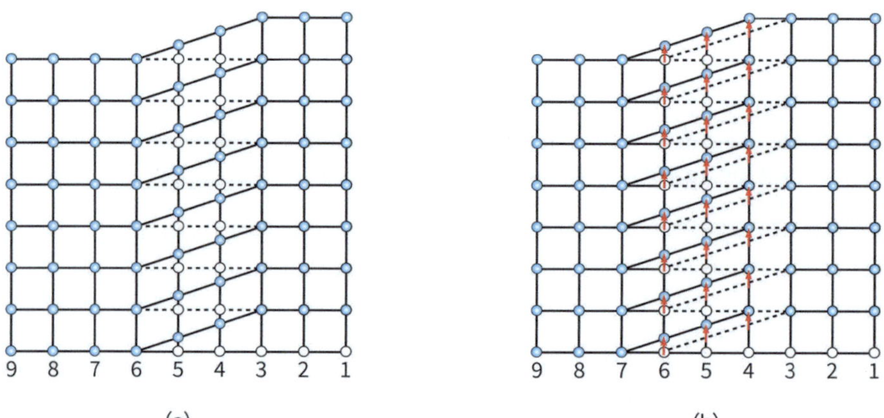

Fig. 5.13 Slip of a screw dislocation

original positions; however, atoms near the dislocation line will produce a small displacement, resulting in shift of screw dislocation to the left with one atomic spacing. As shown in Fig. 5.13, the black dots move up a little distance along the red arrows from the original dashed lines position, and the screw dislocation moves an atomic spacing left. The screw dislocation moves eventually to the left of the crystal and slips out of it, as shown in Fig. 5.14.

For comparison, a slip process of an edge dislocation is schematically plotted in Fig. 5.15. Although the two slip processes occur along Burger's vector, the details of the processes differ. The screw dislocation line moves perpendicular to Burger's vector, which is perpendicular to the direction of shear stress and crystal slip, while the edge dislocation line moves parallel to its Burger's vector, which is parallel to the direction of shear stress and crystal slip.

Screw dislocations are unable to climb due to the lack of an extra half atomic plane. However, they have another type of unique motion mode, cross slip, due to

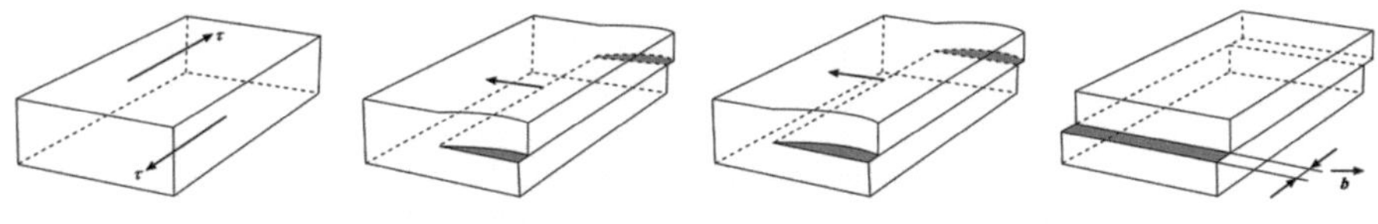

Fig. 5.14 Slip process of a screw dislocation

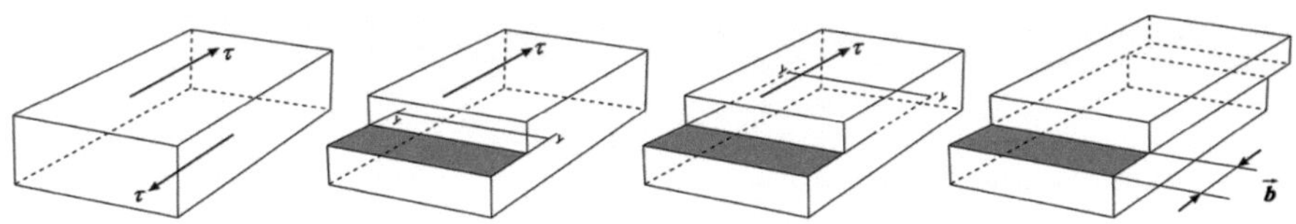

Fig. 5.15 Slip process of an edge dislocation

the fact that their dislocation line is parallel to their Burger's vector. In Fig. 5.16, a pure screw dislocation *BC* moves from right to left under applied shear stress, with its slip plane being *ABCD*. Although *ABCD* is a slip plane for this dislocation, it is not the only one, as any shear stress acting on the prism plane parallel to the dislocation line can also induce slipping. The slip plane is determined by the shear stress direction, passing through the dislocation line and remaining parallel to the plane on which the shear stress is acting. For instance, under shear stress on the upper and lower crystal surfaces, the dislocation line would slip in the y-axis direction, forming a slip plane parallel to *xy*. This results in a step *MN*, as shown in Fig. 5.17a. When the shear stress changes to the left and right crystal surfaces, slip trace *NB* can form. The screw dislocation's ability to slip on multiple planes is linked to its structure. Each point on a helix has the same properties, and moving one circle produces a screw pitch (the Burger's vector). Therefore, the slip plane depends solely on the direction of the external shear stress, enabling screw dislocations to make any slip motion parallel to their Burger's vector in a crystal. Unlike edge dislocations, which have only one slip plane, screw dislocations have numerous slip planes. On the plane perpendicular to the dislocation line, a screw dislocation can produce slip traces of various curved shapes due to different shear stresses (not only the simple trace in Fig. 5.17). However, the number of available slip planes is limited in real crystals, meaning that screw dislocations can only slip on a restricted number of planes. When screw dislocations slide into a different slip plane, the process is referred to as cross slip.

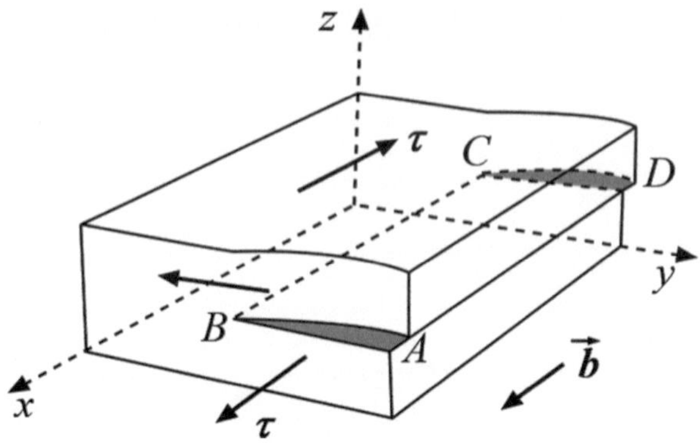

Fig. 5.16 Slip of a screw dislocation under shear stress

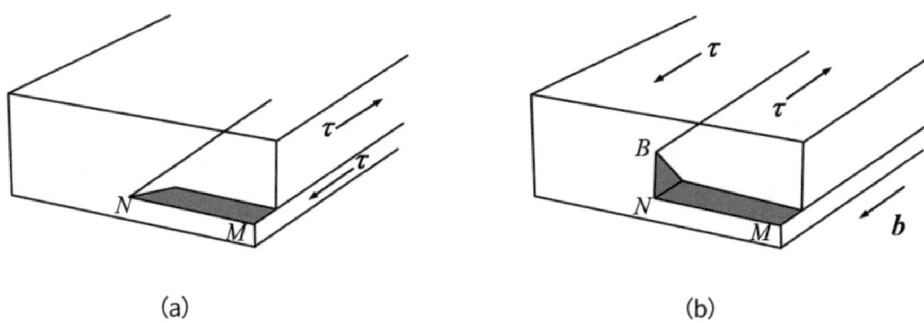

Fig. 5.17 Slip of a screw dislocation on several slip planes

5.4.5 Dislocation Density

To better understand dislocations in crystals, it is important to measure their density. Dislocation density (ρ) refers to the total length of dislocation lines within a unit volume of the crystal, $\rho = l/V$, where l represents the length of dislocation lines within a given volume V and is usually expressed in m^{-2}.

In well-annealed metal crystals, the dislocation density typically falls in the range of 10^{10} to 10^{12} m^{-2}. However, it can increase significantly with plastic deformation, with heavily cold-rolled metals exhibiting values of about 10^{14} to 10^{15} m^{-2}. In non-metallic crystals, dislocation density is typically much lower. For instance, carefully grown semiconductor crystals can have dislocation densities as low as 10^5 m^{-2}.

As plastic deformation increases, dislocation densities also tend to increase. The strength of crystal has a close relationship with dislocation density, and hardening induced by dislocation is given by

$$\sigma_\rho = 0.5MG\boldsymbol{b}(\rho_d)^{1/2} \tag{5.5}$$

where M is the Taylor factor, G is the shear modulus, $\boldsymbol{b}$ is Burgers vector and ρ is the dislocation density.

5.5 Stress Fields of Dislocations

The presence of dislocations in a crystal causes nearby atoms to deviate from their normal positions, resulting in lattice distortion and the formation of a **stress field**. Quantitative analysis of the distribution and energy of distortion caused by dislocations is essential to study the interactions between dislocations and other crystal defects, and ultimately to understand the properties of crystals.

At the core of a dislocation, the arrangement of atoms is highly disordered and beyond the elastic deformation range, rendering Hooke's law inapplicable. However, outside the central region, the elastic stress field created by the dislocation can be modeled using elastic isotropic continuum theory. To analyze the dislocation stress field, it is customary to exclude the central region with a radius of approximately 0.5~1 nm and apply the elastic continuum model to derive the stress field formula for the surrounding region.

5.5.1 Continuum Medium Model for Dislocations

In the continuum medium model, matter is assumed to be continuously distributed in the whole space that it occupies, and the physical parameters of the macroscopic motion of matter are differentiable continuous functions of space and time. For the continuous medium model, modern mathematical tools can be applied.

The continuum model is typically used to calculate the long-range stress field of a dislocation. In this model, the real crystal is replaced by a continuous elastic medium, and the stress field is computed using classical elastic mechanics methods. Although this model has limitations, as it neglects the crystal's lattice structure and cannot be used to analyze the region near the dislocation line with seriously dislocated atoms, its results are reliable and have been confirmed by numerous experiments. Moreover, even in some marginal cases, the continuum model can provide useful results.

When the continuous medium model is applied to calculate the stress field of dislocation, the following steps can be adopted:

(a) It is envisaged that the area with serious dislocated atoms near the dislocation line (the distance from the dislocation is less than r_0) will be completely excavated, and the remaining crystals will be replaced by continuous elastic medium.

(b) According to the general definition of dislocation, the displacement field $\boldsymbol{u}$ (x, y, z) in the medium has a jump in the slip plane along the dislocation line, which is equal to Burger's vector, which can be expressed mathematically as

$$\boldsymbol{u}(M_1) - \boldsymbol{u}(M_2) = \boldsymbol{b} \tag{5.6}$$

where M_1 and M_2 are the two points corresponding to both sides of the S plane; Or use the equivalent Burger's circuit representation, along the arbitrary closed circuit around the dislocation.

$$\oint \mathrm{d}\boldsymbol{u} = \oint \frac{\mathrm{d}\boldsymbol{u}}{\mathrm{d}s}\mathrm{d}s = \boldsymbol{b} \tag{5.7}$$

(c) The elastic field within the medium should meet the elastic equilibrium equation. For the isotropic elastic medium, the equilibrium equation of displacement is

$$(1 - 2v)\nabla^2\boldsymbol{u} + \mathrm{grad}(\mathrm{div}\boldsymbol{u}) = 0 \tag{5.8}$$

where ν is Poisson's ratio. For the plane strain problem, the equilibrium equation can be simplified to the biharmonic equation satisfied by the stress function χ

$$\nabla^4\chi = 0 \tag{5.9}$$

(d) If there is no external binding force, the stress component along the inner and outer surfaces shall be equal to 0.

5.5.2 Stress Components

The stress state at any point in the object can be described by 9 stress components. Figure 5.18 shows the expression of these 9 stress components in rectangular coordinates and cylindrical coordinates, respectively, where σ_{xx}, σ_{yy}, σ_{zz}, (σ_{rr}, $\sigma_{\theta\theta}$, σ_{zz}) are the normal stress components, τ_{xy}, τ_{yz}, τ_{zx}, τ_{yx}, τ_{zy}, τ_{xz} ($\tau_{r\theta}$, $\tau_{\theta r}$, $\tau_{\theta z}$, $\tau_{z\theta}$, τ_{zr}, τ_{rz}) are the shear stress components. The first symbol in the subscript represents the outer normal direction of the plane subjected to the stress, and the second symbol represents the direction of the stress.

In the equilibrium conditions, $\tau_{xy} = \tau_{yx}$, $\tau_{yz} = \tau_{zy}$, $\tau_{zx} = \tau_{xz}$ ($\tau_{r\theta} = \tau_{\theta r}$, $\tau_{\theta z} = \tau_{z\theta}$, $\tau_{zr} = \tau_{rz}$), only six stress components can fully express the stress state of a point. The strain components corresponding to these six stress components are ε_{xx}, ε_{yy}, ε_{zz} (ε_{rr}, $\varepsilon_{\theta\theta}$, ε_{zz}) and γ_{xy}, γ_{yz}, γ_{zx} ($\gamma_{r\theta}$, $\gamma_{\theta z}$, γ_{zr}).

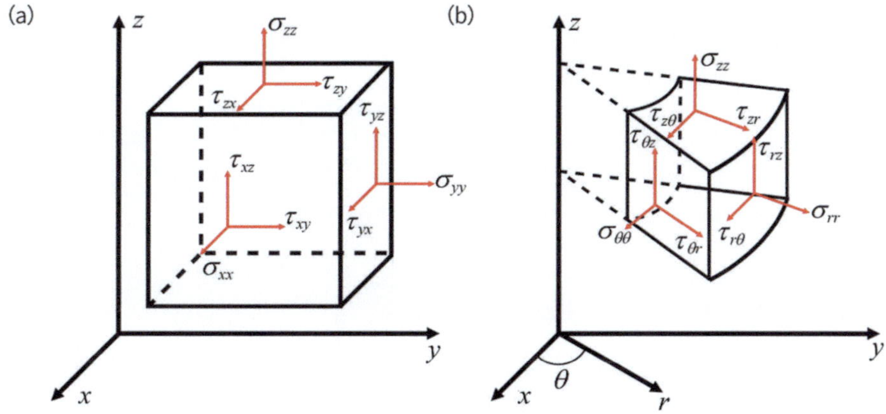

Fig. 5.18 The stress components at a point in the object, which is magnified as a hexahedron in the figure. **a** rectangular coordinates; **b** cylindrical coordinates

5.5.3 Stress Field of a Screw Dislocation

A screw dislocation can be created by cutting a cylinder with radius R made of continuous medium from the surface to the center along the longitudinal direction and then displacing the two sides of the slit relative to each other along the longitudinal direction (z axis) by a distance of b before gluing them together. The Burger's vector $\boldsymbol{b}$ defines the screw dislocation located at the central axis of the cylinder.

To simplify the analysis, a central hole with radius r_0 is made in the cylinder, as shown in Fig. 5.19, to create a continuum medium model of the screw dislocation. In the thick-walled cylinder shown in Fig. 5.19, there is only a relative displacement along the z direction, resulting in two shear strain components and no normal strain components. The cylindrical coordinates can express the two shear strain components, as given in Eq. 5.10, and the corresponding shear stress component is shown in Eq. 5.11.

$$\gamma_{r\theta} = \gamma_{z\theta} = \frac{b}{2\pi r} \tag{5.10}$$

$$\tau_{\theta z} = \tau_{z\theta} = G\gamma_{\theta z} = \frac{G\boldsymbol{b}}{2\pi r} \tag{5.11}$$

where G is the shear elastic modulus.

The rest 7 stress components are all 0. Namely,

$$\sigma_{\mathrm{rr}} = \sigma_{\theta\theta} = \sigma_{\mathrm{zz}} = 0 \tag{5.12}$$

$$\tau_{\mathrm{r}\theta} = \tau_{\theta\mathrm{r}} = \tau_{\mathrm{rz}} = \tau_{\mathrm{zr}} = 0 \tag{5.13}$$

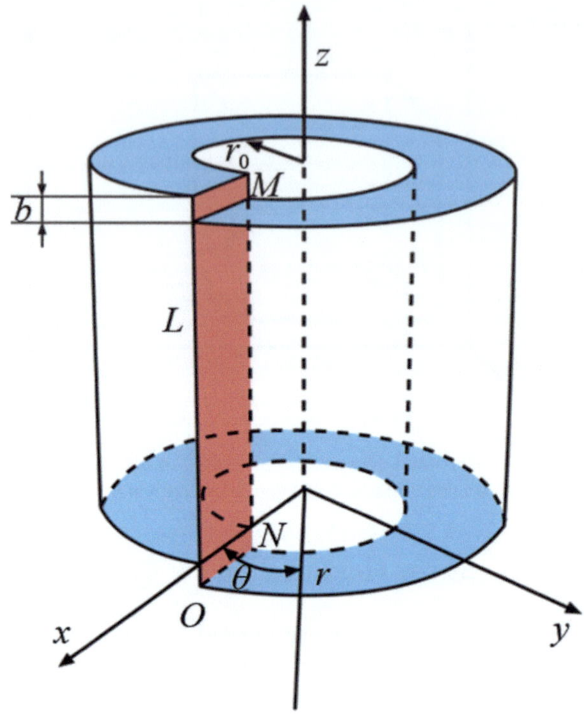

Fig. 5.19 The continuous medium model for a screw dislocation

If converting to stress component expressed in the rectangular coordinates, it will be

$$\begin{aligned} \tau_{yz} &= \tau_{zy} = \frac{Gb}{2\pi} \times \frac{x}{x^2 + y^2} \\ \tau_{xz} &= \tau_{zx} = \frac{-Gb}{2\pi} \times \frac{y}{x^2 + y^2} \\ \sigma_{xx} &= \sigma_{yy} = \sigma_{zz} = \tau_{xy} = \tau_{yx} = 0 \end{aligned} \tag{5.14}$$

From Eqs. 5.11 and 5.14, it can be deduced that the stress field of a screw dislocation has two following characteristics:

1. It has no normal stress component,
2. The shear stress component is only related to the distance r from the dislocation center. The stress state at each point with the same distance from the dislocation center is the same. The farther away from the dislocation center, the smaller the shear stress component.

When r approaches 0, $\tau_{\theta z}$ tends to infinity, which is obviously inconsistent with the real situation. This is the reason why the central part is cut out when building the continuous medium model. Generally, r_0 is taken as 0.5~1 nm.

5.5.4 Stress Field of an Edge Dislocation

The continuous medium thick-walled cylinder is once again utilized for demonstration purposes. Similarly, it is longitudinally cut, but this time the two sides of the cut are joined together after being relatively displaced by the distance b along the radial x axis in rectangular coordinates (or the r axis in cylindrical coordinates). As a result, an edge dislocation with Burger's vector $\boldsymbol{b}$ is formed, as shown in Fig. 5.20. In the figure, *OO'* denotes the location of the dislocation line, *MNOO'* is the slip plane, and the z-y plane corresponds to the additional half atomic plane.

The stress distribution in this thick-walled cylinder, that is, the stress field of the edge dislocation, can be obtained by using elastic mechanics. In the cylindrical coordinate, expressions can be:

$$
\begin{aligned}
&\sigma_{rr} = \sigma_{\theta\theta} = -A\frac{\sin\theta}{r} \\
&\sigma_{zz} = \nu(\sigma_{rr} + \sigma_{\theta\theta}) \\
&\gamma_{r\theta} = \gamma_{\theta r} = A\frac{\cos\theta}{r} \\
&\gamma_{rz} = \gamma_{zr} = \gamma_{\theta z} = \gamma_{z\theta} = 0
\end{aligned}
\tag{5.15}
$$

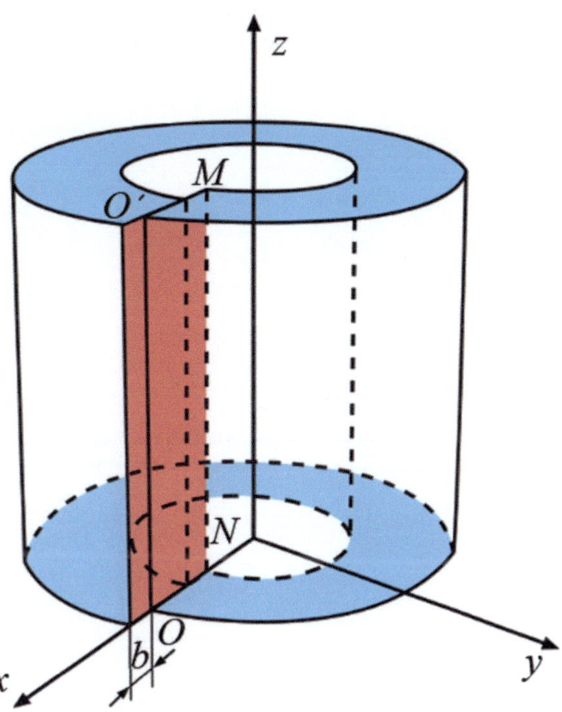

Fig. 5.20 The continuous medium model for an edge dislocation

In the rectangular coordinate, expressions can be:

$$
\begin{aligned}
\sigma_{xx} &= -A\frac{y\left(3x^2 + y^2\right)}{\left(x^2 + y^2\right)^2} \\
\sigma_{yy} &= A\frac{y\left(x^2 - y^2\right)}{\left(x^2 + y^2\right)^2} \\
\sigma_{zz} &= \nu\left(\sigma_{xx} + \sigma_{yy}\right) \\
\gamma_{xy} &= \gamma_{yx} = A\frac{x\left(x^2 - y^2\right)}{\left(x^2 + y^2\right)^2}
\end{aligned}
\tag{5.16}
$$

A can be calculated by

$$A = Gb/2\pi(1 - \nu) \tag{5.17}$$

where ν is Poisson's ratio.

Figure 5.21 shows the stress distribution around an edge dislocation. The stress field of edge dislocation has the following characteristics:

1. The components of normal stress and shear stress coexist.
2. All the stress components are independent of z value, indicating that the stress state at each point on the straight line parallel to the edge dislocation line is the same.

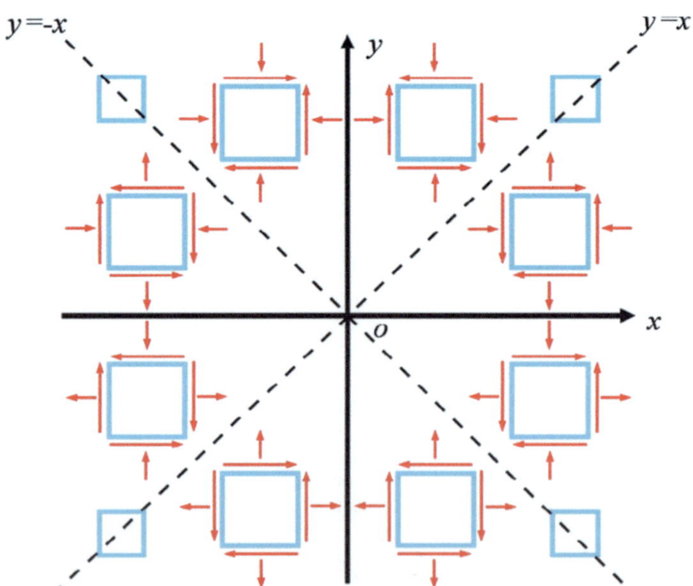

Fig. 5.21 Stress distribution around an edge dislocation

3. The stress field is symmetrical to the y-axis.
4. If $y = 0$, then $\sigma_{xx} = \sigma_{yy} = \sigma_{zz} = 0$, indicating zero normal stress on x–z plane.
5. If $y > 0$, then $\sigma_{xx} < 0$; if $y < 0$, then $\sigma_{xx} > 0$, indicating the upper side of x–z plane in compression and the lower side in tension.
6. If $x = \pm y$, then $\sigma_{yy} = 0$, and $\tau_{xy} = 0$.

5.6 Strain Energy of Dislocations

The existence of the elastic stress field around the dislocation increases the energy of the crystal, which is called strain energy of dislocation. Since the dislocation center region is excluded when calculating the elastic stress field, with regard to a static dislocation, when calculated according to the stress field, the strain energy should include two parts: the strain energy E_0 in the dislocation center region and the elastic strain energy E_e caused by the dislocation stress field calculated by the formula above. That is,

$$E = E_e + E_0 \tag{5.18}$$

The dislocation center region exhibits large lattice distortion, thus E_0 cannot be calculated by linear elastic theory. It is estimated that this part of energy is about $\frac{1}{10} \sim \frac{1}{15}$ of the total strain energy, hence it is usually ignored, instead E_e is employed to desbribe the strain energy of dislocation. The strain energy of dislocation can be calculated by the work done to form the dislocation.

5.6.1 Strain Energy of an Edge Dislocation

It is assumed that the edge dislocation shown in Fig. 5.20 is a dislocation of unit length. In the process of creating this dislocation, the displacement gradually increases from O to b, which is a variable that varies with r, set to be x; At the same time, the forces on the $MNOO$' plane also vary with r. In the process of displacement, when the displacement is x, the shear stress is

$$\tau_{or} = \frac{Gx}{2\pi(1-\nu)} \times \frac{\cos\theta}{r} \tag{5.19}$$

Note that θ = 0, therefore, the work done to overcome the shear stress $\tau_{\theta r}$ is

$$W = \int_{r_0}^{R}\int_{0}^{b} \tau\theta r \mathrm{d}x\mathrm{d}r = \int_{r_0}^{R}\int_{0}^{b} \frac{Gx}{2\pi(1-\nu)} \times \frac{1}{r}\mathrm{d}x\mathrm{d}r = \frac{Gb^2}{4\pi(1-\nu)} \times \ln\frac{R}{r_0} \tag{5.20}$$

This is the stress energy E_e^e of edge dislocation per unit length.

5.6.2 Strain Energy of a Screw Dislocation

Using the shear stress on the screw dislocation $\tau_{\theta z} = \frac{Gb}{2\pi r}$, the strain energy of screw dislocation per unit length can also be obtained through

$$E_{\mathrm{e}}^{\mathrm{s}} = \frac{G\boldsymbol{b}^2}{4\pi} \ln \frac{R}{r_0} \tag{5.21}$$

Comparing Eqs. 5.19 and 5.20, can be seen that when $\boldsymbol{b}$ is the same,

$$E_{\mathrm{e}}^{\mathrm{e}} = \frac{1}{(1-\nu)} E_{\mathrm{e}}^{\mathrm{s}} \tag{5.22}$$

Poisson's ratio of general metals is 0.3~0.4, if $\nu = 1/3$, $E_{\mathrm{e}}^{\mathrm{e}} \approx \frac{3}{2} E_{\mathrm{e}}^{\mathrm{s}}$. Therefore, the elastic strain energy of edge dislocation is about 50% larger than that of screw dislocation.

5.7 Line Tension of Dislocations

Since the energy of a dislocation is proportional to its length, it has a tendency to shorten its length as much as possible. The line tension T of a dislocation is similar to the surface tension of liquid, which is expressed by the energy of the dislocation line per unit length (the energy per unit length is J/m = N·m/m = N, which is the same as the unit of force).

If we want to increase the length of dislocation by dl, we must apply the work $T\mathrm{d}l$ against the linear tension T. Obviously, this work should be equal to the increased energy in dislocation dE, namely, $T\mathrm{d}l = \mathrm{d}E$, $T = \frac{\mathrm{d}E}{\mathrm{d}l}$. This indicates that the value of line tension is equal to the energy of dislocation per unit length, and the energy of dislocation per unit is $E = \alpha G\boldsymbol{b}^2$ Therefore, the dislocation line tension is defined as the increased energy when the dislocation is increased by a unit length, i.e.

$$T = \alpha G b^2 \tag{5.23}$$

For a straight dislocation line, T is about Gb^2, while for bended dislocation, as shown in Fig. 5.22, if the dislocation increases by a unit length, the energy raised in the system will be less than Gb^2. Therefore, in rough estimation, α is often taken as 0.5, so the line tension is

$$T \approx \frac{1}{2} G b^2 \tag{5.24}$$

Line tension is an elastic property of a dislocation. When a dislocation is bent by external force, the line tension will straighten the dislocation line as much as possible, as the energy of the dislocation is proportional to its length, and bending

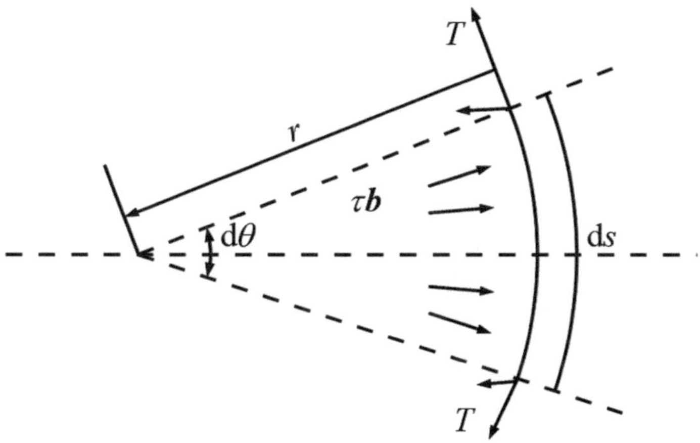

Fig. 5.22 The external force required to bend a dislocation

of the dislocation will increase the dislocation line and thus increase its energy, so it has the tendency to shorten its length and straighten it as much as possible.

Figure 5.22 shows that there is a dislocation line with a length of ds, its radius of curvature is r, and the central angle corresponding to ds is dθ. If there is a shear stress τ, it will try to straighten the dislocation line, and the component of the line tension along the horizontal direction is $2T\sin\frac{\mathrm{d}\theta}{2}$. At equilibrium, these two forces must be equal to each other, that is,

$$\tau \boldsymbol{b}\mathrm{d}s = 2T\sin\frac{\mathrm{d}\theta}{2} \tag{5.25}$$

When dθ is significantly small,

$$\sin\frac{\mathrm{d}\theta}{2} \approx \frac{\mathrm{d}\theta}{2} \tag{5.26}$$

And d$s = r$dθ, hence

$$\tau b = \frac{T}{r} \approx \frac{Gb^2}{2r} \text{ or } \tau = \frac{Gb}{2r} \tag{5.27}$$

From the above discussion, if the shear stress τ produces a force of τb on the dislocation that cannot move freely, the dislocation will bend outwards, and its radius of curvature r is inversely proportional to τ (the smaller the radius of curvature of the bend, the greater the external force required). This will help us understand the motion of dislocation with both fixed ends. The existence of line tension is the reason why the dislocation in the crystal is distributed in a three-dimensional network, because the line tension of each dislocation at the same node in the network automatically tends to the equilibirium state, thus ensuring the relative stability of the dislocation in the crystal.

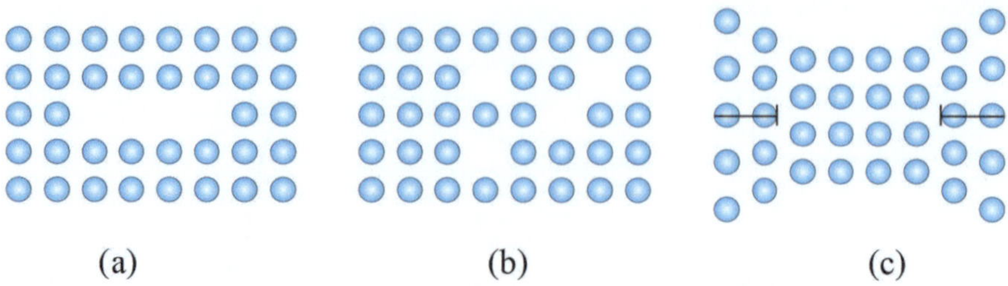

Fig. 5.23 Aggregation of vacancies to form dislocations. **a** vacancies in crystal; **b** vacancies rearrange; **c** form dislocation

5.8 Dislocation Nucleation and Multiplication

The initiation of dislocations in a dislocation-free crystal is referred to as generation of dislocation. On the other hand, the increase in the number of dislocations from the existing ones is termed as multiplication of dislocations. Dislocation kinetics deals with how dislocations move under the influence of external forces. Solving these issues is a crucial requirement for comprehending the origin of dislocations in crystals, the plastic deformation process, and numerous physical characteristics impacted by dislocations.

5.8.1 Dislocation Nucleation

The origin of dislocations can stem from several sources. During solidification, crystals may have slightly different orientations where they meet, leading to the formation of dislocations; component segregation can cause different lattice constants in different parts of the crystal, creating dislocations at the transition region; dislocations may also be initiated by the impact of flowing liquids and local stress concentration during cooling. Stress concentration at the crack tip, interface of precipitate or inclusion, and surface damage can also promote the formation of dislocations. The aggregation of supersaturated vacancies, as illustrated in Fig. 5.23, is also an important source of dislocations. Understanding these sources of dislocations is essential for comprehending the process of plastic deformation, the physical properties impacted by dislocations, and the methods to control their formation.

5.8.2 Dislocation Multiplication

During plastic deformation, it is expected that a significant number of dislocations would slip out of the crystal, leading to a decrease in dislocation density. However, in reality, the dislocation density increases with the level of deformation, and may even increase by several orders of magnitude after severe deformation. This observation suggests that dislocations are being multiplied during the deformation process, and the location at which this occurs is referred to as the dislocation source.

There are many mechanisms of dislocation multiplication; the most important one is the dislocation multiplication mechanism proposed by Frank and Read in 1950 which has been confirmed by experiments, called Frank-Read source, or F-R source for short. Imagine that there is a segment of edge dislocation *AB* on a slip plane in the crystal, and its two ends are pinned by dislocation network nodes, as shown in Fig. 5.24. When the applied shear stress meets the necessary requirements, the dislocation line *AB* will slip under slip force. Under uniform stress field, the slip force $\boldsymbol{F}_t = \boldsymbol{\tau}|\boldsymbol{b}|$ along the whole dislocation line is equal. The dislocation line should have slipped parallel and forward, but because the two ends of dislocation *AB* are fixed and cannot move, it will bend while moving, resulting in the curved dislocation, as shown in Fig. 5.25. Force exerted on the dislocation $\boldsymbol{F}_t$ is always perpendicular to the dislocation itself, even when the dislocation is bent, so if the force continuously exerts on the disocation, each segment of the dislocation should move outward along its normal direction, as shown in Fig. 5.25c and d. When the dislocation line goes forward for a distance, points p and q in Fig. 5.25d meet. From the Burger's vector of dislocations, we can see that the two points p and q should be a left-handed screw dislocation and a right-handed screw dislocation, respectively, so when they meet together, they should offset each other. Therefore, the original whole dislocation line is now divided into two parts, as shown in Fig. 5.25e. Thereafter, the outer dislocation loop continues to expand under the force $\boldsymbol{F}_t$, until reaching the surface of the crystal, while another segment of dislocation inside returns to the original state driven by both the line tension and the force $\boldsymbol{F}_t$. This process is not over, because the stress continues to be applied to the crystal. In fact, the dislocation *AB* will continue to repeat the above-mentioned actions under the action of $\boldsymbol{F}$. In this way, the structure shown in Fig. 5.24 will emit a large number of dislocation loops, resulting in dislocation multiplication.

The following will be the discussion on the amount of stress required to activate the F-R source. When the *AB* dislocation line of F-R source is bent, according to Eq. 5.28, The centripetal restoring force caused by line tension is

$$f = \frac{Gb^2}{2r} \tag{5.28}$$

To remain bending state in the line *AB*, the slip force $\boldsymbol{\tau}|\boldsymbol{b}|$ on the dislocation line per unit length achieves the equilibrium with f, i.e.

$$\tau|\boldsymbol{b}| = \frac{Gb^2}{2r} \tag{5.29}$$

$$\tau = \frac{G|\boldsymbol{b}|}{2r} \tag{5.30}$$

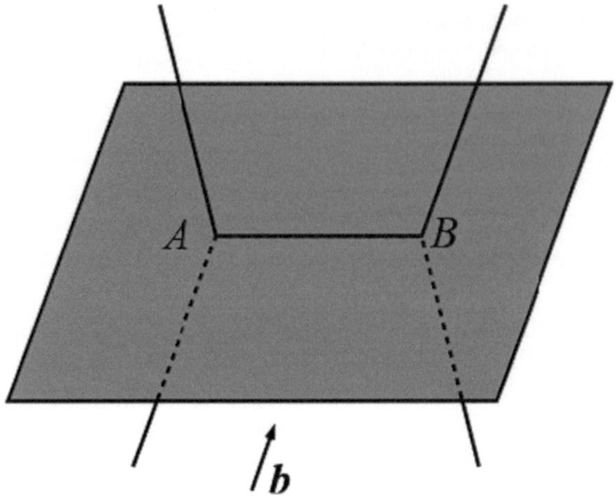

Fig. 5.24 Structure of F-R source

It can be seen that, the applied shear stress τ is inversely proportional to the radius of curvature r of the dislocation line, that is, the smaller r, the bigger the required $\boldsymbol{\tau}$. When the AB dislocation line is bent into a semicircle, its r is the minimum, and $\boldsymbol{\tau}$ is the maximum. After that, if the dislocation continues to extend, its r value will increase instead, $\boldsymbol{\tau}$ thus decreases accordingly. Therefore, the shear stress in equilibrium with the semicircular shaped dislocation is the critical shear stress $\boldsymbol{\tau_c}$ that activates the F-R source. Using Eq. 5.30, we can obtain

$$\tau_c = \frac{Gb}{2r} = \frac{Gb}{l} \tag{5.31}$$

where, l is the length of the edge dislocation AB.

Usually, the Order of magnitude of l is 10^{-4} cm, $|\boldsymbol{b}| \approx 10^{-8}$ cm, then τ_c calculated from Eq. 5.31 will be about 10^{-4} G. Regarding the activation of dislocation source as the yielding of crystal, $\boldsymbol{\tau_c}$ will be the critical shear stress, which is similar with the yield strength of real crystals.

(Supplementary) Other mechanisms of dislocation multiplication:

The above-mentioned F-R source is essentially a segment of slipable dislocation pinned at both ends, so it is called bilateral (or biaxial) F-R source, also known as U-shaped plane source. In addition, there are also unilateral F-R source, double cross slip multiplication and other mechanisms. Figure 5.26 shows the mechanism of double cross slip displacement multiplication. A screw dislocation starts to slip on the (111) plane, due to obstacles or changes in local stress state, a segment of the dislocation crosses slip to $(1\bar{1}1)$ plane, and then returns to another (111) plane parallel to original (111) plane after bypassing the obstacle. At this time, the dislocations at both ends of the $(1\bar{1}1)$ plane are edge dislocations and cannot propagate together with the dislocations on the (111) plane. As a result, the dislocations on the (111) plane will multiplicate in the way as illustrated in Fig. 5.26. The screw

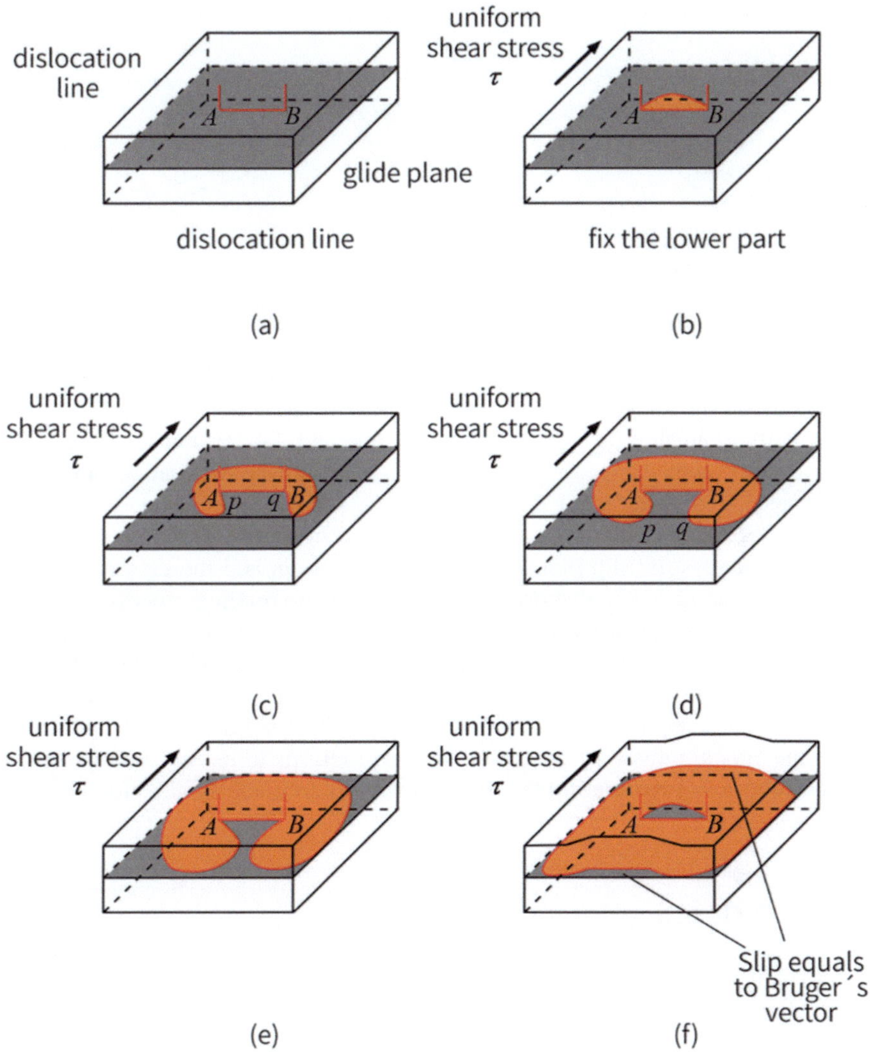

Fig. 5.25 Schematic motion of F-R source

dislocation is usually transferred from the original slip plane to the intersecting slip plane, and then transferred to the slip plane parallel to the original slip plane. Such slip motion of dislocation is called double cross slip. The dislocation multiplication mechanism in such a manner is called double cross slip multiplication. Assuming that, if the dislocation loop on the (111) plane crosses slip again to another parallel (111) plane and becomes a new dislocation source, the dislocation will multiplicate rapidly. It can be seen that double cross slip is a more efficient multiplication mechanism.

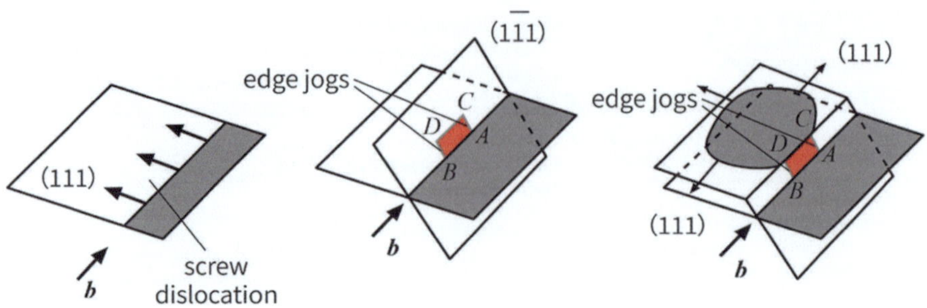

Fig. 5.26 Double cross slip dislocation multiplication mechanism

A unilateral (or uniaxial) F-R source, also known as a L-shaped planar source, is essentially a slipable dislocation pinned at one end. For example, in Fig. 5.27, *CD* is a dislocation on the slip plane, *DE* is a stationary dislocation, and point *D* is pinned. Under the action of shear stress τ, segment *CD* begins to slip, and gradually becomes a curl around point *B*, extending outward continuously. The curl sweeps across the slip plane once every revolution, and the crystal produces a slip of $|\boldsymbol{b}|$. Figure 5.27a~d show several stages of the rotating process. (a) is the initial stage, *DC* is a positive edge dislocation, (b) is after rotating by 90°, Burger's vector is consistent with the direction of the dislocation line (*DC* direction), so it is a right screw dislocation; (c) is after the dislocation line *DC* rotates by 270°, it becomes a left screw dislocation; (d) is after the *DC* rotating by 360°, the upper half of the crystal moves $|\boldsymbol{b}|$, and the dislocation returns to its original position, if the shear stress τ remains constant, the crystal can continuously slip along the slip plane.

5.9 Dislocation Interactions

The theoretical strength value of an ideal crystal is about two orders of magnitude greater than the yield strength of a real material, mainly due to the presence of crystal defects, especially dislocations. The dynamic behavior of dislocations plays a crucial role in determining the characteristics of plastic deformation. The process of dislocation motion across a crystal can be schematically illustrated in Fig. 5.28. Figure 5.28a represents an ideal crystal that does not contain any dislocations. Figure 5.28b shows a distorted crystal with lattice planes marked by vertical lines inside the body. The shear stress $\boldsymbol{\tau}$ initiates a step on the left-hand side surface by breaking bonds in the left outer plane and pushing the upper half-plane into the crystal. The crystal is then sheared along the plane, and the dislocation moves by the vector $\boldsymbol{b}$, permanently altering the crystal's shape due to plastic deformation. The shift $\boldsymbol{b}$ is known as the Burgers vector, which not only describes the magnitude of the shift between two parts of the crystal but also the intensity of the strain field formed around the dislocation. In ductile metallic systems, dislocation motion and

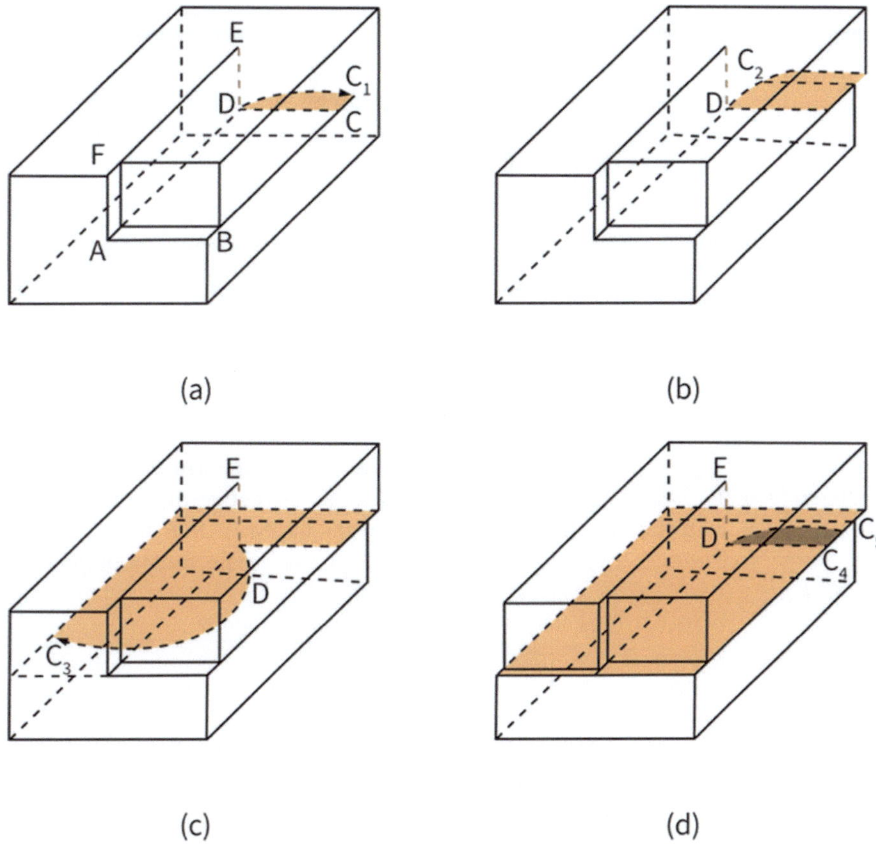

Fig. 5.27 Schematic motion of single-side F-R source

interactions play a crucial role in all deformation and failure mechanisms and can significantly influence the mechanical properties of metals.

Several types of dislocation interactions with different features in crystals are introduced and illustrated, as follows.

5.9.1 Dislocation Interactions with Defect Fields

Two dislocations could generally induce forces on each other, because one dislocation produces a stress field that could cause a force on the other. In this way, two parallel screw dislocations having Burgers vectors of magnitudes $\boldsymbol{b}_1$ and $\boldsymbol{b}_2$ could produce a force along the line joining the dislocation centre, the magnitude of the force can be calculated by

$$F = \frac{\mu b_1 b_2}{2\pi r} \tag{5.32}$$

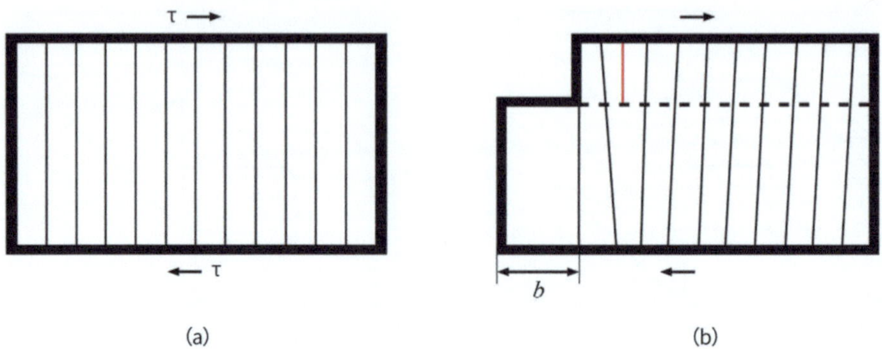

Fig. 5.28 Shearing by the motion of a dislocation through a crystal

The force is repulsive if the two Burgers vectors are in the same direction and is attractive if the two Burgers vectors are opposite. Two edge dislocations with the same sign with parallel Burgers vectors are displayed in Fig. 5.29a; if they have Burgers vectors of magnitude b_1 and b_2, they can create force components F_x and F_y:

$$F_x = \frac{\mu b_1 b_2}{2\pi r}\cos\theta\left(\cos^2\theta - \sin^2\theta\right) \tag{5.33}$$

$$F_y = \frac{\mu b_1 b_2}{2\pi r}\sin\theta\left(1 + \cos^2\theta\right) \tag{5.34}$$

The angle θ of the slip plane to the line joining the dislocations determines the component of force between the dislocations parallel to the slip plane. When 45°

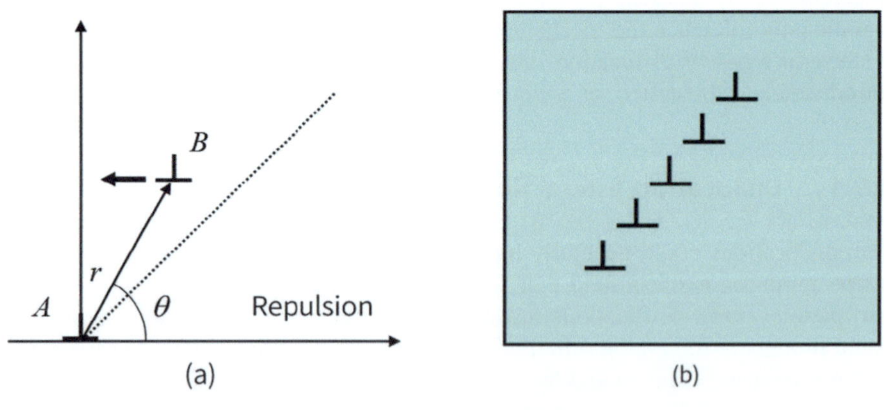

Fig. 5.29 **a** Two edge dislocations of the same sign with parallel Burgers vectors, and **b** the formation of a low-angle grain boundary

$< \theta < 135°$, the dislocations will be attracted to a position directly above each other. A whole series of parallel edge dislocations could lead to the formation of a low-angle grain boundary as shown in Fig. 5.29b.

5.9.2 Dislocation Interactions with Grain Boundaries

Grain boundaries consist of dislocations when the grains are inclined at low angles to each other. When dislocations migrate to a grain boundary, their orientation and type cannot be "lost" without significant atomic rearrangement, and thus they may become arrested at the boundary. New dislocations can be generated through applying continuous strain, and these newly generated dislocations will behave in the same manner as the pre-existing dislocations.

Grain boundaries are highly disordered, which can be regarded to be strained, and it can be considered as a massive dislocation that repels approaching dislocations inside the crystal. As shown in Fig. 5.30, dislocations "pile up" at grain boundaries, causing opposing atomic planes to move steadily in the opposite direction. Eventually, this movement can result in the formation of microscopic cracks on the grain boundaries.

It is known that the presence of grain boundaries leads to the strengthening of polycrystalline materials which is one of the most important strengthening mechanisms. It is the strong interaction between dislocations and grain boundaries during plastic straining that leads to the increment in the material strength, which is grain boundary strengthening. The basic reason is the retard of the dislocation motion by grain boundaries.

5.9.3 Dislocation Interactions with Isolated Obstacles

To strengthen crystalline materials, introducing nanoparticles into the matrix at the nanometer scale is one of the most effective methods. During deformation, stress drives dislocations to these obstacles, and higher stress levels are required to force dislocations to bypass them, inducing plastic deformation and thus strengthening the materials.

If the obstacle is a void, there are two edge dislocation—void interaction mechanisms for mid-plane encounters These mechanisms can be distinguished based on whether the dislocation undergoes a climb process or not when encountering a void, as illustrated in Fig. 5.31. During both mechanisms, the dislocation creates two surface steps with a magnitude of $\boldsymbol{b}$ (Burgers vector) on the incoming and outgoing sides of the void. In Fig. 5.31c1, the dislocation does not climb, and the surface step on the outgoing slip plane forms on the original slip plane. On the other hand, in Fig. 5.31c2, the dislocation climbs by the magnitude of h' relative to the original slip plane by absorbing some vacancies from the void, leaving a surface step on the outgoing slip plane. The occurrence of the situation in Fig. 5.31c2

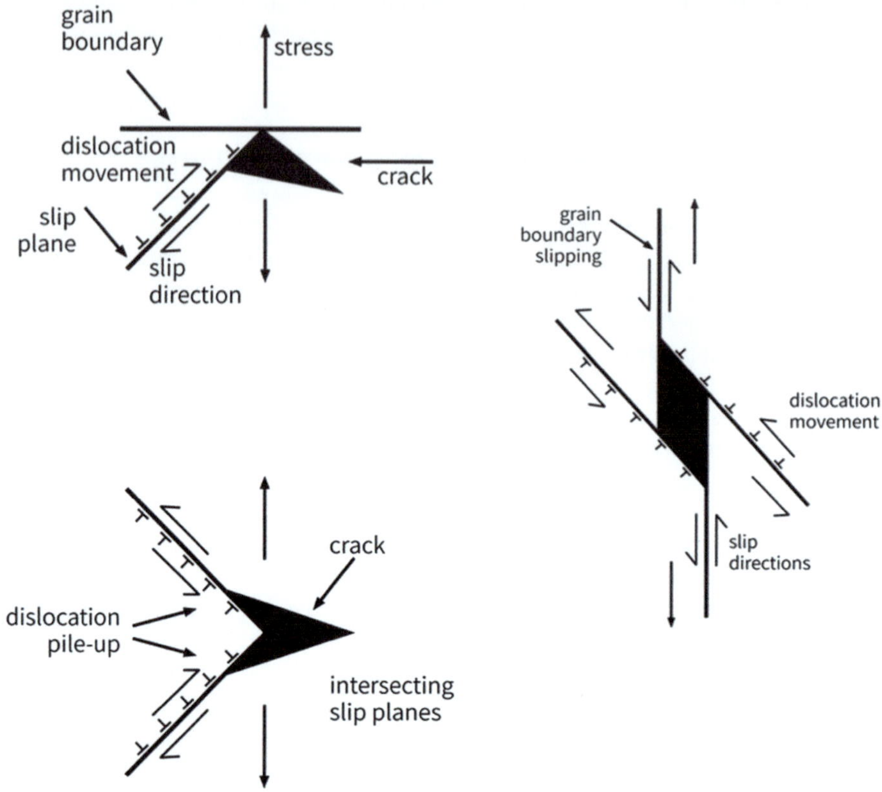

Fig. 5.30 Under stress, dislocations move to grain boundaries and pile up, but the slip thereby generated opens the microscopic cracks which lead to failure

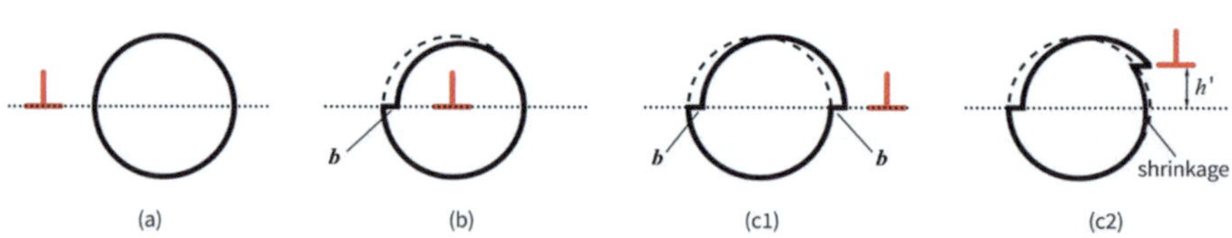

Fig. 5.31 Illustrations of two edge dislocation—void interaction mechanisms, the green dotted line is the original slip plane, and the black solid and blue dashed circles are the current and original void shape, respectively

is due to the formation of jogs along a dislocation line. Consequently, the void shrinks, as shown by the violet area denoted as "shrinkage."

On the other hand, if the obstacle is an impenetrable precipitate other than void, there are two main interaction mechanisms between a dislocation and an impenetrable precipitate: Orowan looping (proposed by Mott and Nabarro and Orowan) and Hirsch looping (proposed by Humphreys and Hirsch), as illustrated

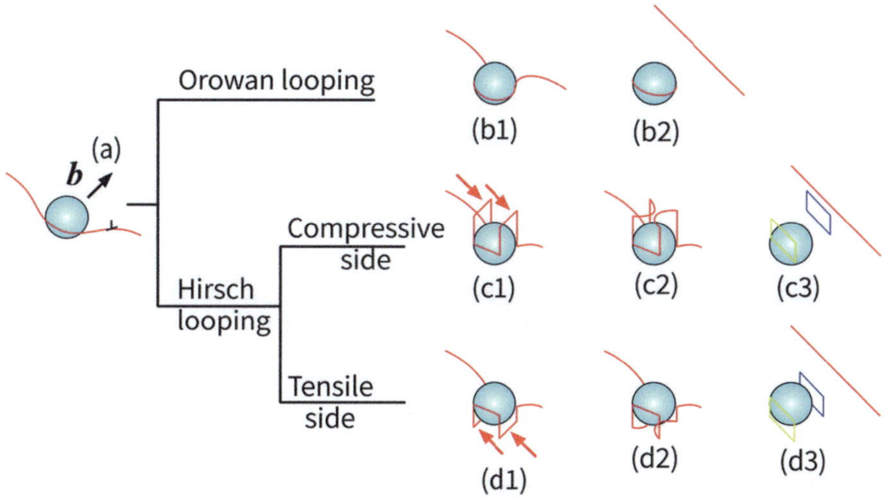

Fig. 5.32 Schematic presentation of two dislocation—impenetrable precipitate interaction mechanisms: Orowan looping and Hirsch looping, where the red curve is the edge dislocation, with its extra half-plane above the slip plane, the compressive and tensile sides of the dislocation stress field lie above and below the slip plane, respectively, and the blue and black square frames represent edge prismatic loops of interstitial and vacancy, respectively

in Fig. 5.32. In Orowan looping, the dislocation bows around the precipitate (Fig. 5.32b1) and a shear loop is then left behind (Fig. 5.32b2) as the dislocation moves forward. In Hirsch looping, the screw components of the dislocation can cross slip either toward the compressive (Fig. 5.32c1) or tensile side (Fig. 5.32d1) of the original slip plane. The two screw dislocation segments having the same Burgers vector but opposite line directions can move together to annihilate each other (Fig. 5.32c2 and d2). From Fig. 5.32c3 and d3, edge prismatic loops of interstitial and vacancy types are formed as the result of Hirsch looping.

5.10 Major Planar Defects

Planar defects play a crucial role in the structural and physical properties of crystalline materials. There are three main types: grain boundaries, twin boundaries, and stacking faults. Their roles in defining the mechanical, thermal, and electrical attributes of materials make them indispensable subjects of analysis. The ability to manipulate these defects presents opportunities for customizing material characteristics for specific applications, from microelectronics to construction materials.

5.10.1 Grain Boundary

Grain boundaries, which are planar defects that separate regions with different crystalline orientations (i.e., grains) within a polycrystalline solid, are a common type of defect. Polycrystalline solids are composed of multiple grains with varying sizes and orientations and lack a periodic arrangement of atoms due to the existence of grain boundaries. The phase transformation of metals begins with small crystals that grow and eventually form a polycrystalline structure. The size of crystallographic grains can range from nanometers to millimeters, and their orientations usually rotate with respect to neighboring grains. Grain boundaries limit the movement of dislocations, impeding dislocation propagation due to the stress field of the grain boundary defect region and the lack of slip planes and slip directions. Thus, reducing crystallite size is a common way to enhance mechanical strength because smaller grains create more obstacles per unit area of slip plane. Therefore, grain size is a critical feature of a metal, and its strength can be controlled by adjusting the grain size. The grain size determines the properties of the metal, where smaller grain sizes increase tensile strength and tend to increase ductility. Figure 5.33 schematically illustrates the presence of grain boundaries between grains in a polycrystalline solid.

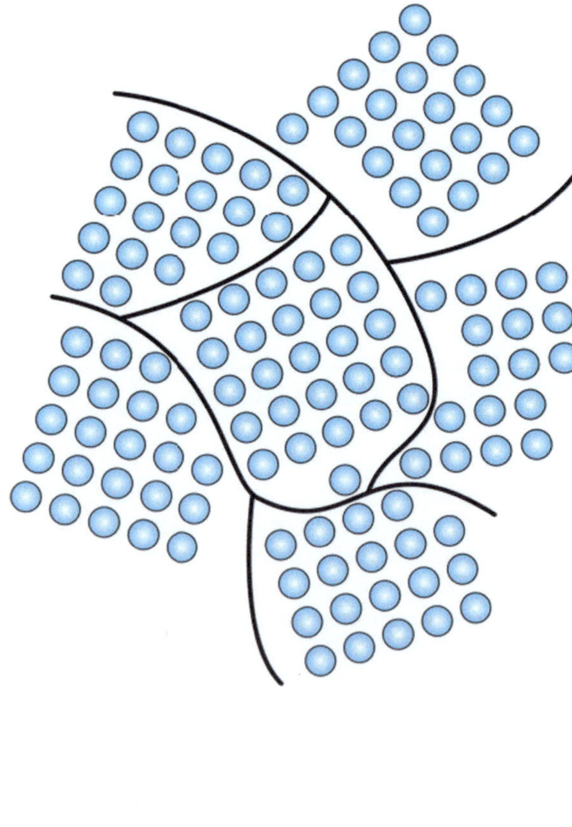

Fig. 5.33 Schematic drawing of grain boundaries

The property of a grain boundary depends on its structure, and this structure mostly depends on the relative orientation of the two grains adjacent to it and the relative orientation of the grain boundary to one of the grains.

① Low Angle Grain Boundary

If the orientation difference between two grains is less than 10°, the grain boundary between them is called a Low-angle grain boundaries (LAGB). Sub-grain boundaries are usually low-angle grain boundaries. There are usually two types of simple low- angle grain boundaries: tilt grain boundaries and twist grain boundaries, the former is composed of edge dislocations, and the latter is composed of screw dislocations. The tilt grain boundaries can be subdivided into symmetric tilt grain boundaries and asymmetric tilt grain boundaries. Specific details will be discussed in the **Interface Chapter**.

② High Angle Grain Boundary

The interface between two grains with an orientation difference greater than 10° is commonly referred to as a High-angle grain boundary (HAGB). Due to its complex structure and irregular atomic arrangement, the dislocation model cannot accurately describe it. Although its structure is not fully understood, it has been suggested that the high-angle grain boundary is similar to the model shown in Fig. 5.34. This model shows that the interface is not a smooth surface, but rather composed of irregular steps, including both a compression zone B and an expansion zone C. This is because the atoms on the grain boundary are influenced by atoms in the two grains with differing orientations. The arrangement of atoms on the high-angle grain boundary is disordered, but also contains some ordered regions, leading to a combination of bad and good regions. As the orientation difference increases, the area of bad regions also increases. The width of high-angle grain boundaries in pure metals is typically no more than 3 atomic distances.

5.10.2 Twin Boundary

A twin refers to a nonparallel regular continuum composed of two or more identical crystals, also known as bicrystal. Between the two monocrystals that constitute twins, there must be some corresponding crystal planes and corresponding crystal edges parallel to each other, but they cannot be all parallel one by one. However, they must be able to coincide with each other or be completely parallel either through a symmetric operation of reflection, rotation by 180° or inversion. Twinning can be regarded as a phenomenon somewhere between a crystallographic defect and a grain boundary. A twin boundary is the common boundary shared by two adjacent twins, as schematically shown in Fig. 5.35.

There are two types of twin boundaries: coherent twin boundaries and incoherent twin boundaries. In the case of a coherent twin boundary, illustrated in Fig. 5.36a, the crystals on both sides take this boundary as the symmetry plane to form a mirror symmetry relationship. The atoms on the boundary are located

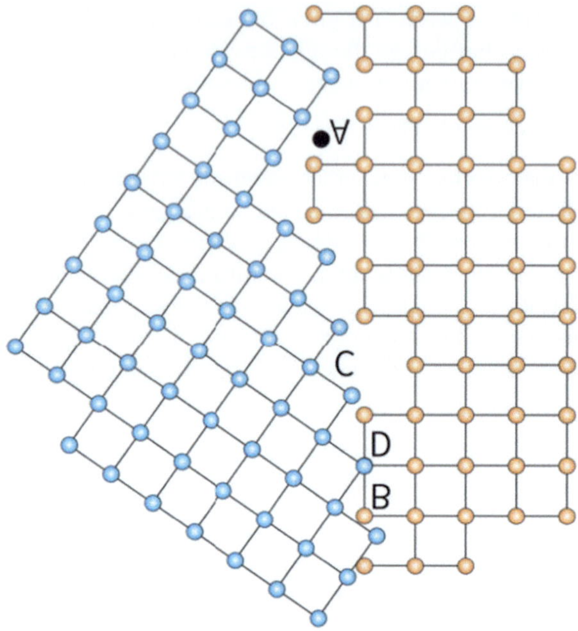

Fig. 5.34 High angle grain boundary model

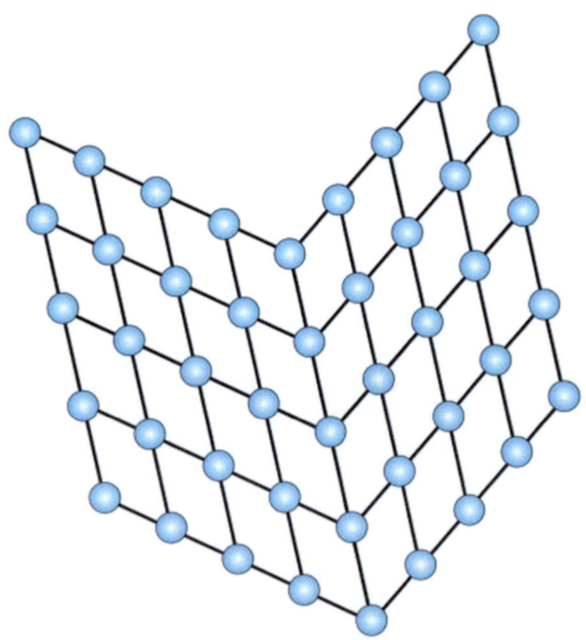

Fig. 5.35 Schematic drawing of a twin boundary

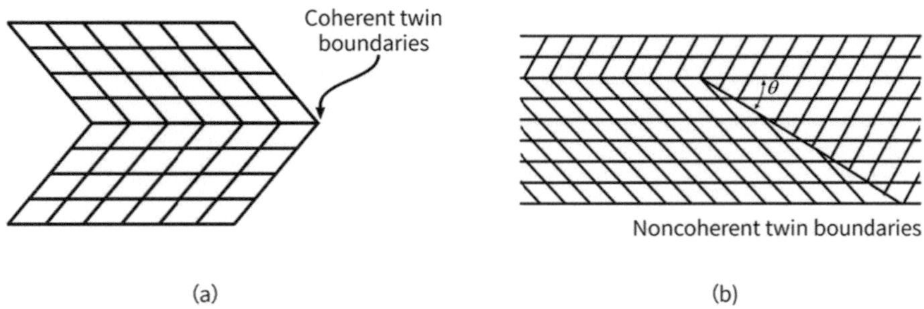

Fig. 5.36 **a** Coherent twin boundaries, and **b** incoherent twin boundaries

at the same positions on the nodes of the two-crystal lattice, which are shared by both crystals, resulting in a completely matched twin plane interface without any distortion. This results in a low energy level and a stable structure.

An incoherent twin boundary is formed when the twin boundary rotates by a certain angle relative to the twin plane. In this case, only a portion of the atoms on the twin boundary are shared by the two crystals, resulting in a severe misarrangement of atoms, as illustrated in Fig. 5.36b. Consequently, the energy of this twin boundary is relatively high.

5.10.3 Stacking Fault

A stacking fault is a type of planar defect that can occur in modular crystal structures, particularly those with layer structures, but not exclusively. In a perfectly periodic sequence of modules, a stacking fault occurs when a module deviates from its expected orientation or position. This results in an out-of-sequence module. Stacking faults can occur in all crystal structures that can be described in terms of layers. These faults can arise in a variety of ways, such as the insertion of an incorrect layer, alteration of the layer sequence, or a different translation between two adjacent layers. Stacking faults can affect either the entire crystal or a specific region within it.

For crystals with ABCABC stacking of the close-packed planes in close-packed structures, there are two possible situations for one layer resting on another. Based on the hard sphere model, a close-packed layer of atoms resting on an A layer could also rest equally well in either a B or C layer, and there is no reason for geometrically selecting a particular position. There are two possible types of stacking faults in a face-centred cubic lattice, which are intrinsic and extrinsic stacking faults. As present in Fig. 5.37, (a) is schematic intrinsic stacking fault and (b) is an extrinsic one, the difference between them is the change in sequence caused by the removal or introduction of an extra layer. The normal stacking sequence of (111) planes in Fig. 5.37 is ABCABC…, and the planes which are in normal relation to one another are separated by Δ, and those with a stacking error by

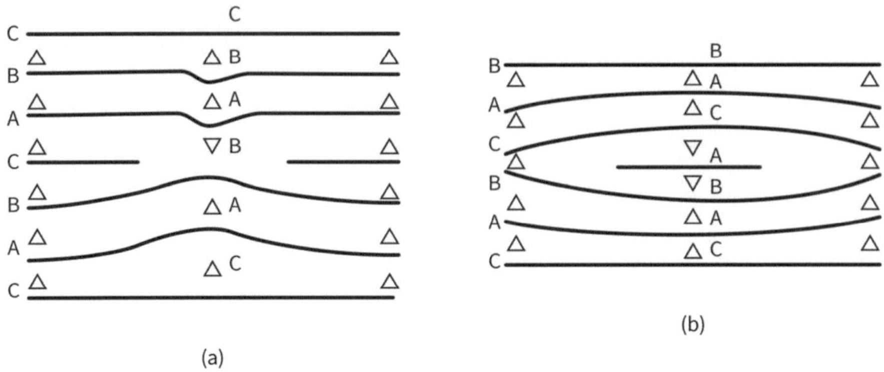

Fig. 5.37 Stacking faults in face centred cubic crystals. **a** Intrinsic stacking fault; **b** extrinsic stacking fault

∇. In Fig. 5.37a, a section of the middle C layer is removed, and this leads to a break in the stacking sequence, such a stacking fault is an intrinsic fault. The stacking sequences below and above the faulted plane are continuous right up to the fault itself. In Fig. 5.37b, an extra A layer has been inserted between a B and a C layer, creating two breaks in the stacking sequence and such a stacking fault is an extrinsic fault. For the extrinsic stacking fault, the inserted layer belongs to none of the continuing patterns of the lattice above or below the fault. In addition, stacking faults are not expected in crystals with ABABAB... stacking sequences in body centred or face centred cubic metals because there is no alternative site for an A layer resting on a B layer.

Stacking faults may form during crystal growth or plastic deformation. For example, the intrinsic stacking fault in the face centred cubic structure can be induced by a sliding process. The presence of stacking faults could significantly influence the plasticity of materials. They can be quantified by the stacking fault energy, which corresponds to the formation enthalpy per unit area. Typical values of stacking fault energy of metals lie in the range 1~1000 mJ/m^2.

5.11 Characterization of Defects and Some Latest Research Progress

5.11.1 Characterization of Defects

The application of electron microscopy in solids has gone through three stages of development: (1) the electron contrast observation of thin crystals emerged in the 1950s and 1960s, especially the observation and analysis of crystal defects; (2) the high-resolution structure images and atomic images of extremely thin crystals was observed for the first time in 1970s; (3) the analytical electron microscopy technology has been developed vigorously after 1980s. These techniques have greatly promoted the development of materials science. The hypothesis of crystals defects was proposed in the early stage to explain certain properties of materials. Due to technical problems, it was difficult to observe these crystal defects directly at that time. It was not until the 1950s that many experimental techniques for directly observing dislocations and other defects appeared, leading to major breakthroughs in the research of crystal defects. The main instruments of defect characterization are optical microscope (OM), scanning electron microscope (SEM), transmission electron microscope (TEM), atom probe tomography (APT), X-ray diffractometer (XRD), etc.

The followings are three commonly used methods for dislocation characterization, including surface method (etch pit), TEM method and XRD method.

① Surface Method

Observation of dislocations can be achieved by using an optical microscope (OM) or scanning electron microscope (SEM) to examine the etching pits that form when the crystal is etched with appropriate chemical or electrolytic etchants, or subjected to heat treatment in vacuum or other atmospheres. These etching pits are formed due to the stress field of dislocations on the crystal surface, or due to the tendency of achieving balance between surface tension and dislocation line tension, which causes the metal to corrode at the position of the dislocation. To verify the relationship between the etch pits and dislocations, thin slices of the single crystal can be produced and corroded to compare the location of the etch pits. If almost identical etch pits are formed on the two opposite surfaces, it indicates that the etch pits are caused by dislocations. Additionally, flat-bottomed pits can form at defects like steps and inclusions, which are easily distinguished from sharp-bottomed pits at dislocation outcrops. Figure 5.38 provides a schematic representation of this method. For example, the morphology of dislocation etch pits on the surface of a Mn-doped yttrium aluminate crystal is shown in Fig. 5.39.

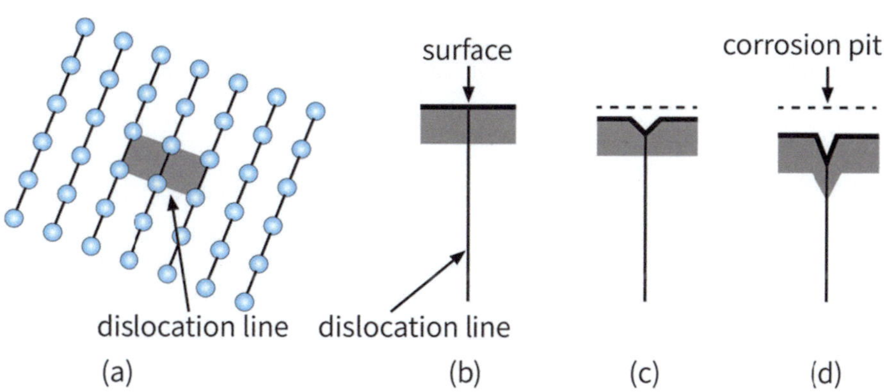

Fig. 5.38 Surface method (etch pit method), **a** the elastic stress field at the outcrop of dislocation, **b** the outcrop of dislocation line, **c** formation of etch pit at the outcrop, and **d** the growth of etch pit

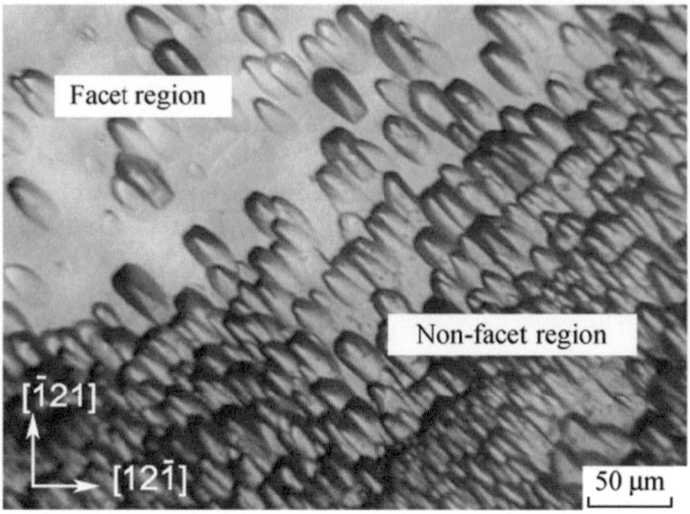

Fig. 5.39 Dislocations in a Mn-doped yttrium aluminate crystal

② TEM Method

High-resolution TEM allows for the direct observation of dislocations in metals. On the other hand, the TEM diffraction method can also be used to observe dislocations in the microstructure of the material.

The initial step in TEM observation involves thinning the sample to a film thickness that allows electron beam transmission. In areas without dislocations, electrons diffract when passing through regularly spaced crystal planes, satisfying Bragg's law with specific diffraction angles, interplanar spacings, and electron wavelengths. However, in regions near the dislocation, lattice distortion causes corresponding changes in the diffraction intensity. Therefore, the image of the region near the dislocation will contrast with the surrounding dislocation-free area, as depicted in Fig. 5.40. Dislocation lines appear as black lines because of electron beam attenuation at the dislocation.

Figures 5.41 and 5.42 show several TEM bright field images of dislocations formed in a high entropy alloy and an aluminum alloy. When observing dislocations with TEM, the magnification is generally between 50,000 and 300,000 times, which is far from the limit of TEM magnification. Some TEMs are also equipped with devices for in-situ heating or deformation of specimens during observation, which can be used to directly observe the real-time movement of dislocations. TEM has now become the most commonly used method for direct defect observations in materials.

③ XRD Method

X-ray analysis of solid materials is an important method for studying crystal defects. It can identify dislocations, stacking faults, grain boundaries, and other

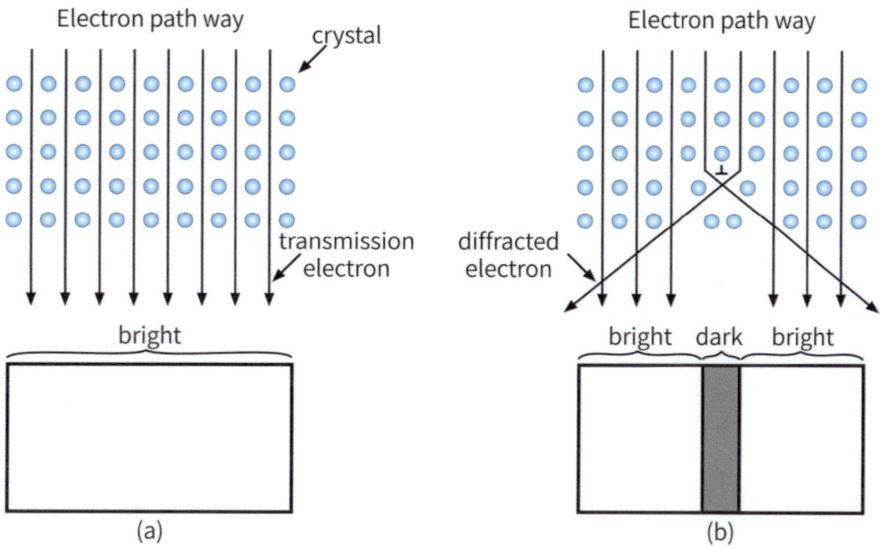

Fig. 5.40 Schematic diagram of the TEM method for dislocation observation

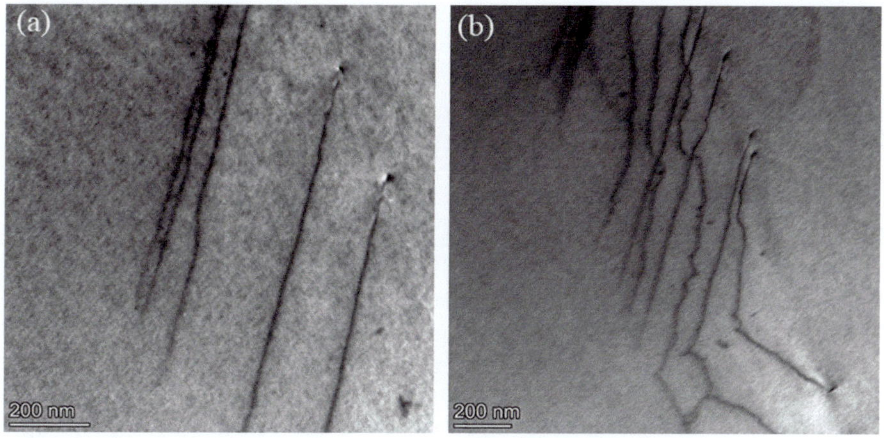

Fig. 5.41 TEM images of dislocations in FCC matrix of $V_{10}Cr_{10}Co_{35}Fe_{45}$ high entropy alloy. (Photographed by Prof. WW Sun, Southeast University, unpublished.)

lattice imperfections in a non-destructive manner. According to X-ray diffraction theory, when X-rays are directed at a perfect crystal, primary extinction occurs due to the interaction between the incident and diffracted waves. However, if there is a defect in the crystal, the normal lattice arrangement is disrupted, causing changes

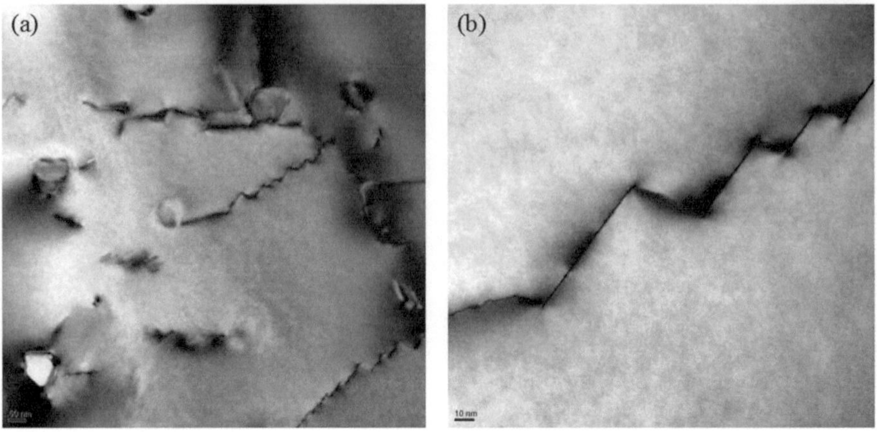

Fig. 5.42 TEM images of dislocations in 2024 commercial aluminium alloy that have been cyclic deformed for 500 cycles at room temperature at the frequency of 0.2 Hz. (Photographed by Prof. WW Sun, Southeast University, unpublished.)

in the lattice plane spacing or local orientation. As a result, the diffraction condition is also altered, and primary extinction is no longer observed. Instead, a kinematic diffraction zone appears.

Field ion microscopy (FIM) and atom probe techniques offer even higher magnification capabilities (typically over 3 million times) than other methods to observe defects. These methods can directly visualize dislocations on material surfaces at the atomic scale.

5.11.2 Research Progress on the High Strength Steels

Strength and ductility are key mechanical properties of metallic materials for developing energy efficient and light weight structural components in a wide variety of industries, including automotive and aerospace. Severe plastic deformation or phase transformation could generate high dislocation densities in metals. These dislocations play an important role on affecting the mechanical properties of super steels. TEM has been used to investigate the motion of dislocations in the deformed and partitioned (D and P) steel. The mobile and immobile dislocation cells in martensite could rearrange to form dislocation cells during deformation. From Fig. 5.43a, the dislocation cells can be observed in martensite. The dislocation cells consist of a cell interior with a low dislocation density and a cell wall with a high dislocation density. Besides, the large austenite grains are decorated with a high number of dislocations as well as stacking faults, whereas the small austenite grains mainly possess dislocations (Fig. 5.43b).

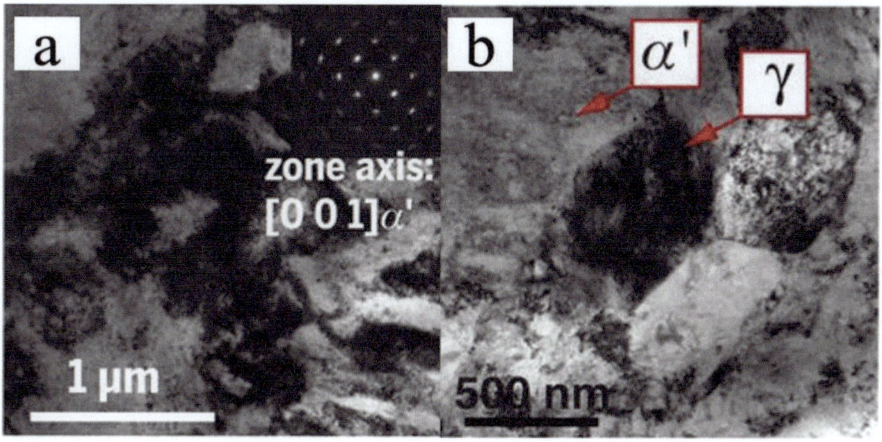

Fig. 5.43 Microstructure of deformed and partitioned steel before the tensile test

5.11.3 Research Progress on the High Entropy Alloys

High-entropy alloys (HEAs), also known as multi-principle element alloys, are materials composed of five or more elements in nearly equiatomic proportions. Due to their unique compositions and chemical structures, they have the potential to achieve exceptional combinations of mechanical properties. Consequently, the investigation of the deformation mechanism of HEAs has become a new area of interest in the field.

The arrangement of atoms during growth or under external forces may deviate from the normal order, resulting in surface defects or stacking faults. These defects have a significant impact on the material's mechanical properties. Stacking faults are usually formed due to the decomposition of full dislocations. Partial dislocations result from the decomposition of full dislocations, and they serve as the boundary between the stacking fault and the crystal around it. Under thermal stress, it is often observed under the electron microscope that full dislocations move and decompose, causing a stacking fault to expand between two partial dislocations, which can be seen as streak contrast. The width between the two partial dislocations, denoted as *d*, represents the width of the extended dislocation. The relationship between the extended dislocation width d and the stacking fault energy γ can be expressed as follows:

$$d = \frac{Gb_1b_2}{8\pi k\gamma} \tag{5.35}$$

where G is shear strain modulus, γ is the energy required to form stacking fault per unit area (stacking fault energy).

It is obvious that the larger the extend dislocation width d, the smaller the stacking fault energy γ.

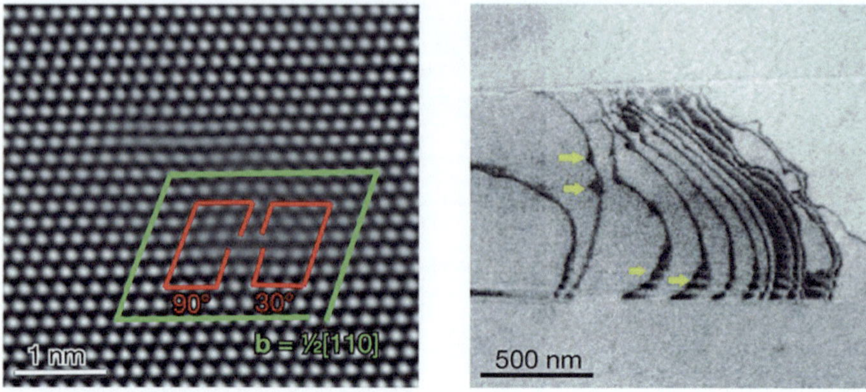

(a) (b)

Fig. 5.44 TEM observation of dislocations in the CrFeCoNiPd HEA

In a coherent structure, an inhomogeneous element distribution inevitably causes the non-uniform distribution of lattice strain due to mismatch of different atoms with various sizes, that is, severe lattice distortion effect. The CrMnFeCoNi HEAs have the low stacking fault energy ($\gamma_{sf} = 30$ mJ·m^{-2}), resulting from 1/2<110>111 full dislocation and 1/2 < 110 > 111 partial dislocations activating at room temperature. The 1/2<110>111 full dislocation is highly active at room temperature in the CrFeCoNiPd HEAs during plastic deformation. Figure 5.44a shows the atomic structure of a full dislocation, with the ***b*** of 1/2[110], in which the full dislocation is dissociated into two partial dislocations. From the measured widths of dislocation core, the γ_{sf} was estimated to be 66 mJ·m^{-2} according to Eq. 5.35. Figure 5.44b shows the sluggish motion of dislocations in a pile-up and the temporary pinning of one of the partials, where the leading dislocation was obstructed by a strong obstacle. Besides, it can be clearly seen the widely separated leading and trailing partials in some of the moving dislocations from the image.

5.12 Questions

Q5.1 Briefly describe the main types of crystal defects.

Q5.2 Which of the following is NOT a type of point defect?

(A) Vacancies (B) Interstitials (C) Dislocations (D) Substitutional impurities

Q5.3 Explain the following basic concepts: vacancy, edge dislocation, screw dislocation, mixed dislocation, Burgers vector, dislocation density.

Q5.4 Point out the types of four dislocations in the figure below, and explain the redundant semi atomic surface of the edge dislocation.

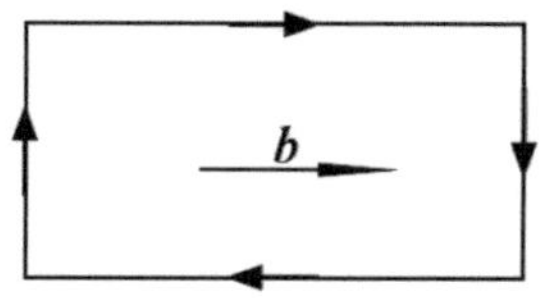

Q5.5 The dislocation capable of cross slip must be ( ).
(A) Edge dislocation (B) Screw dislocation (C) Mixed dislocation

Q5.6 The slip plane of dislocation line is the plane determined by ___and____. Theretically, there are __slip planes for edge dislocation, __slip planes for screw dislocation, both edge and screw dislocations could do __movement, only_____ dislocation could do climbing.

Q5.7 The direction of dislocation motion is perpendicular to the dislocation line everywhere and is variable in the process of motion, the direction of relative sliding of crystals ( ).

(A) changes with the direction of dislocation movement.
(B) is always the direction of the Burger vector.
(C) is always the direction of external force.
(D) is always the direction perpendicular to the Burger vector.

Q5.8 Discuss the Frank-Read mechanism in detail. How does this mechanism contribute to dislocation multiplication?

Q5.9 Using Eq. 5.31, calculate the critical shear stress needed to activate the Frank-Read source given the length of the edge dislocation is 10^{-4} cm and the magnitude of the Burgers vector is approximately 10^{-8} cm.

Q5.10 Discuss the differences and similarities between Low-Angle Grain Boundaries (LAGB) and High-Angle Grain Boundaries (HAGB). What is stacking fault? How does it form?

Q5.11 Describe the roles of grain boundaries, twin boundaries, and stacking faults in determining the mechanical characteristics of polycrystalline materials.

Q5.12 Discuss the methods of defect characterization in crystalline materials, provide their application occasions.

Q5.13 Given the recent advancements in the observation of defects at the atomic scale, discuss the potential future directions and challenges in this area.

Bibliography

Anon (1948) Symposium on internal stresses in metals and alloys. The institute of Metal, London

Darvell BW (2018) Materials science for dentistry. Elsevier, Amsterdam

Ding Q, Zhang Y, Chen X et al (2019) Tuning element distribution, structure and properties by composition in high-entropy alloys. Nature 574(7777):223–227

Feng Z, Fu R, Lin C et al (2020) TEM-based dislocation tomography: challenges and opportunities. Curr Opin Solid State Mater Sci 24(3):100833

He BB, Hu B, Yen HW et al (2017) High dislocation density-induced large ductility in deformed and partitioned steels. Science 357(6355):1029–1032

Humphreys FJ, Hirsch PB (1970) The deformation of single crystals of copper and copper-zinc alloys containing alumina particles-II. Microstructure and dislocation-particle interactions. Proceedings of the Royal Society of London. A. Math Phys Sci 318(1532):73–92

Li HJ, Yang XB, Su LB et al (2008) Growth facets in yttrium aluminum garnet crystals. J Chin Ceram Soc 5

Mott NF, Nabarro FRN (1962) Report of conference on strength of solids. Phys. Soc, London

Ohring M (1995) Engineering materials science. Elsevier, Amsterdam

Diffusion

6

Learning Objectives

1. Differentiate between steady-state and non-steady-state diffusion.
2. Familiarize yourself with Fick's first and second laws of diffusion.
3. Understand the solutions to Fick's second law for diffusion under different conditions.
4. Understand the two atomic mechanisms of diffusion.
5. List the factors that affect diffusion.
6. Identify the types of diffusion.

6.1 Introduction

Diffusion is a physical process where molecules move from an area of high chemical potential to an area of lower chemical potential. An example of diffusion is the scent of Osmanthus fragrans in autumn.

Diffusion is the fundamental method by which particles move in a substance. At temperatures higher than absolute zero (0 K), particles in any system are in thermally induced motion. When there is a chemical potential gradient within a substance, the particles migrate directionally due to thermal motion, which is referred to as diffusion. Therefore, diffusion is a mass transfer process that describes the directional migration of substances on a macroscopic scale. In addition to diffusion, the transfer of substances in gases and liquids can also be accomplished through convection, whereas in solids, diffusion is often the sole method of substance transfer.

Diffusion finds extensive use in advanced industries. For instance, carbon is often diffused into the surface of steel components to enhance their hardness. This

Z. Sun et al., *Fundamentals of Materials Science*,
https://doi.org/10.1007/978-981-96-3904-5_6

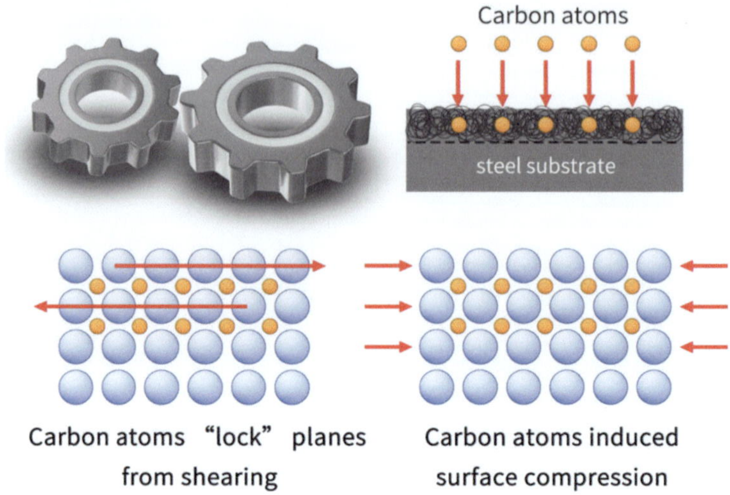

Fig. 6.1 Diffusion of carbon atoms into the steel surface to improve hardness and prevent cracking

is achieved by "locking" planes from shearing through the infiltration of C atoms, thereby rendering the steel difficult to deform. At the same time, C atoms place the surface under compression, making it resistant to cracking, as illustrated in Fig. 6.1. In the process, the steel part is immersed in the carburizing agent, heated, and subjected to an increase in surface carbon content, leading to a concentration difference between the surface and center parts. As a result, carbon on the steel part's surface diffuses into the steel components.

Another example of the applications of diffusion in our lives is in semiconductors. Common semiconductor materials include silicon, germanium, gallium arsenide, etc., which have been widely used in photovoltaics, power conversion, illumination (LED) etc. The conductivity of semiconductors at room temperature lies between that of conductors and insulators and is related to temperature. Silicon, a classic semiconductor, has a conductivity of only 10^{-4} S/m. Doping heteroatoms such as B or P can dramatically enhance the conductivity of Si. Diffusion is a key process to introduce a controlled number of dopants into semiconductors and to alter their properties. There is a monotonic decrease in doping concentration from the surface, whose in-depth distribution can be controlled by the temperature and diffusion time.

Boron is extensively used as a p-type dopant in silicon, while arsenic and phosphorus are common n-type impurities. The high solubility of these three elements (exceeding 5×10^{20} atoms/cm^3 at a diffusion temperature between 800 and 1200 °C) in silicon ensures an effective diffusion process. A series of dopants, such as B, As, P, can be introduced from different sources, including solid sources (BN, As_2O_3, and P_2O_5), liquid sources (BBr_3, $AsCl_3$, and $POCl_3$), and gaseous sources (B_2H_6, AsH_3, and PH_3). For example, the gaseous source is transported to the

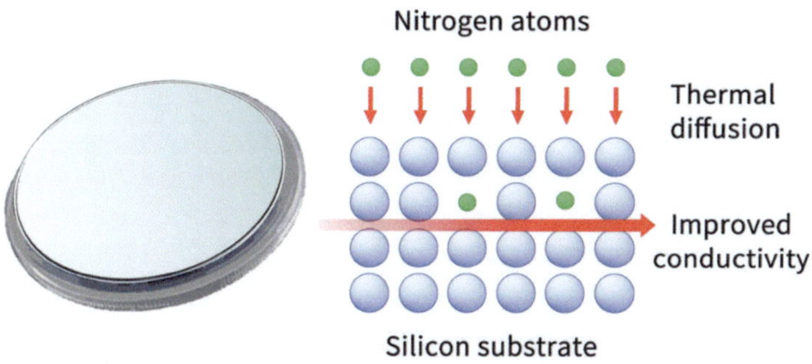

Fig. 6.2 Thermal diffusion of heteroatoms into silicon substrate enhances its conductivity

semiconductor surface by an inert gas (e.g., N_2) and is then reduced at the surface. An effective doping strategy is thermal diffusion technology, as shown in Fig. 6.2: Impurity atoms may enter and occupy vacancies. To produce a large number of lattice vacancies in the crystal, it is necessary to heat the crystal to intensify the thermal motion of the crystal atoms. For Si crystals, the required temperature to form a large number of vacancies is approximately 1000 °C, which is the temperature of thermal diffusion.

6.2 Fick's Laws

Diffusion is a time-dependent process, where the amount of one element diffusing into another is a function of time. To determine the diffusion speed or rate of mass transfer, it is necessary to have the diffusive flux. The diffusive flux, denoted by J, represents the amount of mass (or atoms) M that diffuses perpendicularly through a unit cross-sectional area of a solid per unit time. This can be mathematically expressed as:

$$J = \frac{M}{At} \tag{6.1}$$

where A is the area across which diffusion is occurring and t is the diffusion time.

In a differential form, this expression becomes

$$J = \frac{1}{A}\frac{\mathrm{d}M}{\mathrm{d}t} \tag{6.2}$$

The units for J are kilograms or atoms per square meter per second ($\mathrm{kg/m^2{\cdot}s}$ or $\mathrm{atoms/m^2{\cdot}s}$).

A steady-state condition exists when the diffusion flux J remains constant over time. An example of steady-state diffusion is the diffusion of hydrogen gas through a Pd metal plate, where the concentrations (or pressures) of the diffusing species on both surfaces of the plate remain constant, as shown in Fig. 6.3.

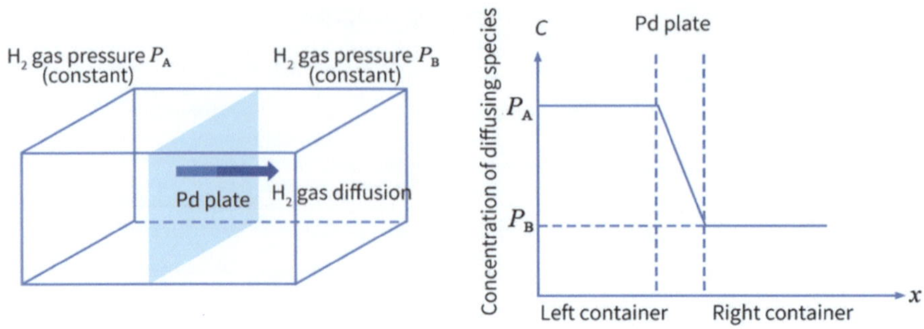

Fig. 6.3 Steady-state diffusion

6.2.1 Fick's First Law

Fick's first Law is used to describe the steady-state diffusion in one direction. At a point of time in diffusion, the diffusion flux J is proportional to the negative value of the concentration gradient dC/dx (Eq. 6.3 or Eq. 6.4), where D represents the **diffusion coefficient**.

$$J = -D\frac{\mathrm{d}C}{\mathrm{d}x} \tag{6.3}$$

$$J\left(\frac{atoms}{m^2 s}\right) = -D\left(\frac{m^2}{s}\right)\frac{\mathrm{d}C}{\mathrm{d}x}\left(\frac{atoms}{m^3} \times \frac{1}{m}\right) \tag{6.4}$$

Here, the net flow of atoms is equal to the diffusion coefficient D times the diffusion gradient $\frac{\mathrm{d}C}{\mathrm{d}x}$. For gas phase, D is constant at a fixed temperature, while in other circumstance, it is variable.

6.2.2 Fick's Second Law

In most practical diffusion situations, the process is non-steady. The diffusion flux and the concentration gradient vary with time, with a net accumulation or depletion of the diffusing species. Figure 6.4 shows the movement in the direction x, taking a volume with a cross-sectional area of unit and a length of Δx, and letting the flux in and out of this volume be J_1 and J_2. For mass balance, we can get Eq. 6.5.

$$J_1 - J_2 = -\frac{\partial J}{\partial x}\Delta x \tag{6.5}$$

According to the definition of flux, M is the mass diffusing through and perpendicular to a unit cross-sectional area of solid per unit of time. ΔM is the amount

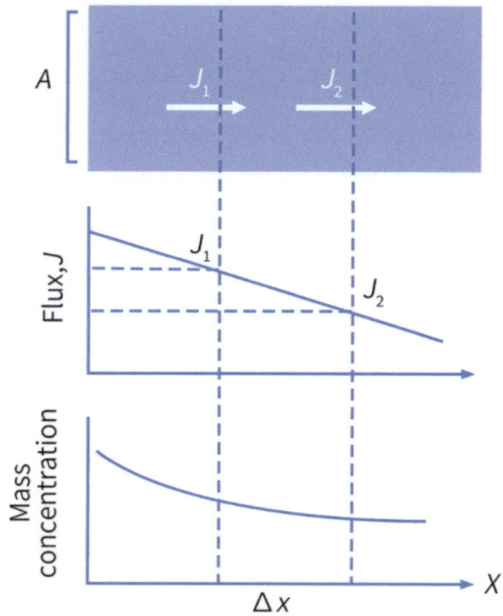

Fig. 6.4 Diffusion flux and mass concentration change with distance

of accumulated diffusion substance across an area of A during a certain amount of time Δt and can be expressed by Eq. 6.6.

$$\Delta M = (J_1 A - J_2 A)\Delta t \tag{6.6}$$

Rearrange:

$$\frac{\Delta M}{\Delta x A \Delta t} = \frac{J_1 - J_2}{\Delta x}$$

When Δx, $\Delta t \to 0$

$$\frac{\partial C}{\partial t} = -\frac{\partial J}{\partial x} \tag{6.7}$$

Merging Eq. 6.7 into Fick's first law Eqs. 6.1–6.4, Fick's second law can be derived as Eq. 6.8.

Fick's second law:

$$\frac{\partial C}{\partial t} = \frac{\partial}{\partial x}\left(D\frac{\partial C}{\partial x}\right) \tag{6.8}$$

Assuming D is independent of concentration, then.

$$\frac{\partial C}{\partial t} = D\frac{\partial^2 C}{\partial x^2} \tag{6.9}$$

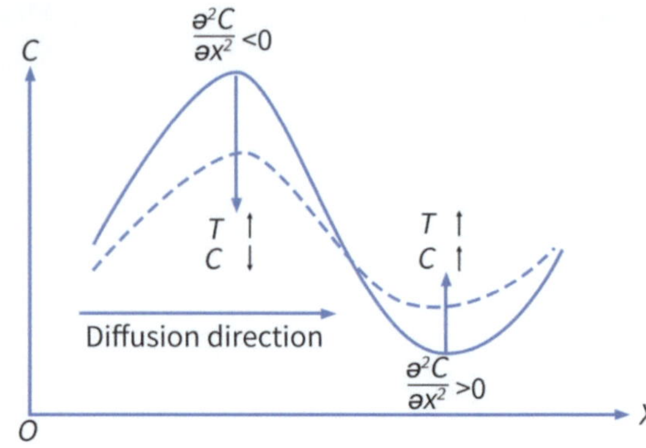

Fig. 6.5 Relationship between Fick's first and second law

Fick's second law states that the rate of concentration change over time at a certain point in the diffusion process is proportional to the second derivative of the concentration distribution curve at that point. As shown in Fig. 6.5, if the second derivative $\frac{\partial^2 C}{\partial x^2}$ of the curve at this point is greater than 0, that is, the curve is concave, the concentration of this point will increase with time, that is, $\frac{\partial C}{\partial t} > 0$; if the second derivative of the curve at this point is less than 0, that is, the curve is convex, the concentration of this point will decrease with the increase of time and $\frac{\partial C}{\partial t} < 0$. Fick's First Law states that diffusion occurs in the direction of decreasing concentration. Consequently, Fick's first law and second law essentially indicate that the process of diffusion always converts an inhomogeneous system to a homogeneous system, eventually reaching an equilibrium.

In three dimensions, assuming that the diffusion coefficient is isotropic, the general formula of Fick's second law is:

$$\frac{\partial C}{\partial t} = D\left[\frac{\partial^2 C}{\partial x^2} + \frac{\partial^2 C}{\partial y^2} + \frac{\partial^2 C}{\partial z^2}\right] \tag{6.10}$$

In a cylindrical coordinate system

$$\begin{cases} x = r\cos\theta \\ y = r\sin\theta \end{cases}$$

Each volume elements are ∂r, $r\partial\theta$, ∂z, and Eq. 6.10 can be transformed as:

$$\frac{\partial C}{\partial t} = \frac{1}{r}\left\{\frac{\partial}{\partial r}\left(rD\frac{\partial C}{\partial r}\right) + \frac{\partial}{\partial\theta}\left(\frac{D}{r}\frac{\partial C}{\partial\theta}\right) + \frac{\partial}{\partial z}\left(rD\frac{\partial C}{\partial z}\right)\right\} \tag{6.11}$$

If the diffusion is in cylindrical symmetry, and D is independent with concentration:

$$\frac{\partial C}{\partial t} = \frac{D}{r}\left[\frac{\partial}{\partial r}\left(r\frac{\partial C}{\partial r}\right)\right] \tag{6.12}$$

In spherical coordinates

$$\begin{cases} x = rsin\,\theta\,cos\,\varphi \\ y = rsin\,\theta\,sin\,\varphi \\ z = rcos\,\theta \end{cases}$$

Each volume elements are ∂r, $r\partial\theta$, $r\sin\theta\partial\varphi$, then

$$\frac{\partial C}{\partial t} = \frac{1}{r^2}\left\{\frac{\partial}{\partial r}\left(Dr^2\frac{\partial C}{\partial r}\right) + \frac{1}{\sin\theta}\frac{\partial}{\partial\theta}\left(D\sin\theta\frac{C}{\partial\theta}\right) + \frac{D}{\sin^2\theta}\frac{\partial C}{\partial\varphi^2}\right\} \tag{6.13}$$

If the diffusion is in spherical symmetry, and D is independent with concentration, then.

$$\frac{\partial C}{\partial t} = \frac{D}{r^2}\frac{\partial}{\partial r}\left(r^2\frac{\partial C}{\partial r}\right) \tag{6.14}$$

6.2.3 Solution to Fick's Second Law and Their Application

For unsteady diffusion, Fick's second law needs to be solved based on the initial and boundary conditions. Here are some simple and practical solutions.

① One-Dimensional Diffusion with Infinite Length

To simplify the calculation, diffusion is restricted to one dimension. Bar A, which has a mass concentration of C_2, and bar B, which has a mass concentration of C_1, are joined together by welding the surface perpendicular to the x-axis, as illustrated in Fig. 6.6. By holding such a diffusion couple at a certain temperature for a period of time, the mass concentration near the welding surface ($x = 0$) is altered. Mass flows from bar A to bar B, resulting in a decreased mass concentration on the left of the welding surface ($x < 0$) and an increased mass concentration on the right ($x > 0$). If the diffusion couple has infinite length, both ends of the bars will maintain their original concentration.

Since the mass flow follows Fick's second law, initial and boundary conditions are figured out to solve the equation of diffusion.

Initial conditions:

$$t = 0\begin{cases} x > 0, C = C_1 \\ x < 0, C = C_2 \end{cases} \tag{6.15}$$

The initial condition states that at time zero, the original concentration on two sides of the interface ($x = 0$) is C_1 and C_2, respectively.

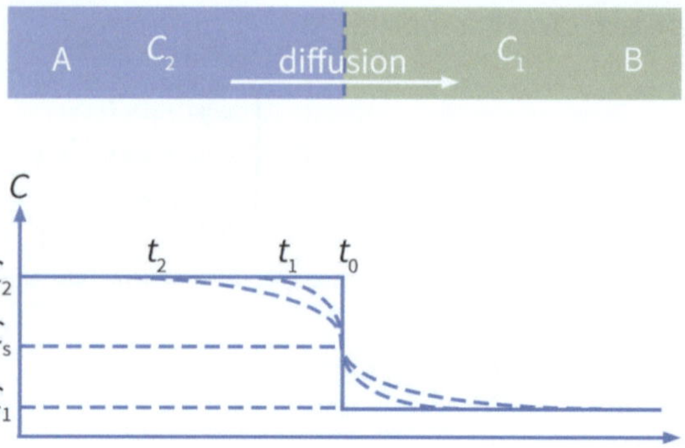

Fig. 6.6 Mass concentration as a function of time in a diffusion coupe with infinite length

The ***Boundary conditions*** are:

$$t \geq 0 \begin{cases} x = +\infty, C = C_1 \\ x = -\infty, C = C_2 \end{cases} \tag{6.16}$$

The boundary condition states that both ends of the diffusion couple always maintain the original concentration in such an infinite-length diffusion couple.

Substitution of intermediate variables makes partial differential equations into ordinary differential equations, by assuming $\beta = \frac{x}{2\sqrt{Dt}}$.

Thus $\frac{\partial C}{\partial t}$ and $\frac{\partial^2 C}{\partial x^2}$ can be expressed in a form of β as follows:

$$\frac{\partial C}{\partial t} = \frac{\mathrm{d}C}{\mathrm{d}\beta}\frac{\partial \beta}{\partial t} = -\frac{\beta}{2t}\frac{\partial C}{\partial t}$$

$$\frac{\partial^2 C}{\partial x^2} = \frac{\partial^2 C}{\partial \beta^2}\left(\frac{\partial \beta}{\partial x}\right)^2 = \frac{\partial^2 C}{\partial \beta^2}\frac{1}{4Dt} = \frac{\mathrm{d}^2 C}{\mathrm{d}\beta^2}\frac{1}{4Dt}$$

Substitute the above two equations into Fick's second law, we can get:

$$-\frac{\beta}{2t}\frac{\partial C}{\partial t} = D\frac{1}{4Dt}\frac{\mathrm{d}^2 C}{\mathrm{d}\beta^2}$$

$$\frac{\mathrm{d}^2 C}{\mathrm{d}\beta^2} + 2\beta\frac{\mathrm{d}C}{\mathrm{d}\beta} = 0$$

$$\frac{\mathrm{d}C}{\mathrm{d}\beta} = A_1 \exp\left(-\beta^2\right)$$

Therefore, ***C* is solved as a function of *β* in** Eq. 6.17, a part of which is described by the Gauss error function erf(β), an integration of Gaussian probability density function as illustrated in Fig. 6.7. A_1 and A_2 are constants that can be derived by boundary conditions later.

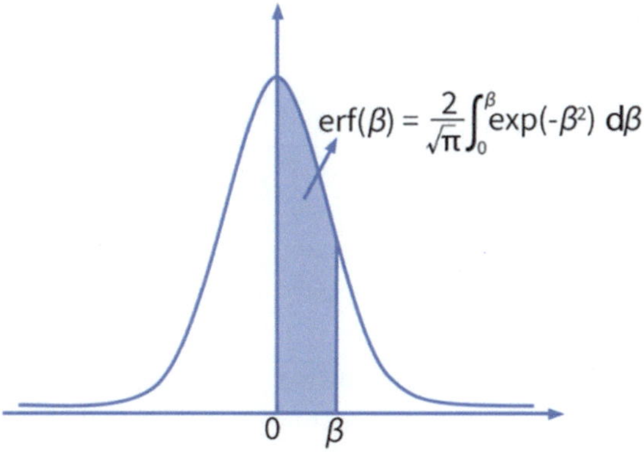

Fig. 6.7 Gauss error function as in integration of Gaussian probability density function

$$C = A_1 \int_0^{\beta} \exp(-\beta^2)\mathrm{d}\beta + A_2 \tag{6.17}$$

$$\mathrm{erf}(\beta) = \frac{2}{\sqrt{\pi}} \int_0^{\beta} \exp(-\beta^2)\mathrm{d}\beta$$

It is therefore proved that:

$$\mathrm{erf}(\infty) = 1, \mathrm{erf}(-\beta) = -\mathrm{erf}(\beta)$$

The error function values corresponding to different β values are shown in Table 6.1.

Since

$$\int_0^{\infty} \exp(-\beta^2)\mathrm{d}\beta = \frac{\sqrt{\pi}}{2}$$

$$\int_0^{-\infty} \exp(-\beta^2)\mathrm{d}\beta = -\frac{\sqrt{\pi}}{2}$$

Based on boundary conditions:

$$C_1 = A_1 \int_0^{\infty} \exp(-\beta^2)\mathrm{d}\beta + A_2$$

Table 6.1 Values of Gauss error function erf(β)

β	0	1	2	3	4	5	6	7	8	9
0.0	0.0000	0.0113	0.0226	0.0338	0.0451	0.0564	0.0676	0.0789	0.0901	0.1013
0.1	0.1125	0.1236	0.1348	0.1459	0.1569	0.1680	0.1790	0.1900	0.2009	0.2118
0.2	0.2227	0.2335	0.2443	0.2550	0.2657	0.1680	0.1790	0.1900	0.2009	0.2118
0.3	0.3286	0.3389	0.3491	0.3593	0.3694	0.3794	0.3893	0.3992	0.4090	0.4187
0.4	0.4284	0.4380	0.4475	0.4569	0.4662	0.4755	0.4847	0.4937	0.5027	0.5117
0.5	0.5205	0.5292	0.5379	0.5465	0.5549	0.5633	0.5716	0.5798	0.5879	0.5959
0.6	0.6039	0.6117	0.6194	0.6270	0.6346	0.6420	0.6494	0.6566	0.6638	0.6708
0.7	0.6778	0.6847	0.6914	0.6981	0.7047	0.7112	0.7175	0.7238	0.7300	0.7361
0.8	0.7421	0.7480	0.7538	0.7595	0.7651	0.7707	0.7761	0.7814	0.7867	0.7918
0.9	0.7969	0.8019	0.8068	0.8116	0.8163	0.8209	0.8254	0.8299	0.8342	0.8385
1.0	0.8427	0.8468	0.8508	0.8548	0.8586	0.8624	0.8661	0.8698	0.8733	0.8768
1.1	0.8802	0.8835	0.8868	0.8900	0.8931	0.8961	0.8991	0.9020	0.9048	0.9076
1.2	0.9103	0.9130	0.9155	0.9181	0.9205	0.9229	0.9252	0.9275	0.9297	0.9319
1.3	0.9340	0.9361	0.9381	0.9400	0.9419	0.9438	0.9456	0.9473	0.9490	0.9507
1.4	0.9523	0.9539	0.9381	0.9400	0.9419	0.9438	0.9456	0.9473	0.9490	0.9507
1.5	0.9661	0.9673	0.9687	0.9695	0.9706	0.9716	0.9726	0.9736	0.9745	0.9755
β	1.55	1.6	1.65	1.7	1.75	1.8	1.9	2.0	2.2	2.7
erf(β)	0.9716	0.9763	0.9804	0.9838	0.9867	0.9891	0.9928	0.9953	0.9981	0.9999

$$C_2 = A_1 \int_0^{-\infty} \exp(-\beta^2)\mathrm{d}\beta + A_2$$

The values of A_1 and A_2 can be derived as:

$$A_1 = \frac{C_1 - C_2}{2}\frac{2}{\sqrt{\pi}}$$

$$A_2 = \frac{C_1 + C_2}{2}$$

The analytical formula of mass concentration C varying with distance x and time t is

$$C(x,t) = \frac{C_1 + C_2}{2} + \frac{C_1 - C_2}{2}\frac{2}{\sqrt{\pi}}\int_0^{\beta} \exp\left(-\beta^2\right)\mathrm{d}\beta = \frac{C_1 + C_2}{2} + \frac{C_1 - C_2}{2}\mathrm{erf}\left(\frac{x}{2\sqrt{Dt}}\right) \tag{6.18}$$

$$C(x,t) = \frac{C_1 + C_2}{2} + \frac{C_1 - C_2}{2}\frac{2}{\sqrt{\pi}}\int_0^{\beta}\exp(-\beta^2)\mathrm{d}\beta = \frac{C_1 + C_2}{2} + \frac{C_1 - C_2}{2}\mathrm{erf}\left(\frac{x}{2\sqrt{Dt}}\right) \tag{6.19}$$

At the boundary $(x = 0), \mathrm{erf}(0) = 0$, thus

$$C_s = \frac{C_1 + C_2}{2}$$

That is, the mass concentration C_s on the interface always remains unchanged due to the assumption that the diffusion coefficient is independent of concentration. As a result, concentration decrease on the left side of the interface is symmetrical to the concentration increase on the right.

② One-dimensional Diffusion with Constant-surface-concentration

In other situations, such as modifying the semiconductor properties of silicon by doping it with heteroatoms to enhance its conductivity, gas sources are used to carry out diffusion. In such a condition, impurity atoms from the vapor source are distributed onto the semiconductor's surface and then diffuse into the semiconductor wafer. The impurity atom concentration remains constant throughout the diffusion process. To obtain a solution for the diffusion equation, initial and boundary conditions are required, such as those for one-dimensional constant-surface-concentration diffusion.

Initial condition:

$$t = 0, x \geq 0, C = C_0 \tag{6.20}$$

The initial condition states that at time zero, the original concentration in the semiconductor wafer is C_0.

Boundary condition:

$$t \geq 0\begin{cases} x = 0, C = C_s \\ x = +\infty, C = C_0 \end{cases} \tag{6.21}$$

The boundary condition states that the concentration of vapor source remained to be C_s regardless of time, and the semiconductor wafer is so long that the end maintains the original concentration of C_0.

The concentration as a function of distance x and time t is derived as:

$$C(x,t) = C_s - (C_s - C_0)\mathrm{erf}\left(\frac{x}{2\sqrt{Dt}}\right) \tag{6.22}$$

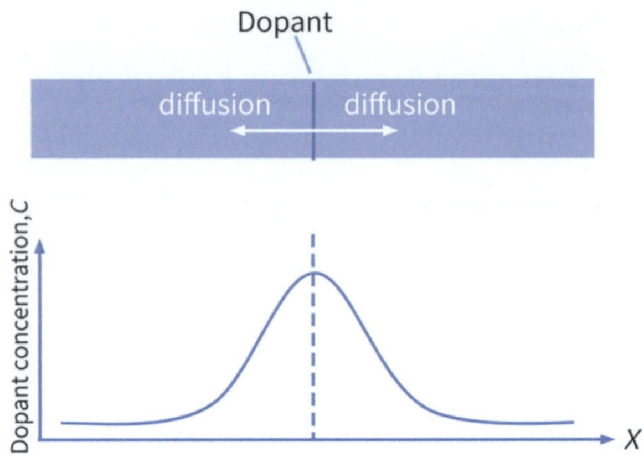

Fig. 6.8 Schematic of a constant-total-dopant diffusion

Assuming the original impurity concentration in semiconductor wafer is zero ($C_0 = 0$), the concentration expression can be simplified as:

$$C(x,t) = C_S\left[1 - \operatorname{erf}\left(\frac{x}{2\sqrt{Dt}}\right)\right] \tag{6.23}$$

③ One-dimensional Diffusion with Limited Source

In this situation, a set amount of dopant is placed between two semiconductor rods by depositing it on the surface of one of the semiconductors. The dopant is then diffused into the semiconductors as illustrated by Fig. 6.8.

Initial condition:

$$t = 0\begin{cases} x = 0, C = C_0 \\ x \neq 0, C = 0 \end{cases} \tag{6.24}$$

The initial condition states that the concentration of solute mass will be symmetrically distributed around the origin, and when $t = 0$, where $|x| > 0$, the mass concentration C are all zero.

Boundary condition:

$$t \geq 0, x = \pm\infty, C = 0 \tag{6.25}$$

Solving the differential equation shows that:

$$C(x,t) = \frac{k}{\sqrt{t}}\exp\left(-\frac{x^2}{4Dt}\right) \tag{6.26}$$

where k is the undetermined constant.

Assuming that the mass of the diffused material is M, the cross-sectional area of the rod is the unit area

$$M = \int_0^t \int_{-\infty}^{\infty} C(x, t)\mathrm{d}x\mathrm{d}t = \int_{-\infty}^{\infty} C(x, 0)\mathrm{d}x \tag{6.27}$$

If $\frac{x^2}{4Dt} = \beta^2$

$$\mathrm{d}x = 2\sqrt{Dt}\mathrm{d}\beta$$

$$M = 2k\sqrt{D} \int_{-\infty}^{\infty} \exp(-\beta^2)\mathrm{d}\beta = 2k\sqrt{\pi D}$$

$$\int_0^{\infty} \exp(-\beta^2)\mathrm{d}\beta = \frac{\sqrt{\pi}}{2}$$

$$\int_0^{-\infty} \exp(-\beta^2)\mathrm{d}\beta = -\frac{\sqrt{\pi}}{2}$$

$$k = \frac{M}{2\sqrt{\pi D}}$$

$$C(x, t) = \frac{M}{2\sqrt{\pi Dt}}\exp\left(-\frac{x^2}{4Dt}\right) \tag{6.28}$$

Equation 6.28 describes a Gaussian distribution of concentration at a certain point of time. The concentration of the attenuated film source after diffusion varies with distance. Changing the time also changes the shape of curves in Fig. 6.9 (numbers indicate different Dt values, Dt = 1/16, 1/4, 1).

If the diffusion is deposited on one end of the metal rod, after diffusion annealing, the mass of the diffusion material (Eq. 6.29) is twice of the above diffusion

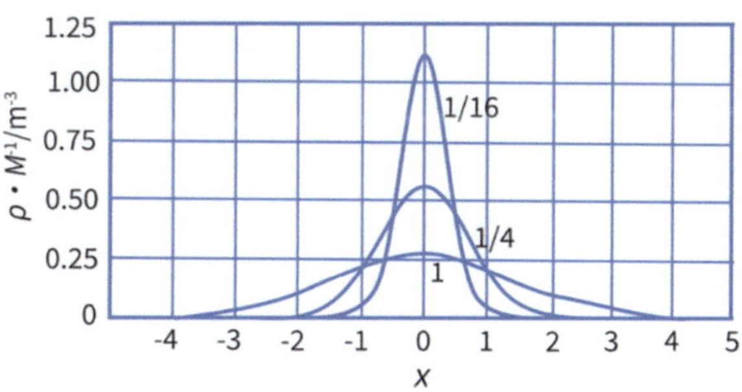

Fig. 6.9 Attenuated film source after diffusion

couple (Eq. 6.28). Because the diffusion material changes from both sides to only one side.

$$C = \frac{M}{\sqrt{\pi D t}} \exp\left(\frac{-x^2}{4Dt}\right) \tag{6.29}$$

④ The Boltzmann-Matano Method

In previous conditions, diffusion coefficient D has been considered as a constant to simplify the calculation. In real conditions, diffusion coefficient D varies with the compositions. The dependence of D on compositions follows Fick's second law.

In a one-dimensional diffusion with infinite length.

Initial condition:

$$t = 0, \begin{cases} C|_{x>0} = C_1 \\ C|_{x<0} = C_2 \end{cases} \tag{6.30}$$

Boundary condition:

$$\left.\frac{dC}{dx}\right|_{x=\pm\infty} = 0 \tag{6.31}$$

This is a non-homogeneous differential equation, which makes solving it extremely complicated. In diffusion metallurgy research, the Boltzmann-Matano analysis is commonly used to calculate D from a concentration profile. This method is based on Boltzmann's solution of Fick's second law in 1884 and was first proposed by Matano in 1933. After preparing and annealing the diffusion couple at a fixed temperature within time t, the concentration-penetration curve $c(x, t)$ is experimentally determined. As the composite curve may not be symmetrical, the error function solution cannot be applied. A solution to Eq. 6.8 may be obtained by applying the so-called λ substitution.

$$\lambda = \frac{x}{\sqrt{t}}$$

which can convert the partial differential Eq. 6.8 into a total differential equation relating C and the variable λ. This transformation results in:

$$\frac{\partial C}{\partial t} = \frac{dC}{d\lambda}\frac{\partial \lambda}{\partial t} = -\frac{\lambda}{2t}\frac{dC}{d\lambda}$$

$$\frac{\partial C}{\partial x} = \frac{dC}{d\lambda}\frac{\partial \lambda}{\partial x} = \frac{1}{\sqrt{t}}\frac{dC}{d\lambda}$$

$$\frac{\partial}{\partial x}\left(D\frac{\partial C}{\partial x}\right) = \frac{1}{\sqrt{t}}\frac{d}{d\lambda}\left(D\frac{dC}{d\lambda}\right)\frac{\partial \lambda}{\partial x} = \frac{1}{t}\frac{d}{d\lambda}\left(D\frac{dC}{d\lambda}\right)$$

Rearrange Eq. 6.8

$$-\frac{\lambda}{2}\frac{\mathrm{d}C}{\mathrm{d}\lambda} = \frac{\mathrm{d}}{\mathrm{d}\lambda}\left(D\frac{\mathrm{d}C}{\mathrm{d}\lambda}\right)$$

$$-\frac{\lambda}{2}\mathrm{d}C = \mathrm{d}\left(D\frac{\mathrm{d}C}{\mathrm{d}\lambda}\right)$$

Integrating dC from C_1 to C:

$$-\frac{1}{2}\int_{C_1}^{C} \lambda \mathrm{d}C = \int_{C_1}^{C} \mathrm{d}\left(D\frac{\mathrm{d}C}{\mathrm{d}\lambda}\right)$$

$$\left(D\frac{\mathrm{d}C}{\mathrm{d}\lambda}\right) = \left(D\frac{\mathrm{d}C}{\mathrm{d}x}\frac{\mathrm{d}x}{\mathrm{d}\lambda}\right) = \left(D\sqrt{t}\frac{\mathrm{d}C}{\mathrm{d}x}\right)$$

It can be rearranged as:

$$-\frac{1}{2}\frac{1}{\sqrt{t}}\int_{C_1}^{C} x\mathrm{d}C = \sqrt{t}\int_{C_1}^{C} \mathrm{d}\left(D\frac{\mathrm{d}C}{\mathrm{d}x}\right)$$

Since

$$\left.\frac{\mathrm{d}C}{\mathrm{d}x}\right|_{C=C_1} = 0$$

$$\frac{1}{2t}\int_{C_1}^{C} x\mathrm{d}C = \left(D\frac{\mathrm{d}C}{\mathrm{d}x}\right)_C - \left(D\frac{\mathrm{d}C}{\mathrm{d}x}\right)_{C=C_1} = \left(D\frac{\mathrm{d}C}{\mathrm{d}x}\right)_C$$

Finally, **the relationship between diffusion coefficient and concentration is derived**:

$$D(C) = -\frac{1}{2t}\left(\frac{\mathrm{d}x}{\mathrm{d}C}\right)_C \int_{C_1}^{C} x\mathrm{d}C \tag{6.32}$$

This equation requires a proper evaluation of $x = 0$ corresponding to the origin interface. The plane at which $x = 0$, so-called Matano interface, is determined by finding the value of x in the diffusion zone (x_0) such that

$$\int_{C_1}^{C_2} \lambda \mathrm{d}C = \frac{1}{\sqrt{t}}\int_{C_1}^{C_2} x\mathrm{d}C = 0 \tag{6.33}$$

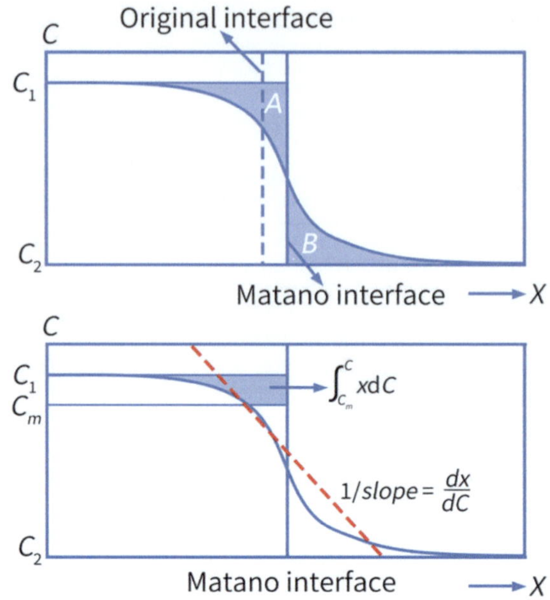

Fig. 6.10 Typical composition profile of an annealed diffusion couple. The Matano interface is positioned to make the hatched areas, and the tangent values employed to calculate the diffusion coefficient at a composition $C = C_{\mathrm{m}}$

$$\left(\int_{C_1}^{C_2} x\mathrm{d}C = \int_{C}^{C_2} x\mathrm{d}C + \int_{C_1}^{C} x\mathrm{d}C = 0\right)$$

Graphically, the Matano interface is the line that makes the two hatched areas A and B in Fig. 6.10 equal.

Practically, to derive the diffusion coefficient at concentration C_{m} from an experimental diffusion curve, one can identify the Matano interface first. The reciprocal of tangent of the curve at $C = C_{\mathrm{m}}$ is $\frac{\mathrm{d}x}{\mathrm{d}C}$. And the area enveloped by $C = C_1$, $C = C_{\mathrm{m}}$ and Matano interface is $\int_{C_{\mathrm{m}}}^{C} x\mathrm{d}C$. Consequently, the diffusion coefficient could be deduced using Eq. 6.32.

The Matano-Boltzmann method is widely used in metallurgical diffusion research as it is very useful for determining the diffusion coefficient (D) of various components in alloys. With the rapid development of electronic analysis technology and computers, this process can now be highly automated, making it even more accessible.

6.3 Theory and Mechanism of Diffusion

The first and second laws of diffusion, along with their solutions under various conditions, reflect the macroscopic laws governing atomic diffusion. These laws form the basis for solving practical problems related to diffusion. The diffusion coefficient is a crucial parameter for measuring an atom's ability to diffuse, but its exact value is still unknown. To determine the diffusion coefficient, we must establish its relationship with other macroscopic and microscopic quantities related to diffusion, which is an important aspect of diffusion theory. In fact, the macroscopic diffusion phenomenon is the statistical result of the irregular hopping of a large number of atoms in the microscopic view. Starting from the microscopic hopping of atoms, the main content of this section is to study the atomic theory of diffusion, the microscopic mechanism of diffusion, and the connection between microscopic theory and macroscopic phenomena.

6.3.1 Atomic Hopping and Diffusion Distance

The parabolic law of diffusion distance and time, derived from the second diffusion equation, shows that atoms in a crystal do not move in a straight line when they hop, but rather along a broken line, similar to Brownian motion. Atoms in a crystal vibrate in place until their energy is high enough to hop to an adjacent position. Since each atom's hopping direction and distance can vary, it is convenient to represent their movement using a displacement vector. If the atom hops *n* times in time (t), the displacement vector of each jump is $\boldsymbol{r}$, then the atom starts from the starting point and reaches the end point after n random hops, The net displacement vector $\boldsymbol{R_n}$ should be the sum of each displacement vector as described by Eq. 6.34 and illustrated in Fig. 6.11.

$$-\boldsymbol{R}_n = \boldsymbol{r}_1 + \boldsymbol{r}_2 + \boldsymbol{r}_3 + \cdots + \boldsymbol{r}_n = \sum_{i=1}^{n} \boldsymbol{r}_i \tag{6.34}$$

When the atom hops, it can hop forward and backward. If the probability of hopping in either direction is equal, the resulting displacement vectors along this axis will cancel each other out. To avoid this, dot product operation in mathematics is applied as shown below.

$$R_n{}^2 = \boldsymbol{R}_n \cdot \boldsymbol{R}_n$$

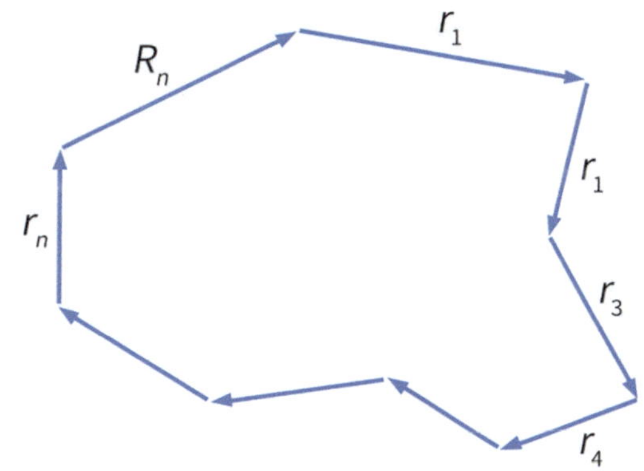

Fig. 6.11 Random walk of atoms

$$
\begin{aligned}
&= \boldsymbol{r}_1 \cdot \boldsymbol{r}_1 + \boldsymbol{r}_1 \cdot \boldsymbol{r}_2 + \boldsymbol{r}_1 \cdot \boldsymbol{r}_3 + \cdots + \boldsymbol{r}_1 \cdot \boldsymbol{r}_n + \\
&\boldsymbol{r}_2 \cdot \boldsymbol{r}_1 + \boldsymbol{r}_2 \cdot \boldsymbol{r}_2 + \boldsymbol{r}_2 \cdot \boldsymbol{r}_3 + \cdots + \boldsymbol{r}_2 \cdot \boldsymbol{r}_n + \cdots + \\
&\boldsymbol{r}_n \cdot \boldsymbol{r}_1 + \boldsymbol{r}_n \cdot \boldsymbol{r}_2 + \boldsymbol{r}_n \cdot \boldsymbol{r}_3 + \cdots + \boldsymbol{r}_n \cdot \boldsymbol{r}_n
\end{aligned} \tag{6.35}
$$

$$
\begin{aligned}
R_n^2 &= \sum_{i=1}^{n} \boldsymbol{r}_i \cdot \boldsymbol{r}_i + 2\sum_{i=1}^{n-1} \boldsymbol{r}_i \cdot \boldsymbol{r}_{i+1} + 2\sum_{i=1}^{n-2} \boldsymbol{r}_i \cdot \boldsymbol{r}_{i+2} + \cdots + 2\sum_{i=1}^{1} \boldsymbol{r}_i \cdot \boldsymbol{r}_{i+n-1} \\
&= \sum_{i=1}^{n} r_i^2 + 2\sum_{j=1}^{n-1}\sum_{i=1}^{n-j} \boldsymbol{r}_i \cdot \boldsymbol{r}_{i+j}
\end{aligned} \tag{6.36}
$$

Since $r_{i+j} = |\boldsymbol{r}_i||\boldsymbol{r}_{i+j}|\cos\theta_{i,i+j}$, $\theta_{i,i+j}$ is the angle between two vectors, so we have:

$$
R_n^2 = \sum_{i=1}^{n} r_i^2 + 2\sum_{j=1}^{n-1}\sum_{i=1}^{n-j} |\boldsymbol{r}_i| \cdot |\boldsymbol{r}_{i+j}| \cos\theta_{i,i+j} \tag{6.37}
$$

In a cubic symmetric crystal, all transition vectors can be assumed to have equal magnitudes.

$$
R_n^2 = nr^2 + 2r^2 \sum_{j=1}^{n-1}\sum_{i=1}^{n-j} \cos\theta_{i,i+j} \tag{6.38}
$$

This refers to the distance between an atom and the origin after n transitions, but diffusion is the result of a large number of atoms after n transitions, so the average value of a large number of atoms after n transitions $\overline{R_n^2}$ is:

$$
-\overline{R_n^2} = nr^2 \left(1 + \overline{\frac{2}{n} \sum_{j=1}^{n-1}\sum_{i=1}^{n-j} \cos\theta_{i,i+j}} \right) \tag{6.39}
$$

If an atom hops an infinite number of times with equal probabilities of hopping forward and backward, then the sum in the above formula (Eq. 6.39) is zero.

So,

$$
\overline{R_n^2} = nr^2 \tag{6.40}
$$

The average diffusion distance is:

$$
\sqrt{\overline{R_n^2}} = \sqrt{n}r \tag{6.41}
$$

Given an atom's hopping frequency Γ (number of hops per unit time), we can calculate Γ by determining the amount of time the atom remains in an equilibrium position, τ seconds, and taking the reciprocal of τ. This gives us the equation $\Gamma = 1/\tau$. We can then use this value of Γ to calculate the number of hops n that the atom makes in a given amount of time, t, using the equation $n = \Gamma$t. Substituting the above formula (Eq. 6.41) to get:

$$\sqrt{\overline{R_n^2}} = \sqrt{\Gamma t} \cdot r \tag{6.42}$$

The formula above establishes a relationship between the macroscopic displacement of diffusion and microscopic quantities such as the hopping frequency and hopping distance of an atom. It shows that the relationship between diffusion distance and time, derived from the microscopic theory of the atom, follows a parabolic law.

6.3.2 Atomic Hopping and Diffusion Coefficient

Based on the analysis above, the microscopic hopping of a large number of atoms determines the macroscopic diffusion distance. This diffusion distance is related to the diffusion coefficient of the atom, establishing an inherent connection between atomic hopping and the diffusion coefficient.

As depicted in Fig. 6.12, due to the irregularity of atomic jumping, solute atoms can jump between crystal planes N_1 and N_2 in either direction. In a solid solution with uniform concentration, the number of solute atoms jumping from plane N_1 to plane N_2 is equal to the number jumping from plane N_2 to plane N_1, resulting in no macroscopic diffusion. However, in a solution with non-uniform concentrations, the solute atoms jump more frequently in one direction than the other, causing a net flow of atoms.

In Fig. 6.12, the areal densities of solute atoms at crystal plane N_1 and crystal plane N_2 are n_1 and n_2, respectively. The distance between the two surfaces is d, and the hopping frequency of the atom is Γ. The probability of atoms jumping from plane N_2 to plane N_1 and from plane N_1 to plane N_2 is P. For example, in a simple cubic crystal, atoms can jump in six directions, but the probability of jumping only in the positive direction of the x-axis is $P = 1/6$. It is assumed here that the probability of the atom hopping in the forward and reverse directions is the same.

In Δt time, the number of solute atoms jumping from crystal plane N_1 to crystal plane N_2 or from crystal plane N_2 to crystal plane N_1 per unit area are respectively:

$$N_{1\to 2} = n_1 P \Gamma \Delta t$$

$$N_{2\to 1} = n_2 P \Gamma \Delta t$$

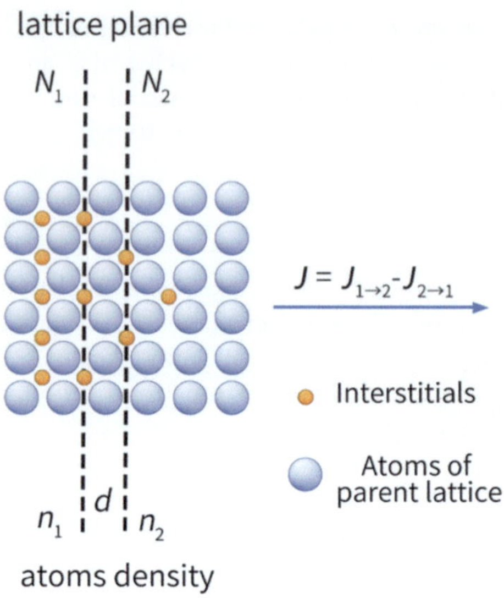

Fig. 6.12 Atom flux from crystal plane N_1 to N_2 and N_2 to N_1

If $n_1 > n_2$, the number of atoms jumping from plane N_1 to plane N_2 is greater than the number of atoms jumping from plane N_2 to plane N_1, resulting in a net transfer of solute atoms

$$N_{1\to2} - N_{2\to1} = (n_1 - n_2)P\Gamma\Delta t \tag{6.43}$$

According to the definition of diffusion flux J, we can get:

$$J = (n_1 - n_2)P\Gamma \tag{6.44}$$

Now the areal density of solute atoms is converted into volume concentration, and the volume concentrations of solute atoms at crystal plane N_1 and crystal plane N_2 are respectively C_1 and C_2.

$$C_1 = \frac{n_1}{1 \times d} = \frac{n_1}{d}$$

$$C_2 = \frac{n_2}{1 \times d} = C_1 + \frac{\partial C}{\partial x}d$$

If the concentration changes over unit distance is $\partial C/\partial x$, then the concentration change over distance d is $(\partial C/\partial x)d$. In fact, C_2 is expanded according to the Taylor series, and only the first-order derivative terms are obtained.

Therefore,

$$n_1 - n_2 = -\frac{\partial C}{\partial x} \cdot d^2$$

$$J = -d^2 P\Gamma \frac{\partial C}{\partial x} \tag{6.45}$$

Compared with the first equation of diffusion, the diffusion coefficient of the atom is:

$$D = d^2 P\Gamma \tag{6.46}$$

d and P depend on the type of crystal structure. In addition to the crystal structure, Γ is closely related to temperature. Its significance lies in establishing the relationship between the diffusion coefficient and the hopping frequency of the atom, the probability of hopping, and the geometric parameters of the crystal.

Substituting Eq. 6.46 into Eq. 6.42,

$$\sqrt{R_n^2} = \frac{r}{d\sqrt{P}}\sqrt{Dt} = K\sqrt{\mathrm{Dt}} \tag{6.47}$$

In the above formula, r is the jumping distance of the atom; d is the distance between adjacent parallel product planes perpendicular to the diffusion direction, that is, the projection value of r in the diffusion direction; K depends on the geometric factor of the structure. The above formula shows that the relationship between atomic diffusion distance and time.

Interstitial Diffusion in Face-centered Cubic and Body-centered Cubic Crystal Taking face-centered cubic and body-centered cubic interstitial solid solutions as examples, the calculation of the jump probability P in Eq. 6.46 is described below. Due to the different structures of the two crystals, the type, number, and distribution of the interstitial sites are also different, which will affect the probability of the hopping of the interstitial atoms. In the face-centered cubic structure in Fig. 6.13a, there are 12 adjacent octahedrons around each interstitial atom, giving it an interstitial coordination number of 12. Since the radius of the interstitial atoms is much larger than the inscribed radius of the interstitial sites, a large elastic distortion will occur in the lattice, and the equilibrium concentration of the interstitial solid solution is very low, so it can be considered that the 12 interstitial sites around the interstitial atoms are empty. When the interstitial atoms located at the body center of crystal plane 1 jump along the y-axis to crystal plane 2, there are 4 possible interstitial sites on crystal plane 2 to jump into, and the jumping probability $P = \frac{4}{12} = \frac{1}{3}$. At the same time, $d = \frac{a}{2}$, and a is the lattice constant. Substituting these parameters into Eq. 6.46, the diffusion coefficient of interstitial atoms in the face-centered cubic structure becomes:

$$D = d^2 P\Gamma = \frac{1}{12}a^2\Gamma \tag{6.48}$$

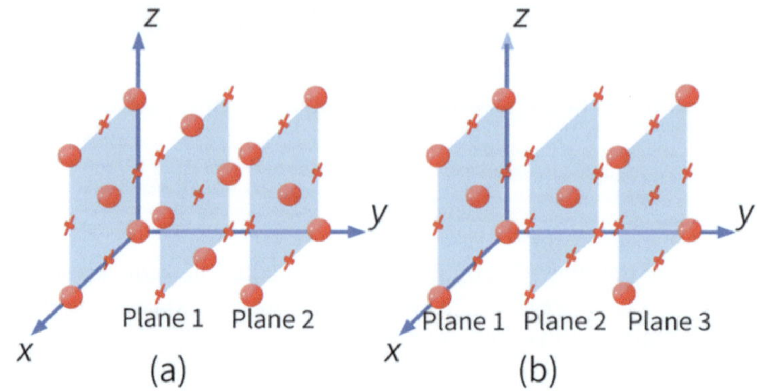

Fig. 6.13 Interstitial diffusion in **a** Face-centered cubic and **b** Body-centered cubic crystal

In the body-centered cubic structure in Fig. 6.13b, the interstitial coordination number is 4. The surrounding environment of each interstitial atom may differ because the interstitial octahedron is asymmetric. When interstitial atoms hop from crystal plane 1 to crystal plane 2, there are two different interstitial sites on crystal plane 1 to consider. If an atom is located at the interstitial sites in the center of the edge and jumps along the y-axis to crystal plane 2, only one of the interstitial sites may be reached. The probability is $\frac{1}{4}$, there are $4 \times \frac{1}{4} = 1$ such interstitial sites on crystal plane 1; if the atom is in the interstitial sites of the face center, when hopping to crystal plane 2, there is no interstitial sites for hopping. The probability is $\frac{0}{4} = 0$. There are $1 \times \frac{1}{2} = \frac{1}{2}$ of such interstitial sites on crystal plane 1. Therefore, the hopping probability is the weighted average of the hopping probability of the interstitial atoms at different positions, that is, $P = \frac{\left(4\times\frac{1}{4}\times\frac{1}{4}+1\times\frac{1}{2}\times 0\right)}{\frac{3}{2}} = \frac{1}{6}$. If the interstitial atom jumps from plane 2 to plane 3, the calculated P value is the same. Similarly, substituting $P = \frac{1}{6}$ and $d = \frac{a}{2}$ into Eq. 6.46, the diffusion coefficient of interstitial atoms in the body-centered cubic structure is:

$$D = d^2 P \Gamma = \frac{1}{24} a^2 \Gamma \tag{6.49}$$

For different crystal structures, the diffusion coefficient can be written in general form

$$D = \delta a^2 \Gamma \tag{6.50}$$

δ is the geometric factor related to the crystal structure; a is the character constant.

6.3.3 Diffusion Mechanism

Diffusion in a solid can be classified into two types: self-diffusion and interdiffusion. Self-diffusion occurs when the chemical potential gradient is zero. If the crystal structure of the pure component is uniform, there will be no concentration gradient, and the net diffusion flux should be zero. In a diffusion annealing experiment where a metal surface is coated with a radioactive isotope, it is observed that the isotope self-diffuses into the metal without depending on a concentration gradient. Figure 6.14 illustrates the self-diffusion of the atoms in the crystal.

Interdiffusion is the process of diffusional exchange of atoms between two materials in contact. This process is driven by the chemical potential gradient across the boundaries. When two materials are tightly joined, the atoms of one material will transfer to the other through diffusion as illustrated in Fig. 6.15.

From an atomic perspective, diffusion refers to the migration of atoms within the lattice. For an atom to move, it must have sufficient energy to overcome the current arrangement of atoms, which typically occurs through defects in the lattice. Although the crystal structure is known for its orderly arrangement of atoms or ions, it is challenging to maintain strict periodicity in practice. During thermal fluctuations, some atoms or ions leave their lattice positions and move into interstitial positions within the crystal lattice or on the crystal surface due to

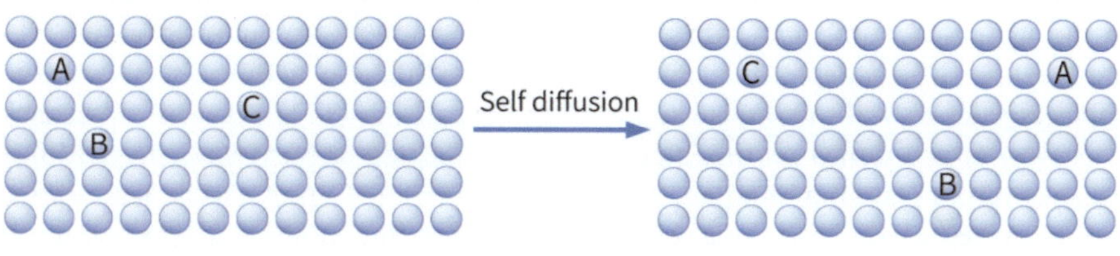

Fig. 6.14 Self diffusion in a solid

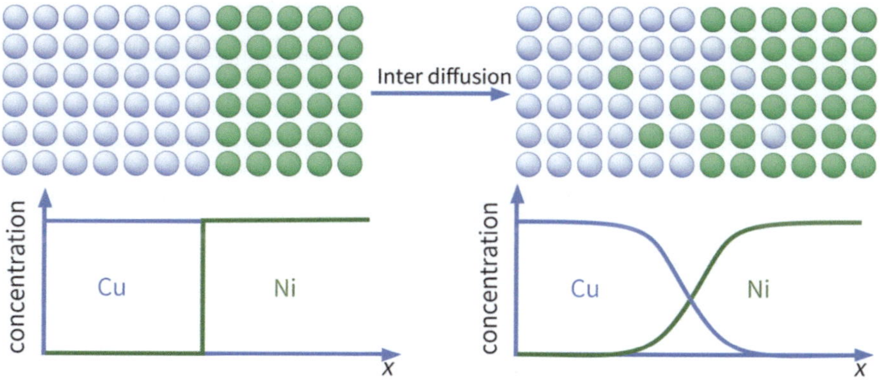

Fig. 6.15 Interdiffusion in a solid

vibration, leaving behind vacancies in the lattice. These atoms in the interstitial positions or the vacancies left on the original lattice points are not permanently fixed and can regain energy from thermal fluctuations, allowing them to change positions within the crystal structure. The mechanism of interdiffusion is divided into vacancy diffusion and interstitial diffusion, which are discussed below.

Vacancy diffusion

At high temperatures, many atoms in the semiconductor can easily move out of their original lattice position, creating vacancies that impurity atoms can fill. Impurity atoms can diffuse through these vacancies and occupy these lattice positions. For example, impurity atoms can replace the silicon atoms in a pure silicon crystal by substitutional diffusion. In detail, impurity atoms move to adjacent positions through vacancies and replace the original silicon atoms. The displacement diffusion mechanism is applicable to most common diffusion species, such as boron, phosphorus, and arsenic. The atomic size of these impurities is too large to squeeze into the voids. Therefore, replacing Si atoms is the only way for them to enter the silicon crystal. Diffusion rate depends on vacancy number and activation energy; more vacancies and lower activation energy lead to a larger diffusion coefficient.

Vacancy diffusion is directly related to the vacancy concentration in the crystal. At any given temperature, there will always be a certain number of vacancies present in the crystal, with higher temperatures resulting in a greater number of vacancies. As a result, atomic diffusion is facilitated at higher temperatures due to the increased number of vacancies that can be found around each atom. Figures 6.16 and 6.17 depict the process of atomic diffusion within a face-centered cubic crystal, as well as the atomic arrangement of the (111) plane. If a vacancy is present at position 0 on this plane, the atom located at position 2 may fill this vacancy through atomic jumping. The diagram in Fig. 6.17 provides a clearer illustration of the relative positional changes of surrounding atoms during hopping. If the atomic diameter is d, the gap between 1 and 3 atoms can be calculated to be $0.73d$. When an atom with diameter d moves through a $0.73d$ gap, it requires enough energy to overcome the obstacles of neighboring atoms and cause local lattice distortion. The denser the crystal structure or the larger the diffused atoms, the more significant the lattice distortion and the higher the activation energy for diffusion. In diffusion through vacancies, energy is required for atoms to jump over the free energy barrier and for vacancies to form, resulting in a higher activation energy for vacancy diffusion compared to interstitial diffusion. In elements like germanium and silicon with a diamond cubic crystal structure that has a lot of free space, atoms may be forced into interstitial positions during diffusion.

Interstitial diffusion

In the interstitial diffusion mechanism, an interstitial atom moves through the crystal lattice jumping from one interstitial site to the next adjacent site. Small interstitial atoms, such as hydrogen, carbon, and nitrogen, can easily diffuse in the crystal in this way.

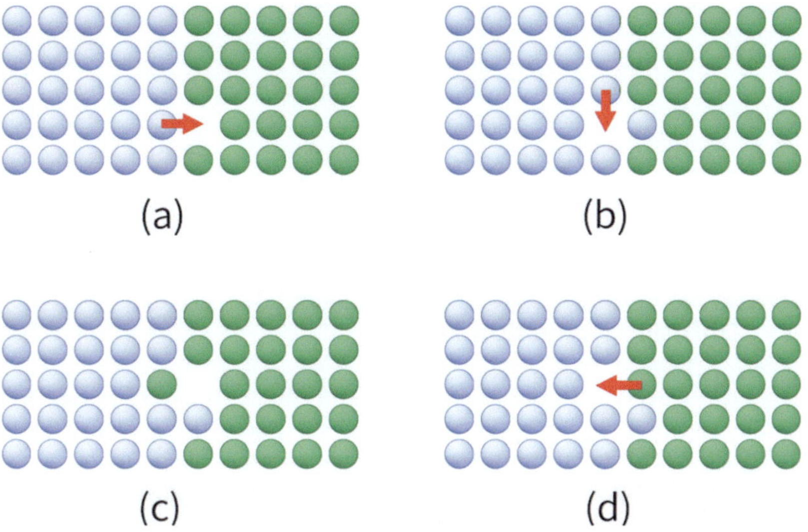

Fig. 6.16 Vacancy diffusion mechanism

Fig. 6.17 Vacancy diffusion in a Face-Center-Cubic (FCC) crystal

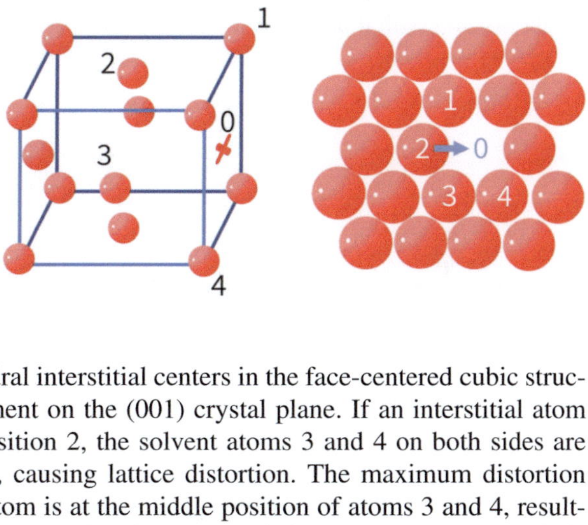

Figure 6.18 shows octahedral interstitial centers in the face-centered cubic structure and the atomic arrangement on the (001) crystal plane. If an interstitial atom moves from position 1 to position 2, the solvent atoms 3 and 4 on both sides are pushed away simultaneously, causing lattice distortion. The maximum distortion occurs when the interstitial atom is at the middle position of atoms 3 and 4, resulting in the highest distortion energy. The free energy is lowest when the atom is in the equilibrium position at the center of the interstitial sites, such as positions 1 and 2. When the atom is in the middle of two adjacent interstitial sites, the free energy is highest, resulting in a free energy barrier that the atom must overcome, known as the diffusion activation energy. The diffusion activation energy is the resistance to the diffusion of atoms. Diffusion can only occur when the free energy of atoms is higher than the activation energy of diffusion. Since interstitial atoms are small, the activation energy of interstitial diffusion is low, making diffusion easier.

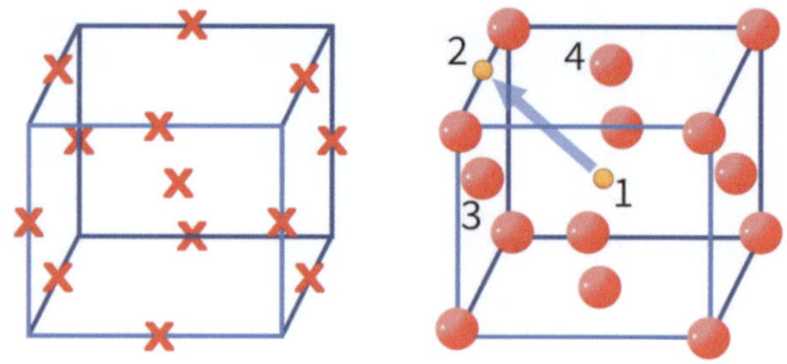

Fig. 6.18 Interstitial diffusion in a Face-Center-Cubic (FCC) crystal

6.3.4 Diffusion Activation Energy

The atoms in the crystal must have enough extra energy to jump from its original equilibrium position. The process of diffusing the atoms to overcome this extra energy is called activation, and this extra energy barrier is called activation energy. Taking interstitial atoms as an example, as shown in Fig. 6.19, the interstitial atoms located at positions 1 and 3 are at the lowest point of potential energy G_1. $G_2 \sim G_1$ is the activation energy.

According to statistical mechanics, the free energy distribution of atoms obeys the Maxwell–Boltzmann distribution when the temperature is T. Suppose the total number of interstitial atoms is N and the number of atoms with molar free energy $G > G_2$ is n_2, then:

$$\frac{n_2(G \geq G_2)}{N} = \mathrm{e}^{-G_2/RT} \tag{6.51}$$

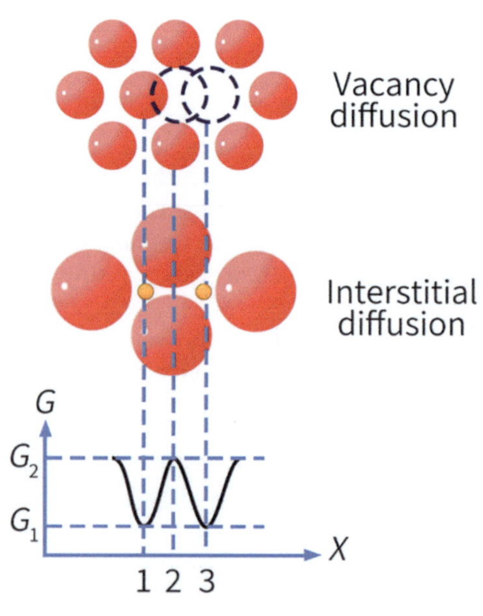

Fig. 6.19 Free energy in vacancy diffusion and interstitial diffusion

Similarly, the number of atoms of molar free energy $G > G_1$ ($G_1 < G_2$), n_1 is:

$$\frac{n_1(G \geq G_1)}{N} = \mathrm{e}^{-G_1/RT} \tag{6.52}$$

So,

$$\frac{n_2(G \geq G_2)}{n_1(G \geq G_1)} = \mathrm{e}^{-(G_2-G_1)/RT} = \mathrm{e}^{-\Delta G^*/RT} \tag{6.53}$$

Note that G_1 is the free energy when the interstitial atom is in the equilibrium position.

$$n_1(G \geq G_1) \approx N$$

At any moment, the percentage of atoms that have enough energy and can change positions to undergo hopping migration (or the hopping probability of any atom) is:

$$p_1 = \frac{n_2(G \geq G_2)}{N} = \mathrm{e}^{-\Delta G^*/RT} \tag{6.54}$$

For diffusion, p is the activation probability of the atom. The larger the ΔG^*, the smaller the p; and the higher the temperature T, the larger the p.

The above analysis helps to clarify the concept of activation energy of diffusion. During diffusion, the energy barrier that an atom must overcome to move from its original equilibrium position to a new one is known as the atom's activation energy. ΔG^* represents the activation energy of diffusion. In the diffusion process, the activation energy required for substituted atoms to overcome the obstacles of surrounding atoms is higher than that required for interstitial atoms.

In metallurgy, many important processes are related to temperature, such as grain growth rate, corrosion rate, etc. If the diffusion coefficient is measured at different temperatures, it is found that the relationship between the diffusion coefficient D and temperature can also be expressed by the Arrhenius equation as follows:

$$D = D_0\mathrm{e}^{-Q/RT} \tag{6.55}$$

D and Q depend on the composition and structure of the substance, but have nothing to do with temperature. D is the diffusion coefficient; Q is the activation energy of diffusion (unit: J/mol).

Calculation of D and Q in the interstitial mechanism

To find out the frequency Γ of the atomic hopping, assuming that the nearest neighbor interstitial positions around interstitial atoms are all empty, then

$$\Gamma = p_1 p_2 \tag{6.56}$$

p_1: is the probability that the atom has the energy to change its position, which is the activation probability; p_2 is the vibration frequency v which is the frequency of vibration toward interstitials. Then in unit time, among all vibrations, the number of hopping, that is, the hopping frequency Γ can be expressed as:

$$\Gamma = \nu \exp\left(-\frac{\Delta G^*}{RT}\right) \tag{6.57}$$

The frequency is ν about 10^{13} s^{-1}.
According to:

$$\Delta G^* = \Delta H^* - T\Delta S^* \tag{6.58}$$

We have:

$$\Gamma = \nu \exp\left(-\frac{\Delta H^*}{RT} + \frac{\Delta S^*}{R}\right) \tag{6.59}$$

Based on Eq. 6.50

$$D = \alpha a^2 \nu \exp\left(\frac{\Delta S^*}{R}\right) \exp\left(-\frac{\Delta H^*}{RT}\right) \tag{6.60}$$

In the formula, ΔS^* is the activation entropy change, which is caused by the change of the vibration frequency of the atom; ΔH^* is the activation enthalpy change, which is constant at a constant temperature and constant pressure solid diffusion condition; $\alpha a^2 \nu \exp\left(\frac{\Delta S^*}{R}\right)$ is the frequency factor D_0;

Based on Eq. 6.55

$$Q = \Delta H^* \approx \Delta E \tag{6.61}$$

ΔE is the change of systematic energy.

Calculation of D and Q in the vacancy mechanism

At a given temperature, there is an equilibrium vacancy concentration within the crystal. The mechanism behind this is the exchange of positions between atoms and vacancies. Within crystals, vacancies diffuse with ease as there are always atoms surrounding the vacancies. However, the diffusion of atoms is not as straightforward since there may not be vacancies present around the atoms. Once an atom completes a transition, it must wait for a new vacancy to move into its neighboring position for the next transition to occur.

Therefore, in contrast to the Eq. 6.56 of the interstitial mechanism, the hopping frequency of atoms in the vacancy mechanism can be expressed as:

$$\Gamma = p_1 p_2 p_3 \tag{6.62}$$

In the formula, p_1, p_2 are the same as in Eq. 6.56; p_3 is the equilibrium concentration of vacancies, that is, the probability that each nearest neighbor atom around the diffused atom exists in the form of vacancies, there is:

$$p_3 = \exp\left(-\frac{\Delta G_v}{RT}\right) = \exp\left(\frac{\Delta S_v}{R}\right)\exp\left(-\frac{\Delta H_v}{RT}\right) \tag{6.63}$$

So, the diffusion coefficient is:

$$D = \alpha a^2 \nu \exp\left(\frac{\Delta S^* + \Delta S_v}{R}\right)\exp\left(-\frac{\Delta H^* + \Delta H_v}{RT}\right) \tag{6.64}$$

In the vacancy diffusion mechanism, the frequency factor

$$D_0 = \alpha a^2 \nu \exp\left(\frac{\Delta S^* + \Delta S_v}{R}\right) \tag{6.65}$$

Diffusion activation energy

$$Q = \Delta H^* + \Delta H_v \approx \Delta E^* + \Delta E_v \tag{6.66}$$

Therefore, the diffusion activation energy of the vacancy mechanism is equal to the sum of the energy required to generate 1 mol of vacancies (ΔE_v) and the energy required to exchange these vacancies with atoms (ΔE^*).

ΔS_v, ΔH_v, are vacancy formation entropy and vacancy formation enthalpy, respectively. ΔS_v is caused by changes in the vibration frequency of surrounding atoms due to the introduction of vacancies. Usually there is $\Delta S < \Delta S^*$.

$$D_0 = \alpha a^2 \nu \exp\left(\frac{\Delta S^*}{R}\right) \tag{6.67}$$

According to the experimental relationship, it is easy to find D and Q by using the graphical method and using single logarithmic coordinates.

Figure 6.20 depicts the experimental results of the diffusion coefficient of a range of low concentration doped impurities in silicon and gallium arsenide. In most instances, plotting the logarithmic relationship between the diffusion coefficient and the absolute temperature yields a linear graph, indicating that the diffusion coefficient can be expressed as:

$$D = D_o e^{-\frac{E_a}{kT}} \tag{6.68}$$

where D_0 describes the diffusion coefficient extrapolated to infinite temperature and E_a is the Arrhenius activation energy.

E_a is linked to the energy required to move dopant atoms from one interstitial position to another. The E_a values of impurity atoms in Si and GaAs are usually within the range of 0.5 to 1.5 eV. When it comes to vacancy diffusion, E_a

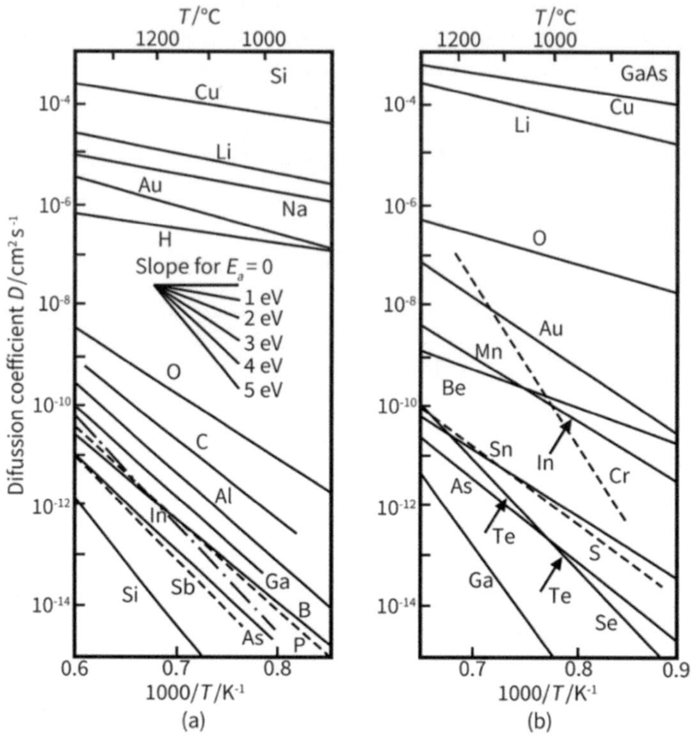

Fig. 6.20 Diffusion coefficient (diffusivity) as a function of the reciprocal of temperature for **a** silicon and **b** gallium arsenide. "S. M. Sze, "VLSI Technology", 2'Ed, McGrow-Hill, (1988)"

relates to both the kinetic energy and vacancy formation energy. Thus, the E_a of vacancy diffusion is usually higher than that of interstitial diffusion, typically ranging between 3 to 5 eV. For materials with a high diffusion rate, such as Cu, the activation energy is less than 2 eV, indicating that diffusion primarily occurs due to the movement of interstitial atoms. In contrast, for slow-diffusing substances such as As, E_a is generally greater than 3 eV, which means that the vacancy diffusion mechanism dominates the diffusion process.

6.3.5 Factors Affecting Diffusion

Temperature

Since,

$$D = D_0 \exp \frac{-Q_d}{RT}$$

Fig. 6.21 Graph of log D versus 1/T

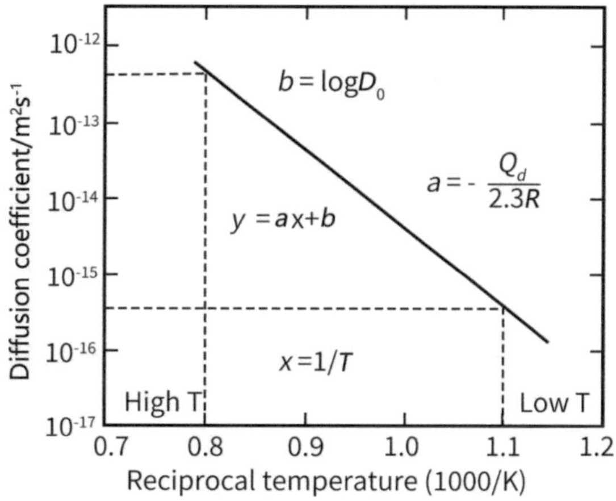

Therefore,

$$\log D = \log D_0 - \frac{Q_\mathrm{d}}{2.3R}\left(\frac{1}{T}\right) \tag{6.69}$$

Here, D_0 is the maximum value of diffusivity, Q_d is the activation energy, R is the Gas constant, and T is temperature. By describing the diffusion coefficient with an Arrhenius relation, at a given temperature T, the ratio of the thermal energy given by $k_\mathrm{B}T$ to the activation energy Q_d sets the value of the diffusion coefficient, and it is maximized at infinite temperature.

The activation energy is defined as that energy required to promote the diffusive motion of one mole of atoms. A small activation energy results in a relatively large diffusion coefficient.

Plot of log D versus $1/T$ has a slope of $-Q_\mathrm{d}/2.3R$ in Fig. 6.21, intercept of log D_0.

$$Q_\mathrm{d} = -2.3R\frac{\log D_1 - \log D_2}{\frac{1}{T_1} - \frac{1}{T_2}} \tag{6.70}$$

If log D is plotted based on the reciprocal of absolute temperature, a straight line should be generated with a slope and intercept of $-Q_\mathrm{d}/2.3R$ and log D_0, respectively. In fact, this is the way to determine the values of Q_d and D_0 through experiments. From the plots of these alloy systems (Fig. 6.22), there is a linear relationship under all displayed conditions.

Chemical composition

When an atom moves in the crystal structure, it must break the bondage of its surrounding atoms, which will partially destroy the atomic bond. Therefore, the diffusion activation energy Q and the diffusion coefficient D must be related to the

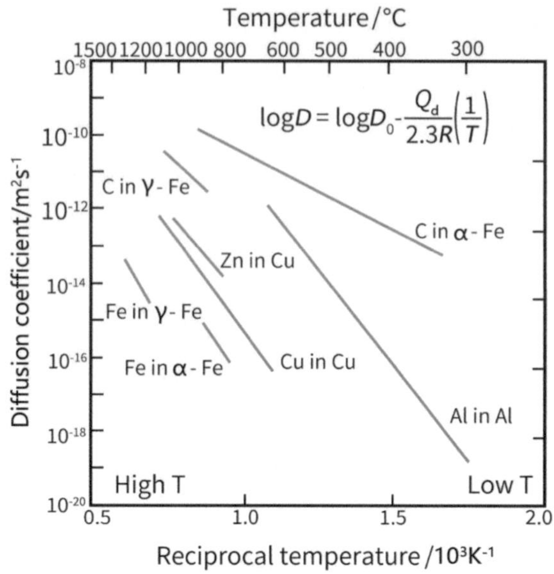

Fig. 6.22 A plot for several alloy systems

macroscopic or the microscopic parameters. In both pure metals and alloys, the strength of atomic bonds determines the size of Q and D. Macroscopic parameters that characterize atomic bond strength include melting point (T_m), latent heat of fusion (L_m), latent heat of sublimation (L_s), coefficient of expansion (α), and coefficient of compression (k). Generally, the smaller the values of T_m, L_m, and L_s, or the larger the values of α and k, the smaller the Q and the larger the D. Empirical formula of diffusion activation and macro parameters are listed in Table 6.2.

Regarding a binary alloy composed of elements A and B, if the addition of element B lowers the alloy's melting point, the interdiffusion coefficient of the alloy will increase. Conversely, if the alloy's melting point is raised, the interdiffusion coefficient will decrease.

In terms of microscopic parameters, any factor that reduces the solubility of a solid solution will also reduce the activation energy required for diffusion of solute atoms and increase the diffusion coefficient. For instance, a greater difference in atomic radius between solid solution components results in larger lattice distortion caused by solute atoms. This, in turn, makes it easier for atoms to diffuse away from the distorted position, leading to a decrease in activation energy (Q) and an

Table 6.2 Empirical formula of diffusion activation and macro parameters

Macro parameters	Melting point (T_m)	Latent heat of fusion (L_m)	Latent heat of sublimation (L_s)	Coefficient of expansion (α)	Coefficient of compression (k)
Empirical formula	$Q = 32T_m$ $Q = 40T_m$	$Q = 16.5L_m$	$Q = 0.7L_s$	$Q = 2.4/\alpha$	$Q = V_0/8k$

increase in the diffusion coefficient (*D*). Table 6.3 lists the diffusion coefficients of some elements in silver.

Structure

(1) Solid solution
Solid solutions are typically classified as either interstitial solid solutions or substitutional solid solutions. These two types of solid solutions have fundamentally different mechanisms for the diffusion of solute atoms. In an interstitial solid solution, solute atoms diffuse through interstitial spaces, and the activation energy for this diffusion is relatively low. On the other hand, in a substitutional solid solution, solute atoms diffuse through vacancies in the crystal lattice. The concentration of vacancies in the crystal is very low, and the activation energy for this type of diffusion is much larger than that for interstitial diffusion. Table 6.4 lists the diffusion activation energy of different solute atoms in γ-Fe.

(2) Crystal structure
The crystal structure reflects the arrangement of atoms in space, and a higher density of the crystal results in a narrower path for atom diffusion, greater lattice distortion, and stronger binding energy between atoms. As a consequence, the diffusion activation energy becomes larger and the diffusion coefficient becomes smaller. This rule is applicable to both pure metal and solid solution diffusion. For example, face-centered cubic crystals are denser than body-centered cubic crystals. Compared with α-Fe, the self-diffusion coefficient of γ-Fe measured experimentally differs by two orders of magnitude at 910 °C, $D_{\alpha\text{-Fe}} \approx 280D_{\gamma\text{-Fe}}$. The diffusion coefficients of solute atoms in different solid solutions are also different. At 910 °C, the diffusion coefficient of C in α-Fe is 100 times larger than that in γ-Fe.

Table 6.3 Diffusion coefficients of elements in silver

Metal	Ag	Au	Cd	In	Sn	Sb
$D/(10^{-10}cm^2/s)$ (1000 K)	1.1	2.8	4.1	6.6	7.6	8.6
Solubility/mole ratio	1.00	1.00	0.42	0.19	0.12	0.05
Radius/nm	0.144	0.144	0.1521	0.1569	0.1582	0.1614

Table 6.4 Diffusion activation energy of different solute atoms in γ-Fe

Type of solid solution	Substitutional						Interstitial		
Solute elements	Al	Ni	Mn	Cr	Mo	W	N	C	H
Q/(kJ/mol)	184	282.5	276	335	247	261.5	146	134	42

The carburizing process for steel typically involves heating it to a temperature range of 900 to 930 °C. For commonly used carburizing steels, this temperature range should fall within the austenite single-phase region. Austenite has a face-centered cubic structure, and while the diffusion rate of C in austenite may seem slow, the higher carburizing temperature actually accelerates the diffusion of C. Additionally, the solubility of C in austenite is much higher than in ferrite.

(3) Crystal anisotropy

In theory, the anisotropy of a crystal inevitably leads to the anisotropy of atom diffusion. However, experiments have shown that in cubic systems with higher symmetry, the diffusion coefficients along different directions are not significantly different. Only in crystals with lower symmetry is there noticeable directional dependence in diffusion. For instance, when copper and mercury diffuse through close-packed hexagonal metals like zinc and cadmium, the diffusion coefficient along the (0001) crystal plane is smaller than the diffusion coefficient along the [0001] direction. This is because (0001) represents the close-packed surface of atoms, and the activation energy of diffusion along this surface is relatively large. Nonetheless, the anisotropy of diffusion gradually decreases with increasing temperature. The three factors that affect crystal diffusion are essentially the same - the lower the crystal density, the faster the diffusion of atoms.

Defects

There are various point, line, and surface defects in solid materials. The energy of the defect is higher than that inside the crystal grain, which can provide a greater driving force for diffusion and make the atoms diffuse faster along the defect. The diffusion along the defect is called Short-circuit diffusion as shown in Fig. 6.23, and the diffusion inside the crystal lattice is called lattice diffusion. Short-circuit diffusion includes surface diffusion, grain boundary diffusion, dislocation diffusion and vacancy expansion. When the temperature is low, the short-circuit diffusion is dominant; when the temperature is higher, the lattice diffusion is dominant.

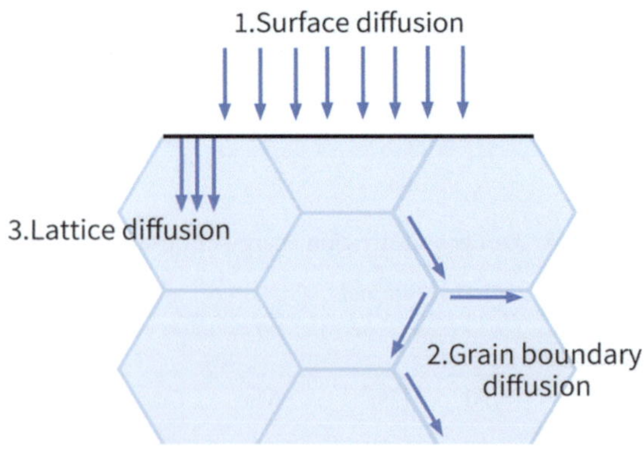

Fig. 6.23 Scheme of short path diffusion

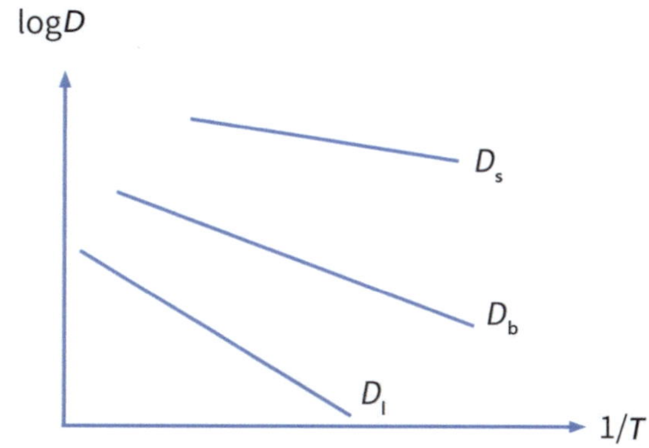

Fig. 6.24 Diffusion coefficient of surface diffusion, grain boundary diffusion and lattice diffusion as a function of temperature

Among all the defects, the energy on the surface is the highest, while the energy on the grain boundary is the second highest, and the energy inside the grain is the lowest. As a result, the activation energy of atom diffusion along the surface is the lowest, followed by the activation energy of diffusion along the grain boundary, with the activation energy of lattice diffusion being the highest. For the diffusion coefficient, $D_s > D_b > D_l$, where D_s, D_b, and D_l are the surface diffusion coefficient, grain boundary diffusion coefficient and lattice diffusion coefficient, respectively, as shown in Fig. 6.24.

In polycrystalline metals, the diffusion coefficient of atoms is a combined result of lattice diffusion and grain boundary diffusion. The smaller the grain size, the larger the grain boundary area of the metal, and the greater the contribution of grain boundary diffusion to the diffusion coefficient. At low temperatures, the activation energy of grain boundary diffusion is much lower than that of lattice diffusion, and grain boundary diffusion plays a dominant role. At higher temperatures, the vacancy concentration in the crystal increases, leading to faster diffusion through the bulk of the material, which then becomes the primary mechanism. Grain boundary diffusion has a significant impact on self-diffusion and inter-diffusion at lower temperatures. However, in interstitial solid solutions, the lattice diffusion activation energy of solute atoms is inherently low, and the diffusion speed is relatively high, making the effect of grain boundary diffusion less pronounced.

Dislocations in a crystal also play a role in promoting diffusion. When solute atoms interact with the elastic stress field around a dislocation, they segregate and form solute atom clusters, known as Cottrell and Snoek air masses, around the dislocation line. When these solute atoms diffuse along the elongated distortion zone centered on the dislocation line, the activation energy is only about half that of bulk diffusion, resulting in a relatively high diffusion rate. However, since dislocations make up only a small proportion of the entire crystal, they do not contribute much to diffusion at higher temperatures but play a significant role at lower temperatures.

6.4 The Kirkendall Effect in Solid State Diffusion

6.4.1 The Kirkendall Effect

The migration of atoms into vacancies, known as vacancy diffusion, is the most prominent among various diffusion mechanisms. Despite initial skepticism due to the perceived low concentration of vacancies, the Kirkendall experiment provided ingenious and simple proof that vacancy diffusion does occur in most metal materials.

Smigeiskas and Kogendal used experiments to demonstrate that the diffusion coefficients of the components involved in the interdiffusion process are not equal, thereby providing evidence for vacancy mechanism of displacement diffusion. The experiments they conducted in 1947 are shown in Fig. 6.25. A very thin molybdenum wire was placed on a rectangular brass rod as a marker. Then, copper was plated onto the brass (Cu + 30%Zn) and the molybdenum wire was wrapped between the two metals. It's important to note that the brass and copper form a diffusion couple and the molybdenum wire, which has a high melting point, is only used as a marker and does not participate in the diffusion process.

The aforementioned samples are heated to 785 °C to promote interdiffusion between zinc and copper. Compared to molybdenum wire, it is evident that zinc diffuses outward while copper diffuses inward. The experiment discovered that, after one day, the two layers of molybdenum wire moved inward by 0.0015 cm and, after 56 days, they moved inward by 0.0124 cm. The data of this displacement over time are listed in Table 6.5. This phenomenon is called the Kirkendall effect.

Fig. 6.25 Kirkendall experiment

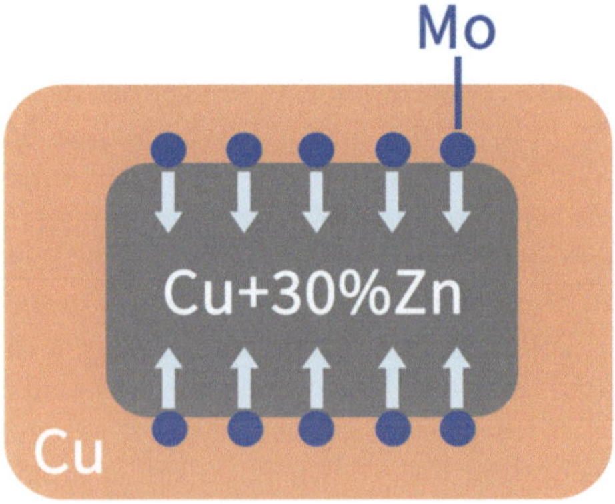

Table 6.5 Displacement of molybdenum wire with time

Time/day	0	1	3	6	13	28	56
Displacement of molybdenum	0	0.0015	0.0025	0.0036	0.0056	0.0092	0.0124

If copper and zinc have equal diffusion coefficients, atomic exchange will occur equally with respect to the molybdenum wire. Zinc has a larger atomic size than copper, so the outer copper lattice constant increases while the internal brass lattice constant decreases after diffusion, causing the molybdenum wire to move inward. However, if lattice constant change were the only reason for the molybdenum wire's movement, it would only move about one-tenth of the observed distance. The experimental results indicate that zinc diffuses much faster than copper, which is the main cause of the molybdenum wire's inward movement. It was also observed that the marker's movement is proportional to the square root of time.

It was later discovered that this phenomenon can occur in substitutional solid solutions such as Cu–Sn, Cu–Ni, Cu–Au, Ag–Au, Ag–Zn, Ni–Co, Ni–Cu, and Ni–Au. Additionally, the marker consistently moves towards the side with a higher concentration of low-melting-point components. The low-melting-point component diffuses faster compared to the high-melting-point component, creating an unequal amount of atomic exchange, resulting in the Kirkendall effect.

6.4.2 The Darken's Equation and the Interdiffusion Coefficient

Darken's mathematical treatment of the Kirkendall effect involves a diffusion couple consisting of a high-melting-point metal A (on the left) and a low-melting-point metal B (on the right), as shown in Fig. 6.26. A high-melting-point marker is placed between the two metals before welding. Darken introduced two parallel coordinate systems: one is a fixed coordinate system (X, Y) relative to the ground, and the other is a dynamic coordinate system (X', Y') that moves with the marker interface. As the high-melting-point metal has a strong atomic bonding force and diffuses slowly, and the low-melting-point metal has a weak atomic bonding force and diffuses quickly, the marker interface shifts to the low-melting-point side at high temperatures. An observer standing on the marker sees that the number of A and B atoms diffusing in opposite directions across the interface is different. However, an observer standing on the ground sees that the number of A and B atoms diffused in both directions is the same.

After the above analysis, the total velocity (V) of the diffusing atom relative to the ground is the sum of the atom's diffusion velocity (V_d) relative to the marker and the marker's movement velocity (V_m) relative to the ground, namely

$$V = V_m + V_d \tag{6.71}$$

The total velocity (V) of atoms can be calculated based on the diffusion system. The number of moles of diffused atoms passing through a unit area in unit time (diffusion flux) is equal to the number of moles of diffused atoms in a volume of V.

$$J = C(V \times 1) = CV \tag{6.72}$$

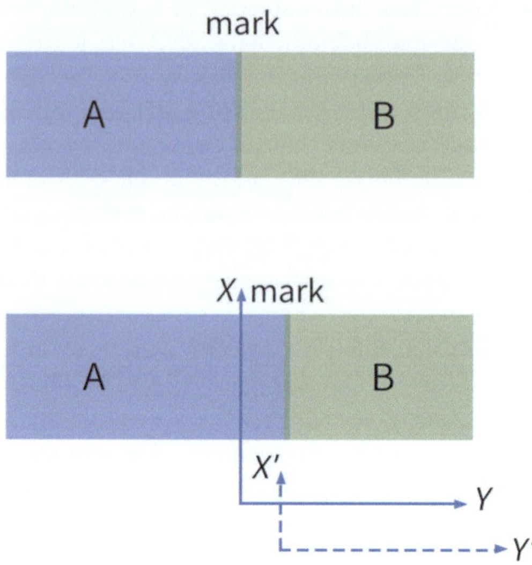

Fig. 6.26 Kirkendall effect in a diffusion couple

C is the molar volume concentration of diffused atoms. Using Eq. 6.71 and Eq. 6.72, the total flux of A and B atoms relative to the fixed coordinate system can be written as

$$\begin{aligned} J_A &= C_A V_A = C_A[V_m + (V_d)_A] \\ J_B &= C_B V_B = C_B[V_m + (V_d)_B] \end{aligned} \tag{6.73}$$

The first term in Eq. 6.73 is the flux of the marker relative to the fixed coordinate system, and the second term is the diffusion flux of the atom relative to the marker. If the diffusion coefficients of A and B atoms are represented by D_A and D_B, respectively, according to the first law of diffusion, the second term caused by diffusion can be written as

$$C_A(V_d)_A = -D_A \frac{\partial C_A}{\partial x}$$

$$C_B(V_d)_B = -D_B \frac{\partial C_B}{\partial x}$$

and

$$J_A = C_A V_A = C_A V_m - D_A \frac{\partial C_A}{\partial x}$$

$$J_B = C_B V_B = C_B V_m - D_B \frac{\partial C_B}{\partial x}$$

The number of A and B atoms crossing a fixed plane should be equal, but in opposite directions

$$J_A = -J_B$$

So

$$V_m(C_A + C_B) = D_A\frac{\partial C_A}{\partial x} + D_B\frac{\partial C_B}{\partial x}$$

The molar volume concentration C_i of the component has the following relationship with the molar density ρ and the **molar fraction x_i**

$$C_i = \rho x_i$$

Since $x_A + x_B = 1$.
The Interface drift speed V_m is

$$V_m = (D_A - D_B)\frac{\partial x_A}{\partial x} = (D_B - D_A)\frac{\partial x_B}{\partial x}$$

The total flux of A and B are

$$\begin{aligned} J_A &= -(D_A x_B + D_B x_A)\tfrac{\partial C_A}{\partial x} = -\tilde{D}\tfrac{\partial C_A}{\partial x} \\ J_B &= -(D_A x_B + D_B x_A)\tfrac{\partial C_B}{\partial x} = -\tilde{D}\tfrac{\partial C_B}{\partial x} \end{aligned} \tag{6.74}$$

$\tilde{D} = (D_A x_B + D_B x_A)$ is the inter-diffusion coefficient of the alloy, and D_A and D_B are the intrinsic diffusion coefficients of the components. Equation 6.74 is called Daken's equations. It can be seen from the deduced results that **the diffusion law is still applicable to the diffusion of substitutional solid solution as long as the diffusion coefficient (*D*) in the first and second laws of diffusion is replaced with the Interdiffusion coefficient (*D*) of the alloy**.

The diffusion coefficients we have encountered so far can be classified into three types:

(1) Self-diffusion coefficient (*D*) refers to diffusion due to thermal motion of atoms in a pure metal or uniform solid solution without a concentration gradient. In experiments, radioisotopes of the same metal are used as tracer atoms to measure the self-diffusion coefficient. When there is a concentration gradient of isotope atoms, observable diffusion occurs. Since the physical and chemical properties of the metal and metal radioisotopes are the same, the measured diffusion coefficient of isotope atoms is the metal's self-diffusion coefficient.
(2) Intrinsic diffusion coefficient (*D*) refers to diffusion caused by the concentration gradient of the components in an alloy. The diffusion driven by the concentration gradient of the component in the alloy is called intrinsic diffusion of the component, and it is described by the intrinsic diffusion coefficient.

(3) Interdiffusion coefficient (D) is the weighted average of the intrinsic diffusion coefficients of each component in the alloy, which reflects the diffusion characteristics of the alloy but does not represent the diffusion properties of a specific component. Both intrinsic diffusion coefficient and interdiffusion coefficient are caused by concentration gradient, so they are collectively referred to as the chemical diffusion coefficient.

6.4.3 The Theoretical and Practical Significance of Kirkendall Effect

The Kirkendall effect demonstrates the internal relationship between the macroscopic diffusion law and the microscopic mechanism. It has universal significance in the development of diffusion theory and its practical applications in production.

The Kirkendall effect provides support for the vacancy mechanism and highlights the unique diffusion coefficient of each component in the diffusion system. As a result of the Kirkendall effect, side effects may arise since $J_{\text{zn}} > J_{\text{ca}}$ and $D_{\text{zn}} > D_{\text{cu}}$ (note that D_{zn} and D_{cu} are displacement coefficients distinct from the diffusion coefficient D used in Fick's law). Complete shrinking of the crystal causes the original interface to shift, whereas incomplete shrinking creates scattered or concentrated vacancies on the side of the low melting point metal, exceeding the equilibrium vacancy concentration and forming Kirkendall holes, as illustrated in Fig. 6.27. This process also reduces the vacancy concentration on the side of the refractory metal below the equilibrium vacancy concentration, altering the crystal density. Studies have also shown changes in the sample's cross-section; for example, after Ni-Cu diffusion couples undergo diffusion, the copper cross-section near the original interface decreases due to atom loss, forming a surface depression. Conversely, the nickel cross-section expands due to atom acquisition, forming surface bumps.

The Kirkendall effect can often have negative effects in practical applications. For example, electronic equipment contains numerous wires, contacts, electrodes, and multilayer structures that must operate at high temperatures for extended periods. These side effects can result in disconnections, breakdowns, performance deterioration, and even complete device failure. To mitigate these issues, proper design and materials selection are crucial.

Fig. 6.27 Cross section of diffusion couple

6.5 Diffusion Kinetics

Fick's first law $J = -D\frac{\partial C}{\partial x}$ expressed by concentration gradient can only describe the downward diffusion of atoms from high concentration to low concentration. When $J = -D\frac{\partial C}{\partial x} \to 0$, that is, when the concentration of alloy becomes more uniform, macroscopic diffusion ceases. However, several diffusion phenomena in alloys involve uphill diffusion, where particles move from areas of low concentration to high concentration. Examples include the amplitude modulation decomposition of solid solution and eutectoid transformation. These uphill diffusion processes demonstrate that the primary driving force for diffusion is not solely concentration gradients.

6.5.1 Driving Force for Diffusion

As per the principles of thermodynamics, the general standard for determining the direction of system change is that, under constant temperature and pressure, the system will change in the direction of decreasing Gibbs free energy. The equilibrium state of the system corresponds to the minimum free energy, and a process is considered spontaneous if it occurs without any external input. The driving force behind system changes is the ability to decrease the value of ΔG, which represents the change in Gibbs free energy.

The same principle applies to diffusion in alloys, where atoms always move from a location with a higher chemical potential to one with a lower chemical potential. If the chemical potential is equivalent (in a single-phase alloy), the system will reach its equilibrium state, and macroscopic diffusion will cease. Thus, the primary driving force behind atomic diffusion is the gradient of chemical potential. If the atom of component i in the alloy moves ∂x distance in the direction due to some external factors (such as temperature, pressure, stress, magnetic field, etc.), its chemical potential decreases $\partial \mu_i$, then the driving force of the atom is

$$F = -\frac{\partial \mu_i}{\partial x} \tag{6.75}$$

The driving force for atomic diffusion is in the same direction as the chemical potential decreases.

6.5.2 Generalized Diffusion Coefficient

Under the influence of chemical potential, diffusing atoms in a solid experience resistance from the solvent atoms as they move in a particular direction. The resistance is proportional to the speed of diffusion. When the diffusion of solute atoms speeds up to the point where the resistance they encounter is equal to the driving force, the diffusion rate of solute atoms reaches its limit, which is the average atomic diffusion rate. The average diffusion speed ν of atoms is proportional to the driving force F:

$$\nu = BF \tag{6.76}$$

The proportional coefficient B is the speed under the unit driving force, which is called the mobility. The diffusion flux is equal to the product of the mass concentration ρ of the diffused atom and its average diffusion speed ν:

$$J = \rho_i \nu_i$$

So we have

$$J = \rho_i B_i F_i = -\rho_i B_i \frac{\partial \mu_i}{\partial x}$$

Derived from Fick's first law

$$J = -D \frac{\partial \rho_i}{\partial x}$$

So we have:

$$D = \rho_i B_i \frac{\partial \mu_i}{\partial \rho_i} = B_i \frac{\partial \mu_i}{\partial \ln \rho_i} = B_i \frac{\partial \mu_i}{\partial \ln x_i}$$

In equation, $x_i = \frac{\rho_i}{\rho}$, In thermodynamics, $\partial \mu_i = kT \partial \mathrm{In} a_i$, a_i is the activity of component i in solid solution, $a_i = r_i x_i$, r_i is activity coefficient. So we have:

$$D = kTB_i \frac{\partial \ln a_i}{\partial \ln x_i} = kTB_i \left(1 + \frac{\partial \ln r_i}{\partial \ln x_i}\right) \tag{6.77}$$

For an ideal solid solution or dilute solid, the factor in the brackets (also known as the Thermodynamic factor) in the above formula is equal to one, so:

$$D = kTB_i \tag{6.78}$$

It can be seen that the diffusion rate of the different component in an ideal solid solution depends only on the mobility B, and the above formula is called the Nernst-Einstein equation.

In the dual system, according to Gibbs–Duhem's relationship

$$x_1 \mathrm{d}\mu_1 + x_2 \mathrm{d}\mu_2 = 0 \tag{6.79}$$

x_1 describes the Mole fraction of component 1 and x_2 describes the mole fraction of component 2.

Since μ_i is a function of x_i, deriving from equation:

$$d_{\mu_i} = RT\mathrm{dln}a \tag{6.80}$$

We have

$$x_i \mathrm{d}\mu_i = RT(\mathrm{d}x_i + x_i \mathrm{dln}r_i)$$

noting $\mathrm{d}x_1 = -\mathrm{d}x_2$, so we have:

$$x_1 \mathrm{dln}r_1 = -x_2 \mathrm{dln}r_2$$

Rearranging:

$$\frac{\mathrm{dln}r_1}{\mathrm{dln}x_1} = \frac{\mathrm{dln}r_2}{\mathrm{dln}x_2} \tag{6.81}$$

$\frac{\mathrm{dln}r_i}{\mathrm{dln}x_i}$ of component 1 and component 2 are equal, and the reason that their diffusion rates D_1 and D_2 are different is due to the difference in mobility B_1 and B_2.

6.5.3 Uphill Diffusion

When $\left(1 + \frac{\partial \ln r_i}{\partial \ln x_i}\right) > 0$ and $D > 0$, it means that the component is "downhill diffusion" from the high-concentration area to the low-concentration area. When $\left(1 + \frac{\partial \ln r_i}{\partial \ln x_i}\right) < 0$ and $D < 0$, the surface component is migrating from the low-concentration area to the high-concentration area.

In summary, it can be seen that the basic factor that determines component diffusion is the chemical potential gradient, whether it is uphill diffusion or downhill diffusion, the result will always lead to a decrease in the chemical potential gradient of the diffusion component until the chemical potential gradient is zero.

Some of the following conditions may also cause uphill diffusion:

(1) The effect of elastic stress. When there is an elastic stress gradient in the crystal, it causes atoms with a larger radius to run to the elongated part of the lattice, and atoms with a smaller radius to run to the compressed part, resulting in uneven content atoms in the solid solution.
(2) Internal adsorption of grain boundaries. The grain boundary energy is higher than that in the crystal, and the regular arrangement of atoms is worse than that in the crystal. If the solute atoms are located on the grain boundary, they will preferentially diffuse to the product boundary and concentrate on the product boundary to reduce the total energy of the system.
(3) The large electric field or temperature field also promotes the diffusion of atoms in the product in a certain direction, resulting in the non-uniformity of the diffused atoms.

6.6 Other Diffusion Types

6.6.1 Reaction Diffusion

All the previously discussed phenomena involve diffusion in a single-phase solid solution, where the concentration of solute atoms does not exceed its solid solubility in the matrix. However, in many phase diagrams, there are often intermediate phases. Therefore, the concentration distribution caused by diffusion and the corresponding solid solubility determined by the alloy system will inevitably generate intermediate phases during the diffusion process. This phenomenon of forming new phases through diffusion is known as heterogeneous diffusion, which is also referred to as phase transition diffusion or reaction diffusion. Reaction diffusion involves two processes: diffusion and reaction. The phase change occurs when the interface reaches a certain concentration during the reaction process.

To illustrate, let's consider the nitriding of pure iron surfaces. At a temperature of 520 °C, pure iron undergoes nitridation, resulting in a reaction and diffusion process. When the nitrogen concentration exceeds approximately 8 wt%, the ε-phase forms on the surface. This iron–nitrogen compound has a broad range of nitrogen content, typically between 8.25 wt% and 11.0 wt% N. Nitrogen atoms occupy the interstitial sites of close-packed hexagons composed of iron atoms, and the γ-phase lies adjacent to the ε-phase. The γ-phase is an interstitial phase compound with a variable composition, existing in a narrow range of N (5.7 wt% to 6.1 wt%). Nitrogen atoms occupy the interstitial position of the face-centered cubic lattice composed of iron atoms.

Similarly, boronizing of steel involves reaction and diffusion processes. If sufficient boron atoms are present, two types of borides can form in the boronizing layer: FeB in the outer layer and Fe_2B in the inner layer.

6.6.2 Diffusion in Ionic Crystal

In metals and alloys, atoms have the ability to transition into adjacent vacancies and interstitial positions. However, in ionic crystals, diffused ions can only enter positions with the same charge. In other words, they cannot enter sites of adjacent heterogeneous ions. Ion diffusion can only rely on vacancies, and the distribution of vacancies also has unique properties. Since separating a pair of heterogeneous ions would greatly increase electrostatic energy, it is necessary to form two types of defects with different charges simultaneously—a cation and an anion—in order to maintain local charge balance. Therefore, Schottky defects and Frenkel defects exist.

Defects Type in Ionic Crystal

(1) Schottky defect

Schottky defects are point defects formed when occupied sites in the crystal lattice become excited. They are named after Walter H. Schottky. In ionic crystals, Schottky defects occur when oppositely charged ions leave their lattice positions

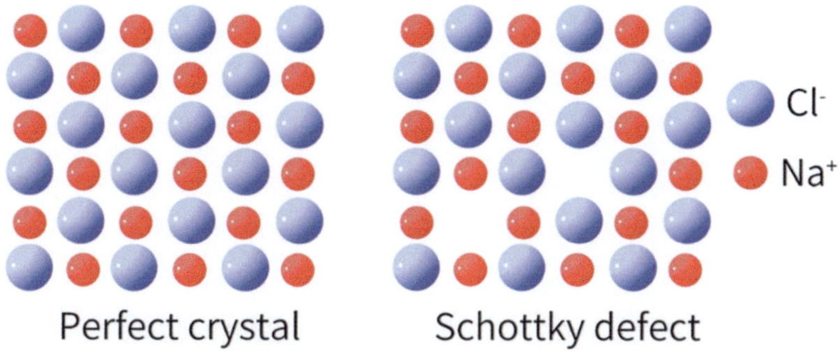

Fig. 6.28 Schottky defect in the NaCl crystal

and bind to the surface, resulting in the formation of oppositely charged vacancies. These vacancies are formed in stoichiometric units, to maintain an overall neutral charge in the ionic solid, but the density reduces due to the presence of vacancies. Schottky defects are commonly observed when there is a small difference in size between the cations and anions that make up a material. Figure 6.28 shows Schottky defect in the NaCl crystal where the Na ion vacancies and Cl ion vacancies formed. Schottky defect can be found in Sodium Chloride (NaCl), Potassium Chloride (KCl), Potassium Bromide (KBr), Silver Bromide (AgBr), Cerium Dioxide (CeO_2), Thorium Dioxide (ThO_2) and so on.

(2) Frenkel defect

A Frenkel defect is a type of point defect in crystalline solids, named after its discoverer Yakov Frenkel. The defect forms when an atom or smaller ion (usually cation) leaves its place in the lattice, creating a vacancy and becomes an interstitial by lodging in a nearby location. Usually, this type of defect is observed in ionic solids, where there is a considerable disparity in the sizes of the ions. Figure 6.29 shows Frenkel defect in the AgCl crystal where the Ag ion vacancies and Cl ion vacancies formed. Common materials where Frenkel defect can be found are: Zinc Sulfide (ZnS), Silver Chloride (AgCl), Silver Bromide (AgBr).

(3) Defects in Non-stoichiometric Compounds

When the valence of ions in a compound change, a situation similar to the two types of defects mentioned above can occur. Two examples are provided here. In Wüstite (FeO), some of the Fe^{2+} ions are oxidized to Fe^{3+} ions, causing some cation positions to become vacant to maintain charge balance in the crystal. This results in cation deficiency, and the resulting compound has an excess of oxygen, forming a non-stoichiometric compound compared to pure FeO. Conversely, in TiO_2, when a part of Ti^{4+} is reduced to Ti^{3+}, oxygen ion vacancies appear to balance the charge, resulting in a lack of oxygen. In addition, replacing ions in a compound with ions of different valences can also lead to hypoxia conditions as mentioned above. For example, adding CaO as a stabilizer for ZrO_2

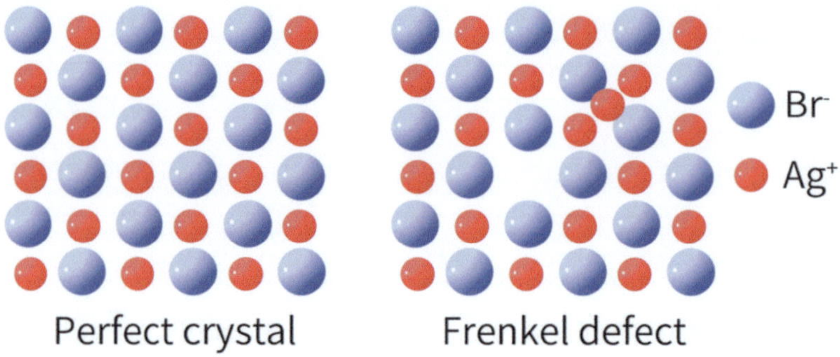

Fig. 6.29 Schottky defect in the AgBr crystal

causes low-chemical state Ca^{2+} ions to replace high-chemical state Zr^{4+}, requiring corresponding oxygen ion vacancies to maintain electrical neutrality.

Diffusion Mechanism in Ionic Crystal

The ionic crystal diffusion mechanism is shown in Fig. 6.30, which mainly includes: (1) vacancy diffusion; (2) interstitial diffusion; (3) sublattice interstitial diffusion. The diffusion of vacancies in MgO is known as vacancy diffusion, while interstitial diffusion refers to the direct movement of interstitial ions between positions. In ionic crystals, interstitial diffusion is typically more difficult than vacancy diffusion due to the large size of the interstitial ions, requiring greater activation energy. Therefore, interstitial-sublattice diffusion can occur, where a nearby lattice ion is replaced by an interstitial ion, resulting in ion movement. This type of diffusion is relatively easy and occurs due to lattice deformation, as seen in the diffusion of Ag in AgBr.

Ionic conductance is the primary form of conductance in ionic crystals. There are two types of ion conductance: intrinsic ion conductance and impurity conductivity. Intrinsic ion conductance, or intrinsic conductivity, results from the

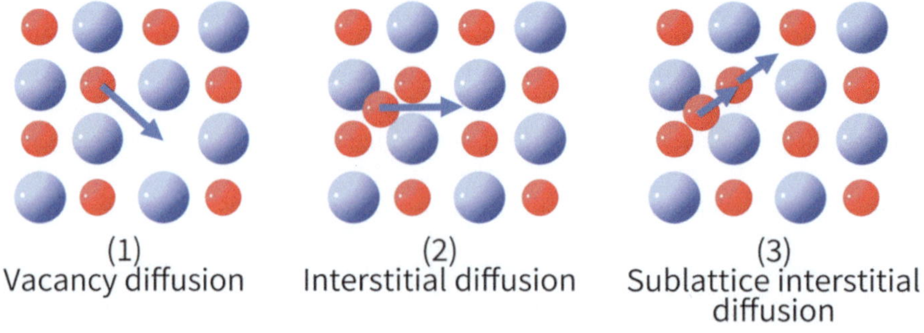

Fig. 6.30 Illustration of diffusion mechanism in ionic lattice

Table 6.6 Diffusion activation energy in ionic materials

Diffusive atoms	$Q/kJ{\cdot}mol^{-1}$	Diffusive atoms	$Q/kJ{\cdot}mol^{-1}$
Fe in FeO	96	Cr in $NiCr_2O_4$	318
Na in NaCl	172	Ni in $NiCr_2O_4$	272
O in UO_2	151	O in $NiCr_2O_4$	226
U in UO_2	318	Mg in MgO	347
Co in CoO	105	Ca in CaO	322
Fe in Fe_3O_4	201		

movement of basic ions in the crystal lattice due to thermal defects. These defects are electrically charged and can act as ion conduction carriers, particularly at high temperatures. Impurity conductivity, on the other hand, is caused by the movement of weakly fixed ions, mainly impurity ions, and is significant at lower temperatures. Both types of conductivity are the result of ion migration and diffusion in the crystal when subjected to an external electric field.

When a solid material is subjected to a constant voltage in an electric field, the electrons and ions within the material will migrate in a specific direction, generating an electric current. In metals and semiconductors, conductance is achieved through the flow of electrons, while in ionic crystals, ions are more mobile than tightly bound electrons, especially at high temperatures, and conductance is achieved through directional ion diffusion. When using isotope atoms to measure the diffusion coefficient D_T, if the number of ions of the original type of diffusion is c and the particle charge is q_i, then the following relation exists between the diffusion coefficient D_T and the electrical conductivity.

$$\frac{\sigma}{D_T} = \frac{cq_i^2}{kT} \tag{6.82}$$

In ionic crystals, since the binding energy of ionic bonds (see Table 6.6) is generally greater than that of metal bonds. As a result, the energy barrier for diffusing ions is much larger than for metal atoms. To maintain local electrical neutrality, paired defects must be created, which limits diffused ions to entering positions with the same charge. The longer migration distance results in a generally lower ion diffusion rate compared to the diffusion rate of metal atoms.

6.6.3 Diffusion in Polymers

When considering polymeric materials, our focus is on the movement of small foreign molecules (such as O_2, H_2O, CO_2, and CH_4) diffusing between molecular chains, rather than the diffusive motion of chain atoms within the polymer structure. In crystals, impurity diffusion occurs through individual atoms or ions within the lattice, whereas in the relatively open structure of amorphous polymer regions, entire molecules pass through the lattice. This diffusion mechanism is similar to interstitial diffusion in metals. The permeability coefficient (P_M) is commonly used to characterize the diffusion properties of polymers, and in the case of steady-state diffusion through a polymer membrane, Fick's first law is modified as

$$J = P_M \frac{\Delta P}{\Delta x} \tag{6.83}$$

In this expression, J is the diffusion flux of gas through the membrane, P_M is the permeability coefficient, Δx is the membrane thickness, and ΔP is the difference in pressure of the gas across the membrane.

Permeability increases with a more open structure, such as low-density polyethylene having higher permeability than high-density polyethylene. Different polymers have varying permeability to different substances; for example, polyethylene has higher permeability to oxygen than to water vapor, while polar-based polymers like vinyl alcohol have higher permeability to water vapor than oxygen. Choosing the appropriate polymer is crucial for different applications based on their permeability to different substances. For instance, polymers used for carrying food require appropriate barrier properties to avoid food spoilage caused by air passing through the film. Similarly, containers for sensitive materials such as ceramic or metal powders require appropriate water and gas barrier properties, with high-density polyethylene being the preferred choice due to its low air permeability. Additionally, to prevent reactions with water vapor in the air, materials like zinc oxide used in rubber, paint, and ceramics must be stored in polyethylene bags.

Polymers can act as a separating agent or filter due to their varying permeability to different substances. When separating two organic gases with differing molecular sizes, a polymer membrane can be selected that allows small molecules to permeate through while larger molecules are retained. As the size of diffusing molecules increases, the diffusion coefficient decreases. Polymer membranes can also separate based on properties such as polarity, with polar membranes allowing the passage of nonpolar molecules while acting as a barrier to polar molecules. Therefore, due to this unique property, polymer membranes can be highly effective in separating different substances. The permeability coefficients of oxygen, nitrogen, carbon dioxide, and water vapor in several common polymers are listed in Table 6.7.

6.7 Questions

Q6.1 If you heat steel in a carbon rich atmosphere, carbon atoms will diffuse into the steel, hardening it. Calculate the diffusivity of a steel plate if the flux of carbon atoms through the plate is 2×10^{-6} g/m^{2}s, the concentration 2 mm under the surface is 300 g/m^3, and the concentration 5 mm under the surface is 100 g/m^3.

Table 6.7 Permeability coefficients PM at 25°C for oxygen, nitrogen, carbon dioxide, and water vapor in a variety of polymers

		PM			
		[$\times 10^{-13}$ (cm^3 STP①)(cm)/(cm^2-s-Pa)]			
Polymer	Acronym	O_2	N_2	CO_2	H_2O
Polyethylene (low density)	LDPE	2.2	0.73	9.5	68
Polyethylene (high density)	HDPE	0.30	0.11	0.27	9.0
Polypropylene	PP	1.2	0.22	5.4	38
Poly (vinyl chloride)	PVC	0.034	0.0089	0.012	206
Polystyrene	PS	2.0	0.59	7.9	840
Poly (vinylidene chloride)	PVDC	0.0025	0.00044	0.015	7.0
Poly (ethylene terephthalate)	PET	0.044	0.011	0.23	/
Poly (ethyl methacrylate)	PEMA	0.89	0.17	3.8	238

Source Adapted from J. Brandrup, E. H. Immergut, E. A. Grulke, A. Abe, and D. R. Bloch (Editors), Polymer Handbook, 4th edition. Copyright @1999 by John Wiley & Sons, New York. Reprinted by permission of John Wiley & Sons, Inc.
① STP, Standard Temperature and Pressure

Q6.2 Phosphorus is introduced into a silicon wafer from a gaseous source at 975 °C for 30 min. Determine the junction depth for a doping concentration of 6.3×10^{16} cm^{-3}. Assume that the diffusion coefficient of phosphorus is 10^{-13} cm^2/s and that its solid solubility is 10^{21} cm^{-3} at 975 °C.

Q6.3 There is a case of one side diffusion with constant total-dopant amount:

(a) A predeposition process is carried out for 15 min on an n-type silicon wafer with a phosphorus dopant concentration of 10^{17} atoms/cm^3 at 950 °C using diborane gas. Assuming intrinsic diffusion, determine the junction depth. At 950 °C the boron saturation concentration is 3.8×10^{20} atoms/cm^3 and the boron diffusion constant is 1.5×10^{-15} cm^2/s.
(b) After the initial predeposition process described in (a), the sample undergoes a drive-in diffusion for 1 h at 1250 °C. What is the final junction depth? At 1250 °C, the boron diffusion constant is 1.2×10^{-12} cm^2 /s.

Q6.4 A 5-mm-thick sheet of palladium with a cross section area of 0.1 m^2 is used as a steady-state diffusional membrane for purifying hydrogen. The hydrogen concentration on the high-pressure (impure gas) side of the sheet is 1.5 kg/m^3 and the concentration on the lower-pressure side is 0.3 kg/m^3, Assume the diffusion coefficient of H_2 in Pd is $D_{pd} = 10^{-9}$ m^2/s.

(a) Calculate the diffusion flux through the membrane.
(b) The mass of hydrogen purified per hour
(c) What would be the effect of making the plate thinner

(d) How can we alter the environment to enhance the performance of this membrane (purify the hydrogen faster?)

Q6.5 A bottle of pop has a CO_2 pressure of about 400 kPa inside the bottle and the CO_2 pressure outside the bottle is 0.4 kPa. Assuming conditions of steady state, calculate the diffusion flux of CO_2 through the wall of the bottle. Assume that each bottle has a surface area of 500 cm^2 and a wall thickness of 0.05 cm.

Bibliography

Callister WD Jr, Rethwisch DG (2020) Fundamentals of materials science and engineering: an integrated approach. John Wiley & Sons, New York

Gale WF, Totemeier TC, eds (2003) Smithells metals reference book. Elsevier, [S.l.]

Glicksman ME (2022) Diffusion in solids: field theory, solid-state principles, and applications. John Wiley & Sons, [S.l.]

Polymer Handbook (1999) Wiley, New York

Shewmon P (1989) Diffusion in solids. Metals Mater Soc 2

Sze SM (1988) VLSI technology, 2nd edn. McGraw-Hill, New York

Material Surfaces and Interfaces 7

Learning Objectives

1. Understand the definition and the importance of material surfaces and interfaces.
2. Learn how to classify interfaces based on state and structure.
3. Understand the formation and effects of surface tension and surface energy and master the calculation methods of surface tension and surface free energy.
4. Describe wetting phenomena at solid–liquid interfaces and understand Young's equations.
5. Explain the phenomenon of adhesion at solid–solid interface and understand the theory of interfacial adhesion.
6. Understand the definition of composites and learn the interface structures and theories in metal matrix, ceramic matrix and polymer matrix composites.
7. Explain the types and mechanisms of interface failure in composites.
8. Learn the basic interface characterization methods to obtain physicochemical properties of interfaces.

7.1 Introduction

The boundary between two immiscible substances that are in contact with each other is referred to as the "interface". When one side of the interface is a gas or vacuum, it is known as a "surface". Many important processes, such as catalysis, molecular recognition, charge transfer, polymerization, take place at these interfaces. The energy of a surface quantifies the disruption of interatomic or intermolecular bonds that occurs when a surface is formed. The interface is typically characterized by a certain thickness, which can be defined as two monolayers,

Z. Sun et al., *Fundamentals of Materials Science*,
https://doi.org/10.1007/978-981-96-3904-5_7

with each monolayer belonging to one side of the interface. However, the atoms or molecules in subsequent layers are also affected by the other side through "longer-ranged" forces.

In certain solid phases consisting of different types of atoms, some atoms tend to accumulate on the interfaces. As a result, interfaces have a significant impact on the properties of materials, such as conductivity, adhesion, friction, and corrosion. The interface of a heterogeneous solid material is especially important because it significantly influences its overall performance.

Solids can generally be categorized into crystalline and amorphous states. Crystalline solids are made up of particles that are arranged in an ordered three-dimensional pattern, and they have a well-defined melting point. Examples of crystalline solids include iron, diamond, silicon, and benzoic acid. Amorphous solids refer to non-crystalline structures in which atoms or molecules are arranged irregularly, more like in a liquid. The distance between two particles tends to vary and these solids are generally isotropic in nature. Examples of amorphous solids include glass, naphthalene, and metallic glasses (amorphous alloys).

Composite materials are made up of a matrix phase and reinforcement phase. The interface between the matrix and reinforcement is critical in determining the performance of composite materials, and as a result, research on interfaces in composite materials has become a distinct subdiscipline.

7.2 Basics Concepts

7.2.1 Surface Energy of Solids

The surface energy of a material is a critical parameter in evaluating its resistance to mechanical stress, also known as its resistance to failure. The process of forming a new surface involves the molecules in the bulk overcoming their intermolecular attraction to become new surface molecules.

The interaction forces between the molecules/atoms/ions in solids are generally strong, and the particles are arranged in a periodic, crystal lattice pattern. As a result, relative motion is much more difficult for these particles in solids. The surface energy of solids is anisotropic due to the anisotropy of the atomic composition and arrangement, and can be varied in different regions of the surface.

7.2.2 Solid–Liquid Interface and Wettability

The wetting or non-wetting behavior of a liquid on a solid is a commonly observed interfacial phenomenon in daily life. This process involves the interaction between three phases: gas, liquid, and solid. The boundary where the liquid, solid, and gas meet is referred to as the "three-phase line" (Fig. 7.1), and the shape of the liquid droplet at this interface is determined by the interfacial tension between the solid and gas (γ_{SG}), the solid and liquid (γ_{SL}), and the gas and liquid (γ_{GL}). The contact angle, denoted as θ, represents the orientation of these vectors and reflects the balance between these three interfacial tensions, with the goal of minimizing the interfacial area. The three interfacial tension vectors acting at the three-phase line are related through Young's equation:

$$\gamma_{SG} = \gamma_{SL} + \gamma_{GL}\cos\theta \tag{7.1}$$

Young's equation is the basis for the study of wettability between liquid and solid. The contact angle can be used to judge wettability as follows.

$\theta = 0°$, perfect wettability, the liquid can spread out on the solid.

$0° < \theta < 90°$, hydrophilic or superhydrophilic, the smaller θ, the better wettability.

$90° < \theta < 180°$, nonwetting.

$150° < \theta < 180°$, superhydrophobic or superoleophobicity when the liquid is water or oil, respectively.

$\theta = 180°$, completely nonwetting, the liquid appears as a ball on the solid surface.

It is important to note that Eq. 7.1 is only valid for ideal "dry" wetting scenarios where the wetting liquid has zero vapor pressure. In reality, most wetting liquids will evaporate above the three-phase line, leading to "wet" wetting, as a result of adsorption of a monolayer or submonolayer of molecules on the solid substrate above the three-phase line. The roughness and chemical composition of a solid surface can also affect the difference between the actual measured contact angles and those calculated using Young's equation.

To better understand the wetting of rough surfaces, Wenzel et al. introduced the concept of "apparent contact angle (APCA)" to account for the effects of surface roughness on wetting behavior. They hypothesized that a liquid can always fill the rough structure as illustrated in Fig. 7.2a. By analyzing the small changes in energy that occur during the wetting process of droplets on rough surfaces, they found that the APCA has a relationship with the intrinsic contact angle, as described in their hypothesis:

$$\cos\theta^* = r\cos\theta \tag{7.2}$$

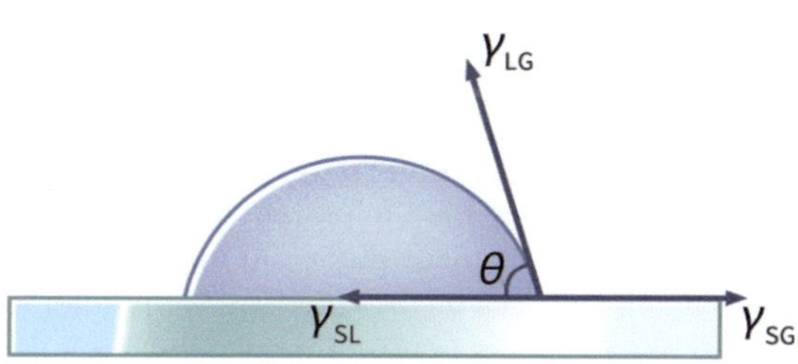

Fig. 7.1 Derivation of Young's equation

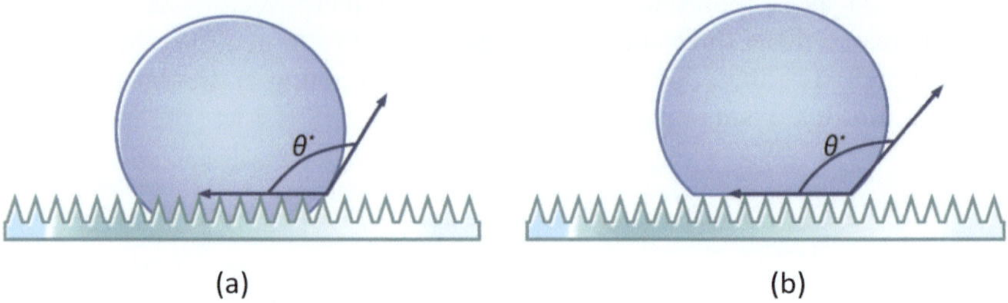

Fig. 7.2 **a** Wenzel's wetting model; **b** Cassie-Baxter's wetting model

This is the famous Wenzel's wetting equation, where r is defined as the roughness factor and is the ratio of APCA and intrinsic contact angle. The behavior of liquids on solid surfaces is influenced by the degree of hydrophilicity or hydrophobicity of the solid. In hydrophilic conditions ($\theta < 90°$), the apparent contact angle (APCA) decreases with an increase in roughness. Conversely, in hydrophobic conditions ($\theta > 90°$), APCA increases with an increase in roughness, leading to improved hydrophobicity. While Wenzel's equation is suitable for describing the wetting behavior of both hydrophilic and part of hydrophobic materials, it does not fully account for the behavior of superhydrophobic materials (APCA > 150°). These materials are thought to form a composite phase boundary containing solid-liquid and gas-liquid interfaces [Fig. 7.2b] rather than a solid-liquid fully infiltrated contact interface. Cassie and Baxter expanded upon Wenzel's equation to further explain the phenomenon of dew droplets rolling on the surface of lotus leaves, while APCA satisfies the following equation:

$$\cos\theta^* = f_1\cos\theta_1 + f_2\cos\theta_2 \tag{7.3}$$

This is the famous Cassie-Baxter equation, where θ_1 and θ_2 are the intrinsic contact angles of the liquid over the solid and the gas, respectively. f_1 and f_2 are the ratio of solid–liquid contact area to apparent contact area and the ratio of gas–liquid contact area to apparent contact area, respectively. They thought that some special microstructure on the surface of superhydrophilic materials can effectively trap part of the gas and makes it impossible for droplets to infiltrate into the interior of the microstructure.

7.2.3 Solid–Solid Interface

A solid–solid interface denotes the boundary between two distinct solid phases, resulting in the formation of interfacial energy. Adhesion is the bonding phenomenon that takes place at a solid interface, arising from the interaction between the two solid phases at the interface. This interaction can be attributed to various van der Waals forces between molecules, such as orientation force (also known as dipole–dipole force), induction force, dispersion force, or chemical bonding, such

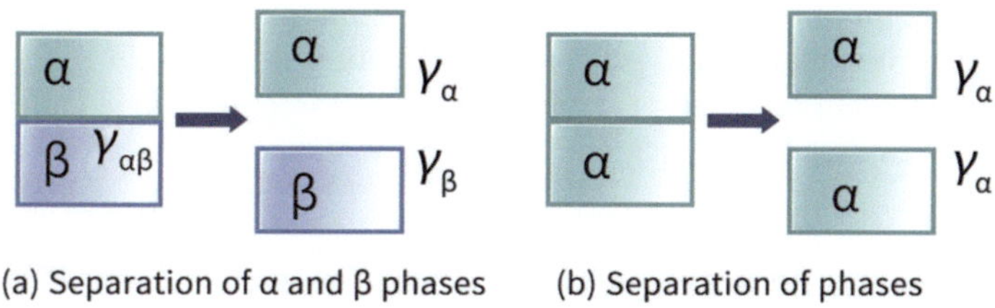

Fig. 7.3 Separation of the phase interface

as ionic bond, covalent bond, metallic bond, etc. In addition, micro-mechanical connections may also play a role in the adhesion process, making it an intricate physical and chemical process.

The interfacial tension between phase α and phase β is denoted as $\gamma_{\alpha\beta}$ [Fig. 7.3a]. The energy required to separate α and β into independent phases is the work of adhesion (W_a), which has the following relationship with the surface tension (γ_α and γ_β) and the interfacial tension $\gamma_{\alpha\beta}$:

$$W_a = \gamma_\alpha + \gamma_\beta - \gamma_{\alpha\beta} \tag{7.4}$$

When the two phases are the same [Fig. 7.3b], the interface disappears and $\gamma_{\alpha\beta} = 0$, $\gamma_\alpha = \gamma_\beta$, where the work of adhesion (W_a) is equal to the cohesive energy (W_c). The greater the cohesive energy of an object, the greater the work required to separate it to create a new surface. Both work of adhesion and cohesive energy are significant physical quantities in surface chemistry. It can be seen from Eq. 7.5 that:

$$W_a = W_c = 2\gamma_\alpha \tag{7.5}$$

7.3 Interface Structures in Crystals

Bulk materials typically exist in a polycrystalline state, where adjacent grains with different orientations in the polycrystalline body form a grain boundary. The existence of grain boundaries significantly affects the properties of materials. In metal smelting and heat treatment processes, controlling grain size is an important method to achieve high-strength, high-ductility materials. In ceramic functional materials, utilizing various physical effects of grain boundaries can produce useful functional components. Even inside the grains, the atomic structure is not always ideal single crystals and some grains can be divided into several sub-units or subgrains, with slight misalignment, besides defects such as vacancies and dislocations. The average diameter of grains is generally between 0.01–0.25 mm, while the average diameter of subgrains is typically around 10^{-3} mm. The atomic structure of the grain boundary is related to the grain orientation, where the structure of a grain boundary with greater misorientation is generally more complex. The problem of grain boundaries is typically divided into low-angle and high-angle grain boundaries based on the extent of misorientation.

7.3.1 Degrees of Freedom of the Interface

The nature of a grain boundary is largely determined by its structure, which in turn depends on the relative orientation of the grains. To determine the orientation of two grains, we can rotate one part of a grain relative to the other by an angle θ around a specific rotation axis μ in a reference coordinate system. In a three-dimensional system, determining the orientation of μ requires 2 variables, so μ and θ together require 3 degrees of freedom to determine the relative orientation of the two grains. The orientation of a grain boundary relative to a crystal can be represented by its normal n, and 2 degrees of freedom are required to determine this orientation in the coordinate system. Therefore, describing a grain boundary geometrically requires 5 degrees of freedom. In the case of heterogeneous interfaces, 5 degrees of freedom are also required to describe the orientation of the interface between them.

7.3.2 Low-Angle Grain Boundaries

When the orientation difference between two grains is less than 10°, the grain boundary between them is called a low-angle grain boundary. There are two types of low-angle grain boundaries: tilted and twisted grain boundaries. Tilted grain boundaries consist of edge dislocations, while twisted grain boundaries consist of screw dislocations.

Symmetrically tilted grain boundaries are single-degree-of-freedom grain boundaries, which can be described as an interface formed by two parts of the same crystal tilted to each other at an angle of $\theta/2$, as shown in Fig. 7.4. The interface can be regarded as composed of a row of parallel edge dislocations. For example, as shown in Fig. 7.5a, if the interface is close to the (100) surface, there is one variable θ. Obviously, the relationship between dislocation distance D and Burgers vector $\boldsymbol{b}$ is:

$$D = \frac{|\boldsymbol{b}|}{2\sin\frac{\theta}{2}} \tag{7.6}$$

When θ is small, $\sin\frac{\theta}{2} \approx \frac{\theta}{2}$, so there is a dislocation spacing on the grain boundary:

$$D = \frac{|\boldsymbol{b}|}{\theta} \tag{7.7}$$

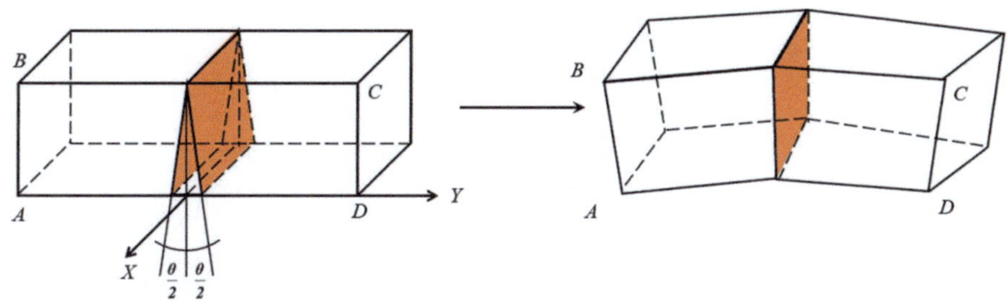

Fig. 7.4 Symmetrical tilted grain boundary structure

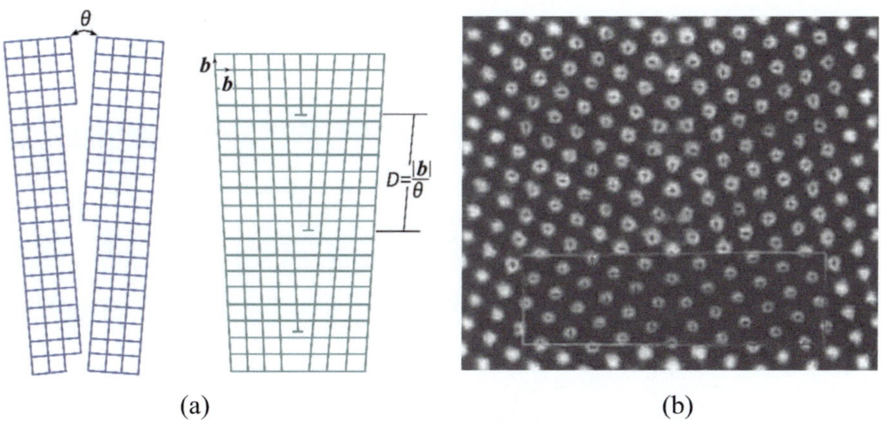

Fig. 7.5 **a** Symmetrical tilted grain boundary model; **b** experimentally observed symmetrical tilted grain boundary

If $|\boldsymbol{b}| = 0.25$ nm, when $\theta = 1°$, put it into the above formula to get $D = 14.2$ nm; when $\theta = 10°$, get $D = 1.4 \text{ nm} \approx 5|\boldsymbol{b}|$ (5 atomic distance). Obviously, at this time, there will be one dislocation at 5–6 atomic distances, and if the dislocation density is too large, the grain boundaries will all be dislocation centers, and this structure is unstable. Therefore, this model does not apply when θ is large. Figure 7.5b are symmetrically tilted grain boundaries observed under a high-resolution electron microscope.

An asymmetric tilted grain boundary is defined as when tilting grain boundary is not close to the (100) plane, but any (hk0) plane, this kind of asymmetric grain boundary needs to be represented by two parallel edge dislocations with Burger's vectors of [100] and [010]. Assuming that the included angle between the (hk0) plane and the [100] direction is φ (see Fig. 7.6), the numbers of two kinds of dislocations in the unit distance along AC are:

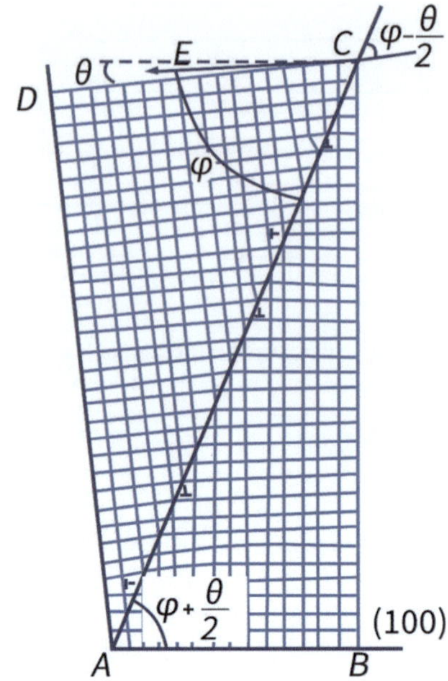

Fig. 7.6 Dislocation model of asymmetric tilted grain boundary (simple cubic)

$$\rho_1 = \frac{EC - AB}{|\boldsymbol{b}| \times AC} = \frac{2}{|\boldsymbol{b}|}\sin\frac{\theta}{2}\sin\varphi \approx \frac{\theta}{|\boldsymbol{b}|}\sin\varphi \tag{7.8}$$

$$\rho_2 = \frac{CB - AE}{|\boldsymbol{b}| \times AC} = \frac{\theta}{|\boldsymbol{b}|}\cos\varphi \tag{7.9}$$

A twisted grain boundary is another type of low-angle grain boundary. Its structure is shown in Fig. 7.7a. A crystal is cut along the middle plane, and then the upper half of the crystal is wound around the axis perpendicular to the cut plane. After turning an angle, it will meet with the lower half of the crystal to form a surface.

Figure 7.7b depicts the structure of a twisted grain boundary between two simple cubic grains. The (001) plane is a common plane (i.e., the drawing surface), and it can be observed that this grain boundary consists of two sets of screw dislocation networks. Although the atomic positions on either side of the twisted grain boundary are not consistent, this mismatch can be concentrated on some atomic positions while the remainder remain consistent. The parts that do not match are screw dislocations, and the entire twisted grain boundary is a lattice composed of two sets of intersecting screw dislocations, with one group parallel to the [100] axis and the other parallel to the [010] direction. The distance D can be expressed with the following equation:

$$D = \frac{|\boldsymbol{b}|}{\theta} \tag{7.10}$$

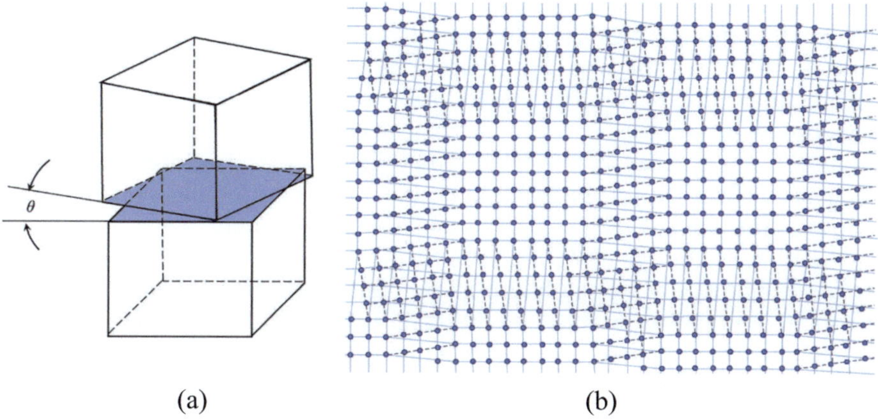

Fig. 7.7 **a** Twisted grain boundary structure; **b** twisted grain boundary model

In general, grain boundaries can have any orientation relationship between the rotation axis and the two adjacent grains, and therefore may require the use of both edge dislocations and screw dislocations to accurately describe their atomic structure. Describing such a grain boundary requires a total of 5 degrees of freedom to be fully defined. Tilted and twisted grain boundaries are special cases of low-angle grain boundaries that can be described using fewer degrees of freedom due to their specific crystallographic orientation relationships.

7.3.3 High-Angle Grain Boundaries

Grain boundaries with orientation angles greater than 10° are commonly referred to as high-angle grain boundaries. These boundaries have a complex atomic arrangement that cannot be described by the dislocation model. It is believed that the structure of high-angle grain boundaries is similar to the model shown in b8(a), where the interface between adjacent grains with different orientations consists of irregular steps. The boundary contains atoms belonging to both crystal grains (atom D), as well as atoms not belonging to any crystal grain (atom A). Additionally, there are compression (region B) and expansion (region C) regions due to the simultaneous influence of atoms from the two grains with different orientations. In general, the arrangement of atoms on high-angle grain boundaries is disordered, but with some ordered regions. As the orientation difference increases, the area of the disordered region increases accordingly. In pure metals, the width of high-angle grain boundaries does not exceed 3 atomic spacings.

In recent years, a "coincident position lattice" model of high-angle grain boundaries has been proposed based on the application of field ion microscopy to study grain boundaries. For example, as shown in Fig. 7.8b, when the orientation difference between two adjacent grains is 28.1° (equivalent to grain 2 being rotated

28.1° around [100] relative to grain 1), some of the atoms on the lattices of the two crystal grains will overlap regularly as the lattices extend to each other through the grain boundary. The new lattice formed by the overlapping positions of these atoms is larger than the original crystal lattice and is generally called the overlapping position lattice. The density of the coincident position lattice, which is 1/17 in the above specific example, reflects the proportion of coincident sites. It should be noted that the density of the coincident position lattice can vary due to differences in crystal structure and the chosen axis and angle of rotation.

According to the coincidence position lattice model, there are a certain number of atoms that overlap the lattice in the high-angle grain boundary structure. The more overlapping positions on the grain boundary, (i.e., the more atoms on the grain boundary shared by two grains), the smaller the distortion of the atomic arrangement, and the lower the grain boundary energy. However, the specific orientations of overlapping lattices of different crystal structures are limited. The model suggests that the higher the density of coincidence positions contained in the grain boundary, the better the two grains match on the interface, resulting in smaller distortion and lower energy. Therefore, on one hand, the grains on both sides of the grain boundary tend to obtain the orientation relationship of the high-density coincident position lattice, and on the other hand, the grain interface tends to coincide with the close-packed plane of the coincident position lattice.

If the grain boundary deviates from the close-packed plane of the coincident position lattice at a certain angle, the grain boundary tends to coincide with the close-packed plane of the coincident position lattice in sections, connected by a small step in the middle, in order to minimize energy. The size of the step is inversely proportional to the deviation angle. For example, in the coincident position lattice of the body-centered cubic crystal shown in Fig. 7.9, the grain boundary

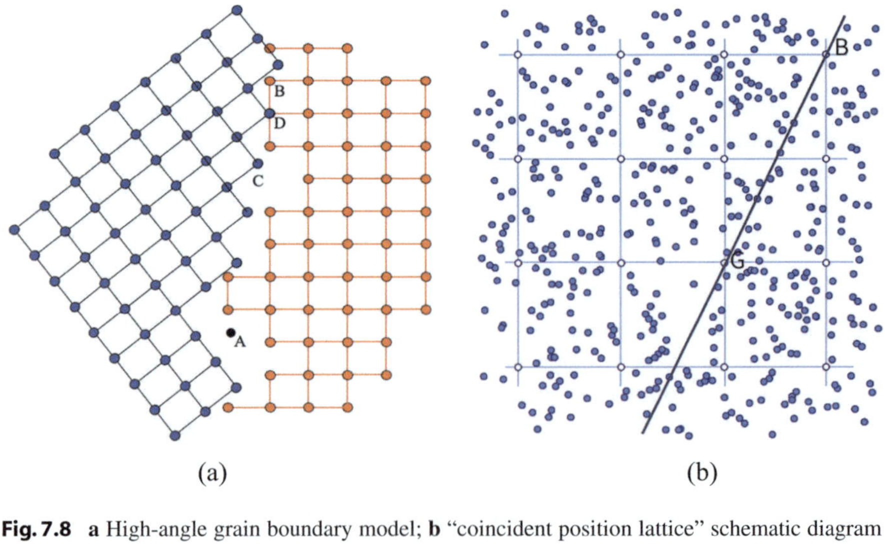

Fig. 7.8 **a** High-angle grain boundary model; **b** "coincident position lattice" schematic diagram

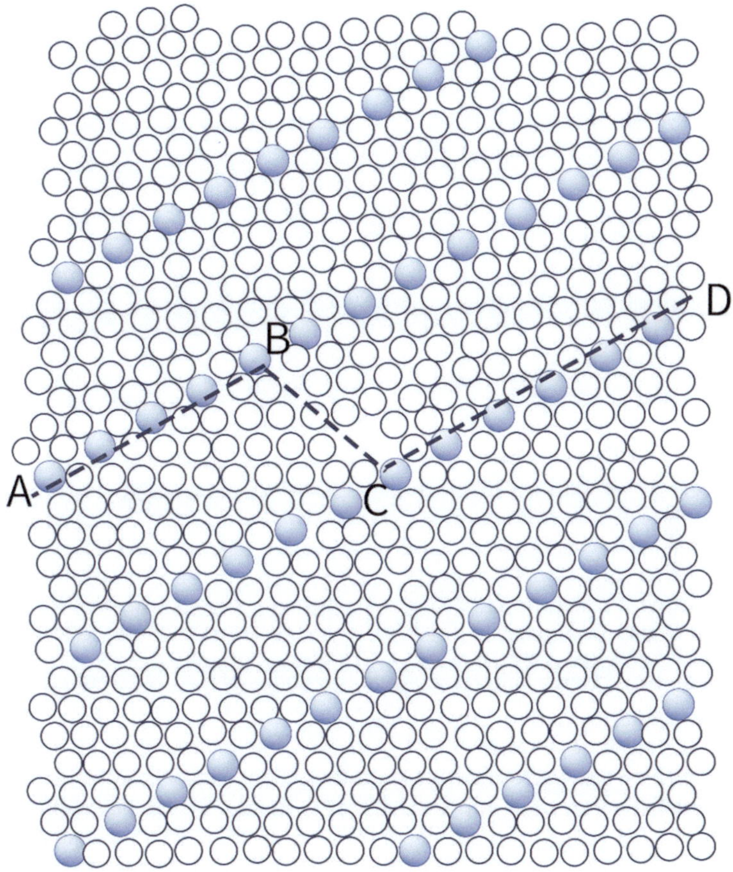

Fig. 7.9 The coincident position lattice in the body-centered cubic crystal

forms AB instead of going directly from A to D, and the two ends of CD coincide with the coincident position lattice (solid circles in the figure), and are connected by a small step BC in the middle. Despite the significant atom misalignment in the BC segment, the total energy of the grain boundary remains low due to the limited size of the steps.

7.3.4 Phase Boundary

If the adjacent grains belong to different phases, the interface in between is called a phase boundary. According to the arrangement of atoms on the interface, phase boundaries in solids can be divided into three types: coherent, semi-coherent and incoherent.

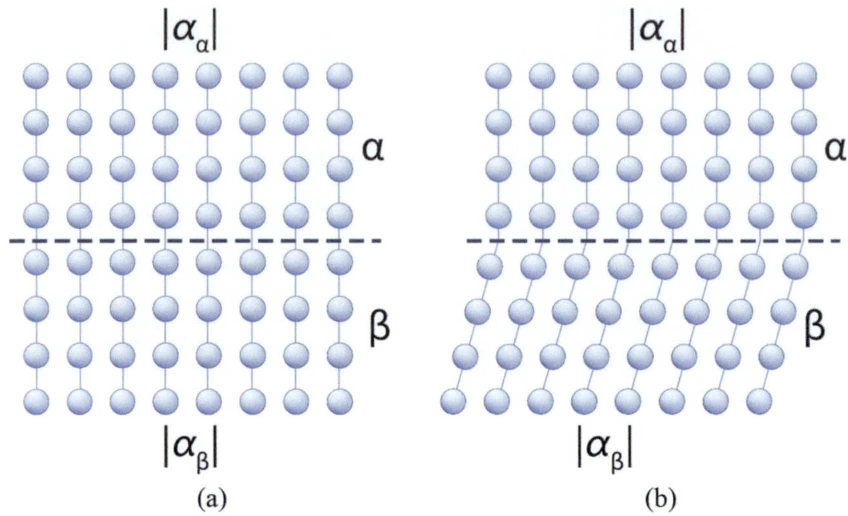

Fig. 7.10 **a** Strain-free coherent phase boundary; **b** strained coherent phase boundary

① Coherent Phase Boundary

The atoms on the coherent interface are at the nodes of two adjacent phase lattices at the same time, that is, the atoms of the two phases completely match each other on the phase interface, as shown in Fig. 7.10a. For example, in a Cu-Si alloy, a coherent interface can be formed between the silicon-rich phase of the hexagonal close-packed lattice and the copper-rich phase matrix of the face-centered cubic lattice. The lattice parameters of the $(111)_{FCC}$ and the $(0001)_{HCP}$ phases are the same, and the atomic spacing is also the same. Therefore, if the two crystals are adjacent along their close-packed planes and the close-packed directions are parallel, the formed interface is completely coherent, and two phases have the following orientation: $(111)_{FCC}//(0001)_{HCP}$.

If the atomic spacing on the interface is not the same, one or both of the two lattices may still remain coherent after a certain distortion, as shown in Fig. 7.10b, which will cause coherent distortion or coherent strain.

② Semi-coherent Phase Boundary

If a_α and a_β are the lattice constants of α and β, respectively, when there is no stress, the mismatch degree δ of the two lattices is defined as

$$\delta = \frac{a_\beta - a_\alpha}{a_\alpha} \tag{7.11}$$

When the δ value is small (<0.05), a coherent interface is formed. The distortion of the coherent interface on both sides increases the total energy of the system. For a relatively large atomic mismatch degree ($0.05 < \delta < 0.25$) or a larger interface area, from an energy point of view, the semi-coherent interface is more favorable than the coherent interface. On the semi-coherent interface, their mismatch can be periodically adjusted and compensated by edge dislocations, as shown in Fig. 7.11a, the mismatch of one-dimensional lattice can be replaced by a group of edge dislocations without long-range strain field to compensate, the spacing D of this group of dislocations should be:

$$D = a_\beta/\delta \tag{7.12}$$

When the value of δ is small, it can be approximated as:

$$D \approx |\boldsymbol{b}|/\delta \tag{7.13}$$

where b is the Burgers vector of the dislocation, $|\boldsymbol{b}| = (a_\alpha + a_\beta)/2$, except for the core part of dislocation on the interface, other places are almost completely matched. It can be seen that when the δ value is small, D is very large, and the interface between the two phases tends to be completely coherent, otherwise, D is small, so that the physical meaning of the dislocation structure is lost, indicating that the phase boundary cannot be described by the dislocation structure, then it should tend to be the incoherent interface.

③ Incoherent Phase Boundary

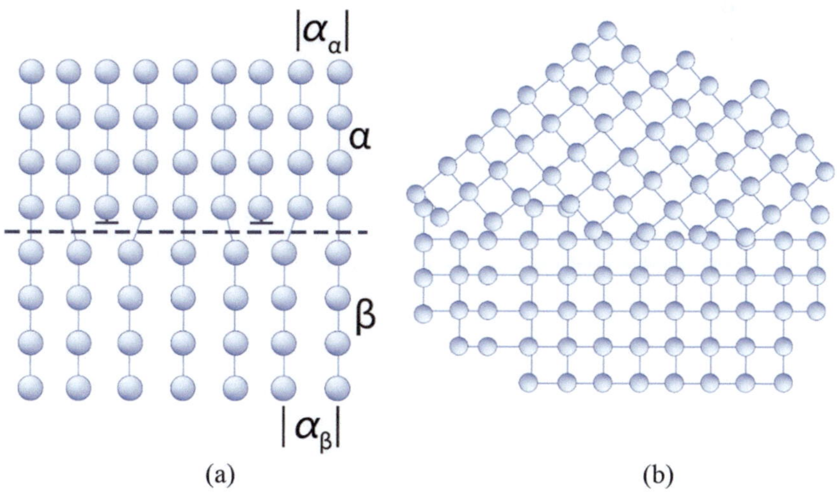

Fig. 7.11 **a** Semi-coherent phase boundary and **b** Incoherent phase boundary

When the atomic arrangement of the two phases at the phase interface is very different, that is, when δ is very large, only an incoherent interface can be formed [Fig. 7.11b]. This kind of phase boundary is similar to the high-angle grain boundary, which can be regarded as a very thin transition layer composed of atoms arranged irregularly.

7.4 Interfaces in Composite Systems

Composite materials are composed of a matrix and reinforcement, and the interfaces between these two phases are considered a new phase. The structure of these interfaces varies depending on the matrix and reinforcement and may be a reaction product layer, a diffusion bonding layer, a composition transition layer, a residual stress layer, an artificially introduced coating, or the gap between the matrix and reinforcement. The interfacial phase has several functions, including transfer effect, blocking effect, and protective effect. However, analyzing the interface can be challenging due to its irregular geometry, complex composition and structure, and uncertain mechanical property. There is no direct, accurate, and quantitative analysis method for its bonding strength, thickness, and stress state. Available experimental techniques for exploring interface characteristics include Raman spectroscopy, electron mass spectrometry, infrared scanning, and other limited methods.

7.4.1 Formation of Composite Interfaces

In composite materials, an "interface" refers to the boundary or contact area between the different constituent phases, like the fibers and the matrix, where they meet and interact with each other; it is a crucial aspect of composite performance as it directly affects how well stress is transferred between the phases, impacting the overall mechanical properties of the composite material. A strong and well-bonded interface is essential for achieving optimal composite properties, while a weak interface can lead to premature failure due to debonding or fiber pull-out.

7.4.2 Interface Structure and Theory of Polymer-Matrix Composites

For polymer-matrix composites, the formation of interfaces can be divided into two stages. The first stage involves the contact and wetting process between the matrix and reinforcement. Due to the different adsorption capacity of the reinforcement to various functional groups on the matrix molecule and other components in the matrix, substances that can reduce the surface energy are adsorbed, resulting in the formation of an interfacial polymerization layer. The second stage involves the curing of the polymer through physical or chemical methods, which leads to the formation of a fixed interface layer. The curing stage is influenced by the first stage and directly determines the structure of the formed interface layer. In the field of polymer composites, the interface layer's structure includes its thickness and microstructure, which plays a critical role in determining the composite's performance. The bonding force at the interface can be categorized into macro and micro bonding forces, where the former is affected by the material's geometrical factors such as surface roughness, cracks, and pores, and the latter is associated with chemical and secondary bonds. The ratio of these bonding forces depends on the material's surface properties and composition. Chemical bonding is the strongest bond and can be achieved through interface chemical reactions. Surface treatments are commonly used to enhance the interface adhesion between the two phases, and the presence of water can weaken the interfacial bonding force. The structure and properties of the interface and its vicinity are distinct from the components themselves, and its thickness and proportion in the interface layer affect the mechanical properties of the composite material. For glass fiber composites, the interface layer also includes a coupling compound generated by a coupling agent. The distance between the reinforcing fiber and the surface of the matrix is affected by several factors such as chemical bonding force, atomic group size, and the curing interface's shrinkage. There is a variety of composite interface theories, each with its own experimental basis, to explain some experimental phenomena. However, due to the interface's complexity, there is no universal theory that can perfectly apply to all composite materials.

The current research has yielded several theories, including wetting and adsorption theory, chemical bonding theory, diffusion theory, electrostatic theory, friction theory, and deformable layer theory. Despite these advancements, there remain experimental observations that these theories cannot fully account for. In some cases, contradictory explanations have emerged for the same phenomenon, while in other cases, multiple theories are required to explain a single phenomenon.

① Wetting and Adsorption Theory

Wetting is a crucial factor in interface formation. When complete wetting occurs between two components, the physical adsorption of the resin on the high-energy surface can provide bonding strength that exceeds the cohesive energy of the matrix, resulting in mechanical bonding and wetting adsorption. Microscopically, even smooth surfaces are uneven, so poor wettability between liquid resin and reinforcing material can limit contact to isolated points, while good wettability can allow the liquid phase to fill surface pits, leading to a large contact area and tight combination through mechanical anchoring. Although good wettability is important for interface contact, it is not the sole condition for interface bonding.

② Chemical Bonding Theory

In order to achieve effective bonding between two phases, the surfaces must possess active groups capable of forming an interface through chemical bonding. Chemical bonding theory is the most widely used and successful approach. An example of this theory in practice is the surface treatment of carbon and organic fibers, which involves processes such as surface oxidation, plasma, and irradiation to generate oxygen-containing active groups (–COOH and –OH) on the surface of the fibers. This improves the compatibility with resins, facilitating chemical bonding. In cases where direct chemical reactions are not possible, coupling agents can mediate chemical bonding between the two phases. For examples, organosilanes are coupling agents that can incorporate both organic and inorganic functionalities into the same molecule; titanate coupling agents can improve the elongation at break and impact strength of mixed plastics; peptide coupling reagents facilitate the formation of peptide bonds between amino acids.

③ Diffusion Theory

The diffusion theory is only applicable to polymer composite systems. The adhesion between two polymers is primarily due to the mutual diffusion of macromolecules at the interface. This process involves the diffusion, penetration, and entanglement of molecular chains, ultimately leading to the formation of an interfacial layer. Factors such as molecular weight, flexibility, temperature, solvent, and plasticizer can affect the diffusion process. The disappearance of the interface during mutual diffusion leads to the formation of a transition region, which can enhance bonding strength. Polymers with similar solubility can easily dissolve and diffuse into each other, resulting in a stronger bond. However, it should be noted that the diffusion theory has limitations and is not applicable for interfaces between polymer binders and inorganic substances.

④ Electrostatic Theory

The electrostatic theory suggests that if two surfaces have different charges, they will bond with each other through electron transfer. Although this theory has some validity, it has not been rigorously proven. Research suggests that the electrostatic force only has a significant effect when the charge density reaches 10^{21} e/cm^3, and cannot be ignored even when the density is as low as 10^{19} e/cm^3. The charge density range varies depending on the material type, preparation method and environmental conditions. For example, the charge density of metals is generally ranging from 10^8 to 10^{10} e/cm^3. Nonetheless, the electrostatic effect has limited contribution to bonding strength, and it cannot explain how temperature, humidity, and other factors affect bonding.

⑤ The Friction Theory

The formation of the interface between the matrix and the reinforcement is primarily attributed to the friction effect. The magnitude of the friction coefficient between the matrix and the reinforcement determines the strength of the composite. This theory can better explain the observed phenomenon that the strength of the composite interface decreases upon exposure to low molecular substances such as water, and the strength can be partially restored after drying. The presence of water and other small molecules on the interface reduces the friction coefficient and consequently weakens its ability to transfer stress, and conversely, the removal of such molecules restores the friction and bonding strength.

⑥ Deformation Layer Theory

Deformation layer theory in the context of composite materials refers to a theoretical approach that analyzes the deformation behavior of a composite laminate by considering the individual deformation of each layer within the laminate, taking into account the different material properties and reinforcement orientations of each layer. Take polymer-matrix composite as an example, during the curing of polymer composites, the polymer will experience shrinkage, leading to the formation of additional stress at the interface between the reinforcement and matrix due to the significant difference in thermal expansion coefficients. As a result, the performance of the composites may decrease. Additionally, external loads generate unevenly distributed stresses in the composites. Observing the microstructure reveals that the interface between the reinforcement and polymer is not uniform, with some areas experiencing higher stress concentrations than the average stress. This stress concentration can break the chemical bond between the fiber and matrix, leading to the formation of micro-cracks that can degrade the performance of the composite.

7.4.3 Interface Structure of Non-polymer-Matrix Composites

Compared to the interface structure of polymer matrix composites, the interface between the matrix and reinforcement of metal matrix composites and ceramic matrix composites exhibits significant differences due to the distinct matrix properties and varied processing techniques.

① Interface of Metal Matrix Composites

In metal matrix composites, the interaction between matrix and reinforcement often leads to the formation of compounds at the interface. The mutual diffusion of matrix and reinforcement creates a diffusion layer. The surface pretreatment coating of the reinforcement makes the shape, size, composition, and structure of the interface very complex. In recent decades, there has been a growing understanding of the importance of the interface in metal matrix composites, and many systematic and detailed studies have been conducted, providing valuable information.

There are mainly three types of interfaces in metal matrix composites. In the first type, where the reinforcement and matrix do not react and dissolve with each other, and the thickness is only a few molecular layers. Besides the original composition, there are almost no other substances on the interface. In the second type, where the reinforcement and matrix do not react but dissolve with each other, the interface is a diffused layer composed of the original composition. In the third type, where the reinforcement and matrix react with each other the interface is a reaction layer that contains submicron interfacial reaction substances.

Interface types are also related to manufacturing methods. Interface bonding of metal matrix composites can be divided into the following forms.

- **Physical bonding**

In the context of material surfaces and interfaces, physical bonding refers to the mechanical interlocking resulting from the rough surface morphology of the material, as well as the friction bonding formed when the reinforcement is wrapped by the shrinkage stress of the matrix. This type of bonding is purely physical and does not involve any chemical reactions. The strength of this bond is largely determined by the roughness of the reinforcement surface.

- **Combination of dissolution and wetting**

This combination corresponds to the second type of interface, characterized by a short-range interaction force between the reinforcement and matrix, typically a few atomic spacings. The presence of an oxide film on the reinforcement surface can hinder the infiltration of liquid metal, necessitating surface treatment of the reinforcement. For example, the oxide film can be removed by ultrasonic mechanical friction, which can reduce the contact angle between the reinforcement and matrix to less than 90° and promote wetting or local mutual dissolution, resulting in improved interfacial adhesion.

- **Reaction bonding**

Reaction bonding is characterized by the formation of a new compound layer, or interfacial reaction layer, between the reinforcement and the matrix, corresponding to the third type of interface. The interfacial reaction layer is typically not a single compound, but a mixture of reaction products. For example, boron-reinforced titanium aluminum alloy has a complex interfacial reaction layer consisting of various reaction products. Generally, as the degree of interfacial reaction increases, so does the bonding strength. However, since many interfacial reaction products are brittle materials, the interfacial layer can become prone to failure due to residual stresses, thereby reducing the bonding strength. Furthermore, some reinforcement surfaces may absorb air and oxidize to form a reaction bonding layer.

② Interface of Ceramic Matrix Composites

In ceramic matrix composites, the uniformity of the reaction layer formed between the reinforcement and the matrix results in good bonding, but it often makes the material brittle. The tensile strength of the composite material begins to decrease when the reaction layer reaches a certain thickness, defined as the first critical thickness. Further increase in the reaction layer thickness leads to a decrease in material strength until it reaches a certain strength, called the second critical thickness.

Silicon nitride is known for its high strength, hardness, corrosion resistance, oxidation resistance, and thermal shock resistance, but its poor fracture toughness limits its performance. Adding fibers or whiskers to silicon nitride can effectively improve its fracture toughness. However, since silicon nitride has a covalent bond structure and is difficult to sinter, sintering aids such as 6% Y_2O_3 and 2% Al_2O_3 are necessary when manufacturing composite materials. The forming process has a significant influence on the interface structure during the preparation of silicon nitride-based carbon fiber composites.

7.4.4 Interface Failure of Composites

The formation of composites relies on the interface between the matrix and reinforcement, which plays a vital role but is also a weak link. Composites are exposed to static, dynamic, and alternating stress, as well as environmental factors such as humidity and heat. In harsh environments and complex working conditions, the interface is often the first to fail, leading to a decline in composite performance. Therefore, it is crucial to comprehend the damage mechanisms of the interface and to develop strategies to delay or prevent interface damage under external loads and harsh environmental conditions.

① Types of Interface failure

- **Cohesive failure**

When the bonding strength between the reinforcing and matrix phases exceeds the cohesive strength of the interface layer, cohesive failure is more likely to occur. For instance, under high temperature conditions, with a thick interface layer, failure is likely to occur within the interface layer itself or in the matrix region adjacent to the interface, or even as cohesive failure of the reinforcement.

- **Adhesive failure**

Adhesive failure occurs when the bonding strength between the two phases is weaker than the cohesive strength of the interface layer, particularly when the interface layer is thin. In this case, the failure is likely to occur at or near the interface.

- **Mixed failure**

In this mode, both cohesive failure and adhesive failure can occur simultaneously. The failure strength of the interface cannot be explained solely by the thermodynamic equilibrium theory with surface tension, as it is not equal to the bonding strength. The failure strength is influenced not only by the bonding force, but also by the mechanical properties, geometry, size, and failure conditions (load and environment) of each phase.

② Interface Damage and Failure Mechanism

The damage and failure mechanism of a composite involves the work of its reinforcement, matrix, and interface. During manufacturing, the interface, reinforcement, and matrix contain inherent micro-damages. These damages nucleate and expand under the influence of external loads and environmental factors, eventually causing the failure of the composite material. For instance, microcracks on the matrix can propagate either perpendicular or parallel to the reinforcement surface, as illustrated in Fig. 7.12a, supposing the reinforcement is a fiber and the matrix is a resin.

As micro-cracks are influenced by external factors, they gradually propagate through the matrix, eventually reaching the fiber surface. During this process, energy is consumed, causing the propagation rate to slow down. The failure mechanism involves energy dissipation during crack propagation, which decelerates the

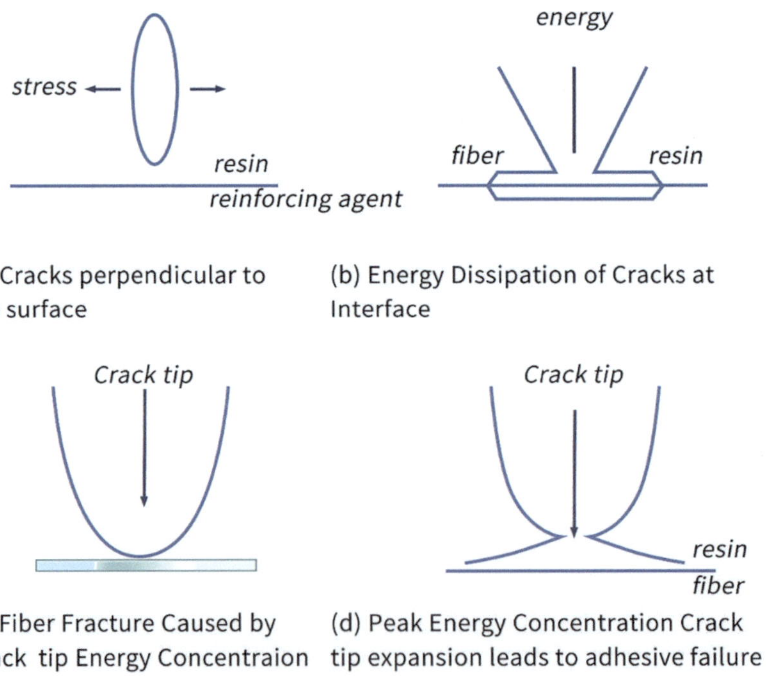

Fig. 7.12 Energy dissipation in interface failure

crack growth rate. The energy is expended to break interfacial bonding, which diminishes the energy concentration at the crack tip. As a result, the gradual failure of the interface occurs instead of the damage of the fiber. Figure 7.12b provides a schematic diagram of energy dissipation at the interface.

When large amount of chemical bonds are formed on the interface, energy dispersion is minimized during crack propagation, resulting in greater energy concentration at the crack tip. This may cause the fiber to break before the concentrated bond fractures, leading to composite material failure, as shown in Fig. 7.12c. Another scenario occurs when the crack propagates without fracturing the fiber, but destroys the interfacial bonding, causing energy dissipation (Fig. 7.12d). If the energy concentration at the crack tip is significant or continues to increase, the interfacial bonding and the fiber may both fracture.

③ Interface Damage Caused by Hydrothermal Environment

The application of composite materials may encounter the challenge of exposure to damp and hot environments, where the interface is a weak link in terms of resistance. During the material manufacturing process, water infiltrates through the channels connected by bubbles, and the osmotic pressure of the solution at the interface increases, leading to the formation of microcracks. The capillary effect of water promotes the expansion of these micro-damages, resulting in swelling of the resin and transverse tensile stress in the interface. When this stress exceeds the bonding strength of the interface, the interface is destroyed, and the fiber is corroded, leading to a deterioration in the performance of the composite material. Although the detrimental effects on the fibers can generally be reversed by drying, the degradation of the resin matrix is irreversible, and the surface of the resin may fall off in fragments, making it necessary to take waterproof measures.

7.5 Composite Principles

Structural and functional composite materials are two types of composite materials that differ in their intended functions. Structural composite materials are designed to primarily exhibit mechanical properties, whereas functional composite materials are intended to exhibit physical or chemical properties such as heat, electricity, magnetism, sound, light, or radiation. The properties of a composite material are determined by various factors including its shape, size, distribution (including continuity and orientation), and interface state. The relationship between a composite material and its components is known as the composite effect, which is influenced not only by the state of the composite but also by the properties of each component. Broadly speaking, composite effects have three characteristics:

- The performance of a composite material, except for some cases, is often not a simple summation of the properties of its components.
- The same component and the same compositing state often have different composite effects on different properties.

- When the same component has different compositing states, it will have different composite effects on a certain property

These three characteristics determine the diversity and flexibility of the design of composite material properties. The composite effect, which is the relationship between a composite and its components, is influenced by the properties of each component as well as the composite state. The composite effect can be classified into two types: linear effects and nonlinear effects. Linear composite effects include average effects, parallel effects, complementary effects, and offsetting effects, while nonlinear composite effects include multiplicative effects, induced effects, system effects, and resonance effects.

The average effect is the most commonly observed composite effect, which follows the well-known mixture law where a certain property of a composite material changes linearly with the volume fraction of its constituent materials. This law is based on the assumption that the interfacial contact is tight and intact, and the formulas used to estimate some mechanical properties of composite materials are corrected by experimental coefficients. For most composite materials, calculating the mechanical properties of laminates with different plies is achieved through the mechanical properties of a single layer, assuming an average effect in the inter-layer compositing.

The commonly used mixture law has two forms, namely

$$A_c = \sum A_i V_i \tag{7.14}$$

$$1/A_c = \sum V_i / A_i \tag{7.15}$$

where V_i is the volume fraction of component i in the composite material; the summation range includes all components that make up the composite material. Equation 7.14 is called the parallel mixture law, which is applicable to the density of composite materials, the elastic modulus of unidirectional fiber composites along the fiber direction, longitudinal Poisson's ratio, etc.; Eq. 7.15 is called the series mixture law, which is applicable to the transverse elastic modulus, longitudinal shear modulus and transverse Poisson's ratio of unidirectional fiber composite materials.

7.6 Surface and Interface Analysis

Compared to bulk materials, only a few techniques can be used to analyze the physical and chemical properties of surfaces and interfaces. The surface or interface layer of a solid contains approximately 10^{15} atoms/cm^2, which limits the number of atoms "seen" by probes. Additionally, the surface and interface are often unstable and their composition can change over time due to factors such as exposure to different atmospheres and temperatures. The surface or interface structure can also significantly impact material properties, with an oxide-covered surface having a completely different structure from the underlying material.

The continuous and rapid development of optoelectronics and microelectronics manufacturing is driving the study of material surfaces and interfaces. Table 7.1 lists most of the technologies that provide surface and interface information. While more than one technique can be used to measure a specific physical property, each technique may have specific limitations, and often multiple technologies need to be combined to obtain sufficient information.

7.6.1 Electron Spectroscopies

Electron spectroscopy comprises a set of powerful techniques for investigating the surface and interface of materials. By measuring the kinetic energies of the electrons emitted when a chemical species is bombarded with X-ray or ultraviolet radiation, the method determines the energy at which electrons are bound. The results provide insights into the structure because variations in atomic arrangement affect the energy required to eject electrons. A number of excitation electron spectroscopy techniques listed in Table 7.1 are vital to surface science research.

X-ray photoelectron spectroscopy (XPS) is an advanced analytical technique for the microscopic analysis of materials and components. The principle of XPS involves radiating samples with X-rays, which excites the inner electrons or valence electrons of atoms or molecules. These excited electrons, known as photoelectrons, have their energy measured to create a photoelectron spectrum, with the kinetic energy/binding energy of the photoelectron plotted on the abscissa and the relative intensity on the ordinate. XPS provides information not only on the atomic or molecular structure and valence state for chemical research, but also on the element composition, chemical state, molecular structure, and chemical bond. It offers general chemical information and surface information, as well as information on the distribution in depth, particularly when analyzing electronic materials. The X-ray beam incident on the sample surface is a photon beam, resulting in negligible damage to the sample. This feature is particularly advantageous for the analysis of organic materials or polymer materials.

Electron Energy Loss Spectroscopy (EELS) is a powerful technique that involves the inelastic scattering of incident electron beam in the sample. The amount of energy lost by electrons during scattering directly reflects the scattering mechanism, as well as the chemical composition and thickness of the sample. This enables analysis of the element composition, chemical bond, and electronic structure of thin films. When the incident electron beam strikes the sample surface, the backscattering phenomenon occurs, resulting in a mixture of elastic and inelastic scattered electrons returning to the surface. By detecting the number of energy loss electrons on the sample according to their energy distribution, a series of spectral peaks can be obtained, which is called the electron energy loss spectrum.

Table 7.1 Surface/interface techniques and physical properties measured

Technique	Surface information
Auger electron spectroscopy (AES)	Chemical composition, depth distribution
X-ray photoemission spectroscopy (XPS)	Chemical composition and bonding states
UV photoemission spectroscopy (UPS)	E_F versus E_C and E_V, Φ_M, valence band
Soft X-ray photoemission spectroscopy (SXPS)	Chemical composition, bonding, E_F versus E_C and E_V, Φ_M, valence band states
Constant initial (CIS) and final (CFS) state spectroscopies	Empty states above E_F
Angle-resolved photoemission spectroscopy (ARPES)	Atomic bonding symmetry, Brillouin zone dispersion
Surface extended X-ray absorption Fine Structure (SEXAFS)	Local surface bonding coordination
Inverse photoemission spectroscopy (IPS)	Unoccupied surface state and conduction band states
Laser-excited photoemission spectroscopy (LAPS)	Band gap state energies, symmetries
Low-energy electron (LEED) and positron (LEPD) diffraction	Surface atomic geometry
X-Ray diffraction (XRD)	Near-surface and bulk atomic geometry
Surface photovoltage spectroscopy (SPV or SPS)	Gap states energies versus E_C and E_V, Φ_M, qV_B
Infrared absorption spectroscopy (IR)	Gap state energies, atomic bonding and coordination
Cathodoluminescence spectroscopy (CLS)	Gap state energies, compound/phase band gap
Photoluminescence spectroscopy (PL)	Near-surface compounds, bandgap
Surface reflectance spectroscopy (SRS)	Surface dielectric response, absorption edges
Spectroscopic Ellipsometry (SE)	Composition, complex dielectric response
Surface photoconductivity spectroscopy (SPC)	Surface/near-surface gap state energies
Static secondary ion mass spectrometry (SSIMS)	Elemental composition, lateral mapping
Ion beam scattering spectrometry (ISS)	Energy transfer dynamics, charge density
Scanning tunneling microscopy (STM)	Atomic geometry, step morphology, filled and empty state geometry
Atomic force microscopy (AFM)	Morphology, electrostatic forces
Kelvin probe force microscopy (KPFM)	Φ_M, qV_B versus morphology
Magnetic force microscopy (MFM)	Magnetic moment and polarization
Scanning tunneling spectroscopy (STS)	Bandgap states, heterojunction band offsets
Field ion microscopy (FIM)	Atomic motion, atomic geometry

(continued)

Table 7.1 (continued)

Technique	Surface information
Low energy electron microscopy (LEEM)	Morphology, atomic diffusion, phase Transformation, grain boundary motion
Total external X-ray diffraction (TEXRD)	Lattice structure, strain
Low-energy electron loss Spectroscopy (LELS)	Chemical reactions, plasmon, interband excitations
Depth-resolved cathodoluminescence spectroscopy (DRCLS)	Sub-surface gap states, interface compound/ phase band gap
Cross-sectional cathodoluminescence spectroscopy (XCLS)	Heterojunction band offsets, interface states, Φ_M, qV_B
Ellipsometry	Sub-surface or interface dielectric response
Raman scattering spectroscopy (RS)	Interface bonding, strain, band bending
Confocal Raman spectroscopy (CRS)	Sub-surface bonding, strain
Rutherford backscattering spectroscopy (RBS)	Near-surface, bulk atomic symmetry, composition, depth distribution
Secondary ion mass spectrometry (SIMS)	Subsurface/interface elemental composition, depth distribution, lateral mapping
Ballistic electron energy microscopy (BEEM)	Barrier heights, heterojunction band offsets, barrier height lateral mapping
Scanning capacitance microscopy	Sub-surface doping, capacitance
Cross-sectional Kelvin probe force microscopy (XKPFM)	Φ_M, qV_B across interfaces (requires UHV)
High resolution transmission electron microscopy (HRTEM)	Interface lattice structure
Electron Energy Loss Spectroscopy (EELS)	Gap state energies, atomic bonding, lattice symmetry

7.6.2 Multi-technique UHV Chambers

Surface science experiments are typically conducted in ultra-high vacuum (UHV) chambers equipped with multiple analytical techniques to maintain and analyze the surface under various conditions. These chambers typically contain at least one analytical technique, such as XPS or AES, to measure surface composition. In addition to excitation sources, these chambers may include devices for in situ surface cleaning, such as ion bombardment guns or focused ion beams (FIBs), which can expose new surfaces for analysis. To study interfaces, these chambers may include metal evaporation or chemical deposition, as well as facilities for measuring thickness. Low-energy electron diffraction (LEED) devices, lasers for annealing, and electrical probes for measuring basic properties may also be included. Figure 7.13 is a typical UHV chamber, illustrating how these different tools are arranged around the manipulator.

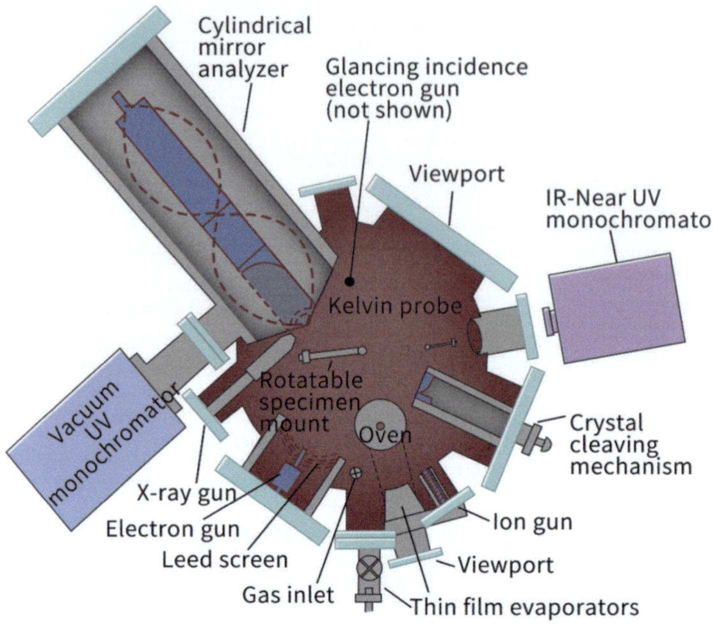

Fig. 7.13 Multi-technique UHV chamber schematic top view slice through the main set of horizontal ports

7.7 Functional Surface

To develop new, reliable, efficient, and multifunctional materials for advanced applications, precise control of their surface properties is essential. Surface functionalization via various biochemical and physical processes enables the modification of material properties without altering their bulk characteristics. For instance, surface modification leads to better interfacial bonding between the matrix and filler; surface functionalization with organic molecules has proven effective in modulating the electronic properties of 2D materials; and surface treatment is an effective approach to stabilize nanoparticles in a nanofluid without the use of surfactants. Three typical surfaces with specific functions will be discussed in the following sections.

7.7.1 Superhydrophobic Surface

Superhydrophobic materials are characterized by a rough surface and a water contact angle exceeding 150°, resulting in poor wettability and water sliding along the surface. The discovery of superhydrophobicity on lotus leaves was a landmark observation, where a high water contact angle and low hysteresis facilitated water droplets rolling off the surface, leading to water repellence and self-cleaning properties. The origin of the superhydrophobic and self-cleaning properties of lotus leaves was found to be rooted in the unique surface structure, which exhibits protruding nub structures coated with rough epicuticular wax crystalloids, thereby leading to the superhydrophobic property. Many researchers have developed hydrophobic coatings and materials by mimicking the structure of lotus leaves. Ge et al. created a lotus leaf-like structure by hydrolysis treatment of a polyacrylonitrile (PAN) membrane. This structure possesses a rough surface with randomly distributed nipples and demonstrates remarkable underwater superoleophobicity with an oil contact angle of 162° (Fig. 7.14), giving it an underwater anti-oil fouling property. Sun et al. prepared biomimetic multiwalled carbon nanotube/polydimethylsiloxane (MWCNT/PDMS) nanocomposites with superior flexibility and electrical conductivity using mechanical stirring and three roll shear grinding. They then used laser-etching to create micropillar arrays on the surface of the MWCNT/PDMS nanocomposites, resulting in excellent hydrophobicity and icephobicity.

Another example of a superhydrophobic surface and interface can be found in the legs of water striders, as shown in Fig. 7.15. Water striders are insects with the remarkable ability to stand and move on water surfaces. The structures of their legs comprise of nanoscale grooved units and a hydrophobic wax layer. To replicate these unique structures, Jiang et al. used PVB electrospun nanofibers to wrap silver wires and create artificial water strider legs. These legs enabled the assembled artificial water strider to float on water, even while carrying heavy glass strips. The maximum weight of the artificial device was 10 g, which is 5 times higher than that of a real water strider.

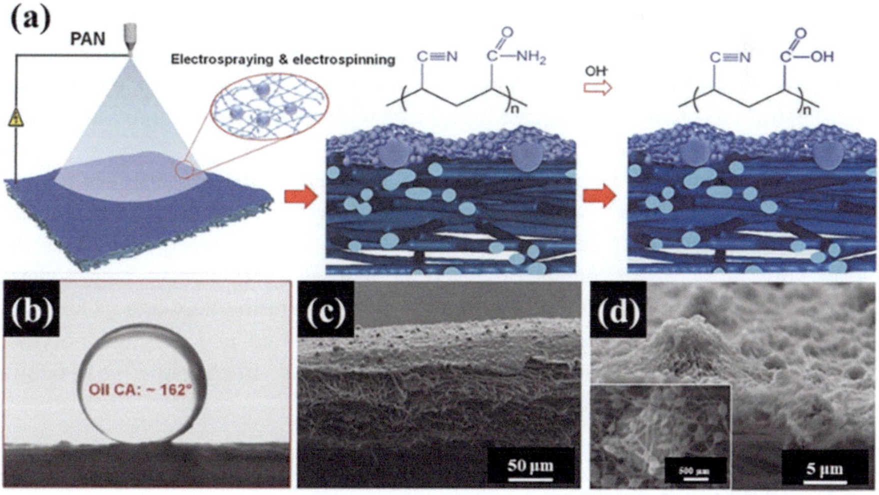

Fig. 7.14 **a** Schematic illustration of fabrication process of lotus-leaf-like surface layer. **b** Photograph of the oil droplet under water on the prepared membrane. **c** Cross-sectional SEM images and **d** high magnification SEM image of the lotus-leaf-like structure of the as-prepared surface layer

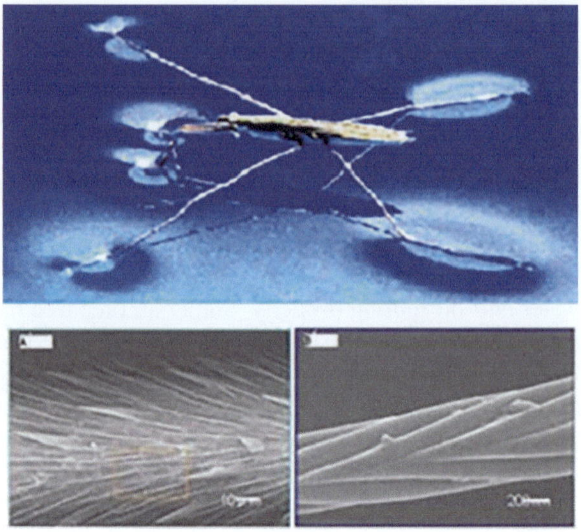

Fig. 7.15 Superhydrophobic surface microstructure of water strider

7.7.2 Wear-Resistant Surface

Friction-induced damage is a persistent challenge for materials under long-term service. Wear can significantly reduce the performance and lifespan of materials, ultimately leading to failure. In China, wear-related losses to industries are estimated to account for approximately 3–4% of GDP. To address this issue, a range of materials and technologies have been developed to minimize wear. Enamel is a well-known example of a wear-resistant material due to its robust outer layer. Qi et al. have developed a hierarchical coating that mimics the hardness of teeth enamel. As illustrated in Fig. 7.16, this involves depositing a thin quasi-linear LBL film containing 2D filler graphene oxide (GO) nanosheets atop a thick exponential LBL substrate to create a protective hard layer with an extremely high stiffness of 31.4 GPa and a hardness of 2.3 GPa.

Another example of an artificial enamel material is the Polymer-Infiltrated-Ceramic-Network (PICN), which encompasses a hybrid structure featuring two interpenetrating networks of ceramic and polymer. This material resembles the interlocking of prismatic bands observed in natural teeth. Unlike single-phase composites, PICN comprises two continuous interpenetrating networks. The interpenetrating/interlocking structure of the feldspar ceramic and acrylate polymer network contributes to this material's remarkable flexural strength relative to dentin and exceptional wear resistance.

7.7.3 Adhesive Surface

The primary requirement of any adhesive bond is to effectively connect two materials while resisting the loads and stresses that they will experience in their intended

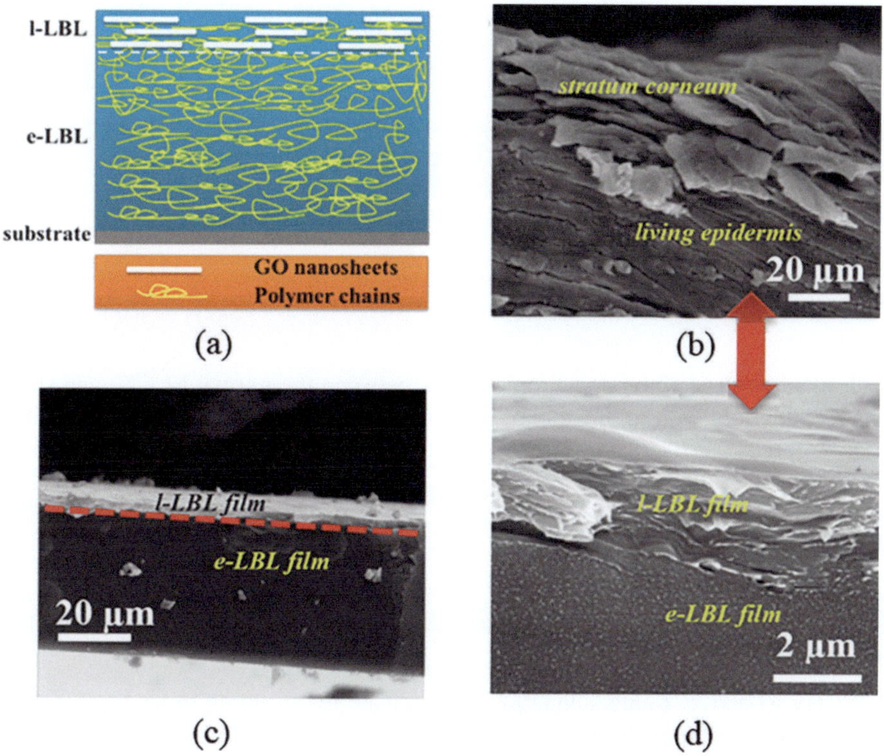

Fig. 7.16 Hybrid LBL films mimicking the structure of epidermis. **a** Schematic illustration of the composition and layout of hybrid LBL films; **b** The cross-sectional SEM images of epidermis from finger skin; **c** A panoramic cross-sectional SEM image of hybrid films showing the deposition of a thin l-LBL film on the top of a thick e-LBL film; **d** A close look at the cross-section of hybrid films showing striking structural similarities to the epidermis shown in (**b**)

use. The adhesion properties of a surface are crucial to designers and engineers, who face the ongoing challenge of enhancing manufacturing processes through the integration of a broader range of materials. Many organisms have evolved specialized adhesive organs and secrete substances to attach themselves to surfaces, either temporarily or permanently.

The gecko is an exceptional climber among animals, leveraging its foot pads to achieve reversible dry adhesion. As illustrated in Fig. 7.17, its attachment pads consist of complex hierarchical structures that include millions of microfibrils (setae) and nano-sized branches (spatula). This intricate topographical design enables physical interactions with virtually any surface, regardless of its texture or presence of contaminants, and at any angle. Researchers have employed a range of materials, from soft polymers to hard ceramics, to replicate this hierarchical fibrillar structure, leading to the development of adhesive surfaces with diverse applications. For instance, wedge-shaped 8.2 cm^2 PDMS fibrillar arrays

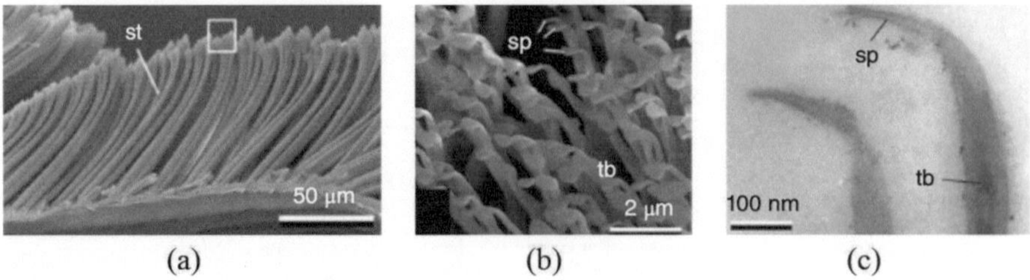

Fig. 7.17 SEM micrographs (at different magnifications) of **a** the setae, and **b**–**c** the spatula of the foot pad of a gecko. (st, seta; sp, spatula; tb, branch)

with tilted angles exhibit anisotropic forces between the normal (2.1 N) and shear (13 N) directions. Approximately 67% of the initial adhesion and 76% of the initial friction forces were retained after 30,000 cycles of attachment and detachment.

7.8 Questions

Q7.1 What are surface energy and surface tension, and what are their similarities and differences?

Q7.2 Try to derive Young's equation and indicate the meaning of each parameter. How to judge the wettability?

Q7.3 Why do interfaces play an important role in the performance of composites?

Q7.4 The surface of the alumina porcelain piece is coated with silver. When it is heated to 1000 °C, can liquid silver wet the surface of the alumina porcelain piece? (As is known that γ (Al_2O_3, solid) = 1 J/m^2, γ (Ag, liquid) = 0.92 J/m^2, γ (Ag, liquid/ Al_2O_3, solid) = 1.77 J/m^2).

Q7.5 The surface energy of periclase is 1000 erg/cm^2(1 erg = 1×10^{-7} J), and if the density is 3.68 g/cm^3, how many calories per gram is required to pulverize it into 1 μm particles? (Assuming that periclase before and after crushing are cubes).

Q7.6 The eutectic of the MgO-Al_2O_3-SiO_2 system is placed on a Si_3N_4 ceramic sheet. At the eutectic temperature, the surface tension of the liquid phase is 900 erg/cm^2, the interface energy between the liquid and the solid is 600 erg/cm^2, and the measured contact angle is 70.52°, calculate the surface tension of Si_3N_4.

Q7.7 The two-dimensional crystal contains second-phase particles and the equilibrium morphology of the particles is a rectangle with side lengths l_1 and l_2. The interface energies on both sides of the rectangle are γ_1 and γ_2, respectively. If the area of the rectangle remains unchanged, prove that the equilibrium shape of the rectangular particle is:

$$\gamma_1 l_1 = \gamma_2 l_2$$

Bibliography

Bai F, Wu J, Gong G et al (2014) ACS Appl Mater Interfaces 6:16237–16242
Gu Y, Zhang W, Mou J (2017) Adv Mech Eng 9(12):1–13
Zhou G, Yang G, Li X et al (2018) ACS Appl Mater Interfaces 10:14276–14280
Persson B, Gorb S (2003) J. Chem. Phys. 119 (21):11437–11444

Phase Diagrams

8

Learning Objectives

1. Create unary phase diagrams, label the various phase regions, and comprehend the significance of lines and triple points.
2. (a) Sketch simple isomorphous, eutectic, and peritectic phase diagrams, (b) label the various phase regions on the diagrams, and (c) identify liquidus, solidus, and solvus lines.
3. Determine the phase(s) present, composition(s) of the phase(s) by tie line, mass fraction(s) of the phase(s) by lever rule, and sketch the microstructure development during the solidification process, given a binary phase diagram, the composition of an alloy, and its temperature, assuming the alloy is at equilibrium.
4. (a) Locate the temperatures and compositions of all eutectic, eutectoid, peritectic, and peritectoid phase transformations for a given binary phase diagram, and (b) write the corresponding reactions for all these transformations during either heating or cooling.
5. (a) Specify whether an iron-carbon alloy containing 0.022 wt% C to 2.11 wt% C is hypoeutectoid or hypereutectoid, (b) name the proeutectoid phase, (c) compute the mass fractions of the proeutectoid phase and pearlite, as well as those of eutectoid ferrite and cementite, and (d) sketch the schematic microstructure development of this alloy during slow cooling from a single austenite-phase region to a temperature just below the eutectoid.
6. Master the compositional triangle, the lever rule, the "center gravity" principle, and the isothermal and vertical sections of ternary phase diagrams.
7. Understand ternary phase diagrams with three/four-phase equilibrium and their applications.

Z. Sun et al., *Fundamentals of Materials Science*,
https://doi.org/10.1007/978-981-96-3904-5_8

8.1 Introduction

A phase diagram is a map that depicts the relationship between different phases in equilibrium in a system, as a function of temperature, pressure, and composition. Understanding phase diagrams for alloy systems is essential because the microstructure and mechanical properties of alloys are strongly correlated, and the development of an alloy's microstructure is related to the features of its phase diagram. Phase diagrams are critical tools for material scientists and engineers to select alloys or compounds with a specific composition, microstructure, and properties, by designing and controlling heat treatment procedures.

The early development of the phase diagram in physics is attributed to prominent scientists such as Gibbs, Maxwell, and van der Waals. The term "phase" was first introduced by J. Willard Gibbs: "In considering the different homogeneous bodies which can be formed out of any set of component substances, it is convenient to have a term which shall refer solely to the composition and thermodynamic state of any such body without regard to its size or form. The word phase has been chosen for this purpose." Based on the work of Gibbs, Maxwell plotted the variables of volume, entropy, and energy in a three-dimensional space for a given temperature and pressure, although he did not construct a phase diagram directly. Van der Waals' work with real gases led to the discovery of coexisting liquid and gas phases below the critical temperature, for which he was awarded the Nobel Prize in physics in 1910.

8.2 Basics of Phase Diagrams

8.2.1 Definitions and Concepts

To start this chapter on phase diagrams, it is important to establish the definitions of several terms that will be commonly used in the subsequent sections.

① System

A **system** can be composed of a single substance or a series of substances that are completely isolated from the surroundings. The system being investigated can undergo changes in overall composition, temperature, or pressure. Such a system can comprise of gases, liquids, and solids made up of pure chemical elements, compounds, or both. For instance, the iron-carbon system consists of the Fe and C elements can be referred to the series of alloys made up of the two elements, regardless of their compositions.

② Component

In phase diagrams, the term **component** refers to the stable individual substances such as pure elements or compounds that make up all the phases present. For instance, in a H_2O state diagram, H_2O is the only component. In the Cu–Zn system, Cu and Zn are the two components that form the terminal solid solutions and the intermediate intermetallic compounds. While in the iron-carbon system, Fe and C are the components, it is sometimes more convenient to select Fe and Fe_3C (iron carbide or cementite) as the components.

③ Solubility Limit

Solubility refers to the ability of a chemical substance, known as the solute, to dissolve in a solvent. The solubility of a substance depends on various physical and chemical properties of the solute and solvent, as well as temperature and pressure. The concentration of a substance dissolved in a particular solvent measures the extent of its solubility. The maximum amount of one component, as solute, that can be dissolved in another component, as solvent, is called the **solubility limit**. The solubility limit is generally temperature dependent. For example, in a sugar-water system, the solubility limit of sugar increases with temperature in water. In the Cu-Ni system, Cu and Ni are completely soluble throughout the compositional range, whereas the solubility of C in Fe varies with temperature.

④ Phases

The term **phase** refers to a physically distinct, chemically homogeneous, and mechanically separable region of a system, where differences in composition, physical state, crystal structure, or properties distinguish one phase from another. For example, ice, water, and steam are different phases despite being identical in composition (H_2O) due to the differences in physical state. Another example is α-Fe and γ-Fe, which are two distinct phases of solid pure iron with different crystal structures (BCC and FCC, respectively). At a given temperature, a sugar-water liquid solution below its solubility limit is considered a single phase, but exceeding the limit causes sugar to become insoluble, forming two separate phases. A Cu-Ni solid solution with unlimited solubility is a single phase.

⑤ Microstructure

The term **microstructure** refers to the number of phases present, their amounts and sizes, and their spatial distribution. The microstructure is influenced by factors such as the constituent elements, their concentrations, and the heat treatment history, which includes the temperature, dwelling time, and cooling rate.

⑥ Phase Diagrams

A **phase diagram**, also known as an equilibrium phase diagram, is a graphical depiction of a material's physical states under varying conditions of temperature and pressure. It shows the solid, liquid, or gaseous states of a substance, such as H_2O, at different conditions. In the case of materials made up of two or more components, a phase diagram can represent the equilibrium states of phases with changes in temperature, pressure, and composition.

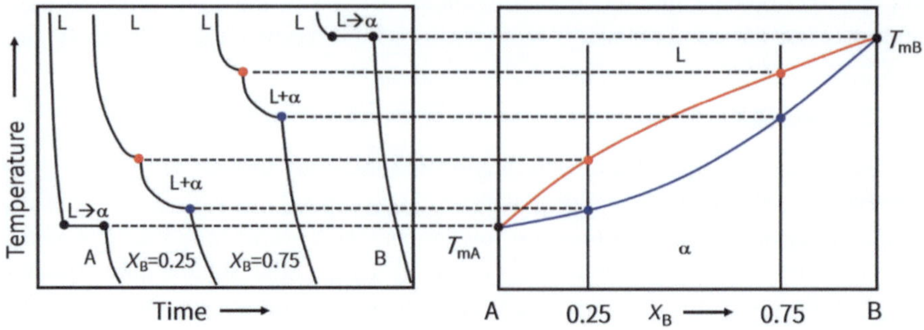

Fig. 8.1 Construction of a hypothetical phase diagram of the A-B system by thermal analysis

8.2.2 Constructions of Phase Diagrams

Either experimental methods or theoretical computations can be used to construct a phase diagram. Experimental methods include thermal analysis, electrical resistance, magnetic analysis, and hardness measurements. Among these methods, thermal analysis is the most widely used and will be used as an illustrative example.

For a given system consisting of components A and B, a set of alloys with different compositions (e.g., 0% B, 25% B, 75% B, 100% B) are chosen. The alloys are then melted at elevated temperatures to ensure the melting of all component elements, and their slow cooling curves are recorded experimentally. The turning points that correspond to the phase change temperatures, namely, the melting and solidification points, are determined from the cooling curves. Note that pure elements have only one turning point, while other compositions have two turning points corresponding to the temperature range for solidification. By connecting these turning points in a composition-temperature plot, the binary phase diagram can be constructed. Figure 8.1 shows an example of such a diagram.

8.2.3 Thermodynamic Basis of Phase Diagrams

① Phase Equilibrium

Equilibrium refers to the state of minimum free energy under a specific set of fixed conditions, including composition, temperature, and pressure. Even a slight alteration in the composition, volume, or physical state of any substance under these conditions can cause an increase in free energy.

- Gibbs free energy

The **Gibbs free energy** G of a system is defined by the following equation:

$$G = H - TS \tag{8.1}$$

where H is the enthalpy, and S is the entropy of the system, T is the absolute temperature.

Enthalpy is a measure of a system's heat content, which can be expressed as:

$$H = E + PV \tag{8.2}$$

where E is the internal energy of the system, P is the pressure, and V is the volume.

Phase equilibrium refers to a state in which the Gibbs free energy of a system is at a minimum, i.e., $\Delta G = 0$. In the case of condensed states of materials such as liquids and solids, the volume change during a phase transformation is small, and $H \approx E$. Therefore, under equilibrium conditions,

$$\Delta G = \Delta H - T\Delta S = 0 \tag{8.3}$$

- Chemical potential

When a small quantity of substance A ($\mathrm{d}n_{\mathrm{A}}$ mol) is added to a large amount of phase under constant temperature and pressure, the system's size increases by $\mathrm{d}n_{\mathrm{A}}$. Consequently, the Gibbs free energy of the system also increases by a small amount, $\mathrm{d}G'$. If $\mathrm{d}n_{\mathrm{A}}$ is sufficiently small, $\mathrm{d}G'$ can be proportional to the amount of A added, as expressed below:

$$\mathrm{d}G' = \mu_{\mathrm{A}}\mathrm{d}n_A(T, P\ \text{constant}) \tag{8.4}$$

$$\mu_{\mathrm{A}} = \frac{\mathrm{d}G'}{\mathrm{d}n_{\mathrm{A}}} \tag{8.5}$$

The proportionality constant μ_{A} is called the partial molar free energy of A or alternatively the **chemical potential** of A in the phase.

The total Gibbs free energy of a system is the sum of the chemical potentials of its components weighted by their amount:

$$G' = \sum n_i \mu_i \tag{8.6}$$

where n_i is the amount of component i.

The molar Gibbs free energy G_m is the Gibbs free energy of one mole of the substance or a mixture:

$$G_{\mathrm{m}} = \sum X_i \mu_i \tag{8.7}$$

where X_i is the molar fraction of component i.

In a one-component (A) system, the chemical potential of the component is equivalent to its molar Gibbs free energy, denoted as $G_{\mathrm{m}} = \mu_{\mathrm{A}}$. For a binary system consisting of A and B at a constant temperature and external pressure, where X_A and X_B vary with $X_A + X_B = 1$, we have the following:

$$G_m = X_A\mu_A + X_B\mu_B \tag{8.8}$$

- Equilibrium criteria

Various criteria can be used to assess the possible equilibrium of a system, including:

(a) A spontaneous phase transformation occurs when the Gibbs free energy of the system decreases.

(b) The equilibrium condition is represented by $dG = 0$, indicating the minimum Gibbs free energy of the system.

In a closed one-component (A) system, substance A may exist in either the α or β phase, which corresponds to G^α and G^β, respectively. The most stable structure corresponds to the lowest Gibbs free energy of the system. If α is more stable than β, $G_A^\alpha < G_A^\beta$; if the system is at equilibrium, $G_A^\alpha = G_A^\beta$, and α and β coexist.

(c) For a two-component or binary (A and B) system, two phases α and β are at equilibrium and correspond to G^α and G^β, respectively.

Total free energy change of the system is $dG = dG^\alpha + dG^\beta$.

At constant T and P,

$$dG^\alpha = \mu_A^\alpha dn_A^\alpha + \mu_B^\alpha dn_B^\alpha \tag{8.9}$$

$$dG^\beta = \mu_A^\beta dn_A^\beta + \mu_B^\beta dn_B^\beta \tag{8.10}$$

As the total amounts of A and B are constant, then $dn_A^\alpha = -dn_A^\beta$, and $dn_B^\alpha = -dn_B^\beta$. Therefore,

$$dG = (\mu_A^\alpha - \mu_A^\beta)dn_A^\alpha + (\mu_B^\alpha - \mu_B^\beta)dn_B^\alpha \tag{8.11}$$

As dn_A^α and dn_B^α are independent, at equilibrium $dG = 0$, then:

$$\mu_A^\alpha = \mu_A^\beta,\ \mu_B^\alpha = \mu_B^\beta \tag{8.12}$$

Specifically, the chemical potentials of each component are equal in all phases, indicating no chemical transfer.

② Gibbs Free Energy-composition Curves

In most applications, the external pressure remains constant, and therefore, the relationships between phase diagrams and Gibbs energies at a constant pressure will be discussed in the following sections.

According to the quasi-chemical approach, the free energy change that occurs when mixing a regular solution can be described by the equation $\Delta G_{\text{mix}} = \Delta H_{\text{mix}} - T\Delta S_{\text{mix}}$, where ΔH_{mix} and ΔS_{mix} represent the differences in enthalpy and entropy between the mixed and unmixed states. For an *ideal solution*, $\Delta H_{\text{mix}} = 0$, while for a *regular solution*, the mixing process can be either endothermic ($\Delta H_{\text{mix}} > 0$) or exothermic ($\Delta H_{\text{mix}} < 0$). Figure 8.2 illustrates the variation in ΔG_{mix} for different values of Ω and temperature, where Ω is calculated using the following equation:

$$\Omega = N_a z \varepsilon \tag{8.13}$$

where

$$\varepsilon = \varepsilon_{\text{AB}} - \frac{1}{2}(\varepsilon_{\text{AA}} + \varepsilon_{\text{BB}}) \tag{8.14}$$

where N_a is Avogadro's constant, z is the number of bonds per atom, ε_{AB} is the A-B bond energy, ε_{AA} is the A-A bond energy, ε_{BB} is the B-B bond energy, and

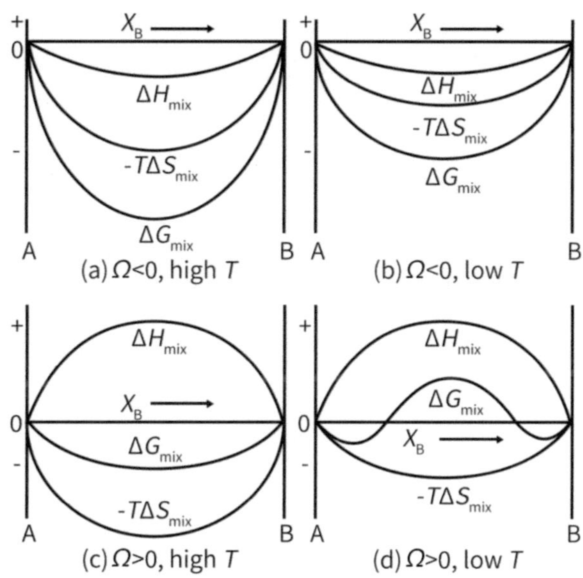

Fig. 8.2 The effect of ΔH_{mix} and T on ΔG_{mix} for a regular solution

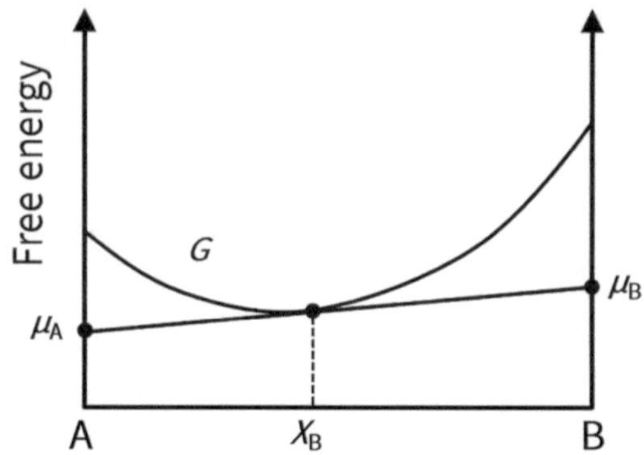

Fig. 8.3 The molar free energy curve variation with alloy compositions

ε is the difference between the A-B bond energy and the average of the A-A and B-B bond energies.

The factor Ω plays a crucial role in determining the arrangement of atoms in solid solutions in the following three cases:

1. Random solid solutions as $\Omega = 0$
2. Ordered solid solutions as $\Omega < 0$
3. Disordered solid solution as $\Omega > 0$

Consequently, when A and B components have the same crystal structure, the molar free energy will vary with the alloy composition, as depicted in Fig. 8.3. According to the definition of chemical potential (Eq. 8.5), the partial molar Gibbs free energies can be obtained from the molar Gibbs energy of the mixture by drawing a tangent line at the alloy composition, as shown in the figure below. The intercepts of the tangent at $X_B = 0$ and 1 correspond to μ_A and μ_B, respectively, for the solution of that composition corresponding to the contact point of the tangent.

③ Common Tangent Construction of Multiphase Equilibrium

If the A and B components have **different crystal structures**, two curves will be present for the two phases α and β. Equilibrium phases and compositions can be determined by graphical representations of the chemical potentials of the two components using the intercept method and common tangent construction, as shown in Fig. 8.4. The intercepts of the tangent of the Gibbs free energy curve of the α phase at composition X_e^{α} represent μ_A^{α} and μ_B^{α}, and similarly for the β phase. A common tangent line can be constructed for both the α and β curves, satisfying the equilibrium conditions for the binary system, i.e., $\mu_A^{\alpha} = \mu_A^{\beta}$ and $\mu_B^{\alpha} = \mu_B^{\beta}$.

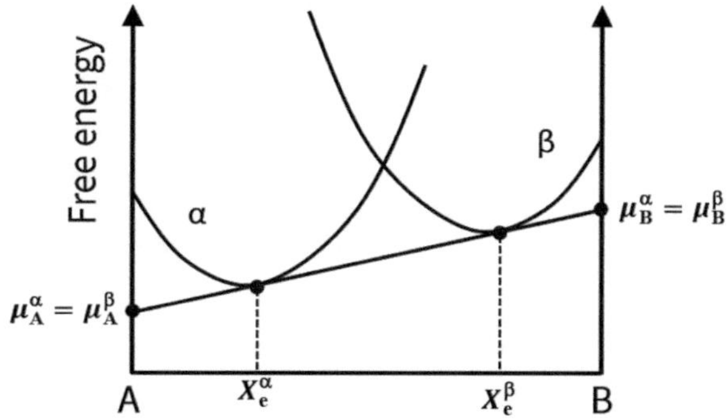

Fig. 8.4 Common tangent construction for a heterogeneous system

8.2.4 The Lever Rule and Gibbs Phase Rule

① The Lever Rule

Using the common tangent construction method, we can determine the equilibrium composition of each phase. For example, the compositions of two equilibrium phases (α and β) are shown as X_e^α and X_e^β, respectively, in Fig. 8.5.

We can calculate the amount of each phase using the **lever rule**. For example, the X_0 alloy is a mixture of two phases (α and β) containing two elements, A and B. The lever rule states that the mass fraction (W_i) of the α and β phases can be calculated as follows:

$$W_\alpha = \frac{X_e^\beta - X_0}{X_e^\beta - X_e^\alpha} \tag{8.15}$$

$$W_\beta = \frac{X_0 - X_e^\alpha}{X_e^\beta - X_e^\alpha} \tag{8.16}$$

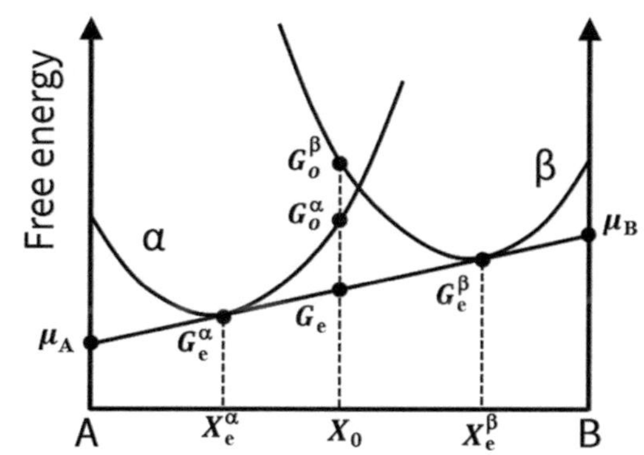

Fig. 8.5 Determination of the amount of each phase by the lever rule

② Gibbs Phase Rule

The Gibbs phase rule defines the mathematical relationship used to determine the stability of phases present in a system under equilibrium conditions. For a heterogeneous system composed of P phases and C components in equilibrium, the Gibbs phase rule assumes that the diffusion of a component from one phase to another doesn't involve any chemical reaction, the system is in equilibrium.

- **The total number of variables**

The total number of variables required to specify the state of a system includes pressure, temperature, and concentration. Each phase has an independent concentration variable of C-1 with respect to the C components. Therefore, the total independent concentration variables for P phases with respect to C components are $P(C\text{-}1)$.

$$\text{Total number of variables} = P(C-1) + 2 \tag{8.17}$$

- **The total number of equilibrium equations**

The condition for equilibrium in a binary system with two phases is given by Eq. 8.12. For systems containing multiple components and phases, a more general requirement for equilibrium is that the chemical potential of each component must be identical in every phase, i.e.,

$$\mu_A^{\alpha} = \mu_A^{\beta} = \mu_A^{\gamma} \cdots$$

$$\mu_B^{\alpha} = \mu_B^{\beta} = \mu_B^{\gamma} \cdots$$

$$\mu_C^{\alpha} = \mu_C^{\beta} = \mu_C^{\gamma} \cdots$$

The number of equilibrium equations for each component is P–1. For C components, the total number of equilibrium equations is $C(P–1)$. Hence,

$$\text{Total number of equilibrium equations involved} = C(P-1) \tag{8.18}$$

The number of degrees of freedom F is given by subtracting Eq. 8.18 from Eq. 8.17, namely

$$F = [P(C-1) + 2] - [C(P-1)] = C - P + 2 \tag{8.19}$$

A consequence of this general condition is the Gibbs phase rule. This expression states that if a system containing C components and P phases is in equilibrium, the number of degrees of freedom F is given by:

$$F = C - P + 2 \tag{8.20}$$

F is the number of externally controlled variables such as temperature, pressure, or composition that can be varied independently while maintaining the system equilibrium. If pressure is maintained constantly, one degree of freedom is lost, and the phase rule becomes:

$$F = C - P + 1 \tag{8.21}$$

The phase rule applies only to systems that are under conditions of true equilibrium without any exceptions and violations.

8.3 Unary Phase Diagrams

The unary phase diagram represents a system with only one component, where the composition is constant. It shows the variation of states of matter as a function of pressure and temperature. Therefore, the unary phase diagram is also known as a **pressure–temperature diagram**.

8.3.1 Phase Diagram of H_2O

① Description of the Phase Diagrams

Taking the H_2O phase diagram as an example, we can learn how to read a unary phase diagram, as shown in Fig. 8.6.

The H_2O system has only one component, H_2O, and the horizontal axis represents temperature (°C), while the vertical axis represents pressure (Pa).

Region (Area): There are three regions, and each corresponds to a single-phase area, i.e., only one phase present when the system is in equilibrium. For instance, *AOC* is a solid-phase region, and *AOB* is a liquid-phase region.

Line (curve): Line is the boundary separating regions, and along the lines two phases coexist in equilibrium, e.g., solid/liquid on *AO*, liquid/gas on *BO*.

Triple Point: Three phases are in equilibrium at the triple point, i.e., solid/liquid/gas phases coexist on point *O*.

Phase rule: It is $F = C - P + 2$ for unary phase diagram. When C equals 1, then the maximum number of phases coexisting is $P = C + 2 - F = 1 + 2 - 0 = 3$, which is consistent with the triple point. For other lines and regions, the phase rule can be applied as well.

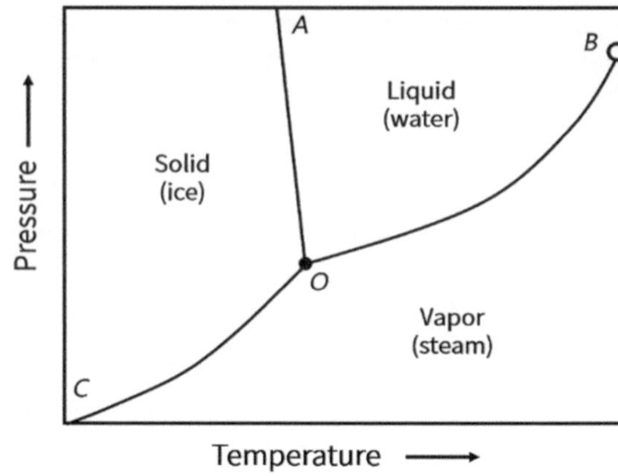

Fig. 8.6 Unary phase diagram of H_2O

② Endpoints of the Coexistence Curves.

Critical Point: The critical point is the point where the liquid/vapor phase equilibrium ceases to exist, such as the critical point *B* of water in Fig. 8.6. At the critical point of water (374.3 °C and 22.1 MPa), the liquid and vapor phases are no longer distinguishable and become a single phase. The phase in the region with temperature and pressure above the critical point is termed **supercritical fluid**. There is no critical point associated with the crystalline solid/liquid curve since those two phases cannot change continuously from one phase to the other.

③ Metastable Extensions of the Coexistence Curves

One feature of pressure–temperature phase diagrams is that the extensions of the two-phase coexistence curves through the triple point must go into the single-phase region of the third phase. In Fig. 8.7a, it can be observed that at a pressure larger than the triple point pressure, the extension of the solid/vapor curve (dashed curve *Ob*) extrapolates into the single-phase liquid region of the diagram. This can be deduced from the Gibbs free energy plots of the three phases at a pressure of P_1 (Fig. 8.7b). The phases in equilibrium at a given pressure are determined by the Gibbs free energy plots of the possible phases versus temperature. The stable phase at point *b* in Fig. 8.7a is the liquid phase as it has the lowest Gibbs free energy, along ac of Fig. 8.7b, and the solid/vapor equilibrium is seen to be **metastable** at this pressure. This rule is one of **Schreinemaker's rules**.

8.3.2 Other Representative Unary Phase Diagrams

① Phase Diagram of Pure Iron

In Fig. 8.8, the system comprises only one component, which is Fe. The unary phase diagram of iron shows **allotropy**, with five regions corresponding to liquid Fe and four solid phases. The α-Fe region is denoted by *AOC*, the γ-Fe region by

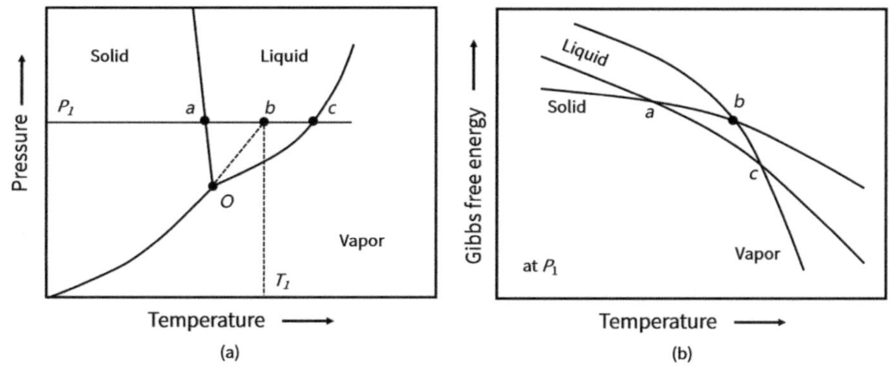

Fig. 8.7 The metastable extensions of the coexistence curves

bPcAOB, and the δ-Fe region by *aPc* as well as the ε-Fe region by *BOC*. Therefore, the unary phase diagram of iron includes five phases.

There are several lines separating the different solid Fe phases. Two solid phases coexist in equilibrium along these lines, such as α-Fe/γ-Fe on *AO* and γ-Fe/δ-Fe on *cP*. There are two triple points in this phase diagram. For instance, δ-Fe/γ-Fe/liquid phases coexist at point *P*, and α-Fe/γ-Fe/ε-Fe phases coexist at point O.

The **Clausius–Clapeyron equation** applies to the first-order transformations in the Fe phase diagram of Fig. 8.8, determining the slope of the two-phase coexisting lines. The equation is:

$$\frac{\mathrm{d}P}{\mathrm{d}T} = \frac{\Delta H}{T\Delta V} = \frac{\Delta S}{\Delta V} \tag{8.22}$$

where ΔH is the molar enthalpy change, ΔS is the molar entropy change, ΔV is the molar volume change, and T is the temperature at which the slope is measured.

Fig. 8.8 The unary phase diagram of pure Fe

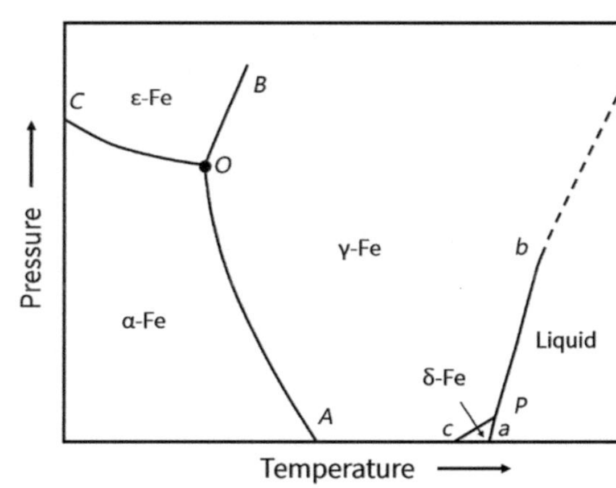

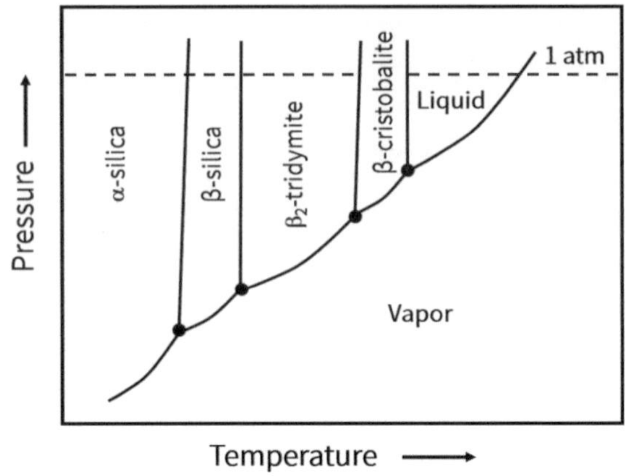

Fig. 8.9 The equilibrium phase diagram of pure SiO_2

As the enthalpy change on melting is usually positive, the sign of the change in volume on melting mostly decides whether the slope of the line is positive or negative. For example, the liquid/γ-Fe line has a positive slope due to the larger volume of liquid than that of γ-Fe. Likewise, the negative slope of γ-Fe/α-Fe is due to the denser atom packaging in the FCC of γ-Fe, i.e., smaller volume.

② Phase Diagram of SiO_2

Silica is the most widely used and abundant raw material, with four crystal structures at different temperatures and pressures: α-quartz, β-quartz, β_2-tridymite, and β-cristobalite, as shown in Fig. 8.9. The analysis method for the triple points, boundary lines, and phase regions of SiO_2 is similar to that of iron.

③ Possible Errors in Unary Phase Diagrams

There are several errors that can arise in the construction of phase diagrams due to violations of thermodynamic principles. These errors include: (1) Showing more than three phases coexisting in equilibrium (Fig. 8.10a), which violates the phase rule. (2) Inconsistent slopes of coexistence curves that don't follow the Clapeyron equation (Fig. 8.10b). (3) Metastable extensions of the two-phase coexistence curves that don't extend into the proper single phase region (Fig. 8.10c), which violates Schreinemaker's rules.

8.4 Binary Phase Diagrams

Binary phase diagrams depict the behavior of two-component systems with respect to changes in temperature and composition, generally keeping pressure constant for the sake of simplicity.

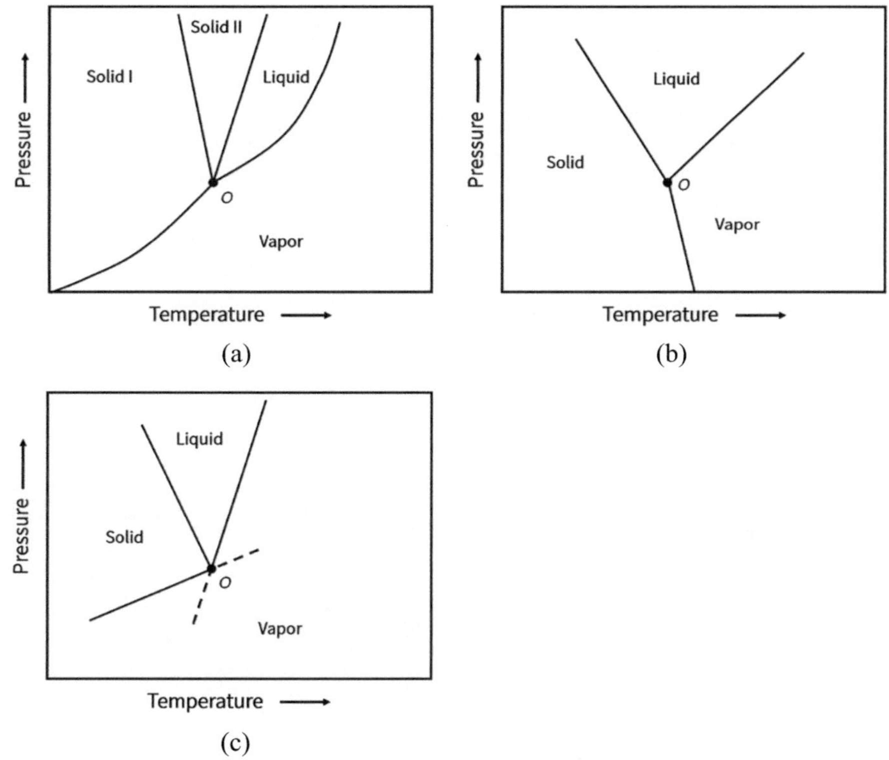

Fig. 8.10 Possible errors during the construction of phase diagrams

8.4.1 Binary Isomorphous Phase Diagrams

An isomorphous phase diagram is a type of binary phase diagram in which the two components are completely soluble across the entire composition range for both solid and liquid phases. Many systems have only isomorphous phase diagrams, such as Cu-Ni, Au–Ag, Au-Pt, NiO-MgO, Al_2O_3-Cr_2O_3, and almost all binary phase diagrams feature at least partially isomorphous transformations.

① Construction of Binary Isomorphous Phase Diagrams.

The equilibrium melting temperatures of the pure components occur when $G_S = G_L$, i.e., at $T_m(A)$ and $T_m(B)$. As the temperature increases, the free energy of both phases decreases. These variations are also crucial for A-B alloys since they determine the relative positions of G_A^S, G_A^L, G_B^S and G_B^L on the molar free energy diagrams of the alloy at different temperatures, as shown in Fig. 8.11.

At a high temperature $T_1 > T_m(A) > T_m(B)$, the liquid is the stable phase for pure A and pure B. In the simple case we are considering, the liquid also has a lower free energy than the solid at all intermediate compositions, as shown in

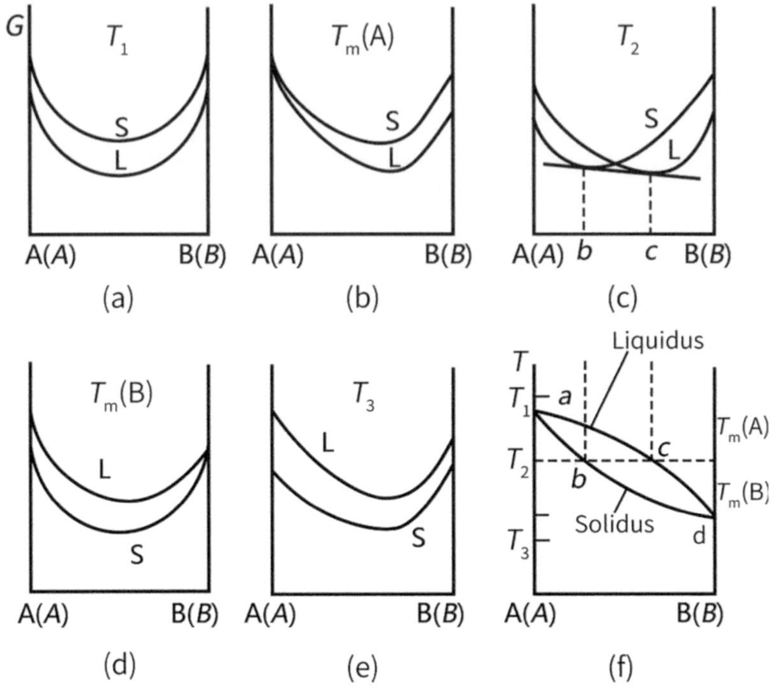

Fig. 8.11 The derivation of a binary isomorphous phase diagram from the free energy curves for the liquid (L) and solid (S)

Fig. 8.11a. Decreasing temperature will have two effects: first, G_A^L and G_B^L will increase more rapidly than G_A^S and G_B^S, and second, the curvature of the G curves will be reduced due to the smaller contribution of $-T\Delta S_{mix}$ to the free energy.

At T_m(A), as shown in Fig. 8.11b, $G_A^S = G_A^L$, corresponding to point a on the A-B phase diagram. At a lower temperature T$_2$, the free energy curves intersect, as shown in Fig. 8.11c. The common tangent construction indicates that alloys between A and b are solid at equilibrium, between c and B they are liquid, and between b and c equilibrium consists of a two-phase mixture (S + L) with compositions b and c. These points are plotted on the equilibrium phase diagram at T_2.

Between T_2 and T_m(B), G_L continues to rise faster than G_S, and points b and c in Fig. 8.11d will both move to the right, tracing out the solidus and liquidus lines in the phase diagram. Eventually, at T_m(B), b and c will meet at a single point, d in Fig. 8.11f. Below T_m(B), the free energy of the solid phase is below that of the liquid, and all alloys are stable as a single-phase solid.

② Description of Binary Isomorphous Phase Diagrams

An example of a binary isomorphous phase diagram is the Cu-Ni phase diagram shown in Fig. 8.12. The two metals in this system form a continuous series of

Fig. 8.12 A **a** The copper–nickel phase diagram. **b** A portion of the copper–nickel phase diagram for which compositions and phase amounts are determined at point *B*

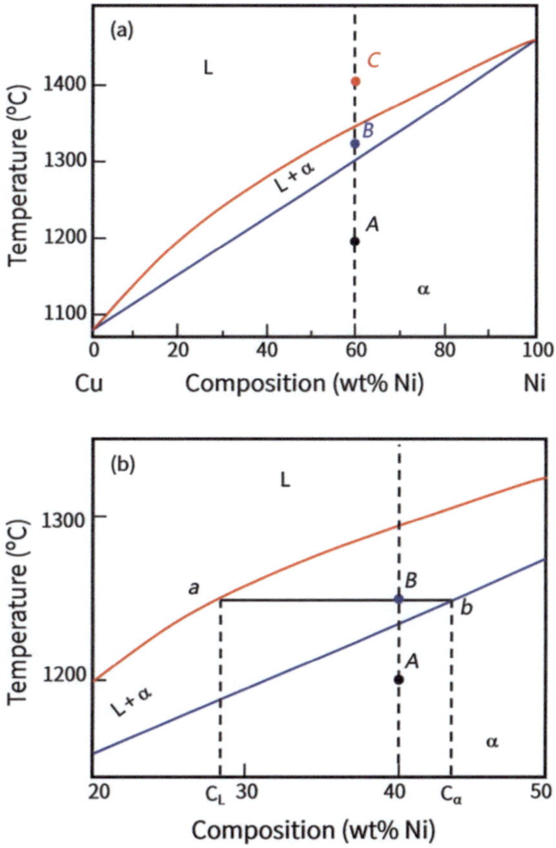

solid solutions with a FCC lattice, displaying **complete solubility** in both liquid and solid states. The close atomic sizes of Cu and Ni, with a difference of less than 15%, and similar atomic valence and electronegativity, allow for complete solubility. The lattice parameters vary almost linearly with atomic radius between copper (3.6153×10^{-8} cm) and nickel (3.5238×10^{-8} cm).

In the Cu-Ni phase diagram, the two components Cu and Ni can be combined to form a series of alloys. The horizontal axis shows the composition of the alloys, varying from 0 to 100% Ni in weight percentage, while the vertical axis represents temperature (°C), with a default pressure of 1 atmosphere.

Points: The phase diagram consists of two end points, which represent the melting points of the pure components Cu (1085 °C) and Ni (1455 °C).

Lines: Two lines separate the diagram into three regions. The upper line, forming the lower boundary of the liquid melt, is known as the **liquidus**, while the lower curve, forming the upper boundary of the solid crystals, is called the **solidus**. The liquidus and solidus indicate the respective compositions of the liquid and solid phases during the equilibrium cooling process.

Regions: Above the liquidus line, there is only the liquid phase (L), and below the solidus line, there is only the solid α phase (α). A two-phase region, or a mixture of liquid and solid phases (L + α), exists between the liquidus and solidus. The melting points of the two pure components are broadened to a melting range in the solid-solution alloys.

③ Interpretation of Binary Isomorphous Phase Diagrams

Phase Rule

In the Cu-Ni system, the Gibbs phase rule follows $F = C - P + 1$, by assuming a constant pressure.

For single-phase regions, $F = 2 - 1 + 1 = 2$, which means that we need to specify two variables to fully describe the characteristics of any alloy. For example, for an alloy with an overall composition of 60 wt% Ni containing only α phase, the temperature can vary freely up to approximately 1315°C without changing the alloy states. Point *A* in Fig. 8.12a has two degrees of freedom to specify this point, i.e., at 1200 °C with 60 wt% Ni.

For two-phase regions, $F = 2 - 2 + 1 = 1$, which means that we only need to stipulate one variable to fully fix the state of the Cu-Ni system. Point *B* in Fig. 8.12a is located in the L + α region, and thus, there is one degree of freedom to describe this alloy, namely, the system temperature or the composition of one phase. If the temperature crossing *B* is 1325 °C, then the composition of both phases is fixed to 52 wt% Ni for the liquid phase and 65 wt% Ni for the α phase by prolonging the **tie line** at this temperature to intercept the liquidus and solidus lines, respectively.

To Determine the Phase Present

To determine the phases present in an alloy, we can locate the corresponding point on the phase diagram by knowing the alloy composition and the system temperature, and read the labeled regions simultaneously. For instance, the alloy with 60 wt% Ni at 1200 °C contains only the α phase (point *A* in Fig. 8.12a). However, this alloy at 1325 °C falls into the L + α region [point *B* in Fig. 8.12a), and at 1400 °C, there is only the liquid phase (point C in Fig. 8.12a).

To Determine the Phase Compositions

To determine the phase compositions, we need to do it separately. In the single-phase region, the phase composition is equal to the overall alloy composition. For instance, a 40 wt% Ni alloy at 1200 °C (point A in Fig. 8.12b) is located in the single α region, and thus, the **phase composition** of α is 40 wt%Ni.

In the two-phase region, the phase composition can be determined with the assistance of a tie line. When an **isothermal** (constant temperature) line passes through the two-phase region on the isomorphous phase diagram, the segment intersecting the two adjacent solubility lines is called a **tie line**. The perpendiculars from the ends of the tie line intersect with the composition axis, showing the compositions of the two phases that exist in equilibrium with each other at this temperature. For example, a 40 wt% Ni alloy at 1250 °C (point B in Fig. 8.12b) is located in the L + α region. By drawing a tie line intersecting the liquidus and solidus at points *a* and *b*, two perpendiculars from points *a* and *b* intersect the horizontal axis. From the intersects, the composition of L and α can be read directly as C_L = 28 wt% Ni and C_α = 44 wt% Ni, respectively.

To Determine the Amount of Phases

To determine the phase amount in both single- and two-phase regions, different methods are used. In a single-phase region, the amount of phase is always 100%. However, in a two-phase region, the amount of each phase can be determined using the tie line and **lever rule**.

First, the compositions of the two phases are obtained, along with the overall composition of the alloy (C_0). The lever rule is then applied to determine the mole fraction or mass fraction of each phase. The following equations can be used to calculate the amount of each phase:

$$W_L = \frac{C_\alpha - C_0}{C_\alpha - C_L} \tag{8.23}$$

$$W_\alpha = \frac{C_0 - C_L}{C_\alpha - C_L} \tag{8.24}$$

For example, in an alloy with 40 wt% Ni at 1250 °C, located in the L + α two-phase region, the compositions of L and α are 28 wt% Ni and 44 wt% Ni, respectively. Using the lever rule, the amount of L and α can be calculated as W_L = 25% and W_α = 75%, respectively.

④ Development of Microstructure in Isomorphous Alloys.

Equilibrium Solidification

During equilibrium cooling, the cooling rate is very slow, allowing the components in the liquid and solid phases to continuously readjust themselves to maintain thermodynamic equilibrium. The slow cooling rate facilitates long-range diffusion of atoms. Using an alloy with a composition of 30 wt% Ni-70 wt% Cu as an example, as shown in Fig. 8.13, we discuss the microstructure development as it slowly cools from 1300 °C to 1200 °C to ensure phase equilibrium at every temperature point.

At point *a* (T_0 = 1300 °C), the alloy is composed entirely of the L phase with a composition of 30 wt% Ni (30Ni), i.e., 100% of L.

Upon cooling to point *b* (T_1 = 1245 °C), which intersects the liquidus, solidification begins. The composition of the first solid phase, point *c*, can be determined by the intersection point of tie line *bc* with the solidus, which is 41 wt% Ni (41Ni). Thus, at this temperature, a two-phase equilibrium forms between the liquid with composition *b* (30Ni) and the solid with composition *c* (41Ni). According to the lever rule calculations, the amount of the first solid phase is negligible.

As the temperature continues to decrease, the composition of the solid phase changes along the solidus, and that of the liquid phase varies along thc liquidus.

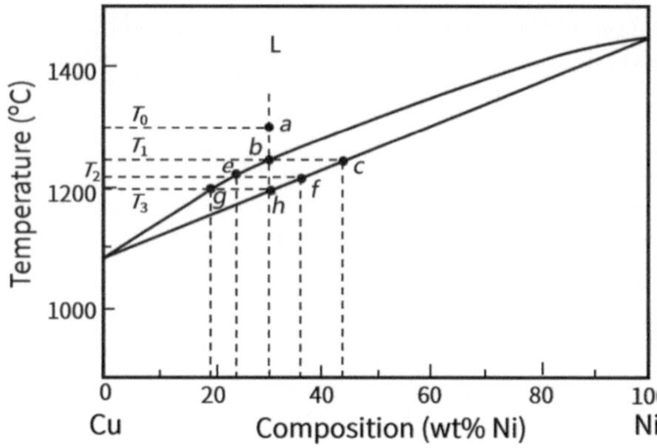

Fig. 8.13 The solidification process of a Cu-30 wt%Ni alloy

At T_2 (1220°C), the composition of liquid is point *e* (24 wt% Ni (24Ni)), and that of solid is point *f* (36 wt% Ni (36Ni)). The amounts of liquid and solid phases are approximately 50% each.

Upon cooling to T_3 (1200 °C), the last drop of liquid with 18 wt% Ni (18Ni) at point *g* reaches equilibrium with the solid phase at point *h*, where the composition of solid solution is equal to the initial alloy composition (30Ni). Further cooling leads to no change in the microstructure, and after solidification, a homogeneous single-phase solid α phase is obtained. The microstructure development of the CuNi alloy during the entire solidification process is shown in Fig. 8.14.

The Diffusion Process During Equilibrium Solidification

During solidification, each nucleus forms a grain. Under equilibrium cooling conditions, sufficient diffusion leads to homogeneous composition within the grains, resulting in a microstructure similar to that of a pure metal. The composition is uniform for all grains.

Solidification of a solid solution occurs over a temperature range, during which the compositions of the liquid and solid phases change continuously as the temperature decreases. This process relies on the diffusion of component atoms and includes three essential processes at each temperature: (1) diffusion in the liquid phase, (2) continuous growth of the solid phase, and (3) diffusion in the solid phase. The diffusion process during the equilibrium solidification from T_1 to T_2 in Fig. 8.13 is illustrated in Fig. 8.15, where *L* and *S* represent respectively the

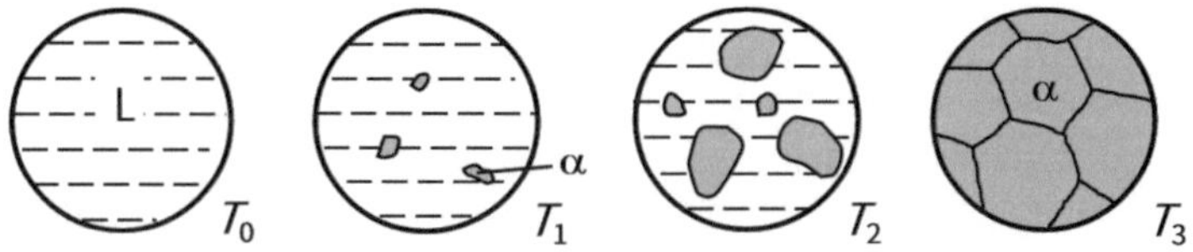

Fig. 8.14 Schematic microstructure development of Cu-30 wt% Ni alloy during the solidification process

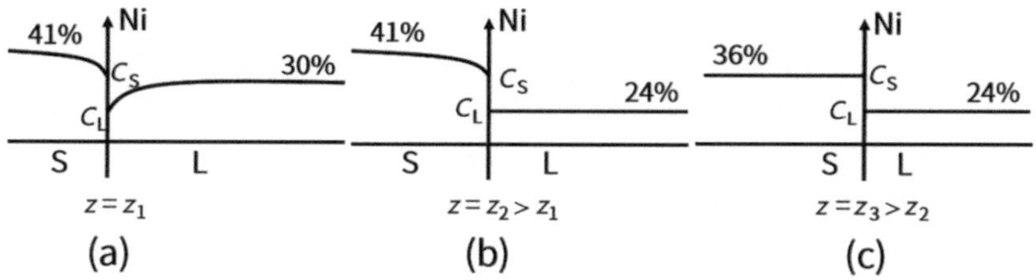

Fig. 8.15 The diffusion process of equilibrium solidification of Cu-30 wt% Ni alloy at T_2

liquid and solid phases, C_L and C_S correspond to the equilibrium compositions of liquid and solid phases at the *L/S* interface at T_2, while z represents the distance from the initial solidification end.

After the crystallization of solid phase (C_S), the content of Ni in the liquid phase around the solid decreases to C_L, while far away from this region, the composition remains unchanged. In the liquid, the diffusion of component atoms continues due to a concentration gradient. Meanwhile, the solid phase grows continuously during the diffusion process until the composition of liquid phase reach C_L (24Ni). Similarly, diffusion occurs in the solid phase due to the concentration gradient, but the slower rate of atomic diffusion in the solid phase prolongs the time required to reach the composition of solid C_S (36Ni).

Nonequilibrium Solidification

In the case of equilibrium cooling, it is assumed that the system is in equilibrium at each temperature point, which is challenging to achieve in practice as it requires an extremely slow cooling rate. Thus, nonequilibrium cooling is a more common scenario. During nonequilibrium solidification, the composition of the solid and liquid phases deviates from the liquidus and solidus lines in the equilibrium phase diagram. The deviation from the solidus line is greater because component diffusion in the solid phase is slower than in the liquid phase. The diffusion in the liquid phase is assumed to be fast enough, so the composition change of liquid phase during cooling remains along the liquidus line.

Figure 8.16a illustrates a schematic diagram of the composition changes of the liquid and solid phases during nonequilibrium solidification. At temperature T_1, a solid phase with composition α_1 crystallizes from liquid L_1.

Upon cooling to T_2, the equilibrium composition of the solid is α_2, and the liquid changes to L_2. The rapid cooling rate results in insufficient diffusion in the solid phase, resulting in a **cored structure** with an internal composition of α_1 and an external composition of α_2. The **average composition** α_2' of the solid solution is expected to be between α_1 and α_2. Upon further cooling to T_3, the equilibrium composition of the crystallized solid is α_3, and the actual composition of solid α_3' is the average of α_1, α_2 and α_3. The solidification process is complete when the alloy is cooled to T_4, where the composition of the solid solution becomes α_4', which is the initial alloy composition. The dotted line in the figure connecting the

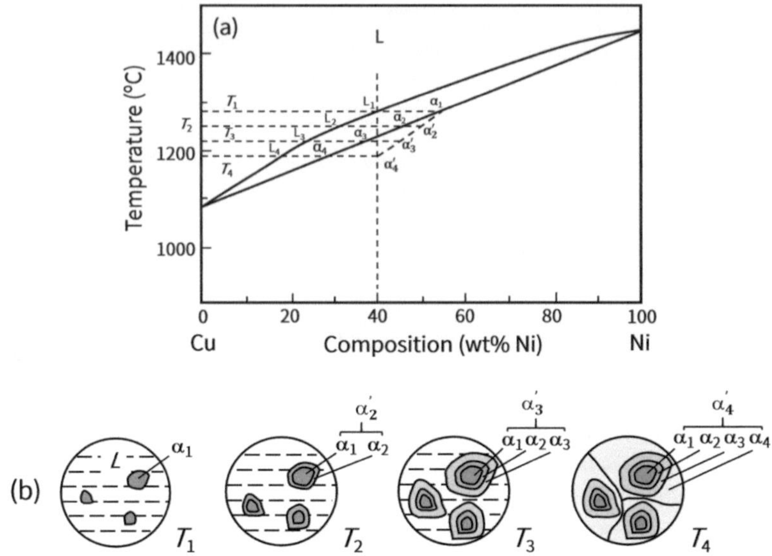

Fig. 8.16 **a** Schematic phase diagram and the change in the composition of the liquid and **b** solid phases during nonequilibrium solidification

average composition points of the solid phase at each temperature is called **the average composition line of solid**. The composition and structural changes of the liquid and solid phases are shown in the figure. The first solidified part is rich in Ni, and the latter part is rich in Cu, resulting in the average concentration of α always being higher than the equilibrium concentration at the α/L interface.

Because solid solution crystallization occurs in a dendritic manner, the inhomogeneous composition is distributed along the dendrites. The center of the dendrites is rich in the high-melting element Ni, while the region between the dendrites contains an increased amount of low-melting element Cu, forming **dendritic segregation**, as shown in Fig. 8.16b.

The extent of dendritic segregation is influenced by the following factors: (1) The higher the cooling rate, the more inhomogeneous the diffusion process, resulting in more pronounced segregation. (2) A larger crystallization range in the phase diagram leads to a greater difference in the segregation composition.

Dendritic segregation results in the presence of inhomogeneous properties in materials. However, such inhomogeneities can be eliminated through diffusion annealing at high temperatures. This process ensures sufficient atom diffusion, which results in a homogenized composition.

8.4.2 Binary Eutectic Phase Diagrams

Binary eutectic phase diagrams represent two components that are fully soluble in the liquid state, but have limited solubility in the solid state. This leads to the formation of a two-phase mechanical mixture through the eutectic reaction. Examples of phase diagrams belonging to this type include Al-Si, Pb–Sn, Ag–Cu, MgO-CaO.

① Construction of Binary Eutectic Phase Diagram

Figure 8.17 illustrates the free energy curves of a system where the liquid phase is approximately ideal, but the solid phase has a positive enthalpy of mixing ($\Delta H_{mix} > 0$), indicating that the A and B atoms 'dislike' each other. At low temperatures (T_3), the free energy curve for the solid has a negative curvature in the middle, forming a miscibility gap (Fig. 8.17c), where the solid solution is most stable as a mixture of two phases α' and α'' with compositions *e* and *f*. As the temperature increases and $-T\Delta S_{mix}$ becomes larger, the compositions *e* and *f* approach each other, and the miscibility gap disappears, as shown in the phase diagram in Fig. 8.17d. The positive ΔH_{mix} in the solid is also responsible for a minimum melting point mixture at higher temperatures. The fact that alloys can melt at temperatures below the melting points of their individual components can be attributed to the repulsion between atoms in the alloy, allowing for the disruption of the solid into a liquid phase at lower temperatures than in either pure A or pure B.

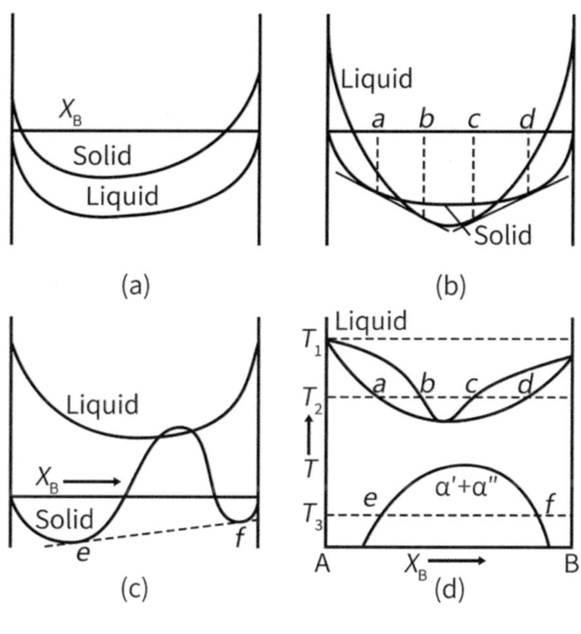

Fig. 8.17 The derivation of a phase diagram where $\Delta H^S_{mix} > \Delta H^L_{mix} = 0$. Free energy vs. composition curves for **a** T_1, **b** T_2, and **c** T_3. **d** The derived phase diagram with miscibility gap

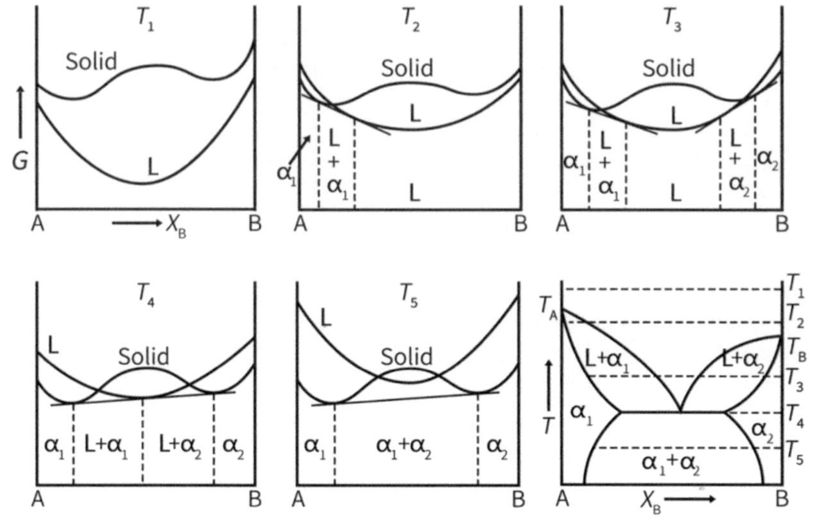

Fig. 8.18 Derivation of a eutectic phase diagram where both solid phases have the same crystal structure

When ΔH_{mix} is significantly positive, the miscibility gap shown in Fig. 8.17d can expand into the liquid phase, resulting in eutectic phase diagram patterns as displayed in Fig. 8.18.

A comparable phase diagram can be produced when A and B possess distinct crystal structures, depicted in Fig. 8.19.

② Description of Binary Eutectic Phase Diagrams

The Pb–Sn binary eutectic phase diagram in Fig. 8.20 will be discussed below as an example. The Pb–Sn phase diagram can be utilized to form a range of alloys with two components, Pb and Sn. Sn–Pb solder is one such alloy that is used to permanently join metal workpieces together. In electrical and electronics applications, solder is used to fuse connections and devices, creating conducting pathways. The properties of solder can vary depending on the composition of metals in the alloy.

Phases: The system consists of three phases: liquid (L) and two solid solutions (α and β). The α phase is a solid solution rich in Pb with Sn as the solute, while the β phase solid solution is rich in Sn with Pb as the solute.

Points: The two end points on the vertical axes represent the respective melting points of pure components Pb and Sn. Points *B* and *G* are the maximum solubilities of the two solid phases (α and β), respectively. Point *E* is the **eutectic point** (or **invariant point**) of the alloy, which represents the lowest melting point of the alloy.

Lines: The liquidus line is *AEF*, and the solidus line is *ABEGF*. The solvus line separates a single solid solution into a solid two-phase region and changes as a

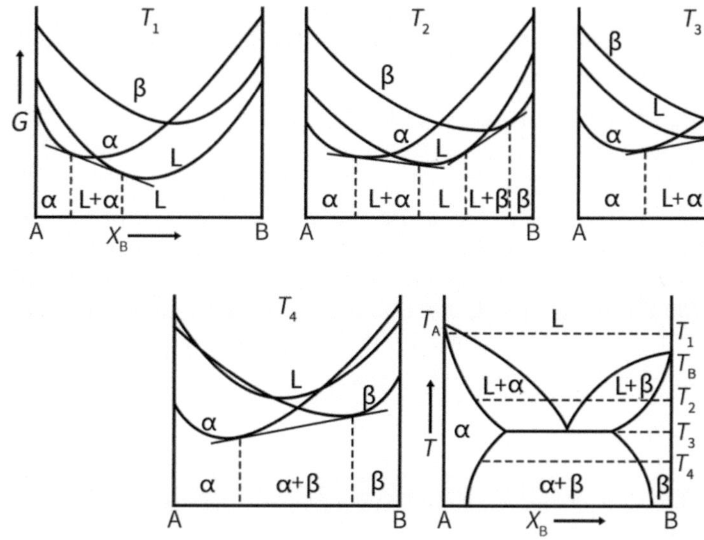

Fig. 8.19 Derivation of a eutectic phase diagram where each solid phase has a different crystal structure

Fig. 8.20 Pb–Sn binary eutectic phase diagram

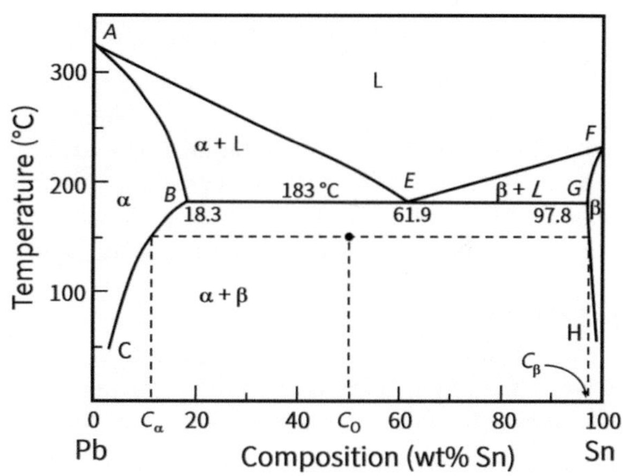

function of temperature. The **solvus line** of α is *BC*, and the solvus line of β is *GH*, indicating a decrease in solubility limit with decreasing temperature. The **eutectic line** is *BEG* and corresponds to a **eutectic isotherm** $\boldsymbol{T_E}$, where a eutectic reaction occurs. On the eutectic line, three phases coexist at equilibrium.

$$\mathrm{L} \rightarrow \alpha + \beta \tag{8.25}$$

Regions: The system contains three single-phase regions, namely, the two solid solutions α and β, as well as the liquid phase L. There are also three two-phase regions, namely L + α, L + β, and α + β.

③ Interpretation of Binary Eutectic Phase Diagrams

Phase Rule

The Pb–Sn system follows the Gibbs phase rule $F = C - P + 1$, by assuming a constant pressure. The phase rule can be used to calculate the degree of freedom, F, for single-phase and two-phase regions, similar to the Cu-Ni system. For alloys located on the *BEG* line, the eutectic reaction occurs at a fixed temperature and the compositions of the three phases involved are constant, according to the calculation $F = C - P + 1 = 2 - 3 + 1 = 0$.

To Determine the Phase Present

Similar to the isomorphous phase diagrams, we can locate the point corresponding to the alloy composition and temperature on the phase diagram and read the labeled regions directly. For example, a Pb-50 wt% Sn alloy cooled from 300 °C to room temperature successively passes through the single-phase L region, (α + L) two-phase region, three-phase coexisting region (line *BEG*) and (α + β) two-phase region.

To Determine the Phase Compositions

To determine the phase compositions, we need to consider each region separately. In the single-phase regions (L, α, β), the phase composition is equal to the alloy composition. For the two-phase regions (L + α, L + β, and α + β), we can use a tie line to determine the phase compositions. For alloys with compositions in the range of *BEG* and at T_E, the L phase with composition of *E* will be transformed into the α of point *B* and β of point *G* according to the eutectic reaction.

To Determine the Amount of Phases

In the single-phase regions, the phase amount is equal to 100%. In the two-phase regions, we can use the lever rule to calculate the amount of each phase after determining the phase compositions and the known composition of the alloy.

Example problem For a Pb-50 wt% Sn alloy at 150°C, determine (a) what phases are present, (b) the compositions of the phases, and (c) the relative amount of each phase present in terms of mass fraction.

Solution: (a) The α and β phases coexist at these conditions. (b) The compositions of the α and β phases are 11 wt% Sn and 98 wt% Sn, respectively. (c) The amount of each phase present can be determined using the lever rule. The mass fraction of the α phase is 55%, and the mass fraction of the β phase is 45%.

$$W_\alpha = \frac{C_\beta - C_0}{C_\beta - C_\alpha} = \frac{98 - 50}{98 - 11} = 0.55$$

$$W_{\beta} = \frac{C_0 - C_{\alpha}}{C_{\beta} - C_{\alpha}} = \frac{50 - 11}{98 - 11} = 0.45$$

④ Development of Microstructure in Eutectic Alloys

Equilibrium Solidification

The microstructure development of the binary eutectic phase diagram was demonstrated by examining four representative alloy compositions.

Alloy C_1 has a composition in the range of 0 to the minimum solvus of α. When the liquid phase with C_1 composition cools down to the temperature range between the liquidus and solidus lines, the microstructure development is similar to the isomorphous cases, where the **primary α phase** crystallizes gradually from L. When the temperature is below the solidus line, only α phase exists. Upon further cooling to room temperature, no change in the microstructure was observed. Figure 8.21 shows a schematic plot of the equilibrium solidification process of a C_1 Pb–Sn alloy.

Alloy C_2 has a composition in the range of the minimum solvus to the maximum solubility of α. Upon cooling, as shown in Fig. 8.22, the liquid phase of C_2 to the temperature range between the liquidus and solidus lines, the microstructure gradually transforms from L to α, similar to the isomorphous cases. When the temperature drops below the solidus line, the solidification of L completes, and only α remains. Upon further cooling to the solvus line, no change occurs in the microstructure of solid α. As the temperature decreases below the solvus line, small β particles, called the **secondary β phase** or β_{II}, precipitate along the boundaries or defects of α grains. With a further decrease in temperature, the amount and size of β particles increase. The equilibrium composition of α and β_{II} changes along the

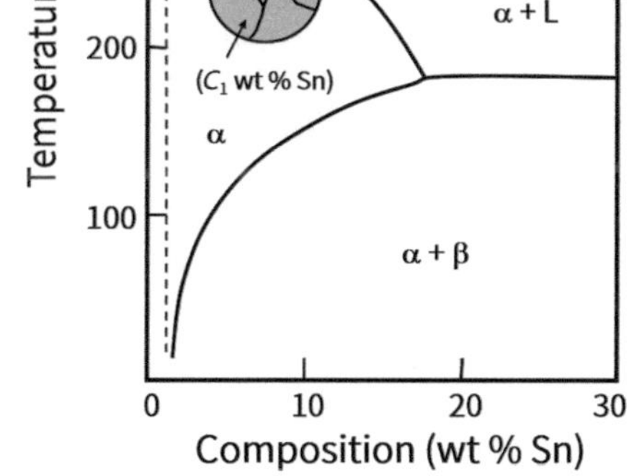

Fig. 8.21 Schematic plot of the equilibrium solidification process of a C_1 Pb–Sn alloy

Fig. 8.22 Schematic plot of the equilibrium solidification process of a C_2 Pb–Sn alloy

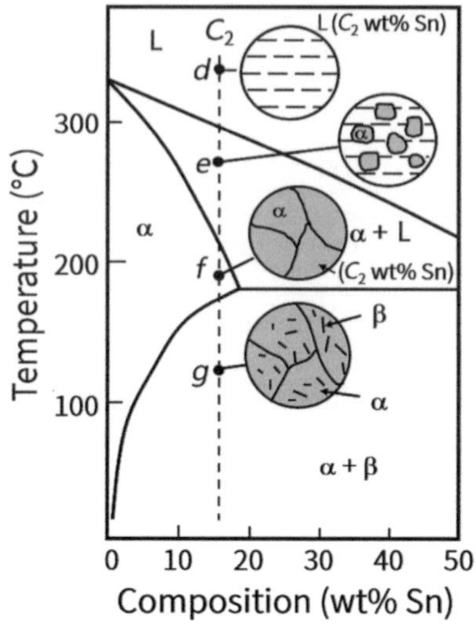

solvus lines. The phase amount in the two-phase regions, such as L + α and α + β_{II}, can be determined by creating tie lines and applying the lever rule. Figure 8.22 shows the schematic plot of the equilibrium solidification process of a C_2 Pb–Sn alloy.

Alloy C_3 has the eutectic composition. Upon cooling the liquid phase with C_3 composition to the eutectic temperature (T_{E}), a **eutectic structure** forms according to the eutectic reaction, where the corresponding cooling curve is also present (Fig. 8.23). The temperature of the system will not decrease further until the completion of the reaction. Upon further cooling to room temperature, the equilibrium compositions of α and β will change along the solvus lines, resulting in the precipitation of secondary β_{II} and α_{II}. However, it is difficult to distinguish the secondary phases from primary phases, and the change in microstructure below T_{E} is negligible. The phase amount in the α + β regions can be determined using the lever rule.

Alloy C_4 is a **hypoeutectic alloy**, ranging from the maximum solid solubility of α to the eutectic composition, as shown in Fig. 8.24. During cooling, primary α precipitates from the liquid phase via isomorphous transformation as the temperature drops below the liquidus line. The compositions of α and L change along the solidus and liquidus, respectively. At the eutectic isotherm T_{E}, the remaining liquid completely transforms into a eutectic structure (α + β) as the eutectic reaction occurs. The final microstructure consists of a primary α solid solution and eutectic structure, where the eutectic structure is also known as a **microconstituent**. There is no change in the microstructure upon further cooling to room temperature.

The amount of primary α and eutectic α must be determined separately. The lever rule can be applied in the L + α two-phase region just above T_{E} to determine the amount of **primary α phase**. The amount of L phase is equal to that of the eutectic structure (α + β). The amount of **total α phase** can be determined by

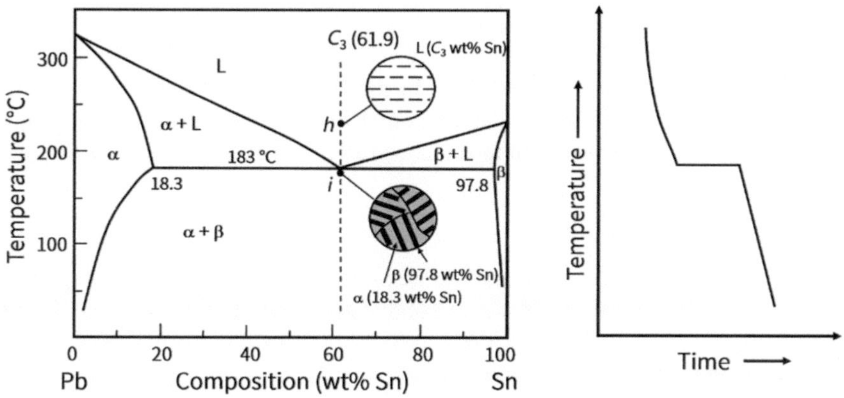

Fig. 8.23 Schematic plot and the cooling curve of the equilibrium solidification process of a C_3 Pb–Sn alloy

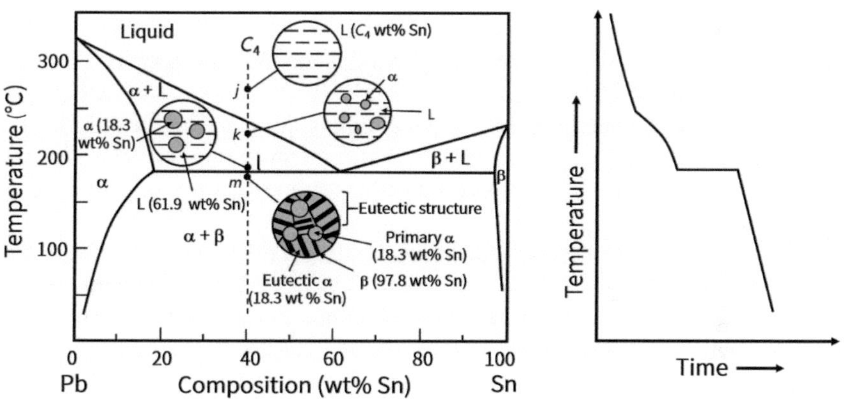

Fig. 8.24 Schematic plot and the cooling curve of the equilibrium solidification process of a C_4 Pb–Sn alloy

applying the lever rule in the (α + β) two-phase region just below T_E. Then, the amount of eutectic α can be obtained by subtracting the amount of primary α from that of total α.

$$\text{Primary } \alpha \text{ phase: } W_{\alpha P} = \frac{C_E - C_0}{C_E - C_\alpha} \tag{8.26}$$

$$\text{Eutectic structure: } W_E = \frac{C_0 - C_\alpha}{C_E - C_\alpha} \tag{8.27}$$

$$\text{Total } \alpha \text{ phase: } W_\alpha = \frac{C_\beta - C_0}{C_\beta - C_\alpha} \tag{8.28}$$

$$\text{Eutectic } \alpha \text{ phase: } W_{\alpha E} = W_{\alpha} - W_{\alpha P} \tag{8.29}$$

Conversely, alloys with compositions in the range of the eutectic composition to the maximum solid solubility of β are called **hypereutectic alloys**, and the final microstructure of the alloy is composed of a primary β solid solution and a eutectic structure (α + β).

Nonequilibrium Solidification

Under equilibrium solidification conditions, only alloys with eutectic compositions can form eutectic structures. However, under nonequilibrium solidification conditions, some hypoeutectic or hypereutectic alloys can also form 100% eutectic structures, which are called **pseudoeutectic**.

Alloy I with eutectic transformation forms a eutectic structure in the hatched area surrounded by two liquidus lines (point *r* in Fig. 8.25a) as the liquid is supercooled. The hatched area is called **pseudoeutectic region**. Outside this area, the microstructure has eutectic structure and dendrites. The degree of supercooling affects the size of this region. Pseudoeutectic regions are symmetrically distributed when the alloy components have similar melting points, but when they differ, the pseudoeutectic region is biased towards the high-melting component, as seen in the Al-Si alloy (Fig. 8.25b).

Some alloys form a single-phase solid solution under equilibrium solidification, but under high cooling rates, small amounts of nonequilibrium eutectic structures may appear, as seen for alloys with compositions to the left of *a* or to the right of *c* in Fig. 8.25a.

In Fig. 8.25a, the average composition of alloy II deviates from the solidus under nonequilibrium solidification conditions. The dotted line represents the average composition of the solid during quick cooling. As alloy II cools down to the solidus, a small amount of liquid remains. Upon further cooling to the eutectic

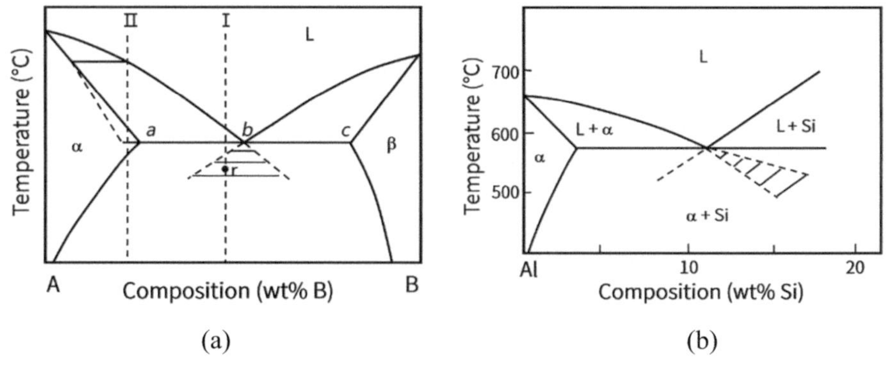

Fig. 8.25 **a** Representative nonequilibrium eutectic regions in a hypothetical phase diagram and **b** the pseudoeutectic region in Al-Si alloys

temperature, the composition of the remaining liquid reaches the eutectic composition, and the eutectic reaction occurs. The resulting **nonequilibrium eutectic structure** is distributed on grain boundaries and among the dendrites of the α phase.

Because only a small fraction of nonequilibrium eutectics are present, the α phase of eutectics grows adhesively to the primary α, pushing the β phase to solidify to grain boundaries. Consequently, the microstructural features of eutectics, i.e., alternating layers of two phases, disappear. These eutectics with phase separation are called **divorced eutectic structures**.

8.4.3 Binary Peritectic Phase Diagrams

The peritectic phase diagram consists of two components that are fully soluble in the liquid state but only partially soluble in the solid-state. In binary phase diagrams, the peritectic transition is an isothermal transformation in which the solid phase reacts with the remaining liquid phase to form the other solid phase. Examples of binary phase diagrams with peritectic transformation include Fe–C, Cu–Zn, Ag–Sn, and Ag–Pt.

① Description of Binary Peritectic Phase Diagrams

Figure 8.26 depicts a peritectic transition in the Pt–Ag system, with end points *A* and *B* being the melting points of pure components Pt and Ag, respectively. Points *D* and *P* denote the maximum solubilities of α and β, respectively. Point *P* is called the **peritectic point**.

The figure shows three phases: L and two solid solutions (α and β). *ACB* represents the liquidus line, *ADPB* is the solidus line, *DE* is the solvus line of the α phase, and *PF* is the solvus line of the β phase. The **peritectic line** *DPC* is a horizontal line, corresponding to a **peritectic isotherm** T_P, on which a peritectic reaction will occur for the alloys with compositions in the *DC* range:

$$L + \alpha \rightarrow \beta \tag{8.30}$$

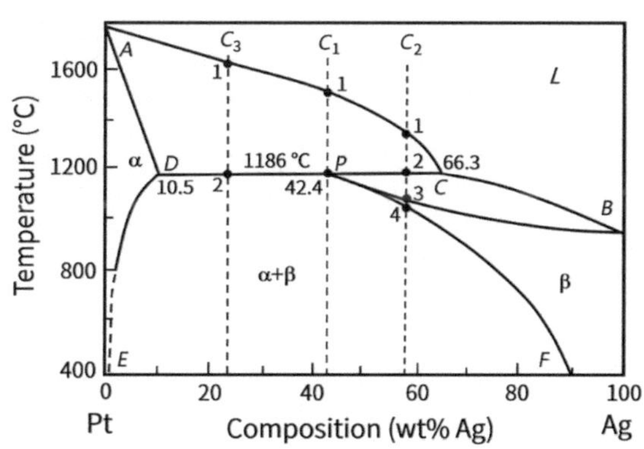

Fig. 8.26 Pt–Ag binary peritectic phase diagram

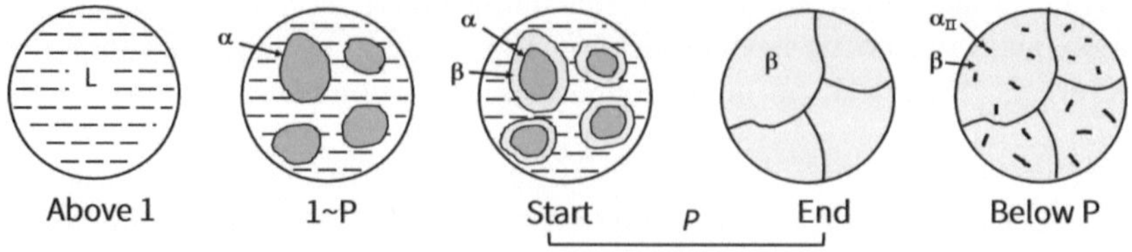

Fig. 8.27 Schematic drawings of the equilibrium microstructures for Pt–Ag alloy C_1 during the solidification process

The peritectic reaction is an isothermal transformation.

② Development of Microstructure in Peritectic Alloys

The microstructure development of the binary peritectic phase diagram is demonstrated by examining three representative alloy compositions.

Alloy C_1 (42.4 wt% Ag): Alloy C_1 was found to begin with the crystallization of primary α as the alloy was cooled down from high temperatures to point 1 intersecting with the liquidus line. During the continuous cooling process, the amount of α increased gradually, and the fraction of the liquid phase decreased steadily, with the compositions of α and liquid phases along solidus *AD* and liquidus *AC*, respectively. Upon cooling to the peritectic reaction temperature T_P (point *P*), the composition points of the α phase and liquid phase in the alloy reached points *D* and *C*, respectively. At the onset of the peritectic reaction, the relative fraction of the two phases is determined by the lever rule. Below T_P, the liquid phase reacted with the α phase to form β entirely. As the temperature continued to decrease, the solubility of Pt in the β phase decreased with temperature along the *PF* line, leading to continuous precipitation of α_{II} from the β solid solution. The equilibrium microstructure of the alloy at room temperature is $\beta + \alpha_{II}$, and the corresponding solidification process is illustrated in Fig. 8.27.

Alloy C_2 (42.4 wt%-66.3 wt% Ag): This alloy follows the same crystallization process as alloy C_1 in the temperature range of point 1 to point 2 during slow cooling. However, because the relative amount of liquid phase is greater than what is required for peritectic reaction, the remaining liquid phase will crystallize into the β phase in the same way as an isomorphous transformation during further cooling. The compositions of the liquid and β phases vary along the liquidus lines *CB* and *PB*, respectively. At point 3, the solidification is completed, and the composition of the β phase is the same as that of the initial alloy. The β phase remains single-phase in the range of point 3 to point 4. At temperatures below point 4, the α_{II} phase will continuously precipitate from the β phase. The equilibrium microstructure of the alloy at room temperature is $\beta + \alpha_{II}$, and the corresponding solidification process is shown in Fig. 8.28.

Alloy C_3 (10.5 wt%-42.4 wt% Ag): The crystallization process of the alloy in the temperature range of point 1 to point 2 during slow cooling is the same as

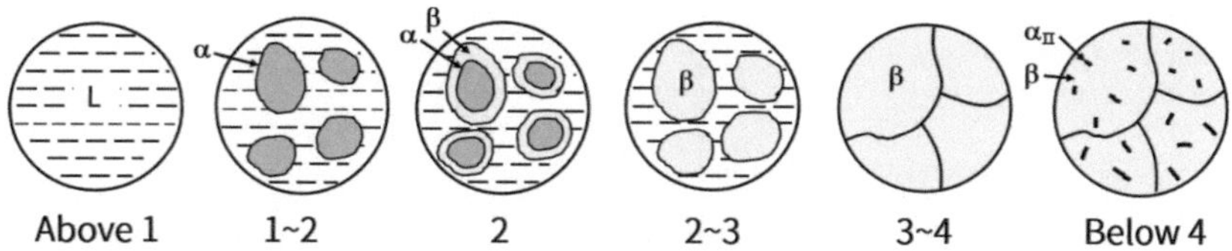

Fig. 8.28 Schematic drawings of the equilibrium microstructures for Pt–Ag alloy C_2 during the solidification process

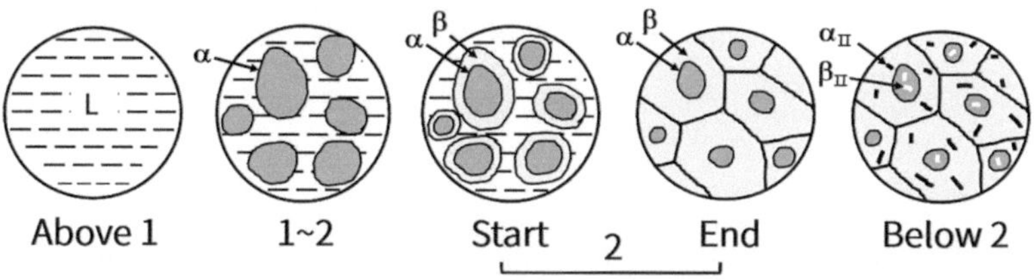

Fig. 8.29 Schematic drawings of the equilibrium microstructures for Pt–Ag alloy C_3 during the solidification process

that of alloy C_2. The relative amount of the α phase is greater than that required for peritectic reaction. After the peritectic reaction, there are some remaining α phases in addition to the newly formed β phase. Below the peritectic temperature, the α_{II} phase precipitates from the β phase, and the β_{II} phase precipitates from the α phase. The equilibrium microstructure of the alloy at room temperature is $\alpha + \beta + \alpha_{II} + \beta_{II}$, and the corresponding solidification process is shown in Fig. 8.29.

8.4.4 Other Types of Binary Phase Diagrams

① Binary Phase Diagrams Having Iintermediate Phases

In some binary phase diagrams, one or more compounds may form, which are called **intermediate phases** because they lie in the middle of the phase diagrams. If the compound has no solubility, it appears as a vertical line in the phase diagram, which can be regarded as an independent component dividing the phase diagram into two separate parts. These compounds are also called line compounds. Figure 8.30 shows the Mg–Si phase diagram, in which Mg_2Si is a line compound, an intermediate phase. The stable compound Mg_2Si can be considered a component dividing the phase diagram into separate Mg-Mg_2Si and Mg_2Si-Si eutectic phase diagrams for analysis. Additionally, there are many binary alloy systems (Cu–Mg, Fe–P, Mn–Si, etc.) and ceramic systems (SiO_2–MgO, BeO–Al_2O_3, etc.) that form stable compounds.

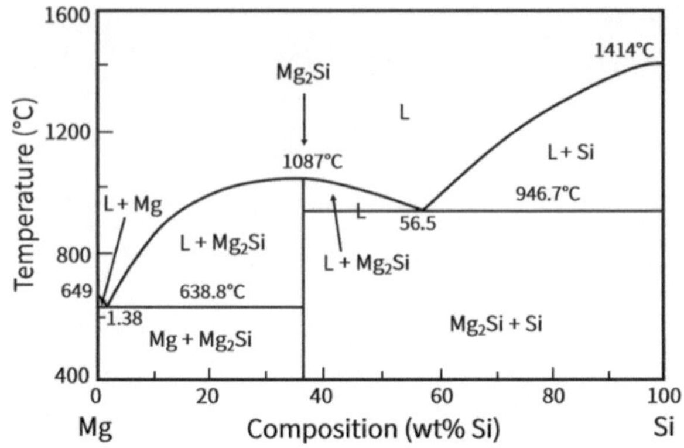

Fig. 8.30 Mg-Si binary phase diagram with an intermediate phase

② Binary Monotectic Phase Diagrams

A **monotectic reaction** refers to that one liquid phase L_1 decomposes isothermally into one solid phase and another liquid phase L_2. In the Cu–Pb binary phase diagram (Fig. 8.31), a monotectic reaction takes place at 955 °C.

$$L_{36} \rightleftharpoons Cu + L_{87} \tag{8.31}$$

Binary systems with monotectic transitions include Cu–S, Mn-Pb, Cu–O, etc.

③ Binary Syntectic Phase Diagrams

There are only a few alloy systems, such as Na-Zn and K-Zn, that exhibit **syntectic transitions** where two liquid phases L_1 and L_2 react to form a solid phase. The schematic diagram in Fig. 8.32 shows the syntectic reaction occurring at a certain temperature:

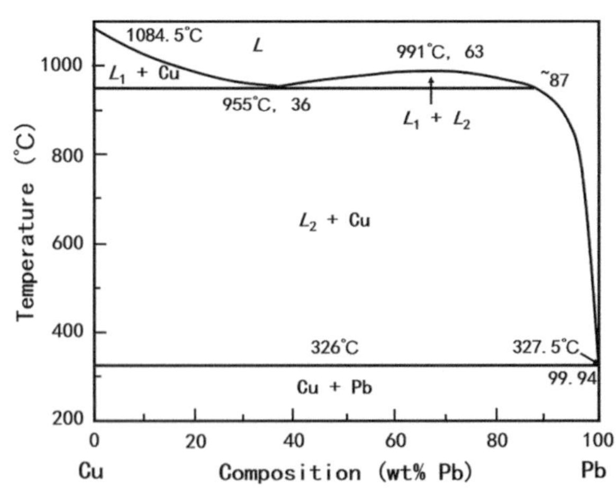

Fig. 8.31 The Cu–Pb binary monotectic phase diagram

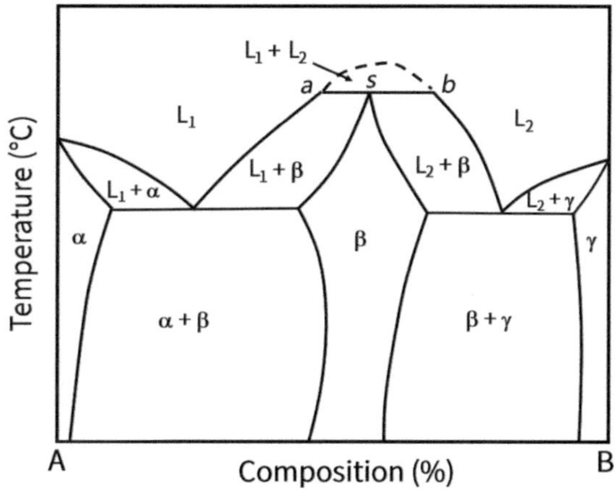

Fig. 8.32 A hypothetical binary phase diagram containing syntectic reaction

$$L_{1a} + L_{2b} \rightleftharpoons \beta_s \tag{8.32}$$

④ Binary Metatectic Phase Diagrams

Figure 8.33 shows the Fe-B binary phase diagram containing the metatectic transition. At 1381 °C, the Fe-B alloy with trace amounts of B decomposes isothermally into one liquid phase and another solid phase in a reaction called the **metatectic transition**. There are also very few alloy systems with metatectic reactions, such as Fe-S and Cu-Sb.

$$\delta \rightleftharpoons \gamma + L \tag{8.33}$$

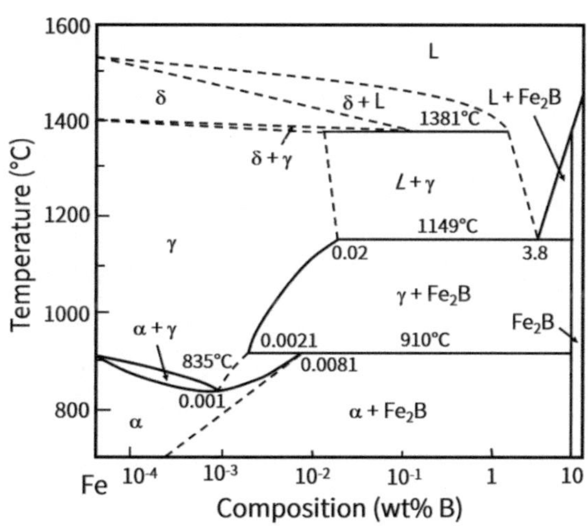

Fig. 8.33 The Fe-B binary metatectic phase diagram

⑤ Binary Phase Diagrams Having Solid-State Transformations

Binary Eutectoid Phase Diagrams

The **eutectoid** (or eutectic-like) reaction has a similar form to the eutectic reaction, where a solid phase is transformed into two other solid phases at a constant temperature. For example, the Cu-Sn phase diagram in Fig. 8.34 shows four eutectoid reactions:

$$\begin{aligned} &\text{IV: } \beta \rightleftharpoons \alpha + \gamma \\ &\text{V: } \gamma \rightleftharpoons \alpha + \delta \\ &\text{VI: } \delta \rightleftharpoons \alpha + \varepsilon \\ &\text{VII: } \zeta \rightleftharpoons \delta + \varepsilon \end{aligned} \tag{8.34}$$

Cu–Zn and Fe–C systems are two examples of binary phase diagrams with eutectoid transitions. The Fe–C system is particularly significant, and it will be further discussed in Sect. 8.4.6.

Binary Peritectoid Phase Diagrams

The **peritectoid (or peritectic-like) reaction** is similar to the peritectic reaction, in which a solid phase reacts with another solid phase to form a third solid phase at a constant temperature. In the Cu-Sn phase diagram (as shown in Fig. 8.34), there are two peritectoid reactions:

$$\begin{aligned} &\text{VIII: } \gamma + \varepsilon \rightleftharpoons \zeta \\ &\text{IX: } \gamma + \zeta \rightleftharpoons \delta \end{aligned} \tag{8.35}$$

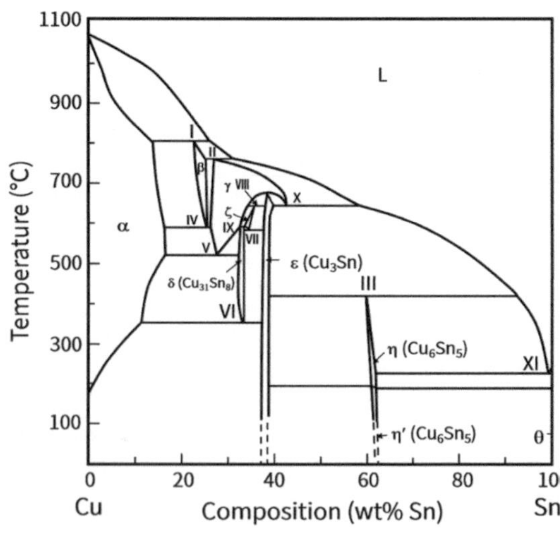

Fig. 8.34 The Cu-Sn binary phase diagram

⑥ Summary of Binary Phase Diagrams

Complex binary phase diagrams are composed of the basic phase diagrams described above. Understanding the characteristics and transition rules of basic phase diagrams can simplify the analysis of complex diagrams. The following methods are commonly used for analysis:

- Determine whether there is a stable compound in the phase diagrams.
- Distinguish each phase region according to **the rule of contact**, which states that the difference in the phase number of adjacent phase regions is one (except in cases of point contact).
- Identify the horizontal lines for three-phase coexistence and analyze the types of isothermal transitions using schematic diagrams like those shown in Table 8.1.
- Analyze phase transitions and microstructural developments of specific alloys with temperatures using the phase diagram, tie line, and lever rule.
- Phase diagrams provide information on the phases present in the system at equilibrium and their relative amounts, but not their shape, grain size, and spatial distribution. Therefore, special attention should be paid to the possible phases and microstructures that may be present under nonequilibrium conditions during actual production.
- Evaluate possible errors and mistakes when constructing phase diagrams using the Gibbs phase rule.

Table 8.1 Schematic diagrams of the isothermal transitions among three phases in binary systems

Isothermal transition type		Reaction type	Diagram feature
Eutectic type	Eutectic transition	$L \rightleftharpoons \alpha + \beta$	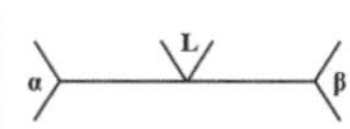
	Eutectoid transition	$\gamma \rightleftharpoons \alpha + \beta$	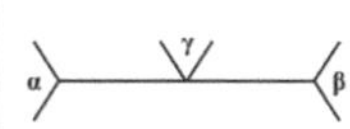
	Monotectic transition	$L_1 \rightleftharpoons \alpha + L_2$	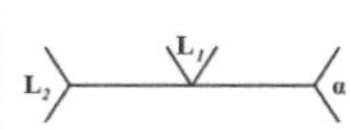
	Metatectic transition	$\delta \rightleftharpoons L + \gamma$	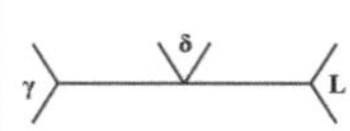
Peritectic type	Peritectic transition	$L + \beta \rightleftharpoons \alpha$	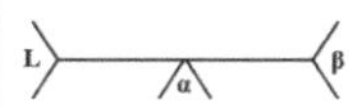
	Peritectoid transition	$\gamma + \beta \rightleftharpoons \alpha$	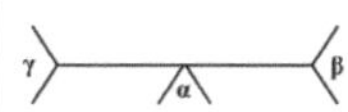
	Syntectic transition	$L_1 + L_2 \rightleftharpoons \alpha$	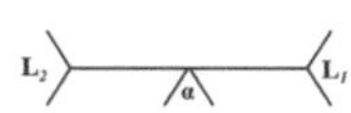

8.4.5 Examples of Practical Binary Phase Diagrams

SiO_2-Al_2O_3 is the most important binary system in ceramics, and therefore, it is often analyzed as an example. The binary phase diagram for SiO_2–Al_2O_3 is shown in Fig. 8.35, and it is a critical tool in the study of ceramics and refractories. It is important to note that the composition of mullite, which is an intermediate compound, with oxygen concentration varying between 72 wt% and 78 wt%, corresponding to the formula between $2SiO_2$–$3Al_2O_3$ and SiO_2-$2Al_2O_3$.

The SiO_2–Al_2O_3 phase diagram features three complex structures of compounds. The component α-Al_2O_3, also known as corundum, belongs to the rhombohedral crystal structure, while SiO_2 has several allotropic structures (as shown in Fig. 8.9). The intermediate compound mullite has a monoclinic structure. There are two three-phase isothermal transitions in the SiO_2–Al_2O_3 phase diagram. The first is the eutectic transformation at 1587 °C (L → SiO_2 + mullite), and the second is the peritectic transition that occurs at 1828 °C (L + Al_2O_3 → mullite).

Hypoeutectic ceramic (<10 wt% Al_2O_3). The ceramic melt starts to crystallize into SiO_2 (cristobalite) as the temperature decreases to liquidus. With decreasing temperature, the amount of SiO_2 increases, while that of Al_2O_3 in the liquid phase increases. When the temperature drops to 1587 °C, the composition of the liquid phase reaches the eutectic point, and a eutectic reaction occurs, resulting in a mechanical mixture of SiO_2 and mullite. The microstructures after the eutectic reaction contain primary cristobalite and eutectic structures. As the temperature decreases, the primary SiO_2 and SiO_2 in the eutectic structure gradually transform into other allotropes. When the eutectic reaction is completed, there is no precipitation during the cooling process because SiO_2 and mullite are insoluble.

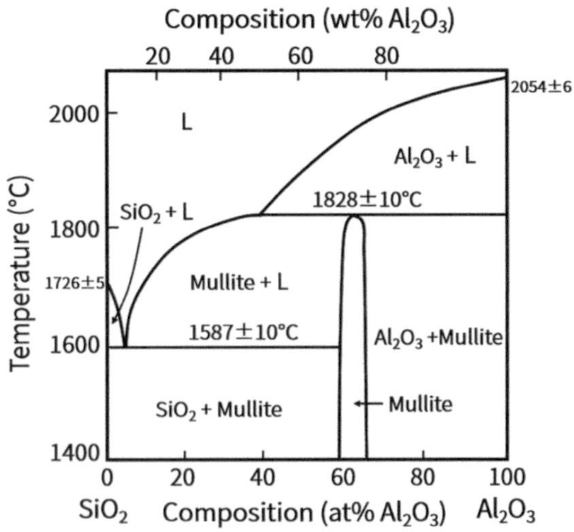

Fig. 8.35 Binary SiO_2–Al_2O_3 phase diagram

Eutectic ceramic (10 wt% Al_2O_3). The eutectic ceramic melt undergoes a eutectic reaction at 1587 °C to form a eutectic structure of SiO_2 and mullite. The relative amounts of SiO_2 and mullite can be calculated by the lever rule to be 86% and 14%, respectively. SiO_2 transforms into one of the three low-temperature allotropes after the end of the eutectic reaction according to the cooling rate.

Hypereutectic ceramic (10 wt%~55 wt% Al_2O_3). The ceramic melt of this composition starts to crystallize into mullite as the temperature decreases to the liquidus. With decreasing temperature, the amount of mullite increases and that of Al_2O_3 in the liquid phase decreases. At 1587 °C, a eutectic reaction occurs, resulting in primary mullite and eutectic structures. Within this composition range, the maximum amount of primary mullite can be calculated by the lever rule to be 72.5%. After the eutectic reaction, the SiO_2 in the eutectic structure undergoes an allotropic transition.

Ceramic composition in 55 wt%~72 wt% Al_2O_3. The ceramic melt starts to crystallize into Al_2O_3 as the temperature decreases to liquidus. With decreasing temperature, the amount of Al_2O_3 increases, while that of the liquid phase decreases. At 1828 °C, a peritectic reaction occurs, leading to the combination of peritectic mullite and the liquid phase into mullites in an isomorphous way. The composition of the liquid phase is changed along the liquidus line. Finally, at 1587 °C, as the composition of the liquid phase reaches $w(Al_2O_3) = 10\%$, a eutectic reaction occurs, resulting in mullite and eutectic structures.

Ceramic composition in 72 wt%~78 wt% Al_2O_3. The ceramic melt starts to crystallize into Al_2O_3 as the temperature decreases to liquidus. When the temperature drops to 1828 °C, a peritectic reaction occurs. After the peritectic reaction, it enters the single-phase region of mullite, which is maintained at room temperature.

Ceramic composition above 78 wt% Al_2O_3. This ceramic melt begins to crystallize into Al_2O_3 when the temperature declines to liquidus. When the temperature decreases to 1828 °C, the peritectic reaction arises. After the peritectic reaction at 1828 °C, the microstructures contain primary Al_2O_3 and the peritectic product mullite, which do not change during the further cooling process because they are completely insoluble.

In the SiO_2-Al_2O_3 binary system, varying the Al_2O_3 content leads to several commonly used refractory products. Refractories with Al_2O_3 content above 10% show improved high-temperature resistance with increasing Al_2O_3 content. Refractory bricks are typically composed of mullite or mullite and Al_2O_3, resulting in significantly improved refractoriness. The highest refractoriness temperature can be achieved with pure Al_2O_3. Sintered Al_2O_3 is commonly used as lab glassware, while fused cast Al_2O_3 is typically used as a refractory for glass tank furnaces.

8.4.6 The Iron-Iron Carbide (Fe-Fe_3C) Phase Diagrams

Fe–C alloys, which are composed of iron and carbon, are exemplified by carbon steels and cast irons widely used in engineering. The Fe–C phase diagram provides the basis for analyzing the microstructure, formulating processing, and predicting properties of Fe–C alloys. It has significant practical value in engineering, making it essential to study and understand. The Fe–C phase diagram includes solid lines (Fe-Fe_3C) and dashed lines (Fe–C), representing the two forms of carbon in the alloys, graphite and compound Fe_3C. Fe_3C is also referred as **cementite**. This **dual phase diagram** is depicted in Fig. 8.36.

① Description of Fe-Fe_3C Phase Diagram

In this discussion, we will only analyze the solid line phase diagram consisting of Fe and Fe_3C. Solid iron has different crystal structures (allotropes) in different temperature ranges: a BCC structure termed α-Fe from room temperature to 912 °C, a FCC structure termed γ-Fe in the temperature range of 912~1394 °C and a high-temperature BCC structure termed δ-Fe between 1394~1538 °C. Fe_3C is considered an independent component to construct the Fe-Fe_3C phase diagram.

Phases: The Fe-Fe_3C phase diagram consists of five phases: one liquid phase (L), three solid solutions (α, γ and δ) and one intermediate phase Fe_3C. Carbon atoms act as interstitial impurities in iron crystals, forming solid solutions. The α phase, also known as **ferrite**, has a BCC crystal structure with a maximum carbon solubility of 0.022 wt% at 727 °C. The γ phase, also known as **austenite**, has a FCC crystal structure with a maximum carbon solubility of 2.11 wt% at 1148 °C. The δ phase, similar to the α phase, has a BCC crystal structure with a maximum carbon solubility of 0.09 wt% at 1495 °C. The intermediate phase Fe_3C, termed

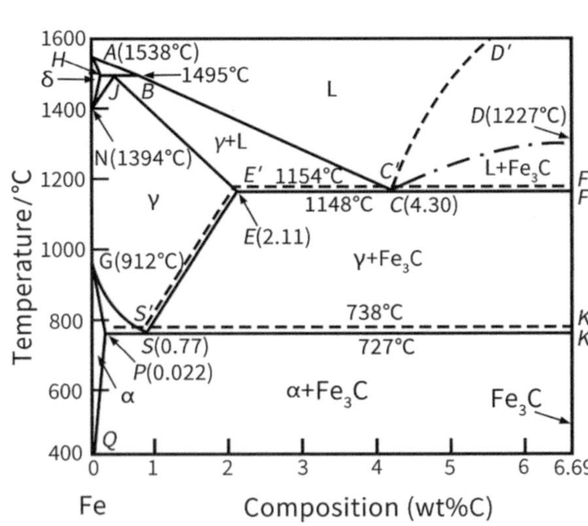

Fig. 8.36 Binary Fe–C and Fe–Fe_3C dual phase diagram

Table 8.2 Temperature, composition and meaning of each characteristic point in the Fe-Fe_3C phase diagram

Point	Temperature/°C	C concentration	Statements
A	1538	0	The melting point of pure Fe
B	1495	0.53 wt%	Composition of liquid during peritectic reaction
C	1148	4.30 wt%	Eutectic point $L_C \rightleftharpoons \gamma_E + Fe_3C$
D	1227	6.69 wt%	The melting point of cementite (calculated)
E	1148	2.11 wt%	The maximum solubility of C in γ-Fe
F	1148	6.69 wt%	Cementite
G	912	0	The allotropic transition from α-Fe to γ-Fe (A_3)
H	1495	0.09 wt%	The maximum solubility of C in δ-Fe
J	1495	0.17 wt%	Peritectic point $L_B + \delta_H \rightleftharpoons \gamma_J$
K	727	6.69 wt%	Cementite
N	1394	0	The allotropic transition from γ-Fe to δ-Fe (A_4)
P	727	0.022 wt%	The maximum solubility of C in α-Fe
S	727	0.77 wt%	Eutectoid point $\gamma_S \rightleftharpoons \alpha_P + \mathrm{Fe_3C}$
Q	Room Temp	0.0008 wt%	The solubility of C in α-Fe

cementite, has an orthogonal crystal structure with a fixed carbon concentration of 6.69 wt%.

Points: The temperature, composition and meaning of each characteristic point in the Fe-Fe_3C phase diagram are shown in Table 8.2. The symbols of the characteristic point are internationally used as standards.

Lines: The Fe–Fe_3C phase diagram includes a liquidus line, *ABCD*, and a solidus line, *AHJECF*, which contains three isothermal transition lines. On line *HJB* at 1495 °C, a peritectic reaction occurs, forming an **austenite** microstructure (A):

$$L_B + \delta_H \rightleftharpoons \gamma_J \tag{8.36}$$

On line ECF at 1148 °C, a eutectic reaction takes place, resulting in the **ledeburite** microstructure (L_d), which is a mixture of austenite and cementite:

$$L_C \rightleftharpoons \gamma_E + \mathrm{Fe_3C} \tag{8.37}$$

On line PSK at 727 °C, a eutectoid reaction occurs, leading to the **pearlite** microstructure (P), which is a mixture of ferrite and cementite:

$$\gamma_S \rightleftharpoons \alpha_P + \mathrm{Fe_3C} \tag{8.38}$$

This eutectoid reaction temperature T_E is also marked as A_1. In addition, the phase diagram also includes several important solid-state transition lines. Line *GS*, also called the A_3 line, is the starting line of the precipitation of ferrite from austenite during cooling or the ending line of the complete dissolution of ferrite into austenite upon heating. Line *ES* is the solid solubility curve of carbon in austenite, also known as the A_{cm} line, below which **secondary cementite** (Fe_3C_{II}) precipitates from austenite. This is different from the **primary cementite** (Fe_3C_I) precipitated from liquid along line *CD*. Line *PQ* is the solid solubility curve of carbon in ferrite, below which **tertiary cementite** (Fe_3C_{III}) precipitates from ferrite.

Regions: The diagram also contains five single-phase regions, including three solid solutions (α, γ and δ), one compound Fe_3C and one liquid phase (L) as well as seven two-phase regions, i.e., L + δ, L + γ, L + Fe_3C_I, δ + γ, α + γ, γ + Fe_3C_{II}, α + Fe_3C_{III}.

Development of microstructure in Fe-Fe_3C alloys.

Seven representative compositions of Fe-Fe_3C alloys are examined to demonstrate the microstructural development, as shown in Fig. 8.37.

C_1: Industrial Pure Iron (<0.022 Wt% C)

Between the temperature points 1 and 2, the δ solid solution crystallizes from the liquid alloy via isomorphous transition. Between points 2 and 3, only a single δ solid solution is present. The polymorphic transition of δ→γ occurs between points 3 and 4, and the γ phase nucleates and grows first on the grain boundary of the δ phase. The alloy remains entirely γ phase until point 5. Between points 5 and 6, another polymorphic transition of γ→α occurs, where the α phase preferentially nucleates and grows on the grain boundary of the γ phase. Tertiary cementite Fe_3C_{III} precipitates from α below point 7. The crystallization process of industrial pure iron is shown in Fig. 8.38.

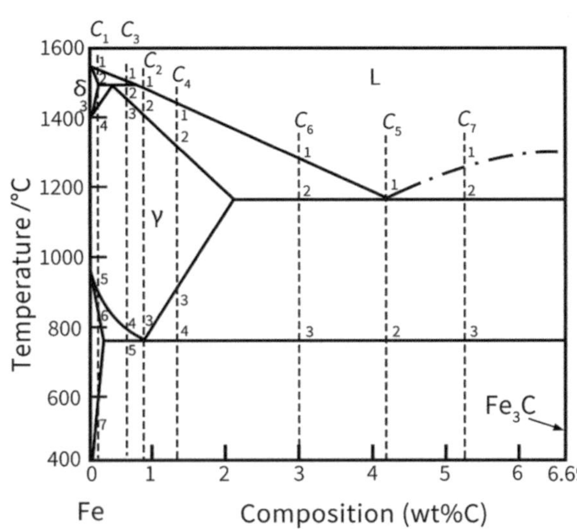

Fig. 8.37 The cooling process of several representative Fe–C alloys

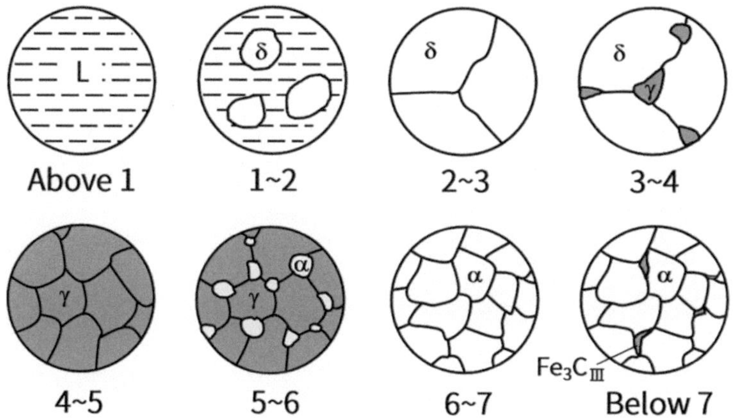

Fig. 8.38 Schematic drawings of the equilibrium microstructures for Fe–C alloy C_I during the solidification process

C_2: Eutectoid Steel (0.77 Wt% C)

Between the temperature points 1 and 2, the γ solid solution forms by the isomorphous transition during the solidification of the liquid alloy. Below point 2, the alloy is solidified completely into a single phase γ. This state is maintained until point 3 when the eutectoid reaction occurs at 727°C, and the γ phase transforms completely into pearlite by the end of the transition. **Pearlite** is composed of alternating lamellae of **eutectoid ferrite** (α) and **cementite** (Fe_3C). As the temperature continues to decrease, a small amount of Fe_3C_{III} precipitates from the α phase and becomes unidentifiable from the microstructure due to its preferential growth with eutectoid cementite. The crystallization process of eutectoid steel is shown in Fig. 8.39.

At room temperature, the relative amounts of α and Fe_3C in pearlite can be obtained by the lever rule.

$$W_\alpha = \frac{6.69 - 0.77}{6.69 - 0.0008} \times 100\% = 88\%$$

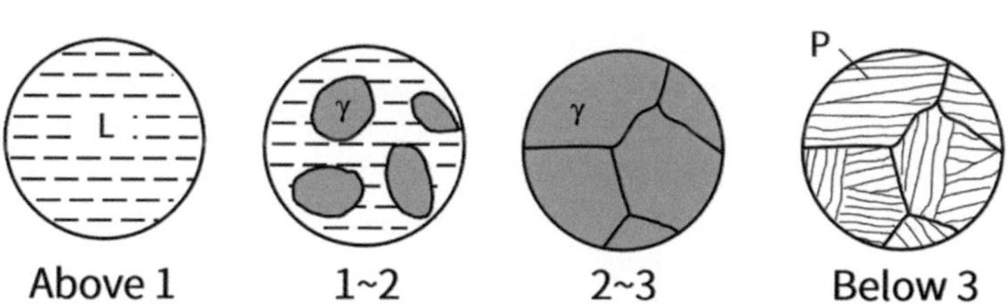

Fig. 8.39 Schematic drawings of the equilibrium microstructures for Fe–C alloy C_2 during the solidification process

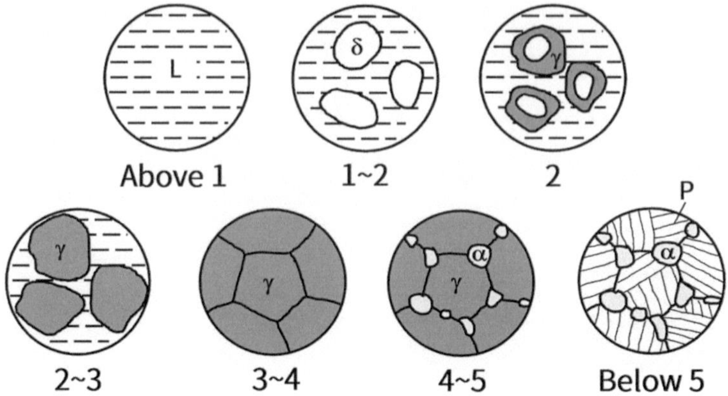

Fig. 8.40 Schematic drawings of the equilibrium microstructures for Fe–C alloy C_3 during the solidification process

$$W_{Fe_3C} = 100\% - 88\% = 12\%$$

Here, the solubility limit carbon of ferrite at room temperature is set to 0.0008 wt%.

The lamellar spacing of pearlite decreases with increasing cooling rate, resulting in finer pearlite with higher strength, toughness, and plasticity. With appropriate annealing treatment, eutectoid cementite in pearlite can have a spherical shape, known as **spheroidal pearlite**, which has lower strength but better plasticity and toughness than lamellar pearlite.

C_3: Hypoeutectoid Steel (0.022wt%-0.77 Wt% C)

For **hypoeutectoid** (less than eutectoid) **steel**, the δ solid solution crystallizes from the liquid alloy by an isomorphous transition between points 1 and 2. At point 2, the peritectic reaction occurs, leaving a residual liquid phase due to the carbon concentration of the alloy being greater than the composition of the peritectic point (0.17 wt% C). From point 2 to 3, this liquid phase continuously solidified into the γ phase. The alloy remains entirely γ phase from point 3 to 4. Below point 4, the α phase, also called **proeutectoid** (meaning "pre- or before eutectoid") **ferrite**, starts to precipitate from the grain boundary of the γ phase. As the temperature decreases, the amount of proeutectoid ferrite increases, and the carbon concentrations of the α phase and the remaining γ change along the *GP* and *GS* lines, respectively. At point 5, the carbon concentration of the remaining γ reaches 0.77 wt%, where a eutectoid transition takes place to form pearlite. Below point 5, tertiary cementite precipitates from the proeutectoid ferrite, but its amount was negligible. The microstructure of alloy C_3 at room temperature consists of proeutectoid α and pearlite, as shown in Fig. 8.40.

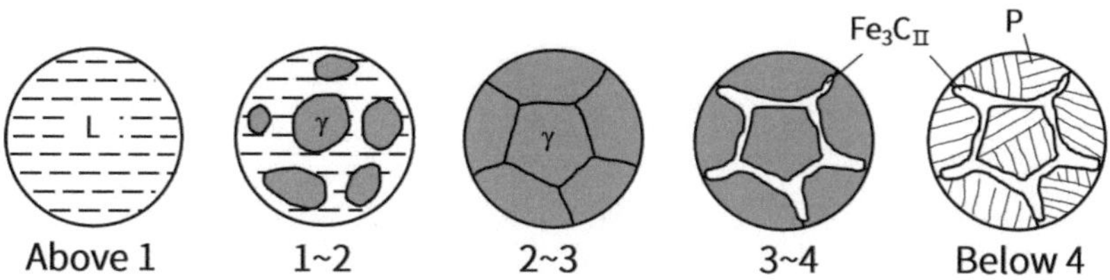

Fig. 8.41 Schematic drawings of the equilibrium microstructures for Fe–C alloy C_4 during the solidification process

C_4: Hypereutectoid Steel (0.77 Wt%-2.11 Wt% C)

For a **hypereutectoid** (more than eutectoid) **steel**, as the temperature decreases from point 1 to 2, the γ solid solution forms gradually according to the isomorphous transition. The alloy remains entirely γ phase from point 2 to 3. As the alloy is cooled to point 3, secondary cementite Fe_3C_{II}, also called **proeutectoid cementite**, precipitates from the grain boundary of the γ phase until point 4, during which the composition of the γ phase varies along the *ES* line. The precipitates of Fe_3C_{II} are generally located along the grain boundaries of the γ phase, exhibiting a network distribution. When the temperature decreases to point 4, the carbon concentration of γ decreases to a eutectoid composition, and then an isothermal eutectoid transition arises. The final microstructure consists of network Fe_3C_{II} and pearlite, as schematically shown in Fig. 8.41.

C_5: Eutectic White Cast Iron (4.30 Wt% C)

When alloy C_5 is cooled to point 1, a eutectic reaction occurs, resulting in a eutectic structure called ledeburite (L_d). Continuous cooling into the range of point 1 to 2, the secondary cementite Fe_3C_{II} continuously precipitates from the γ phase in the eutectic structure, which usually attaches to the eutectic cementite and cannot be identified. As the temperature decreases to point 2, the carbon concentration of the eutectic γ phase decreases to the eutectoid point (0.77 wt%). At this temperature, the γ phase transforms into pearlite by an isothermal eutectoid reaction. The resulting microstructure is called room-temperature ledeburite, or **transformed ledeburite** (L_d'), which maintains the morphology of eutectic ledeburite but transforms the eutectic γ into pearlite, as schematically shown in Fig. 8.42.

C_6: Hypoeutectic White Cast Iron (2.11 Wt%-4.30 Wt% C)

As alloy C_6 cools into the temperature range of point 1 to 2, the primary γ phase precipitates from the liquid phase, and the composition of both phases changes along the JE and BC lines, respectively. At point 2, the liquid phase with eutectic composition transforms into ledeburite via eutectic transition, and secondary cementite Fe_3C_{II} precipitates from the primary γ and the eutectic γ. The composition of the γ phase varies along the *ES* line. Upon cooling to point 3, all γ phases transform into pearlite through eutectoid reactions. The resulting microstructure is

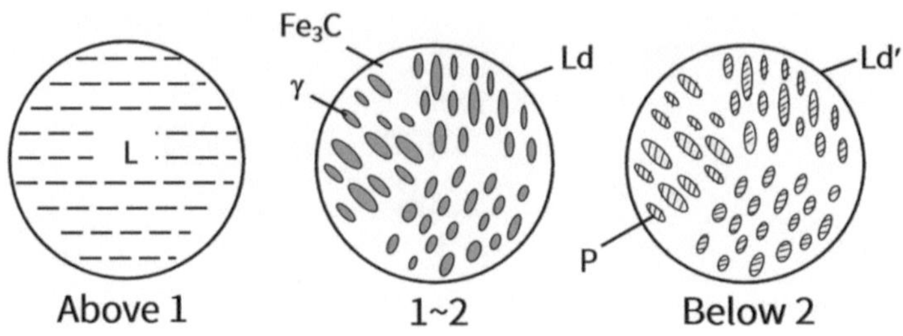

Fig. 8.42 Schematic drawings of the equilibrium microstructures for Fe–C alloy C_5 during the solidification process

composed of pearlite, Fe_3C_{II} and transformed ledeburite, as schematically shown in Fig. 8.43.

C_7: Hypereutectic White Cast Iron (4.30 Wt%-6.69 Wt%C)

As alloy C_7 is cooled from point 1 to 2, primary cementite Fe_3C precipitates from the liquid phase, forming a strip-shaped morphology. The subsequent transition follows the process of eutectic white cast iron. At room temperature, the resulting microstructure of hypereutectic white cast iron consists of Fe_3C_I and transformed ledeburite, as shown in Fig. 8.44.

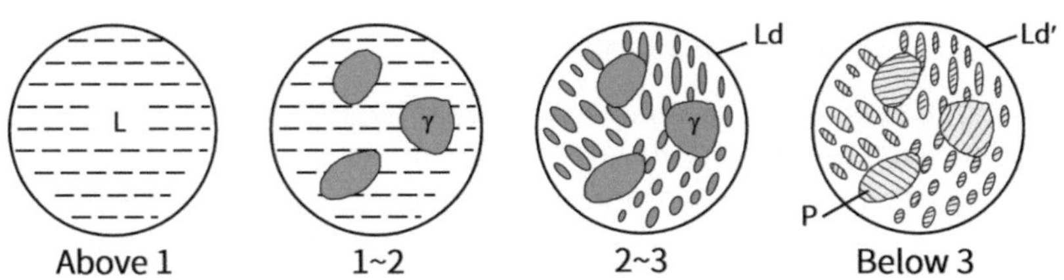

Fig. 8.43 Schematic drawings of the equilibrium microstructures for Fe–C alloy C_6 during the solidification process

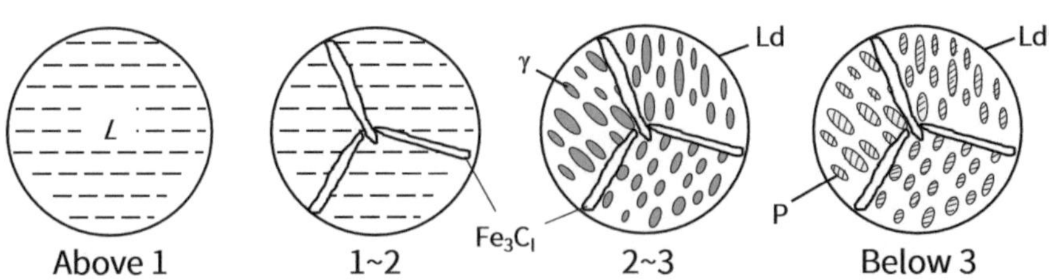

Fig. 8.44 Schematic drawings of the equilibrium microstructures for Fe–C alloy C_7 during the solidification process

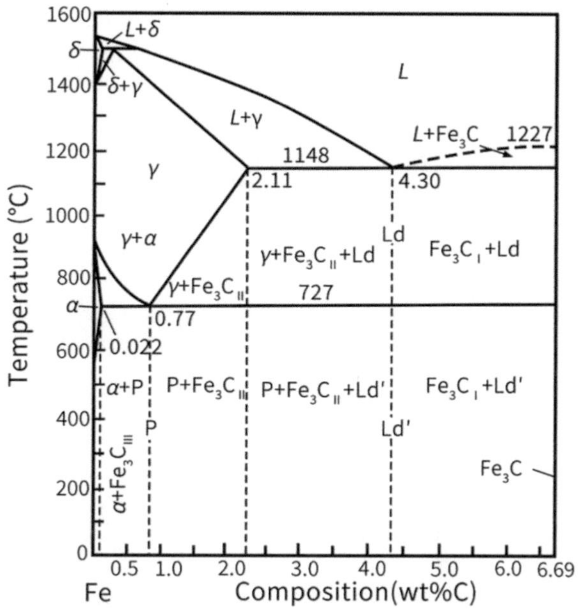

Fig. 8.45 The phase regions in the Fe-Fe_3C phase diagram labeled according to the resultant microstructures

② Effects of Carbon Content on the Microstructure and Properties of Fe–C Alloys

The phase regions in the Fe-Fe_3C phase diagram can be identified based on the resulting microstructures, as shown in Fig. 8.45, according to the above analysis of the transition process of various Fe–C alloys.

With increasing carbon content, the microstructure of the Fe–C alloy changes as follows:

$$\alpha + Fe_3C_{III} \rightarrow \alpha + \text{P (pearlite)} \rightarrow \text{P} \rightarrow \text{P} + Fe_3C_{II} \rightarrow \text{P} + Fe_3C_{II} + {L_d}' \rightarrow {L_d}' \rightarrow {L_d}' + Fe_3C_I$$

In addition, from the perspective of the phase constituent, the equilibrium structure of the Fe–C alloy at room temperature is composed of ferrite and cementite phases. With increasing carbon content, the morphology of cementite changes as follows:

$$Fe_3C_{III}\text{(thin plate)} \rightarrow \text{eutectoid } Fe_3C \text{ (lamella)} \rightarrow Fe_3C_{II}\text{(network)} \rightarrow \text{eutectic } Fe_3C \text{ (continuous matrix)} \rightarrow Fe_3C_I\text{(thick flake)}$$

The effect of the carbon content on the mechanical properties is mainly related to the final microstructure and the relative amount of the phases present. The equilibrium microstructure of Fe–C alloys at room temperature consists of α and Fe_3C phases. As mentioned earlier, α has low hardness and strength but good ductility, while Fe_3C is very hard and brittle. Pearlite is a mixture of α and Fe_3C, where fine lamellar Fe_3C is distributed on the α matrix to strengthen the alloy. The content of pearlite has a great influence on the properties of Fe–C alloys.

The amount of pearlite in hypoeutectoid steel gradually increases with increasing carbon content, leading to increased strength and hardness but decreased ductility and toughness. In hypereutectoid steel, the microstructure contains Fe_3C_{II} and pearlite. At low carbon content (below 1.0 wt%), the Fe_3C_{II} on the grain boundaries has little effect on the mechanical properties since it has not yet formed a continuous network. When the carbon content is close to 1.0 wt%, strength is highest. However, when the carbon content exceeds 1.0 wt%, the Fe_3C_{II} phase increases to form a continuous network, greatly increasing brittleness and lowering toughness and strength. White cast iron is brittle and weak due to its high cementite content.

8.5 Ternary Phase Diagrams

Ternary phase diagrams depict the systems containing three components, showing possible phases present in equilibrium at specified composition, temperature and pressure. As discussed in binary phase diagrams, the pressure is typically fixed at 1 atmosphere. Due to the addition of the third component, these diagrams are represented as a 3-dimensional triangular prism. The base of the prism is known as a compositional triangle, with each side of the prism analogous to a binary phase diagram. The lines and regions in binary systems are expanded into surfaces and spaces in ternary systems, respectively.

8.5.1 Basics of Ternary Phase Diagrams

① Ternary Space Model

To fully represent the phase equilibrium of a ternary system under constant pressure, a three-dimensional model called a "space model" is necessary. This model includes a triangular prism, as shown in Fig. 8.46, where the vertical axis represents temperature and the base of the prism corresponds to composition, typically displayed as an equilateral triangle. The vertical sides of the prism represent the three binary systems AB, BC, and AC, which combine to form the ternary system ABC.

② Compositional Triangle

The triangular base of the prism corresponds to the compositional triangle (as shown in Fig. 8.47), in which each corner represents one pure components, points along the sides correspond to binary alloys and points within the triangle represent ternary alloys.

Weight percentages are often used to express the proportions of each component in a ternary alloy. For example, in Fig. 8.47, the weight percentage of each

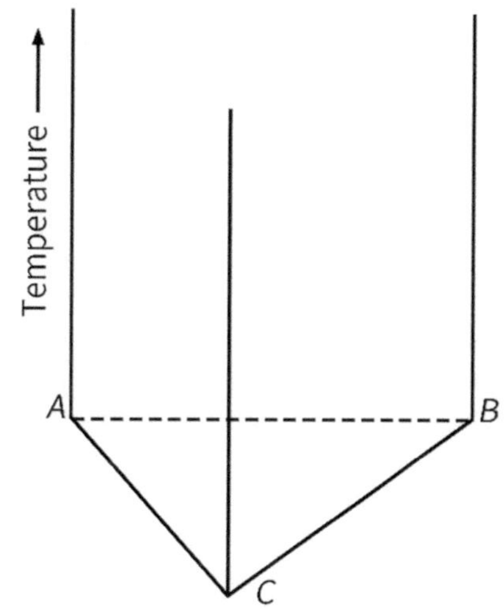

Fig. 8.46 Space model used for representing ternary systems

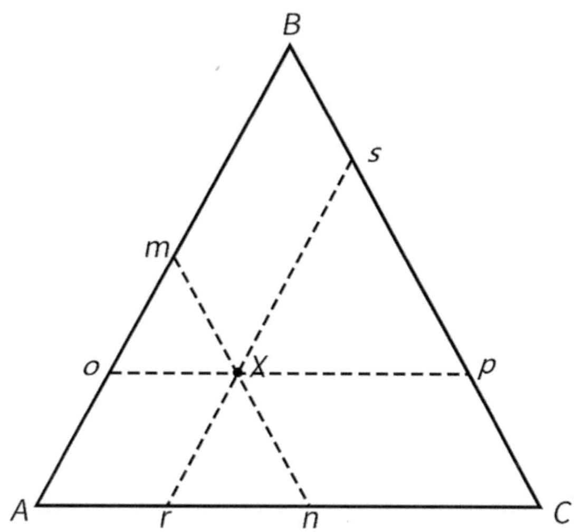

Fig. 8.47 Alloy compositions represented by equilateral triangle

component in alloy X can be determined by the following procedure. Drawing a line *mn* parallel to the side of *BC*, it can give the percentage of A in terms of the lengths *mB* or *nC*. The lengths *oA* or *pC* obtained by drawing the line *op* parallel to the side of *AC* give the percentage of B. In the same way, the percentage of C can be obtained by drawing *rs* parallel to the *AB* side and determining the lengths *r*A or *s*B.

Therefore,

$$\%\text{A} = mB = nC = sX = pX$$

$$\%B = oA = pC = rX = nX$$

$$\%C = rA = sB = oX = mX$$

As an example, let us consider plotting the composition point of an alloy Y that contains 30% C and 50% B. Starting from point *A* on the compositional triangle, the length *Ab* corresponding to 50% B could be marked on the *AB* side, and the length *Ac* corresponding to 30% C on the *AC* side. Thus, the composition point of the alloy is located at the intersection of two lines *by* and *cy*, as shown in Fig. 8.48. Alternatively, the alloy compositions can be represented on the triangle by taking the height of the triangle as 100%, and the percentages of each component in a ternary alloy are given by the length of the vertical height line from the composition point to the opposite sides of the triangle. However, this representation is less convenient than the previous method.

It is sometimes useful to represent a composition using a right-angled triangle. For example, in Fig. 8.49, the percentage of A is obtained by subtracting the percentage of B + C from 100. This representation is particularly useful when the interest is focused on alloys of metal A with relatively small amounts of B and C, which is often the case in aluminum-based alloys where an A-rich corner of the system needs to be represented.

The compositional triangles of ternary systems include the following orientation lines:

- When a line is drawn from one corner of the triangle to intersect the opposite side, the ratio of the other two components in the alloy remains constant. For

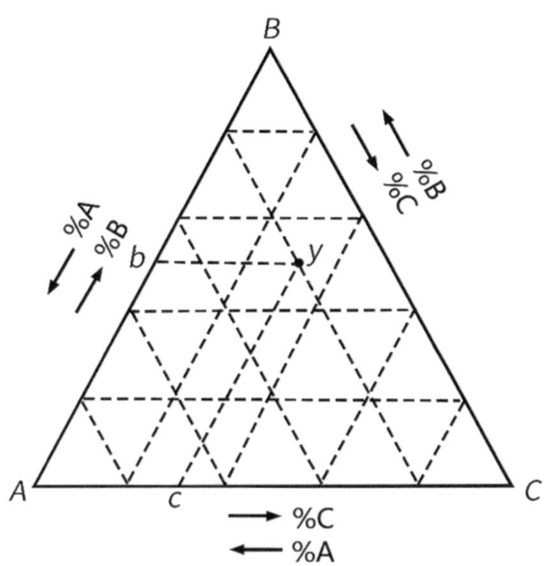

Fig. 8.48 Alloy composition points drawn on the compositional triangle

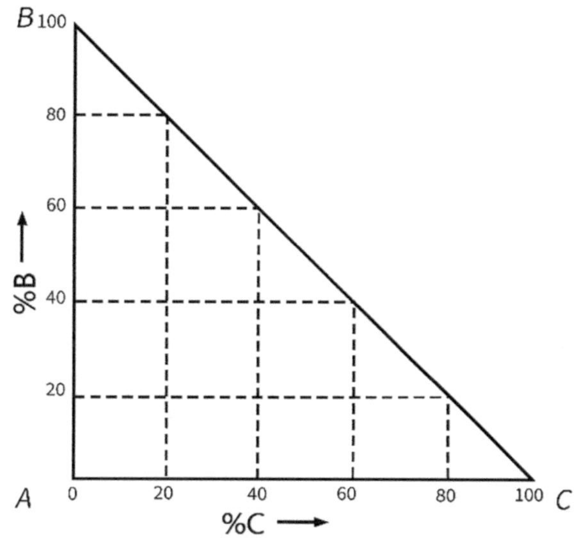

Fig. 8.49 The right-angled triangle to represent compositions

example, as shown in Fig. 8.50, taking the alloy compositions along *Am* into account, the percentage of A present in the alloy decreases as the composition point is gradually moved away from *A*, while the relative amounts of B and C remain the same.

- Along a line parallel to one side of the triangle, the alloys contain a constant percentage of one component. For instance, the percentage of A will remain the same along *no* in Fig. 8.50.

③ Lever Rule in Ternary Systems

If a mixture f is composed of two alloys d and e (Fig. 8.51), then the composition of f lies on the straight line connecting the original alloy compositions (**tie line**).

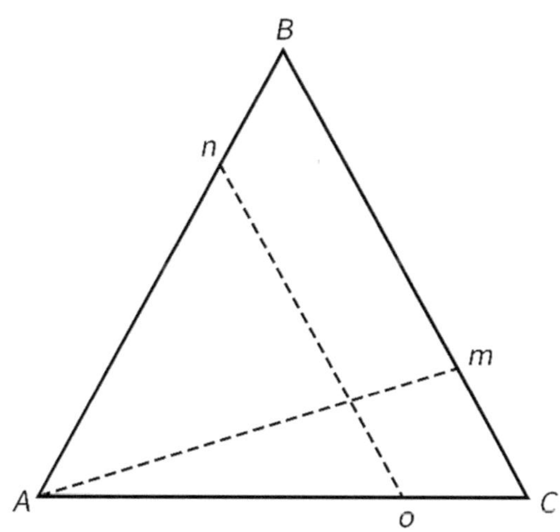

Fig. 8.50 Compositional aspects of ternary system

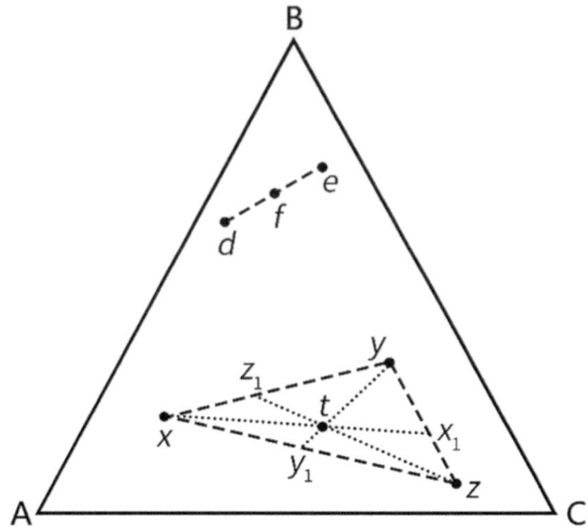

Fig. 8.51 Mixtures including two or three alloys or phases

The composition point *f* divides the tie line into two segments, and the length of each segment is inversely proportional to the weight percent of the corresponding alloy. The following formulas can be used to calculate the weight percentages:

$$\frac{\text{Weight of } d}{\text{Weight of } e} = \frac{ef}{df} \times 100\% \quad (8.39)$$

$$d = \frac{ef}{de} \times 100\% \quad (8.40)$$

$$e = \frac{df}{de} \times 100\% \quad (8.41)$$

This lever rule is virtually the same as stated in the binary phase diagrams, which has an important application in determining the composition of a two-phase mixture resulting from the decomposition of a single alloy along a tie-line.

④ The 'center of gravity' principle

As the three alloys x, y, and z are mixed together to form the mixture **t**, its composition (Fig. 8.51) is located inside a triangle formed by connecting the alloy compositions with straight lines, which is called the tie triangle. The principle of the center of gravity states that the triangle can be balanced on the fulcrum at point *t*, and the quantity of each alloy is proportional to the length from the fulcrum to the opposite side. Therefore, the weight percentage of the three alloys in the mixture at equilibrium is given by the following formula:

$$x(\%) = \frac{x_1 t}{x x_1} \times 100\% \quad (8.42)$$

$$y(\%) = \frac{y_1 t}{y y_1} \times 100\% \tag{8.43}$$

$$z(\%) = \frac{z_1 t}{z z_1} \times 100\% \tag{8.44}$$

In addition, this principle will be applied to 'tie-triangles' representing a three-phase equilibrium in a system.

8.5.2 Ternary Phase Diagrams with Two-Phase Equilibrium

① The Space Model and Projected View

The ternary isomorphous system depicted in Fig. 8.52a consists of three completely soluble components A, B, and C in both the liquid and solid states. The liquid + solid solution two-phase space, which represents a bundle of tie-lines with varying directions, is enclosed by liquidus and solidus curved surfaces. Each tie-line connects the composition points of the coexisting liquid and solid at a specific temperature, and no two tie-lines at the same temperature intersect. Experimental or thermodynamic methods are necessary to verify the location and direction of tie-lines in specific systems.

The projected views on the base of the space model of paths that represent the change in liquid and solid compositions are illustrated in Fig. 8.52b. The equilibrium solidification process in the system can be illustrated by using the alloy composition point X as an example. Upon cooling through the liquid region to liquidus temperature t_1, solidification starts, and solid solution crystals form, which contain more B than the original liquid. The compositions of the liquid and solid phases are given by the endpoints of tie line L_1S_1, and the positions of multiple tie

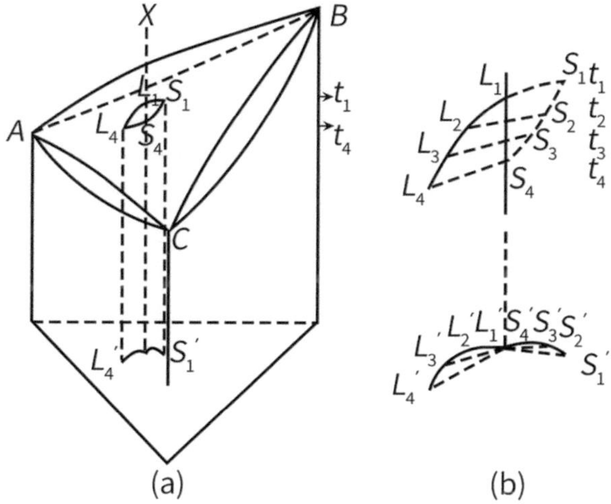

Fig. 8.52 **a** Space model of the ternary system showing complete solid solubility of the components and **b** The compositions of L and S phases and their projected views on the base of the space model during the solidification process of typical alloy X

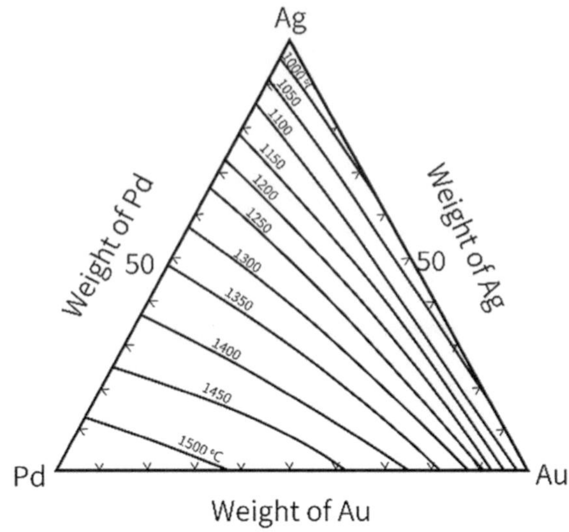

Fig. 8.53 Liquidus projection for the Ag-Au–Pd system (From Metals Handbook, 7th Edn., 1948)

lines passing through the alloy composition point X at low temperatures (t_2 -t_4) are also shown. Therefore, L_1S_1 tie-line represents the beginning of solidification, and L_4S_4 indicates the end of solidification. The endpoints of these tie lines locate on the liquidus and solidus surfaces, and the paths of $L_1L_2L_3L_4$ and $S_1S_2S_3S_4$ represent the gradual changes in the liquid and solid phase compositions during the solidification process, respectively.

A series of experimental measurements of the liquidus and solidus temperatures of an alloy can be used to draw isotherms on the liquidus and solidus surfaces. These isotherms can be conveniently represented by projecting a view onto the base of the space model. For example, Fig. 8.53 shows a projection of the liquidus plane for the Ag-Au–Pd system.

② Isothermal Sections and the Application of Lever Rule

Tie lines for alloy X are indicated on the isothermal sections of the space model at temperatures t_2 and t_3 (Fig. 8.54). These sections also show other typical tie-lines for the liquid + solid regions, and some of them may point coincidently toward a corner of the system. The lever rule can be applied to the appropriate tie line at specified temperature to calculate the relative amounts of liquid and solid, such as at t_2 (Fig. 8.54a):

$$\frac{\text{Weight of liquid}}{\text{Weight of solid}} = \frac{X_2S_2}{X_2L_2} \tag{8.45}$$

$$\%\text{Liquid} = \frac{X_2S_2}{L_2S_2} \times 100\% \tag{8.46}$$

$$\%\text{Solid} = \frac{X_2L_2}{L_2S_2} \times 100\% \tag{8.47}$$

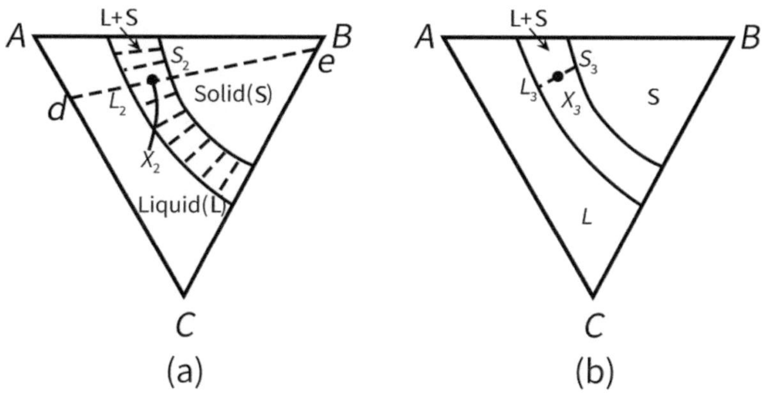

Fig. 8.54 Isothermal sections through the space model of Fig. 8.52 for temperatures **a** t_2, and **b** t_3

As the temperature continues to decrease to T_3 (Fig. 8.54b), the amount of solid solution increases due to the increased length of X_3L_3 relative to the corresponding tie lie. With further cooling to lower temperature, the composition of the solid solution eventually reaches the composition of alloy X, indicating no remaining liquid and the completion of solidification.

Vertical Section

Figure 8.55 depicts the vertical section taken along the direction of tie-line L_2S_2 on the isothermal section at temperature t_2. Except the tie-line L_2S_2, the tie lines at other temperatures for the liquid + solid regions do not lie in the plane of the vertical section. As a result, the vertical section of a ternary system differs significantly from a binary phase diagram. The compositions of the liquid and solid, coexisting in equilibrium during the solidification of alloys whose compositions lie in the vertical section, vary out of the plane of this section. Therefore, it is incorrect to assume that a horizontal line cutting the liquidus and solidus surfaces in this plane defines the compositions of the liquid and solid in equilibrium.

8.5.3 Ternary Phase Diagrams with Three-Phase Equilibrium

Although there are several types of ternary phase diagrams with three-phase equilibrium, this section will focus on the following two cases:

- Ternary phase diagram containing one eutectic three-phase region: Liquid (L) + two primary solid solutions (α and β) forming a three-phase eutectic reaction $L \rightarrow \alpha + \beta$;
- Ternary phase diagram containing one peritectic three-phase region: Liquid (L) + two primary solid solutions (α and β) forming a three-phase peritectic reaction $L + \alpha \rightarrow \beta$;

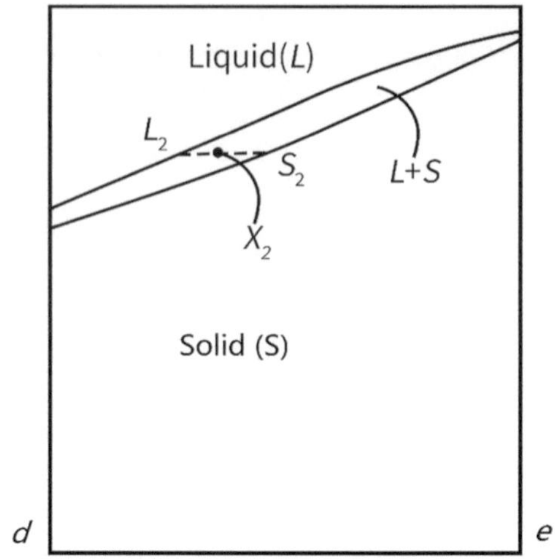

Fig. 8.55 Vertical section taken along the direction of tie-line L_2S_2 of Fig. 8.54a

① Systems with a Eutectic Reaction L → α + β

The Space Model of the System

Figure 8.56 shows the space model of a hypothetical system ABC, where complete liquid solubility is assumed. The solid solutions are called α and β, with complete solid solubility in binary system AC and partial solubility in AB and BC of the eutectic type. The eutectic point *M* representing the reaction L → α + β in system AB is assumed to have a higher temperature than that of point N representing the reaction L → α + β in system BC.

In the space model, the liquidus consists of *AMNC* and *BMN* surfaces, which correspond to the solidification of primary α and β, respectively. They intersect along a eutectic curve or 'valley' *MN* that falls smoothly from side *AB* to side *BC*. Curves *DG* and *EF* connect the compositions of the α and β phases formed in binary systems during eutectic reactions, represented by the horizontal straight lines *DME* and *GNF*. The solidus of the system is composed of surfaces *ADGC*, *DEFG*, and *BEF*. Each of the liquid + solid regions (i.e., liquid + α and liquid + β) is surrounded by respective liquidus and solidus surfaces, similar to the two-phase region of the solid solution system.

Curves *M*N, *DG*, and *EF* are not in the same plane. Curve *MN* is located above surface *DEFG*, and there are three curved surfaces (*DMNG*, *MEFN*, and *DEFG*) that enclose a three-phase space, representing the coexistence of liquid, α, and β. Each surface consists of tie lines that represent the coexistence of two phases among liquid, α, and β. Surfaces *DMNG* and *MEFN* separate the three-phase space from the L + α and L + β regions, respectively. *DEFG* separates the three-phase space from the α + β region and is therefore the solidus surface for the α + β portion of the system.

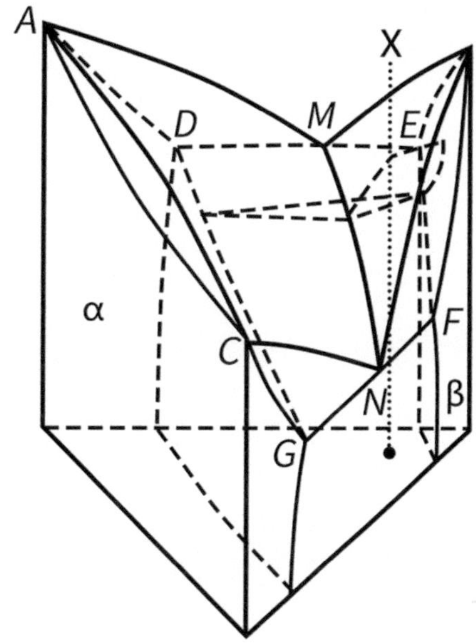

Fig. 8.56 Space model of system showing a eutectic reaction: L → α + β

The Eutectic Reaction and the Application of Lever Rule

The three-phase space in the ternary system is related to the eutectic reaction L → α + β, which occurs over a temperature range. The three-phase region can be considered as a collection of isothermally tie triangles in the form shown in Fig. 8.57,where the corners represent the compositions of the coexisting phases at that particular temperature. As the eutectic reaction temperature decreases, the liquid composition follows curve *MN*, while the α and β compositions vary along curves *DG* and *EF*, respectively. Therefore, the composition of each phase is restricted to a curve that varies with temperature when three phases coexist.

The eutectic solidification process in equilibrium conditions is illustrated by alloy X in Fig. 8.56, which is composed of primary β crystals and α and β eutectic mixtures as solidified. As the temperature decreases, the liquid composition follows the path on liquidus surface, and the β composition changes along the path on solidus surface. At temperature T_1, the liquid composition reaches a point on curve *MN* before complete solidification, and the β composition moves to a point on curve *EF*, forming a tie triangle, as shown in Fig. 8.56 and Fig. 8.58(a). The alloy composition is located on tie-line $L_1\beta_1$, forming one of the sides of $L\alpha\beta$ tie-triangle at T_1, and the relative amounts of L and β can be obtained using the lever rule.

As the temperature drops to T_2, the alloy composition point is now within the tie triangle (Fig. 8.58b), entering the three-phase region. The liquid, α, and β compositions can be represented by points L_2, α_2, and β_2, respectively, at the triangle corners. Therefore, the "center of gravity" principle can be applied to

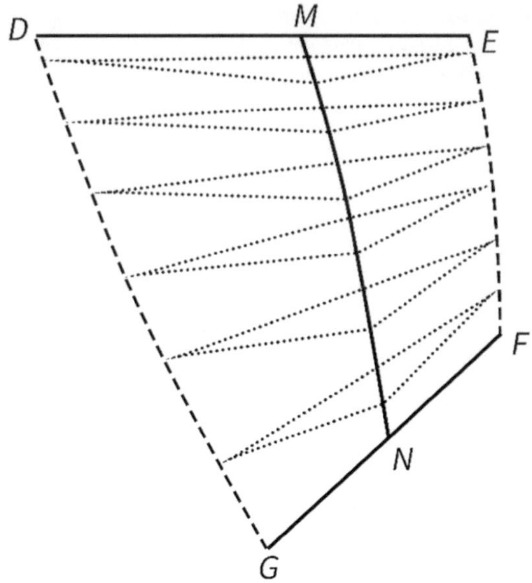

Fig. 8.57 Three-phase region (liquid, α and β) of Fig. 8.56

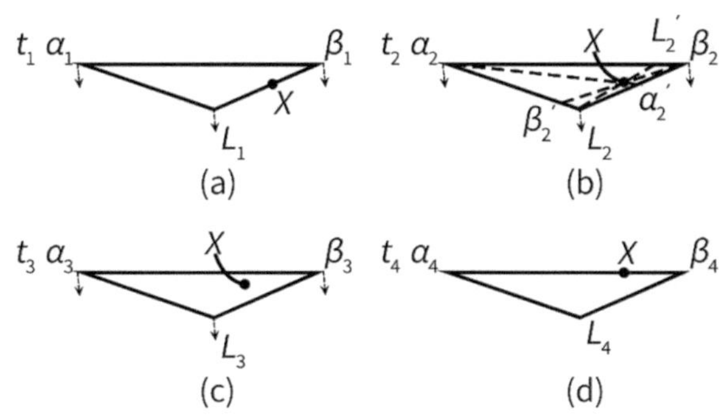

Fig. 8.58 Tie triangles at various temperatures illustrate the progress of the eutectic reaction in alloy X of Fig. 8.56

determine the amount of each phase present.

$$\%Liquid\,(L_2) = \frac{XL_2'}{L_2L_2'} \times 100\% \tag{8.48}$$

$$\%\beta_2 = \frac{X\beta_2'}{\beta_2\beta_2'} \times 100\% \tag{8.49}$$

$$\%\alpha_2 = \frac{X\alpha_2'}{\alpha_2\alpha_2'} \times 100\% \tag{8.50}$$

In this example, it is assumed that the size and shape of tie triangle are roughly constant throughout the entire solidification range of alloy X, but in an actual system, these characteristics of the triangle at different temperatures may vary significantly.

Upon cooling to a lower temperature T_3, the compositions of the liquid, α and β vary along the *MN*, *DG*, and *EF* curves, respectively. The tie triangle corresponding to temperature T_3 is shown in Fig. 8.58c, and the position of the alloy composition X relative to the triangle is different from that at T_2. Using the "center of gravity" principle shows that the amount of liquid at T_3 is less than that at T_2, which is consistent with the eutectic property of the $L \rightarrow \alpha + \beta$ reaction.

As the alloy is further cooled, the compositions of α and β change continuously until the eutectic reaction is completed at the solidus temperature T_4. The tie triangle in Fig. 8.58d describes this situation. At this temperature, the alloy composition is located on the $\alpha + \beta$ side of the tie triangle, indicating that there is no liquid. The relative quantity of α and β can be determined by using the lever rule to the αβ tie line.

The eutectic solidification process can be observed through the movement of the tie triangle relative to the alloy composition point. The completion of solidification is indicated by the presence of alloy composition points on the αβ side of a tie-triangle.

Projected View of the System

The solidification process of the alloy in the given system can be tracked using the data for the liquidus and the solid state regions. The data can be represented as a system projected view on the base of the space model, as shown in Fig. 8.59. The curve mn represents the eutectic valley *MN*. The solidus lines *DG* and *EF* is depicted by curves *dg* and *ef*, respectively, which represent the boundaries between the α and β solid-solution regions and the $\alpha + \beta$ region as the solidification ends. The liquidus areas of α and β are represented by the surfaces *AmnC* and *Bmn*, respectively, while the areas of α, β and $\alpha + \beta$ solidus are denoted by *AdgC*, *Bef*, and *defg*, respectively. Figure 8.59 also displays the position of a typical tie-triangle. Note that even though a tie-triangle is essentially isothermal, the curves *mn*, *dg*, and *ef* are projected lines that extend with decreasing temperature, as shown by the arrows.

In cases where the solubility remains unchanged after solidification, the boundary positions of the solid state regions in the projected view (Fig. 8.59) will be the same as those shown in the isothermal section through the solid state part of the system (Fig. 8.61b). However, after the completion of solidification in practice, the boundary positions in the two types of diagrams will generally be different due to changes in solid solubility, but for discussion purposes, these changes can be conveniently ignored.

The system's projected view can be used to track the solidification processes of various alloys as follows. The position of the alloy composition point in either region *AmnC* or *Bmn* determines whether α or β will be the primary phase during solidification. Solidification occurs directly by the eutectic reaction without the separation of primary α or β if the alloy composition lies on curve *mn*. In the absence of experimental data for a given system, it is impossible to estimate

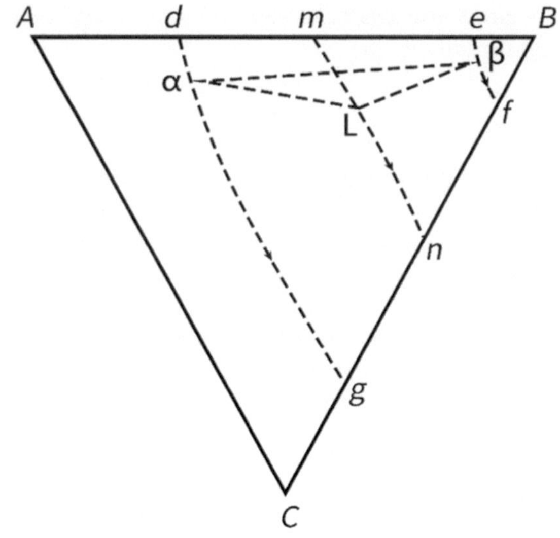

Fig. 8.59 Projected view of the base of the space model of the system shown in Fig. 8.56, including a typical tie-triangle

the composition of the primary phase when solidification begins or the accurate changes of liquid and solid compositions at the initial stage of solidification. Nevertheless, the final changes in liquid composition can be estimated by the procedure discussed later if the primary β solid solubility is very limited.

If the alloy composition is within the single-phase regions *AdgC* or *Bef*, solidification is completed according to the way interpretated for a system having complete solubility. When the alloy is within the two-phase region *defg*, the liquid composition reaches curve *mn* during solidification, and the eutectic reaction continues until all liquid is consumed upon further cooling.

As mentioned earlier, the eutectic curve may exhibit a maximum or minimum value. The projected view distinguishes these cases through the direction of the arrows on the projection of the eutectic curve, as illustrated in Fig. 8.60a and b. At the maximum or minimum points, the tie triangles reduce to a tie line that connects the compositions of L, α, and β under equilibrium conditions.

Isothermal Sections

Figure 8.61a shows a representative isothermal section through the space model between the eutectic temperatures in the AB and BC systems. The tie triangle of L + α + β region is related to the surrounding single-phase and two-phase regions. Typical tie lines are shown crossing the two-phase regions. An isothermal section in the solid-state with a temperature slightly lower than that of eutectic BC is shown in Fig. 8.61b. As mentioned previously, if the changes in solid solubility after solidification are ignored, the positions of the boundary curve are the same as the solidus projection in Fig. 8.59.

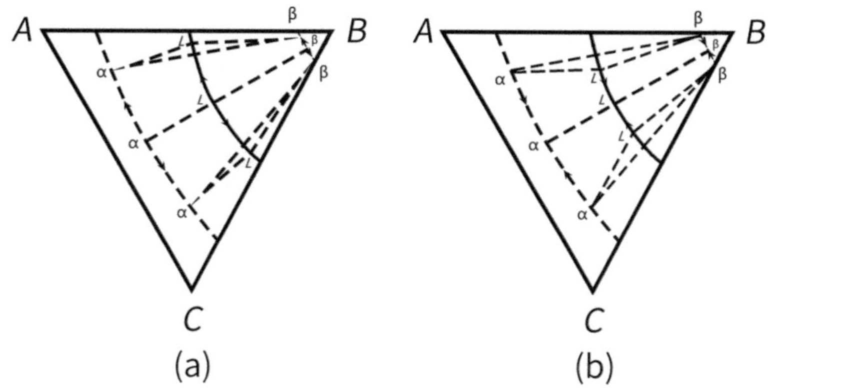

Fig. 8.60 Projected view of the system in which the eutectic valley shows **a** a maximum and **b** a minimum

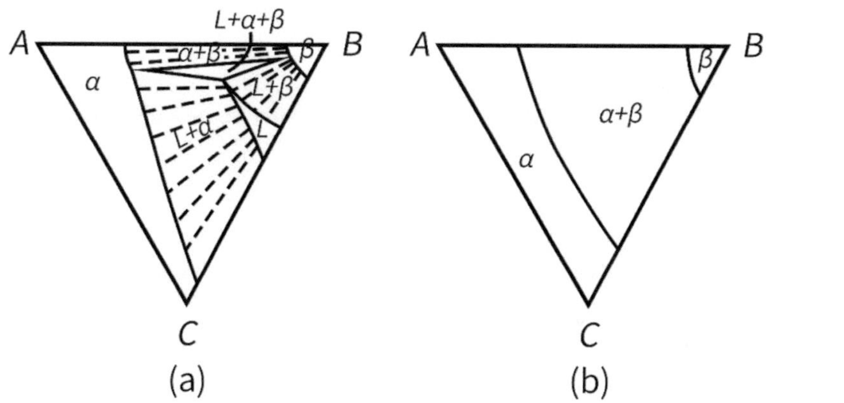

Fig. 8.61 Representative isothermal sections through the space model of Fig. 8.56 **a** at a temperature between that of the eutectics in systems AB and BC; **b** at a temperature below that of eutectic BC

Vertical Section

Figure 8.62 depicts a vertical section through corner *B* to the midpoint *r* of side *AC*, highlighting the three-phase region, L + α + β, where eutectic reactions occur within a specific temperature range. However, it should be noted that the boundaries of this area are not necessarily straight lines because they are sections of the surface formed by a series of tie lines, which are generally not on the plane of the section. Additionally, in the two-phase region, the tie lines are generally not on the plane of the section either.

While this type of section does not provide information on the composition of the coexisting phases in the equilibrium state, it shows the temperature at

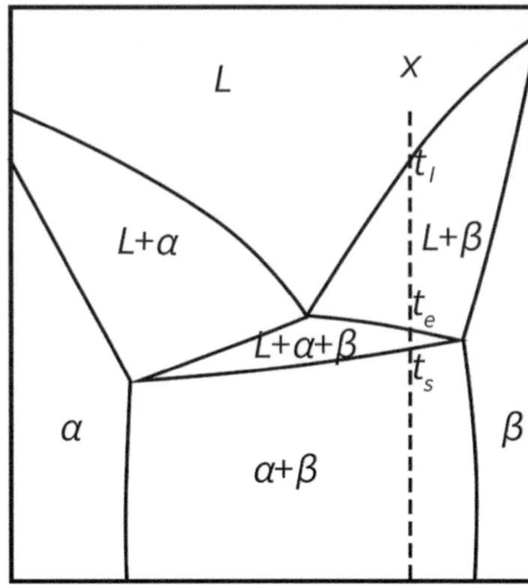

Fig. 8.62 Vertical section of the system shown in Fig. 8.56 from corner B to the midpoint r of side AC

which phase transformation occurs in a specific alloy and the stage of solidification. For instance, t_l represents the liquidus, t_e denotes the temperature at which eutectic reaction begins, and t_s indicates the temperature at which solidification is completed.

② Systems Containing a Peritectic Reaction L + α→ β.

The system depicted in Fig. 8.63 is a hypothetical ternary system ABC with complete liquid solubility, while the binary system BC exhibits complete solid solubility. Systems AB and AC have partial solid solubility and are of the peritectic type. The solid solutions are labeled as α and β, respectively, with the latter extending entirely across the *BC* side of the system. The temperature of the peritectic reaction L + α→ β in system AB is higher than that of the same reaction in system AC.

The liquidus comprises two surfaces, *AMN* and *MBCN*, corresponding to the primary separation of α and β, respectively. These surfaces intersect along the peritectic curve *MN*, which is not a valley as in the eutectic system, but more like a "shoulder." Curves *EF* and *DG* connect the compositions of α and β participating in the peritectic reactions in the binary systems, indicated by the horizontal straight lines *EDM* and *FGN*. Surface *AEF* represents the solidus of the liquid + α region, while *DBCG* is the solidus surface of the L + β region. Curves *MN*, *DG*, and *EF* do not lie in the same plane, with curve *DG* located below the surface *EMNF*. The three-phase space, which represents the coexistence of liquid, α, and β during the peritectic reaction, is bounded by the three surfaces *EMNF*, *DMNG*, and *EDGF*. Each of these surfaces is composed of tie lines, that is Lα, Lβ, and αβ tie-lines respectively.

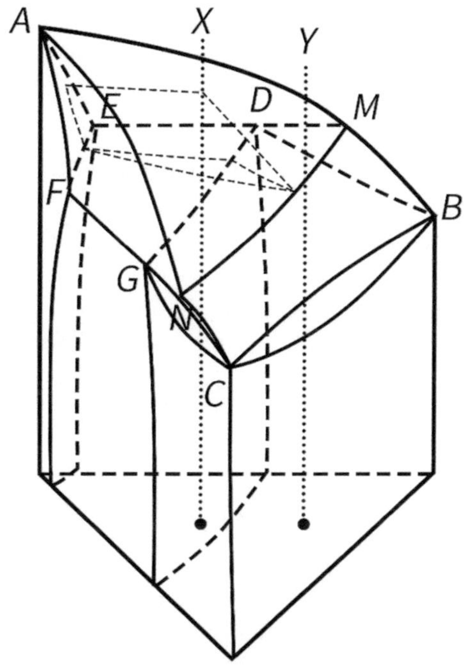

Fig. 8.63 Space model of the system showing a peritectic reaction: L + α→ β

8.5.4 Ternary Phase Diagrams with Four-Phase Equilibrium

When four phases are in equilibrium, invariant reactions take place in ternary systems. Two types of reactions will be discussed here, each involving a liquid (L) and three solid solutions (α, β, and γ):

(1) Ternary eutectic reaction:

$$\mathrm{L} \rightarrow \alpha + \beta + \gamma$$

(2) Ternary peritectic reaction:

$$\textbf{Type a}: \mathrm{L} + \alpha \rightarrow \beta + \gamma$$

$$\textbf{Type b}: \mathrm{L} + \alpha + \beta \rightarrow \gamma$$

The term "peritectic" is sometimes used only for the **Type b** reaction, while **Type a** is considered to have intermediate properties between eutectic and peritectic. However, in this discussion, the term "peritectic" applies to both types.

Each reaction involves the intersection of three liquidus curves, which represent the path of the liquid phase composition changes in a three-phase reaction. The fundamental characteristics of each invariant reaction can be understood by examining the liquidus curves and associated tie triangles.

① Systems with a Ternary Eutectic Reaction

Figure 8.64 illustrates the ternary system ABC, where each binary system is eutectic. The primary solid solutions are α, β, and γ respectively and there is no intermediate phase. Eutectic valleys (*DG*, *FG*, and *EG*) run from the binary systems to lower temperatures and intersect at the ternary eutectic point *G*. For each eutectic reaction, a three-phase region consisting of tie triangles exists. For the L → α + β reaction, the curve *DG* represents the composition of the liquid phase, the curve M_1M represents the composition of α, and N_1N represents the composition of β. Similarly, for the L → β + γ reaction, the liquid composition is located on *FG*, and the β and γ compositions are located in N_2N and O_2O, respectively. Finally, for the L → α + γ reaction, the liquid composition lies on *EG*, and the α and γ compositions are located in M_2M and O_1O, respectively.

As temperature decreases, the positions of tie triangles change due to the variation in compositions of different phases during eutectic reactions. The **ternary eutectic point** (*G* in Fig. 8.64) is the temperature at which the corners of the liquid compositions of the three tie triangles coincide and the ternary eutectic reaction occurs. At a constant eutectic temperature, the Lα sides of two tie triangles overlap, as do the Lβ and Lγ sides. Figure 8.65a shows the relative positions of tie triangles just above the ternary eutectic point, and Fig. 8.65b shows their combination at the ternary eutectic temperature. The ternary eutectic reaction is illustrated by

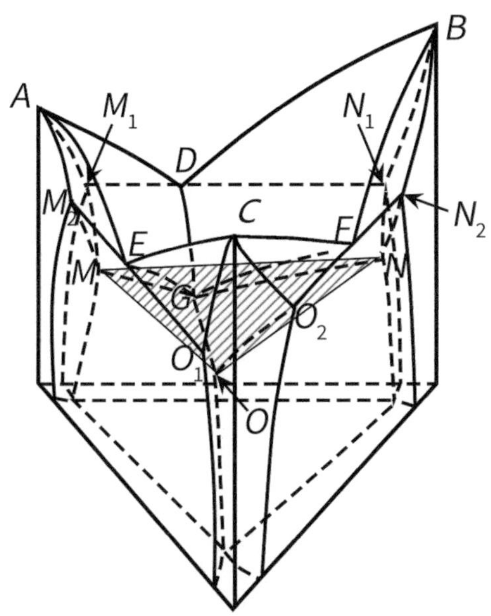

Fig. 8.64 Space model of the system showing a ternary eutectic reaction L → α + β + γ

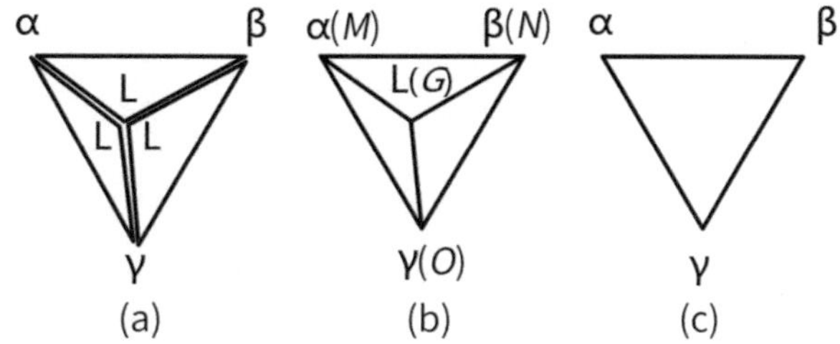

Fig. 8.65 Tie-triangles **a** at a temperature just above that of the ternary eutectic; **b** at the ternary eutectic temperature; **c** below the ternary eutectic temperature

a plane in the space model (*MNO* in Fig. 8.64). The compositions of the four phases involved in the reaction are fixed and represented by points on the plane. The liquid composition point lies within the triangle forming by the compositions of the three solid phases. Under equilibrium conditions, the temperature remains constant until all the liquid is consumed during the eutectic reaction.

There is a three-phase region below the ternary eutectic plane, namely, α + β + γ, and the equilibrium state at this temperature is represented by the αβγ tie triangle shown in Fig. 8.65c. Thus, the ternary eutectic reaction consists of three three-phase regions from higher temperatures to be approaching the eutectic plane and one three-phase region located below the eutectic plane. For a specified alloy, the present phases, their compositions and fractions can be determined by using the 'center of gravity' principle in the corresponding tie triangles, respectively.

② Systems Containing a Ternary Peritectic Reaction (Type a)

A system in which one binary system belongs to the peritectic type and two binary systems are of the eutectic type, for which the constant equation is:

$$\mathrm{L} + \alpha \rightarrow \beta + \gamma$$

In the ternary system ABC shown in Fig. 8.66, the AB binary system is peritectic, while the BC and AC binary systems are eutectic. The system forms terminal solid solutions α, β, and γ, and there are no intermediate phases.

Curve *DG* represents the liquid composition path for the L + α→ β peritectic reaction, while curve *EG* represents the liquid composition curve for the eutectic reaction L→ α + γ. The eutectic valley *FG* corresponds to the liquid composition curve for the eutectic reaction L→ β + γ. Point *G* is the composition of the liquid involved in the invariant peritectic reaction L + α→ β + γ.

Each curve is associated with a three-phase region composed of tie triangles. The α-phase composition and the composition of β are represented by curves M_1M and N_1N in the Lαβ region. In the Lαγ region, the α and γ compositions are located on curves M_2M and O_1O, respectively. In the Lβγ region, the β and γ compositions follow curves NN_2 and OO_2, respectively. The ternary peritectic reaction can be analyzed based on the tie triangles that form the three-phase region. Just above the ternary peritectic temperature, the positions of the Lαβ and Lαγ tie triangles are shown in Fig. 8.67a. As the temperature decreases, the positions of the tie triangles

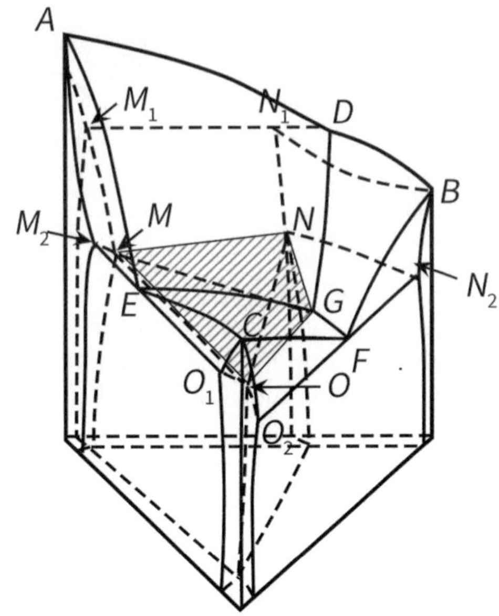

Fig. 8.66 Space model of the system showing a ternary peritectic reaction: L + α → β + γ

change due to variations in the phase compositions. At the ternary peritectic temperature, the liquid composition reaches point *G*, and the α composition reaches point *M*, causing the *Lα* sides of the triangles to overlap, forming a trapezium composed of the two triangles (Fig. 8.67b).

At the peritectic temperature, the four phases L, α, β, and γ coexist, and their compositions form the corners of a trapezium called the peritectic plane (*MNGO* in Fig. 8.66). The liquid composition point *G* is located outside the tie triangle formed by the solid phase compositions α, β, and γ (i.e., points *M*, *N*, and *O* in

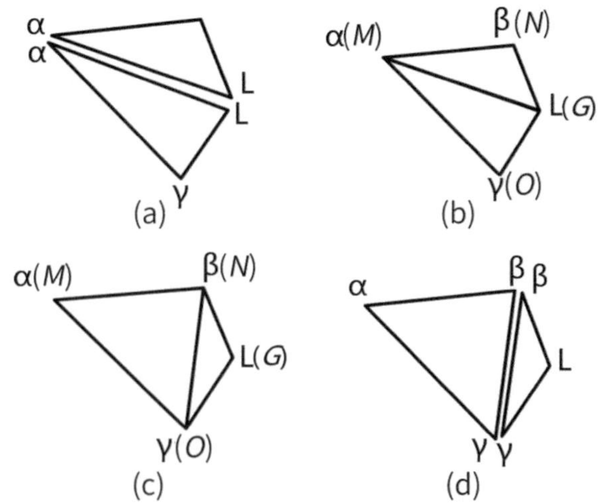

Fig. 8.67 Tie-triangles **a** at a temperature just above that of the ternary peritectic; **b** and **c** at the ternary peritectic temperature; **d** below the ternary peritectic temperature

Fig. 8.67c). The peritectic reaction occurs between L and α, located at opposite corners of the trapezium, to form a mixture of the two phases β and γ, located at the other opposite corners. After completion of the peritectic reaction, the trapezium is divided into two triangles along the *βγ* diagonal, representing the phase equilibria between αβγ and Lβγ (Fig. 8.67c). Above the peritectic plane, there are two three-phase regions: Lαβ and Lαγ. Below the plane, there are two regions: αβγ and Lβγ. As the temperature decreases, the tie triangles αβγ and Lβγ separate due to changes in phase composition (Fig. 8.67d). The Lβγ triangle is displaced along the eutectic curve *GF* as the liquid composition changes, while the β and γ compositions change along curves NN_2 and OO_2, respectively.

At the peritectic temperature, alloys with compositions located in the αβγ triangle (*MNO*) complete their equilibrium solidification process through the ternary peritectic reaction. However, for alloys with compositions located in the Lβγ triangle (*GNO*) at the peritectic temperature, the α-phase is exhausted in the peritectic reaction. As the temperature continues to decrease, solidification eventually occurs, and the Lβγ triangle changes position until the alloy composition point lies on the βγ side of the triangle. An alloy whose composition corresponds to the intersection of the Lα and βγ diagonals of the trapezium at the practice temperature consists entirely of L and α just above this temperature and wholly of β and γ just below it.

To illustrate the nature of the invariant reaction, the phase percentages just above and below the peritectic temperature in a typical alloy can be calculated. For example, in alloy X (Fig. 8.68), assuming the peritectic temperature remains constant at the start of the invariant reaction:

$$\%Liquid = \frac{L_1X}{LL_1} \times 100\% \tag{8.51}$$

$$\%\alpha = \frac{\alpha_1X}{\alpha\alpha_1} \times 100\% \tag{8.52}$$

$$\%\gamma = \frac{\gamma_1X}{\gamma\gamma_1} \times 100\% \tag{8.53}$$

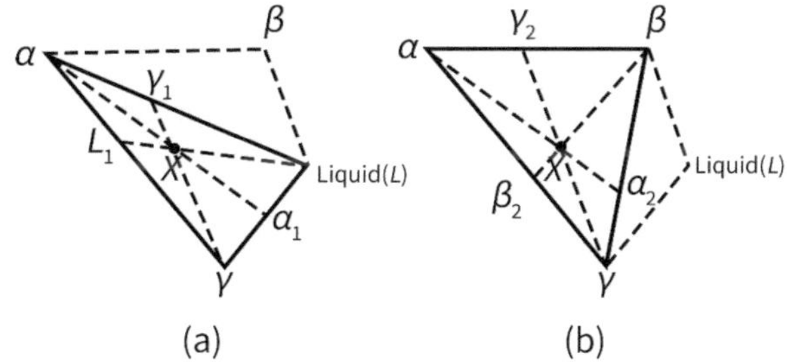

Fig. 8.68 Tie triangles at the ternary peritectic temperature used to calculate the percentages of the phases present in alloy X before and after the occurrence of the invariant reaction

After the invariant reaction is accomplished:

$$\%\alpha = \frac{\alpha_2 X}{\alpha\alpha_2} \times 100\% \tag{8.54}$$

(i.e., less amount than at the beginning of the reaction)

$$\%\gamma = \frac{\gamma_2 X}{\gamma\gamma_2} \times 100\% \tag{8.55}$$

(i.e., more amount than at the beginning of the reaction)

$$\%\beta = \frac{\beta_2 X}{\beta\beta_2} \times 100\% \tag{8.56}$$

(no liquid is present)

Therefore, the percentages of β and γ have increased at the expense of the L and α phases during the peritectic reaction process of L + α → β + γ.

③ Systems Containing a Ternary Peritectic Reaction (Type b)

The ternary system ABC shown in Fig. 8.69, is an example of a system with one binary system of the eutectic type and two of the peritectic type, exhibiting an invariant reaction of the form L + α + β → γ. The terminal solid solutions in the system are α, β, and γ, without intermediate phases present. The AB system is of the eutectic type, while the BC and AC systems are essentially peritectic.

The eutectic valley *DG* (L → α + β) extends into the ternary system. The point *G* represents the composition of the liquid participating in the invariant peritectic

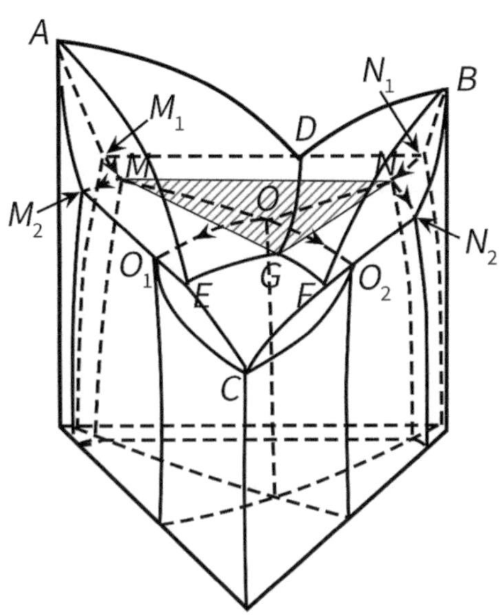

Fig. 8.69 Space model of the system showing a ternary peritectic reaction: L + α + β → γ

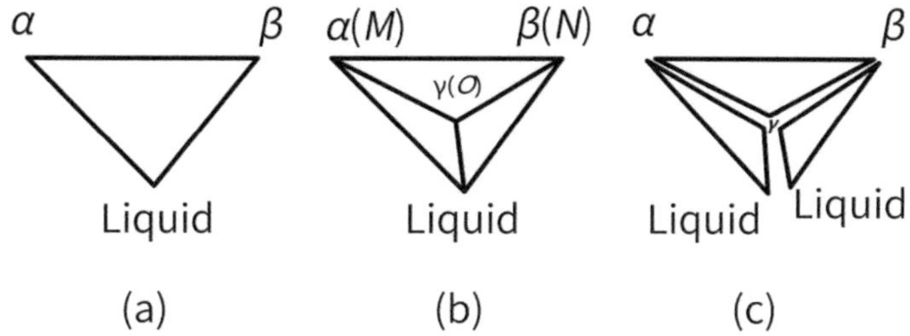

Fig. 8.70 Tie-triangles **a** at a temperature just above that of the ternary peritectic, **b** at the peritectic temperature, **c** below the peritectic temperature

reaction L + α + β → γ. Below *G*, there are two peritectic curves: *GE* (L + α → γ) and *GF* (L + β → γ). During the eutectic reaction L → α + β, as the liquid composition changes along *DG*, the compositions of α and β move along M_1M and N_1N, respectively. In the case of the peritectic reaction L + α → γ, the compositions of liquid, α, and γ change along the curves *GE*, MM_2, and OO_1, respectively. While in the case of the reaction L + β → γ, the compositions of liquid, β, and γ vary following *GF*, NN_2, and OO_2, respectively.

The three-phase region of L + α + β extends down to the plane corresponding to the invariant reaction. At temperatures just above the peritectic, Fig. 8.70a shows a typical Lαβ tie triangle in this region. The triangle (*GMN*) at the invariant reaction temperature determines the liquid, α, and β compositions that participate in the reaction. One characteristic of the peritectic reaction is that the liquid composition *G* is located outside the triangle (*MNO*) formed by three solid compositions, with the composition of γ located at point *O*. The reaction creates three three-phase regions below the peritectic surface, as shown in Fig. 8.70b for tie triangles αβγ (*MNO*), Lαγ (*GMO*), and Lβγ (*GNO*). As the temperature decreases, the change in composition leads to the separation of the three triangles mentioned above (Fig. 8.70c). The triangles Lαγ and Lβγ correspond to L + α → γ and L + β → γ, respectively (Fig. 8.70).

At the peritectic temperature, alloys located in the αβγ region will solidify completely through the invariant reaction, consuming all the liquid. However, for alloys located on the peritectic plane for the Lαγ or Lβγ regions, some liquid will remain after the completion of the invariant reaction, and solidification will continue through the L + α → γ or L + β → γ reactions, respectively.

8.6 Questions

Q8.1 Read the unary phase diagram of carbon and determine the following information:

(1) How many phase(s) presents in this diagram?
(2) Determine the temperature and pressure for the triple point.

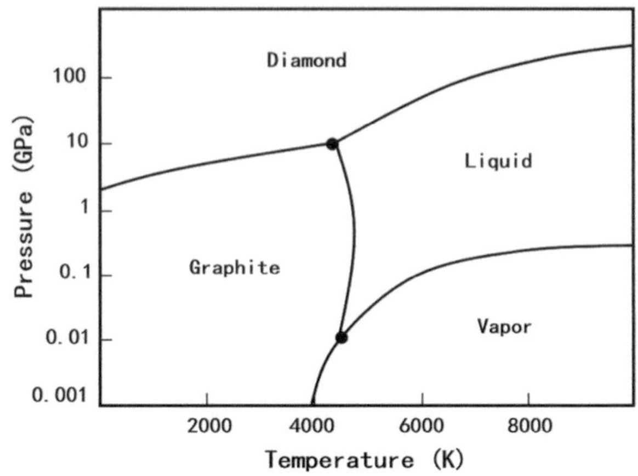

Fig. 8.71 The unary phase diagram of carbon

(3) Comparing graphite and diamond, one is the stable phase and one is the metastable phase. Under what condition will the metastable condition transform into a stable one.

Q8.2 A copper–nickel alloy (Fig. 8.12) of composition40 wt% Cu–60 wt% Ni is slowly heated from a temperature of 1300 °C. (a) At what temperature does the first liquid phase form? (b) What is the composition of this liquid phase? (c) At what temperature does complete melting of the alloy occur? (d) What is the composition of the last solid remaining prior to complete melting?

Q8.3 Read the following Cu-Ag binary phase diagram and determine the following information:

(1) How many phase(s) presents in this diagram?
(2) Point out the two-phase regions.
(3) What are the liquidus, solidus and solvus lines in this phase diagram?
(4) Write down the possible isothermal reaction(s) in this phase diagram.
(5) For a Cu-40 wt% Ag alloy, please schematically draw the microstructure development during the slow cooling from 1200°C to room temperature.
(6) Determine the amount of the primary α, the total α and the eutectic α phases for the Cu-40 wt% Ag alloy during slow cooling from 1200 °C to room temperature.

Q8.4 According to the Cu-Ag phase diagram (Fig. 8.72), a 90 wt% Cu–10 wt% Ag alloy is heated to a temperature within the liquid phase region. Upon cooling, if the composition of the liquid phase becomes 20 wt% Ag, determine: (a) The temperature of the alloy, (b) The composition of the phase and (c) The mass fractions of both phases.

Q8.5 For a copper–silver alloy (Fig. 8.72) of composition Cu-25wt% Ag and at temperatures just below 779 °C (i.e. eutectic isotherm T_E), do the following: (a)

Fig. 8.72 The binary Cu-Ag phase diagram

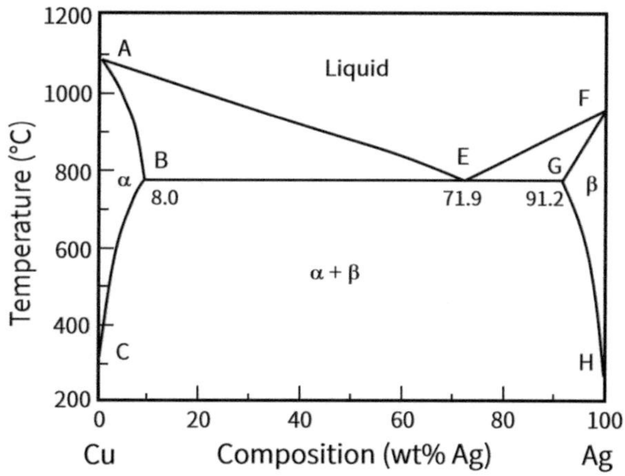

Determine the present phases. (b) Determine the mass fractions of primary α and eutectic microconstituents. (c) Determine the mass fraction of eutectic α.

Q8.6 The following Fig. 8.73 is the Cu-Mg phase diagram for which only single-phase regions are labeled. Specify all temperature-composition points at which eutectics, eutectoids, peritectics, and congruent phase transformations occur and write down the reaction upon cooling for each point.

Q8.7 The following Fig. 8.74 is the Al-Li phase diagram for which only single-phase regions are labeled. Specify all temperature-composition points at which eutectics, eutectoids, peritectics, and congruent phase transformations occur and write down the reaction upon cooling for each point.

Q8.8 Read the following Fe-Fe_3C binary phase diagram and determine the following information:

Fig. 8.73 The binary Cu-Mg phase diagram

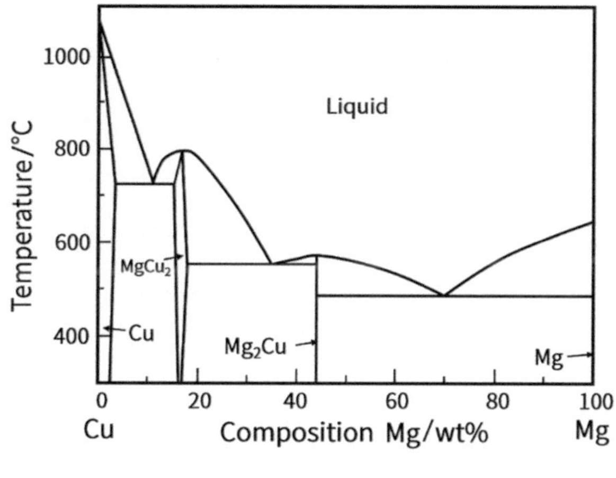

Fig. 8.74 The binary Al-Li phase diagram

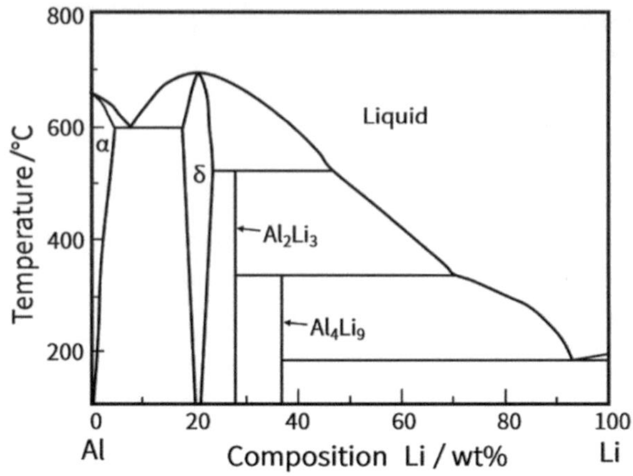

(1) How many phase(s) presents in this diagram?
(2) Label all of the two-phase regions.
(3) Write down the possible isothermal reactions in this phase diagram.
(4) For a steel with a composition of 0.4wt% C, please schematically draw the microstructure development during the slow cooling from 1000 °C to room temperature. Determine the amount of the primary α, the total α and the eutectic α phases.
(5) For a steel alloy with a composition of 1.5wt% C, please schematically draw the microstructure development during the slow cooling from 1000 °C to room temperature. Determine the amount of the primary Fe_3C, the total Fe_3C and the eutectic Fe_3C phases.

Q8.9 Read the following Fe-Fe_3C binary phase diagram in Fig. 8.75 and determine the following information:

(1) At what temperature and carbon concentration does austenite transform to ferrite and cementite during cooling?
(2) At what temperature and carbon concentration range does the primary cementite (Fe_3C_I), the secondary cementite (Fe_3C_{II}) and the tertiary cementite (Fe_3C_{III}) precipitates from austenite?

Q8.10 In a Fe-Fe_3C binary system, (1) Why does the pearlite phase have a layered structure? (2) What is the difference between proeutectoid ferrite and pearlite in terms of their microstructure and carbon concentration? (3) Explain the difference between hypoeutectoid and hypereutectoid steels in terms of their composition and microstructure.

Q8.11 In a system of ABC, the ternary alloy composed of 30 wt% B and 30 wt% C consists of the following three-phase equilibrium at a specific temperature:

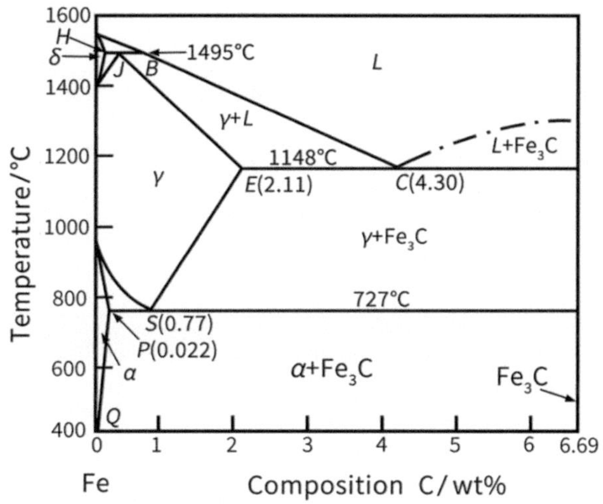

Fig. 8.75 The binary Fe-Fe_3C phase diagram

Liquid phase: 50% A, 40% B, 10% C.

α solid solution: 85% A, 10% B, 5% C.

γ solid solution: 10% A, 20% B, 70% C.

(a) Calculate the weight proportions of liquid, α and γ present in this alloy.
(b) The composition of the alloy was deduced at the same temperature, which consisted of equal proportions of the α and γ phases of the compositions mentioned before but without the liquid phase.

Q8.12 The melting points of A, B and C are 1000 °C, 900 °C and 750 °C, respectively. The constituent binary systems of ternary system ABC each indicate complete solubility in both the liquid and solid-states. The following data refer to three binary alloys:

Composition/wt %			Temperature/°C	
A	B	C	Liquidus	Solidus
50	50	–	975	950
50	–	50	920	850
–	50	50	840	800

For the ternary system, diagrams consistent with the above data are drawn and labeled to represent the possible forms of (a) a liquidus projection, showing isotherms for 950 °C and 850 °C, (b) a solidus projection showing isotherms for 950 °C and 850 °C, (c) an isothermal section at 950 °C, and (d) a vertical section connecting A to the midpoint of BC. Assume that there is no maximum or minimum value on the surface of the liquid and solid.

Q8.13 With reference to Fig. 8.56:

(a) State what phase regions are separated by the following surfaces:

(i) ADGC, (ii) BMN, (iii) DMNG, (iv) DEFG.

(b) State what surfaces separate the following phase regions: (i) L + β/β, (ii) L + β/α + β + L.

Q8.14 In system ABC, a ternary alloy containing 40% B and 8% C undergoes a eutectic reaction $L \rightarrow \alpha + \beta$ over the temperature range 550 ~ 500 °C as part of its solidification sequence. The compositions of the phases coexisting in equilibrium at 540 °C and 510 °C are as follows:

Temperature	α	β	L
540 °C	85%A, 10%B, 5%C	5%A, 93%B, 2%C	55%A, 30%B, 15%C
510 °C	82%A, 11%B, 7%C	6%A, 89%B, 5%C	48%A, 32%B, 20%C

Calculate:

(a) The proportions by weight of α and β phases present in the alloy at 540 °C.
(b) The ratio of the proportions of liquid phase present at 540 °C and 510 °C.

Q8.15 With reference to Fig. 8.64:

(a) State what phase regions are enclosed by the following surfaces: (i) BDN_1, $BDGF$, DN_1NG, FN_2NG, BN_1NN_2, BFN_2, (ii) M_2MGE, O_1OGE, M_2MOO_1, MGO.
(b) State what surfaces separate the following phase regions: (i) L + α + β/L + α; (ii) γ/L + γ.

Q8.16. With reference to Figure 8.66:

(a) State what surfaces enclose the following phase regions: (i) L + α + β, (ii)L + α.
(b) State what phase regions are separated by the surfaces: (i) $OGFO_2$; (ii) MNO; (iii) NGO.

Bibliography

Callister WD Jr, Rethwisch DG (2010) Materials science and engineering: an introduction, 8th edn. John Wiley & Sons, New York

Hu GX, Cai X, Rong YH (2010) Fundamentals of materials science, 3rd edn. Shanghai Jiao Tong University Press, Shanghai

Pan JS, Tong JM, Tian MB (2011) Fundamentals of materials science. Tsinghua University Press, Re.ed. Beijing

Porter DA, Easterling KE, Sherif MY (2009) Phase transformations in metals and alloys. 3rd ed.[S.l.]:Taylor & Francis Group, LLC

Shi DK (2003) Fundamentals of materials science, 2nd edn. China Machine Press, Beijing

Tao J, Yao ZJ, Xue F (2018) Fundamentals of materials science, 2nd edn. Chemical Industry Press, Beijing

West DRF, Saunders N (2002) Ternary phase diagrams in materials science. 3rd ed.[S.l.]:Maney Publishing

Yu YN (2012) Fundamentals of materials science, 2nd edn. Higher Education Press, Beijing

Solidification and Crystallization 9

Learning Objectives

1. Derive the relationship between the thermodynamic driving force for the solidification of pure metals and supercooling.
2. Cite equations to describe the relationships between the Gibbs free energy change and the nucleation radius for both homogeneous and heterogeneous nucleation.
3. Derive expressions for the critical nucleus radius and nucleation barrier for homogeneous nucleation during solidification of pure metals.
4. Describe the differences in nucleation barriers between homogeneous and heterogeneous nucleation during solidification of pure metals.
5. Distinguish between smooth and rough solid-liquid interfaces.
6. Understand the growth mechanisms of solidification of pure metals.
7. Describe the classical ingot structures.
8. Understand the solidification processes in sing-phase alloys, eutectic alloys, ceramics, and polymers.

9.1 Introduction

Solidification is the process by which a liquid material transforms into a solid through a phase transformation. This transformation, also known as crystallization, occurs when atoms in the material arrange themselves into a periodic structure, forming a crystal lattice in the solid state.

Except for certain metal products produced by powder metallurgy, nearly all metal products undergo the process of smelting and solidification. Smelting is the process of producing a melt with the required composition, which is then poured

Z. Sun et al., *Fundamentals of Materials Science*,
https://doi.org/10.1007/978-981-96-3904-5_9

into a mold and allowed to solidify, resulting in the formation of a casting or an ingot. The ingot is subsequently subjected to hot and/or cold deformation to manufacture various profiles, bars, plates and wires. The microstructure formed during solidification decisively affects the properties of both castings and profiles. Although heat treatment can improve the structure and performance of the ingot, macro-defects such as macro-segregation, non-metallic inclusions, shrinkage cavities, and cracks cannot be fully eliminated and directly affect the final properties of the product. The crystallization or solidification process is closely related to the structure of an ingot or a casting. Parameters such as nucleation rate and growth rate play a critical role in determining the characteristics of the product. Therefore, controlling the solidification process is crucial for enhancing the performance of engineering materials and realizing their full potential.

This chapter delves into the principles and laws governing the solidification process and the resulting material structures. Understanding these concepts is crucial for guiding the preparation, processing, and performance optimization of materials. The chapter provides detailed discussions on the crystallization behavior of various materials, including pure metals, single-phase alloys, eutectic alloys, and the formation of ingot structures. It also covers the solidification of ceramics and crystallization of polymers.

9.2 Solidification and Crystallization of Metals

Solidification or crystallization of metals is the process by which metal atoms transition from a disordered liquid to an ordered solid. This fundamental process is crucial for various fabrication techniques, including ingot casting, single-crystal growth, and, more recently, the rapid solidification of alloys and glasses. This section will focus on exploring these solidification processes and the underlying principles that govern them.

9.2.1 Structure of Molten Metal

Real gases can be modeled using the ideal gas model with corrections introduced through perturbation theory. Similarly, a real crystal can be represented by incorporating disorder as a perturbation to the ideal crystal, which consists of atoms or molecules arranged in a periodic lattice. However, our understanding of liquids remains limited, and the atomic structure of liquids has yet to be fully determined.

It is widely recognized that atoms in a liquid state are disordered and lack long-range periodicity of crystalline solids. The atoms in liquid can move freely, while remaining in contact, and thus, the liquids have a definite volume but no definite shape. For molten metals, research has shown that many of their thermal properties are comparable to those of solid metals, suggesting that the atoms in liquid metals are closely packed, much like in solids (see Table 9.1).

Table 9.1 Thermophysical properties of common metals

Element	Melting temperature /K	Volume change from liquid to solid /%	Heat of fusion L_m / (kJ/mol)	Heat of vaporization L_b /(kJ/mol)	L_m/L_b (%)	Entropy of fusion ΔS / (J/mol·K)
Al	933	6	10.46	291.2	3.6	11.51
Au	1336	5.1	12.80	342.3	3.7	9.25
Cu	1356	4.15	13.01	304.6	4.3	9.62
Zn	693	4.2	7.20	115.1	6.3	10.67
Mg	923	4.1	8.70	133.9	6.5	9.71
Cd	594	4.0	6.40	99.6	6.4	10.29
Fe	1809	3.0	15.19	340.2	4.5	8.37
Sn	505	2.3	6.97	295.0	2.4	13.78
Ga	302.9	−3.2	5.57	256.0	2.2	18.4
Bi	544.5	−3.35	10.84	179.5	6.1	19.95
Sb	903	−0.95	19.55	227.0	8.6	21.65

It is worth noting that the heat of fusion is significantly lower than the heat of vaporization (L_m/L_b: 2.2%~8.6%). Metallic bonds persist up to the boiling point, which helps explain the minimal change in volume between solid and molten metals (approximately 1% to 6%). Although molten metals lack long-range order or periodicity, they do have short-range ordering. X-ray and neutron diffraction measurements have proven that the local arrangement of atoms in molten metals closely resembles that in solid metals. However, compared to solids, the average distance between atoms in liquids is slightly larger, and the coordination number is lower (see Table 9.2).

Table 9.2 Comparison of structural data obtained by X-ray diffraction

Element	Liquid		Solid	
	Average atomic distance /nm	Coordination number	Average atomic distance/nm	Coordination number
Al	0.296	10–11	0.286	12
Zn	0.294	11	0.265	6
			0.294	6
Cd	0.306	8	0.297	6
			0.330	6
Au	0.286	11	0.288	12

9.2.2 Solidification of Pure Metals

① Thermodynamics

Defining the enthalpy (H) as zero for a pure metal in its most stable state at 298 K, the enthalpy can be calculated by

$$H = \int_{298}^{T} C_P \, dT \tag{9.1}$$

Since the entropy of a system is generally considered to be zero at absolute zero (0 K), the entropy (S) is given by

$$S = \int_{0}^{T} \frac{C_P}{T} dT \tag{9.2}$$

Based on the above, the Gibbs free energy for a pure metal, whether in liquid or solid state, can be expressed as a function of temperature.

$$G_L = H_L - TS_L \tag{9.3}$$

$$G_S = H_S - TS_S \tag{9.4}$$

where H_L and H_S are the enthalpies of the liquid and solid phases, respectively; S_L and S_S refer to the corresponding entropies.

The Gibbs free energy can be used to understand the solidification behavior of pure metals, i.e., the liquid–solid phase transformation. Figure 9.1 shows the empirical variation in Gibbs free energy with temperature. At the equilibrium melting temperature (T_m), where $G_L = G_S$, the solid and liquid phases are in equilibrium and coexist at this temperature. T_m is also referred to as the solidification temperature. When the temperature is above T_m and $G_L < G_S$, the liquid phase is thermodynamically stable. Conversely, when the temperature drops below T_m and $G_L > G_S$, the solid phase becomes stable, and the liquid–solid phase transformation, also known as solidification, occurs.

The difference in Gibbs free energy between the solid and liquid phases serves as the driving force for solidification. The larger the driving force, the stronger the tendency for solidification to occur. At a given temperature, the Gibbs free energy change is given by

$$\Delta G = G_S - G_L = \Delta H - T\Delta S \tag{9.5}$$

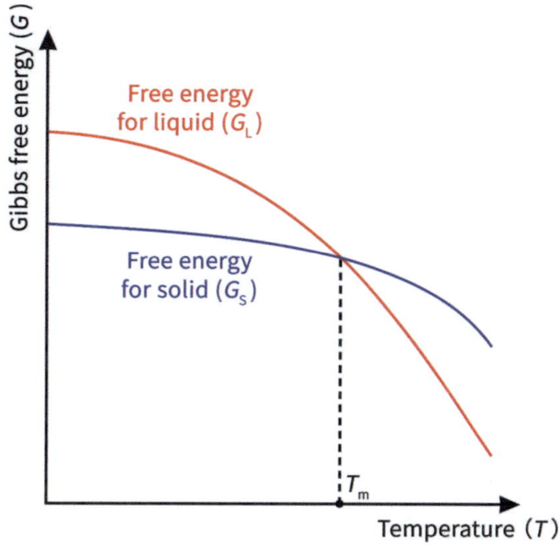

Fig. 9.1 The Gibbs free energy for the liquid and solid phases of a pure metal. T_m denotes the melting temperature

At the melting temperature T_m, $\Delta G = 0$. Equation 9.5 can be converted to

$$\Delta S = \frac{\Delta H}{T_m} = \frac{L_m}{T_m} \tag{9.6}$$

where L_m is the heat of fusion, and ΔS is the entropy of fusion. For most metals, the entropy of fusion is nearly constant, typically $\Delta S \approx 8.314$ J/mol·K. For small supercooling, namely, when the temperature is close to the melting temperature, the heat capacities at constant pressure for the liquid and solid phases are approximately equal ($C_P^S \approx C_P^L$), marking ΔH effectively independent of temperature. Substituting Eqs. 9.5 and 9.6 yields

$$\Delta G = \Delta H - T\Delta S = L_m - T\frac{L_m}{T_m} = \frac{L_m(T_m - T)}{T_m} = \frac{L_m \Delta T}{T_m} \tag{9.7}$$

where ΔT=*Tm-T* is the degree of supercooling. Therefore, the driving force for the liquid–solid phase transformation of pure metals is directly proportional to the degree of supercooling: The greater the supercooling, the larger the driving force for solidification.

② Nucleation

There are two types of nucleation: homogeneous nucleation and heterogeneous nucleation. In homogeneous nucleation, a new phase forms randomly and spontaneously within the parent phase. In contrast, heterogeneous nucleation occurs at preferential sites, such as container surfaces, impurities, or phase boundaries in the parent phase. In practice, homogeneous nucleation rarely occurs, and the solidification of metals typically proceeds via heterogeneous nucleation. Nevertheless, studying the theory of homogeneous nucleation remains important, as it serves as the fundamental basis for the solidification of metals and solid-state phase transformations.

Homogeneous Nucleation

Homogeneous nucleation is described by classical nucleation theory. Considering a liquid at constant pressure below its melting temperature, atom clusters called embryos may form within the melt owing to random thermal fluctuations. These embryos are thermodynamically unstable and may either grow or shrink and redissolve into the liquid.

For the sake of simplicity, an embryo is assumed to be spherical with a radius r; then, the total Gibbs free energy change due to the formation of the embryo in the melt is given by

$$\Delta G = \frac{4}{3}\pi r^3 \Delta G_V + 4\pi r^2 \sigma \tag{9.8}$$

The first term is the energy released by the liquid–solid phase transformation, and ΔG_V is the free energy change per unit volume. The second term refers to the energy required to form the solid–liquid interface, and σ is the interfacial free energy per unit area. At a given temperature, both ΔG_V and σ are definite values, so the free energy change ΔG depends only on r. Figure 9.2 illustrates the variation of ΔG as a function of r. The radius, r^*, corresponds to the maximum free energy change, ΔG^*. If the radius of the embryo r is smaller than r^*, the system can lower its free energy by redissolution of the embryo, making the embryo unstable and likely to shrink and disappear. Conversely, if $r > r^*$, the growth of the embryo will lead to a decrease in the free energy of the system, and the embryo becomes a stable nucleus that will continuously grow. r^* is referred to as the critical nucleus size.

The critical nucleus size can be easily obtained by differentiation of Eq. 9.8 that

$$r^* = -\frac{2\sigma}{\Delta G_V} \tag{9.9}$$

and the corresponding free energy change (known as the nucleation barrier)

$$\Delta G^* = \frac{16\pi\sigma^3}{3(\Delta G_V)^2} \tag{9.10}$$

For a given degree of supercooling ΔT, ΔG_v is given by Eq. 9.7 as

$$\Delta G_V = \frac{L_m \Delta T}{T_m} \tag{9.11}$$

Fig. 9.2 The change in Gibbs free energy, ΔG, associated with homogeneous nucleation of a sphere with radius r

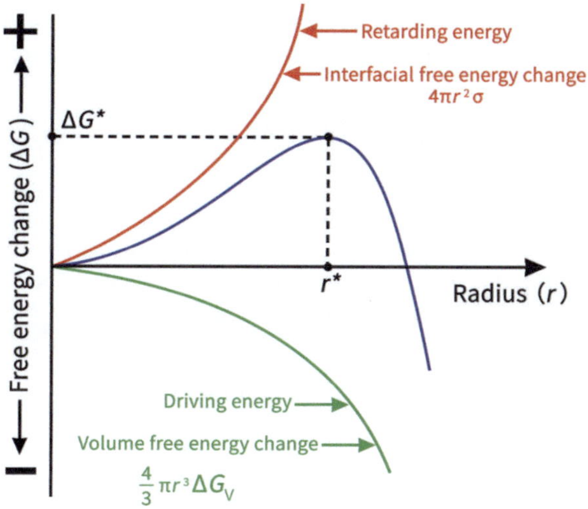

Therefore, we can get

$$r^* = \frac{2\sigma \cdot T_m}{L_m \cdot \Delta T} \tag{9.12}$$

and

$$\Delta G^* = \frac{16\pi\sigma^3 T_m{}^2}{3(L_m \cdot \Delta T)^2} \tag{9.13}$$

It can be concluded that the critical nucleus size is inversely proportional to the degree of supercooling, with a smaller ΔT resulting in a larger r*. Specifically, at the melting temperature ($\Delta T = 0$), the critical nucleus size becomes infinite, indicating that all embryos will redissolve and solidification cannot occur. Moreover, as the supercooling decreases, the nucleation barrier increases. It is important to note that forming a critical nucleus requires overcoming a nucleation barrier equal to one-third of the interfacial free energy ($4\pi r^{*2}\sigma = \frac{16\pi\sigma^3 T_m{}^2}{(L_m \cdot \Delta T)^2}$), indicating that the reduced volume free energy can only compensate for two-thirds of the interfacial free energy, and the other one-third of which depends on the energy fluctuations in the liquid.

The rate of homogeneous nucleation is defined as the number of nuclei generated per unit time per unit volume of the liquid, depending on both the nucleation barrier and the atomic mobility. According to classical nucleation theory, the number of stable nuclei n^* (i.e., those with radii greater than r^*) is related to the nucleation barrier ΔG^* by

$$n^* = K_1 \exp\left(-\frac{\Delta G^*}{kT}\right) \tag{9.14}$$

where the constant K_1 is related to the total number of potential nucleation sites e, k is Boltzmann's constant, and T is temperature. For the exponential term in Eq. 9.14, it is noted that the change in temperature has a greater impact on the ΔG^* term than on the T term in the denominator. Therefore, as the temperature drops below T_m, ΔG^* becomes smaller, making the exponential term n^*, as depicted in Fig. 3a.

Another important step in nucleation is the clustering of atoms, which occurs through diffusion of atoms during the formation of nuclei. Taking the diffusion during nucleation into consideration, the frequency (ν_d) at which atoms attach to a solid nucleus from the surrounding liquid is:

$$\nu_d = K_2 \exp\left(-\frac{Q_d}{kT}\right) \tag{9.15}$$

where Q_d is the activation energy for diffusion, and K_2 is a temperature-independent constant. Thus, a decrease in temperature leads to a reduction in ν_d (see Fig. 3b).

The nucleation rate $\dot{N}$ (number of nuclei per unit volume per second), which is proportional to the n^* and ν_d, is expressed as:

$$\dot{N} = K_3 n^* \nu_d = K_1 K_2 K_3 \exp\left(-\frac{\Delta G^*}{kT}\right)\exp\left(-\frac{Q_d}{kT}\right) \tag{9.16}$$

where K_3 is the number of atoms on the nucleus surface. The temperature dependence of nucleation rate is illustrated in Fig. 9.3c. Note that the energy barrier for nucleation depends on the degree of supercooling. As the temperature decreases below T_m, the nucleation rate increases very rapidly from essentially zero to a maximum, then decreases with further cooling. This behavior can be explained as follows: At low supercooling, ΔG is greater than Q_d, and the formation of nuclei is inhibited by the high energy needed for nucleation. When the supercooling is high and the temperature is significantly reduced, the diffusion becomes difficult, thereby suppressing the nucleation rate. Consequently, the nucleation rate reaches a maximum at an intermediate degree of supercooling. It is worth mentioning that excessive supercooling can result in complete suppression of nucleation, resulting in the formation of an amorphous metal.

Heterogeneous Nucleation

Homogeneous nucleation rarely occurs in most metals. During the solidification process of metals, the wall of the mold, impurities, and nucleating agents in the melt can serve as nucleation sites to promote heterogeneous nucleation. Consider an embryo forming in contact with a flat mold wall, as illustrated in Fig. 9.4. Assuming γ_{SL} is isotropic, it can be proven that the system reaches its minimum interfacial free energy when the embryo takes the form of a spherical cap and the interfacial free energies of the solid-wall, solid–liquid, and liquid-wall interfaces satisfy the condition

Fig. 9.3 Schematic plots of the relationship between temperature and **a** the number of stable nuclei, **b** the frequency of atomic attachment, and **c** the nucleation rate

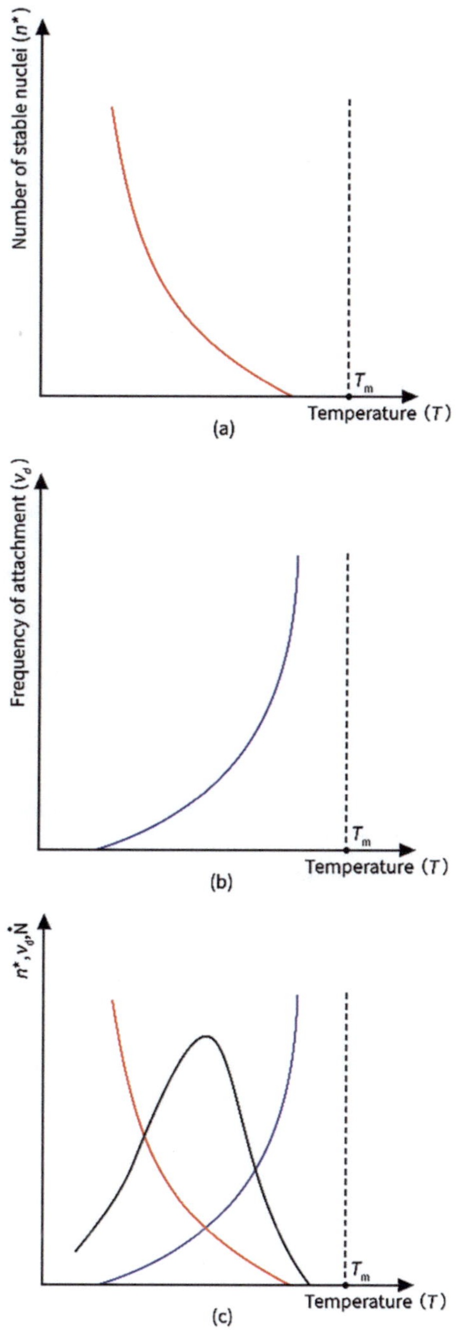

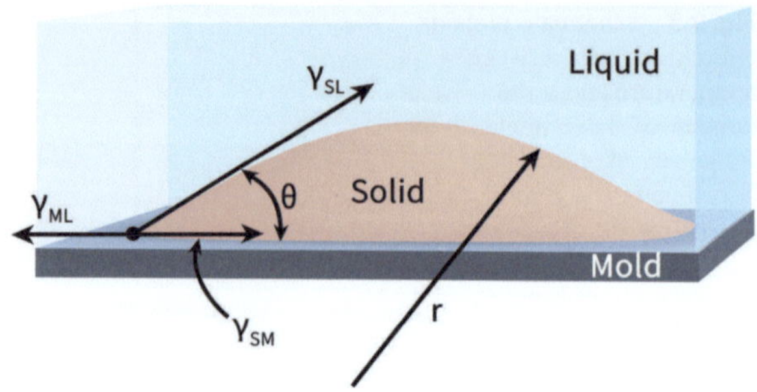

Fig. 9.4 Heterogeneous nucleation of a spherical cap on a flat mold

$$\gamma_{SM} + \gamma_{SL}\cos\theta = \gamma_{ML} \tag{9.17}$$

where γ_{SL}, γ_{SM}, and γ_{ML} are the interfacial free energies of the solid–liquid, solid-wall, and liquid-wall interfaces, respectively. θ is called as the wetting angle.

Accordingly, the free energy change of the system is given by

$$\Delta G_{het} = V_S \cdot \Delta G_V + A_{SL} \cdot \gamma_{SL} + A_{SM} \cdot \gamma_{SM} - A_{SM} \cdot \gamma_{ML} \tag{9.18}$$

where V_S is the volume of the spherical cap and is calculated by $V_S = \frac{\pi r^3}{3}(2 - 3\cos\theta + (\cos\theta)^3)$; A_{SL} and A_{SM} are the areas of the solid–liquid and solid-wall interfaces, i.e., $A_{SL} = 2\pi r^2(1-\cos\theta)$ and $A_{SM} = \pi r^2(\sin\theta)^2$. Note that the fourth term refers to the decrease in interfacial free energy due to the elimination of the liquid-wall interface.

Substituting Eq. 9.17 into Eq. 9.18, one can obtain

$$\Delta G_{het} = \left(\frac{4\pi r^3}{3}\Delta G_V + 4\pi r^4 \cdot \gamma_{SL}\right)\left(\frac{2 - 3\cos\theta + \cos^3\theta}{4}\right) = \left(\frac{4\pi r^3}{3}\Delta G_V + 4\pi r^4 \cdot \gamma_{SL}\right) \cdot f(\theta) \tag{9.19}$$

where $f(\theta)$ depends on the shape of the embryo and varies from 0 to 1. Figure 9.5 shows the variation in **Gibbs free energy** for both homogeneous and heterogeneous nucleation. By differentiation of Eq. 9.19, it can be shown that

$$r^*_{het} = -\frac{2\gamma_{SL}}{\Delta G_V} \tag{9.20}$$

and

$$\Delta G^*_{het} = \frac{16\pi {\gamma_{SL}}^3}{3(\Delta G_V)^2} \cdot f(\theta) \tag{9.21}$$

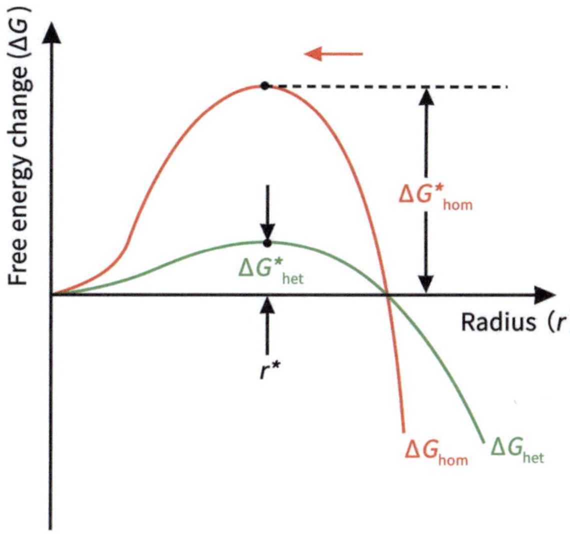

Fig. 9.5 The change in Gibbs free energy for homogeneous and heterogeneous nucleation

Consequently, the nucleation barrier for heterogeneous nucleation ($\Delta G_{\text{het}}{}^*$) is smaller than $\Delta G_{\text{hom}}{}^*$ by the shape factor $f(\theta)$. Furthermore, the critical nucleus size remains unaffected by the presence of the mold wall and solely relies on the degree of supercooling. Combing Eqs. 9.13 and 9.21 gives

$$\Delta G^*_{\text{het}} = \Delta G^*_{\text{hom}} \cdot f(\theta) \tag{9.22}$$

Due to the lower value of ΔG^* for heterogeneous nucleation, the nucleation rate versus temperature curve is shifted to higher temperatures for the heterogeneous case, as shown in Fig. 9.6. In other words, heterogeneous nucleation occurs at much lower degrees of supercooling than homogeneous nucleation.

③ Crystal Growth

Microscopically, crystal growth is a process in which atoms of the liquid phase become attached to the surface of the solid nuclei, i.e., crystal growth occurs at the solid–liquid interface. Therefore, the growth mechanism depends on the atomic structure of the solid–liquid interface.

Solid–Liquid Interface

In terms of atomic structure, there are basically two types of solid–liquid interfaces at the equilibrium state, namely, rough and smooth interfaces, as shown schematically in Fig. 9.7. In the case of a rough interface, the transformation from the liquid to the solid phase occurs over several irregular atomic layers. For a smooth interface, the transformation from the liquid to the solid phase occurs over a single layer of atoms, and such interface can also be described as sharp, faceted, or planar.

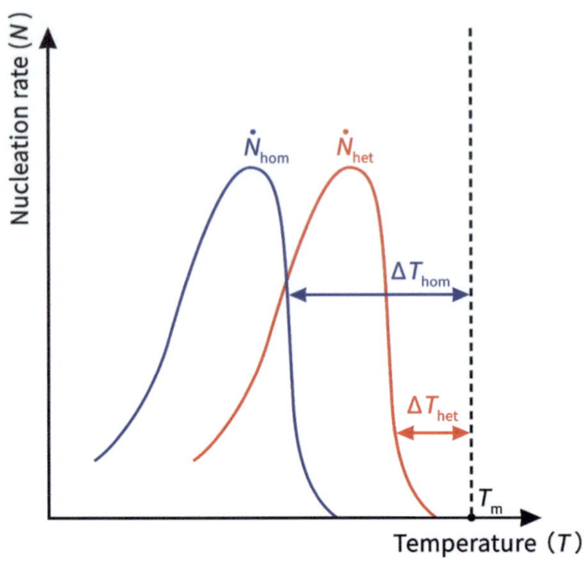

Fig. 9.6 Temperature dependences of nucleation rates for homogeneous and heterogeneous nucleation

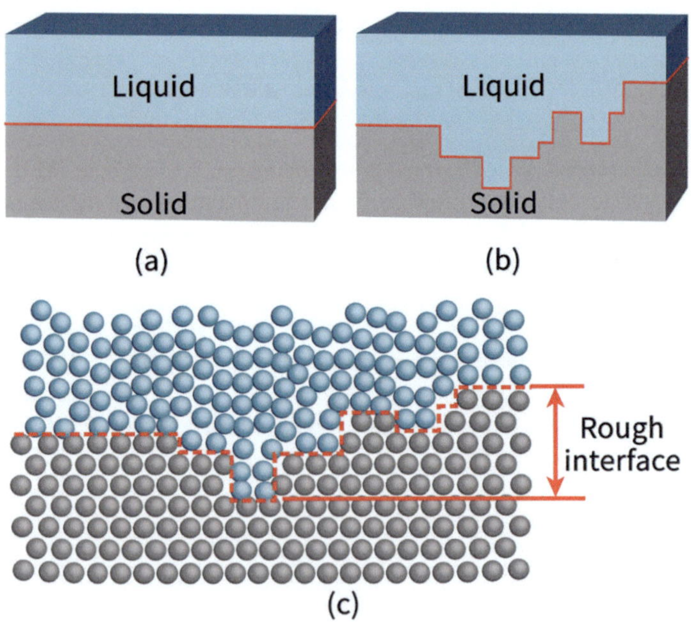

Fig. 9.7 Solid–liquid interface: **a** smooth interface, **b** rough interface. **c** The packing of atoms near a rough interface

In order to understand the atomic arrangement quantitatively at the solid–liquid interface, K.A. Jackson proposed a parameter α to discuss the type of a solid–liquid interface by assessing its atomic configuration at the lowest energy state $\alpha = \frac{\xi L_m}{RT_m}$, where L_m is the heat of fusion, T_m the melting temperature, R the gas constant, and ξ a crystallographic factor (typically less than 1). According to Jackson's theory, atomically rough interfaces should have $\alpha \leq 2$ and smooth interfaces should have $\alpha > 2$. For $\alpha \leq 2$, the system's energy is minimized when 50% of the sites at the interface are occupied by solid atoms and the other 50% are left vacant. This results in a rough interface with several atomic layers forming a transformation region, although the interface may appear macroscopically flat. Most metals are predicted to exhibit rough interfaces due to the significant lattice mismatch between their solid and liquid phases. For $\alpha > 2$, the solid–liquid interface is predicted to be smooth, and the lowest energy state is achieved when either only a few sites are occupied or most of the sites are occupied by solid atoms. In this case, the interface is composed of a one-atom-thick transformation layer, which is highly ordered and smooth. Smooth interfaces are typically observed in materials such as semiconductors and ceramics, which have a low degree of lattice mismatch between their solid and liquid phases.

Due to the different atomic structures of the interface, the growth mechanisms are different. For a rough interface, the continuous growth mechanism is assumed; for a smooth interface, the lateral growth mechanism is assumed.

Growth Mechanisms

(a) Continuous Growth

Considering a rough solid–liquid interface, such an interface always contains a large number of vacancies, which allows individual atoms of liquid to be statistically added to the growing solid surface, and the interface advances vertically along its normal direction. The process is continuous, and the growth rate is proportional to the degree of supercooling and is expressed by

$$\nu = K_1 \Delta T_i \tag{9.23}$$

where K_1 is a constant of proportionality in units of cm/sec·K, and ΔT_i is the supercooling of the interface. For a continuous growth mechanism, the attachment of liquid atoms to the solid surface does not require additional energy, and the migration of the interface is continuous, resulting in a very fast growth rate. This mechanism is applicable to most metals.

(b) Lateral Growth

Considering a smooth solid–liquid interface, if a single liquid atom becomes attached to the solid surface, the interfacial free energy will increase. Thus, the atom will most likely jump back into the liquid. However, if the interface contains steps, the liquid atoms will be added to the steps with a much lower increase in interfacial free energy, allowing for advancement of the interface by the lateral growth of the steps. In this case, crystal growth relies very much on how such steps are supplied. Two common mechanisms are shown below.

Two-Dimensional Nucleation Growth Mechanism: Through homogenous nucleation, a large number of atoms come together to form a two-dimensional nucleus with the thickness of a single atomic layer on the surface of the solid crystal. Then, liquid atoms will be attached to the step created by the nucleus continuously and make the nucleus expand laterally to the surroundings until cover the entire surface, thereby creating a new solid-liquid interface (Fig. 9.8). A new two-dimensional nucleus will then forms on the new interface and expand laterally to form another atomic layer. This process is repeated, with the interface advancing by the continuous formation and lateral expansion of two-dimensional nuclei. The process is discontinuous, and the growth rate depends on the nucleation rate of the two-dimensional nuclei by

$$\nu = K_2 \exp(-\frac{b}{\Delta T_i}) \tag{9.24}$$

where K_2 and b are constants. The growth rate increases with increasing supercooling. However, when the supercooling of the interface is relatively small, the growth rate is extremely low due to the high nucleation barrier for the formation of two-dimensional nuclei.

Spiral Growth Mechanism: When dislocations intersect the solid–liquid interface, they can create steps on the surface of the solid crystal, which can eliminate the

Fig. 9.8 Schematic diagram of two-dimensional nucleation growth on the smooth solid–liquid interface

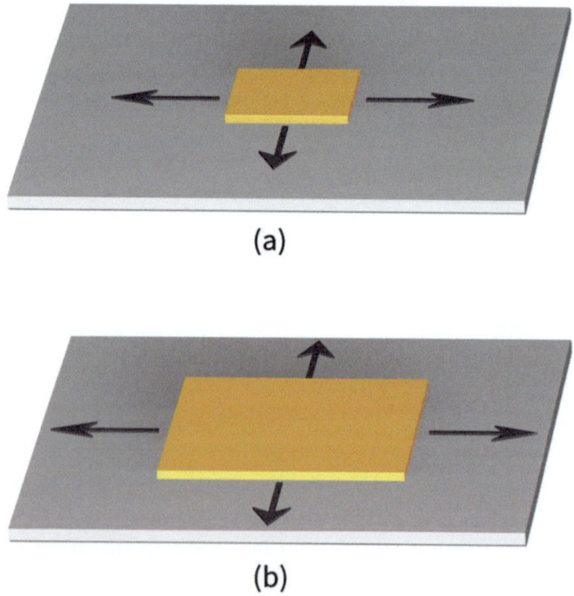

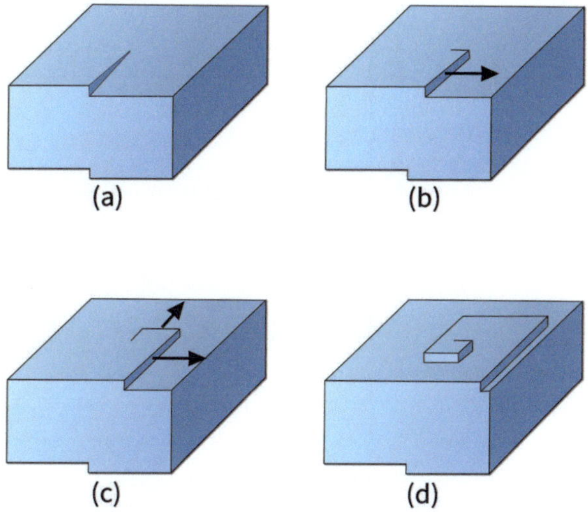

Fig. 9.9 Schematic diagram of spiral growth

need for two-dimensional nucleation. Figure 9.9 shows a schematic diagram of the formation and development of spiral growth.

Initially, the dislocation creates a step on the solid surface. Then, liquid atoms will attach to the step, causing it to advance and generate a second step. As the second step moves forward, it creates a third step, which in turn produces a fourth step, and so on, resulting in a spiral pattern around the dislocation core. In the case of spiral growth, the growth rate is given by

$$v = K_3(\Delta T_i)^2 \tag{9.25}$$

where K_3 is a constant. Due to the limited attachment sites provided by dislocations, the growth rate in spiral growth is very small.

For comparison, Fig. 9.10 shows the growth rate variations of continuous growth, two-dimensional nucleation growth, and spiral growth as a function of supercooling of the interface. Compared to a rough interface, crystal growth on a smooth interface requires larger supercooling. For a given degree of supercooling, the migration of a smooth interface is more difficult, resulting in a lower growth rate.

9.2.3 Solidification of Single-Phase Alloys

Solidification of pure metals involves no change in composition, and the crystal growth is only related to the temperature gradient of the liquid. However, in the solidification of single-phase solid solution alloys, compositional changes occur. During this process, the compositions of the liquid and solid phases are different from that of the parent liquid, leading to solute redistribution. This redistribution

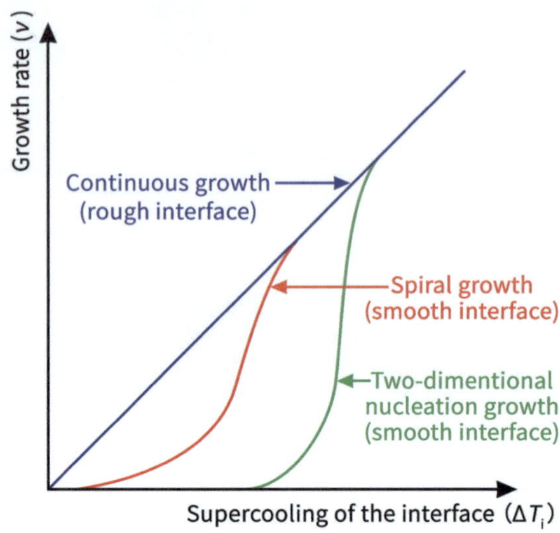

Fig. 9.10 Growth rate as a function of supercooling of the interface

is influenced by cooling conditions, which alter the degree of liquid supercooling and affect the crystal growth morphology at the solid–liquid interface front.

Consider an alloy with composition X_0 in Fig. 9.11, where both the solidus and liquidus lines are assumed to be straight. Assuming that at any temperature, the ratio of the solute concentration in the solid to that in the liquid at equilibrium is constant, that is,

Fig. 9.11 Phase diagram for a single-phase alloy

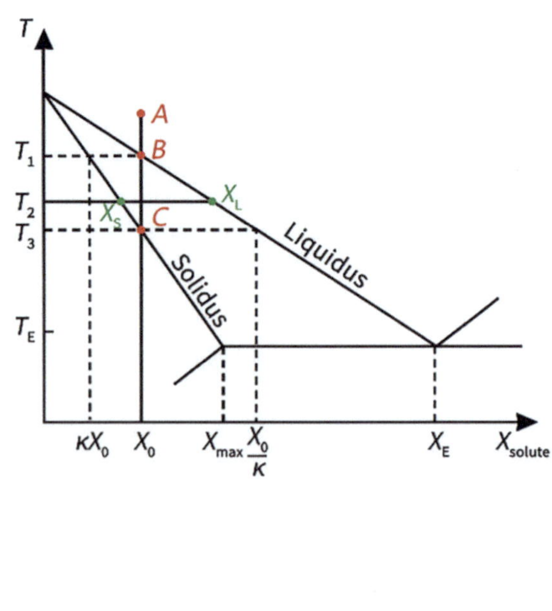

$$k = \frac{X_S}{X_L} \tag{9.26}$$

where k is the partition coefficient and describes the extent to which solute is ejected into the liquid during solidification. X_S and X_L are the mole fractions of solute in the solid and liquid, respectively. In most cases, the solidus and liquidus are not straight lines, but they are often close enough that we can assume that k is independent of temperature.

In practice, the solidification process of such alloys is very complex and influenced by various factors, including temperature gradients, cooling rates, and growth rates. However, to simplify the analysis, let us consider the movement of a planar solid–liquid interface along a bar of alloy as shown in Fig. 9.12.

① Equilibrium Solidification

According to Fig. 9.11, an alloy with composition X_0 is cooled from the liquid state (point A), and as the temperature decreases and reaches T_1 (Point B), solidification begins with the formation of a solid with composition $kX_{0,}$ while the composition of the remaining liquid remains at X_0. As the temperature decreases further, more solid forms. If the cooling rate is slow enough to allow sufficient time for diffusion, the solid and liquid will remain homogeneous, and their compositions can be calculated based on the lever rule (Fig. 12b). Because the solidification process is one-dimensional, to satisfy the conservation of solute, the shaded areas in Fig. 9.12b must be equal. At temperature T_3, the compositions of the liquid and solid are X_0/k and X_0, respectively.

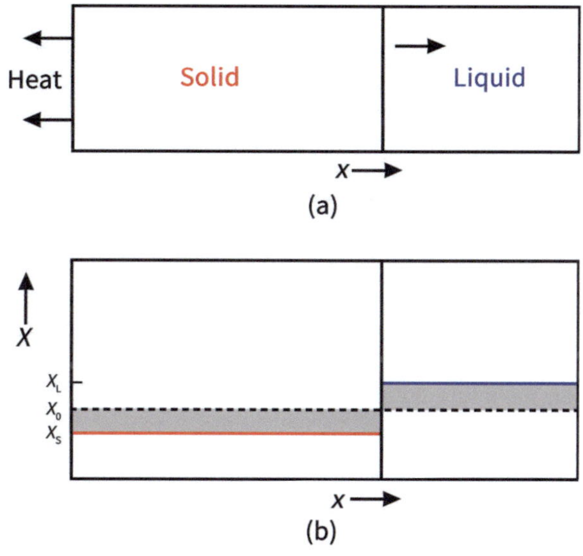

Fig. 9.12 Unidirectional solidification of alloy X_0. **a** A planar solid–liquid interface and heat flow, **b** Composition profile at temperature T_2 assuming complete equilibrium. Two shaded areas must be equal to ensure the conservation of the solute

② Non-equilibrium Solidification

The solidification of single-phase alloys relies on the diffusion of solute in both the liquid and solid phases. Equilibrium solidification occurs when there is sufficient time for complete diffusion in both phases, and the composition of the alloy can be predicted using the phase diagram. However, in industrial production, the cooling rate of the alloy melt is relatively fast, and enough diffusion time cannot be maintained at each temperature, so that the solidification process deviates from the equilibrium condition, which is called non-equilibrium solidification. The faster the cooling rate, the greater the deviation from equilibrium.

In the following, we will discuss two limiting cases and introduce the corresponding quantitative analysis of the compositions of the solid and liquid phases during solidification.

No Diffusion in the Solid, Complete Mixing in the Liquid

Usually, the cooling rate during solidification is so fast that there is no substantial diffusion in the solid phase. To study the solidification process under this condition, we can assume that there is no diffusion in the solid, and that the liquid remains homogenous by convection or mechanical mixing. For simplicity, we consider unidirectional solidification. As shown in Fig. 13a, an alloy with initial composition X_0 starts to solidify at temperature T_1 with the formation of a solid with a composition of kX_0. Since k is less than 1, the first formed solid is purer than the liquid, so the solute concentration of the remaining liquid will increase and become larger than X_0 (Fig. 13b). Therefore, for further solidification, the temperature of the interface must be lowered to a temperature below T_1, and the next layer of solid has a higher solute concentration than that of the first. As this process continues, the solute concentration of the liquid becomes progressively higher, and solidification occurs at a gradually decreasing temperature (Fig. 13c). It is assumed that at any stage of solidification, there is a local equilibrium at the solid–liquid interface; that is, at a given interface temperature, the compositions of the solid and liquid in contact with each other can be obtained by the phase diagram. However, due to the assumption of no diffusion in the solid, the different layers of solid will maintain the original compositions. Therefore, the average composition of the solid ($\overline{X_S}$) is always lower than that at the solid–liquid interface (see the dashed line in Fig. 13a). At a given interface temperature, the compositions of the solid and liquid ($\overline{X_S}$ and X_L) are given by the lever rule. In addition, the solute concentration of liquid can become larger than X_0/k, even reaching a eutectic composition X_E, at which solidification will terminate with the formation of a eutectic structure ($\alpha + \beta$). Figure 9.13d shows the solute distribution along the length of the solidified bar, including the relationship of $\overline{X_S} = X_0$.

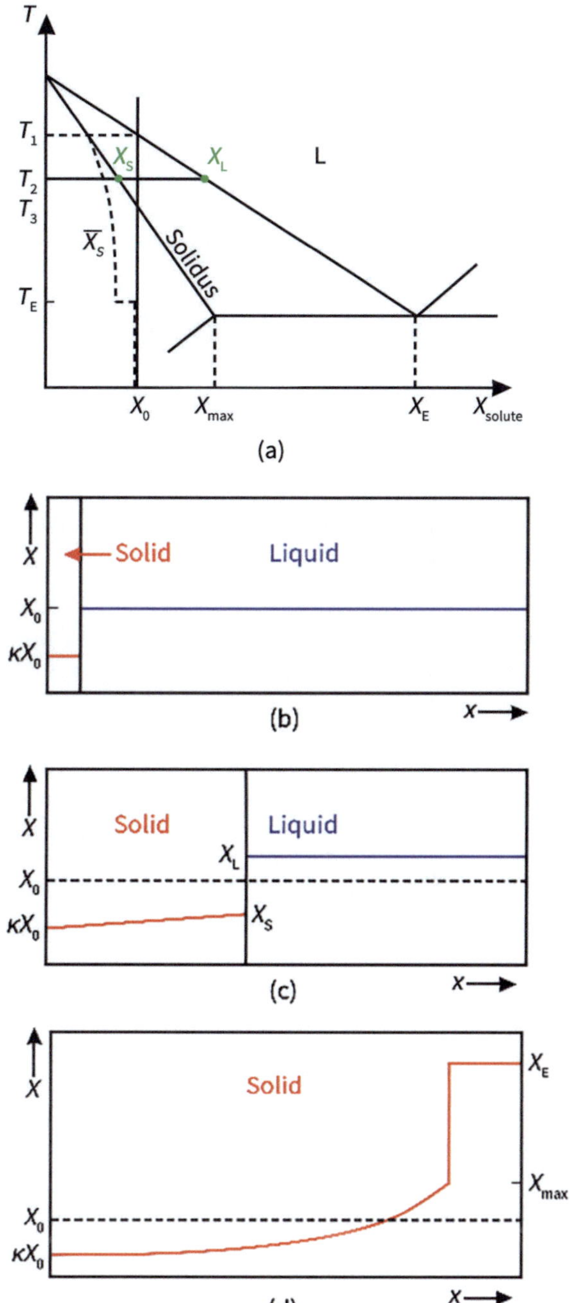

Fig. 9.13 Planar front solidification of alloy with composition X_0 assuming no diffusion in the solid but complete mixing in the liquid. **a** Phase diagram with the average composition of the solid. **b** Composition profile at temperature just below T_1. **c** Composition profile at temperature T_2. **d** Composition profile at temperature T_E and below

The variation in X_S along the length of the solidified bar can be calculated by equating the solute rejected into the liquid at the early stage of solidification with the resulting increase in solute concentration in the liquid:

$$(X_L - X_S)df_S = (1 - f_S)dX_L \tag{9.27}$$

where f_S is the solidified volume fraction.

Applying the boundary condition $X_S = kX_0$ at $f_S = 0$, and integrating Equation 9.27 gives

$$X_S = kX_0(1 - f_S)^{(k-1)} \tag{9.28}$$

and

$$X_L = X_0 f_L{}^{(k-1)} \tag{9.29}$$

where $f_L = 1 - f_S$. Equation 9.29 is known as the Scheil equation.

No Diffusion in the Solid, Diffusional Mixing in the Liquid

If there is no convection or mechanical mixing in the liquid, solute rejected from the solid will only be transferred away by diffusion. Therefore, the solute will quickly accumulate ahead of the solid-liquid interface, and the composition of the solid formed will increase rapidly (see Fig. 14a). This is the so-called initial transient. If solidification proceeds with a constant rate, v, it can be noted that a steady state is reached when the temperature of the interface reaches T_3 in Fig. 9.11, where the composition of the liquid adjacent to the interface is X_0/k, and the solid has composition X_0.

During steady-state growth, the concentration profile in the liquid must be such that the rate at which the solute diffuses down the concentration gradient away from the solid–liquid interface is balanced by the rate at which the solute is rejected from the solidifying liquid, i.e.,

$$-DC_L' = v(C_L - C_S) \tag{9.30}$$

where D is the diffusivity in the liquid, C_L' refers to dC_L/dx at the interface, and C_L and C_S are the equilibrium solute concentrations of the liquid and solid at the interface, respectively. If the diffusion equation for steady-state solidification is solved, it can be shown that the concentration profile in the liquid is given by

$$X_L = X_0 + \frac{X_0(1 - k)}{k}\exp(-\frac{x}{D/v}) \tag{9.31}$$

Note that the solute concentration of the liquid, X_L, decreases exponentially from X_0/k at $x = 0$ (i.e., the interface) to X_0 far from the solid–liquid interface. This concentration profile has a characteristic width of D/v.

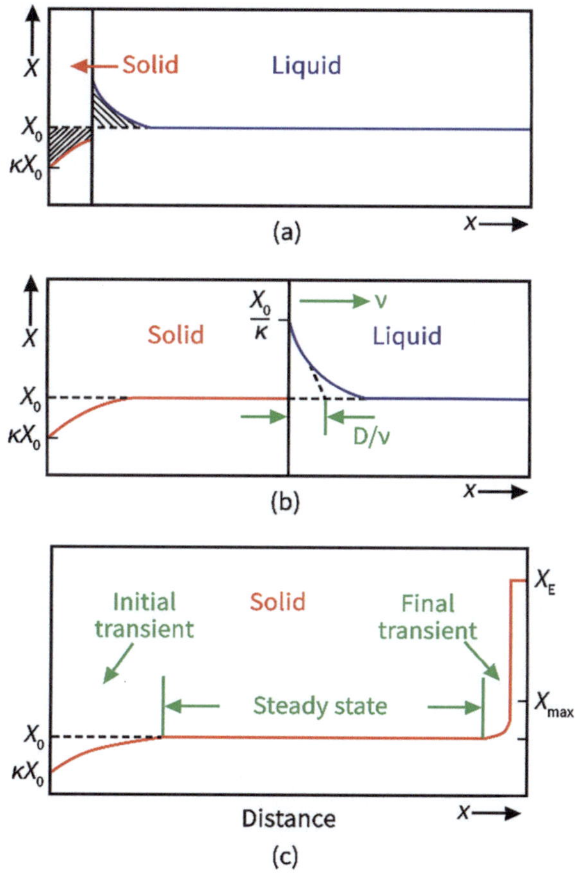

Fig. 9.14 Planar front solidification of alloy with composition X_0 assuming no diffusion in the solid and diffusional mixing in the liquid. **a** Concentration profile when the temperature of the solid–liquid interface is between T_2 and T_3 in Fig. 9.11. **b** Steady-state solidification at temperature T_3. The composition solidifying equals the composition of the liquid far ahead of the solid (X_0). **c** Concentration profile at temperature T_E and below

At the final stage of solidification, the solid–liquid interface will be within ~ D/v of the end of the bar, the bow-wave of the solute is compressed into a very small volume, and the composition of the interface rises rapidly, leading to a final transient and eutectic formation (Fig. 14c).

③ Cellular and Dendritic Solidification

The previous discussion on solidification pertains to a planar interface, where the solid–liquid boundary is assumed to be flat. Consider steady-state solidification at such a planar interface, as shown in Fig. 9.15. The solute distribution in the liquid phase is shown as the curve X_L, which determines the equilibrium solidification temperature T_e. However, the actual temperature of the liquid can vary along any line, such as T_L, depending on the specific conditions of the solidification process. When the temperature gradient in the liquid is below a certain critical value, the temperature of the liquid ahead of the solid–liquid interface may drop below its equilibrium freezing temperature. This occurs due to compositional effects, and the phenomenon is referred to as constitutional supercooling, which can result

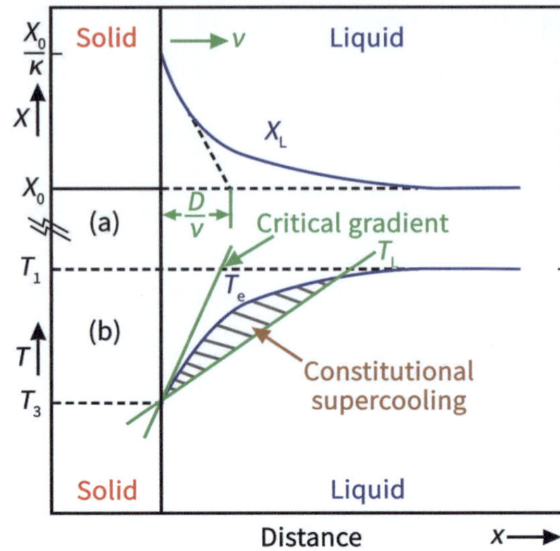

Fig. 9.15 The diagram of constitutional supercooling

in the formation of non-equilibrium microstructures and defects in the final solid products.

The morphology of the solid–liquid interface is highly influenced by the constitutional supercooling. Under steady-state growth, the critical gradient, which represents the minimum temperature gradient required to avoid constitutional supercooling and maintain equilibrium solidification, can be given as $(T_1 - T_3)/(D/v)$, where T_1 and T_3 are the liquidus and solidus temperatures for the bulk composition X_0, respectively.

(1) If the temperature gradient of the liquid phase $dT_L/dx > (T_1—T_3)/(D/v)$, the planar interface will remain.
(2) If the temperature gradient ahead of interface decreases gradually below the critical value, a narrow zone of constitutional supercooling is created. Within this zone, protrusions can grow, but they are closely spaced due to the small amount of constitutional supercooling. Eventually, this process leads to the formation of a cellular structure.
(3) When the temperature gradient of the liquid phase is sufficiently low, the primary arms of solid, or cells, develop secondary arms due to the increased constitutional supercooling range. This results in dendritic growth, characterized by of a large growth rate and continuous branching of the crystal into complex dendritic structures.

9.2.4 Solidification of Eutectic Alloys

During the solidification of a binary eutectic alloy, two solid phases form simultaneously from the liquid, that is, $L \rightarrow \alpha + \beta$. There are different types of eutectic solidification. In this section, the formation and growth of lamellar eutectics are discussed.

Considering a binary eutectic alloy consisting of components A and B, its eutectic is made up of an A-rich α phase and a B-rich β phase, and the growth of α and β at the solidification front (i.e., solid–liquid interface) involves short-range diffusion of A and B atoms in the liquid. The growth rate of the eutectic, R, is governing by atomic diffusion and is strongly influenced by the interlamellar spacing, λ. Generally, smaller interlamellar spacings result in a higher growth rate.

There is a lower limit for λ of a binary eutectic alloy, which is determined by thermodynamic constraints. The Gibbs free energy change associated with the formation of the eutectic includes two parts: the volume free energy change and the interfacial free energy change. The latter is related to the interlamellar spacing. When one mole of liquid solidifies into a two-phase eutectic with interlamellar spacing λ, the Gibbs free energy change of the system can be expressed as

$$\Delta G(\lambda) = \Delta \dot{G}(\infty) \cdot V_{\mathrm{m}} + \frac{2\gamma_{\alpha\beta} \cdot V_{\mathrm{m}}}{\lambda} \tag{9.32}$$

where $\Delta \dot{G}(\infty)$ is the decrease in free energy per unit volume for very large values of λ and is related to the supercooling ΔT_0 by $\Delta \dot{G}(\infty) = \frac{\Delta H \cdot \Delta T_0}{T_{\mathrm{E}}}$ (ΔH is an enthalpy term, ΔT_0 is the supercooling of the interface), V_{m} refers to the molar volume of the eutectic, $\gamma_{\alpha\beta}$ is the α/β interfacial free energy, and there is a total of $(2/\lambda)$ m^2 of α/β interface per unit volume of eutectic, and T_{E} is the eutectic temperature.

Using the relationship $\Delta G(\lambda^*) = 0$, the minimum spacing is obtained by

$$\lambda^* = -\frac{2\gamma_{\alpha\beta} \cdot T_{\mathrm{E}}}{\Delta H \cdot \Delta T_0} \tag{9.33}$$

The growth rate of the eutectic R is determined by the diffusion coefficient D and concentration gradient and is approximately given by

$$R = K_1 \cdot D \cdot \frac{\Delta X}{\lambda} \tag{9.34}$$

where K_1 is a proportionality constant. The term $\Delta X = X_{\mathrm{B}}^{\mathrm{L}/\alpha} - X_{\mathrm{B}}^{\mathrm{L}/\beta}$ varies with λ: when $\lambda = \infty$, $\Delta X = \Delta X_0$; when $\lambda = \lambda^*$, $\Delta X = 0$. If $\lambda^* < \lambda < \infty$, $\Delta X = \Delta X_0(1 - \frac{\lambda^*}{\lambda})$, as shown in Fig. 9.16. Accordingly, smaller interlamellar spacings result in a higher growth rate.

For small supercoolings, $\Delta X_0 \propto \Delta T_0$, thus

$$R = K_2 \cdot D \cdot \Delta T_0 \cdot \frac{1}{\lambda}(1 - \frac{\lambda^*}{\lambda}) \tag{9.35}$$

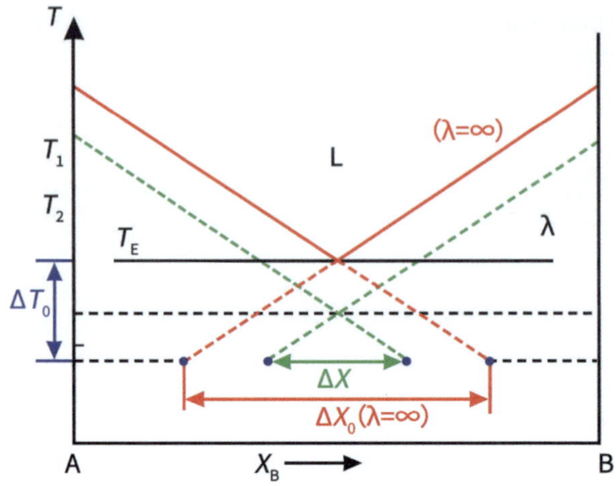

Fig. 9.16 Eutectic phase diagram showing the relationship between ΔX and ΔX_0

where K_2 is a proportionality constant. By differentiation of Eq. 9.35 and assuming $\frac{dR}{dr} = 0$, the maximum growth rate is

$$R_0 = K_2 \cdot D \cdot \frac{\Delta T_0}{2\lambda^*} \tag{9.36}$$

and the corresponding spacing $\lambda_0 = 2\lambda^*$. According to Eq. 9.33, $\Delta T_0 \propto \frac{1}{\lambda^*}$ so the following relationships are predicted:

$$R_0 \cdot {\lambda_0}^2 = K_3 \tag{9.37}$$

$$R_0 = K_4 \cdot (\Delta T_0)^2 \tag{9.38}$$

where K_3 and K_4 are constants. Therefore, both a decrease in the minimum interlamellar spacing and an increase in interface supercooling can promote the growth of the eutectic.

9.2.5 Ingot Microstructure and Solidification Technology

Metal casting is a manufacturing process in which a liquid melt is introduced into a mold and then allowed to cool to form a solidified part. Generally, if the solidified part retains its shape without undergoing plastic deformation, it is called a casting; if the solidified part undergoes subsequent plastic deformation, it is known as an ingot. The underlying principles of solidification are similar in both cases. In this section, we will focus on ingots, beginning with a description of their classical microstructure, followed by a discussion of solidification technologies.

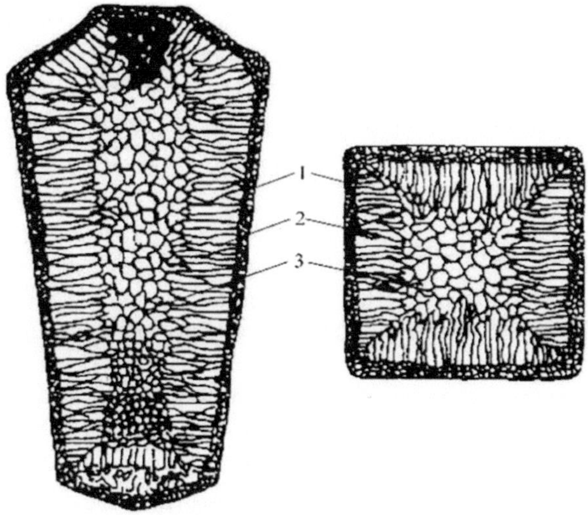

Fig. 9.17 Schematic illustration of the vertical and horizontal cross section of the cast structure of an ingot; 1: chill zone, 2: columnar zone, 3: equiaxed zone

① Description of Classical Ingot Structures

The classical microstructure of an ingot consists of three distinct zones: a chill zone adjacent to the mold wall and composed of fine equiaxed grains, a columnar zone composed of elongated grains, and a central equiaxed zone composed of randomly oriented round grains (Fig. 9.17). While it's common to observe all three regions in ingots, it's also not unusual for one or more regions to be absent. For instance, stainless steel ingots are often characterized by a microstructure composed entirely of columnar grains without any chill or equiaxed zones.

Chill Zone

During pouring the melt into the mold, the melt is rapidly cooled below the liquidous temperature in the vicinity of the cold mold, heterogeneous nucleation takes place at the mold wall and produces a large quantity of solid nuclei. Since these nuclei formed almost simultaneously, their growth is restricted by the effect of impingement, and thus, the chill zone is characterized by fine grains with uniform size and random orientations. In addition, the chill zone's high crystallization rate prevents the release of latent heat of crystallization, which causes the temperature at the interface between the melt and solid crystals to rise significantly and halts further nucleation. Thus, the chill zone is a thin layer with fine grains.

Columnar Zone

Once the chill zone is formed, the temperature of the mold wall increases and the cooling rate of the remaining melt decreases. At the same time, the release of latent heat during crystallization decreases the degree of supercooling of the melt at the front of the chill zone, making further nucleation increasingly difficult. As a result, solidification proceeds mainly through the growth of pre-existing crystals formed in the chill zone. Due to competitive growth, these crystals grow in the direction of the temperature gradient, i.e., perpendicular to the mold wall, advancing from the chill zone to the center of the melt. This leads to the formation of elongated, columnar grains, giving rise to the columnar zone. The growth process plays a more significant role than nucleation in the formation of the columnar zone.

Equiaxed Zone

As the columnar grains continue to grow, the solidification front moves farther from the mold wall, making heat dissipation increasingly difficult. Thus, the cooling rate of the melt decreases, and the temperature gradient in the melt diminishes, thereby preventing the rapid growth of columnar grains. When the temperature of the melt drops below the melting temperature, many nuclei appear and grow in various directions, eventually forming a central equiaxed zone with random oriented grains.

The equiaxed zone, which comprises equiaxed grains of random orientation, is typically considered more desirable than the columnar zone regarding their mechanical properties. In the columnar zone, the grains are elongated and have a long axis that is perpendicular to the growth direction, creating long, parallel grain boundaries. This structure can lead to segregation, voids, and internal cracks. In contrast, the equiaxed zone has a more isotropic grain structure with random orientations, which can help to prevent the formation of impurity-rich regions and reduce the risk of voids and internal cracks. Additionally, the equiaxed zone often exhibits a finer grain size, which can lead to improved strength and toughness.

② Solidification Technologies

To obtain an ingot with the desired properties, it is critical to tailor its microstructure. This can be done by selecting and controlling an appropriate technique, each of which has its own advantages and disadvantages, and the selection of a particular technique depends on the material and the desired properties.

Grain Refinement in Solidification

In many practical cases, the final grain structure is the result of solid-state recrystallization following mechanical deformation or grain growth. However, controlling the grain structure during solidification is equally crucial for achieving optimal material performance.

Grain size plays an important role in determining the mechanical properties of materials. Finer grains generally strength, hardness, and toughness. Therefore, grain refinement, the process of reducing grain size and suppressing the formation of coarse columnar grains in favor of equiaxed grains, is of significant industrial importance.

There are several methods to achieve grain refinement during solidification, including:

Increase Supercooling

According to the Johnson-Mehl equation, the relationship between the number of nuclei formed at time *t*, the nucleation rate *N*, and the growth rate v_g is given by

$$P(t) = k\left(\frac{N}{v_g}\right)^{3/4} \tag{9.39}$$

where k is a constant related to the shape of the nuclei, and $P(t)$ is inversely proportional to the grain size. Based on Eq. 9.39, an increase in the nucleation rate N leads to a decrease in grain size, whereas an increase in the growth rate v_g tends to increase grain size. Experiments have shown that for the solidification of metals, both N and v_g increase with increasing supercooling. However, the increase in N is generally faster than the growth rate v_g, leading to an overall decrease in grain size as supercooling increases.

Add Nucleating Agents or Inoculants

Normally, the solidification of metals occurs via heterogeneous nucleation. During the actual production process, nucleating agents are commonly added in the melt as nucleation sites to promote heterogeneous nucleation and grain refinement. This is the so-called inoculation treatment. The choice of nucleating agent or inoculant depends on the material being produced and the desired microstructure. Several nucleating agents used in metals are listed in Table 9.3.

Vibration and Stirring

Grain refinement can be achieved by vibration or stirring in the process of casting. The mechanism is to refine the microstructure by promoting nucleation.

Table 9.3 Typical nucleating agents used in metals

Metals	Nucleating agents
Mg–Al	C
Mg–Al-Mn	$FeCl_3$
Mg-Zn	$FeCl_3$; Zn-Fe; NH_3
Cu-Al_2Cu eutectic	Ti
Cu-7%Al	Mo; Nb; W; V
Cu-9%Al	Bi
Al-Si hypereutectic alloy	Cu-P; $PNCl_2$; P
Low alloy steels	Ti
Silicon steel	TiB_2
Lead alloys	S; Se; Te
Tin alloys	Ge; In
Fe–C-Si	C
Aluminum alloys	Nb; Al-B; Al-Ti-B; Al-Ti; Ti salt

By carefully selecting and combining these methods, it is possible to achieve a microstructure with the desired grain size and properties. Overall, grain refinement during solidification is a critical process for achieving optimal mechanical properties in materials.

Single Crystal

A single crystal is defined as a crystal consisting of a noninterrupted repetition of the unit cell in three dimensions, with no grain boundaries. In producing a single crystal, the generation of multiple nuclei must be avoided during solidification, and only one nucleus is allowed to form and grow. The absence of grain boundaries and defects means that the physical and chemical properties of a single crystal are homogeneous throughout the entire crystal. This uniformity can result in unique and exceptional properties, such as high mechanical strength, improved electrical conductivity, and high optical transmission, making single crystals ideal for use in various applications. Single crystals have important theoretical significance for the study of the intrinsic properties of materials, and their applications in industry, especially in optics and electronics, are becoming increasingly widespread.

However, producing large single crystals can be difficult and expensive, as it requires careful control of the crystal growth process to ensure the formation of a single nucleus and the absence of defects. Nevertheless, advances in crystal growth techniques and equipment have enabled the production of larger and higher-quality single crystals, opening up new possibilities for their use in a wide range of applications.

Metallic Glass

Metallic glass, also known as amorphous alloy, is generally produced by rapidly cooling a molten metal to suppress crystallization, resulting in a liquid-like metastable structure. Unlike conventional crystalline alloys, in which dislocations and grain boundaries play vital roles in physical, mechanical, and chemical properties, metallic glasses lack these crystalline defects and possess unique properties, including high elastic limits and strength, excellent thermoplasticity, excellent soft magnetic properties, high resistivity, and good corrosion and wear resistance. Owning to these characteristics, metallic glasses offer significant advantages for both structural or functional applications, and have attracted widespread attention in materials science and engineering.

However, metallic glasses also face certain limitations, such as low fracture toughness and limited formability. Researchers are currently working to overcome these limitations by developing new compositions, refining processing techniques, and exploring new applications for these unique materials.

9.3 Solidification and Crystallization of Ceramics

The solidification and crystallization process of ceramics is more complex than that of metals, although the functionally principles of crystallization remain the same. Therefore, this section provides a brief overview of ceramic solidification.

Like metals, the solidification and crystallization process of ceramics requires a certain degree of supercooling and is the process of nucleation and crystal growth. According to thermodynamic theory, when the temperature of the melt is lower than the melting temperature, the Gibbs free energy of the melt is higher than that of the stable crystalline phase. Crystallization will occur, and finally, a polycrystal with different grain sizes and shapes will form. However, crystallization does not typically occur exactly at the melting temperature; instead, the melt must be supercooled below a certain threshold to initiate nucleation. Between this threshold and the melting temperature, the melt exists in a metastable state, in which a crystalline phase cannot yet form. Once the temperature of the melt drops below this threshold, crystallization takes place readily, and supercooling acts as the driving force for the liquid–solid phase transformation.

Crystallization of ceramics involves two stages, i.e., the formation of stable nuclei and the growth of these nuclei into crystals. The crystallization rate depends on both the nucleation rate and the growth rate, which are influenced by factors such as composition, cooling rate, and processing conditions. The viscosity of the ceramic melt in ceramics can vary greatly depending on the composition, which directly affects the solidification behavior. The solidification of ceramics is often considered a nonequilibrium process, meaning that it does not reach thermodynamic equilibrium during cooling. This can lead to the formation of amorphous glasses, characterized by disordered atomic structures rather than the ordered lattice structures found in crystalline materials.

In daily life, we are surrounded by different kinds of ceramic products. One common example is the ceramic vase (Fig. 9.18). Ceramic products are often coated with glazes, whose purposes are protection and decoration. Glaze is a thin vitreous silicate coating to a ceramic product by firing. In general, raw materials of glazes consist of glass former (silica), flux (various metal oxides, such as Na, K, and Ca), and stabilizer (such as kaolin, clay, and feldspar). In structure, glaze is mainly composed of an amorphous phase but may also include bubbles, cracks, and even crystalline phases (Fig. 18a). Crystalline glaze is a specialist type of glaze, in which visible and distinct crystals form in the matrix of the glaze (Fig. 18b). The formation of such crystals is facilitated by slower cooling rates, which promote crystallization during the solidification of the glaze.

Glass–ceramic materials are composite materials that consist of both an amorphous phase and one or more crystalline phases. They are produced by controlling the crystallization of glass, typically through a two-step process. In the first step, glass is formed by a glass manufacturing process, and in the second step, the glass is reheated and partially crystallized in a heat treatment. The degree of crystallinity depends on the number of crystal nuclei, treatment time, and temperature. The type

(a) (b)

Fig. 9.18 Ceramic vases with **a** glaze and **b** crystalline glaze

of nucleation that occurs in the material, whether homogeneous or heterogeneous, is also important.

The properties of glass–ceramic materials depend on the type and volume fraction of crystalline phases, as well as the composition and characteristics of the remaining glass phase. Glass–ceramics offer a combination of desirable properties, such as high mechanical strength, low conductivity, high dielectric constant, good machining performance, chemical resistance, and thermal stability. Because of these attributes, glass-ceramics are applied in various fields, including machinery manufacturing, optics, electronics and microelectronics, aerospace, chemistry, industry, biomedicine, and construction. One example is bio-ceramics, a subclass of glass–ceramic material that developed for medical applications such as dental implants (see Fig. 9.19) and bone substitutes.

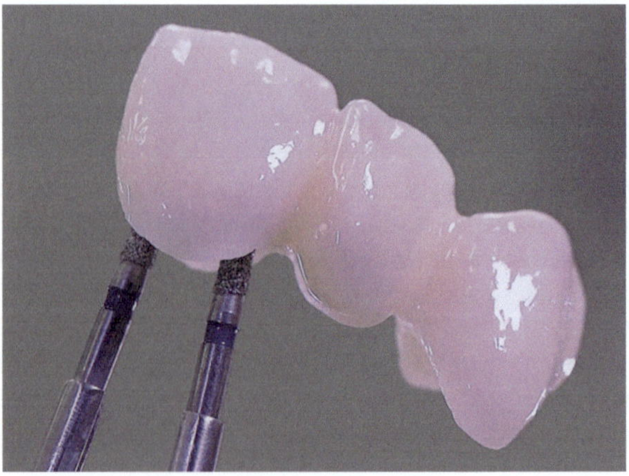

Fig. 9.19 A typical glass–ceramic used in dental work

9.4 Solidification and Crystallization of Polymers

Crystallization of polymers can occur from either molten polymers or polymer solutions ; the latter is discussed in Chap. 4. This section focuses on the crystallization of molten polymers. While there are similarities between the solidification and crystallization of polymers, metals, and ceramics, the long-chain structure of macromolecules in polymers makes it much more challenging to arrange the chains of high molecules into a lattice with a high degree of regularity compared to low molecules. As a result, polymer crystallization is often characterized by incompleteness, imperfection, and slow crystallization rates.

9.4.1 Common Rules Affecting Crystallization

Grain Size Affected by Supercooling

Crystallization of molten polymers occurs when they are cooled to a temperature between their melting point and glass transition temperature. This process requires supercooling, and the degree of supercooling is directly proportional to the nucleation rate. Upon cooling and crystallization, polymers typically form spherulites, the size of which can be controlled by adjusting the cooling rate. A faster cooling rate results in greater supercooling and the formation of smaller spherulites.

Nucleation and Growth

Nucleation is divided into two types: homogeneous and heterogeneous nucleation. Homogeneous nucleation occurs when the crystallization process starts spontaneously without the presence of foreign particles or surfaces to initiate the process. This type of nucleation is based on the orderly arrangement of polymer chains by thermal motion as crystal nuclei. Heterogeneous nucleation, on the other hand, occurs when the presence of foreign particles, surfaces, or imperfections initiates the nucleation process. This type of nucleation occurs at a lower supercooling compared to homogeneous nucleation and is more common in practice. To accelerate crystallization, nucleating agents are often added to promote heterogeneous nucleation, increasing the overall crystallization rate of polymers.

Avrami Equation

The isothermal crystallization of polymers is characterized by a time-dependent sigmoidal curve when the crystallized volume fraction is plotted against time. The Avrami equation provides a mathematical expression for this relationship

$$y = 1 - \exp(-kt^n) \tag{9.40}$$

where k is the crystallization rate constant, and n is the so-called Avrami exponent, which is related to the crystallization mechanism. Usually, the transformation rate

Table 9.4 Typical values for the Avrami exponent

Dimensionality of growth	Homogeneous nucleation	Heterogeneous nucleation
3	$n = 3 + 1 = 4$	$n = 3 + 0 = 3$
2	$n = 2 + 1 = 3$	$n = 2 + 0 = 2$
1	$n = 1 + 1 = 2$	$n = 1 + 0 = 1$

is represented by

$$rate = \frac{1}{t_{0.5}} \tag{9.41}$$

where $t_{0.5}$ refers to the time required to reach 50% crystallization. In practice, the rate is dependent on the temperature and the molecular weight of polymers.

The typical values of the Avrami exponent n are shown in Table 9.4. In the case of homogeneous nucleation, where nuclei form from the regular arrangement of macromolecular chains, n equals the dimensionality of growth + 1. For heterogeneous nucleation, where nuclei originate from system impurities, n equals the dimensionality of growth.

The Avrami equation is commonly used to describe the crystallization process of polymers and has been successfully applied in many studies. However, there are cases where the experimental data deviates from the equation, such as when the value of n is not an integer or the data does not fit a straight line. These deviations suggest that the crystallization process of polymers is more complex than what the Avrami model can fully describe.

9.4.2 Chain Structure Affecting Crystallization

The crystallization of polymers is typically incomplete. Even for highly crystallizable polymers such as polyethylene, the maximum crystallinity is around 95%, while most general polymers reach only about 50%. This is due to the characteristics of the macromolecular chain structure. The factors that affect the crystallization ability of a polymer include:

Chain Symmetry

The higher the structural symmetry of the polymer chain, the easier it is to crystallize. Polyethylene and polytetrafluoroethylene are good examples. They have main chains composed of carbon atoms with either hydrogen or fluorine atoms bonded around the carbon atoms, resulting in high symmetry that makes them easy to crystallize. In contrast, the introduction of asymmetry in the chain structure can decrease or even eliminate the polymer's crystallization ability. For example, chlorination of polyethylene can destroy the symmetry of the molecular chain, disrupting the regular packing of the chains and reducing the polymer's ability to crystallize.

Regularity of the Chain

Polymers with a completely random main chain and lacking a symmetrical center typically cannot crystallize. Free radical polymerized polystyrene and polymethyl methacrylate are examples of such amorphous polymers that cannot crystallize. However, directional polymerization can create a regular configuration for the asymmetric center on the main chain, such as isotactic or syndiotactic macromolecules. This leads to a higher degree of order and enables the polymer to gain crystallization ability.

Copolymerization Effect

Copolymers are formed by two or more different monomer molecules. In general, random copolymerization disrupts symmetry and regularity of the chain, thereby reducing or even eliminating crystallization ability. However, if the homopolymers of the individual copolymerized units share the same type of crystal structure, the copolymer can still crystallize. This occurs because the copolymer chains have regions with similar chain conformations as the homopolymers and can still align to form crystals in those regions. For example, a copolymer of polyethylene and polypropylene can still form crystals if the content of each monomer unit is in a specific range, and the resulting copolymer has a crystalline structure similar to those of its parent homopolymers.

Chain Flexibility

The flexibility of the chain is necessary for the diffusion and alignment of chain segments at the crystal surface during crystallization. Polymers with more flexibility, like polyethylene, can easily align their chain segments to form crystals, resulting in higher crystallinity. However, structural features, such as bulky side groups, rigid rings, or branching, impede chain alignment and reduce crystallization. In addition, cross-linking can greatly limit the mobility of the polymer chain, preventing it from aligning to form crystals, which can also reduce the polymer's crystallization ability.

9.5 Questions

Q9.1 Show that differentiation of Eq. 9.8 leads to Eqs. 9.9 and 9.10.

Q9.2 Show that Eq. 9.19 follows from Eq. 9.18 by the following relationships for a spherical cap:

$$V_S = \frac{\pi r^3}{3}(2 - 3\cos\theta + (\cos\theta)^3)$$

$$A_{SL} = 2\pi r^2(1 - \cos\theta)$$

$$A_{SM} = \pi r^2(\sin\theta)^2$$

Q9.3 Define the following terms: embryo, nucleation, homogeneous nucleation, and heterogeneous nucleation.

Q9.4 What is the difference between homogenous nucleation and heterogeneous nucleation?

Q9.5 From the thermodynamic point of view, what two conditions must be satisfied for solidification from liquid to solid to proceed?

Q9.6 What are the characteristics and differences between rough and smooth solid–liquid interfaces?

Q9.7 Describe the relationship between growth mechanisms and solid–liquid interface.

Q9.8 What is an ingot? Describe the classical microstructure of an ingot.

Q9.9 Describe the common solidification technologies used for tailoring the microstructure of an ingot.

Q9.10 In cast materials, why does solidification almost always begin at the mold walls?

Q9.11 What are the typical values of the Avrami exponent found in the isothermal crystallization of polymers, and how do they relate to the nucleation and growth mechanisms involved in the process?

Bibliography

Avrami M (1939) Kinetics of phase change. I General Theory. J Chem Phys, 7:1103–1112

Avrami M (1940) Kinetics of phase change. II Transformation-time relations for random distribution of nuclei. J Chem Phys, 8:212–224

Avrami M (1941) Kinetics of phase change. III Granulation, phase change, and microstructure. J Chem Phys, 1941, 9:177–184

Becker R, Döring W (1935) Kinetische Behandlung der Keimbildung in übersättigten Dämpfen. Ann Phys 24:719–752

Frenkel J (1939) A general theory of heterophase fluctuations and pretransition phenomena. J Chem Phys 7:538–547

Jackson KA (1958) Liquid metals and solidification. Am Soc Metals, Cleveland, Ohio

Johnson WA, Mehl RF (1939) Reaction kinetics in processes of nucleation and growth. Trans Metall Soc AIME 135:416–458

Kolmogorov AN (1937) On the statistical theory of the crystallization of metals. Bull Acad Sci USSR, Math Ser 1:355–359

Volmer M, Weber A (1926) Keimbildung in übersättigten Gebilden. Z Phys Chem 119:277–301

Solid-State Phase Transformations 10

Learning Objective

1. Define first-order and second-order phase transformations.
2. Cite equations to describe the relationships between Gibbs free energy change and nucleation radius for homogeneous and heterogeneous nucleation during solid-state phase transformation; derive the critical nucleus radius and nucleation barrier for homogeneous nucleation.
3. Draw a schematic transformed fraction versus time plot for a typical solid-state phase transformation and cite an equation to describe the behavior.
4. Describe the characteristics of martensitic transformation.
5. Distinguish and understand the difference between diffusional, diffusionless, and transitional phase transformations.
6. Draw schematic diagrams of the microstructure for typical phases in steel alloys: pearlite, bainite, and martensite.
7. Know the mechanical characteristics for each of the following phases: pearlite, bainite, and martensite; briefly explain these behaviors in terms of composition and microstructure.
8. Given the TTT or CCT diagram for an iron-carbon alloy of eutectoid composition, design a heat treatment to obtain a specific phase in steel alloys.

10.1 Introduction

Solid-state phase transformation is a process in which the internal phase structure of a solid material undergoes changes due to alterations in temperature and pressure. Unlike other transformations like solidification or evaporation, solid-state phase transformation occurs within solid materials, where both the initial and

Z. Sun et al., *Fundamentals of Materials Science*,
https://doi.org/10.1007/978-981-96-3904-5_10

the new phases are solids. These transformations typically result in changes to the composition, crystal structure, microstructure, and properties of the material. Solid-state phase transformation is often accompanied by at least one of the following changes: (1) changes in crystal structure, such as allotropic transformation of metals, pleomorphic transformation of solid solutions, and martensitic phase transformation; (2) changes in chemical composition, such as spinodal decomposition; and (3) changes in order degree, such as magnetic transformation and superconductive transformation.

Changes in the properties of solid materials result from alterations in their microstructure caused by solid-state transformations. These transformations are typically influenced by the heating or cooling temperature, which is constrained by thermodynamic conditions and affects phase transformation kinetics. To maintain its microstructure and properties, the material is usually subjected to a higher temperature during the final heat treatment process than its service temperature. As a result, the properties conferred by the structure can persist for an extended period, often indefinitely.

Solid-state phase transformations follow the principle of minimum free energy, like the solidification process of a liquid. The volume free energy difference between the new phase and the parent phase drives the transformation, which typically involves two stages of nucleation and growth. The degree of supercooling influences the mechanism and rate of nucleation and growth. The parent phase, which is a crystal with ordered atoms, has stronger interatomic bonds than the liquid state and contains defects like vacancies, dislocations, and grain boundaries. This results in new characteristics of solid-state transformation as compared to solidification.

In this chapter, we will cover the fundamental concepts of solid-state phase transformations. Despite the diversity of phase transformations, the common fundamental processes and the unique characteristics of each transformation will be explored and explained.

10.2 Classification and Characteristics of Solid-State Phase Transformations

10.2.1 Classification of Solid-State Phase Transformations

Solid-state phase transformations play a crucial role in material fabrication, as they can change the microstructure and composition distribution. These transformations can be categorized based on various criteria such as thermodynamics, kinetics, microstructure, or mechanism. This section will focus on the classification of solid-state phase transformations through thermodynamics and kinetics.

① Classification through Thermodynamics

In 1933, Ehrenfest proposed that specific thermodynamic variables could be employed to distinguish different types of phase transformations. The order of a phase transformation is determined by the derivative of the relevant thermodynamic potential, such as the Gibbs free energy. When the first derivatives exhibit a discontinuity, the phase transformation is considered to be of first order transformation.

$$\left(\frac{\partial G}{\partial T}\right)_{\mathrm{P}} = -S; \left(\frac{\partial G}{\partial P}\right)_{\mathrm{T}} = V; \left(\frac{\partial\left(\frac{G}{T}\right)}{\partial\left(\frac{1}{T}\right)}\right)_{\mathrm{P}} = H \tag{10.1}$$

In first-order phase transformations, the entropy (S), volume (V), and enthalpy (H) exhibit discontinuity, indicating volume changes and the release or absorption of latent heat. Many solid-state phase transformations, such as the crystallization of alloys and the martensite phase transformation, are first-order transformations. On the other hand, second-order transformations are characterized by discontinuous second derivatives:

$$\left(\frac{\partial^2 G}{\partial T^2}\right)_{\mathrm{P}} = -\frac{C_{\mathrm{P}}}{T}; \left(\frac{\partial^2 G}{\partial P^2}\right)_{\mathrm{T}} = -\beta V; \left(\frac{\partial H}{\partial T}\right)_{\mathrm{P}} = C_{\mathrm{P}} \tag{10.2}$$

where the isothermal compressibility $-\frac{1}{V}\left(\frac{\partial^2 G}{\partial P^2}\right)_{\mathrm{T}} = \beta$ and the specific heat capacity C_{P} exhibit discontinuities. Unlike first-order transformations, there is no latent heat involved in second-order transformations at the transformation temperature. Despite being rare in nature, some transformations, such as magnetic transitions and superconductive transitions, are considered to be second-order transformations.

② Classification through Kinetics

Solid-state phase transformations take place through the movement of atoms. As a result, these transformations can be grouped into three categories based on the kinetics of this atomic migration: diffusional, diffusionless, and transitional transformations.

Diffusional transformations typically occur at elevated temperatures, and involve long-distance atomic migration and movement of the interface during the nucleation and growth of the new phase. At the start of nucleation, the interface between the new and parent phases is usually coherent, but this changes to an incoherent interface during the growth process.

In diffusionless transformations, also known as displacive transformations, the new phase is formed through slip or twining. The interface between the two phases is coherent, and the chemical composition remains unchanged. Martensitic transformation is the most common example of this type of transformation.

Bainite transformation is a common type of transitional transformation. Bainite is a non-lamellar structure composed of ferrite and carbide. The carbide grows through carbon diffusion, while ferrite is formed through individual atom movement less than one interatomic spacing, like a martensitic transformation.

Both diffusional and diffusionless characteristics can be observed in bainite transformation.

10.2.2 Characteristics of Solid-State Phase Transformations

Atoms in crystalline solids are arranged in an orderly and symmetrical pattern, repeating throughout the crystal, exhibiting anisotropy. However, there are always structural defects with uneven distribution. These properties of solids indicate that solid-state phase transformations are distinct from solidification and the following are the characteristics of solid-state phase transformations.

① Diffusion in Materials

The diffusion of atoms in solids is slower than in liquids; for example, near the melting point, the diffusion coefficient of a liquid is around 10^{-5} cm^2 s^{-1}, much higher than that of a solid (10^{-10} cm^2 s^{-1}). As a result, the rate of diffusion can play a significant role in determining the rate of transformation.

At elevated temperatures, atomic motion is fast enough that the diffusion process is not typically the limiting step. Instead, the rate of transformation is determined by the driving force, which is influenced by the amount of supercooling. This means that as supercooling increases, the transformation rate will also increase. However, when temperatures are low enough, atomic diffusion becomes the determining factor for the transformation rate, and as temperatures decrease, the overall transformation rate will also decrease, even with a larger driving force. Rapid cooling can therefore reduce high-temperature phase transformations, and different cooling rates can provide a range of possibilities for diverse transformations. For example, rapidly cooling a Fe–C alloy can result in the formation of martensite instead of pearlite.

② Nucleation

Heterogeneous and Homogeneous Nucleation

Nucleation is the initial step in the formation of a new thermodynamic phase, and typically dominates the kinetics of the process. In solid-state phase transformations, heterogeneous nucleation is the predominant mechanism, facilitated by various types of defects such as point, line, surface, and volume defects in the parent solid phase. These defects, with their distinct energies, are unevenly distributed and favorable for nucleation. The aggregation of new phase particles at these defects weakens the formation energy barrier, thereby promoting nucleation kinetics. Interface defects, which possess high formation energies, are particularly effective nucleation sites. Thus, surfaces, grain boundaries, phase boundaries, twin planes, and subcrystal planes of crystals are the preferred nucleation positions. This explains why the new phase typically forms along these regions under low supercooling.

In contrast, homogenous nucleation is rare in solid-state phase transformations due to the demand for a large driving force and high supercooling. However, homogeneous nucleation can still occur in some cases, such as when the driving force is very large, and the resistances, such as interfacial energy and strain energy, are small. Additionally, a low concentration of defects in the material can also promote homogeneous nucleation by reducing the energy barrier for nucleation. One example of a solid-state phase transformation where homogeneous nucleation can occur is in Zn-Cu alloys during the formation of transition phases θ" from G.P. Zones. The high driving force, low resistances, and low defect concentration in the alloy promote the homogeneous nucleation of the new phase throughout the grain interiors, rather than only at pre-existing nucleation sites.

Interface

In a solid-state phase transformation, a new interface forms between the parent and new phases, which are described in detail in Chap. 7.

The degree of atomic matching at the interface between two phases in a material can be described by the relative strain (δ), which can be defined as $\delta = \frac{a_\alpha - a_\beta}{a_\alpha}$, where a_α and a_β are the unstressed interplanar spacings of the matching planes in the α and β phases, respectively. When the value of δ is less than 0.05, the interface is said to be coherent, meaning that the matching plane in the two phases are perfectly aligned with each other. When the value of δ is between 0.05 and 0.25, the interface is semicoherent, indicating that there is some degree of lattice mismatch between the two phases. When the value of δ is greater than 0.25, the interface is incoherent, suggesting that there is significant lattice mismatch between the two phases.

The relative strain (δ) is a measure of the compatibility between the new phase and the parent phase in a solid-state phase transformation. The value of δ can be used to categorize the interface into three types. A coherent interface demonstrates excellent atomic matching and possesses lowest interface energy. An incoherent interface, on the other hand, exhibits poor atomic matching and the highest interface energy. The semicoherent interface falls between the coherent and incoherent interfaces in terms of both atomic matching and interface energy. To minimize the nucleation work during the transformation, the most effective way is to form nuclei with the lowest interfacial energy, similar to the nucleation process in liquid metals. The resistance at the coherent or semicoherent interface is lower than that of an incoherent interface. Therefore, in the initial stages of transformation, coherent or semicoherent interfaces are often observed.

It is important to note that a completely coherent interface only occurs at the coherent twin boundary, as there will always be some differences in the crystal structure or lattice constant between the two phases in other cases. Therefore, when a coherent interface forms, a certain amount of elastic strain is generated near the interface to maintain lattice matching and coherence. This elastic strain energy is known as the coherency strain energy, which is different from the volume strain energy caused by the specific volume difference between the two phases. In the case of a coherent interface, with a relatively small lattice mismatch, the lattice continuity requires a significant amount of lattice distortion to accommodate the mismatch, resulting in a higher elastic strain energy compared to semicoherent and incoherent interfaces. As the degree of mismatch increases, the strain energy also increases. However, despite the higher strain energy, coherent interface has the lowest interface energy among the three types of interfaces. The semicoherent interface exhibits intermediate value of strain energy and interface energy, while the incoherent interface has the lowest strain energy but the highest interface energy.

Orientation Relationship

When a new solid-state phase forms, there is usually a specific orientation relationship between the new formed phase and its parent phase. This relationship is characterized by the alignment of crystallographic planes and directions between the two phases. One such observed orientation relationship is known as the Kurdjumov–Sachs relationship (K-S relationship). For example, in the case of the transformation from γ-Fe to α-Fe, the K-S relationship manifests as follows: the {110} crystallographic plane of the α-Fe phase aligns parallel to the {111} crystallographic plane of the γ-Fe phase, while the <111>crystallographic direction of the α-Fe phase aligns parallel to the <110> crystallographic direction of the γ-Fe phase. During the nucleation process, the orientation of the new phase is constrained by the parent phase, which can affect the formation and growth of crystal nuclei. The formation energy is the lowest when the crystal planes or directions of the new and parent phases are parallel to each other. This alignment can facilitate the formation of crystal nuclei by minimizing the energy required for the new phase to form.

The orientation relationship between the new and parent phases can also affect the properties and behavior of the material. For example, it can influence the mechanical properties, such as the strength and ductility, and can affect the behavior of dislocations and other defects. Therefore, understanding the orientation relationship between phases is important for predicting and controlling the properties of materials undergoing solid-state phase transformations.

③ Growth Characteristics

Habit Plane

During a solid-state phase transformation, new phases tend to grow in a certain direction and along specific crystal planes of the parent phase. The crystal plane and direction that determine the growth of the new phase are collectively called the habit plane and habit direction, respectively. These are usually described by the crystal plane and crystallographic direction indices of the parent phase. The selection of the habit plane and habit direction is influenced by several factors, including the interfacial and strain energies, as well as the growth conditions of the new phase.

If the interfacial energy changes with the growth direction of the new phase, then the new phase will grow along the direction that minimizes the interfacial energy, resulting in a fully developed phase interface with the lowest interfacial energy. Similarly, if the strain energy changes with the growth direction of the new phase, then the new phase will grow along the direction with the smallest strain energy, thereby minimizing the strain energy. Additionally, the specific habit plane and habit direction can vary depending on the phase transformation conditions.

Coherent and Incoherent Growth

For diffusional phase transformation, the growth of the new phase occurs through the migration of atoms across an incoherent interface between the parent and new phases. Even if a coherent interface forms in the nucleation stage, it is difficult to maintain the coherence due to the large strain energy associated with the misfit between the two crystal lattices. As a result, the coherent interface gradually disappears and evolves into an incoherent interface.

In contrast, diffusionless phase transformation involves the shear and shuffling of atomic planes, without significant atomic migration. The coherent interface between the parent and new phases can be maintained during nucleation and growth, as long as the lattice misfit between the two phases remains below a critical value. However, if the new phase grows to a certain extent with a large strain energy, the coherence can be destroyed, and the growth may stop.

The mechanisms of diffusional and diffusionless phase transformation have important implications for the microstructure and properties of materials. For example, diffusional phase transformation can lead to the formation of complex microstructures with a high density of dislocations and defects, while diffusionless phase transformation can result in highly ordered structures with low defect densities. Understanding these mechanisms is essential for designing and engineering advanced materials with tailored microstructures and properties.

④ Transitional Phase

One notable feature of solid-state phase transformations is the occurrence of transitional phases, which are metastable phases that have crystal structures or compositions intermediate between the parent and new phases. In some cases, the transitional phase may be the final product of the transformation, and multiple transitional phases may appear in sequence. For example, in the aging of an Al-Cu alloy, two transitional phases (θ'' and θ') form before the stable phase θ ($CuAl_2$).

Transitional phases have a crystal structure and composition that are closer to those of the parent phase, and this can facilitate the phase transformation by reducing the resistance to the transformation. However, the formation of the transitional phase is energetically unfavorable from a thermodynamic perspective, but beneficial in terms of kinetics. Understanding the formation and behavior of transitional phases is important for designing materials with tailored properties and microstructures.

10.3 Thermodynamics of Solid-State Phase Transformations

Thermodynamics plays a fundamental role in the study of phase transformations, as it provides a theoretical framework for understanding the thermodynamic properties and behavior of materials during phase transformations. Thermodynamics can be used to analyze and calculate various thermodynamic phenomena in the process of phase transformation, such as phase equilibrium, phase stability, phase transformation direction, and driving force. Thermodynamics can also be used to predict the stability of different phases under different conditions, such as changes in temperature, pressure, or composition. The driving force for phase transformation can be calculated using thermodynamic quantities such as the free energy, enthalpy, and entropy.

Overall, the application of thermodynamics to phase transformations is a powerful tool for predicting and controlling the behavior of materials under different conditions and is an important area of study in materials science and engineering.

10.3.1 Nucleation in Solid-State Phase Transformations

Classical solid-state phase transformations involve nucleation and growth. Nucleation can be classified into two types based on time sensitivity: thermally activated and nonthermally activated. In thermally activated nucleation, both temperature and time affect the nucleation process, and crystal nuclei can form through isothermal processes. Diffusion-type phase transformations typically occur in higher temperature ranges where thermally activated nucleation is common. When supercooling is slight, crystal nuclei usually appear at defects due to low driving force requirements, which is known as heterogeneous nucleation. In contrast, if supercooling is significant, homogenous nucleation may occur due to the increased driving force. Non-thermally activated nucleation occurs through rapid cooling, and it is not time-sensitive. Crystal nuclei generally do not form through an isothermal process in this type of nucleation. Heterogeneous nucleation is usually observed in this type, and it requires a large degree of supercooling for fast nucleation, as in the case of martensitic transformation.

To initiate the nucleation-growth solid-state phase transformation, nucleation is a crucial step. There are two types of nucleation: homogenous and heterogeneous nucleation. Homogenous nucleation occurs when nuclei form spontaneously without any foreign surface acting as a catalyst. On the other hand, heterogeneous nucleation occurs when nuclei form at defects such as grain boundaries, dislocations, or other imperfections in the parent phase. Heterogeneous nucleation is typically preferred because the energy barrier for nucleation is much smaller compared to homogenous nucleation.

① Homogenous Nucleation

Both interfacial energy and strain energy pose resistance to phase transformation, which sets it apart from solidification. In classical nucleation theory, an embryo is assumed to be a sphere with radius *r*, and the Gibbs free energy change due to its formation is expressed as:

$$\Delta G = V\Delta G_V + V\omega + S\sigma = \frac{4}{3}\pi r^3 \Delta G_V + \frac{4}{3}\pi r^3 \omega + 4\pi r^2 \sigma \quad (10.3)$$

Here, ΔG_V, *w*, and σ are the free energy change per unit volume, the strain energy per unit volume, and the interfacial free energy per unit area, respectively. By differentiating Eq. 10.3 with respect to *r* and setting it to zero ($\frac{\partial \Delta G}{\partial r} = 0$), the critical nucleus size r^* can be determined:

$$r^* = \frac{-2\sigma}{(\Delta G_V + \omega)} \quad (10.4)$$

The critical activation energy for nucleation is then given by:

$$\Delta G^* = \frac{16\pi \cdot \sigma^3}{3(\Delta G_V + \omega)^2} \quad (10.5)$$

If the radius of the embryo is smaller than r^*, it will shrink and disappear; otherwise, it will grow, as described in Fig. 10.1. The larger the critical radius, the fewer nuclei with critical size will be. Compared with solidification process, under the same conditions, the additional strain energy resistance exists in the solid-state phase transformation increases the nucleation barrier and the critical nucleus size.

The homogeneous nucleation rate per unit volume per second, $\dot{N}$, can be described by

$$\dot{N} = K\exp\left(-\frac{\Delta G^*}{kT}\right)\exp\left(-\frac{Q}{kT}\right) = K\exp\left(-\frac{16\pi\sigma^3}{3kT(\Delta G_V + \omega)^2}\right)\exp\left(-\frac{Q}{kT}\right) \quad (10.6)$$

where *k* is Boltzmann's constant, *K* is the constant, and *Q* denotes the activation energy for the diffusion of atoms. Equation 10.6 is very similar to the expression of the nucleation rate of solidification. In the case of solid-state phase transformation, the nucleation barrier is often associated with the formation of strain energy, which is a measure of the deformation or distortion of the crystal lattice as the new phase nucleates. The strain energy arises due to the difference in lattice parameters between the parent and the new phases, which results in a local distortion of the crystal lattice. The magnitude of the strain energy is proportional to the elastic modulus of the material and the square of the elastic strain. Therefore, the greater the rigidity of the solid, the larger the strain energy, which is not conducive to

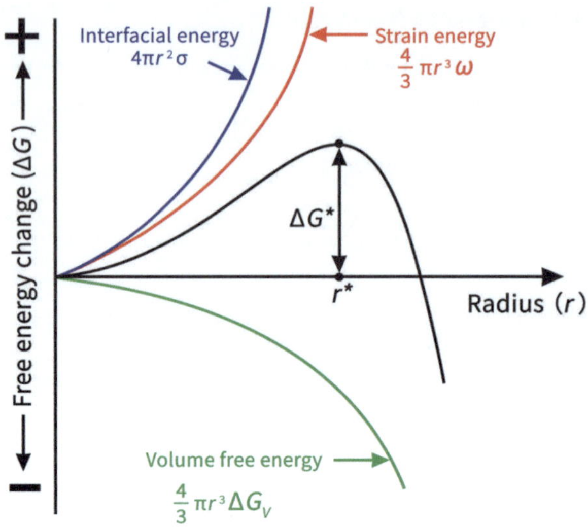

Fig. 10.1 Schematic diagram of the total changes in free energy as a function of nucleus radius for homogeneous nucleation

the phase transformation. Moreover, the activation energy for diffusion in solids is generally much larger than that in liquids, which further reduces the nucleation rate.

The Driving Force

The driving force of the solid-state phase transformation is the volume free energy change ΔG_V, which depends on several factors, including temperature, composition, and crystal structure.

1. Temperature Dependences of Driving Force

If a new phase β precipitates from α phase with decreasing temperature, the free energy change associated with the transformation can be calculated as follows:

$$\Delta G_V^{\alpha\to\beta}(T) = \Delta H_V^{\alpha\to\beta}(T) - T\Delta S_V^{\alpha\to\beta}(T) \tag{10.7}$$

At the temperature T_0, the free energies of both phases are the same and can be expressed as:

$$\Delta G_V^{\alpha\to\beta}(T_0) = \Delta H_V^{\alpha\to\beta}(T_0) - T\Delta S_V^{\alpha\to\beta}(T_0) = 0 \tag{10.8}$$

Therefore, $\Delta S_V^{\alpha\to\beta}(T_0) = \frac{\Delta H_V^{\alpha\to\beta}(T_0)}{T_0}$, and when the supercooling ($T_0 - T$) is small, $\Delta H_V^{\alpha\to\beta}(T) \approx \Delta H_V^{\alpha\to\beta}(T_0)$, and $\Delta S_V^{\alpha\to\beta}(T) \approx \Delta S_V^{\alpha\to\beta}(T_0)$. Substitute those equations into Eq. 10.7, the free energy change can be derived:

$$\Delta G_V^{\alpha\to\beta}(T) \approx \frac{\Delta H_V^{\alpha\to\beta}(T_0)\Delta T}{T_0} \tag{10.9}$$

This shows that the free energy change is directly proportional to the degree of supercooling, and the magnitude of the driving force increases with decreasing temperature. This means that the transformation is more likely to occur at lower temperatures, where the driving force is greater.

2. Composition Dependences of Driving Force

Under constant pressure, the Gibbs free energy of the unary phase system is only a function of temperature. The composition in alloys can also affect the free energy, as demonstrated by the following example. When the alloy X_0 is cooled rapidly from T_1 to T_2, the new phase β precipitates from the supersaturated solution as shown in Fig. 2a. In Fig. 2b, the free energy of the alloy decreases by an amount ΔG_0, which is the total driving force for the transformation, not nucleation, as the first nuclei do not significantly change the composition of α. If a small number of nuclei with composition X_B^β precipitate, the total free energy of the system will change by ΔG_1, as represented by point P:

$$\Delta G_1 = \mu_A^\alpha X_A^\beta + \mu_B^\alpha X_B^\beta \tag{10.10}$$

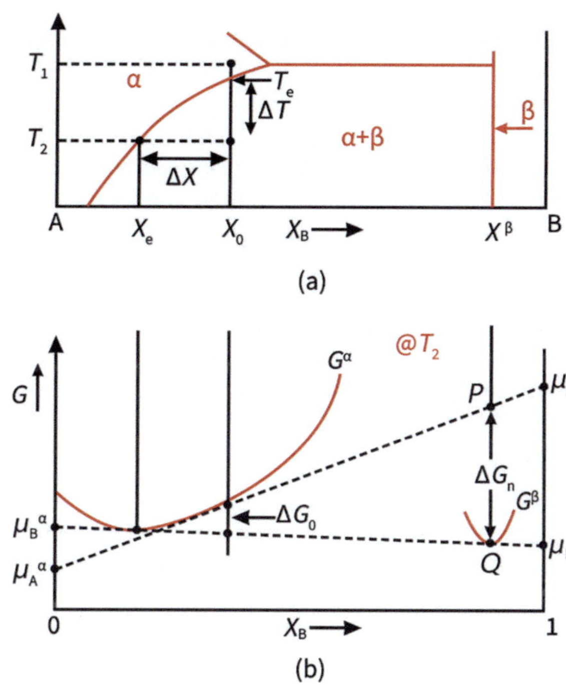

Fig. 10.2 Free energy changes during precipitation

If the removed atoms are crystalized into the β phase, the total free energy of the system will increase by ΔG_2, as given by point *Q*:

$$\Delta G_2 = \mu_A^\beta X_A^\beta + \mu_B^\alpha X_B^\beta \tag{10.11}$$

Therefore, the driving force for the transformation can be express as:

$$\Delta G_n = \Delta G_2 - \Delta G_1 \tag{10.12}$$

where ΔG_n represents the line segment *PQ*, and the driving force of nucleation $\Delta G_V = \Delta G_n / V_\beta$ (V_β is molar volume of β). For dilute solution, it can be show that approximately as $\Delta G_V \propto \Delta X$ and $\Delta X = X_0 - X_e$.

Strain Energy

Strain energy plays a vital role in nucleation and can be divided into two types. The first type is coherency strain energy caused by the lattice mismatch between the two coherent phases. The greater the mismatch, the higher the strain energy will be. The second type is caused by specific volume difference between the parent and the new phase, which can result in a volume change and elastic strain when the new phase forms, as seen in the transformation from austenite (FCC) to ferrite (BCC).

While there are some systems in which the equilibrium phase can nucleate homogeneously, such as Co precipitating from a Cu-Co solution due to its similar crystal structure and 2% difference in lattice parameter to Cu, most examples of homogeneous nucleation are limited to metastable phases or zones, such as the G. P. zone in Al-Cu alloys.

② Heterogenous Nucleation

Solid-state phase transformations mostly take place through heterogeneous nucleation due to the uneven distribution of various defects in the material. The free energy change of the system during heterogeneous nucleation is given by:

$$\Delta G = V\Delta G_V + V\omega + S\sigma + \Delta G_d \tag{10.13}$$

This equation differs from that of homogeneous nucleation only by the inclusion of ΔG_d which is the formation energy of the defect. Nucleation at the defect can release the formation energy of the defect, reducing the energy barrier and promoting the nucleation process. The following discussion will be based on the contribution to nucleation in descending order.

Nucleation on Grain Boundary

Due to the high formation energy, grain boundaries are favorable nucleation sites for new phases. When a new phase nucleates at a grain boundary, the interface between the new and parent phases requires only partial reconstruction. Usually, one side of the crystal nucleus forms a coherent or semicoherent interface with lower interface energy, while the other side forms an incoherent interface. As the nuclei grow, these interfaces eventually become incoherent.

Nucleation on Dislocation

Electron microscopy observations have confirmed that dislocations can also be advantageous for nucleation due to the following main reasons:

1. Dislocations can interact with solute atoms to form clusters, concentrating solute atoms near the dislocation line, which is beneficial for nucleation.
2. The nucleation barrier can be reduced when the original dislocation vanishes.
3. Dislocations can act as atomic diffusion channels, reducing the activation energy for atom diffusion.
4. New phases with different specific volumes can nucleate in the tensile/compressive zone around the dislocation, reducing strain energy.

Nucleation on Vacancy

The vacancies indirectly affect nucleation and can be concluded below:

1. When a vacancy group reaches a certain size, it can collapse into a dislocation ring, promoting nucleation as a dislocation.
2. The resistance to phase transformation increases with an increase in the specific volume difference. By absorbing or releasing vacancies, strain energy can be reduced, facilitating nucleation.
3. Vacancies increase the diffusion coefficient, which is conducive to nucleation.

10.3.2 Metastable Versus Equilibrium

There are three thermodynamic states: stable, metastable, and unstable, as shown in Fig. 10.3. In thermodynamics, the stable state of a material refers to the equilibrium state with the lowest Gibbs free energy. As shown in Fig. 10.3, state A is a stable equilibrium state, corresponding to a global minimum in the free energy. On the other hand, state B is a metastable state with a local minimum in free energy, and the system is stable to small fluctuations, but will eventually evolve to a more stable state.

Phase transformations in materials may be driven by varying temperature, composition, and external pressure. However, temperature changes by heat treatments are most commonly employed to promote a phase transformation, which corresponds to crossing a phase boundary on the phase diagram. During a phase transformation, the material will typically reach an equilibrium state, which is characterized by the phase diagram. However, most phase transformations occur

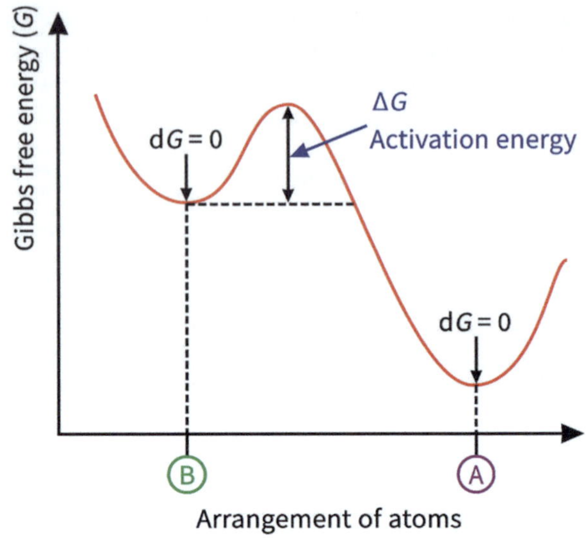

Fig. 10.3 Gibbs free energy as a function of the arrangements of atoms in phases. **A**: stable configuration, **B**: metastable configuration

within a finite time frame, and the rate of approach to equilibrium for solid systems is so slow that it is rarely possible to obtain a structure in true equilibrium. Equilibrium states can only be retained if heating or cooling is carried out at extremely slow rates. Otherwise, the transformations are shifted to lower or higher temperatures than displayed by the phase diagram, which are called supercooling and superheating, respectively.

For many materials, the preferred state is the metastable one between the initial and equilibrium states. In some cases, a far-from-equilibrium state is actually desired. In such cases, it is crucial to investigate the time dependence of phase transformations, which can be more valuable than knowledge of the equilibrium state.

Under certain conditions, a material may have only one stable equilibrium state, but non-equilibrium metastable states can appear in multiple forms, including:

1. *Microstructural refinement*: Finer scale distributions of phases lead to an increase in Gibbs free energy by providing more interfacial area per unit volume.
2. *High defects concentrations*: Crystallographic defects cause the atoms to deviate from their equilibrium positions, leading to a decrease in the regularity of the atomic arrangement and an increase in the free energy of the system.
3. *Formation of oversaturated solid solutions*: A stable crystalline phase may be found with solute levels beyond the solubility limit at the ambient temperature or beyond the equilibrium limit at any temperature.
4. *Formation of metastable phases*: A non-equilibrium transformation occurs, and a new metastable phase with a different structure from the original one forms.

5. *Crystalline-to-amorphous phase transformation*: The crystalline-to-amorphous transformation is a process in which the atomic arrangements change from order to disorder, leading to an increase in free energy.

10.4 Kinetics of Solid-State Phase Transformations

Although the temperature dependences of transformation were discussed above, it is also important to consider the time dependence of the rate. The growth of a new phase occurs through interface movement, which can be classified into two types based on the mechanism of movement: diffusion and diffusionless.

10.4.1 Diffusional Growth

The growth process of the new phase is attributed to the movement of its interface. The rate of interface movement is related to the structure of the interface itself. In the case of incoherent interfaces, the movement of the interface is controlled by volume diffusion, which involves the diffusion of atoms through the bulk of the material. In contrast, for coherent or semicoherent interfaces, the movement of the interface is controlled by interface diffusion, which involves the exchange of atoms across the interface between the two phases. The rate of interface diffusion is usually much faster than volume diffusion, which leads to faster interface movement and therefore faster phase growth in the case of coherent or semicoherent interfaces.

① Volume Diffusion

Take the precipitation of the β phase (spheroidal particles) from a supersaturated solution α as an example. The alloy with composition x_0 is heated to a single-phase zone and then rapidly cooled to a temperature below the solubility curve. Then, the β phase will precipitate from the parent phase α. If β particles of radius r precipitate homogenously in α, the interface between the two phases becomes incoherent.

As the β particle grows by the distance dr over time dt, the increased volume can be approximated as $S\mathrm{d}r \approx 4\pi r^2\mathrm{d}r$. Therefore, the growth rate of β can be expressed as:

$$\frac{\mathrm{d}m_\beta}{\mathrm{d}t} = \frac{S\mathrm{d}r}{V_\mathrm{m}^\beta \mathrm{d}t}\left(x^{\frac{\beta}{\alpha}} - x^{\frac{\alpha}{\beta}}\right) = \frac{SD^\alpha}{V_\mathrm{m}^\alpha}\frac{\mathrm{d}x^\alpha}{\mathrm{d}r} \tag{10.14}$$

where S and V_m^β are the surface and molar volumes, respectively. D^α denotes the diffusion coefficient of β in the α solid, $x^{\frac{\beta}{\alpha}}$ and $x^{\frac{\alpha}{\beta}}$ refer to the compositions of β and α at the interface, respectively. For a solid solution with low supersaturation, this process approximates a steady-state diffusion process in which the diffusion flow rate is independent of time. Therefore, the growth rate of β can be calculated as:

$$\frac{\mathrm{d}m_\beta}{\mathrm{d}t} = 4\pi r^2 D^\alpha \frac{x_0 - x^{\frac{\alpha}{\beta}}}{r} \tag{10.15}$$

Through the use of Eqs. 10.14 and 10.15, we can represent the radius of the β particle as:

$$r^2 = \frac{2D^\alpha (x_0 - x^{\frac{\alpha}{\beta}})}{x^{\frac{\beta}{\alpha}} - x^{\frac{\alpha}{\beta}}} \frac{V_\mathrm{m}^\beta}{x V_\mathrm{m}^\alpha} t \tag{10.16}$$

This equation shows that the growth rate of the β particle over time follows a parabolic relationship. Similarly, if the β phase precipitates in the form of flakes, the thickness of the β phase also exhibits a parabolic relationship with time.

② Interface Diffusion

Aaronson proposed the ledge growth mechanism, which has been repeatedly proved in various transformations, such as discontinuous precipitation, austenite to pearlite, and massive transformations. As shown in Fig. 10.4, during nucleus growth, atoms move perpendicularly to the incoherent interface to make the step move laterally. If the horizontal movement speed of the steps is v, the step height is a, and the distance between the steps is b, then the number of steps passing along the direction of v per unit time is expressed as $n = v/b$. The thickening growth rate of the new phase can be obtained as:

$$u = na = \frac{av}{b} \tag{10.17}$$

This equation demonstrates that the growth rate of the new phase increases with the step density on the interface. The horizontal movement of the steps relies on the motion of dislocation, where the Bermansch vector of edge dislocations is parallel to the slip surface. As the dislocation moves along the slip surface, the new phase grows along the normal direction.

Fig. 10.4 Schematic diagram of the ledgewise growth mechanism

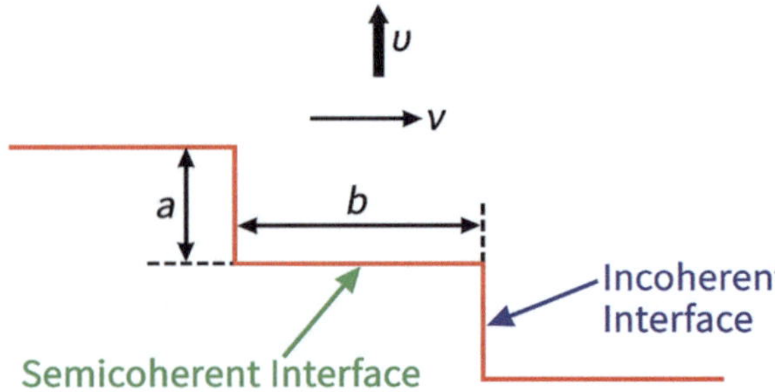

10.4.2 Kinetic of Phase Transformations

The kinetics of solid-state phase transformation involves the study of how the transformation progresses over time and temperature, and the fraction of the phase that has transformed at a given time and temperature. The nucleation rate and growth rate are the two main factors that affect the kinetics of phase transformations. The nucleation rate is a measure of how frequently new nuclei form, while the growth rate determines how fast the newly formed nuclei grow. Together, these two rates determine the overall rate and extent of the phase transformation.

① Kinetic Equations with Constant Nucleation Rate

It is assumed that the new phase forms through homogenous nucleation, with unchanged composition in the parent phase. Both the nucleation rate $\dot{N}$ and the growth rate u are independent of time t. After derivation, the transformed volume fraction and time satisfy the following equation called the Johnson-Mehl formula:

$$x_t = \frac{V}{V_0} = 1 - \exp\left(\frac{-\eta \dot{N} u^3 t^4}{4}\right) \tag{10.18}$$

where V is the material volume before transformation, V_0 refers to the transformed phase volume, and η is the shape-dependent constant of the new phase ($\eta = \frac{4\pi}{3}$).

② Kinetic Equations with Time-dependent Nucleation Rate

Solid-state phase transformations usually take place through heterogeneous nucleation. In this case, the growth rate u of the new phase can be assumed to be constant, while the nucleation rate $\dot{N}$ is a function of time. If there are N_0 available nucleation sites per unit volume in the parent phase, the nucleation rate at a constant temperature over time t can be expressed as

$$\dot{N} = N_0 v \exp(-vt) \tag{10.19}$$

where v is the frequency of nucleation, which is a constant. The volume fraction of the new phase can be described by the Avrami equation, which is similar to the Johnson-Mehl formula. The volume fraction of the new phase is given by:

$$x_\mathrm{t} = 1 - \exp\left(-bt^n\right) \tag{10.20}$$

where b and n are time-independent constants for the reaction, which can be determined experimentally.

A schematic diagram of the transformation rate from the above equation is shown in Fig. 10.5. At the beginning of the transformation, the rate is low because the nucleation of new phases has not yet begun. As the transformation progresses, the rate increases as more and more nuclei are formed and begin to grow. The rate reaches a maximum when the growth of the new phase is fastest, and then slows down as the available reactants are consumed or as the new phase reaches the

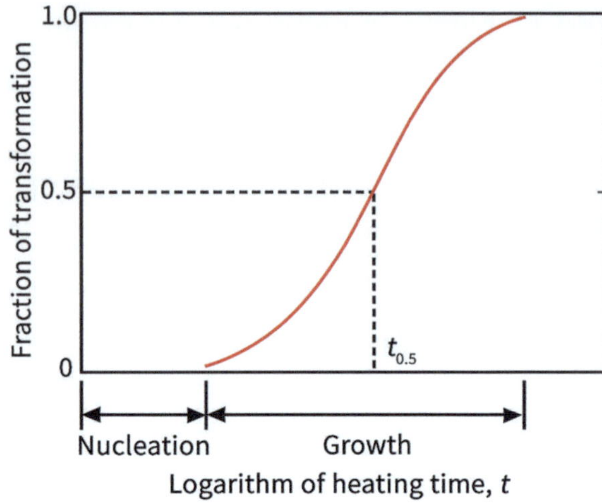

Fig. 10.5 Plot of transformed volume fraction versus the logarithm of time, typical of many solid-state transformations occuring at a constant temperature

limits of the existing matrix. Finally, the transformation rate reaches a plateau as the transformation is completed and the system reaches its equilibrium state. The S-shaped curve shown in the figure reflects the typical kinetic behavior observed in most solid-state phase transformations.

10.5 Diffusional Phase Transformations

10.5.1 Precipitation from Solid Solution

Solid-state phase transformations, such as precipitation, can occur in a parent phase that is a solid solution, compound, or even a solute atom-rich zone. For precipitation to occur, certain conditions must be met, including the supersaturation of the solid solution, a decrease in solubility with decreasing temperature, and sufficient atomic diffusion for precipitation.

In the case of continuous precipitation, the precipitated phase forms homogeneously in the matrix as small, isolated particles, and the composition change of the parent phase around the new phase is continuous. At the initial stage, the precipitated phase is coherent with the matrix due to its similar crystal structure to the parent phase, resulting in a disc or flake shape. However, in the later stage, the interface becomes incoherent due to the different crystal structures, and the precipitated phase is usually equiaxed.

Discontinuous precipitation occurs when the precipitated phase nucleates at the grain boundary of the parent phase and grows into the grain in the form of a cellular structure. At the interface between the parent and precipitated phases, the composition changes abruptly. This discontinuous change in composition gives rise to the name of discontinuous precipitation.

① Continuous Precipitation

Continuous precipitation is a solid-state phase transformation phenomenon where a precipitate phase forms continuously within a material. This gradual and continuous process leads to the homogeneous distribution of fine particles throughout the material's matrix. In continuous precipitation, the composition change between the parent phase and the precipitate phase is continuous, meaning there is no abrupt change in composition at the interface. The precipitate phase grows and develops within the parent phase, maintaining coherence with the matrix in the initial stages. However, as the process continues, the interface between the parent and precipitate phases may become incoherent due to the differences in crystal structures.

One prominent example of continuous precipitation is observed in aluminum-copper (Al-Cu) alloys during the aging process. In Al-Cu alloys, continuous precipitation involves the formation of a continuous network or distribution of precipitate phases within the aluminum matrix. This precipitation occurs as a result of the decreasing solubility of copper in aluminum with decreasing temperature. During aging, the alloy is held at a specific temperature for a defined period, allowing the formation and growth of the precipitates within the material.

Precipitation in Aluminum-Copper Alloys

The aging of Al-Cu alloy is a widely studied and exemplary instance of continuous precipitation. Figure 10.6 shows the Al-rich end of the Al-Cu phase diagram. At the outset, the Al-4%Cu alloy has a mixed α and θ ($CuAl_2$) phase composition. Once the alloy is heated to roughly 540 °C, all the copper present dissolves in the aluminum, leading to the creation of a stable α phase, which is subsequently subjected to rapid water cooling. This process results in the formation of a supersaturated solid solution. Afterward, the alloy undergoes a period of aging at a fixed temperature (such as room temperature or below roughly 180 °C), leading to the emergence of various transitional phases that impact the mechanical properties of the Al-Cu alloy.

The first precipitation is not the θ phase, but rather the coherent Cu-rich Guinier–Preston (GP) zones. These zones were independently detected in 1938 by Guinier and Preston from X-ray diffraction patterns, and they homogeneously nucleate in the matrix. Specifically, in the (110) planes of the matrix, the Cu atoms are enriched, and the average concentration of Cu atoms is approximately 90%. In addition, the crystal structure of the enrichment zone is identical to that of the α phase. GP zones are fully coherent with the matrix accompanied by a very low interface energy. Therefore, the zones minimize their strain energy by adopting a disc shape that is approximately 2 atomic layers thick (0.3~0.6 nm) and 10 nm in diameter.

The second transitional phase, θ'', forms through either homogenous nucleation or by deriving from a GP zone. The θ'' phase has a tetragonal unit cell ($a = b =$ 0.404 nm, $c = 0.78$ nm), which is different from that of the GP zones. It is noted that the atomic structure of the (001) planes is identical to that of the matrix. Similar to the GP zones, the θ'' phase is also in the form of discs, with thicknesses

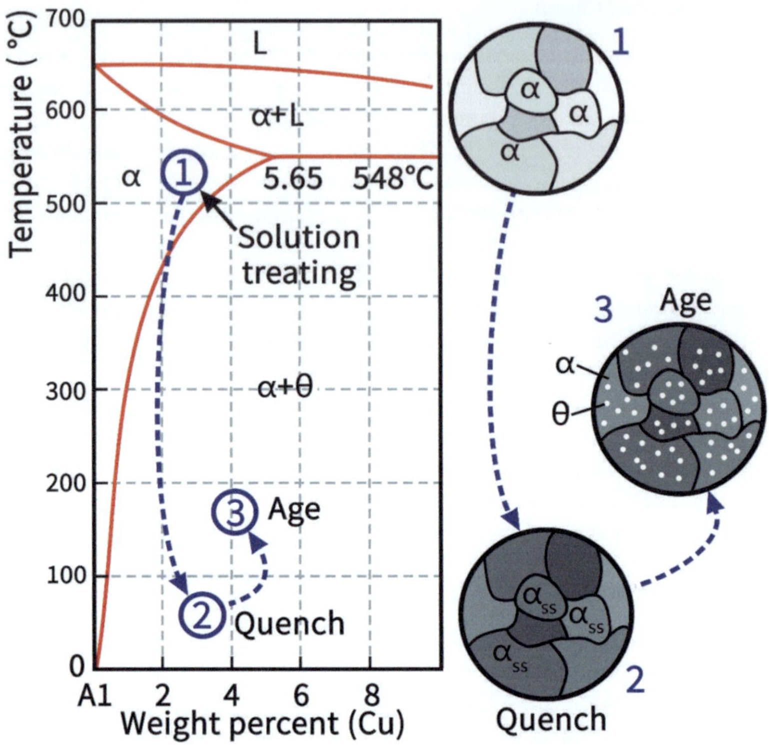

Fig. 10.6 The Al-rich end of the Al-Cu phase diagram showing the three steps in the age-hardening heat treatment and the microstructures that are produced

of 2 nm and diameters of 30 nm. The θ'' phase is fully coherent with the matrix, and its habit plane is $\{001\}_{\alpha}$.

The third transitional phase, θ', appears with increasing aging time, and its composition is approximately $CuAl_2$. The structure of θ' is also disc-shaped. The broad faces of the plates are coherent or semicoherent with the matrix, while the edges of the plates are incoherent.

The equilibrium phase of θ ($CuAl_2$) has a complex body-centered tetragonal structure, which is incoherent with the matrix due to lattice mismatching.

The formation of transitional phases with lower activation energy barriers is due to their crystal structures being intermediate between those of the matrix and the equilibrium phase. This results in a high degree of coherence with a low interfacial energy contribution. In contrast, the equilibrium phase with its complex crystal structure is incompatible with the matrix, resulting in high-energy interfaces.

The formation of GP zones and subsequent transitional precipitates depends on the composition and temperature, and they are only observed when the alloy is aged at a temperature below the GP zone solvus. If the aging temperature is above the θ'' solvus but below the θ' solvus, the first precipitate will be θ', not the GP zone. It is important to note that these transitional precipitates can undergo reversion; when an alloy with GP zones is heated to above the GP zone solvus, the zones will dissolve.

Age Hardening

Transitional phase precipitation has attracted much attention due to its ability to improve the mechanical properties of Al-Cu alloys. Figure 10.7 shows the dependence of the hardness of Al–Cu alloys on time and temperatures (130 °C and 190 °C) and how different transitional phases, such as the GP zones θ'' and θ', affect the hardness. After quenching the Al-4.5%Cu alloy, the primary contribution to hardness is solid solution hardening. As GP zones and θ'' form at 130 °C, the hardness increases due to the coherent interphases. Eventually, the hardness decreases with the formation of θ' due to dislocation bowing. Peak hardness is associated with the existence of θ'' and θ'. Upon further aging, the average distance between the precipitates increases, and the dislocations bow more easily, resulting in a decreased hardness. Aging beyond peak hardness is referred to as overaging.

When the Al-4.5%Cu alloy is aged at 190 °C, the first precipitate turns to θ' due to the instability of GP zones at high temperatures. The fraction of θ'' increases with time, leading to an increase in hardness as shown in Fig. 7b. However, at 190 °C, the driving force for θ'' is smaller compared to 130 °C, leading to a coarser dispersion of precipitates and a reduction in θ'' precipitates, which weakens the peak hardness. Additionally, the time to reach peak hardness is shorter due to faster atomic diffusion. In commercial alloy product processes, the heat treatment temperatures are usually restricted by economic considerations. Therefore, a double aging treatment is employed in some high-strength alloy products, where a fine dispersion of GP zones is obtained at a relatively low temperature below the GP zone solvus. These GP zones act as heterogeneous nucleation sites for precipitation at higher temperatures, leading to a finer distribution of precipitates in the subsequent aging process.

② Discontinuous Precipitation

In some alloys, the supersaturated solid solution decomposes through a discontinuous cellular reaction, resulting in the formation of a new phase β and a solute-depleted zone (α phase) that exhibits the same crystal structure as the initial phase α_0 ($\alpha_0 \rightarrow \beta + \alpha$). This reaction is called "discontinuous precipitation" due to the sharp change in composition between the initial phase of α_0 and the solute-depleted α phase across the reaction front. When the precipitated phase nucleates on the grain boundary, it does not grow along the grain boundary or into the grain as a needle-like or lamellar Widmanstatten structure; instead, it forms a cellular structure, as depicted in Fig. 10.8. The front of the cellular precipitation is almost spheroidal, and it is separated from the matrix by a clear interface. Like pearlite, cellular precipitation is a lamellar structure distributed by two phases.

Discontinuous precipitation and continuous precipitation are both diffusional phase transformations, but the atomic diffusion distance is significantly different.

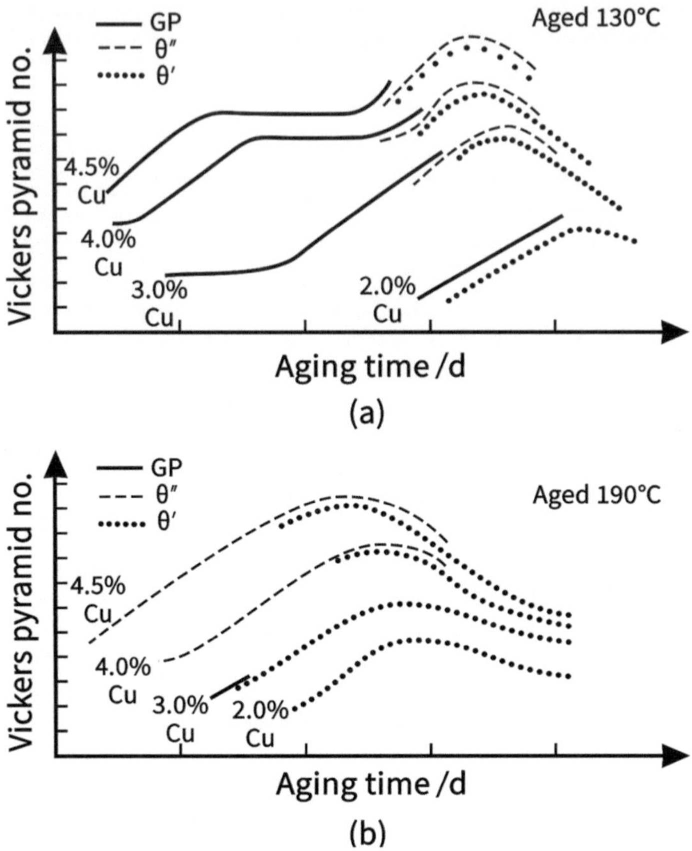

Fig. 10.7 Time dependence of hardness for various Al–Cu alloys at **a** 130 °C and **b** 190 °C

In continuous precipitation, the transformation rate is limited by volume diffusion, while in the case of discontinuous precipitation, short-range interfacial diffusion dominates the transformation rate. Consequently, the growth in the discontinuous precipitation process is rapid due to the short-range interfacial diffusion. However, the uneven phase distribution in discontinuous precipitation is not conducive to material strength.

10.5.2 Eutectoid Transformation

The eutectoid transformation is a type of solid-state phase transformation where a single solid phase decomposes into two different solid phases: $\gamma \rightarrow \alpha + \beta$. This transformation is similar to the eutectic transformation, but it occurs in a single solid phase instead of a liquid–solid mixture. Due to the solid-state nature of the

Fig. 10.8 A schematic diagram of discontinuous (left) precipitation

transformation, atomic diffusion is much more difficult, leading to a slow transformation rate. Furthermore, in cases of fast cooling rates or low temperatures, the transformation can be partially or completely suppressed.

The most studied and illustrative example of eutectoid transformation is pearlitic transformation in Fe–C system. When austenite (γ or A) with a carbon content of 0.77% is cooled to the eutectoid temperature A_{r1}, it decomposes into a mixture of ferrite (α or F) and cementite (Fe_3C), with the reaction formula of $\gamma_{0.77\%C} \rightarrow \alpha_{0.0218\%C} + Fe_3C_{6.69\%C}$, where γ represents the high-temperature phase austenite with a face-centered cubic structure, α represents the low-temperature phase ferrite with a body-centered cubic structure, and Fe_3C represents the high-carbon phase cementite with a complex orthogonal structure.

The eutectoid mixture of ferrite and cementite is called pearlite (represented by P). The formation of pearlite involves two simultaneous processes: one is the composition redistribution to form high-carbon Fe_3C and low-carbon α phase through the diffusion of carbon atoms; the other is lattice reconstruction, in which the face-centered cubic structure of austenite converts into a body-centered cubic structure

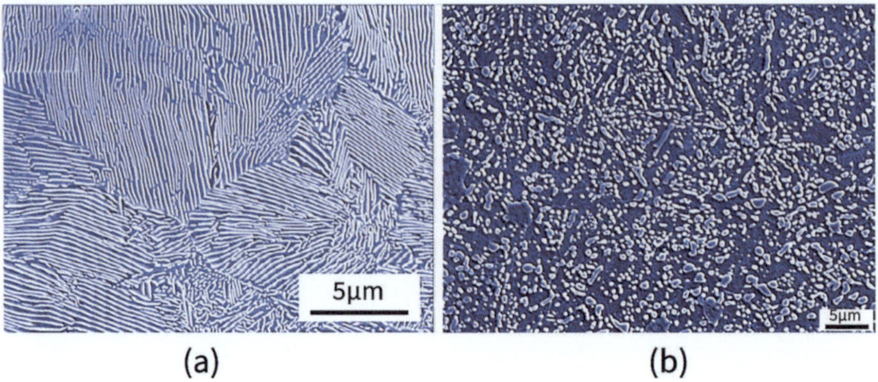

Fig. 10.9 Typical microstructure of pearlite: **a** lamellar pearlite and **b** spheroidal pearlite

α and Fe_3C with a complex orthogonal structure. There are two basic types of pearlite-type microstructures: lamellar pearlite and spheroidal pearlite, which are illustrated in Fig. 10.9. Lamellar pearlite can be obtained through isothermal cooling at elevated temperatures or through continuous cooling at a slow rate. Moreover, lamellar pearlite can convert to spheroidal pearlite (also known as granular pearlite) after spherical annealing.

Lamellar Pearlite

The lamellar pearlite structure may consist of various regions with different orientations of layers, where each region with the same orientation is called a pearlite cluster. The interlamellar spacing (λ) is the most significant characteristic of the pearlite structure, and it is defined as the distance between adjacent layers of ferrite and cementite. There are three types of pearlite structures based on the interlamellar spacing: ordinary pearlite (P), sorbite (S), and troostite (T), which have different interlamellar spacings ranging from 150~450 nm, 80~150 nm, and 30~80 nm, respectively. The size of the interlamellar spacing is influenced by the isothermal transformation temperature during the formation of pearlite. Lower transformation temperatures result in thinner interlamellar spacings, which can lead to higher strength but lower toughness due to the increased resistance to dislocation movement. Therefore, controlling the interlamellar spacing of pearlite is important in tailoring the mechanical properties of the steel for specific applications.

Spheroidal Pearlite

Spheroidal pearlite is a dual-phase structure consisting of spheroidal or granular carbides, typically cementite, that are dispersed and precipitated within a matrix of ferrite. The structure is usually obtained by a process called spheroidizing annealing, which involves heating the steel to a temperature just below the eutectoid point, holding it at that temperature for a period of time, and then cooling it slowly.

The resulting structure provides improved mechanical properties, such as increased strength, ductility, and toughness, compared to the lamellar pearlite structure. The size, quantity, shape, and distribution of carbides in the spheroidal pearlite structure are important factors that affect its structure and performance.

① Nucleation of Pearlite

In the nucleation process of pearlite, the driving force is the change in chemical free energy per unit volume, which can be represented by $\Delta G_V^{\gamma \to P} = G_V^P - G_V^\gamma$. To initiate nucleation, the steel must be cooled below the eutectoid temperature T_c, and the supercooling is defined as $\Delta T = T_c - T$, where T is the actual transformation temperature. The relationship between the driving force and supercooling can be expressed as:

$$\Delta G_V^{\gamma \to P}(T) = \frac{\Delta H_V^{\gamma \to P}(T_c)\Delta T}{T_c} \tag{10.21}$$

The term $\Delta H_V^{\gamma \to P}(T_c)$ refers to the latent heat of the phase transformation. The resistances to transformation are the elastic strain energy and interfacial energy.

Pearlite nuclei typically form on the original austenite grain boundaries due to fluctuations in energy, microstructure, and composition. However, if there is significant supercooling (ΔT) or the austenite composition is homogeneous, nucleation can occur inside the grains. The first phase to form during pearlite nucleation is known as the leading phase. Electron microscopy studies suggest that in the transformation of hypoeutectoid steel into pearlite, the leading phase is ferrite. In contrast, for eutectoid and hypereutectoid steels, the leading phase is cementite.

When Fe_3C is the leading phase in pearlite formation, it typically takes the form of small flakes to minimize elastic strain energy. As shown in Fig. 10.10, when the Fe_3C sheet forms, the carbon concentration in the γ at the Fe_3C/γ interface decreases due to the formation of high carbon concentration in Fe_3C. This decrease in carbon concentration in the γ phase triggers the formation of α sheets, leading to the development of the pearlite nucleus.

When a steel is cooled rapidly or at lower transformation temperatures, carbon diffusion is restricted, which favors the formation of lamellar pearlite. However, this type of pearlite has a high surface area and total interface energy. To reduce the total interface energy, the lamellar pearlite can transform into spheroidal pearlite by keeping the steel at an elevated temperature just below the A_{r1} line for an extended period of time. During this process, the carbides become spheroidized and distribute more uniformly, resulting in a lower total interface energy and improved mechanical properties.

② Growth of Pearlite

In the growth process of pearlite, carbon diffuses from the edges of ferrite lamellae to neighboring cementite lamellae, as well as from austenite to the tips of the

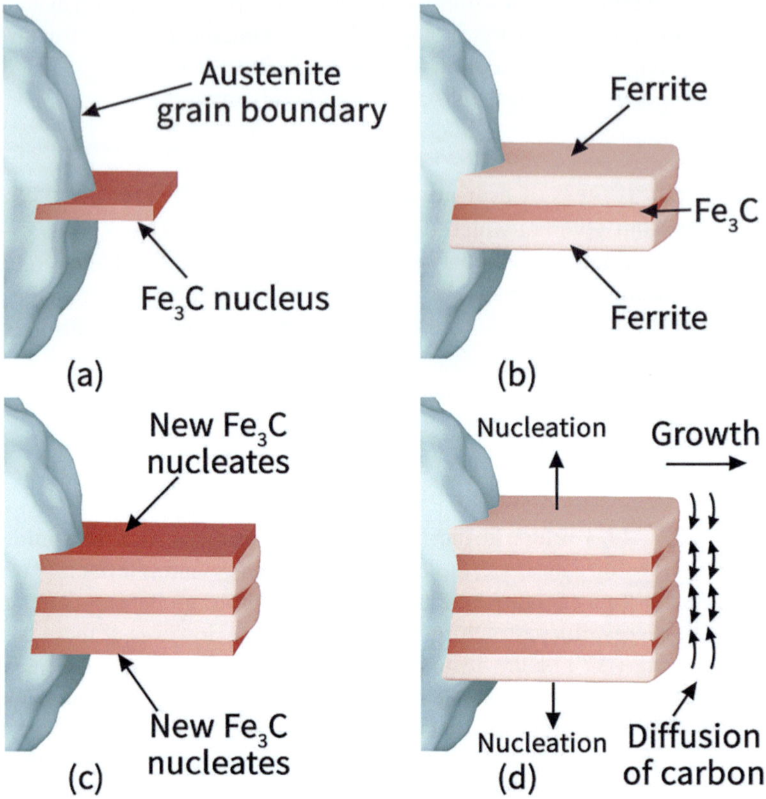

Fig. 10.10 Schematic diagram of nucleation and growth of pearlite

pearlite region. The driving force behind this diffusion is the concentration difference in the austenite around the interfaces ($C_{\gamma/\alpha}$—$C_{\gamma/\text{cem}}$), where $C_{\gamma/\alpha}$ and $C_{\gamma/\text{cem}}$ are the concentrations in austenite that are in equilibrium with ferrite and cementite, respectively. Figure 10.11 illustrates this process. In addition, because of the concentration difference of C_{γ}—$C_{\gamma/\text{cem}}$ and $C_{\gamma/\alpha}$—C_{γ} (C_{γ} is the carbon concentration far from the interface in austenite solution), carbon is also distributed in austenite.

The pearlite phase transformation is controlled by diffusion, and the rate of transformation is completely dependent on the nucleation rate and growth rate. The transformation process over time can be described by the Avrami equation.

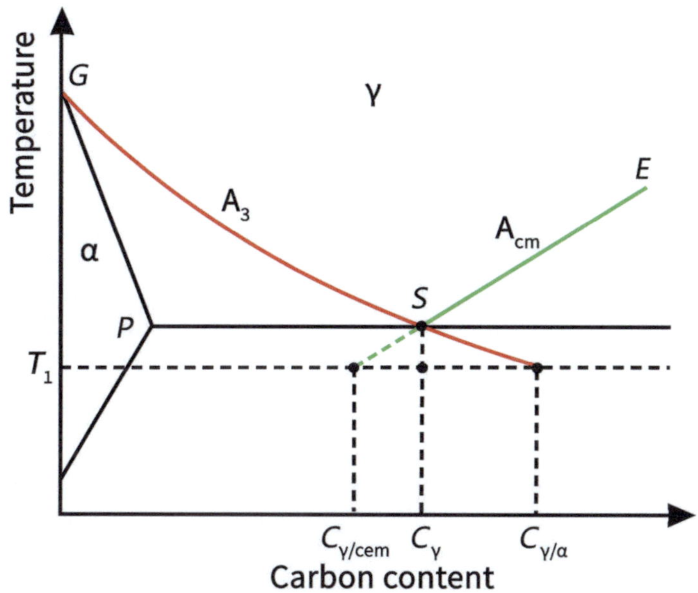

Fig. 10.11 Phase equilibrium concentration in iron-iron carbide phase diagram

③ Mechanical Properties of Pearlite

Lamellar Pearlite

Slow cooling of eutectoid steel leads to the formation of the lamellar pearlite structure, which can be divided into ordinary pearlite, sorbite, and troostite based on their layer spacing (λ). The performance of pearlite depends on λ, where decreasing λ by decreasing the transformation temperature leads to increased strength and hardness, along with improved plasticity and toughness. Among these three types of structures, troostite with the smallest λ exhibits the highest hardness, which can reach HRC38-43.

The matrix of pearlite is ferrite, which exhibits favorable plasticity and toughness, while the strength of pearlite mainly depends on cementite. The strengthening effect of cementite is not due to its own high hardness but rather by the phase interface. The interfaces between cementite and ferrite resist the movement of dislocations, thereby improving the strength and hardness of pearlite. Decreasing λ increases the phase interface area per unit volume and enhances the strengthening effect. At the same time, small λ (thin cementite sheet) produces a certain elastic deformation under stress, which helps improving the plasticity and toughness of pearlite.

Spheroidal Pearlite

Spheroidal pearlite has lower strength and hardness compared to lamellar pearlite, but it has higher plasticity and toughness. The mechanical properties of spheroidal pearlite are determined by the degree of spheroidization of the cementite. The more spheroidal the cementite, the better the mechanical properties of the spheroidal pearlite will be.

10.6 Diffusionless Phase Transformation

Diffusionless phase transformation, also known as displacive transformation, occurs when there is no diffusion of atoms during the phase transformation or diffusion is not the main process. Instead, the transformation happens through coordinated atomic displacements, where the atoms move less than one interatomic spacing and maintain their relationship with their neighbors. Examples of such transformations include β to ω transformation in Ti-Zn alloy and martensite transformation in steels.

In this section, the focus is on martensite transformation, which is a type of diffusionless phase transformation that occurs in steels when austenite is rapidly cooled below the critical quenching rate V_c, the minimum cooling rate required to obtain complete martensite. This transformation is crucial for material strengthening and is widely used in production, with the heat treatment process to obtain martensite structures known as martensite quenching.

Martensite is a term that describes any diffusionless transformation product in metals and alloys where the individual atoms move less than one interatomic spacing. If the cooling rate is fast enough to avoid diffusional phase transformation, martensitic transformation can occur in various metals and alloys. This type of transformation has been observed in many materials, including the $\beta \rightarrow \beta'$ transformation in Cu-Al-Ni alloy and the FCC → FCT transformation in In-Ti alloy.

10.6.1 Characteristics of Martensitic Transformation

① Diffusionless

The martensitic transformation can occur at low temperatures, such as in eutectoid steel where the transformation temperature ranges from 230 °C to −50 °C. Since atoms cannot diffuse at such low temperatures, it has been confirmed that the carbon content and ordered structure of martensite remain the same as that of the parent austenite phase. Despite the lack of diffusion, the transformation rate is extremely fast, even at low temperatures. For instance, Fe–C and Fe–Ni alloys can form martensite in as little as 0.05 to 0.5 ms at temperatures ranging from −20 °C to −195 °C.

The fast transformation rate observed during martensitic transformation cannot be explained by diffusion, as diffusion requires the movement of atoms over relatively large distances, which is impossible at low temperatures. Therefore, the diffusionless nature of martensitic transformation is considered its fundamental characteristic. Unlike diffusion-based transformations, the atoms migrate collectively into the martensite phase at the interface, with the migration distance not exceeding the interatomic spacing. This migration process is a result of the generation and movement of large numbers of defects, such as dislocations and stacking faults, which allow the crystal lattice to change rapidly and without the need for atomic diffusion. The fast transformation rate and the absence of atom diffusion are two key features that make martensitic transformation unique from other phase changes. While some experiments have shown that smaller carbon atoms can diffuse over short distances, this is not a common occurrence during martensitic transformation.

② Surface Relief Morphology

Surface relief is a distinct characteristic of the martensitic transformation that is closely related to the shape alteration during the transformation. This feature can be detected by examining the displacement of scratches made on a flat surface prior to the transformation. During the transformation, the initial horizontal surface of the austenite phase undergoes a shearing process that results in a change in its orientation, indicating that a shape alteration is associated with the transformation to martensite (see Fig. 10.12). The degree of surface relief can vary depending on factors such as the cooling rate, the composition of the material, and the crystallographic orientation of the grains. As a result, the surface relief feature can provide important information about the microstructure and properties of the material, making it a useful tool for understanding the martensitic transformation and its effects on materials.

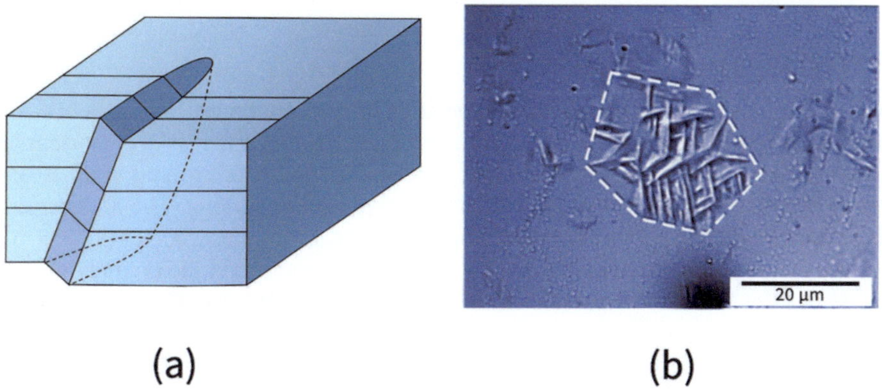

Fig. 10.12 **a** Schematic of surface relief where a martensite plate intersects the surface **b** Surface relief formed by transformed martensite plates during isothermal transformation at 233 K on a polished surface of a 12Cr–9Ni–4Mo–2Cu steel

③ Habit Plane and Orientation Relationship

The habit plane of martensite refers to the crystallographic plane that separates the transformed martensite phase from the parent austenite phase during the martensitic transformation in steels.

The habit plane is determined by the crystal structure of both the parent austenite and the martensite phase, as well as the transformation temperature and composition of the alloy. In general, the habit plane has a specific orientation relationship with the parent austenite crystal, and this relationship can vary depending on the specific alloy composition. For low carbon steels with carbon content less than 0.6%, the habit plane of martensite is usually $\{111\}_\gamma$. For medium carbon steels with carbon content between 0.6% and 1.4%, the habit plane is typically $\{225\}_\gamma$. And for high carbon steels with carbon content greater than 1.4%, the habit plane can change to $\{259\}_\gamma$. However, it's important to note that the specific habit plane can also be influenced by factors such as cooling rate, alloying elements, and deformation history.

Due to the coherent interface, martensite and the austenite matrix exhibit a definite orientation relationship. When the close-packed planes or close-packed directions of the atoms in both phases are either parallel or nearly parallel to each other, the phase boundary energy is low. The observed orientation relationships include the following:

1. Fe-1.4%C alloy exhibits the K-S (Kurdjumov–Sachs) orientation relationship: $\{111\}_\gamma // \{110\}_{\alpha'}$, $<110>_\gamma // <111>_{\alpha'}$.
2. Below -70°C, Fe-30%Ni alloy exhibits the N-W (Nishiyama-Wassermann) orientation relationship: $\{111\}_\gamma // \{110\}_{\alpha'}$,$<112>_\gamma // <110>_{\alpha'}$.
3. Fe-0.8%C-22%Ni alloy displays the G-T (Greniger-Troiano) orientation relationship, which is similar to the K-S relationship but with a slight deviation of 1°~2 °.

④ M_s Temperature

Martensite formation occurs within a specific temperature range. The austenite-to-martensite transformation starts at a specific temperature, known as M_s (martensite transformation start temperature), after significant supercooling, and ends at M_f (martensite transformation end temperature) upon further cooling. The M_s temperature can vary over a wide temperature range from as high as 500 °C ($\approx$ 539 °C for nearly pure iron) to below room temperature, depending on the composition of the steel. Most elements decrease the M_s, except for Co and Al. The large specific volume of martensite results in volume expansion, which stabilizes untransformed austenite even if the temperature falls below M_f. This phenomenon is referred to as the incompleteness of martensite transformation.

⑤ Reversibility

The martensitic transformation is theoretically reversible. As shown in Fig. 10.13, quenching austenite to M_s initiates the martensite transformation, which ends at

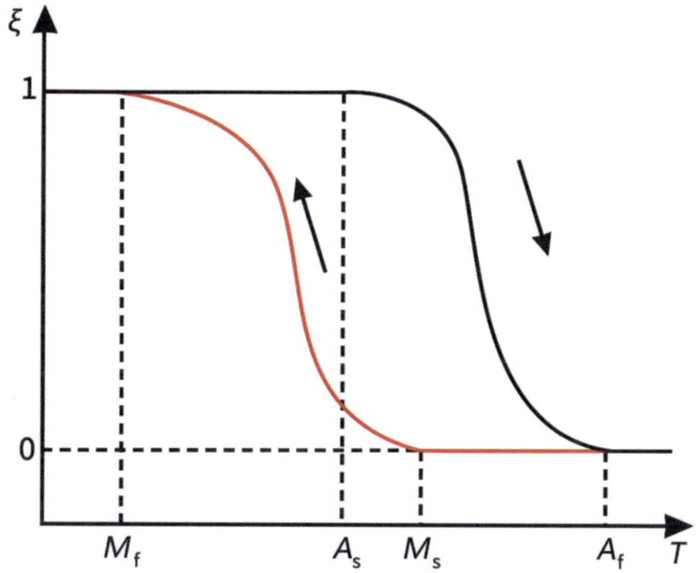

Fig. 10.13 Temperature dependence of the martensite fraction (ξ)

M_f. Conversely, heating martensite causes a transformation from martensite to austenite when the temperature is above A_s (austenite transformation start temperature). As the temperature continues to rise, all martensite will be transformed into austenite when it reaches the austenite transformation end temperature (A_f). This phenomenon is reversible and observed in shape-memory alloys, such as copper–aluminum-nickel and nickel-titanium alloys. Notably, these reversible phase transformations are both reversible diffussionless transformations.

The shape memory effect is a phenomenon in which a material exhibits a temperature-induced phase transformation with reverse deformation. At low temperatures, martensite is thermodynamically favored, while austenite is stable at elevated temperatures. The transformation from austenite to martensite creates internal strain energy in the interface due to different lattice sizes and symmetry, resulting in the formation of many twins in the martensitic phase, a process called "self-accommodating twinning." These twins provide an easy path for deformation when the martensite is under stress, and all atoms are in the same position relative to nearby atoms. When the temperature is raised, all atoms rearrange themselves to the austenite structure, which has the same macroscopic shape as the pre-deformation shape, resulting in the observation of shape-memory behavior.

10.6.2 Thermodynamics of Martensitic Transformation

① The Driving Force

The driving force for martensite transformation is the chemical free energy change, which is expressed as $\Delta G_V^{\gamma \to \alpha'} = G_V^{\alpha'} - G_V^{\gamma}$. As the temperature decreases (supercooling increases), the driving force for the transformation also increases. The temperature at which the free energy becomes equal is known as the equilibrium temperature T_0. However, the formation of martensite is hindered by various resistances, including interfacial energy, strain energy, and defect formation energy. Due to the low yield strength of the austenite surrounding the martensite, a small amount of plastic deformation occurs in the austenite, leading to an increase in strain energy.

Martensite transformation is a process that occurs only when the temperature is below the martensite start temperature (M_s). The M_s is the critical temperature at which the free energy change reaches the minimum driving force required for the austenite-to-martensite phase transformation to take place. The lower the M_s value, the larger the transformation resistance, and the more driving force is required for the transformation to occur. It's worth noting that the equilibrium temperature (T_0) is a key factor in determining the M_s value. When T_0 is constant, a lower M_s value indicates a greater transformation resistance, which means more energy is required to overcome the resistances. The driving force can be expressed as:

$$\Delta G_V^{\gamma \to \alpha'} = \Delta S_V^{\gamma \to \alpha'} (T_0 - M_s) \tag{10.22}$$

② Factors Affecting M_s

Chemical Composition

The composition of the alloy plays a significant role in determining the M_s and M_f temperatures. The presence of interstitial atoms, such as carbon and nitrogen, has a significant effect on the M_s temperature. As the carbon content increases, the M_s temperature decreases due to the increased resistance of the martensitic transformation. On the other hand, some alloying elements such as Co and Al tend to increase the M_s temperature. However, most alloying elements, except for these few, reduce the M_s temperature. These replacement atoms usually have less impact on the temperatures than the interstitial atoms, which have a more significant impact on the martensitic transformation process.

Plastic Deformation

Some materials exhibit a stress-induced martensitic transformation when subjected to plastic deformation above the M_s temperature, resulting in an increase in the M_s temperature to become the M_d temperature. The M_d temperature is the start temperature of the stress-induced martensitic transformation. The degree of plastic deformation and the temperature at which it occurs influence the stress-induced transformation.

Austenitizing

The martensitic transformation process is influenced by the heat treatment procedure. Specifically, austenitizing is a process in which the alloy is subjected to temperatures above the upper critical temperature (A_{c3} or A_{cm}), marking the crucial point at which the microstructure of the alloy changes from ferrite and pearlite to austenite, and held to ensure complete transformation. Generally, the higher the austenitizing temperature or the longer the holding time, the more carbon and alloying elements are dissolved into the austenite, which lowers the M_s temperature. However, other factors such as the cooling rate, homogenization, and soaking time also affect the austenitizing process and, in turn, the M_s temperature.

③ Nucleation of Martensite

It has been experimentlly confirmed that nuclei in martensite form through heterogeneous nucleation. To measure the fraction of transformed martensite, Fe–Ni–C alloy wafers with different size are austenitized and cooled to low temperature. Results are shown below:

1. The fraction of transformed martensite increases with increasing wafer size, indicating heterogeneous nucleation. Because the large wafers contain more defects.
2. If the wafer size is the same, the lower transformation temperature leads larger fraction of transformed martensite. Nucleation occurs on defects with high energy, whereas decreasing the temperature, defects with low energy can also serve as nucleation sites due to increased driving force.

Martensite nucleation occurs at various places in the parent phase, including grain boundaries, subgrain boundaries, and dislocations. Zener explains that within the FCC structure, a full dislocation occurring on the close-packed plane undergoes a decomposition into two partial dislocations, and the stacking fault region situated between these partial dislocations can transform into a BCT (body-centered tetragonal) structure under appropriate conditions, thus providing an explanation for the FCC → BCT martensitic transformation. This mode of nucleation has been confirmed in some alloys.

10.6.3 Crystal Structure of Martensite in Steel

When steel is rapidly cooled through quenching, the carbon atoms that were previously in equilibrium at elevated temperatures become solid-dissolved in α-Fe, resulting in a supersaturated state. The resulting microstructure is known as martensite, which consists of supersaturated α-Fe with interstitial carbon. As shown in Fig. 10.14, there are various possible tetrahedral and octahedral positions in α-Fe, but most carbon atoms occupy the center of octahedral holes, even though their radius is larger than the hole size. This is because the elastic strain energy caused by the carbon atoms in the oblate octahedron is relatively small. However, the introduction of carbon inevitably leads to a strong asymmetric distortion of the α-Fe lattice, which causes the short axis of the octahedral hole to extend by 36% and the long axis to shrink by 4%. As a result, martensite with a BCT (body-centered tetragonal) structure is formed. The distorted Fe–C lattice in martensite has been confirmed by X-ray diffraction, and the *c*/*a* ratio of the lattice increases with increasing carbon content. These measurements suggest that there is a certain long-range order in the distribution of carbon interstitials.

10.6.4 Microstructure of Martensite

The microstructure of martensite is related to the chemical composition and transformation temperature. Based on their microstructure, martensite can be categorized into two types: lath martensite and plate martensite.

① Lath Martensite

Lath martensite is produced when low or medium carbon steel is quenched, and it is characterized by a substructure with high-density dislocations, also known as low-carbon martensite. Figure 10.15 illustrates the microstructure of lath martensite, which is composed of packets, blocks, and laths. The microstructure is largely influenced by the primary austenite grains, which typically consist of 3 ~ 5 packets with various orientations. A packet is made up of one or several blocks separated by a high-angle boundary, with each block containing a mass of laths that are approximately parallel to each other. The laths contain numerous high-density dislocations with densities ranging from approximately $(0.3 \text{ to } 0.9) \times 10^{12}\ \text{cm}^{-2}$. Furthermore, a continuous retained austenite film can be observed between the martensite laths.

② Plate Martensite

Plate martensite is a common microstructure that forms when medium or high carbon steel is quenched. Unlike lath martensite, the internal substructure of plate martensite mainly consists of twins. Plate martensite formation is often accompanied by a phenomenon known as "burst", where the nucleation of one martensite plate induces a sequence of plates. This burst may occur due to stress concentrations that develop when the first plate reaches a grain boundary or another martensite plate. Figure 10.16 show that each grain of plate martensite consists of lenticular martensite and retained austenite. The martensite plate that forms first generally crosses the entire austenite grain, while the sizes of the martensite plates that form subsequently decrease. Each plate has a clear "midrib" in the center, which is fully twinned as thin plate martensite. The surrounding region is the "twinned region" and contains many twins spaced approximately 5 nm apart. The edge of the plate has a complex dislocation configuration that helps relieve some of the elastic stress generated by twin deformation.

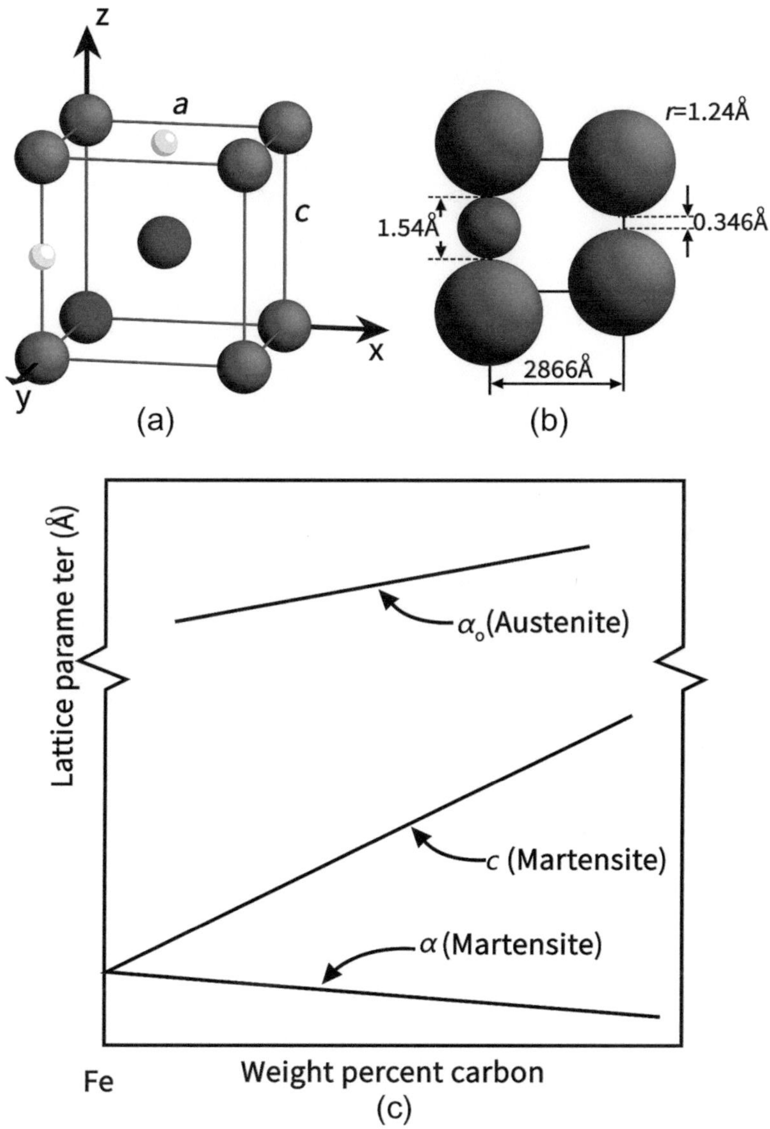

Fig. 10.14 **a** Possible positions for interstitial atoms in the BCC lattice. **b** The distortion caused by carbon atom. **c** Carbon content dependence of a and c

10.6.5 Mechanisms of Martensitic Transformation

The martensite transformation is diffusionless without any interchange in the position of neighbor atoms. Several mechanisms and shear models have been

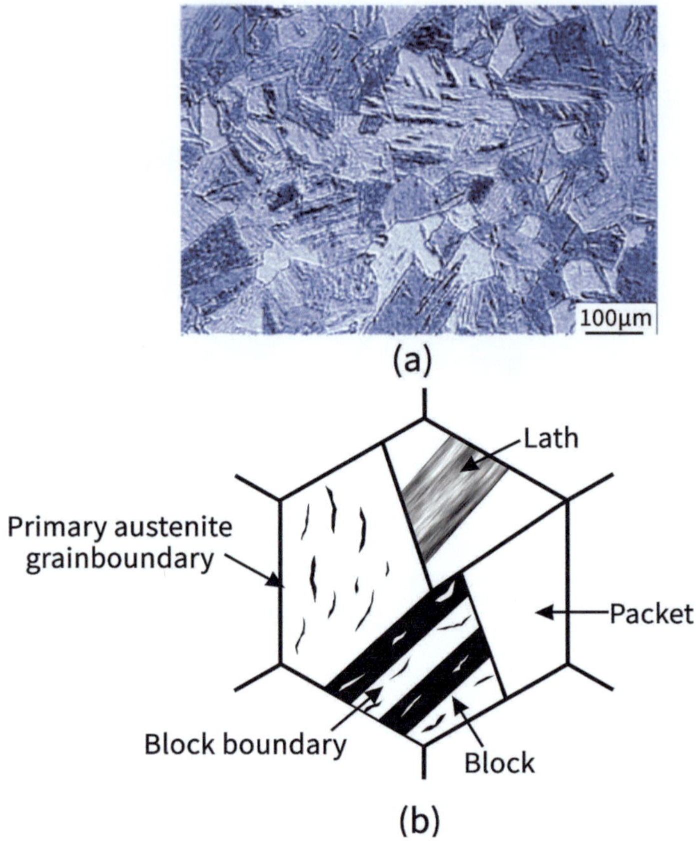

Fig. 10.15 **a** Microstructure of lath martensite. **b** Sketch diagram of lath martensite

developed to explain this transformation, including the B (Bain), K-S (Kurdjurmor-Sachs), G-T (Greninger-Troiano) and N (Nishiyama) models.

To explain the martensite transformation, models must meet several requirements, including the crystal structure, orientation relationship between martensite and austenite, habit plane, and surface relief. This section does not discuss specific shear models but instead qualitatively analyzes the transformation mechanism in terms of slip and twinning, as shown in Fig. 10.17. The internal substructure of martensite contains various defects such as dislocations and twins. These defects are the products of slip or twin during the transformation process, which may be caused by the following two kinds of shear processes:

1. The crystal structure of the matrix transforms into the martensite phase crystal structure of BCT, which is the first shear process of the martensite phase transformation. However, upon calculation, it is found that the phase transformation

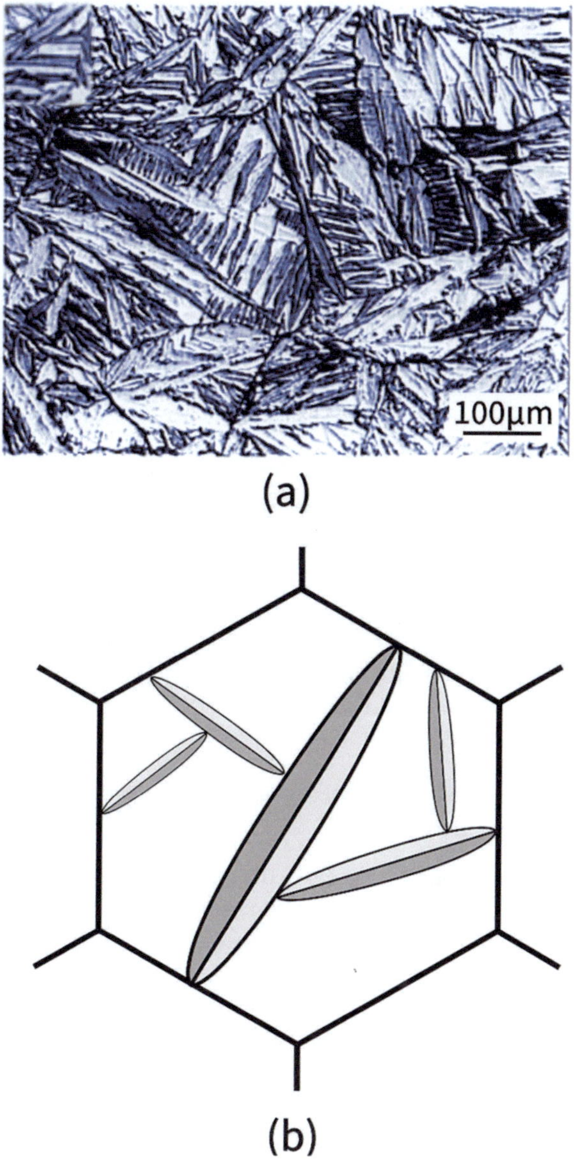

Fig. 10.16 **a** Microstructure of plate martensite. **b** Sketch diagram of plate martensite

strain produced by one shear process is inconsistent with the actual value, and fails to meet the requirements of the martensite substructure.

2. Employing crystal slip or twining as an auxiliary strain, the transformation strain can be consistent with the actual measured value without changing the formed martensite structure. This is the secondary shear of the martensite

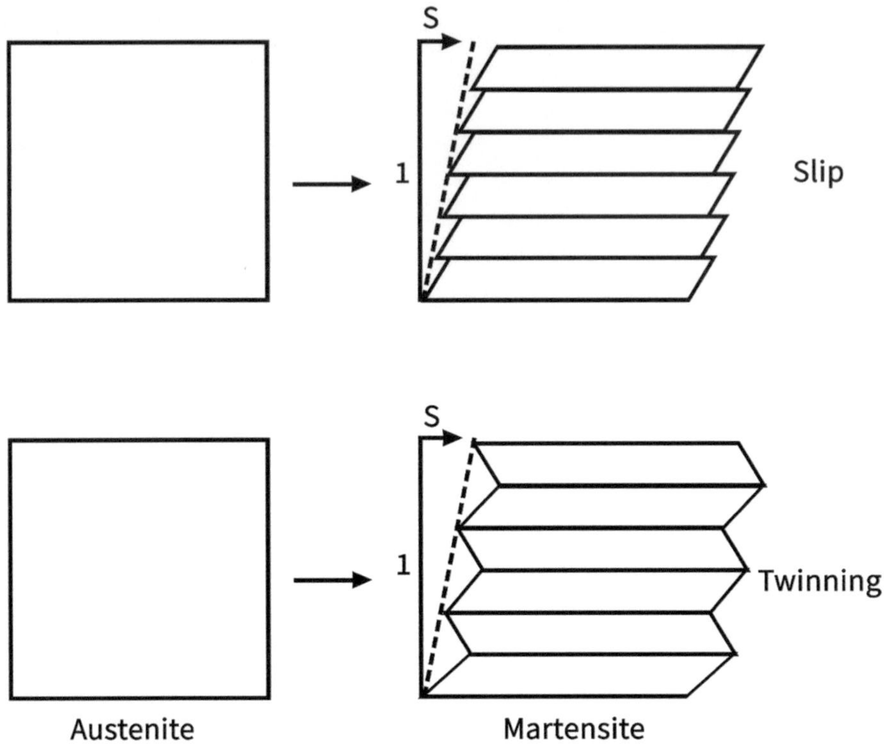

Fig. 10.17 Schematic diagram of slip and twinning in martensitic transformation

transformation process. If the lattice slips, the resulting substructure is a dislocation, and the resulting martensite is lath martensite. If the lattice movement is twinning, the resulting substructure is a twin, resulting plate martensite.

10.6.6 Mechanical Properties of Martensite

Martensite's high strength and hardness depend largely on the content of interstitial solute atoms, such as C and N. Figure 10.18 shows the hardness response of quenched steel with varying carbon contents. At low carbon content, the hardness of martensite (line 1) increases rapidly with an increase in carbon content. However, the hardness tends to plateau beyond 0.4% carbon content. In the actual quenching process (line 2), as the carbon content increases, the M_S point of the steel decreases, resulting in an increase in retained austenite after quenching, which leads to a decrease in the hardness of the quenched steel.

To understand how martensite gains its strength, we can look at the substructure of the material and the role of interstitial solute atoms. Firstly, interstitial atoms such as C and N are partially or completely dissolved in martensite, resulting in a distorted dipole stress fields around the heteroatoms. This stress field

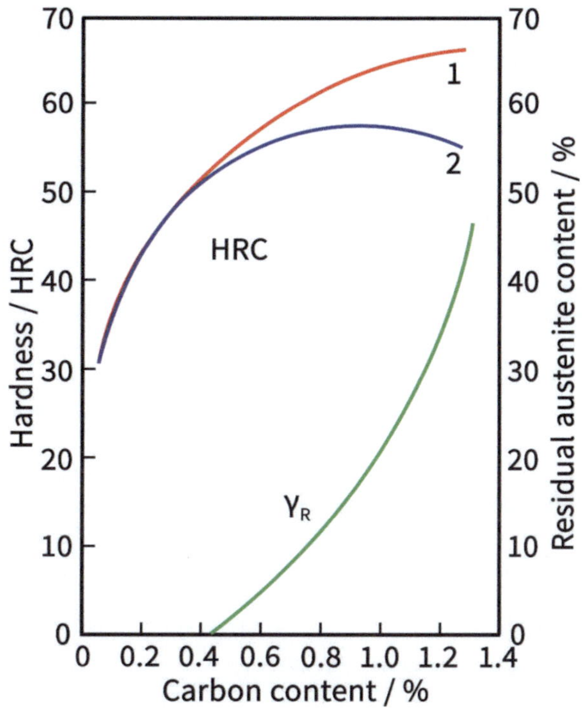

Fig. 10.18 Carbon content dependence of hardness of a quenched steel, 1: hardness of Martensite, 2: hardness of a quenched steel

interacts strongly with dislocations, leading to a significant reduction in dislocation mobility due to the formation of numerous solute atom clusters. This is known as solid solution strengthening, an important mechanism for martensite strengthening. However, after the carbon content exceeds 0.4%, the effect of further strengthening the martensite significantly weakens. This may be because the C atoms are too close together, resulting in the mutual cancellation of the distorted dipole stress fields and a reduction in stress. Secondly, high-density defects, such as dislocations, twinned crystals, and stacking faults, further contribute to the high strength and hardness of martensite. However, due to the mobility of dislocations, lath martensite has lower strength and hardness compared to plate martensite. On the other hand, lath martensite exhibits better plasticity and toughness than plate martensite.

10.7 Transitional Phase Transformation

Transition phase transformation, also known as semi-diffusional phase transformation, includes massive transformation and bainite transformation and mainly occurs in the medium temperature range. Unlike diffusional phase transformation, where atoms move through long-range diffusion, or diffusionless phase transformation, where atoms move cooperatively and homogeneously between phases, transitional phase transformation involves atom movement at the phase interface over a short distance without changing chemical composition as in massive transformation, or where larger atoms do not diffuse with smaller atoms as in bainite transformation.

The bainite transformation in steel is a typical transitional phase transformation. When austenitized steel is cooled, either through continuous cooling or isothermal cooling, to the middle temperature range between pearlite and martensite transformation, bainite transformation occurs. During this transformation, iron atoms cannot diffuse, but smaller carbon atoms can still diffuse. The transformation can be expressed by the reaction formula: $\gamma \rightarrow \alpha + Fe_3C$, and although it has a similar chemical composition to pearlite, its microstructure is distinct.

10.7.1 Basic Features of Bainite Transformation

In 1930, E. C. Bain discovered the bainite transformation, which was later divided into upper bainite and lower bainite by R. F. Mehl in 1939. Over the last few decades, various types of bainite have been discovered in different steels, all of which exhibit certain common characteristics, as listed below:

1. The temperature range in which the bainite transformation takes place has two boundary points: the starting transformation temperature, known as B_s, and the end transformation temperature, called B_f. Generally, the temperature range above the bainite transformation is known as the pearlite transformation zone, while the temperature range below it is called the martensite transformation zone. Within the temperature range of B_s to B_f, the upper bainite transformation occurs in the upper part, and the lower bainite transformation occurs in the lower part. In the case of eutectoid steels, the temperature range for bainite transformation falls between 240 °C and 550 °C.
2. Bainite is a mixed structure comprising supersaturated ferrite and carbides. The morphology of bainite can be classified as either upper or lower bainite, depending on the transformation temperature. Upper bainite forms at higher transformation temperatures, and has a finer lath-like ferrite morphology with a lower supersaturation of carbon. The carbides tend to precipitate along the boundaries of the ferrite laths due to their high diffusion ability. Lower bainite forms at lower transformation temperatures, and has a coarser, flake-like morphology with a higher supersaturation of carbon. The reduced carbon diffusion rate at lower temperatures means that carbides are more likely to precipitate inside the ferrite flakes.
3. Bainite transformation occurs in an intermediate temperature range between the austenite-ferrite transformation temperature and the austenite–martensite transformation temperature. At this temperature range, the diffusion of iron and alloying elements atoms is significantly suppressed, and only small carbon atoms can diffuse through the austenite matrix. This leads to a slower transformation rate compared to the martensitic transformation. The rate of upper bainite transformation is mainly dependent on the slower diffusion of carbon

in austenite, while the rate of lower bainite transformation mainly depends on the slower diffusion of carbon in ferrite. The denser FCC structure of austenite results in a much smaller diffusion coefficient for carbon compared to the BCC structure of ferrite. At higher temperatures, carbon can diffuse from ferrite into austenite, resulting in the formation of finer microstructures for upper bainite, resulting in the formation of upper bainite with a finer microstructure. Conversely, at lower temperatures, the slower diffusion of carbon in ferrite leads to the formation of lower bainite with a coarser microstructure.

4. Bainite transformation follows a process of nucleation and growth. During the incubation period, concentration fluctuations caused by carbon diffusion in austenite lead to the formation of ferrite nuclei in carbon-poor regions. Once the ferrite nucleus exceeds the critical nucleus size, it grows, and during the growth process, supersaturated carbon atoms continue to diffuse from the ferrite to the austenite, resulting in the precipitation of carbide particles in the ferrite bars or within the ferrite.
5. In bainite transformation, the redistribution of iron atoms takes place through homogeneous shear that is parallel to the habit plane.

10.7.2 Microstructure of Bainite

The structure of bainite depends on the chemical composition and transformation temperature. There are two basic types of bainite: upper bainite and lower bainite.

① Upper Bainite

At high temperatures ranging from 350~550 °C, upper bainite forms, consisting of needle- or lath-shaped ferrite with carbides present between the laths, as shown in Fig. 19a. With increasing carbon content in the steel, the carbides between the ferrite laths increase, and the shape gradually changes from granular or short rods to continuous rods. In cases where the carbon content is high enough, carbides can also precipitate inside the ferrite lath. The substrate structure of ferrite in upper bainite consists of high-density dislocations, with densities ranging from approximately 10^8~10^9 cm^{-2}, which are lower than those of lath martensite with the same carbon content.

Due to the high transformation temperature, the driving force for the bainite transformation is small. Therefore, ferrite nuclei first form at the austenite grain boundaries and then grow into the grains side by side. The slower diffusion rate of carbon in austenite compared to ferrite leads to the gradual accumulation of carbon between the ferrite laths. As the carbon concentration in the enrichment region increases, carbides eventually precipitate between the ferrite laths, forming an upper bainite structure. This microstructure is characterized by a fine and uniform distribution of ferrite and carbides (Fig. 19c), which leads to a good combination of strength and toughness in the steel.

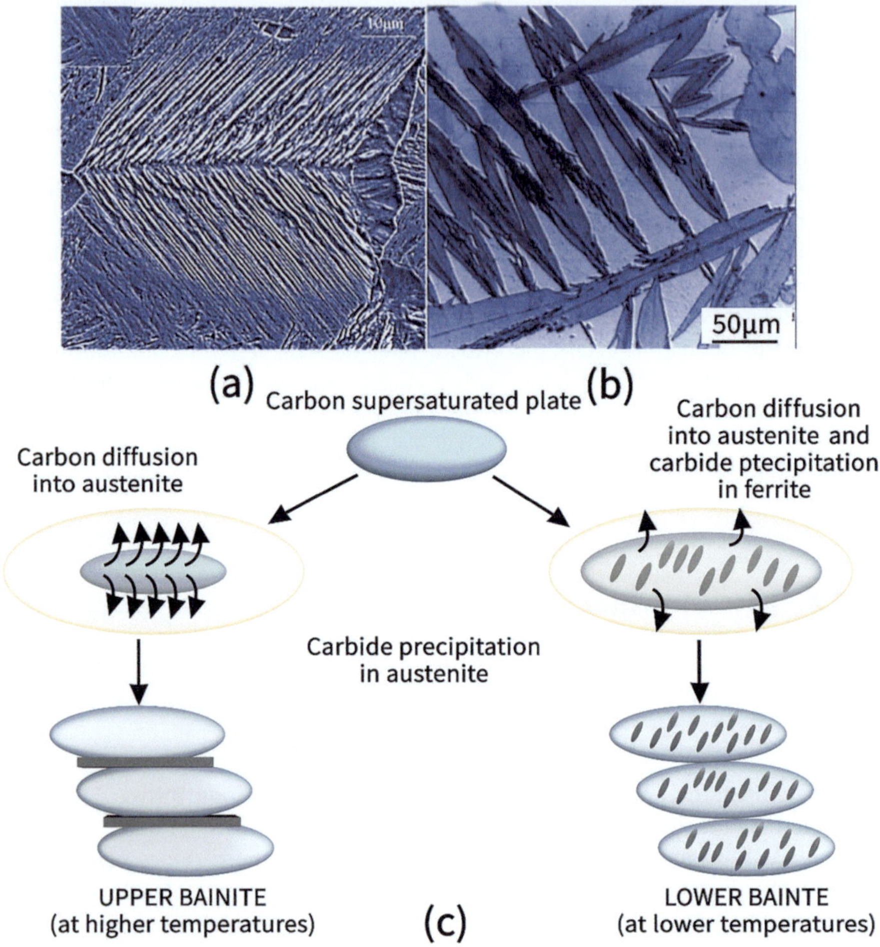

Fig. 10.19 **a** Microstructure of upper bainite. **b** Microstructure of lower bainite. **c** Sketch diagram of forming of bainite

② Lower Bainite

The formation of lower bainite occurs in the lower part of the bainite transformation temperature range, which typically ranges from approximately 240~350 °C for medium- and high-carbon steels. The microstructure of lower bainite is characterized by lenticular or plate-like ferrite with carbides precipitated inside (Fig. 19b), and the substrate structure of ferrite is also high-density dislocations, with a higher density than that of upper bainite.

The relatively low transformation temperature of lower bainite makes it difficult for carbon atoms to diffuse in austenite. However, they are able to move in ferrite, causing carbon atoms to segregate and disperse on certain crystal planes in the ferrite plate and ultimately forming a lower bainite structure, as depicted in Fig. 10.19c.

10.7.3 Mechanical Properties of Bainite

The mechanical properties of bainite are largely determined by its microstructure. The relative amounts and distribution of ferrite and carbide in the microstructure can have a significant impact on the strength, ductility, toughness, and other mechanical properties of the bainite. For example, a finer and more uniform distribution of carbides in the ferrite matrix can lead to improved strength and toughness, while large, irregular carbides can act as stress concentrators and reduce the material's toughness. Similarly, a higher proportion of ferrite can lead to increased ductility, while more carbide can result in a higher hardness and strength.

① Influence of Ferrite

The mechanical properties of bainite are influenced by the size, shape, and distribution of ferrite.

Tiny ferrite improves the strength, hardness, toughness, and ductility of bainite. The size of ferrite is affected by the austenitizing temperature, with lower austenitizing temperature leading to smaller ferrite sizes during bainite transformation. The shape of ferrite also plays a role in the properties of bainite, with ferrite strips or plates exhibiting higher strength and hardness compared to bulk-shaped ferrite. The morphology of ferrite changes from bulk and strip to flake as the temperature decreases. Additionally, lowering the bainite transformation temperature increases the supersaturation of ferrite and dislocation density, which in turn increases the strength and hardness.

② Influence of Carbide

When the carbide size is constant, increasing the carbon content in the steel results in higher strength and hardness of the bainite, but lower ductility and toughness. This is because higher carbon content leads to more carbide precipitation, which impedes the movement of dislocations and makes it more difficult for the material to deform plastically, resulting in reduced ductility and toughness.

On the other hand, when the carbon content is constant, lower temperature leads to more homogeneous carbide dispersion, which enhances the strength and hardness of the material. Additionally, the presence of granular carbide can improve both plasticity and toughness, likely due to the fact that the fine carbide particles can act as barriers to dislocation motion without producing significant stress concentrations.

In terms of the microstructure of bainitic steel, upper bainite exhibits coarser ferrite and carbides, particularly short rod carbides that are discontinuous between ferrite laths. This microstructure is susceptible to the formation of microcracks along the ferrite laths, which can lead to a significant decrease in plasticity and toughness. In contrast, lower bainite has a finer carbide dispersion that precipitates on the ferrite matrix. This microstructure results in high strength, hardness, plasticity, and toughness, likely due to the more homogeneous distribution of carbides, which impedes dislocation motion more uniformly and reduces the likelihood of stress concentrations.

10.8 Isothermal and Continuous Cooling Transformation Diagram

10.8.1 Isothermal Transformation Diagram

Isothermal transformation diagrams, also known as time–temperature-transformation (TTT) diagrams or C curve, are a graphical representation of temperature as a function of time. They were first developed by Davenport and Bain in 1930 to investigate the transformation kinetics of steel alloys. These diagrams were a significant breakthrough as they provided a simple and easily understandable way to visualize phase transformation processes. TTT diagrams are constructed experimentally by measuring the transformed volume fraction as a function of time at various temperatures, and they are useful for understanding the isothermal phase transformations that occur in materials and for controlling their microstructure and properties.

Consider the austenite-to-pearlite transformation in iron-carbon alloy as shown in Fig. 10.20, the upper portion is the S-shaped curve representing the transformed fraction of pearlite as a function of the logarithm of time at 675 °C. By obtaining S-shaped curves at different temperatures, we can present the time and temperature dependence of the transformed fraction of pearlite in a single diagram as shown in the bottom part. The diagram shows two solid C-shaped curves that indicate the time required for the initiation and completion of the transformation at each temperature. The dashed line refers to 50% of transformation completion. According to Fig. 10.20, at temperatures above the eutectoid temperature (727 °C), only austenite will exist. Only if the alloy is cooled to below 727 °C, the austenite-to-pearlite transformation takes place. The left side of the initiation curve represents the austenite zone, where only austenite is present. The right side of the finish curve represents the pearlite zone, where only pearlite exists. The region between the initiation and finish curves is the transformation zone, where austenite and pearlite coexist. The time required for the initiation and completion of the transformation is temperature-dependent, and the shorter the time, the higher is the rate of transformation.

Figure 10.21 is the complete isothermal transformation diagram for an iron-carbon alloy of eutectoid composition, in which the austenite-to-pearlite and austenite-to-bainite transformations are both presented. In contrast to the iron-iron carbide phase diagram, the TTT diagram can also represent the nonequilibrium phase, martensite. The transformed fraction of the austenite-to-martensite transformation is indicated by the horizontal lines labeled M (start), M (50%), and M

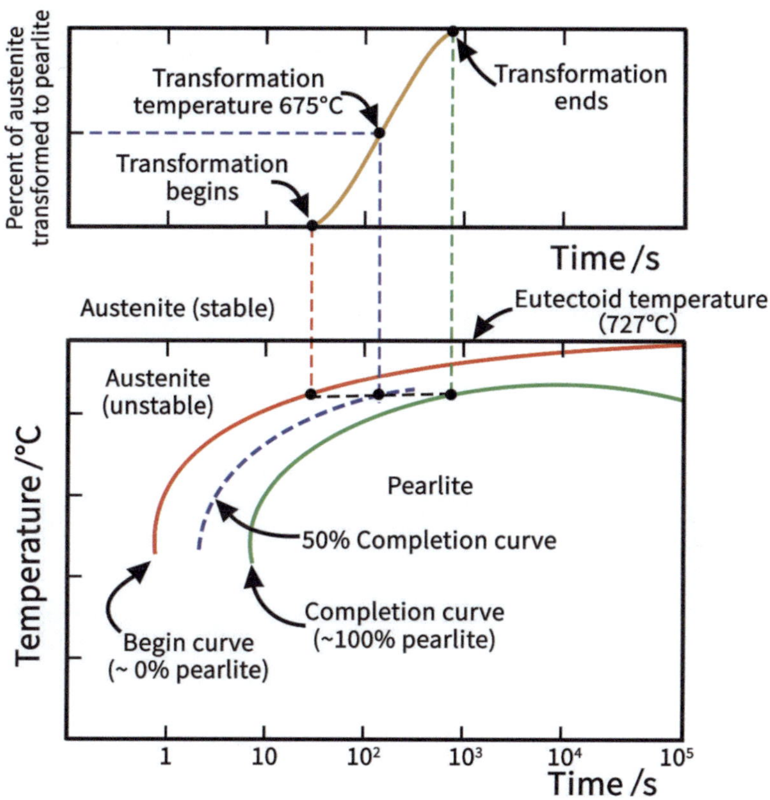

Fig. 10.20 The TTT diagram for austenite-to-pearlite transformation

(90%). These lines demonstrate that the martensitic transformation is not dependent on time, only on temperature. The instantaneous and diffusionless nature of the martensitic transformation suggests that it is an athermal transformation.

10.8.2 Continuous Cooling Transformation Diagram

Heat treatments for steel alloys often involve continuous cooling a specimen to room temperature, which can be described using a continuous cooling transformation (CCT) diagram.

TTT diagrams are valid for isothermal conditions, and to obtain CCT diagrams, TTT diagrams should be modified. In the case of continuous cooling, the start and end times for transformation are delayed, causing the curves on the TTT diagram to shift to lower temperatures and longer times, as depicted in Fig. 10.22. A figure containing such modified curves is called CCT diagram. It is noted that

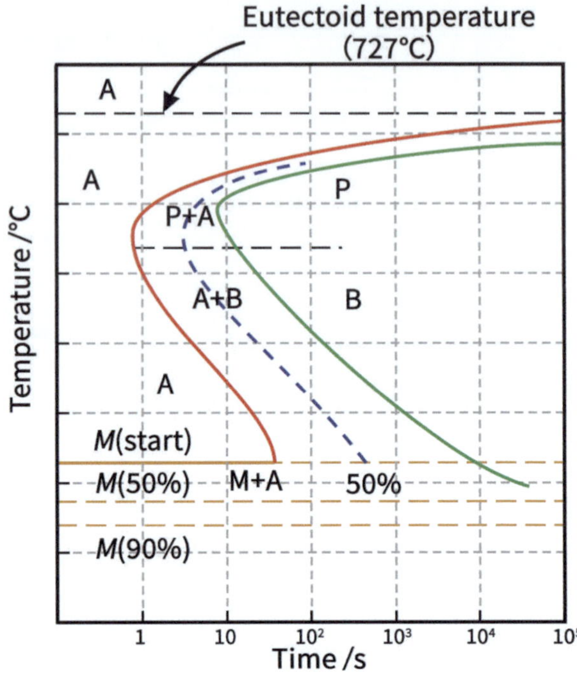

Fig. 10.21 The complete isothermal transformation diagram for an iron-carbon alloy of eutectoid composition. A: Austenite; B: Bainite; M: Martensite; P: Pearlite

the martensitic transformation occurs at the same temperatures for both isothermal and continuous cooling conditions.

The CCT diagram for an iron-carbon alloy of eutectoid composition is depicted in Fig. 10.23, which shows the transformation from austenite to pearlite as occurring between the initiation and finish curve, similar to the TTT diagram. The diagram also highlights the existence of a critical cooling rate, marked with a red dashed line, which represents the minimum cooling rate required to transform austenite to 100% martensite. If the applied cooling rate is greater than the critical rate, only martensite will exist. Conversely, low cooling rates, which are less than 35 °C/s, will result in 100% pearlite. In between, both pearlite and martensite will coexist.

Carbon and other alloying elements (chromium, nickel, molybdenum, manganese, etc.) can shift the curves of CCT diagram to the right side, thus decreasing the critical cooling rate, which promote the formation of martensite and totally martensite structure can be obtained over a larger range of cooling rates. Conversely, it is difficult to form martensite in iron-carbon alloys with less than 0.25% carbon, due to the need for a high cooling rate.

In summary, isothermal and continuous cooling transformation diagrams can be regarded as phase diagrams with introduction of time. These diagrams can predict the evolution of phase and microstructure during constant-temperature and continuous-cooling heat treatments.

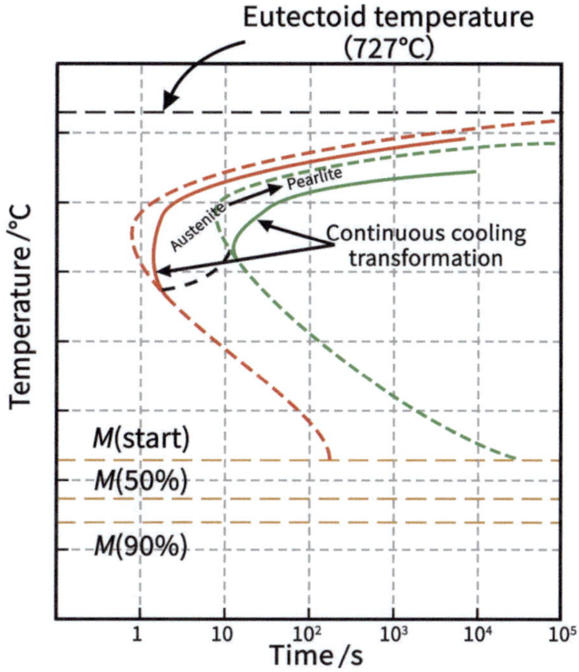

Fig. 10.22 Superimposition of TTT and CCT diagrams for an iron-carbon alloy of eutectoid composition

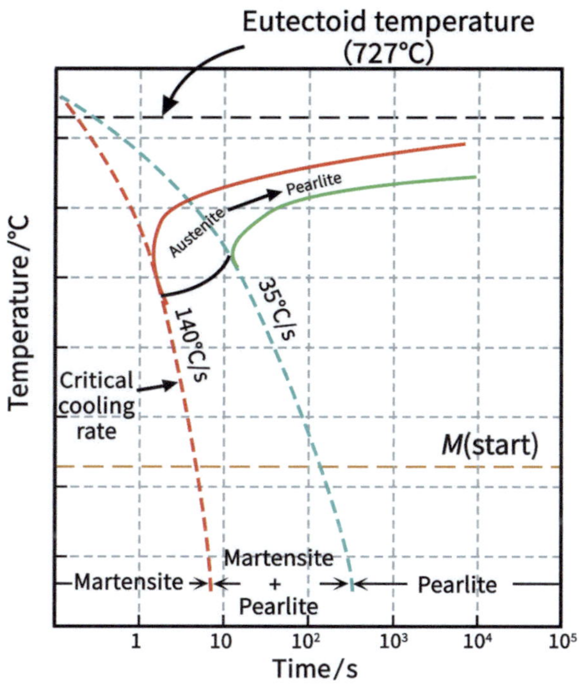

Fig. 10.23 CCT diagram and two superimposed cooling curves for an iron-carbon alloy of eutectoid composition

10.9 Questions

Q10.1 What is the difference between first and second order phase transformation?

Q10.2 Show that differentiation of Eq. 10.3 leads to Eqs. 10.4 and 10.5.

Q10.3 Derive the Eq. 10.9.

Q10.4 What is the difference between Johnson-Mehl formula and Avrami equation?

Q10.5 Define the eutectoid transformation.

Q10.6 What is the difference between the lamellar pearlite and the spheroidal pearlite in terms of the structure and mechanical properties?

Q10.7 Give an exact definition of the habit plane of martensite.

Q10.8 Describe the characteristics of martensitic transformation.

Q10.9 Define M_S and present the influencing factors of M_S.

Q10.10 What is the driving force of martensite transformation? And explain why the driving force for the nucleation of martensite at the M_S temperature is independent of the carbon concentration in Fe–C steels.

Q10.11 Draw a schematic diagram of bainite formation and describe the difference between upper bainite and lower bainite in terms of microstructure and mechanical properties.

Q10.12 Which is more stable, the pearlitic or the spheroiditic microstructure? Why?

Q10.13 Cite two major differences between martensitic and pearlitic transformations.

Q10.14 Briefly cite the differences between pearlite and bainite relative to microstructure and mechanical properties.

Q10.15 Make a copy of the TTT diagram for an iron–carbon alloy of eutectoid composition (Fig. 10.21) and then sketch and label on this diagram a time–temperature path that will produce 100% pearlite.

Bibliography

ANDRÉ GUINIER (1938) Structure of Age-Hardened Aluminium-Copper Alloys. Nature 142: 569–570

Aaronson H, Enomoto M, Lee JK (2010) Mechanisms of diffusional phase transformations in metals and alloys. CRC Press, Taylor & Francis Group

Porter D, Easterling K, Sherif K (2021) Phase Transformations in Metals and Alloys. 4th ed. CRC Press, Taylor & Francis Group
Shiflet GJ, Mangan MA, Meng WG (1998) Growth by ledges. Interface Sci 6:133–154
Zener C (1948) Theory of strain interaction of solute atoms. Phys Rev 74:639–648

Functional Character of Materials 11

Learning Objectives

1. Define the concept of heat capacity, specific heat, and thermal conductivity.
2. Explain the phenomenon of thermal expansion from an atomic perspective using a potential energy-atomic spacing curve of the diatomic model.
3. Calculate the electrical conductivities of metals, semiconductors and insulators given their charge carrier density and mobility.
4. Explain the two sources of magnetic moments in materials.
5. Understand the nature and sources of diamagnetism, paramagnetism and ferromagnetism.
6. Define the index of refraction.
7. Understand why materials are opaque to visible light.

11.1 Introduction

Solid materials can be divided into two categories based on their performance: structural materials and functional materials. Structural materials prioritize strength and toughness, while functional materials prioritize special functional properties such as thermal, electrical, magnetic, and optical properties. Functional materials are widely used in energy resources, information technology and other fields. Unlike structural materials, the performance of functional materials depends on the electronic structures and movements of electrons within atoms (scattering, excitation, transition, etc.). This chapter will focus on the physical basis of the functional properties of materials, including the thermal, electrical, magnetic, and optical behaviors, and the factors that influence them.

Z. Sun et al., *Fundamentals of Materials Science*,
https://doi.org/10.1007/978-981-96-3904-5_11

11.2 Thermal Properties

Thermal properties describe how materials respond to heat and are related to the energy changes of atoms and free electrons in solids. When a solid absorbs heat energy, there are three key thermal effects: heat absorption, thermal expansion, and heat transfer. When temperature increases, materials expand and their temperature gradient leads to energy transfer from the high to the low temperature regions. These three thermal effects can be characterized by three parameters: heat capacity, thermal expansion coefficient, and thermal conductivity. In practical applications, thermal stress is also a critical aspect of material thermal properties. To cater to specific needs, materials with varying thermal properties have been developed, such as low thermal conductivity ceramics for space-shuttle tiles, high heat capacity polymers for building walls, and high thermal conductivity metals used for heat sinks in high-power appliances.

11.2.1 Heat Capacity

The temperature of a solid material increases as it absorbs heat. Heat capacity is a measure of a material's ability to absorb heat from the environment and is defined as the amount of energy required to raise the temperature by 1 K, as expressed by

$$C = \frac{\Delta Q}{\Delta T} \tag{11.1}$$

where ΔQ is the energy required to raise the temperature by ΔT. In engineering calculation, the specific heat capacity (C) is used, which represents the heat capacity per unit mass of the material, and the unit is $J \cdot (kg \cdot K)^{-1}$. For one mol of material, the energy required to raise the temperature by 1 K is called molar heat capacity and is expressed in $J \cdot (mol \cdot K)^{-1}$.

There are two ways to determine the specific heat capacity of a solid material, depending on the environmental conditions for heat transfer. One way is to keep the volume constant, so that the absorbed heat is equal to the change in internal energy, and the slope of the internal energy curve with respect to temperature is C_V. The other way is to keep the pressure constant, so that the absorbed heat is equal to the change in enthalpy, and the slope of the enthalpy curve with respect to temperature is C_P. C_V and C_P are thermodynamically related, with C_V always being less than C_P but becoming equal at absolute zero. C_V is more easily related to the basic properties of solids, however, it is more difficult to determine experimentally than C_P for condensed phases. Nonetheless, at room temperature or lower, C_V and C_P in the condensed phase are very close.

For many solid materials with relatively simple structures, the change in heat capacity with temperature is shown in Fig. 11.1. At low temperature C_V increases rapidly as temperature increases, and the relationship between C_V and the absolute temperature is given by Eq. 11.2.

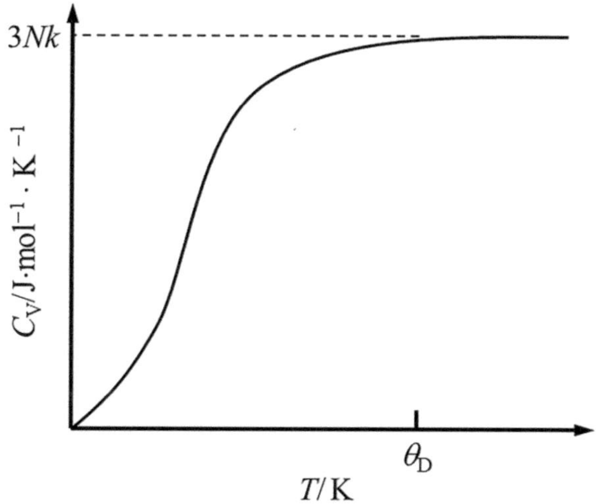

Fig. 11.1 Temperature dependence of the heat capacity at constant volume. θ_D is the Debye temperature

$$C_V = AT^3 \tag{11.2}$$

where A is a constant independent of temperature. Above the Debye temperature θ_D, the energy required to increase the temperature by 1 K remains the same, despite the total energy of the material increasing with temperature. We know that in solids, atoms vibrate around their equilibrium positions, which is called lattice thermal vibration. At low temperature, such vibration can be approximated as "harmonic oscillators". The lattice vibration theory uses the harmonic oscillator model to represent the vibration of each atom in one degree of freedom. The energy is partitioned based on the degrees of freedom, with the average potential and kinetic energy of each degree of freedom being $kT/2$. Each atom in a solid has three degrees of freedom of vibration, leading to a sum of 3 kT for the average kinetic energy and potential energy. The total energy per mole of solid with N atoms can be calculated as:

$$E = 3NkT = 3RT \tag{11.3}$$

where N is Avogadro's constant and k is Boltzmann's constant. At constant volume, the C_V is

$$C_V = \left(\frac{dE}{dT}\right)_V = 3Nk = 3R \approx 25J \cdot (mol \cdot K)^{-1} \tag{11.4}$$

Equation 11.4 shows that C_V is about 25 J · (mol · K)$^{-1}$ which is independent of temperature. Figure 11.1 presents experimental specific heats for a few materials.

11.2.2 Thermal Expansion

Most solid materials expand with increasing temperature, and this expansion is characterized by the coefficient of thermal expansion, α, sometimes referred to as the CTE. The length (or volume) of a material as a function of temperature is expressed as

$$\frac{\Delta l}{l_0} = \alpha_l \cdot \Delta T$$

$$\frac{\Delta V}{V_0} = \alpha_V \cdot \Delta T \tag{11.5}$$

where l_0 and V_0 are the initial length and volume, respectively. When the temperature is increased by ΔT, the increment of material length and volume is Δl and ΔV, respectively; the linear and volume thermal expansion coefficients, α_l and α_V, have units of reciprocal temperature [°C^{-1} or K^{-1}]. In many materials, the value of α_V is anisotropic, depending on the crystallographic direction of measurement. For materials whose thermal expansion is isotropic, α_V is approximately $3\alpha_l$.

At the atomic scale, solid materials expand because the energy of the atoms increases with heating, causing vibrations that deviate from the equilibrium position and increasing the distance between the atoms. This results in an increase in volume or line size at the macro level. The potential energy-atomic spacing curve of the diatomic model provides further explanation of solid materials. For instance, at 0 K, assuming one of the two atoms is fixed at the origin and the equilibrium position of the other atom is r_0, the corresponding potential energy curve is shown as a solid curve in Fig. 11.2. As temperature increase (T_1, T_2, T_3), the amplitude of atomic vibration also increases, resulting in increased vibration energy to (E_1, E_2, and E_3). Furthermore, the atoms vibrate asymmetrically around the equilibrium position. This asymmetry causes the average position of the atoms to move rightward, leading to an increase in the average interatomic distances and thus exhibiting the property of thermal expansion at the macroscopic level. If the potential energy curve is symmetric (as shown by the dotted line in Fig. 11.2), the temperature increases only increases the atomic amplitude, and the amplitudes on both sides remain equal, leading to no thermal expansion.

The coefficient of thermal expansion mainly depends on the binding force between atoms (or molecules, chain segments). Stronger binding forces result in a smaller change in equilibrium distance between atoms with increasing temperature, leading to a lower coefficient of thermal expansion. Thus, ceramics, with strong covalent and ionic bonds, have the lowest thermal expansion coefficient, followed by metallic materials. Polymer materials, with weaker van der Waals forces, have the highest thermal expansion coefficient. The thermal expansibility of materials must be considered in engineering applications. For example, in the diesel engine, the gap between the piston and the cylinder liner is small, which not only allows the piston to reciprocate in the cylinder liner, but also ensures air tightness, which requires the thermal expansion of the cylinder liner and the piston

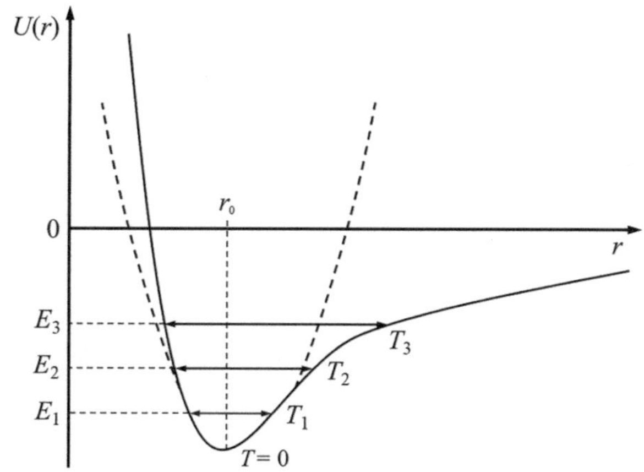

Fig. 11.2 Potential energy curves for interatomic interactions

Table 11.1 Tabulation of the thermal properties for a variety of materials

Materials		C_p/ J · (kg · K)$^{-1}$	α_l/ × 10^{-6} K^{-1}	κ/ W · (m · K)$^{-1}$
Metal	Gold	130	13.8	315
	Silver	235	19.0	428
	Copper	386	16.5	398
	Aluminum	900	23.6	247
	316 Stainless steel	502	16.0	16.3
Ceramics	Alumina (Al_2O_3)	775	8.8	30.1
	Spinel ($MgAlO_4$)	790	7.6	15.0
	Magnesia (MgO)	940	13.5	37.7
	Soda–lime glass	840	9.0	1.7
	Diamond	6.12	1.0~1.5	2000~2320
Polymers	Polyethylene	2100	60~220	0.38
	Polypropylene	1880	80~100	0.12
	Polystyrene	1360	50~85	0.13
	Polyisoprene	–	220	0.14

materials to be almost the same to avoid sticking or leakage. Likewise, materials with small expansion coefficient are used in precision instruments and equipment to maintain their precision and reliability, under varying temperature. Table 11.1 lists the linear coefficients of thermal expansion for several materials.

11.2.3 Thermal Conductivity

Heat conduction is the spontaneous transfer of heat from hot to cold regions of a solid. Thermal conductivity, κ, characterizes the heat transfer ability and can be expressed with the following formula:

$$Q = -\kappa \frac{dT}{dx} \tag{11.6}$$

where Q is heat flux, passing through a unit cross-sectional area, perpendicular to the direction of heat flow, in unit time, and the unit is $W \cdot m^{-2}$. The minus sign indicates that the direction of heat flow is from high temperature to low temperature; κ is the thermal conductivity in $W \cdot (m \cdot K)^{-1}$; dT/dx is the temperature gradient along the direction of heat flow. It should be noted that Eq. 11.6 is only applicable to steady-state heat flow, that is, the heat flow does not change with time.

Heat conduction occurs when the energy is transferred between adjacent parts of a material due to temperature difference, and it can happen through two mechanisms: free electron conduction and lattice vibration conduction, which corresponds to electron thermal conductivity and lattice thermal conductivity respectively, and the thermal conductivity of the material is equal to the sum of the two. Free electron conduction occurs when the free electrons in the heated region absorb heat, gain kinetic energy, and migrate to the cooler region, colliding with atoms or defects to transfer energy. The kinetic energy is then transformed into vibration energy of atoms. Lattice vibration conduction happens because the atomic amplitude in the heated region increases, leading to stronger thermal vibration of the lattice, while the adjacent lattice has weaker thermal vibration. Due to the interaction force (such as chemical bonds, etc.) between the lattices, the vibration of the lattice with weak vibration is intensified and driven by the lattice with strong vibration, leading to heat transfer.

Metals have strong metallic bonds, allowing a large number of free electrons to migrate freely throughout the lattice, resulting in excellent thermal conductivity. The thermal conductivity of metals typically ranges from 20 to 400 $W \cdot (m \cdot K)^{-1}$, and the thermal conductivity of electrons is far greater than that of the lattice. Since free electrons dominate electrical conductivity (σ) and thermal conductivity (κ) in metals, the Wiedemann–Franz Law is used to describe the relationships between them:

$$L = \frac{\kappa}{\sigma T} \tag{11.7}$$

where the theoretical value of L is 2.44×10^{-8} $(\Omega \cdot W) \cdot K^{-2}$, which is independent of temperature, and remains the same for all metals when heat is completely transferred by free electrons. In alloys, especially in solid solutions, impurity atoms will hinder the migration of free electrons, resulting in the reduction of thermal and electrical conductivity.

Ceramic materials are mainly bonded via ionic bonds and covalent bonds, resulting in fewer electrons that can freely move. As a result, the thermal conductivity of lattice is greater than that of electrons. Additionally, phonons are less effective as free electrons in heat transfer because they are more susceptible to scattering by lattice defects. The thermal conductivity of ceramic materials is low, typically ranging from 2 to 50 $W \cdot (m \cdot K)^{-1}$ at room temperature, making them excellent thermal insulators. When atoms are arranged in a highly disordered manner, phonons scatter more strongly, leading to lower thermal conductivity in amorphous ceramics compared to crystal ceramics.

Polymers primarily conduct the heat through the phonon heat conduction between molecules or chain segments. The thermal conductivity of polymers is generally very low, about 0.3 $W \cdot (m \cdot K)^{-1}$ at room temperature, making them good thermal insulation materials. As temperature increases, the crystallinity and thermal conductivity of polymers increase. The thermal conductivity reaches the maximum value when the temperature rises to the glass transition temperature. When the temperature is higher than the glass transition temperature, the thermal conductivity becomes smaller as the molecular arrangement becomes looser.

11.3 Electrical Properties

The electrical conductivity of a solid is one of the most concerned properties. In 1782, Alessandro Volta, an Italian physicist, utilized the electroscope to distinguish metals, insulators, and semiconductors. Nowadays, the theory of electrical conduction and transmission based on energy band theory plays a vital role in information technology. In this section, we discuss the conductance and its mechanism based on electronic band structure. Furthermore, we delve into the conductance of metals, semiconductors, and insulators, with a focus on the characteristics of semiconductors. For electrical properties, metals with high conductivity are often used as wires in precision devices, while ceramics with different band gaps and doping types can realize switching effects in field-effect transistors. Polymers like polymethyl methacrylate are often used as dielectric materials in micro-processors.

11.3.1 Electrical Conductivity

One of the most important electrical properties of solid materials is their ability to conduct electric current. This property is described by Ohm's law, which relates the current to the applied voltage:

$$I = \frac{V}{R} \tag{11.8}$$

where R is the resistance of current through the material; the units of V, I and R are Volt (V), Ampere (A) and Ohm (Ω), respectively. Resistance depends on the sample's shape and is usually independent of current. The electrical resistivity, ρ, is independent of the size or geometry of the sample, and is related to the resistance through the following formula:

$$\rho = \frac{RA}{L} \tag{11.9}$$

where L is the distance between the two ends in measurement, A is the section area perpendicular to the direction of the current, and the unit of ρ is $\Omega \cdot m$. From Ohm's law and the formula above, we have:

$$\rho = \frac{VA}{IL} \tag{11.10}$$

Sometimes the electrical conductivity, σ, is used to describe the electrical properties of a material, which is the inverse of the resistivity, i.e.

$$\sigma = \frac{1}{\rho} \tag{11.11}$$

Conductivity measures a material's ability to conduct electric current, expressed in $(\Omega \cdot m)^{-1}$, thus it is equivalent to discussing electrical performance with either resistivity or conductivity. In addition to Eq. 11.10, Ohm's law can also be expressed as:

$$J = \sigma \xi \tag{11.12}$$

where J is the current density, *i.e.*, the current per unit area of the sample (I / A), and ξ is the electric field strength or the voltage between two points divided by the distance:

$$\xi = \frac{V}{L} \tag{11.13}$$

The equivalence of two expressions of Ohm's law can be easily proved.

Electric current is the result of charged particles responding to an external electric field, causing their movement, as illustrated in Fig. 11.3. Positively charged particles accelerate in the direction of the electric field, while negatively charged particles accelerate in the opposite direction. In the majority of materials, the movement of electrons is responsible for the flow of electric current, a phenomenon known as electronic conduction. However, in some materials such as electrolytes or ionic solids, electric current is carried by the movement of ions, a process known as ionic conduction. The subsequent section will delve into the details of electronic conduction and its mechanisms.

11.3.2 Conductivity of Metals

A typical metal possesses several properties, including luster, ductility, excellent abilities of conduction and heat transfer, among others. These characteristics stem from the free electrons moving within metal crystals. Metal atoms are held together by metal bonds, in which positively charged ions are embedded in a 'sea' of delocalized electrons. This non-directional bonding allows atoms to slide past

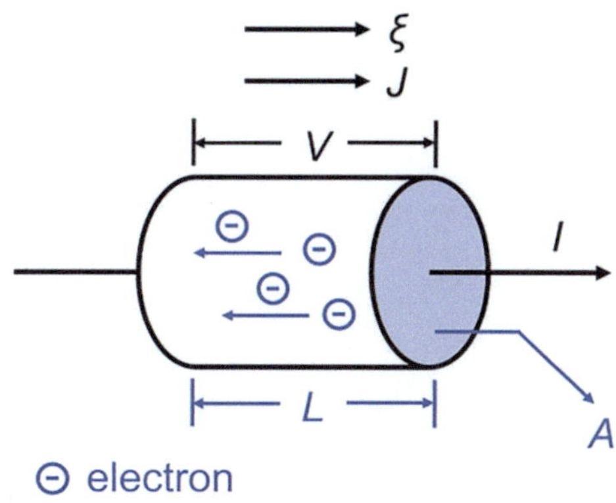

Fig. 11.3 Current flowing through a uniform cylindrical conductor with area A and length L under voltage V and electric field ξ

one another under stress without breaking the bond structure, giving metals their characteristic ductility. Commonly used metals include gold, silver, copper, iron, aluminum, tin, manganese, and zinc. Most metals are solid at one atmospheric pressure and room temperature, except for mercury, which is liquid. Most metals are good electrical conductors, with several commonly used materials' conductivities listed in Table 11.2. Metals exhibit high conductivity due to the presence of many electrons that can be excited to become free electrons. The siemens (S) is the unit of electrical conductance in the International System of Units (SI), which is defined as one ampere per volt (A/V). This implies that when a potential difference of one volt is applied across a material, and it permits a current of one ampere to flow through it, the material is said to have a conductance of one siemens. Conductivity values are commonly expressed in siemens per meter (S/m), which takes into account the length of the material.

The electrical conduction in metals can be conveniently discussed in terms of their electrical resistivity. Crystal defects in metal act as scattering center for conduction electrons and increasing the number of defects can increase the resistivity. Several factors, including temperature, composition, and degree of cold working of metals, largely affect the defect concentration. Electrons in metals are mainly scattered by thermal vibrations, impurities and grain boundaries as depicted in Fig. 11.4. This can be expressed mathematically as follows:

$$\rho_{\mathrm{total}} = \rho_{\mathrm{t}} + \rho_{\mathrm{i}} + \rho_{\mathrm{d}} \tag{11.14}$$

Table 11.2 The electrical conductivities of several metals under room temperature

Metal	Copper	Silver	Gold	Aluminum	Tungsten	Zinc	Iron
Conductivity /(S/m)	6.0×10^7	6.3×10^7	4.1×10^7	3.8×10^7	2.8×10^7	2.7×10^7	2.0×10^7

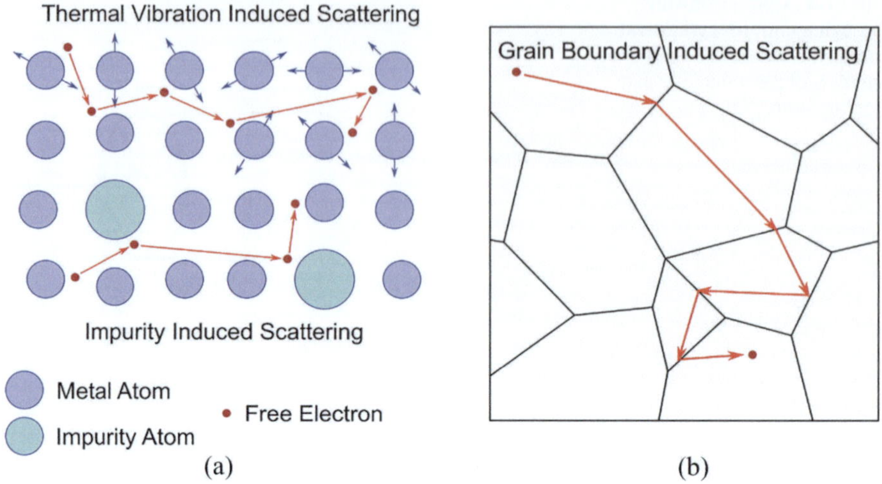

Fig. 11.4 Three different scattering processes induced by **a** thermal vibration, impurity and **b** grain boundary in metal. The red arrows indicate the scattering pathways for free electrons, and blue arrows indicate thermal vibrations of atoms

In the formula, ρ_t, ρ_i, and ρ_d represent the contributions of temperature, impurities, and lattice deformation induced by grain boundaries to the resistivity, respectively. Figure 11.5 illustrated the impact of each variable ρ on the total resistivity for annealed and deformed copper and copper-nickel alloys. Their resistivity changes with temperature, and the arrows indicate the resistivity contributions at 200 K.

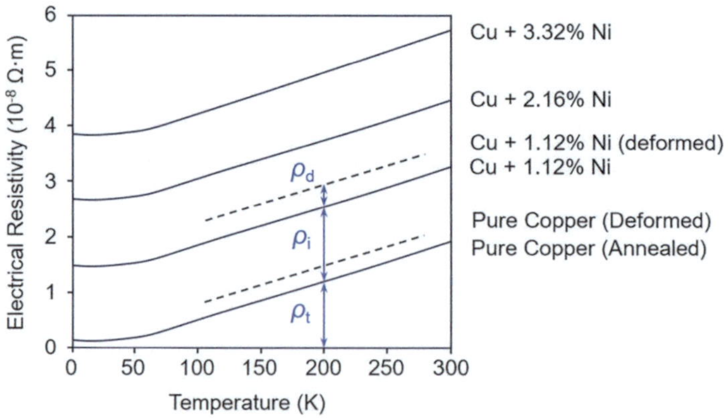

Fig. 11.5 The relationship between electrical resistivity and temperature of copper and three Cu-Ni alloys. The dashed lines indicate deformed pure copper and Cu-Ni alloy

Table 11.3 Intrinsic conductivities and band gap energies of ceramic materials

Material	Type	Temperature/°C	Conductivity/(S/m)	Band Gap/eV
Graphite[1]	Conductor	20	2.0×10^5	0.00
Graphite[2]	Conductor	20	3.3×10^2	0.00
Graphene[3]	Conductor	601	2.0×10^5	0.00
Graphene[4]	Conductor	525	2.0×10^4	0.00
Graphene[5]	Conductor	507	8.0×10^3	0.00
Si	Semiconductor	20	4.4×10^{-4}	1.11
Ge	Semiconductor	20	2.2	0.67
GaAs	Semiconductor	20	$10^{-8} \sim 10^3$	1.35
MoS_2 multilayer	Semiconductor	25	$10^{-9} \sim 10^{-7}$	0.32
MoS_2 layer	Semiconductor	25	$10^{-10} \sim 10^{-8}$	1.80
Alumina	Insulator	20	$10^{-10} \sim 10^{-12}$	–
Porcelain	Insulator	20	$10^{-10} \sim 10^{-12}$	–
Mica	Insulator	20	$10^{-11} \sim 10^{-15}$	–

11.3.3 Conductivity of Ceramics

Ceramic materials are typically characterized by high melting point, high hardness, and high strength, and can exhibit conductive, insulating, and semiconducting properties. The electrical conductivities of various ceramics are listed in Table 11.3.

Semiconductors are a type of material with conductivity values between metals and insulators, and play a vital role in information technology. Generally, the first-generation semiconductor refers to silicon, the second-generation semiconductor refers to compound semiconductor such as GaAs, and the third-generation semiconductor (wide-bandgap semiconductor) refers to SiC. The fourth-generation semiconductors, such as Ga_2O_3 and some antimonides, have lower energy consumption and can be well used in optoelectronic devices. The electrical properties of semiconductors are extremely sensitive to even minute impurity concentrations. Intrinsic semiconductors have electrical behavior determined by their pure material's inherent electronic structure. When the electrical characteristics are determined by impurity atoms, the semiconductor is extrinsic.

Intrinsic semiconductors can be characterized by the electronic energy band structure, which possess a filled valence band and an empty conduction band

[1] Parallel to basal plane.

[2] Perpendicular to basal plane.

[3] Prepared by hydrogen arc discharge exfoliation.

[4] Prepared by argon arc discharge exfoliation.

[5] Prepared by thermal exfoliation.

separated by a narrow band gap. Silicon and germanium are the two basic semiconductors with band gap energies of approximately 1.1 eV and 0.7 eV, respectively, and belong to IVA group in the periodic table. Gallium arsenide (GaAs) and compounds composed of elements in Groups IIB and VIA, such as cadmium sulfide (CdS) and zinc telluride (ZnTe), also exhibit semiconductor behavior. Table 11.3 lists the band gap energies and conductivities of some common semiconductors.

In semiconductors, a valence electron becomes a free electron after excitation with a hole left behind as illustrated in Fig. 11.6a and b. According to energy band theory, this leads to the appearance of a vacant electronic state in the valence band. Under the influence of the electric field, the escaped electron (hole) can move within the crystal by continuously filling incomplete bonds with other valence electrons Fig. 11.6c. A hole is considered to have a positive charge ($+1.6 \times 10^{-19}$ C), opposite to that of an electron. Therefore, the electrons and holes excited under the electric field move in opposite directions and are both scattered by defects in the semiconductor.

The conductivity of an intrinsic semiconductor is different from that in metal because there are both free electrons and holes in the crystal:

$$\sigma = n|e|\mu_e + p|e|\mu_h \tag{11.15}$$

where, n (p) is the number of electrons (holes) per unit volume, μ_e and μ_h are the electron and hole mobilities, respectively. For an intrinsic semiconductor, each excited electron crosses the band gap and leaves a hole in the valence band. Therefore, $n = p$:

$$\sigma = n|e|(\mu_e + \mu_h) = p|e|(\mu_e + \mu_h) \tag{11.16}$$

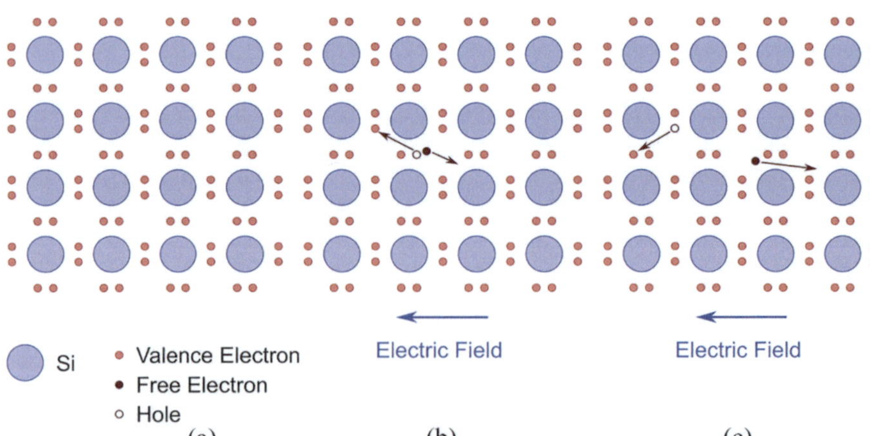

Fig. 11.6 Electron bonding model for explaining the electronical conduction in an intrinsic semiconductor Si **a** before and **b**, **c** after electron excitation

In fact, commercial semiconductors are all extrinsic, whose electrical behaviors are determined by impurities. These impurities exist in small concentrations and can introduce additional electrons or holes into solids. Take *n*-type silicon semiconductor as an example. One silicon atom has 4 valence electrons, and each of them is covalently bonded to 4 adjacent atoms. Now assume one silicon atom in crystal is replaced by an impurity atom with 5 valance electrons, such as atoms in the VA group (P, As, Sb, etc.) in the periodic table. They have 5 valence electrons and only 4 can participate in covalent bonding. The remaining electron is weakly attracted by the impurity atom (as shown in Fig. 11.6a), and can easily break away from impurity atoms to become a free electron or conduction electron due to the small binding energy (in the order of 0.01 eV) as shown in Fig. 11.6b and c. Similarly, p-type semiconductor could be constructed by replacing silicon atoms with III A group atoms (such as Al, B and Ga). Meanwhile, p-type semiconductor is mainly conducted via holes.

The energy state of this weakly bonded electron can be explained by the electron energy band model. The energy level of this electron state is in the band gap just below the bottom of the conduction band Fig. 11.7a, with the corresponding electron binding energy being the energy required to excite an electron from the impurity state to the conduction band. These impurities are called donors, and the impurity states are donor states. Notably, there are no corresponding holes in the valence band since donor electrons are excited from the impurity states. Similarly, in p-type semiconductors, acceptors are formed that accept electrons from valence band and leave holes. The energy state caused by such impurities in band gap is called the acceptor state as shown in Fig. 11.7b.

For *n*-type extrinsic semiconductors, the number of electrons in the conduction band is larger than the number of holes in the valence band, i.e., $n \gg p$, then the

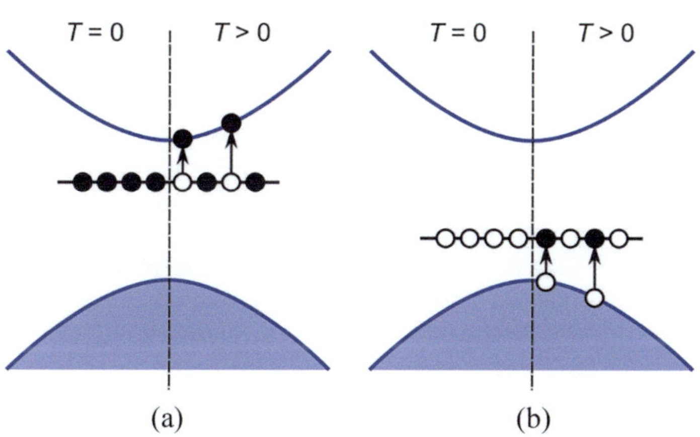

Fig. 11.7 **a** Donors in an n-type semiconductor and **b** acceptors in a p-type semiconductor

Table 11.4 Conductivities of several polymers at room temperature

Polymer	Type	Conductivity /(S/m)
Polyacetylene[6]	Conductor	10^6
Polypyrrole[7]	Conductor	10^5
Polythiophene[8]	Conductor	10^5
Polyaniline[9]	Conductor	10^4
Phenolic aldehyde	Insulator	$10^{-9} \sim 10^{-10}$
Nylon	Insulator	$10^{-9} \sim 10^{-12}$
Polyethylene	Insulator	$10^{-13} \sim 10^{-17}$
Polystyrene	Insulator	$< 10^{-14}$

first term of Eq. 11.15 is greater than the second term, thus:

$$\sigma \approx n|e|\mu_{\mathrm{e}} \tag{11.17}$$

Similarly, for p-type extrinsic semiconductors, $n \ll p$, and thus the conductivity:

$$\sigma \approx p|e|\mu_{\mathrm{h}} \tag{11.18}$$

In extrinsic semiconductors, many charge carriers, such as electrons or holes depending on the impurity type, are generated by thermal activation at room temperature. As a result, these materials have relatively high conductivity at room temperature, making them suitable for electronic devices.

11.3.4 Conductivity of Polymers

A polymer is a substance composed of long chains of repeating structural units, which are convalently bonded to form large macromolecules. Polymers can be **natural**, such as cellulose, proteins, and DNA, or **synthetic**, such as polyethylene, nylon, and polystyrene, depending on their origin and method of production. Table 11.4 lists the electrical conductivity of various polymers at room temperature. Some conductive polymers possess antistatic properties and are suitable for organic solar cells, organic light-emitting diodes, biosensors and supercapacitors. Conductive polymers are classified into composite conductive polymers and structural (or intrinsic) conductive polymers. Composite conductive polymers comprise various conductive substances filled in polymer matrix via different processing techniques.

---

[6] Doped with I_2, Br_2Li, Na or AsF_5.
[7] Doped with BF_4^-, ClO_4^- or tosylate.
[8] Doped with BF_4^-, ClO_4^-, tosylate or $FeCl_4^-$.
[9] Doped with HCl.

Structural conductive polymers with a conjugated long chain structure exhibit a unique property whereby the delocalized π electron residing on the double bonds can migrate along the molecular chain, facilitating the flow of electric current. This migration of π electrons, often referred to as π-electron delocalization, enables the polymers to exhibit intrinsic conductivity.

Furthermore, the conductivity of these conductive polymers can be enhanced through the utilization of doping technology. Doping involves introducing dopant molecules or ions into the polymer matrix, which results in the modulation of its electronic structure. The dopants can either accept or donate electrons, thereby altering the charge carrier concentration in the material and consequently affecting its conductivity. This process effectively improves the overall electrical conductivity of the polymer. Doping can significantly enhance the conductivity of conductive polymers by facilitating the formation of charge carriers, such as positively charged holes or negatively charged electrons, within the polymer matrix. The doped conductive polymers exhibit improved electrical properties, making them suitable for a wide range of applications, including organic electronics, sensors, and energy storage devices.

In fact, most covalently bonded polymers are insulators at room temperature. In actual life, many materials are used based on their insulating properties. The electrical conductivity of the insulator may increase with the increase of temperature; this would lead to the transition from an insulator to a semiconductor.

11.3.5 Superconductivity

Superconductivity is a physical phenomenon in which a material conducts electric current with zore electrical resistance when cooled below a certain critical temperature. The discovery of superconductors has been a significant accomplishment in physics and materials science. Especially, the discovery of high-temperature superconductors has challenged traditional theories and has great application prospects in the future. A material's conductivity is related to its band structure and sensitive to temperature. The resistivity of conductors usually decreases as temperature decreases. However, normal conductors (silver, copper, aluminum, etc.) still exhibit resistance even close to absolute 0 K, because the relaxation time and concentration of carriers in the Fermi surface are not infinite, and the effective mass of electrons cannot be zero. Theoretically, the conductivity at absolute zero cannot be infinite, and the resistivity cannot be zero.

In 1911, the Dutch physicist, H. K. Onnes, discovered that the DC resistance of mercury suddenly dropped to zero near 4.2 K (i.e., liquid helium temperature). He believes that when temperature $T < 4.2$ K, mercury enters a new physical state with superconductivity. This experiment found that the temperature zone where the resistance drops suddenly from a normal value to zero is extremely narrow. Furthermore, this superconductivity phenomenon is reversible, meaning that as the temperature rises above 4.2 K, the resistance of mercury returns to its normal value. Therefore, resistance is solely determined by temperature and is independent

Fig. 11.8 The relationship between the resistance of mercury and temperature

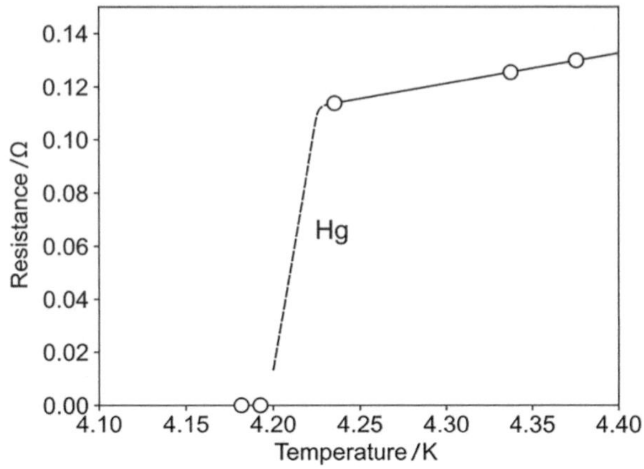

of the process. This transition is considered a phase transition, and the superconducting transition temperature (or critical temperature) is T_c. Figure 11.8 shows the experimental results of Onnes's mercury measurement.

In 1957, American physicists John Bardeen, Leon Cooper and Robert Schrieffer proposed the BCS theory to explain the microscopic mechanism of superconductivity. The theory states that the vibration of the lattice, known as phonon, leads to the formation of a Cooper pair with zero momentum, comprising two electrons with opposite spin and momentum, which is called electro-phonon interaction. According to the theory of matter waves in quantum mechanics, the wavelength of the Cooper pair is so long that it can flow freely around the defects of the lattice and create an electric current without resistance. For this work, Bardeen, Cooper and Schrieffer were awarded the Nobel Prize in Physics in 1972.

Since the discovery of superconductivity in 1911, for a long time physicists believed that the upper limit temperature of superconductivity would not exceed 30 K. Later discovered superconductors with a critical temperature higher than 30 K are called high-temperature superconductors. Currently, it has been discovered that 28 elements and thousands of alloys and compounds are superconductors under normal pressure, and more substances are superconductors under high pressure or structural disorder. In 1986, Karl Müller, a scientist at the IBM Laboratory in Zurich, Switzerland, discovered superconductivity with $T_c = 35$ K in lanthanum-barium-copper-oxide system. At the end of 1986, Chinese scientists Zhong-xian Zhao in the Chinese Academy of Sciences, Ching-wu Chu and Maw-kuen Wu in the University of Houston and University of Alabama have all carried out works on yttrium-barium-copper-oxygen system and obtained a superconductor with $T_c > 90$ K, which triggered a global research upsurge on superconductors.

Zero resistance is one of the basic characteristics of superconductors. When $T < T_c$, the material enters a superconducting state and its resistance suddenly drops to zero. This property of the superconductor is called zero resistivity. Since there is no heat loss caused by resistance below the critical temperature, the current in

Table 11.5 The classification of superconductors

Classification	Examples (T_c)
Metals, alloys, and compounds	Tl (2.39 K), Nb (9.25 K), Ta (4.48 K), Pb (7.20 K), β-phase La (5.98 K), Nb_3Sn (18.1 K), NbTi (9.5 K), Nb_3Ge (23.2 K), MgB_2 (39 K)
Heavy-fermion-based superconductors	$CeCu_2Si_2$ (0.7 K), $CeCoIn_5$ (2.3 K), $CePt_3Si$ (0.75 K), $CeIn_3$ (0.2 K), UBe_{13} (0.85 K), UPt_3 (0.48 K)
High-temperature superconductors	$La_{1.85}Sr_{0.15}CuO_4$ (39 K), $Bi_2Sr_2CaCu_2O_8$ (89 K), $YBa_2Cu_3O_7$ (92 K), $Tl_2Ba_2Ca_2Cu_3O_{10}$ (125 K), $HgBa_2Ca_2Cu_3O_{10}$ (134 K)
Organic superconductors	K_3C_{60} (18 K), Rb_3C_{60} (29 K), $Cs_2Rb_3C_{60}$ (33 K)

superconductor can be maintained for more than one year without external voltage. This current is called super current or continuous current. Superconductors can be extensively applied in several areas, particularly in strong current applications where their zero-resistance characteristics and high current densities are advantages, such as superconducting strong magnets and superconducting motors. Superconductors can be classified based on their material properties as shown in Table 11.5.

11.4 Magnetic Properties

Magnetism is a physical phenomenon produced by electric currents or the magnetic moments of elementary particles, resulting in attractive or repulsive forces between magnetic materials, electric currents, or moving charged particles. Many devices rely on magnetic materials, such as mobile phones, computers, televisions, headphones, radios and so on, which have a significant impact on our daily lives. As early as the Warring States Period, China invented Sinan, the predecessor of the compass, to indicate directions. However, the basic principle and mechanism of magnetic phenomena are complex and subtle, which has puzzled scientists for a long time. This section provides a brief introduction to the origin of magnetism, indicators characterizing magnetism, categories of magnetism and common magnetic materials. For magnetic properties, soft magnetic materials and hard magnetic materials play a significant role in aviation, medical equipment, new energy sources and artificial intelligence.

11.4.1 Magnetic Field

The magnetic force can be conveniently analyzed using the magnetic field lines of force, which are often used to describe the magnetic field distribution (Fig. 11.9a). The direction of tangent at each point on the magnetic field lines of force is consistent with the magnetic field direction at this point. In magnetic materials the magnetic moment can be considered as a small magnet consisting of north and south poles oriented from south to north (Fig. 11.9b). The magnetic moment is affected by the torque of the external magnetic field in the magnetic field and aligns with the direction of the magnetic field to minimize the static magnetic energy of the system. The compass mentioned in the brief introduction works based on this mechanism.

The applied external magnetic field can be expressed by $\boldsymbol{H}$, which is called the magnetic field strength. Magnetism and electricity are inseparable. If the external magnetic field is generated by a solenoid composed of N closely spaced turns with length L and current I (Fig. 11.10a), then

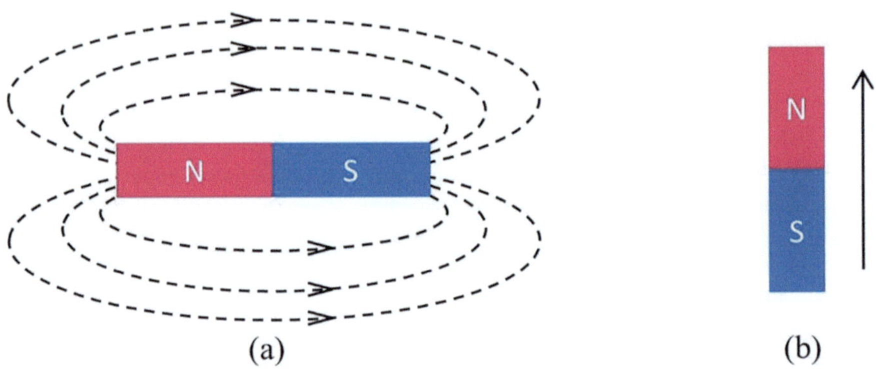

Fig. 11.9 **a** Magnetic field lines of force around a bar magnet. **b** The magnetic moment as designated by an arrow

Fig. 11.10 **a** The external magnetic field $\boldsymbol{H}$ is generated by a solenoid composed of N closely spaced turns with length L and current I. The magnetic induction density $\boldsymbol{B}_0$ in the presence of a vacuum is equal to $\mu_0\boldsymbol{H}$. **b** The magnetic induction density $\boldsymbol{B}$ within a solid material is equal to $\mu\boldsymbol{H}$

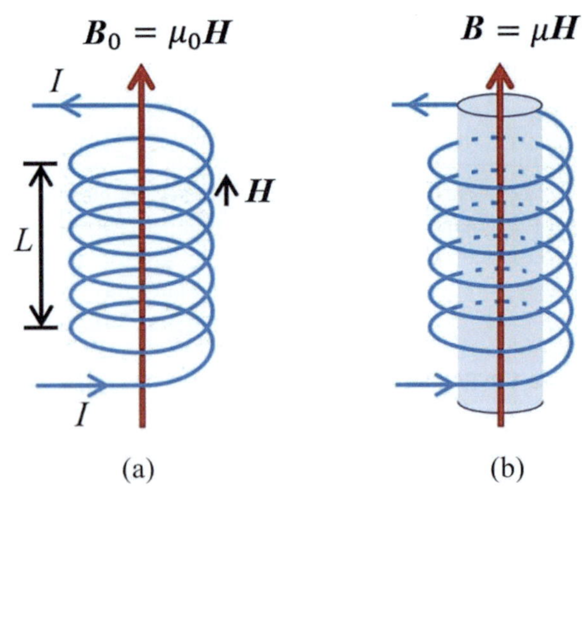

$$H = \frac{NI}{L} \tag{11.19}$$

The magnetic field generated by the current circuit and the strip magnet is ***H***, and its direction can be determined by the right-hand screw law, in the unit of A/m.

Under the influence of the external magnetic field, the magnitude of the magnetic field inside a substance is represented by the magnetic induction, or magnetic flux density (***B***), as shown in Fig. 11.10b. ***B*** is measured in teslas (Wb/m^2). The direction of ***B*** corresponds to the tangent direction of the magnetic line of force at a specific point on the magnetic induction line, and its magnitude is directly proportional to the density of the magnetic line of force. Similar to ***H***, ***B*** is a vector. The relationship between the two can be expressed by:

$$\boldsymbol{B} = \mu \boldsymbol{H} \tag{11.20}$$

where μ represents the permeability and is a characteristic constant of the material that reflects the magnetic induction strength inside the material under the influence of an external magnetic field. The unit of μ is Wb/(A•m) or H/m.

When in a vacuum, the magnetic induction is denoted as B_0:

$$\boldsymbol{B}_0 = \mu_0 \boldsymbol{H} \tag{11.21}$$

where μ_0 represents the permeability of a vacuum, a universal constant with a value of $4\pi \times 10^{-7}$ Wb/(A•m).

The magnetic properties of solid materials parameters are described by several parameters, including relative permeability, magnetization, and magnetic susceptibility, which are defined as follows:

Relative permeability, μ_r, is the ratio of a material's permeability to that of a vacuum, which is expressed as

$$\mu_r = \frac{\mu}{\mu_0} \tag{11.22}$$

It characterizes a substance's susceptibility to magnetization under the influence of an external magnetic field ***H***. The relative permeability of common materials is shown in Table 11.6.

The magnetization ***M*** is defined as.

$$\boldsymbol{B} = \mu_0 \boldsymbol{H} + \mu_0 \boldsymbol{M} \tag{11.23}$$

M is a measure of the change in magnetic field strength within a substance due to the alignment of the magnetic moments in the direction of the external magnetic

Table 11.6 Relative permeability of common materials

Material	Relative permeability	Material	Relative permeability
Vacuum	1	Air	1.00000037
Nickel	100 ~ 600	Bismuth	0.999834
Water	0.999992	Copper	0.999994
Aluminum	1.000022	Platinum	1.000265
Permalloy	100,000	Sapphire	0.99999976

field ***H***. The value of ***M*** is equal to the induced magnetic moment per unit volume of the material and is proportional to the external magnetic field ***H***.

$$\boldsymbol{M} = \chi_m \boldsymbol{H} \tag{11.24}$$

where χ_m is the magnetic susceptibility, which is unitless. The relationship between the magnetic susceptibility and the relative permeability is

$$\chi_m = \mu_r - 1 \tag{11.25}$$

11.4.2 Origins of Magnetic Moments

The nineteenth century French scientist Ampère proposed the hypothesis of molecular currents, suggesting that these currents existed within the molecules of matter and were responsible for its magnetism. Recent advances in physics have confirmed that the origin of magnetism in matter lies in the currents within molecules, and the macroscopic magnetism of a material arises from the magnetic moment of atoms. The magnetism of an atom is derived from the magnetic moment of the nucleus and the magnetic moment of the electrons, as the atom comprises the nucleus and the electrons moving around it. The magnetic moment of the electron comes from two sources: the magnetic moment of the orbital motion of the electrons around the nucleus, which is in the direction of the spin axis; and the spin of the electrons. Each electron spins around its own axis, producing a magnetic moment along the spin axis, which can only be in either the upward or downward direction. The magnetic moment of the nucleus, resulting from the movement of the atomic nucleus, is negligible for macroscopic magnetism, about 1/2000 times the magnitude of the magnetic moment of the electron.

The motion of an electron around the nucleus in orbit corresponds to a closed current, and the resulting magnetic moment is called the orbital magnetic moment m ($\mathrm{A{\cdot}m^2}$) of the electron (Fig. 11.11a):

$$m = l \times \frac{eh}{4\pi m_e} = l \times \mu_B \tag{11.26}$$

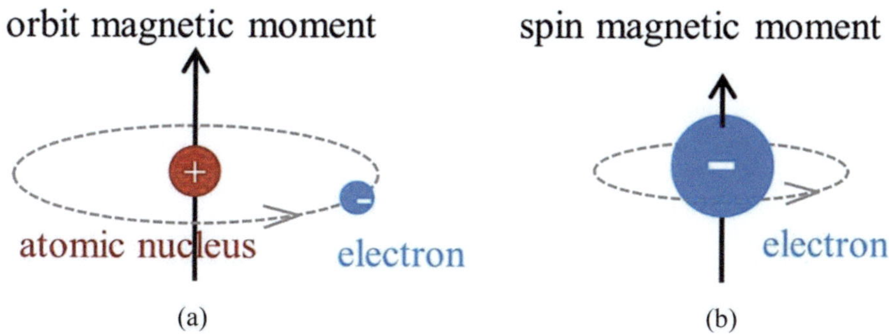

Fig. 11.11 Schematic diagram of orbital magnetic moment formed by electron around nucleus and spin magnetic moment generated by electron spin

where e is elementary charge, m_e is electron mass, h is Planck constant, and l is the angular momentum of the electron orbit in $h/2\pi$. Bohr magneton μ_B is the most basic magnetic moment, with a value of 0.927×10^{-23} A·m^2. When the orbital angular momentum of an electron is a quantum unit, the orbital magnetic moment of the electron is equal to one Bohr magneton. In a shell filled with electrons, the spin angular momentum of electrons also counteracts each other, as the electrons can only have positive and negative spin directions and are always paired in the full shell. As a result, there is a net spin magnetic moment only when the number of electrons in the unfilled shell is unpaired.

In addition to orbital motion, electrons also have spin motion, producing a spin magnetic moment m_s (Fig. 11.11b):

$$m_s = s \times \frac{eh}{2\pi m_e} = 2s\mu_B \tag{11.27}$$

where s represents the electron spin angular momentum in $h/2\pi$. When the shell filled with electrons, their spin angular momentum of electrons also counteracts each other. This is because electrons can only have positive or negative spin directions, with these spin directions always being paired in a full shell. Therefore, there is only a net spin magnetic moment when there are unpaired electrons in the unfilled shell.

In summary, an atom's magnetic moment depends on its structure. In the case of inert gas free atoms such as He, Ne and Ar, their 3 s and 3p shells are fully filled with electrons, causing the orbital and electron spin magnetic moments of all electron pairs to cancel each other, resulting in no intrinsic magnetic moment. On the other hand, the Fe free atom has only 6 electrons on the 3d shell, giving it an intrinsic magnetic moment.

11.4.3 Magnetic Classification

In general, the magnetic properties of substances can be classified as diamagnetism, paramagnetism, ferromagnetism, and antiferromagnetism. Each material exhibits at least one of these magnetic properties, which is determined by its composition and structure.

In the absence of an external magnetic field ***H***, the atoms that make up a material do not possess an intrinsic magnetic moment and therefore show no magnetism. When an external magnetic field is applied, a weak, non-permanent diamagnetism in the material is induced. The diamagnetism is caused by changes in electron orbital motion due to the external magnetic field, inducing a small magnetic moment in the opposite direction to the external magnetic field. The relative permeability μ_r of diamagnetism is therefore less than 1 and the magnetic susceptibility χ_m is negative (-10^{-5} orders of magnitude), resulting in a lower magnetic induction ***B*** inside the diamagnetism material than in a vacuum. While all materials exhibit diamagnetism, it is weak and typically only observable when other types of magnetism are absent. Typical representatives are inert gases He, Ne, Ar, etc.). Figure 11.12a depicts the change in the atomic magnetic moment of a diamagnetic material in the absence or presence of an external magnetic field, with the direction of the arrow indicating the direction of the atomic magnetic moment.

If atoms have intrinsic magnetic moments, the constituent substances may not exhibit macroscopic magnetism because the thermal motion of the atom disturbs the alignment of the atomic intrinsic magnetic moment, causing disorder. While the application of an external magnetic field increases the probability that the atomic magnetic moment tends to align with the external magnetic field, resulting in paramagnetism. Due to the enhanced magnetic field, the relative susceptibility is greater than 1 and the magnetic susceptibility is positive ranging from 10^{-5} to 10^{-2}. Figure 11.12b shows the change in atomic magnetic moment of paramagnetic material in the absence or presence of an external magnetic field ***H***. In some metals, the spin magnetic moment of free electrons aligns with the external magnetic field, resulting in a greater number of free electrons parallel to the external magnetic field than electrons in reverse spin. Therefore, the paramagnetism of

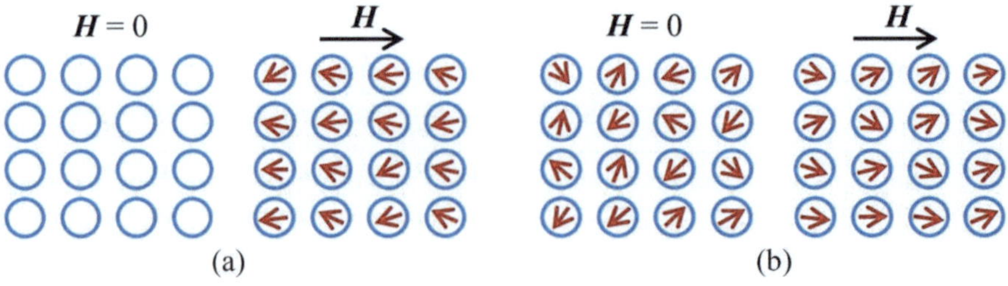

Fig. 11.12 Schematic diagram of atomic magnetic moment arrangement of **a** diamagnetic material, and **b** paramagnetic material, without external magnetic field and with magnetic field, respectively

free electrons is greater than diamagnetism. For example, rare metals with partially filled 4f and 5d orbital electrons, exhibit high paramagnetism, due to a net spin magnetic moment. Transition metals such as V and Ti also have paramagnetism. However, metals such as Au, Ag and Cu display diamagnetism when the diamagnetism of metal ions is greater than the paramagnetism of electrons. Both diamagnetic and paramagnetic materials are considered non-magnetic materials because they require an external magnetic field to exhibit magnetic properties, and their magnetic induction intensity is almost equal to that of a vacuum.

Ferromagnetism is a property of certain martials in which they exhibit a permanent magnetic moment without external magnetic field, which can produce a significant magnetization under external magnetic field, and maintain considerable permanent magnetism after the external magnetic field is removed. Figure 11.13a shows the change in atomic magnetic moment of ferromagnetic materials without magnetic field $\boldsymbol{H}$. The magnetic susceptibility of ferromagnetic materials is much higher than that of paramagnetic materials (Fig. 11.13b), up to 10^6. Like paramagnetic materials, ferromagnetic materials have a net magnetic moment contributed by the spin magnetic moment of unpaired electrons. For example, Fe, Co and Ni are the three most common ferromagnetic materials, each with $4\mu_B$, $3\mu_B$ and $2\mu_B$ net magnetic moment due to their 4, 3 and 2 unpaired 3d electrons. However, paramagnetic materials such as Ti and V also have net magnetic moment, which shows that ferromagnetic materials should meet other conditions beyond having unfilled electron shell in the atom. At room temperature, metals Fe, Co and Ni exhibit the tendency of spontaneous magnetization, forming a small region in which the magnetic moments of adjacent atoms are arranged in one direction, known as domain (Fig. 11.14a).

The size of domains varies, but is typically smaller than the grain size, around 10^{-15} m^3 on average and each magnetic domain contains about 10^9~10^{15} atoms. Due to the varying magnetic moment directions of each domain, the material does

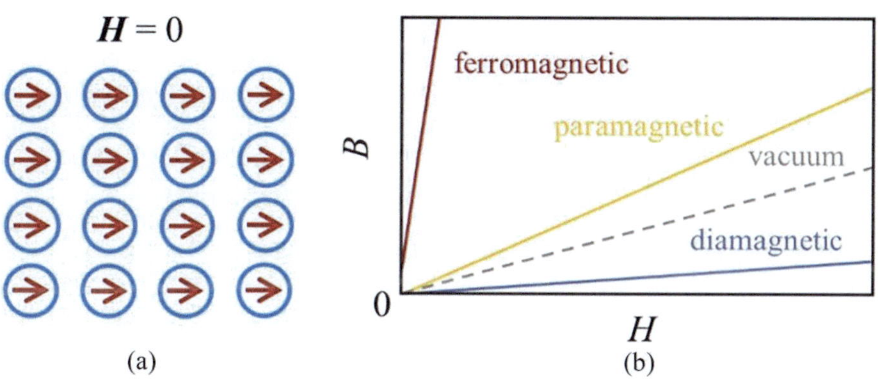

Fig. 11.13 **a** Schematic diagram of atomic magnetic moment arrangement in ferromagnetic materials. **b** Relationship between magnetic induction intensity $\boldsymbol{B}$ and external magnetic field $\boldsymbol{H}$ of ferromagnetic, paramagnetic, and diamagnetic materials

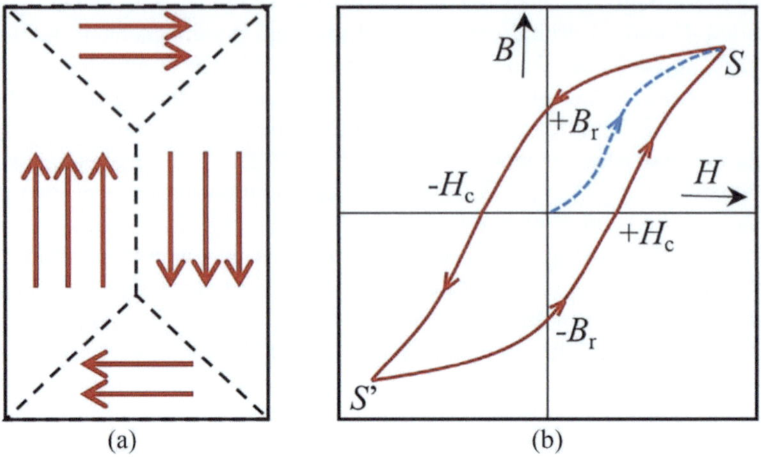

Fig. 11.14 **a** Schematic diagram of domains in a ferromagnetic material. Arrows represent atomic magnetic moment. Within each domain, all dipoles are aligned, whereas the direction of alignment varies from one domain to another. **b** Magnetic induction intensity $\boldsymbol{B}$ versus the external magnetic field $\boldsymbol{H}$ for a ferromagnetic material. The hysteresis loop is represented by the solid red curve; the dashed blue curve indicates the initial magnetization. The remanence $\boldsymbol{B}_r$ and the coercive force $\boldsymbol{H}_c$ are also shown

not exhibit macroscopic magnetism. The domains will be orderly oriented with the direction of the external magnetic field, resulting in strong magnetization ability. The tendency of a material to exhibit spontaneous magnetization and form magnetic domains is directly proportional to the atomic spacing within the lattice and the diameter of the 3d orbitals within the atom. If the ratio between the lattice atomic spacing and the diameter of 3d atomic orbitals in a material is between 1.4 and 2.7, it is considered the ferromagnetic to form magnetic moment. Materials outside of this range, although there are unpaired 3d electrons, do not exhibit the tendency of spontaneous magnetization to form magnetic moment, and are classified as non-ferromagnetic materials.

When an external magnetic field is applied, magnetic induction intensity $\boldsymbol{B}$ initially increases slowly with the increase of magnetic field intensity $\boldsymbol{H}$, then increases faster, and eventually reaches saturation. Upon removal of magnetic field, the magnetic induction intensity of the material does not weaken along the original route. When the external magnetic field is completely removed, there is remanent magnetism ($\boldsymbol{B}_r$). This residual magnetism can be eliminated only by applying reverse magnetic field of a certain intensity, called coercivity ($\boldsymbol{H}_c$). Figure 11.14b shows a hysteresis loop representing the saturation magnetization of the material.

Whether there is an external magnetic field or not, increasing temperature always intensifies the atomic thermal motion, which leads to disruption of the order of magnetic moment arrangement and subsequent decrease in saturation magnetization. The saturation magnetization of ferromagnetic materials reaches the maximum at 0 K. Once heated to a certain temperature, the coupling between magnetic moments is destroyed, resulting in a drop in saturation magnetization to 0. This characteristic temperature is known as Curie temperature, T_C. The Curie temperatures of Fe, Co, and Ni are 1041, 1393, and 608 K, respectively.

In certain atomic interaction, the magnetic moments of adjacent atoms or ions may be arranged in reverse parallel, resulting in ferrimagnetism or antiferromagnetism. Ferrimagnetism occurs when two magnetic moments are reverse parallel, but different in magnitude, creating a net magnetic moment as shown in Fig. 11.15a. Although this property exhibits strong magnetism similar to ferromagnetism, the saturation magnetic induction intensity is lower than that of ferromagnetic materials. Ferrite, a commonly used ferrous material such as Fe_3O_4, demonstrates this characteristic. The molecular formula of Fe_3O_4 can be expressed as $Fe^{2+}O^{2-}$—$(Fe^{3+})_2(O^{2-})_3$, where Fe ions exist in + 2 and + 3 valence states at the ratio of 1:2. While O^{2-} has no net magnetic moment, Fe^{2+} and Fe^{3+} have 4 μ_B and $5\mu_B$, respectively. The spin directions of Fe^{3+} in octahedron and Fe^{3+} in tetrahedron are opposite, causing their magnetic moments to cancel each other out. Thus, the net magnetic moment is generated by Fe^{2+} ions in octahedron.

When two magnetic moments are parallel and have the same magnitude, there is no net magnetic moment and the magnetic susceptibility is very small, resulting in antiferromagnetism, as shown in Fig. 11.15b, as demonstrated by NiO. The magnetic moments of adjacent Ni^{2+} in NiO are antiparallel, while the spin and orbital magnetic moments of O^{2-} ions completely cancel out, yielding no net magnetic moment in the material.

Both ferrimagnetic and antiferromagnetic materials exhibit a critical temperature known as Néel temperature, below which ferrimagnetic materials behave similarly to ferromagnetic materials, displaying high magnetic susceptibility. However, above the Néel temperature, they become paramagnetic, like the Curie temperature of ferromagnetic materials. Antiferromagnetic materials are antiferromagnetic below the Néel temperature. With increasing temperature, the atomic

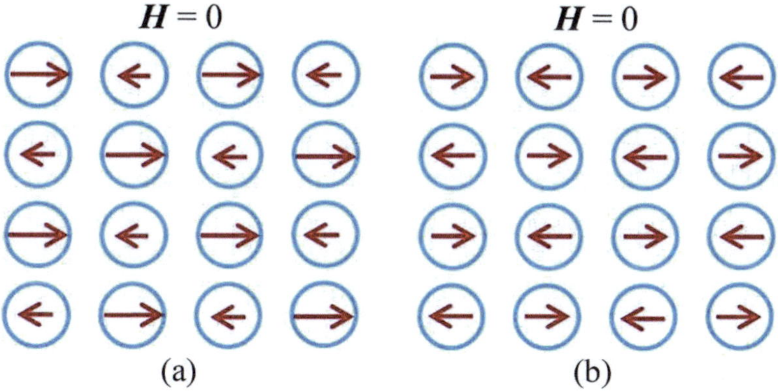

Fig. 11.15 Schematic diagram of atomic magnetic moment arrangement of ferrimagnetic material **a** and antiferromagnetic material **b** without external magnetic field

thermal motion leads to a rise in the irregular arrangement trend of spin magnetic moment, causing the effect of reverse parallel arrangement to weaken, resulting in an increase in magnetic susceptibility. Above the Néel temperature, the magnetic susceptibility decreases like paramagnetic materials. Thus, ferrimagnetic material exhibits the largest magnetic susceptibility at Néel temperature.

11.4.4 Magnetic Materials

Magnetic materials primarily refer to substances composed of transition elements such as iron, cobalt, nickel, and their alloys, which can directly or indirectly generate magnetism. These materials can be roughly divided into two categories: soft magnetic materials and hard magnetic materials, based on the shape and size of hysteresis loops. In 2018, China ranked first in the world in the annual production of various categories of magnetic materials.

Soft magnetic materials are characterized by high magnetic induction, high permeability, low coercivity and low hysteresis loss. They are easy to magnetize and demagnetize in the external magnetic field, and are commonly used in electrical and electronic equipment. The hysteresis loop of soft magnetic materials is characterized by thin and narrow shape, such as the one shown in Fig. 11.16a.

Soft magnetic materials are commonly used in devices that operate under alternating magnetic fields, where low energy loss is required, such as transformer cores. The area of hysteresis loop represents the magnetic energy loss per unit volume of material in each magnetization and demagnetization cycle, which is expressed in the form of heat. Therefore, the thin and narrow hysteresis loop of soft magnetic materials meets the needs of transformer cores.

The resistance of soft magnetic materials is generally high because the current generated by the changing magnetic field in the soft magnetic material causes energy loss. To reduce the energy loss caused by current, high resistivity is required

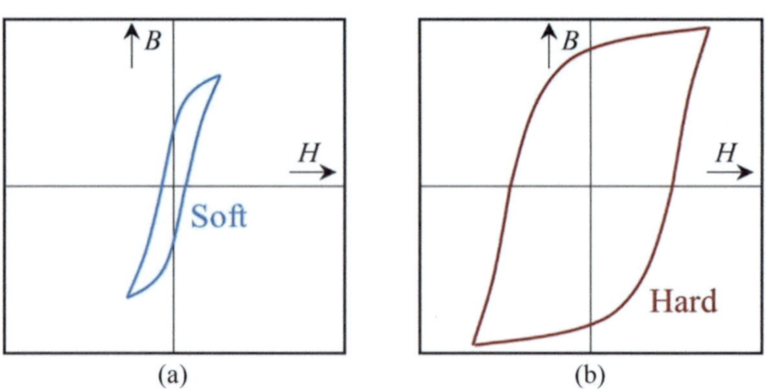

Fig. 11.16 Schematic magnetization curves for **a** soft and **b** hard magnetic materials

in magnetic materials. Therefore, ceramic ferrites and solid-solution alloys are often used as soft magnetic materials due to their high resistivity.

When a soft magnetic material is used as disk material in computers, it is required that the material can store information. In this case, the material should have low remanence, low magnetic saturation strength and low coercivity. A steep magnetization curve is designed to meet the needs of removing information. These materials can be magnetized under a small external magnetic field, so their saturation magnetization and coercivity are very low, allowing for fast transmission of information.

Hard magnetic materials, also known as permanent magnetic materials, are materials that are difficult to magnetize and demagnetize. These materials are characterized by high coercivity (greater than 1×10^4 A/M), high residual magnetic induction and larger magnetic energy product ($\boldsymbol{B} \times \boldsymbol{H}$). The hysteresis loop of hard magnetic materials is characterized by being wide and high (Fig. 11.16b). Hard magnetic materials are widely used and can be divided into two categories: one uses the magnetic field generated by hard magnetic alloys, while the other uses the hysteresis characteristics of hard magnetic alloys to generate rotating torque to convert electrical energy into mechanical energy.

The hysteresis behavior of hard magnetic materials is related to the difficulty of domain wall movement and rotation. All structural factors that hinder domain movement can improve the hard magnetic properties. For example, fine precipitates can be introduced to improve the hard magnetic properties, and the properties of hard magnetic materials can be improved by powder metallurgy.

Due to their characteristics of energy saving and high efficiency, hard magnetic functional devices are widely used in office equipment, communication, automobiles and other fields. In particular, hard magnetic materials are indispensable in the fields of new energy vehicles and wind power generation.

11.5 Optical Properties

Electrons can be excited by absorbing photons with a specific energy. When electrons are in an excited state, they can also transition to a lower energy level and release energy in the form of light radiation. The optical performance of a material represents its response under electromagnetic radiations, especially in visible light. Materials with desired optical responses can be used in all aspects of daily lives and productions. This section will first discuss some basic principles and concepts related to electromagnetic radiation, as well as the possible interaction between electromagnetic radiation and solids. We will then explore the optical behaviors of materials by describing their reflection, refraction, transmission, and absorption characteristics. Some metals have good reflection capacity and can be used as mirrors in radio telescopes. Ceramics and polymers with proper band gaps, can absorb light and have been widely used in photocatalytic domains for water splitting and pollutant degradation.

11.5.1 Electromagnetic Radiation

In the classical sense, electromagnetic radiation can be conceptualized as waves. It arises from the interplay of electric and magnetic fields, which are matually perpendicular to each other and to the direction of wave propagation. Various forms of radiation, such as light, heat, radio waves, and X-rays, are all electromagnetic radiation. Radiation is mainly characterized by a specific range of wavelengths. The electromagnetic spectrum covers a wide range, with wavelength increasing in the direction of γ-ray (10^{-12} μm) → X-ray → ultraviolet light → visible light → infrared light → radio wave (10^5 μm).

Visible light occupy a very narrow region in the radiation spectrum, with wavelengths ranging from 0.4 ~ 0.7 μm. The wavelength of light determines the perceivable color. For instance, radiation with a wavelength of about 0.4 μm appears purple, while green and red occur at roughly 0.5 μm and 0.65 μm, respectively. The spectral ranges of several colors are depicted in Fig. 11.17. White light is a simple mix of all colors. The discussion in this section mainly concerns visible light.

Electromagnetic radiations travel through the vacuum at a constant speed, *i.e.*, 3×10^8 m/s. The relationship between the vacuum permittivity ε_0, vacuum permeability μ_0 and the light speed c is:

$$c = \frac{1}{\sqrt{\varepsilon_0 \mu_0}} \tag{11.28}$$

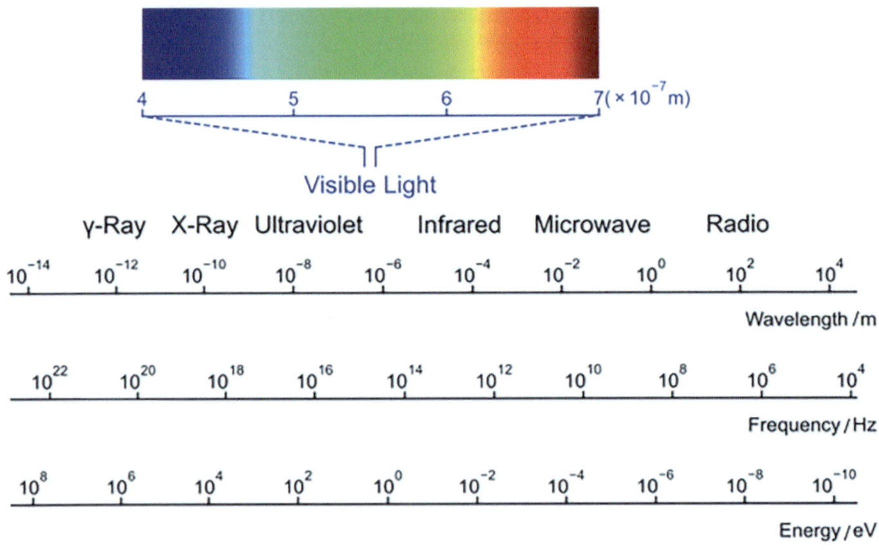

Fig. 11.17 The electromagnetic radiation spectrum

Hence, there is a correlation between the electric and magnetic constants and the speed of light. The frequency ν and wavelength λ of electromagnetic radiation are a function of the light speed:

$$c = \lambda\nu \tag{11.29}$$

The unit of frequency is hertz (Hz). Electromagnetic radiation can also be considered as a photon due to the wave-particle duality. The energy of a photon can be quantized and defined by:

$$E = h\nu = \frac{hc}{\lambda} \tag{11.30}$$

In this formula, h is the Planck constant (6.63×10^{-34} J•s). Therefore, the energy of a photon is proportional to the frequency of radiation and inversely proportional to the wavelength.

When considering the optical phenomenon of the interaction between radiation and matter, it is often convenient to explain it from a photon point of view. In other cases, the wave point of view is more appropriate. Therefore, both methods are used in the discussion in this section.

11.5.2 Light Interaction with Solid

When light travels from one medium to another, it undergoes several events. Some light radiation can pass through the medium, some is absorbed, and some is reflected at the interface between the two media. From the perspective of energy conservation, the radiation beam intensity incident onto the surface of solid medium (I_0) must be equal to the sum of the transmission intensity (I_T), absorption intensity (I_A) and reflection intensity (I_R):

$$I_0 = I_T + I_A + I_R \tag{11.31}$$

The unit of intensity is W/m^2, which corresponds to the energy per unit area and unit time in the perpendicular direction of propagation. This formula can also be expressed as:

$$T + A + R = 1 \tag{11.32}$$

where T, A and R represent transmittivity (I_T/I_0), absorptivity (I_A/I_0), and reflectivity (I_R/I_0), respectively.

Materials with light transmission and less absorption, reflection are transparent. Translucent materials scatter light inside the material and diffuse light through, making it difficult to distinguish objects viewed through them. Any material that cannot transmit light is called opaque. For example, metals are opaque to visible light, meaning that all light radiation are either completely absorbed or reflected. On the other hand, all electrically insulating materials can be made transparent.

11.5.3 Reflection of Light

When light (electromagnetic waves) hit the medium interface, reflection and refraction inevitably occur. A portion of the light is reflected from the interface while another portion is transmitted into the medium. Reflectivity R represents the fraction of light reflected by the interface:

$$R = \frac{I_R}{I_0} \tag{11.33}$$

where I_R and I_0 are incident beam intensity and reflected beam intensity, respectively. Light is a transverse wave, and the electric vector can oscillate in any direction within a plane perpendicular to the propagation direction. Therefore, it can be decomposed into two orthogonal linear polarization components: S component or S wave, whose vibration direction is perpendicular to the incident surface of the light, and the P component or P wave, whose vibration direction is parallel to the incident surface. For polarized light that vibrates perpendicular to the incident surface, reflectivity can be derived as:

$$R_S = \frac{\sin^2(\alpha - \gamma)}{\sin^2(\alpha + \gamma)} \tag{11.34}$$

where α and γ are the incident and refraction angles, respectively. For polarized light that vibrates parallel to the incident surface, the reflectivity is:

$$R_P = \frac{\tan^2(\alpha - \gamma)}{\tan^2(\alpha + \gamma)} \tag{11.35}$$

If the incident light is perpendicular to the interface, $\alpha = \gamma = 0$, the above two equations are equal:

$$R = \left(\frac{n_2 - n_1}{n_2 + n_1}\right)^2 \tag{11.36}$$

where n_1 and n_2 are the refractive indices of the two media. The above formula shows that the greater the difference in refractive indices between the two media, the greater the reflectivity. When light enters the solid from vacuum or air, then:

$$R = \left(\frac{n_s - 1}{n_s + 1}\right)^2 \tag{11.37}$$

Since the refractive index of air is close to 1, that is, $n_s \approx 1$, the greater the refractive index of the solid (n_s), the greater the reflectivity.

11.5.4 Refraction of Light

When light enters a transparent material, its speed is reduced, causing it to bend at the interface. This phenomenon is known as refraction. The refractive index n is defined as the ratio of the light speed in vacuum c to its speed in medium ν, that is:

$$n = \frac{c}{\nu} \tag{11.38}$$

The refractive index n is dependent on the wavelength of light, and this effect can be demonstrated by the dispersion of white light by glass prisms. The light of different colors is deflected at different degrees when entering and leaving the prism, causing the separation of different colors. The refractive index not only reflects the optical path of light, but also the fraction of light that is reflected on the surface.

The light speed in a medium is defined as:

$$\nu = \frac{1}{\sqrt{\varepsilon\mu}} \tag{11.39}$$

where ϵ and μ are permittivity and permeability of medium, respectively. Then the refractive index is:

$$n = \frac{c}{\nu} = \frac{\sqrt{\varepsilon\mu}}{\sqrt{\varepsilon_0\mu_0}} = \sqrt{\varepsilon_r\mu_r} \tag{11.40}$$

where ϵ_r and μ_r are relative permittivity (also known as dielectric constant) and relative permeability, respectively. For materials with weak magnetism, $\mu_r \equiv 1$, then:

$$n \equiv \sqrt{\varepsilon_r} \tag{11.41}$$

Therefore, for transparent materials, there exists a relationship between refractive index and dielectric constant. The electronic component of the dielectric constant can be determined by measuring the refractive index using the above formula.

For ceramics and glasses with cubic crystal structure, the refractive index is isotropic, meaning it is not dependent on the crystal orientation. On the other hand, the refractive index of non-cubic crystals is anisotropic, with the highest refractive index occurring along the orientation with the highest density of ions.

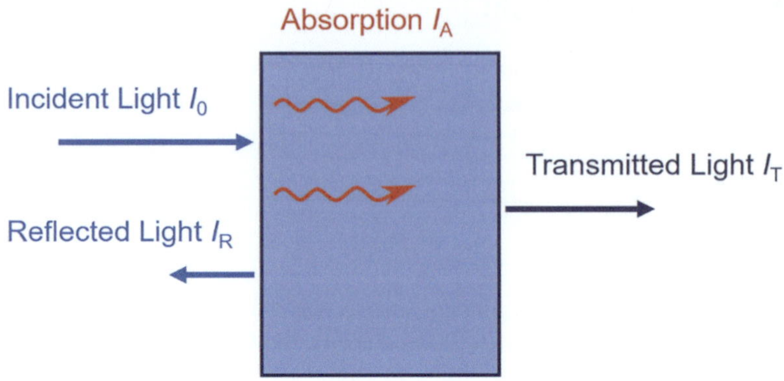

Fig. 11.18 The transmitted intensity of light emitted from the back surface

11.5.5 Transmission of Light

The relationship between transmission, absorption and reflection phenomena can be investigated by passing light through a transparent solid, as shown in Fig. 11.18. When light with an incident intensity I_0 hits the front surface of the transparent sample, the transmitted intensity of light emitted from the back surface is:

$$I_T = I_0(1 - R)^2 e^{-\beta l} \tag{11.42}$$

where R denotes the reflectivity, and β and l correspond to the absorption coefficient and the specimen thickness, respectively. This equation assumes that the media outside the front and back surfaces are homogeneous.

Therefore, the fraction of incident light that passes through the transparent material depends on the losses caused by absorption and reflection. Note that according to Eq. 11.24, the sum of the reflectivity R, absorptivity A and transmissivity T is equal to 1, and each variable in R, A and T depends on the wavelength of light. Figure 11.19 illustrates the incident light transmission, absorption and reflection fractions varying with wavelength through green glass. For example, for light with a wavelength of 0.4 μm, the transmission, absorption, and reflection fractions are about 0.90, 0.05 and 0.05, respectively. However, for light with a wavelength of 0.55 μm, their fractions are 0.50, 0.48, and 0.02, respectively.

11.5.6 Absorption of Light

Light propagating through a conductive medium is subject to attenuation, which is commonly known as light absorption. This absorption occurs when the radiant energy of light is absorbed by the material, causing the transition of electrons from a lower energy level to a higher energy level. Various experiments have

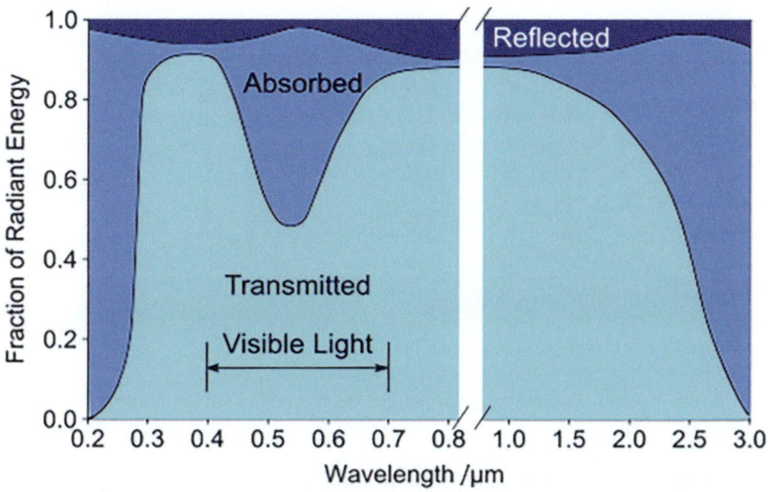

Fig. 11.19 The variations of incident light transmission, absorption, and reflection fractions with wavelength through a green glass

demonstrated that the electronic transition is the primary absorption process in research. When a semiconductor is illuminated by light of a certain wavelength, electrons will absorb enough energy to transition from the valence band to the conduction band. As a result, nonmetallic materials can be either transparent or opaque, depending on the band gaps and impurity levels in materials.

Exciting electrons in the almost filled valence band to cross the band gap and enter the empty states in the conduction band generates free electrons in the conduction band and holes in the valence band, as shown in Fig. 11.20a. Additionally, the excitation energy ΔE is dependent on the frequency of the absorbed photon. These excitations, accompanied by absorption, can only occur when the photon energy is greater than the band gap energy E_g, expressed as:

$$h\nu > E_g \tag{11.43}$$

or:

$$\frac{hc}{\lambda} > E_g \tag{11.44}$$

The minimum wavelength of visible light is approximately 0.4 μm. Since $c = 3 \times 10^8 \mathrm{m/s}$ and $h = 4.13 \times 10^{-15} \mathrm{eV \cdot s}$, the maximum band gap energy corresponding to visible light absorption is:

$$E_{\mathrm{gmax}} = \frac{hc}{\lambda_{\min}} = 3.1\mathrm{eV} \tag{11.45}$$

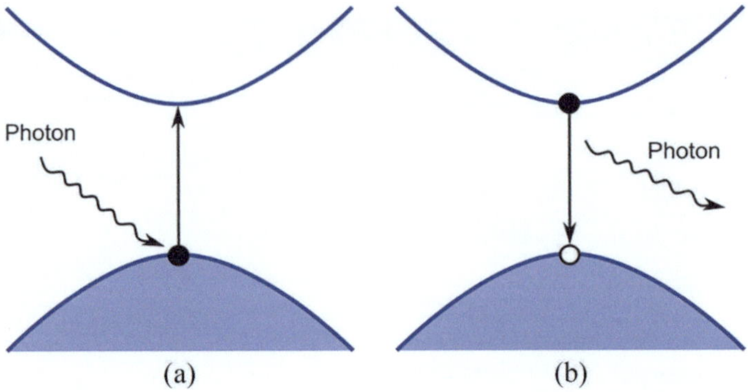

Fig. 11.20 **a** The photon absorption and **b** the emission mechanisms in nonmetallic materials

Thus, nonmetallic materials with band gap energy greater than 3.1 eV will not absorb visible light. If these materials are highly pure, they will appear colorless and transparent.

Conversely, the maximum wavelength of visible light is approximately 0.7 μm, which corresponds to the minimum band gap energy:

$$E_{\text{gmin}} = \frac{hc}{\lambda_{\text{max}}} = 1.8\text{eV} \tag{11.46}$$

This result suggests that semiconductor materials with a band gap energy lower than 1.8 eV have the capability to fully absorb visible light through electronic transitions from the valence band to the conduction band, making them opaque. Conversely, materials with a band gap energy ranging from 1.8 to 3.1 eV can absorb certain portions of the visible spectrum, thereby displaying various colors.

Each nonmetallic material becomes opaque at a certain wavelength, which is determined by the material's band gap energy E_g. For example, diamond has a band gap energy of 5.6 eV, which means that it becomes opaque to radiation with a wavelength of less than 0.22 μm.

11.6 Questions

Q11.1 Please prove the equivalence between two Ohm's law expressions, i.e., $\boldsymbol{I} = \frac{\boldsymbol{V}}{\boldsymbol{R}}$ and $\boldsymbol{J} = \sigma \xi$.

Q11.2 The conductivity of intrinsic Si is 4.4×10^{-4} $(\Omega{\bullet}\text{m})^{-1}$ as listed in Table 11.3. Assume that the electron and hole mobilities are 0.15 and 0.05 $\text{m}^2/(\text{V·s})$, respectively. Estimate the carrier concentration of Si.

Q11.3 Assume that the electron and hole mobilities of intrinsic Ge are 0.38 and 0.18 $\text{m}^2/(\text{V}{\bullet}\text{s})$, respectively. Estimate the carrier concentration of Ge.

Q11.4 Summarize the conducting behavior difference between intrinsic and extrinsic semiconductors.

Q11.5 Summarize the conducting behavior difference among metals, semiconductors, insulators and superconductors.

Q11.6 Write down the relationship among transmission intensity ($\boldsymbol{I_T}$), absorption intensity ($\boldsymbol{I_A}$) and reflection intensity ($\boldsymbol{I_R}$).

Q11.7 Assume a light with wavelength 0.8 μm hits the surface of green glass, read the transmission, absorption and reflection fractions from Fig. 11.19.

Q11.8 TiO_2 is a common photocatalytic material with a band gap about 3.2 eV. Determine whether TiO_2 can absorb visible light.

Q11.9 Summarize the difference between light absorption and emission mechanisms in a nonmetallic material.

Q11.10 Summarize the influence of impurity in light absorption and emission procedures.

Q11.11 Estimate the energy required to raise the temperature of 5 kg of the following materials from 20 to 150 °C: aluminum, aluminum oxide (alumina), and polypropylene.

Q11.12 A bimetallic strip is constructed from strips of two different metals that are bonded along their lengths. Explain how such a device may be used in a thermostat to regulate temperature.

Q11.13 Briefly explain why the thermal conductivities are higher for crystalline than nanocrystalline ceramics.

Q11.14 Calculate the heat flux through a sheet of copper 7.5 mm (0.30 in.) thick if the temperatures at the two faces are 150 and 50 °C; assume steady-state heat flow.

Q11.15 For cooling and heating, what are the nature of the surface stresses?

Q11.16 Summarize the differences between ferromagnetic and antiferromagnetic materials from a macroscopic and microscopic point of view, respectively.

Q11.17 The metals Iron, Cobalt, Nickel, Titanium and Vanadium all have a net magnetic moment, but why is it that only Iron, Cobalt and Nickel are ferromagnetic materials, while Titanium and Vanadium are paramagnetic materials?

Q11.18 A 10-m-long wire coil with 500 turns carries 5A current. What is the magnitude of the magnetic field strength and flux density if the coil is in a vacuum?

Q11.19 What happens to the magnetic properties of ferromagnetic materials (antiferromagnetic materials) when their temperature exceeds the Curie temperature (Néel temperature)?

Q11.20 What are the characteristics of the different shapes and sizes of hysteresis lines, and which devices are suitable for each?

Bibliography

Bertram HN (1994) Theory of magnetic recording. Cambridge University Press, Cambridge

Callister WD (2000) Fundamentals of materials science and engineering. Wiley, London

Ewing JA (1900) Magnetic induction in iron and other metals. D. Van Nostrand Company, London

Fossheim K, Sudbø A (2004) Superconductivity: physics and applications. John Wiley & Sons, New York

Green MA (1990) Intrinsic concentration, effective densities of states, and effective mass in silicon. J Appl Phys 67(6):2944–2954

Hubert A, Schäfer R (2008) Magnetic domains: the analysis of magnetic microstructures. [S.l.]: Springer Science & Business Media

Matula RA (1979) Electrical resistivity of copper, gold, palladium, and silver[J]. J Phys Chem Ref Data 8(4):1147–1298

Spaldin NA (2010) Magnetic materials: fundamentals and applications. Cambridge University Press, Cambridge

Stuchebryukov SD, Selektor SL, Silantieva DA et al (2013) Peculiarities of the reflection-absorption and transmission spectra of ultrathin films under normal incidence of light. Prot Met Phys Chem Surf 49:189–197

Weltner W (1989) Magnetic atoms and molecules.Chelmsford: Courier Corporation

Yoshio S (1954) The electrical resistivity of Cu-Ni alloys and Matthiessen's rule. Science Reports of the Research Institutes, Tohoku University. Ser. A, Physics, Chemistry and Metallurgy, 6:199–206

Zel'Dovich YB (1975) Interaction of free electrons with electromagnetic radiation. Soviet Physics Uspekhi 18(2):79

Deformation and Stress–Strain Behavior of Solid Materials

12

Learning Objectives

1. Understand the measurement of mechanical properties and the meanings of related parameters.
2. Describe the typical deformation and mechanical behavior of some different materials, e.g. metals, ceramics and polymers. Understand the difference of their typical stress–strain curves.
3. Define slip, twinning and kinking in single crystals and know the corresponding atomic mechanisms. Determine the slip system (slip plane and slip direction) for typical crystal structures, such as FCC, BCC and HCP.
4. Describe the effect of grain orientation and grain boundaries of a polycrystalline metal on the plastic deformation behavior.
5. Explain how grain boundaries hinder dislocation motion and know the grain refinement strengthening mechanism.
6. Describe solid-solution strengthening during the plastic deformation of substitutional multiphase alloys.
7. Explain the phenomenon of work hardening from the interaction between dislocations and the strain field.
8. Describe the recovery and recrystallization in terms of both the change of microstructure and mechanical properties of the material.
9. Describe the grain growth after recrystallization from both macroscopic and atomic viewpoints.
10. Explain the underlying mechanisms for the recrystallization texture and annealing twins.
11. Describe the molecular motion of polymers and cite the WLF equation.
12. Build the relationship between molecular motions and mechanical behaviors.

Z. Sun et al., *Fundamentals of Materials Science*,
https://doi.org/10.1007/978-981-96-3904-5_12

13. Understand the meaning of glass transition and the influencing factors for glass transition.
14. Describe the characteristics and analyze the thermodynamics of high elasticity of polymers.
15. Understand the mechanical behaviors and models of viscoelasticity of polymeric materials.
16. Describe the time–temperature equivalence principles for polymeric materials.
17. Draw the typical stress–strain curves and understand the fracture of polymers.

12.1 Introduction

The study of how materials respond to applied forces and their properties has been emphasized by humans since the beginning of our existence. It is essential to understand the measurement of mechanical properties and the meaning of related parameters to design critical components and structures. The deformations (such as elastic or plastic) and mechanical characteristics (including strength, ductility and toughness) must match the designed lifespan of solid material components.

Many engineering materials, such as steel alloys in cable-stayed bridges, aluminum alloys in airplane wings, and ceramics in automobile axles, are also subjected to loads. In these situations, it is crucial to identify the material's characteristics and design the member that can endure the load or deformation before final fracture, such as stiffness, strength, ductility, and toughness.

To investigate the mechanical properties of materials, experiments should be conducted in a laboratory, ideally designed to replicate service conditions. Testing factors that should be considered include the applied load and environmental conditions. Components or structures may endure different static or dynamic loads, such as tensile, compressive, shear, torsion, fatigue, and creep. Service temperature and lifespan are also significant factors for the experiments' design.

Theoretical and mathematical stress analyses can be used to determine stresses and stress distributions within structure members. To design and produce materials that meet service requirements, experiments or simulations may be conducted. Understanding the relationship between microstructure features and mechanical properties of materials is also essential.

This chapter discusses the mechanical behavior of various materials, including metals, ceramics, and polymers. These materials have different stress–strain behaviors and primary mechanical properties.

12.2 Basic Definitions and Classical Mechanical Properties

12.2.1 Basic Definitions

① Stress and Strain

To evaluate a material's deformation behavior under static or quasi-static loads applied uniformly across its cross-section or surface at room temperature, corresponding tests can be performed. These tests typically involve applying different modes of loads, such as tension, compression, shear, and bending (as shown in Fig. 12.1).

When a tensile load is applied along the length axis of a specimen, the material experiences an extension, which is referred to as tensile deformation (as shown in Fig. 12.1a). As the tensile force (F) increases, the specimen's length (l) also increases, which can be plotted on a graph (Fig. 12.2a). However, it's important

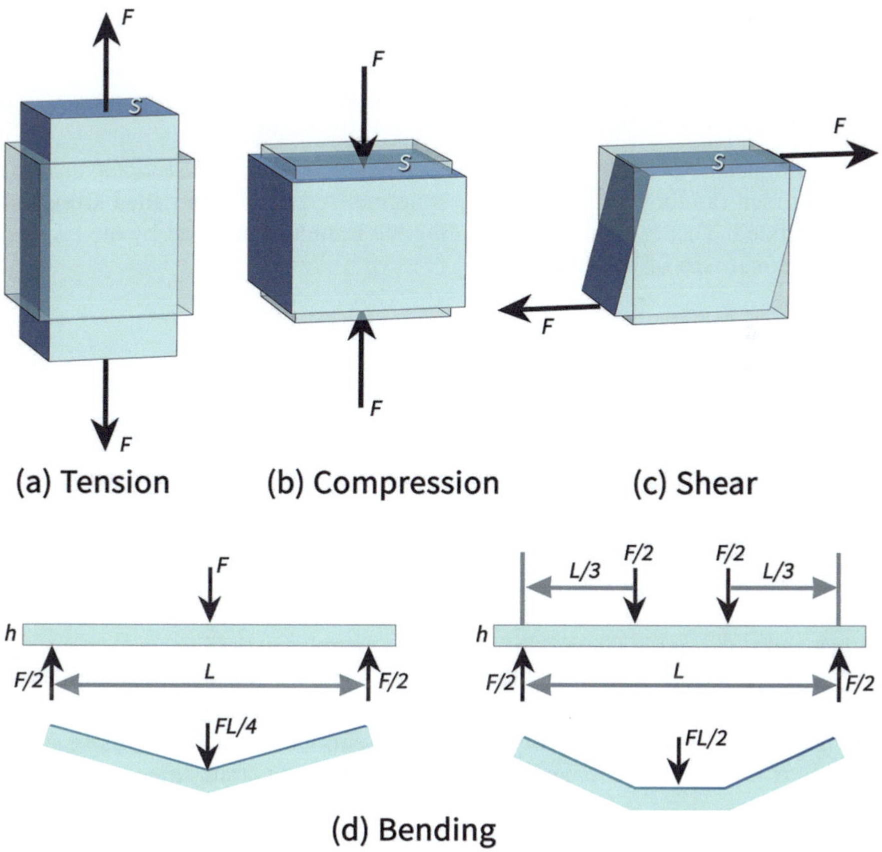

Fig. 12.1 Basic uniaxial deformation modes: **a** tension, **b** compression, **c** shear and **d** bending

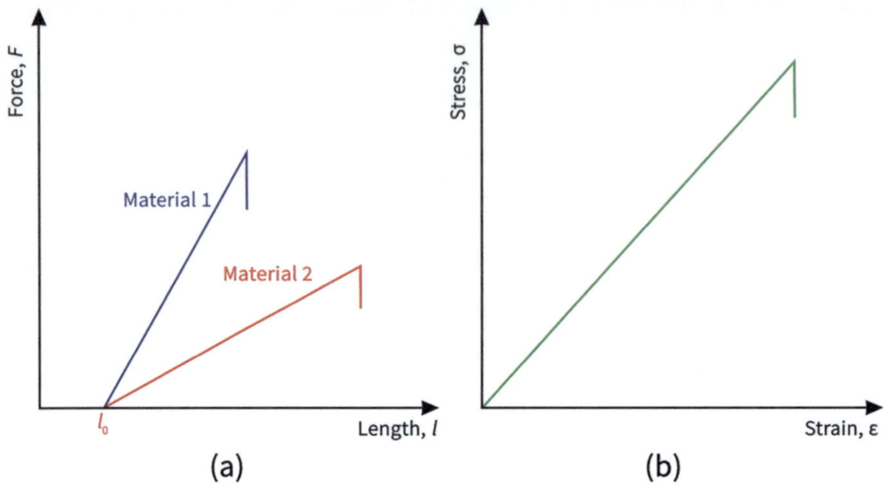

Fig. 12.2 Force-extension diagram **a** and corresponding stress–strain curve **b** for two differently sized, iso-shaped bodies

to note that the relationship between force and deformation can vary with material types.

When a tensile force (F) is applied perpendicular to the cross-sectional area (S) of a specimen (as shown in Fig. 12.1a), a normalized parameter called **stress** (σ) can be defined. This is obtained by dividing the instantaneous load by the original cross-sectional area of the specimen.

$$\sigma = F/S \tag{12.1}$$

In a similar manner, the length change in a specimen can be normalized by defining the parameter **strain** (ε). This is the ratio of the length change ($l_1 - l_0$) to the original length of the specimen (l_0).

$$\varepsilon = (l_1 - l_0)/l_0 \tag{12.2}$$

A *stress–strain curve* can be plotted as shown in Fig. 12.2b to eliminate the size effect of specimens (Fig. 12.1a). This line is obtained by plotting stress (σ) against strain (ε) and is commonly referred to as the constitutive curve or constitutive relation in mechanics. It's important to note that the stress and strain are not dependent on the size of the specimen, but rather on the material type (as shown in Fig. 12.2b). Additionally, it's worth mentioning that **compressive stress** is given as a negative value, and thus, the corresponding strain in compression is also negative.

There are two additional points to consider when examining the stress–strain relationship:

1. The stress need not be uniform across a cross-sectional area S within an inhomogeneous specimen composed of polycrystalline or multiple grains. This is because the various phases may require different stresses to achieve the same extension (of the specimen).
2. The strain is a probability normalized result along the length of the specimen. In an inhomogeneous specimen, the various grains may experience different strains along the length axis of the specimen due to their different orientations or deformation characteristics.

These points imply that the σ-ε constitutive relation is a line consisting of the *average* stress and *average* strain. Therefore, it is important to understand that Eqs. 12.1 and 12.2 are normalized by the initial cross-sectional area S and initial length l_0, respectively. Such defined *average* stress and *average* strain are also known as **engineering stress** and **engineering strain**.

As shown in Fig. 12.1c, when shear forces F act tangentially and in opposite directions on the top and bottom faces of area S of a cube, the **shear stress** τ can be defined as follows:

$$\tau = F/S \tag{12.3}$$

Shear stresses induce changes in the angle between two lines within the body, causing a rotation of planes perpendicular to the top and bottom surfaces on which the shear stresses act. This rotation leads to a displacement, "a", of the top face with respect to the bottom face. The **shear strain**, "γ", is defined as the ratio of the displacement "a" to the height "h" of the cube.

$$\gamma = a/h \tag{12.4}$$

Additionally, the shear strain, γ, can be described as a rotation angle, α, where $\tan\alpha = a/h = \gamma$. To overcome the challenges of conducting tensile tests on brittle materials, bending or flexural tests are often utilized instead. These tests involve specimens with rectangular geometries and can be performed using two testing geometries: three-point and four-point bending tests, as shown in Fig. 12.1d. The four-point bending test is preferred because it creates an extended region with a constant bending moment between the inner supporting rollers. The maximum tensile stress in the beam at the point of fracture is referred to as the **modulus of rupture** (MOR), σ_r. For a beam under low stress, MOR is related to the maximum moment of the beam, M_r. The MOR for ceramics under three- and four-point bending can be expressed using the following simple equations:

$$\sigma_r = \frac{6M_r}{BW^2} \tag{12.5}$$

3-point bending:

$$\sigma_r = \frac{3PL}{2BW^2} \tag{12.6}$$

4-point bending:

$$\sigma_r = \frac{3PD}{BW^2} \tag{12.7}$$

Here, W and B are the height and thickness of the beam, respectively. It should be noted that these equations are only applicable when the distance between the inner rollers is much greater than the specimen height.

In literature, the terms flexural strength, fracture strength, and bending strength are often used interchangeably with MOR. The bending test is also referred to as a flexural test, and the resistance of a beam to bending is known as its flexural rigidity. However, one major disadvantage of the bending test is that the stress distributions is complex and non-uniform. This can result in an overestimation of the strength, particularly if the largest flaws in the sample are located in the interior of the specimen.

② Modulus and Poisson's ratio

At relatively low levels of stress, the σ-ε curve of a specimen under tension or compression can be described by a linear relation by Hooke's law:

$$\sigma = E\varepsilon \tag{12.8}$$

where the proportionality constant, E, is called the **elastic modulus** or Young's modulus. Note that E has the same dimension as stress.

Similarly, at the relatively low levels of shear stress, a linear relation between shear stress and shear strain can hold according to Hooke's law:

$$\tau = G\gamma \tag{12.9}$$

where the constant G is called the **shear modulus** or modulus of rigidity. It has the same dimension as stress.

When a tensile stress is applied to a metal specimen, the specimen is elastically elongated in the direction of the applied stress, as shown in Fig. 12.3, resulting in a strain ε_z. At the same time, there are section reduction in the lateral (x and y) directions perpendicular to the applied stress, resulting in compressive strains ε_x and ε_y. For an isotropic material subjected to uniaxial stress (only in the z direction), the compressive strains ε_x and ε_y are equal (i.e., $\varepsilon_x = \varepsilon_y$). The ratio of the lateral and axial strains is called **Poisson's ratio** (v). The minus sign in v indicates the compressive deformation of lateral (x and y) reduction and axial (z) elongation.

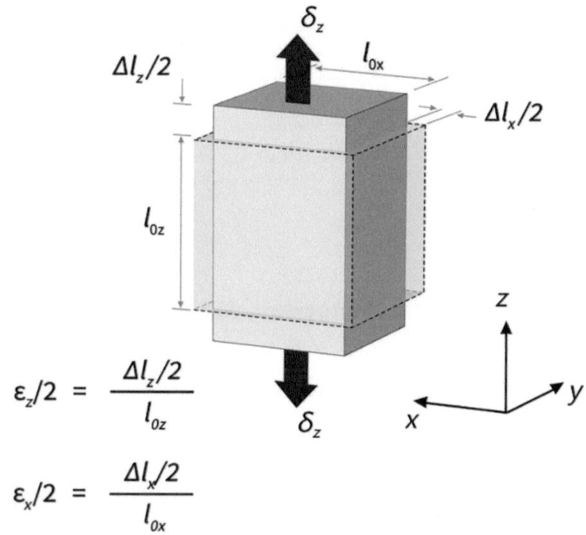

Fig. 12.3 Axial (z) elongation (positive strain) and lateral (x and y) contractions (negative strains) in response to an imposed tensile stress. Solid lines represent dimensions after stress application; dashed lines, before

$$\nu = -\frac{\varepsilon_x}{\varepsilon_z} = -\frac{\varepsilon_y}{\varepsilon_z} \tag{12.10}$$

In general, elastic deformation is characterized by three elastic constants: E, G, and ν. Of these three elastic constants, only two are independent. A relation between them is defined by:

$$E = 2G(1 + \nu) \tag{12.11}$$

Young's modulus is a property that is directly related to the interatomic bonding forces.

In the case of a composite or a material with multiple phases, it should be noted that the overall modulus of the material is a combination of the moduli of each constituent phases. There are two analytical expressions that can be used to estimate the upper and lower bounds for Young's modulus, for a two-phase material, with E_1 and E_2 being respectively the modulus for phase 1 and phase 2, and V_1 and V_2 being the volume fraction of the respective phases.

Voigt Model

Assumption: strain in each constituent is the same. Represents upper bound of Young's modulus.

$$E = V_1E_1 + V_2E_2 \tag{12.12}$$

Reuss Model

Assumption: stress in each phase is the same. Represents lower bound of Young's modulus.

$$\frac{1}{E} = \frac{V_1}{E_1} + \frac{V_2}{E_2} \tag{12.13}$$

12.2.2 Classical Mechanical Properties

① Tensile Stress–Strain Curve

The strength of a material can be determined by analyzing the stress–strain curve obtained from mechanical tests. Characteristics such as elastic modulus, yield strength, and fracture strength obtained from these curves provide important information for the design of structures. A schematic representation of a tensile stress–strain curve is presented in Fig. 12.4. In this plot, engineering stress is shown on the *y*-axis, while engineering strain is on the *x*-axis. Both stress and strain are defined with respect to the initial cross-sectional area and length of the specimen.

As shown in Fig. 12.4, the determination of the elastic limit σ_0 from the stress–strain curve can be difficult for some materials that do not exhibit clear yielding behavior. Therefore, for engineering purposes, the **yield strength** of these materials is taken as the stress that produces a certain amount of strain offset, typically 0.2%, and the yield is then denoted by the symbol $\sigma_{0.2}$. In Fig. 12.4, unloading after reaching the stress level $\sigma_{0.2}$ leaves a residual strain of 0.2% that cannot be recovered.

The maximum stress on the stress–strain curve under uniaxial tension is referred to as the **ultimate tensile strength** (UTS), denoted by σ_{UTS}. Apparently, the yield strength is a more reliable parameter than the ultimate strength for the design of

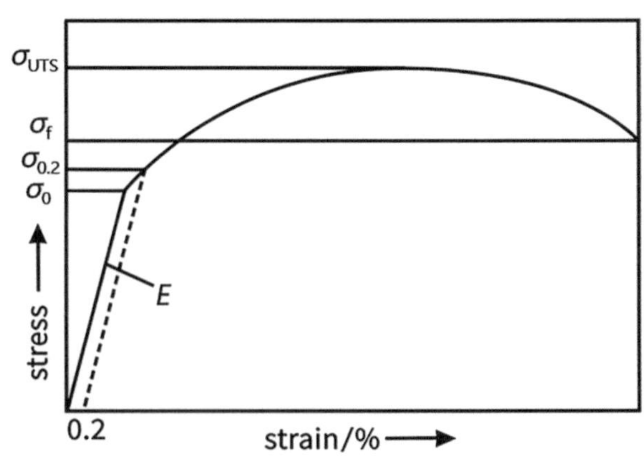

Fig. 12.4 Schematic stress–strain curve for uniaxial tensile loading

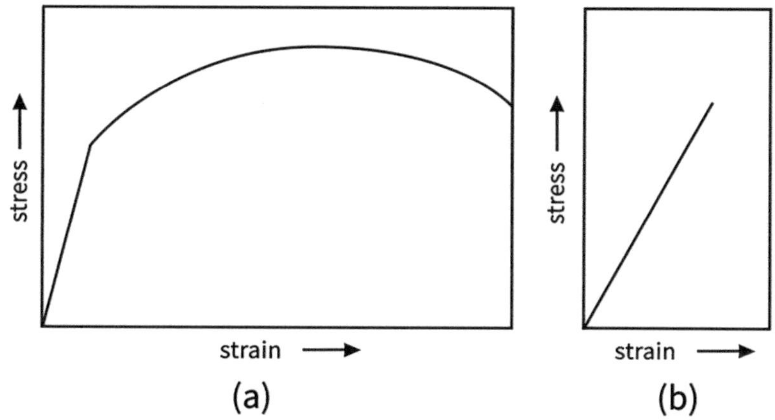

Fig. 12.5 Schematic stress–strain curves for uniaxial tensile loading of **a** a ductile and **b** a brittle material

structures subjected to loading. When the stress surpasses σ_{UTS}, it decreases with a corresponding increase in strain, accompanied by necking on the specimen. Eventually, the specimen fractures at the engineering stress level σ_{f}, which is defined as the **fracture strength**.

The first linear part of the stress–strain curve represents the **elastic deformation**, which follows Hooke's law and the specimen recovers to the original dimension after the removal of loading. The slope of the straight line in this region is the elastic modulus, E (see Eq. 12.8). The second curved part is dedicated to the **plastic deformation** for ductile materials until fracture, which corresponds to permanent and non-recoverable deformation. On the other hand, brittle materials generally do not show this plastic deformation behavior and may fracture within the linear part. Schematic stress–strain curves for ductile and brittle materials are shown in Fig. 12.5.

The transition from the elastic deformation region to the plastic deformation region occurs at the **elastic limit**, corresponding to the σ_0 in Fig. 12.4. When the load is released at a stage of modest plastic deformation (see point $\sigma_{0.2}$ in Fig. 12.4), only the reversible elastic strain as part of the total strain is recovered, while the plastic strain remains as a permanent strain. The unloading curve is almost a straight line with a slope equal to the initial elastic slope (Fig. 12.6). Renewed application of the load follows the straight line recorded upon unloading. The stress required to initiate plastic deformation now increases to a value higher than the point $\sigma_{0.2}$, and this is called strain-hardening effect.

Ductility is a measure of the extent of plastic deformation that a material undergoes before fracture. One of the major phenomena in materials science is the change in strength of a material upon plastic deformation at room temperature, which is also referred to as "work hardening" or "strain hardening" in practice. It can be expressed quantitatively as either percent elongation or percent reduction

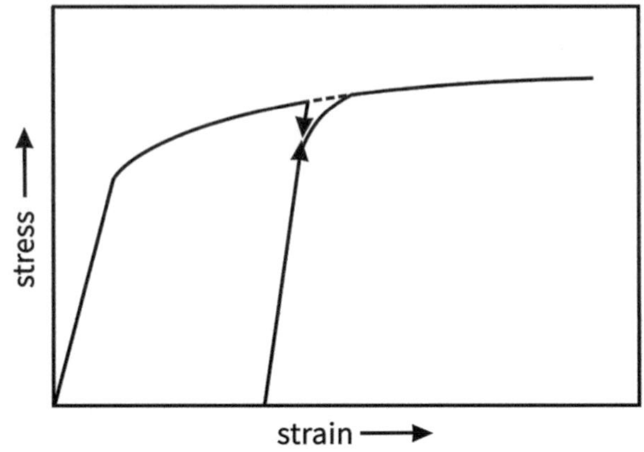

Fig. 12.6 The strain-hardening behavior of material upon unloading and subsequent reloading at a moderate stage of plastic deformation

in cross-sectional area. Percent elongation *%EL* is the percentage of plastic strain at fracture and can be calculated using the following formula:

$$\%EL = \left(\frac{l_f - l_0}{l_0}\right) \times 100 \tag{12.14}$$

where l_f is the fracture length and l_0 is the original length as given earlier.

Percent reduction in area *%RA* is defined as

$$\%RA = \left(\frac{A_0 - A_f}{A_0}\right) \times 100 \tag{12.15}$$

where A_0 is the original cross-sectional area and A_f is the cross-sectional area at the point of fracture.

It is customary to assume that the deformable material is plastically isotropic without volume change, which leads to the incompressibility relation that the sum of the three normal strains is zero:

$$\varepsilon_x P + \varepsilon_y P + \varepsilon_z P = 0 \tag{12.16}$$

These three normal strains are independent components. Thus, for *ideal plastic deformation,* Poisson's ratio, *v*, is normally assumed to be 0.5 (see Eq. 12.10).

Metals are known for their high ductility, which means they can be subjected to severe plastic deformation without breaking. Industrial processes such as rolling, forging, drawing, and hot iso-static pressing rely on such plastic deformation to achieve their desired shapes. To explain plastic deformation, much of the plasticity theory is based on the concept of dislocations.

The main difference between elastic and plastic deformation is that elastic deformation preserves the local atomic arrangements, while plastic deformation requires the breaking and re-establishment of atomic bonds. Consequently, the plastic deformation mechanisms in crystalline and amorphous solids are essentially different: slip of dislocations is the primary plastic deformation mechanism for crystalline solids, while viscous flow is the primary mechanism for amorphous solids.

② True stress and strain

During a loading experiment, the dimensions of the specimen change, which affects the calculation of the engineering strain, defined as:

$$\varepsilon = (l - l_0)/l_0 \quad (12.17)$$

where l_0 and l are the initial and current lengths of the specimen, respectively. For a more accurate measure of strain, the true strain ε_{true} should be defined with respect to the instantaneous length of the specimen:

$$\begin{aligned} \varepsilon_{\text{true}} &= (l_1 - l_0)/l_0 + (l_2 - l_1)/l_1 + (l_3 - l_2)/l_2 \\ &= \sum (l_{i+1} - l_i)/l_i \\ \int (1/l)\, \mathrm{d}l &= \ln(l/l_0) \end{aligned} \quad (12.18)$$

Combining Eq. 12.17 and Eq. 12.18, we get:

$$\varepsilon_{\text{true}} = \ln(\varepsilon + 1) \quad (12.19)$$

Since no volume change occurs during ideal plastic deformation, we can use the symbols S and S_0 to represent the cross-sectional areas at different loading stages.

$$S_0 l_0 = Sl \text{ or } S = (S_0 l_0)/l$$

Then, we can express the true stress σ_{true} as:

$$\sigma_{\text{true}} = F/S = (F/S_0)(l/l_0) = \sigma(1 + \varepsilon) \quad (12.20)$$

where F is the applied force.

Figure 12.7 provides a schematic presentation of the engineering stress–engineering strain curve and the corresponding true stress–true strain curve.

The equations presented for $\varepsilon_{\text{true}}$ and σ_{true} are only valid up to the ultimate tensile strength. Beyond this point, the loading is no longer uniaxial, and **necking** occurs in a complex manner at the site of the specimen. The true stress and true strain curve, from the point of yielding until necking starts, is essential for the theory of plasticity. A commonly used, purely phenomenological, but often inadequate representation of this curve is:

$$\sigma_{\text{true}} = K'(\varepsilon_{\text{true}})n \quad (12.21)$$

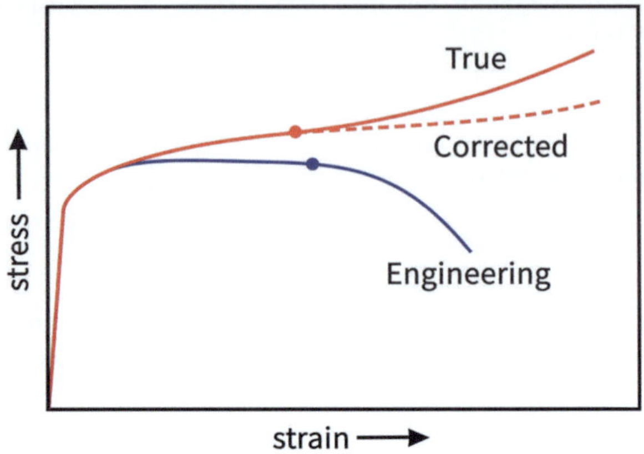

Fig. 12.7 Schematic engineering stress–strain curve for uniaxial tensile loading and the corresponding true stress–strain curve

K' is a parameter that represents the hypothetical true stress required to achieve a hypothetical true strain of 100%. The exponent $\boldsymbol{n}$, also known as the "**strain-hardening exponent**" indicates the degree of strain hardening exhibited by the material. For values of n between 0 and 1, a higher value of n corresponds to a higher true stress required to achieve the same true strain. During tensile loading, the cross-sectional area of the material may gradually decrease, leading to an increase in true tensile stress. If the material exhibits strain hardening, its intrinsic capacity to carry a load can increase as plastic deformation progresses. This can result in a delay in the definitive occurrence of necking if the local initiation of necking can be repaired.

The true stress at which necking occurs corresponds to the maximum load. Therefore, the instability criterion (also described by Hoffman and Sachs in 1953) indicates that no further changes in load can occur beyond this point.

$$\mathrm{d}F = \mathrm{d}(\sigma_{\mathrm{true}}S) = \sigma_{\mathrm{true}}\mathrm{d}S + S\mathrm{d}\sigma_{\mathrm{true}} = 0 \tag{12.22}$$

and thus

$$\mathrm{d}\sigma_{true}/\sigma_{true} = -\mathrm{d}S/S \tag{12.23}$$

Because of the constant volume that holds for ideal plastic deformation

$$\mathrm{d}S/S = -\mathrm{d}l/l = -\mathrm{d}\varepsilon_{\mathrm{true}} \tag{12.24}$$

Combining ($dF = d(\sigma_{true}S) = \sigma_{true}\,dS + S d\sigma_{true} = 0$ Eq. 12.21) and ($dS/S = -dl/l = -d\varepsilon_{true}$ Eq. 12.23) it follows

$$d\sigma_{true}/d\varepsilon_{true} = \sigma_{true} \tag{12.25}$$

In summary, **Considère's criterion** states that definitive necking occurs at the point on the true stress-true strain curve where the slope equals the true stress. It should be noted that this criterion applies specifically to uniaxial tensile loading. For other types of loading, such as bi-axial or tri-axial loading, alternative strength parameters may be defined that more closely relate to the strength of the material under the specific applied loading conditions.

③ Typical Moduli and Poisson's Ratio of Materials

The elastic properties of a material are represented by the modulus of elasticity (E), shear modulus (G), and Poisson's ratio (v). The E and G have the same dimensions as stress and are typically expressed in gigapascals (GPa), or 10^9 pascals. For example, the elastic modulus of steel is approximately 200 GPa, while ceramics often have higher values of E, such as silicon carbide with an elastic modulus of approximately 450 GPa. In contrast, polymers exhibit much smaller values of E, such as Nylon 6,6 and Polystyrene with E values of approximately 3 GPa. It's worth noting that the testing methods for determining E are different for metals, ceramics, and polymers. Table 12.1 shows the values of E, G, and v tested at room temperature for three different typical materials.

④ Anelastic and Viscoelastic Deformation

As mentioned earlier, while the strain in elastic materials like many metallic solids can disappear instantaneously after the release of the applied load, other materials exhibit different behavior. For example, many polymers at temperatures above the *glass transition temperature*, T_g, behave as rubbery solids. When subjected to loading, the strain induced develops as a function of time, and a part of the strain may not be immediately recovered but requires some time to release completely after the load is released. This time-dependent elastic behavior is known as **anelasticity**, and it is due to time-dependent microscopic and atomic processes that occur during deformation. For metals, the anelastic part is typically small and is often neglected. However, for some polymeric materials, its magnitude is significant, and in this case, it is referred to as **viscoelasticity**. This type of deformation can be broken down into two parts: linearly elastic (time-independent) and viscous (time-dependent), as shown in Fig. 12.8.

⑤ Fracture Mechanics

The intrinsic fracture toughness of materials can be determined by introducing artificial defects such as notches in the test material. Parameters of fracture dynamics include Crack Tip Opening Displacement test (CTOD), which takes into account plastic behavior, and the values G and K, which treat fracture as linear elastic behavior. Since most ceramic materials can be treated as elastic, only the value K

Table 12.1 Room-Temperature elastic and shear moduli and poisson's ratios for various materials

Materials	Modulus of elasticity, E/GPa	Shear modulus, G/GPa	Poisson's ratio, ν
Metal Alloys			
Tungsten	407	160	0.28
Steel	207	83	0.30
Nickel	207	76	0.31
Titanium	107	45	0.34
Copper	110	46	0.34
Brass	97	37	0.34
Aluminum	69	25	0.33
Magnesium	45	17	0.35
Ceramic materials			
Aluminum oxide (Al_2O_3)	393	–	0.22
Silicon carbide (SiC)	345	–	0.17
Silicon nitride (Si_3N_4)	304	–	0.30
Spinel ($MgAl_2O_4$)	260	–	–
Magnesium oxide (MgO)	225	–	0.18
Zirconia (ZrO_2)	205	–	0.31
Mullite ($3Al_2O_3$-$2SiO_2$)	145	–	0.24
Glass–ceramic (Pyroceram)	120	–	0.25
Fused silica (SiO_2)	73	–	0.17
Soda–lime glass	69	–	0.23
Polymers			
Phenol–formaldehyde	2.76–4.83	–	–
Poly(vinyl chloride) (PVC)	2.41–4.14	–	0.38
Poly(ethylene terephthalate) (PET)	2.76–4.14	–	0.33
Polystyrene (PS)	2.28–3.28	–	0.33
Poly(methyl methacrylate) (PMMA)	2.24–3.24	–	0.37–0.44
Polycarbonate (PC)	2.38	–	0.36
Nylon 6,6	1.59–3.79	–	0.39
Polypropylene (PP)	1.14–1.55	–	0.40

(continued)

Table 12.1 (continued)

Materials	Modulus of elasticity, E/GPa	Shear modulus, G/GPa	Poisson's ratio, ν
Polyethylene—high density (HDPE)	1.08	–	0.46
Polytetrafluoroethylene (PTFE)	0.40–0.55	–	0.46
Polyethylene—low density (LDPE)	0.17–0.28	–	0.33–0.40

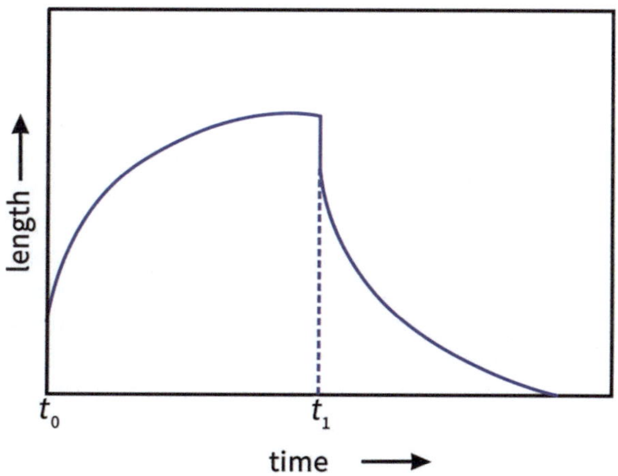

Fig. 12.8 Schematic length–time curve of a viscoelastic material under tensile loading and unloading

needs to be considered. The value G is seldom used and can be calculated from the value K. The ASTM standard has been adopted internationally as the method of measuring the linear fracture toughness value K_{IC} for metals (ASTM Standard Book). It is also desirable to determine the fracture toughness of ceramics based on this standard. Figure 12.9 shows the bending test specimen's shape for which this standard seems applicable for ceramics.

According to the ASTM standard for measuring K_{IC} in metals, fatigue cracks are induced at the tip of the notch. The strength of the material at the point of fracture is greater than its resistance to slipping, so there is typically some slipping deformation that occurs before the material fractures at the crack tip.

$$K_{\mathrm{IC}} = \left(\frac{Pl}{BW^{\frac{3}{2}}}\right) f\left(\frac{a}{W}\right) \tag{12.26}$$

$$f(a/W) = \frac{3(a/W)^{1/2}\left[1.99 - (a/W)(1 - a/W)(2.15 - 3.93a/W + 2.7a^2/W^2)\right]}{2(1 + 2a/W)(1 - a/W)^{3/2}} \tag{12.27}$$

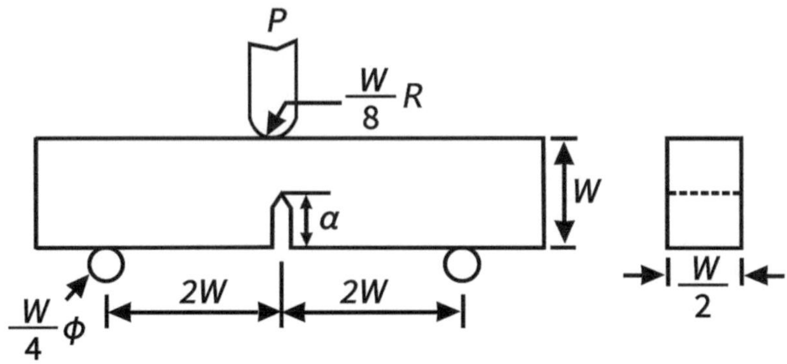

Fig. 12.9 Fracture toughness test specimen

where P, a, W and B are the fracture load and the notch length, width and depth, respectively. The value K_{IC} derived is the intrinsic value for the material tested.

Equation 12.26 provides the value of K_{IC} based on the fracture load (P), and the dimensions of the notch (a, W, and B). The intrinsic value of K_{IC} is a material property. On the other hand, the value of K, expressed as $K = f(a/W)$, depends on the dimensions of the test specimen. Standard references provide $f(a/W)$ values for various sample geometries. The calculated K_{IC} value can then be used to compare the toughness of different materials and to determine the maximum permissible load or defect size.

⑥ Fatigue

In materials science, fatigue is the failure of a material caused by cyclic loading that results in progressive, brittle and localized structural damage. Once a crack has initiated, each loading cycle will grow the crack a small amount, even when repeated alternating or cyclic stresses are of an intensity considerably below the normal strength. The stresses could be due to vibration or thermal cycling. Fatigue damage is caused by: (a) simultaneous action of cyclic stress, (b) tensile stress (whether directly applied or residual), and (c) plastic strain.

The majority of engineering failures are caused by fatigue. Although the fracture is of a brittle type, it may take some time to propagate, depending on both the intensity and frequency of the stress cycles. Nevertheless, there is very little, if any, warning before failure if the crack is not noticed. The number of cycles required to cause fatigue failure at a particular peak stress is generally quite large, but it decreases as the stress is increased. For some mild steels, cyclical stresses can be continued indefinitely provided the peak stress (sometimes called fatigue strength) is below the endurance limit value. A good example of fatigue failure is breaking a thin steel rod or wire with your hands after bending it back and forth several times in the same place. The type of fatigue of most concern in nuclear power plants is thermal fatigue. Thermal fatigue can arise from thermal stresses produced by cyclic changes in temperature. Large components like the pressurizer, reactor vessel, and reactor system piping are subject to cyclic stresses caused by temperature variations during reactor startup, change in power level, and shutdown.

The S–N Curve

Fatigue life, N_f, is the number of stress cycles of a specified character that a specimen sustains before failure of a specified nature occurs. Fatigue life is affected by cyclic stresses, residual stresses, material properties, internal defects, grain size, temperature, design geometry, surface quality, oxidation, corrosion, etc. For some materials, notably steel and titanium, there is a theoretical value for stress amplitude below which the material will not fail for any number of cycles, called a fatigue limit, endurance limit, or fatigue strength.

Engineers use a number of methods to determine the fatigue life of a material. One of the most useful is the stress-life method, which is commonly characterized by an S–N curve, as illustrated in the Fig. 12.10. It plots applied stress (S) against number of cycles to failure (N). The following terms are defined for S–N curve.

Fatigue limit (also called the endurance limit) is the stress level, below which fatigue failure does not occur. This limit exists for some ferrous alloys, for which the S–N curve becomes horizontal at higher N values (Fig. 12.10a). Typical values of the fatigue limit for steels are 35~60% the tensile strength.

Other structural metals, such as aluminium and copper, do not have a distinct limit and will eventually fail even from small stress amplitudes. For these materials, the following two terms are defined. **Fatigue strength** is the value of stress at which failure occurs after some specified number of cycles (e.g., 10^7 cycles). **Fatigue life** is the number of cycles to cause failure at a specified stress level. Both of them are taken from the S–N curve, as shown in Fig. 12.10b.

Because fatigue like brittle fracture has such a variable nature, the data used to plot the curve will be treated statistically. The scatter in results is a consequence

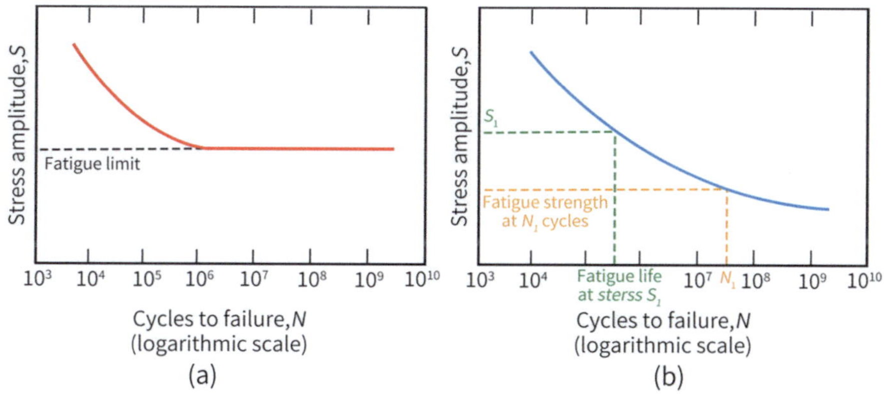

Fig. 12.10 The typical fatigue S–N curve for a material **a** with fatigue limit and **b** without fatigue limit

of the fatigue sensitivity to a number of test and material parameters that are impossible to control precisely.

Fatigue Failure

The process of fatigue failure is characterized by three distinct stages:

1. Crack initiation: a small crack forms at some point of high stress concentration.
2. Crack propagation: the crack advances incrementally with each stress cycle. Most of the fatigue life is generally consumed in the crack propagation stage.
3. Ultimate failure: it occurs very rapidly once the advancing crack has reached a critical size.

Cracks associated with fatigue failure almost always nucleate on the surface of a component at some point of stress concentration. Anything which leads to stress concentration, and the development of cracks, will reduce fatigue life. Therefore, increasing the degree of surface finish, polishing as compared to grinding, improves fatigue life. Increasing the strength and hardness of the surface layers of metal components will also improve fatigue life.

High Cycle Fatigue Versus Low Cycle Fatigue

Fatigue has been separated into regions of high cycle fatigue (HCF) and low cycle fatigue (LCF). The main difference between HCF and LCF is the number of cycles to failure. Transition between LCF and HCF is determined by the stress level, i.e. transition between plastic and elastic deformations.

HCF require more than 10^4 cycles to failure where stress is low and primarily elastic, while LCF is characterized by repeated plastic deformation in each cycle and hence the small number of cycles to failure. Large changes in strain can be produced by small changes in stress during plastic deformation. Fatigue failures, both for HCF and LCF, all follow the same basic steps process of crack initiation, crack propagation, and ultimate failure.

⑦ Creep Behavior (Fatigue is even more important a property than creep. We must introduce FATIGUE)

Many metal parts, such as steam boilers, turbines, and aircraft engines, are exposed to high temperatures for extended periods. To account for the effects of high temperature and continuous loading over time, engineers use the concept of creep. Creep refers to the slow plastic deformation of a material under continuous stress at a given temperature. The stress that causes creep is known as the creep stress. As the material is subjected to this continuous stress, creep deformation gradually increases and can ultimately lead to fracture, which is called creep fracture as shown in Fig. 12.11.

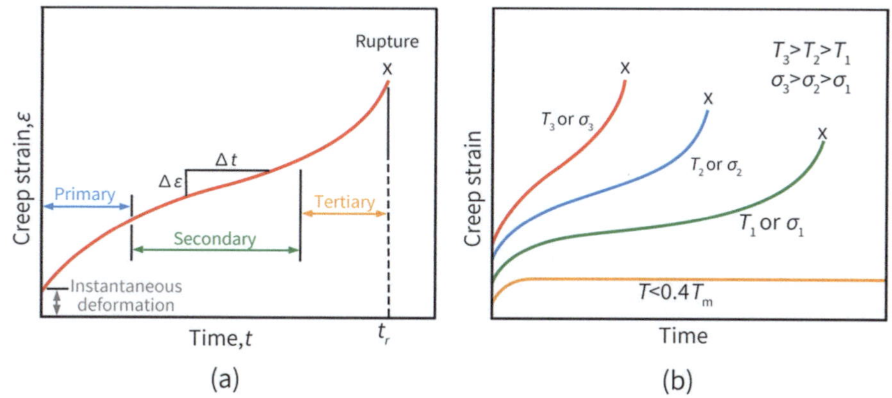

Fig. 12.11 **a** The typical creep curve and **b** the influence of stress and temperature on creep behavior

Creep Curve

As seen from Fig. 12.11, creep is time-dependent and undergoes through several stages:

Primary Creep: In the initial stage, known as primary or **transient creep**, the strain rate is relatively high but decreases with time as the material experiences an increase in creep resistance or strain hardening.

Secondary Creep: During the secondary stage, sometimes referred to as **steady-state creep**, the strain rate becomes nearly constant and is balanced by work hardening and annealing (thermal softening). This stage of creep generally lasts for the longest time span and is the most well-understood stage. No material strength is lost during the primary and secondary stages of creep.

Tertiary Creep: In the tertiary stage, the strain rate accelerates exponentially with stress due to the formation of internal cracks, cavities, or voids. Necking phenomena occur, leading to a decrease in the effective cross-sectional area and an increase in the strain rate. Strength is rapidly lost in this stage, and the material's shape is permanently changed. The acceleration of creep deformation in the tertiary stage eventually leads to failure, often referred to as rupture, and is caused by microstructural and/or metallurgical changes.

The slope of the secondary portion of the creep curve ($\Delta\varepsilon/\Delta t$ in Fig. 12.11) is the most fundamental parameter in a creep test and is known as the minimum or **steady-state creep rate**. This parameter is the engineering design consideration for long-life applications, such as nuclear power plant components that are used for several decades and where failure or excessive strain is unacceptable. Additionally, time to rupture or the **rupture lifetime**, t_r, is an important design consideration for many relatively short-life creep cases, such as turbine blades in military aircraft and rocket motor nozzles. Creep rupture tests must be conducted to the failure point to determine t_r. The creep properties of a material allow the design engineer to determine its suitability for a particular application.

Stress and Temperature Effect on Creep

Creep characteristics are influenced by both temperature and the applied stress, as shown in Fig. 12.11. At temperatures well below $0.4T_m$, after initial deformation, the strain is essentially independent of time. However, with either increasing stress or temperature, the following effects are observed: (1) the instantaneous strain at the time of stress application increases, (2) the steady-state creep rate increases, and (3) the rupture lifetime decreases.

Empirical relationships have been developed in which the steady-state creep rate as a function of stress and temperature is expressed. Its dependence on stress can be written:

$$\dot{\varepsilon_S} = K_1\sigma^n \tag{12.28}$$

where K_1 and n are material constants. When the influence of temperature is included,

$$\dot{\varepsilon_S} = K_2\sigma^n \exp\left(-\frac{Q_c}{RT}\right) \tag{12.29}$$

where K_2 and Q_c are constants and Q_c is the activation energy for creep.

Various theoretical mechanisms have been suggested to explain the creep behavior of different materials, including stress-induced vacancy diffusion, grain boundary diffusion, dislocation motion, and grain boundary sliding. Each mechanism results in a different **stress exponent** (n) value. By comparing the experimental n value of a specific material with the predicted values of the different mechanisms, it is possible to elucidate the creep mechanism for that material. Additionally, correlations have been established between the activation energy for creep (Q_c) and the activation energy for diffusion (Q_d).

12.3 Deformation and Strength of Metals

Metals inevitably undergo deformation under external forces in the fabrication and application processes. Deformation can be elastic or plastic depending on the magnitude of the external force. At the macro level, there are various deformation modes, such as tension, compression, torsion, and complex ones. However, at the microscopic level, plastic deformation of single crystals occurs primarily through two basic ways: slip and twinning. Both of these mechanisms involve the movement of atoms and dislocations under applied stress. This section will demonstrate the fundamental role of dislocations in plastic deformation.

12.3.1 Plastic Deformation of Single Crystals

When a material is loaded, it undergoes deformation. If the applied force is removed, the material may recover its original dimensions, which is called elastic deformation. Elastic deformation follows Hooke's law and is reversible, which means that it does not alter the structure of the material. When the stress exceeds the elastic limit, the material undergoes plastic deformation, resulting in irreversible or permanent deformation. Plastic deformation is crucial in the processing and application of crystalline materials, especially metals. Although most commonly used metals are polycrystalline, the deformation of polycrystalline materials is based on that of each single crystal. Therefore, it is essential to understand the basic process of single crystal deformation, which is mainly affected by slip, twinning, and kinking at room temperature.

① Slip

Slip Line and Slip Band

When stress surpasses the crystal's elastic limit, slips occur between the sheets of the crystal, leading to macroscopic plastic deformation of the crystal. The slip lines are spaced approximately 100 atoms apart, and the amount of slip along each slip line can reach around 1000 atom distances, as shown in Fig. 12.12. The observation of slip lines also indicates that the plastic deformation of the crystal is non-uniform. Slip occurs only on certain crystal planes, while the crystal layers between slip bands or slip lines do not deform but only move relative to each other.

Slip System

As mentioned earlier, during plastic deformation, dislocations move only along certain crystal planes and crystal directions, called **slip planes** and **slip directions**,

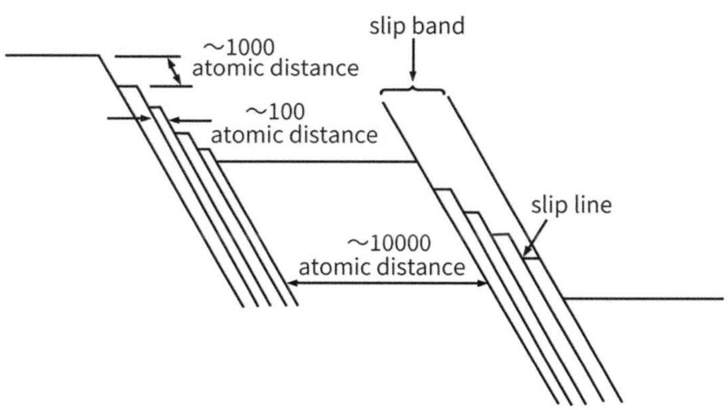

Fig. 12.12 Formation mechanism of slip bands

respectively (as shown in Fig. 12.13). A **slip system** is a set of symmetrically identical slip planes and the associated family of slip directions on which dislocation motion can easily occur and lead to plastic deformation. The magnitude and direction of slip are represented by the Burgers vector. The slip plane and slip direction are typically the crystal plane and crystal direction with the densest packing of atoms. The resistance to dislocation motion primarily arises from the lattice resistance, known as the Peierls-Nabarro force (P-N force), which is equivalent to the critical shear stress required for edge dislocation movement in simple cubic crystals. The P-N force is given by the expression:

$$\tau_{P-N} = \frac{2G}{1-\nu}\exp\left[-\frac{2\pi a}{b}\right] \tag{12.30}$$

where a is the planar spacing of the slip planes, b is the lattice spacing in the slip direction, and v is Poisson's ratio, and G is the shear modulus. The spacing of the densest packing plane is the largest, resulting in the least lattice resistance and hence the easiness of slip. The spacing of the lattice along the direction of the densest packing is the smallest, leading to the smallest Burgers vector of the slip.

Different crystal structures have different slip systems. In metal crystals, the typical slip systems in common structures such as face-centered cubic (FCC),

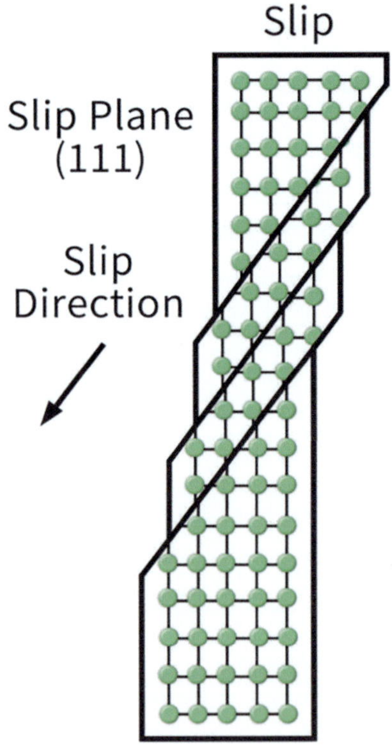

Fig. 12.13 Slip deformations in a face-centered cubic crystal

body-centered cubic (BCC), and hexagonal close-packed (HCP) structures are as follows.

Slip systems in FCC crystals: FCC crystals have a {111} slip plane and <110> slip direction, leading to 12 slip systems.

Slip systems in BCC crystals: As the BCC structure is not a close-packed structure, the slip plane of the BCC structure is unstable and is generally {112} at low temperature, {110} at medium temperature, and {123} at high temperature. However, the slip direction of the BCC structure is stable at <111>. Therefore, there are 12 to 48 possible slip systems.

Slip systems in HCP crystals: In HCP crystals, the slip direction is generally <1120>, but the slip plane depends on the axial ratio. When c/a is close to or greater than 1.633, the most dense-packed plane is {0001}, providing three slip systems of {0001} <1120>. When c/a is less than 1.633, the slip plane becomes {1010} or {1011}, corresponding to 3 or 6 slip systems, respectively. The number of operating slip systems varies with temperature, and HCP crystals have poor ductility due to their limited slip systems.

Not all slip systems may be activated under external force, and the number of operating slip systems may increase with temperature.

Critical Resolved Shear Stress

Crystal slip occurs under the effect of shear stress, but not all slip systems are involved simultaneously. The shear stress on the slip plane along the slip direction is what truly contributes to the slip. Only when the shear stress reaches a critical value, the slip process can start. **Critical resolved shear stress (CRSS)** is the component of shear stress, resolved in the direction of slip, required to initiate slip in a grain.

Assuming a cylindrical single crystal with cross-sectional area A, φ is the included angle between the normal line of the slip plane and the central axis. λ is the angle between the slip direction and the external force F, as shown in Fig. 12.14. The component force of the external F in the slip direction is $F\cos\lambda$, and the area of the slip surface is $A/\cos\varphi$. Then, the shear stress in the slip direction is given by:

$$\tau = \frac{F\cos\lambda}{A/\cos\varphi} = \frac{F}{A}\cos\lambda\cos\varphi = \sigma\cos\lambda\cos\varphi \tag{12.31}$$

When the shear stress in Eq. 12.31 reaches the critical value, slip between the crystal planes begins, which corresponds to the macroscopic yield. At this point, F/A should be equal to σ_S, and the critical resolved shear stress τ_S can be determined by:

$$\tau_S = \sigma_S(\cos\phi\cos\lambda) \tag{12.32}$$

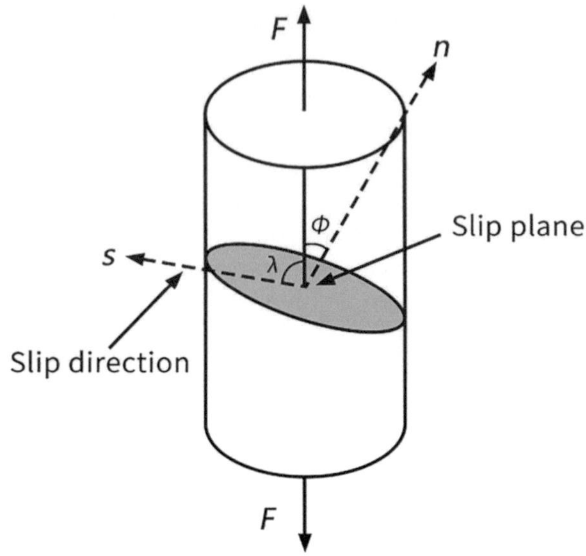

Fig. 12.14 Analysis diagram for calculating the resolved shear stress

The CRSS is related to the nature of the material, the test temperature, and the loading speed and is independent of the loading direction. The orientation factor or **Schmid factor**, $\cos\phi\cos\lambda$, represents the likelihood that a particular slip system will operate. The critical shear stress required for the initiation of the slip system is a constant and independent of the orientation of the external force, known as Schmid's law or Critical-shear-stress law. If the normal stress corresponding to the initiation of the slip plane is regarded as the yield strength, the yield strength is related to the orientation of the applied force. A material with a large value of $\cos\phi\cos\lambda$ can reach τ_S and slip under the application of a small σ_S, known as a soft orientation. Conversely, a hard orientation requires a larger σ_S to initiate slip.

Rotation of Crystal Plane During Slip

Figure 12.15a illustrates the process of crystal slip, showing the assumed slip plane and slip direction. When the axial tension F is large enough, each part of the crystal will undergo delamination movement, i.e., slip. If both ends of the crystal are free, the slip will shift the axis of the crystal (Fig. 12.15b). However, if the two ends of the crystal cannot move horizontally or the tensile axis remains fixed, then the orientation of a single crystal must be rotated accordingly. As a result of the rotation, the slip plane gradually tends to be parallel to the stress axis. Near the clamping head, the crystal plane will bend to a certain extent due to the limitation of the clamping head (Fig. 12.15c). The rotation results in the slip plane and slip direction tending to be parallel to the tensile direction.

Similarly, when the crystal is under compressive deformation, the crystal plane rotates correspondingly, and the result of rotation makes the slip plane roughly perpendicular to the axis of compression, as shown in Fig. 12.16.

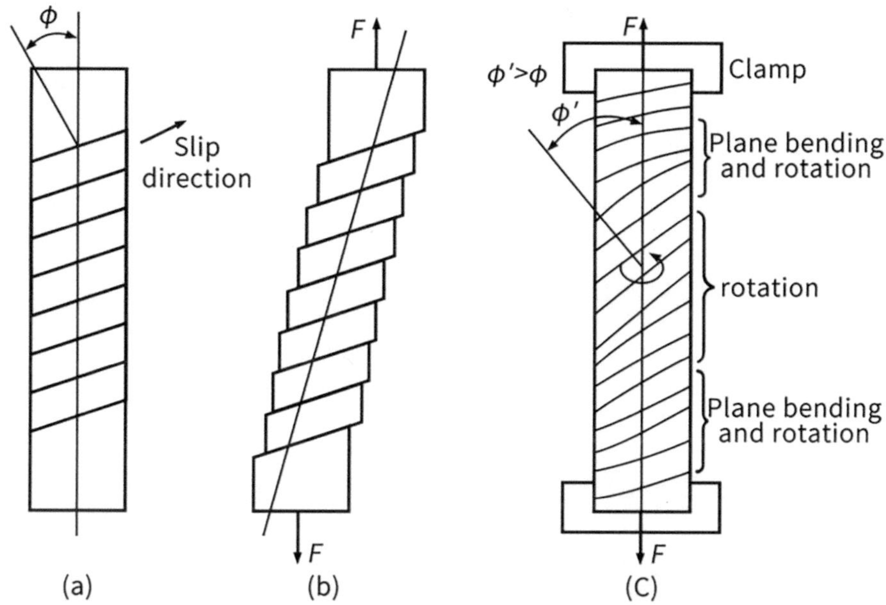

Fig. 12.15 Tensile deformation of a single crystal with slip deformation

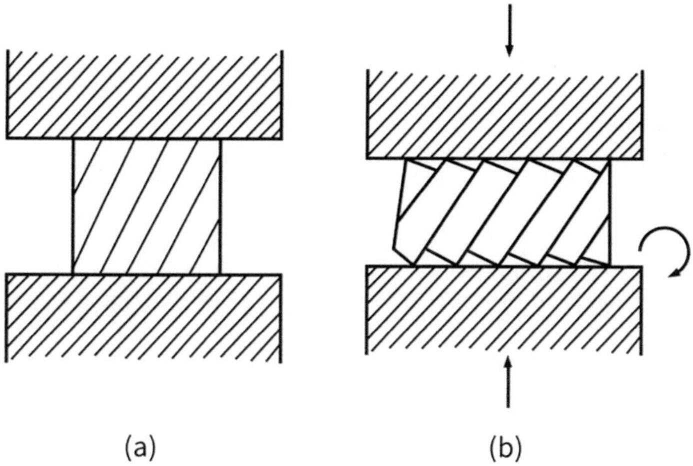

Fig. 12.16 Rotation of the crystal planes during compression

Taking the case of uniaxial tension as an example, Fig. 12.17 shows three layers adjacent to the two typical slip planes in the crystal. Before slip, the force acting on the center layer of the crystal is located at the O_1 and O_2 points. When slip begins, due to the relative movement of layers, the application points of force

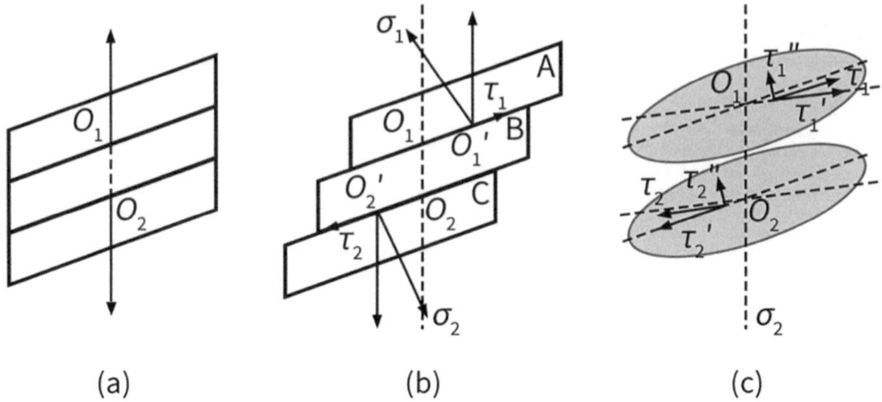

Fig. 12.17 The role of the force couple originating from crystal rotation during uniaxial tension

move to O_1' and O_2', respectively. Then, the acting forces perpendicular to the slip plane can be decomposed into σ_1 and σ_2, and those parallel to the slip plane can be decomposed into τ_1 and τ_2. The force couple of σ_1 and σ_2 causes the rotation of the slip plane toward the tensile axis. The two components τ_1 and τ_2 in the slip plane can be further decomposed along the direction parallel to and perpendicular to the slip direction. The component parallel to the slip direction ($\tau_1^{'}$ and $\tau_2^{'}$) is the shear stress causing the slip, and the other two components ($\tau_1^{''}$ and $\tau_2^{''}$) form a couple that causes the slip direction to rotate to the direction of the maximum shear stress.

Multi-slip System

When a crystal has multiple slip systems, slip initially occurs in the system with the best orientation (highest shear stress). However, due to crystal rotation during deformation, the resolved shear stress on other slip planes may gradually increase over the critical value, resulting in slip. As a result, crystal slip may occur simultaneously or alternately in two or more sets of slip planes, resulting in a **multi-slip system**.

For crystals with multiple slip systems, **cross slip** is often observed, where two or more slip planes slip simultaneously or alternately along a common slip direction. Cross slip involves the screw dislocation transitioning from one slip plane to another at the junction, without changing the slip direction. Cross slip makes slip more flexible, but dislocation movement becomes more challenging in the case of a multi-slip system because the dislocations from different slip systems interact, making this an essential strengthening mechanism.

Dislocation Mechanism of Slip

Crystals do not slip as a single rigid body, but through dislocation motion on the slip plane. A dislocation serves as the boundary between the slip and nonslip zones. Therefore, the critical shear stress required for crystal slip should be equal to the external force needed to overcome the resistance of dislocation motion at the micro level. In pure metals, the resistance to dislocation motion mainly arises from four factors: (1) Lattice resistance, (2) Interaction resistance with other dislocations, (3) Jogs and kinks formed when dislocations intersect, and (4) Interaction of dislocations with point and planar defects.

② Twinning

Twinning is a common method for inducing plastic deformation in crystals. It involves subjecting one part of a crystal to uniform shear along a specific plane (**twinning plane**) and direction (**twinning direction**) with respect to another part, under the influence of shear stress. Twinning occurs when slip becomes difficult. For instance, crystals with HCP structures, such as Zn and Mg, often deform by twinning as they have few slip systems.

When comparing the structural and shape changes resulting from slip and twinning under shear stress, it is evident that twinning is a uniform shear process, whereas slip is an inhomogeneous shear process. The twinned part forms mirror symmetry with the primary crystal, while the slipped part undergoes no orientation change.

The Deformation Process of Twinning

During twinning, each plane of atoms moves through a specific distance in the same direction, proportional to its distance from the twinning plane. The extent of movement increases for successive planes. Under shear stress, the crystal will twin along the twinning plane, and the atomic arrangements on either side of the twinned plane are mirror images of each other.

For example, in the FCC structure, Fig. 12.18a shows twinning planes and directions, and Fig. 12.18b shows the atomic arrangement of $(1\bar{1}0)$ where layers of (111) planes stack in the sequence of ABCABC. A uniform shear of $\frac{a}{6}[11\bar{2}]$ between several layers of (111) in the local area along the direction $[11\bar{2}]$ results in twinning deformation. The sheared region maintains the FCC structure, but its orientation changes to be mirror symmetric with the initial region. The two regions of the crystal are called twins, and the interface between them is the twin boundary. The **twinning plane** is the uniform shear plane, and the **twinning direction** is the moving direction of the twinning plane.

Formation of Twins

Crystals can form twins through three main ways. Mechanical deformation produces deformation twins or **mechanical twins**, which have lenticular or flaky microstructural features. **Growth twins** form during crystal growth from a gas, liquid or solid state. **Annealing twins** form during recrystallization annealing of deformed metals and traverse entire grains with twinning planes as parallel boundaries. Twin growth can be categorized into **nucleation and growth**, where the crystal first undergoes nucleation of thin sheet twins at a fast rate and then widens

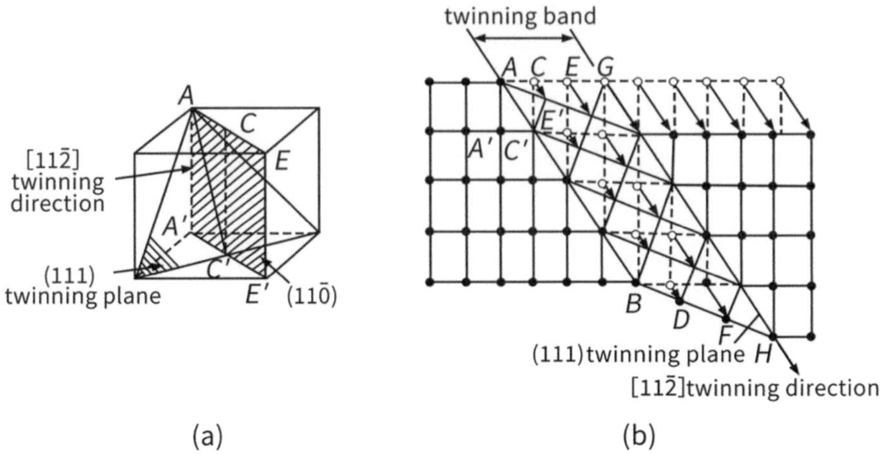

Fig. 12.18 Schematic twinning deformation in a FCC crystal

the twins by extending twin boundaries. Although twinning itself has a smaller direct contribution to crystal deformation than slip, it alters the crystal orientation to favor slip systems and stimulates further slip and deformation. Slip and twinning alternate to achieve a significant deformation ability for the crystals.

Dislocation Mechanism of Twinning

During twinning deformation, uniform shear occurs in the entire twin region, where the relative displacement of each plane is achieved by the movement of one **partial dislocation**, known as the Shockley partial dislocation. For instance, in a FCC crystal, when a full dislocation $\frac{a}{2}\langle 110\rangle$ sweeps on a slip plane $\{111\}$, the crystals on both sides of the slip plane will produce a relative slip displacement of one interatomic distance ($\frac{\sqrt{2}}{2}a$), while the stacking order of the $\{111\}$ plane remains the same, i.e., ABCABC (Fig. 12.19a). However, if there is a Shockley partial dislocation $\frac{a}{6}\langle 11\bar{2}\rangle$ sweeping through a group of parallel and adjacent $\{111\}$ planes, then the relative displacement between slip planes is not an atomic distance, but $\frac{\sqrt{6}}{6}a$. . This causes stacking faults to form in crystal planes, resulting in a change of the stacking sequence of planes from the original ABCABCABC… to ABCACBACB…, which forms a sheet of twin in the top half of the crystal (Fig. 12.19b–d).

The polar axis mechanism, proposed by A.H. Cottrell and B.A. Bilby, explains the formation of deformation twins through dislocation multiplication. Figure 12.20 is a schematic diagram of the **polar axis mechanism.** Dislocation lines *OA*, *OB*, and *OC* intersect at node *O*, where *OA* and *OB* are **stationary** polar axis dislocations out of the slip plane, and *OC* is a **sweeping dislocation** on the slip plane with Burgers vector b_3 that can rotate around node *O*. When the Burgers

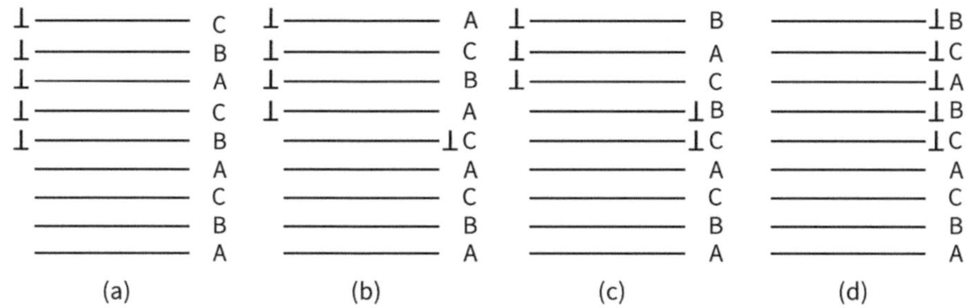

Fig. 12.19 The formation of twins in FCC crystals. The top half {111} planes swept by **a** a full dislocation, and **b–d** a Shockley partial dislocation

vectors b_1 and b_2 of *OA* and *OB* have a component perpendicular to the **sweeping** plane equal to the planar spacing of the sweeping plane, the sweeping plane becomes a continuous curl or helical surface. With each rotation of the sweeping dislocation *OC*, a single atomic layer is produced, and *OC* itself climbs one atomic distance upward to the adjacent crystal plane. This continuous sweeping of the dislocation causes the dislocation line *OC* and node *O* to rise continuously, which generates uniform shear in a fairly wide area of the crystal, resulting in the formation of deformation twins.

③ Kinking

In some cases, a crystal may exhibit very different deformation behaviors at deferent parts, making slip or twinning impossible. In such cases, the crystal undergoes

Fig. 12.20 Schematic diagram of the polar axis mechanism

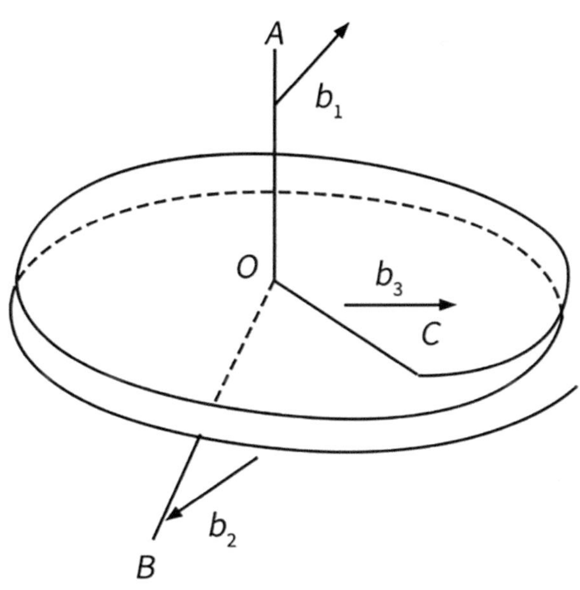

plastic deformation by other means. For instance, when an external force is parallel to the basal plane (0001) of an HCP crystal (i.e., the slip plane), the shear stress on the slip plane is zero, and the crystal cannot deform by slip or twinning. Under such circumstances, the crystal bends locally when the external force exceeds a critical value, resulting in a deformation mode known as **kinking**, and the deformation region is called the **kink band**, as depicted in Fig. 12.21.

Kinking changes the crystal orientation of the kink band asymmetrically, with lattice distortion occurring in area *ABCD*, and bending taking place on its left and right sides. The upper and lower interfaces (*CD*, *AB*) are composed of two kinds of edge dislocations with opposite signs. The dislocation signs on the left and right sides of the bending region are opposite, indicating that the kink band was originally a converging region of dislocations from other areas. The convergence of dislocations produces bending stress, which causes the crystal lattice to bend and form a kink band. Kinking is a coordinated deformation that causes stress relaxation and prevents the crystal from breaking. After kinking, the slip system in this region becomes favorable, resulting in the occurrence of slip.

The kink band does not only occur under the conditions described above but can also be formed in conjunction with twin formation. When a crystal deforms by twinning, the adjacent regions of the crystal experience a large strain due to

Fig. 12.21 Schematic diagram of kinking

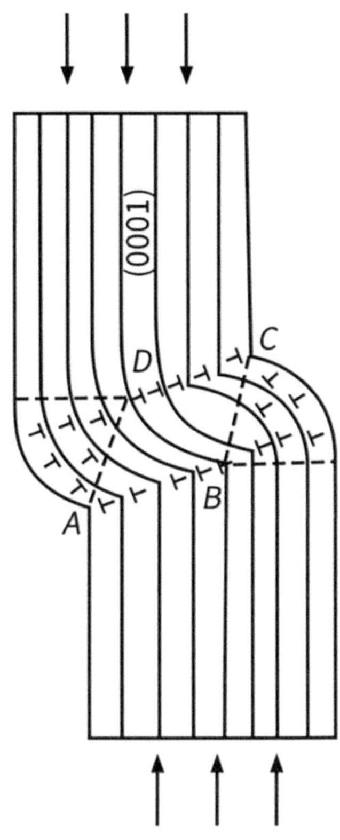

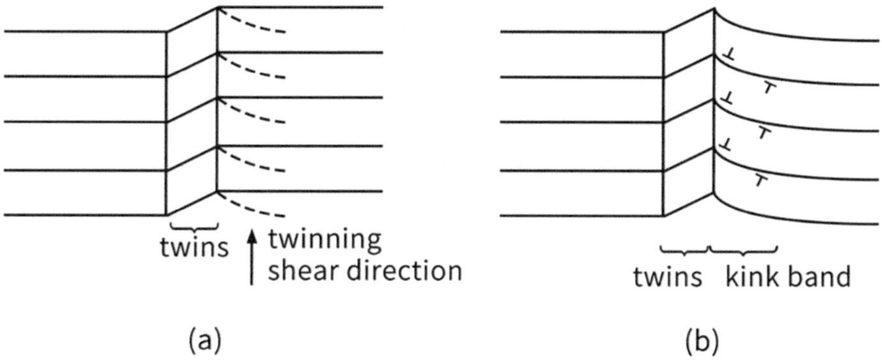

Fig. 12.22 Formation of the kink band with twinning deformation

the shear displacement of the twin region. This effect is particularly pronounced when the crystal is constrained at both ends, as in the case of a tensile clamp. To alleviate the influence of the shear strain and accommodate the constraints, a kink band will form adjacent to the twin boundary region to facilitate the strain transition, as shown in Fig. 12.22.

12.3.2 Plastic Deformation of Polycrystals

The materials commonly used are typically composed of polycrystals. At room temperature, the deformation behavior of each grain in the polycrystal is essentially the same as that of a single crystal. However, due to differences in adjacent grain orientations and the presence of grain boundaries, plastic deformation in polycrystals is relatively complex.

① Effect of Grain Orientation

The plastic deformation of polycrystals is affected by the orientation of each grain, which can cause mutual restriction and coordination during deformation. When external force is applied, crystals with different orientations endure different levels of stress. This results in differences in the resolved shear stress acting on the grains. Thus, the grains do not deform simultaneously, and those in favorable orientations slip first while others in disadvantageous orientations have not yet started to slip.

Furthermore, the slip systems are not uniform for all grains, making it difficult for the slip to extend directly from one grain to another. However, because each grain is surrounded by other grains, its deformation must coordinate with that of its neighbors to maintain continuity between grains. This coordination is achieved by each grain slipping not only on the single slip system with the most favorable orientation but also on several slip systems, including those with relatively favorable orientations. Theoretical analysis indicates that each grain can slip on at least five independent slip systems during polycrystalline plastic deformation. This is because the deformation process can be represented by six strain components (three normal strains and three shear strains). Because the plastic deformation volume is constant (the sum of the three normal strains is zero), there are five independent strain components, each of which needs an independent slip system to generate.

The plastic deformation of polycrystals is coordinated by the multi-slip system of each grain. The ability of a polycrystal to deform plastically depends on whether it has at least five independent slip systems to meet the coordination requirements. FCC and BCC crystals have many slip systems and can meet this requirement, which results in good plasticity of their polycrystals. In contrast, the inferior plastic deformation ability of HCP crystals is due to the lack of sufficient slip systems and poor strain coordination between grains.

② Effect of Grain Boundary

After tensile tests, samples with only two grains exhibit bamboo-like joints at grain boundaries, as shown in Fig. 12.23a. This indicates that the slip near the grain boundaries is blocked, resulting in relatively smaller deformation extent near the boundary, while the deformation within the grain is relatively large, leading to non-uniform deformation across the grain.

The slip band in each grain terminates near the grain boundary after the polycrystalline specimen is stretched. Electron microscopy reveals that dislocations cannot pass through the grain boundaries during the deformation process, as shown in Fig. 12.23b. As a result, the **pile-ups of dislocations** generated near the grain boundary produce a reaction force on the dislocation source within grains. The reaction force increases with the number n of dislocation pile-ups, which can be expressed as:

$$n = \frac{k\pi \tau_0 L}{Gb} \tag{12.33}$$

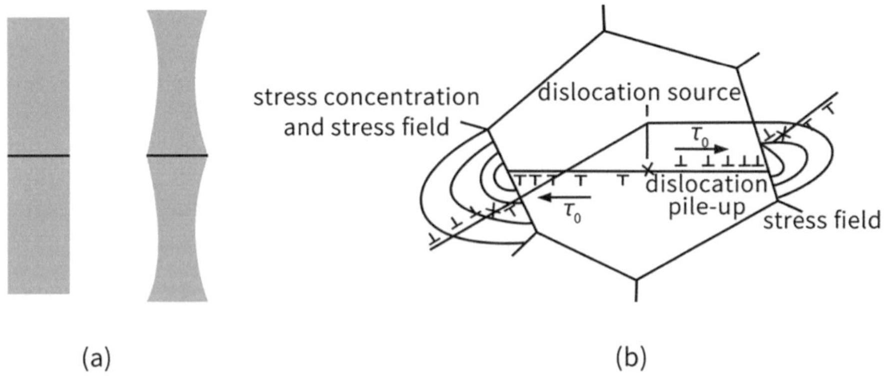

Fig. 12.23 **a** Bamboo joints formed in sample with two grains during tensile deformation and **b** dislocation pile-ups at grain boundaries

where τ_0 is the resolved shear stress acting on the slip plane; L is the distance from the dislocation source to the grain boundary; and k is a coefficient for screw dislocation $k = 1$ and edge dislocation $k = 1 - \nu$ (ν is the Poisson ratio). When n increases to a certain value, the dislocation source can stop operating, and subsequently, the crystal is significantly strengthened. Therefore, the applied stress for polycrystals must be large enough to trigger the action of dislocation sources in a large number of grains, resulting in slip, before macroscopic deformation can be detected.

Since the number of grain boundaries is directly dependent on the grain size, the effect of grain boundaries on the initial resistance for plastic deformation of polycrystals can be directly reflected by the grain size. It has been proven that the strength of polycrystals increases with grain refinement. The relationship between the yield strength σ_y and average grain diameter d of polycrystals can be expressed by the famous **Hall–Petch equation**:

$$\sigma_y = \sigma_o + Kd^{-\frac{1}{2}} \tag{12.34}$$

where σ_o reflects the resistance to deformation within the grain, which is equivalent to the maximum yield strength of a single crystal, and K reflects the influence coefficient of the grain boundary on deformation, which is related to the structure of the grain boundary. Therefore, **grain refinement** is an effective way to strengthen materials, as it is the only strengthening method to improve both material strength and toughness in most situations.

12.3.3 Plastic Deformation and Strengthening of Alloys

Most metallic materials used in engineering are alloys, which deform similarly to pure metals but exhibit new features due to the presence of alloying elements. Based on their phase constituents, alloys can be classified as single-phase solid solution alloys or multiphase alloys, each displaying different characteristics during plastic deformation.

① Plastic Deformation of Solid Solutions

The addition of solute atoms to a solvent metal results in **solid solution strengthening**, which improves the deformation resistance of alloys. Figure 12.24 illustrates the variation of the strength, hardness, and plasticity of Cu-Ni solid solution alloys with composition. The strength (σ_b) and hardness (*HB*) increase to peak values and then decrease gradually with increasing solute content, while the plasticity (δ) shows reverse tendency.

Different solute atoms have varying effects on solid solution strengthening. Several factors influence this process, including:

1. Solute concentration: The strengthening effect generally increases with an increase in solute concentration, especially at low concentrations.

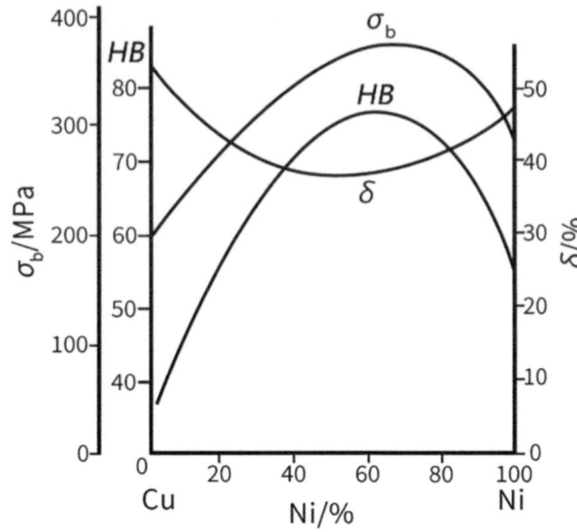

Fig. 12.24 The relationship between the strength and ductility of the Cu-Ni solid solution and composition

2. Atomic size difference: The strengthening effect is better when there is a larger size difference between the solute and solvent atoms. The solubility limit of the solute atom decreases with increasing size difference.
3. Type of solute atoms: Interstitial solute atoms have a greater strengthening effect than substitutional ones, especially in BCC crystals. However, the solubility of interstitial atoms is low, so the actual strengthening effect is limited.
4. Role of electrons: The strengthening effect is more significant when there is a larger difference in the number of valence electrons between the solute atom and the solvent metal.

Solid solution strengthening results from various aspects, including elastic interactions, chemical interactions, and electrostatic interactions between solute atoms and dislocations. Moreover, as solid solutions undergo plastic deformation, dislocation movement alters the distribution of solute atoms in the solid solution structure, leading to an increase in system energy and a consequent increase in slip resistance to deformation.

② Plastic Deformation of Multiphase Alloys

A multiphase alloy, which contains one or more additional phases other than the matrix phase, are more complex to deform than single-phase solid solution alloys due to the different characteristics of the additional phase(s), such as the size, shape, and distribution. The behavior and strengthening effects of two-phase alloys depend on the size of the second phase particles, with aggregated two-phase alloys having a second phase with a grain size similar to that of the matrix, and dispersed two-phase alloys having a small-grain-sized second phase. The plastic deformation behavior and strengthening effect differ between the two types.

Plastic Deformation of Aggregated Dual Phase Alloy

If the grain sizes of the two phases in the alloy are of the same order of magnitude, the deformation behavior of the alloy is determined by the relative proportion of the two phases. As a first-order approximation, it can be assumed that the strains or stresses experienced by the two phases are the same during deformation of the alloy. Therefore, the average flow stress of the alloy at a certain strain and the average strain at a certain stress can be expressed by the rule of mixtures:

$$\sigma = f_1\sigma_1 + f_2\sigma_2 \tag{12.35}$$

$$\varepsilon = f_1\varepsilon_1 + f_2\varepsilon_2 \tag{12.36}$$

where f_1 and f_2 are the volume fractions of the two phases, σ_1 and σ_2, ε_1 and ε_2 are the stresses and strains of the two phases, respectively.

Plastic Deformation of Dispersed Dual Phase Alloy

When the second phase in the alloy is uniformly distributed in the matrix phase as fine particles, it can have a significant strengthening effect. This effect is manifested through the hindrance of the movement of dislocations by the second-phase particles. In general, the second-phase particles can be divided into two types: "non-deformable" and "deformable".

The hindering effect of non-deformable particles on dislocations is depicted in Fig. 12.25. When a moving dislocation encounters the particles, it is blocked, causing the dislocation line to bend around the particle. As the applied stress increases, the blocked portion of the dislocation line bends more sharply, leading the dislocation lines on either side of the particle to meet, which results in the annihilation of the positive and negative dislocations and the formation of a dislocation loop surrounding the particle. The remaining part of the dislocation line continues to move through the particle. Obviously, the resistance to dislocation movement is significant using this mechanism. Each remaining dislocation loop exerts a reverse stress on the dislocation source, so the applied stress must be increased to overcome this reaction and continue the deformation. This increase in stress results in a rapid increase in flow stress. This dislocation movement mechanism was first proposed by E. Orowan and is therefore commonly referred to as the **Orowan mechanism**, which has been confirmed by experiments.

When the second phase particles in an alloy are deformable, they can be cut through by dislocations, a phenomenon known as the **Friedel effect**, as shown in Fig. 12.26. In this case, the strengthening effect of the particles is primarily determined by the nature of the particle and its combination with the matrix. The strengthening mechanism is complex and varies between alloys. The following are some of the main effects:

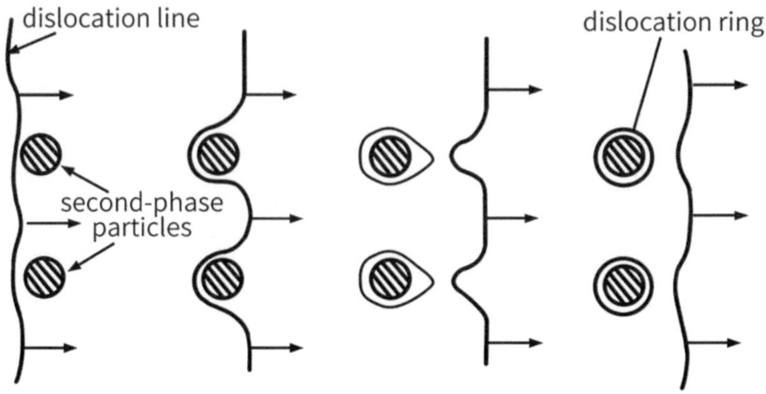

Fig. 12.25 Schematic diagram of dislocations bypassing second phase particles

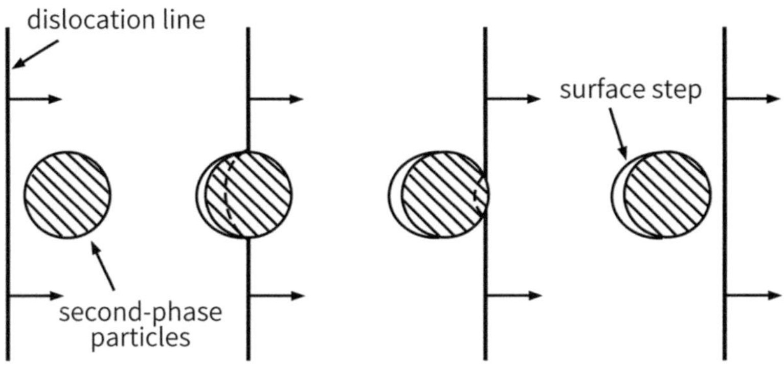

Fig. 12.26 The mechanism of dislocation cutting through particles (Friedel effect)

1. When the dislocation cuts through the particles, the particles produce surface steps with a width of *b*. Because of the increase in surface area, the total interface energy increases.
2. If the particle has an ordered structure, when the dislocation cuts through the particle, it will disrupt the orderly arrangement around the slip plane and produce antiphase boundaries, leading to an increase in energy.
3. Due to the different crystal lattices or lattice parameters between the second-phase particle and the matrix, when the dislocation cuts through the particle, atomic misfits are introduced on the slip plane, which requires additional energy to trigger further dislocation movement.
4. Because of the difference in specific volume between the particle and the matrix and the coherent or semicoherent interface between the particle and parent phase, an elastic stress field is generated around the particle, which interacts with the dislocation and hinders the movement of dislocations.

5. The orientations of slip planes in the matrix and the particle are usually not consistent. When a dislocation cuts through a particle, it produces a jog that hinders the movement of the entire dislocation line.
6. Because the stacking fault energy of the particle is different from that of the matrix, when the extended dislocation passes through, its width changes, leading to an increase in energy.

The combined effect of these strengthening factors improves the strength of the alloy.

In summary, the two mechanisms mentioned above can explain not only the strengthening effect but also the plasticity of multiphase alloys. However, both mechanisms are influenced by the nature, size, and distribution of the particles, among other factors. Therefore, by controlling these parameters, we can adjust the strength and plasticity of precipitation-strengthened and dispersion-strengthened alloys within a certain range.

12.3.4 Effect of Plastic Deformation on Microstructure and Properties

Plastic deformation not only changes the shape and size of a material but also affects its internal structure and various properties.

① Changes in Microstructure and Substructure

After undergoing plastic deformation, the microstructure of metal materials changes significantly. As the degree of deformation increases, the original equiaxed grains gradually elongate along the deformation direction, and a large number of slip bands or twinning bands appear inside each grain. When the amount of deformation is large, the crystal grains become blurred and difficult to distinguish, forming a fibrous structure. The distribution direction of the fibers is the stretching direction of the material.

Plastic deformation of a crystal is caused by the movement and multiplication of dislocations under stress. As the degree of deformation increases, the **dislocation density** in the crystal increases rapidly. Accordingly, the dislocation configuration and distribution and other substructure changes in deformed crystals also change. After severe cold deformation, the dislocation density can increase from10^6~10^7 cm^{-2} in the original annealed state to 10^{11}~10^{12} cm^{-2}. The substructure evolution caused by plastic deformation plays an essential role in the mechanical properties of materials.

② Changes in Properties

In the process of plastic deformation, the mechanical, physical, and chemical properties of the material undergo significant changes as the internal microstructure and structure vary.

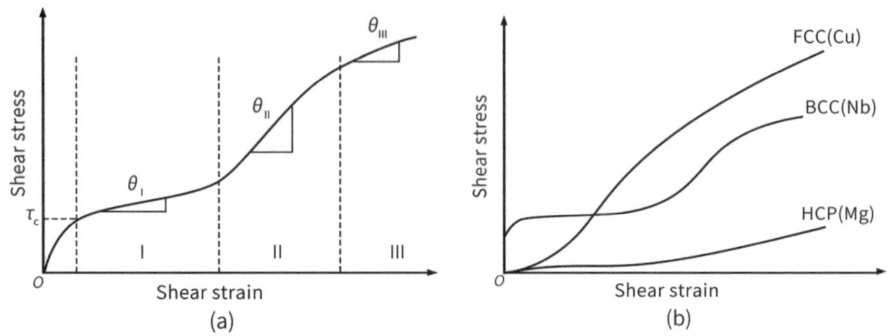

Fig. 12.27 Typical stress–strain curves of **a** a single crystal and **b** several single crystals with different structures

After the metal material undergoes cold working, its strength (hardness) is significantly improved, while its ductility drops rapidly, a phenomenon known as **work hardening**. Work hardening is an important feature of metal materials and can be used as a way to strengthen metals. This is particularly important for materials that cannot be strengthened by heat treatment, such as pure metals and certain alloys like austenitic stainless steel.

Figure 12.27a shows a typical stress–strain curve (also called the work hardening curve) of a metal single crystal, and its plastic deformation is composed of three stages.

Stage I (Easy-slip stage): When the shear stress reaches the critical resolved shear stress (τ_c) of the crystal, the stress does not increase much, and considerable deformation can occur. This section is close to a straight line, its slope θ_I or the work hardening rate is low, and generally around $\theta_I \approx 10^{-4}G$ (where G is the shear modulus of the material).

Stage II (Linear hardening stage): As the strain increases, the stress increases linearly. This section also shows a straight line with a large slope. Work hardening is very significant, θ_{II} is almost a constant of $G/300$.

Stage III (Parabolic hardening stage): As the strain increases, the stress rises slowly, showing a parabolic shape, and θ_{III} gradually decreases.

The actual stress–strain curves of different crystals can vary due to factors such as their crystal structure, crystal orientation, impurity content, and testing temperature. However, they share the same fundamental characteristics, although the length of each stage may differ due to the specifics of the material. The different stages of the curve are determined by dislocation movement, multiplication, and interaction. In some cases, a certain stage may even disappear, as shown in Fig. 27b.

③ Deformation Texture

The process of plastic deformation causes the slip plane and direction of each crystal grain to rotate towards the main deformation direction, resulting in a certain degree of spatial orientation of each crystal grain. This phenomenon is known as preferred orientation, and the resulting microstructure is referred to as **deformation texture**.

The existence of a texture structure has a significant impact on the processing formability and application performance of the material due to its anisotropic nature. There are two primary types of deformation textures that are formed through different processing and deformation methods. The texture formed during wire drawing is referred to as the fiber texture, where each grain has a specific crystal orientation that is roughly parallel to the wire drawing direction. In contrast, the texture formed during plate rolling is called the plate texture, where each grain has a certain crystal plane and orientation that tends to be parallel to the rolling surface and direction.

④ Residual Stress

During plastic deformation, a significant portion of the work done by the external force is converted into heat, while a small portion is stored in the deformed material in the form of distortional energy, which is known as stored energy. The magnitude of this energy varies with the deformation extent, mode, temperature, and material properties and accounts for only a few percent of the total deformation work. The specific forms of stored energy include macroscopic residual stress, microscopic residual stress, and lattice distortion.

1. **The first type of internal stress**, also known as the macroscopic residual stress, is caused by the unevenness of the macro-deformation in different parts of the workpiece, resulting in a stress-balanced range that encompasses the entire workpiece. The distortion energy corresponding to this type of residual stress accounts for only about 0.1% of the total stored energy.
2. **The second type of internal stress**, also known as microscopic residual stress, is caused by the unevenness of deformation between crystal grains or subgrains, with the stress acting over an area equivalent to the grain size. This internal stress can sometimes reach very large values and may even cause microcracks and damage to the workpiece.
3. **The third type of internal stress**, also known as lattice distortion, is caused by a large number of lattice defects, such as vacancies, interstitial atoms, and dislocations, formed during plastic deformation. Its action range is about tens to hundreds of nanometers. The majority of the stored energy (80% to 90%) in the deformed metal is used to form lattice distortion, which increases the energy of the deformed crystal to a thermodynamically unstable state. This energy has a spontaneous tendency to restore the deformed metal to a stable structure state with the lowest free enthalpy, causing recovery and recrystallization processes during heating.

The residual stress in metal materials after plastic deformation is inevitable and can affect the deformation, cracking, and stress corrosion of the workpiece. Therefore, it is necessary to eliminate it through various measures, such as stress relief annealing treatment. However, under certain conditions, the existence of residual stress can be beneficial. For example, treating parts subjected to alternating loads with surface rolling and shot peening can produce a strain layer with compressive stress on the surface of the parts, achieving the purpose of surface strengthening and doubling the fatigue life.

12.4 Annealing of Plastically Deformed Metals

Plastic deformation of metals and alloys results in changes to their internal structure and properties, such as the increase in the density of structural defects like vacancies and dislocations, and the accumulation of distortion energy. As a result, these materials become thermodynamically unstable and possess a high free energy state. Therefore, there is a natural tendency for the plastically deformed material to spontaneously return to the low free energy state that existed before deformation. When a cold deformed metal is heated, several processes such as **recovery, recrystallization, and grain growth** occur. Recovery refers to the stage where substructure and performance changes occur before the appearance of new undistorted grains. Recrystallization denotes the process of gradually replacing deformed grains with undistorted equiaxed grains, and grain growth refers to the continuing enlargement of grains after recrystallization.

12.4.1 Changes in Structure and Properties of Cold-Deformed Metals During Heating

Figure 12.28 illustrates the process of recovery, recrystallization, and grain growth that occurs in plastically deformed materials. During the recovery stage, there is no high-angle grain boundary migration, so the shape and size of the grains remain the same as the deformed morphology, which may be fibrous or flat. In the recrystallization stage, new undistorted grain cores are generated in areas of large distortion and gradually consume the surrounding deformed matrix, resulting in the formation of fine equiaxed grains. Finally, driven by the interfacial energy of the grain boundary, the new grains swallow each other and grow, resulting in a relatively stable size under these conditions, which is referred to as the grain growth stage.

During the recovery stage, the hardness change is minimal, accounting for approximately 1/5 of the total change, whereas it drops significantly during the recrystallization stage. The strength exhibits a similar changing tendency to the hardness. This situation is mainly related to the dislocation mechanism. In the recovery stage, the deformed metal maintains a high dislocation density. However, after recrystallization, the strength and hardness significantly decrease because the dislocation density is substantially reduced.

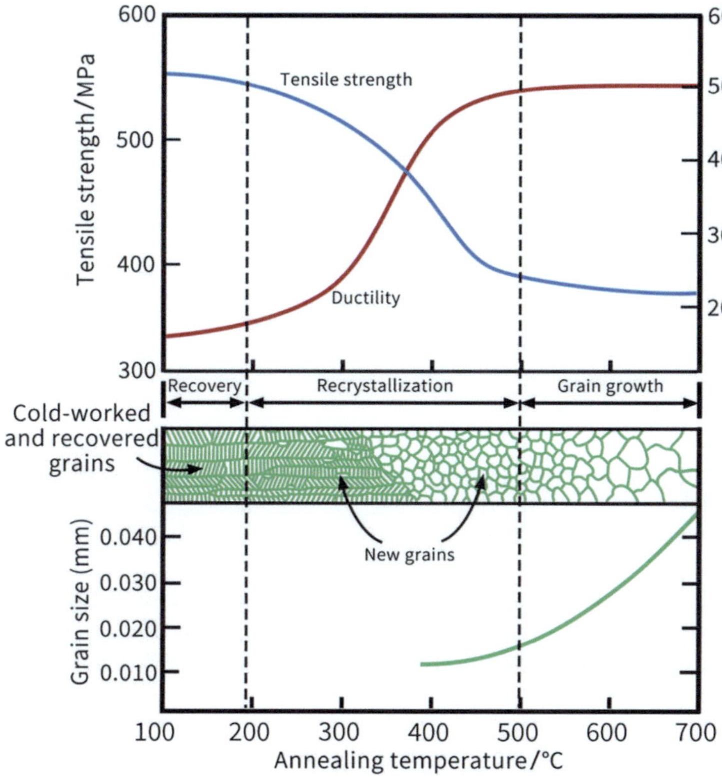

Fig. 12.28 The variation in microstructure and properties of cold deformed metals during annealing

12.4.2 Recovery of Cold-Worked Metals

The process of returning a cold-worked metal to a strain-free state by heating it without significant changes in its microstructure is called recovery or stress relieving. This process is carried out below the recrystallization temperature and is essential for releasing internal stresses in forgings, welding-fabricated equipment, and boiler tubes without affecting the strength achieved during cold working.

During recovery, some of the stored internal strain energy is relieved by dislocation motion in the absence of externally applied stress due to enhanced atomic diffusion at elevated temperatures. The number of dislocations is reduced, and dislocation configurations with low strain energies are produced. In addition, physical properties such as electrical and thermal conductivities return to the states before cold working.

① Recovery Kinetics

The kinetic curves of recovery show that it is a relaxation process characterized by: (1) no incubation period, (2) a very large initial recovery rate at a certain temperature that gradually slows down until it approaches zero, and (3) a limit value for the degree of recovery at each temperature. The higher the annealing temperature, the higher the limit value, and the shorter the time required to reach this limit value. (4) The initial recovery rate is faster for larger pre-deformations, and smaller grain sizes also accelerate the recovery process.

② Recovery Mechanisms

The mechanisms of recovery in cold deformed metals differ with annealing temperature as follows:

1. At low temperatures, recovery is mainly related to the migration of point defects. A large number of point defects such as vacancies and interstitial atoms are generated during cold deformation. The thermal activation energy required for the movement of point defects is relatively low, making it feasible at lower temperatures.
2. At slightly higher temperatures, dislocation movement and redistribution occur. The mechanism of recovery is mainly related to the slip of dislocations, such as the attraction and elimination of dislocations with different signs on the same slip plane and the counterbalance of two dislocation lines of a dislocation dipole.
3. At high temperatures (~0.3 T_m), the edge dislocations can obtain enough energy to cause climbing, which results in two important consequences: (i) redistribution of irregular dislocations on the slip plane, where the edge dislocations are arranged vertically into walls. This distribution can significantly reduce the elastic distortion energy of the dislocations, resulting in a large strain-energy release in this temperature range. (ii) The dislocation walls (small-angle subgrain boundaries) arrange in the direction perpendicular to the slip plane and with a certain misorientation, forming **polygonal domains**.

The driving force of the high-temperature recovery polygonization process mainly comes from the decrease in strain energy. The conditions for the polygonization process are as follows: (i) Plastic deformation causes the crystal lattice to bend. (ii) There are pile-ups of edge dislocations of the same sign on the slip plane. (iii) The metal needs to be heated to a higher temperature so that the edge dislocation can produce climbing motion. The arrangement of edge dislocations after the polygonization process is shown in Fig. 12.29, where subgrain boundaries are formed.

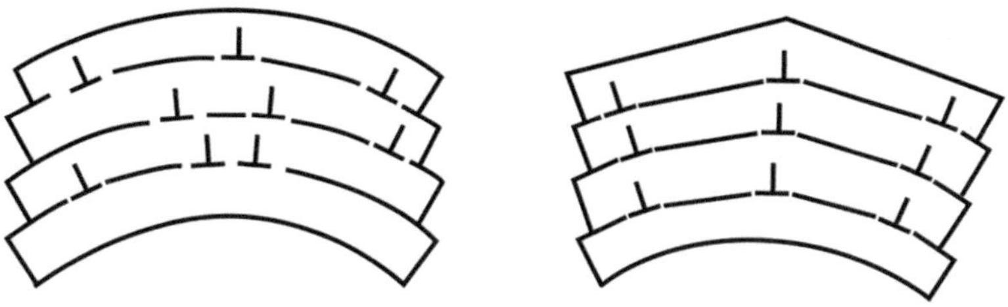

Fig. 12.29 The redistribution of dislocations during the polygonization process

12.4.3 Recrystallization of Cold-Worked Metals

When a cold-worked metal is heated to a certain temperature, a new set of undistorted grains are generated in the original deformed microstructure, and the performance changes significantly, returning to the state before deformation. This process is called recrystallization, which is a process of microstructure reorganization, unlike the changes observed during recovery. The driving force of recrystallization is fundamentally different from that of phase transformation, and hence their nature is not the same. While the driving force of phase transformation is the molar Gibbs free energy difference between the new phase and the parent phase, the driving force of recrystallization is the mechanical storage energy of the deformed metal.

① The Process of Recrystallization

The nucleation process can be divided into two types according to the deformation extent.

Recrystallization is a process of microstructure reorganization that generates a set of new undistorted grains in the original deformed microstructure, resulting in a significant change in the material's performance. This process involves **nucleation and growth**, which means producing new undistorted nuclei on the matrix of the deformed microstructure and gradually growing to form equiaxed grains to replace the entire deformed microstructure. However, unlike other solid phase transformations, the recrystallized nucleus is not a new phase, and its crystal structure remains unchanged.

According to many experimental observations, the recrystallization nucleus is preferentially formed at the primary grain boundary and the region with uneven local deformation, which is prepared during the recovery process. The driving force of recrystallization is fundamentally different from that of phase transformation. The critical nucleation core size for recrystallization is very large (>0.1 μm) or requires a local elastic strain of the order of 20% within a few nanometers, based on the classical nucleation theory. Therefore, the so-called recrystallization nucleus is just a small volume that exists before recrystallization in the deformed structure. The nucleation process can be divided into two types according to the deformation extent.

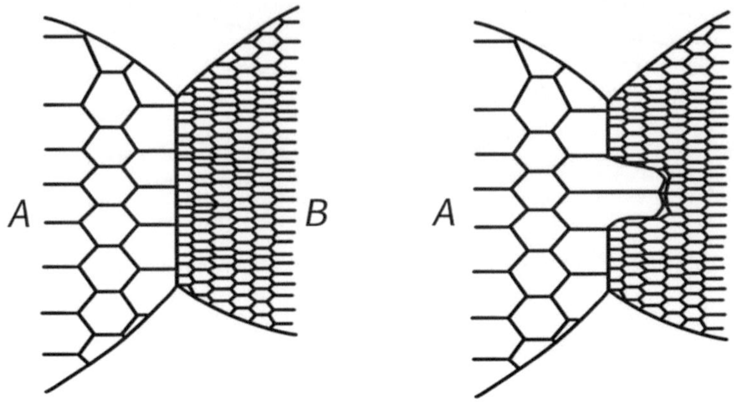

Fig. 12.30 Grain boundary bowing nucleation

Grain boundary bowing nucleation. For metals with small deformation (usually less than 20%), the recrystallization core is mainly formed through grain boundary bowing mechanisms. This involves the movement of the grain boundaries induced by strain or bowing nucleation. When the degree of deformation is low, the dislocation density is unevenly distributed among crystal grains due to non-uniform deformation. As shown in Fig. 12.30, if two adjacent grains (A and B) have different dislocation densities due to varying degrees of deformation, the size of the subcrystals formed after polygonization in grain B will be relatively small. To reduce the free energy of the system, at certain temperature conditions, some subcrystals from grain A adjacent to the grain boundary will migrate through the grain boundary and penetrate into grain B, causing the subcrystals in grain B to be consumed. This process initiates the formation of distortion-free recrystallized nuclei.

Subcrystalline nucleation. This mechanism generally occurs under a large degree of deformation. When the degree of deformation is large, the dislocations in the crystal will continue to multiply, and the cell structure composed of dislocation entanglements will easily occur in the heating process to form subcrystals. With subcrystals as the recrystallization core, its nucleation mechanism can be divided into subcrystalline combination and subcrystalline migration mechanisms.

After the recrystallized nucleus is formed, it grows into the surrounding distortion area by the movement of the interface. The driving force of the interface migration is the strain energy difference between the undistorted new grain and the surrounding distorted matrix. The grain boundary always deviates from its center of curvature and advances toward the distorted area. Recrystallization is complete until all distorted grains transform into equiaxed grains.

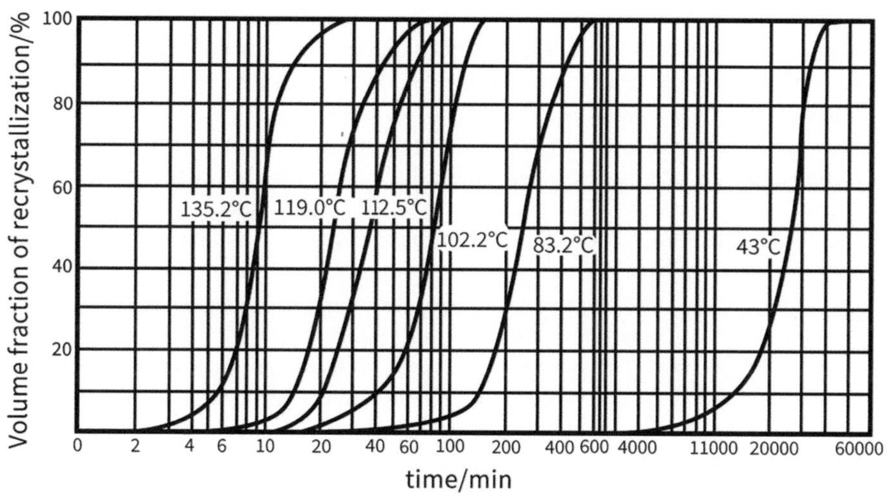

Fig. 12.31 Isothermal recrystallization curves of 98% cold-rolled pure copper at different temperatures

② Recrystallization Kinetics

The kinetics of recrystallization are determined by the nucleation rate N and the growth rate G. When the volume fraction of recrystallization is plotted on the y-axis and time on the x-axis, the constant temperature kinetic curve obtained from experiments has typical "S" curve characteristics, as shown in Fig. 12.31. The figure illustrates that the recrystallization process has an incubation period, during which the rate of recrystallization is slow and gradually accelerates. When the volume fraction of recrystallization reaches approximately 50%, the recrystallization rate reaches its maximum and then gradually slows down. The volume fraction of recrystallization (y) is dependent on time (t) during constant-temperature recrystallization, which is usually described by the Avrami equation:

$$y = 1 - \exp(-kt^n) \tag{12.37}$$

where k and n represent constants that are independent of time and specific to the reaction being considered.

③ Recrystallization Temperature and the Influencing factors

The **recrystallization temperature** is the temperature at which recrystallization just reaches completion in 1 h. Generally, the recrystallization temperature is about 0.3–0.5 times melting point in degrees Kelvin. The recrystallization temperature is not a fixed value and varies depending on the material as well as factors such as the degree of cold deformation and original grain size.

1. Degree of deformation: With an increase in cold deformation, the stored energy also increases, resulting in a higher driving force for recrystallization. Therefore, the recrystallization temperature is lower, and the recrystallization rate during isothermal annealing is faster. However, the recrystallization temperature becomes stable when the degree of deformation is beyond a certain limit.
2. Original grain size: Finer grains have higher resistance to deformation and higher energy storage after cold deformation. This leads to lower recrystallization temperatures. Moreover, grain boundaries serve as favorable nucleation sites for recrystallization, thereby increasing the nucleation rate N and growth rate G of fine-grained metals. As a result, newly formed grains are smaller, and the recrystallization temperature is further reduced.
3. Trace solute atoms: The presence of trace solute atoms significantly increases the recrystallization temperature due to their interaction with dislocations and grain boundaries. Solute atoms tend to segregate at these locations and hinder the slip and climb of dislocations and the migration of grain boundaries, thereby hampering the nucleation and growth of recrystallization cores and hindering the recrystallization process.
4. Second-phase particles: The effect of second-phase particles on recrystallization depends on the size and distribution of the particles in the matrix metal. They can either promote or hinder the recrystallization of the matrix.
5. Recrystallization annealing process parameters: The parameters of the annealing process, such as heating rate, heating temperature, and holding time, have some influence on the recrystallization of deformed metals.

12.4.4 Grain Growth After Recrystallization

When the new strain-free grains of the worked metal obtained after recrystallization are heated at a temperature higher than that required to cause crystallization, a progressive increase in grain size is achieved, known as grain growth. Grain growth can be classified into two types: normal growth and abnormal growth.

Normal grain growth suggests that the grain size increases gradually and simultaneously. From the macroscopic perspective, the driving force of grain growth is the reduction of interfacial energy. For a single grain, the driving force is mainly caused by the interface curvature of the grain boundaries. Grain growth is actually a process of grain boundary migration toward the curvature center. Grain growth is sometimes referred to as coarsening. For many polycrystalline materials, the grain diameter d varies with time t according to the following relationship:

$$d^n - d_0^n = Kt \tag{12.38}$$

where d_0 is the initial grain diameter at $t = 0$, K and n are time-independent constants, and the value of n is generally equal to or greater than 2.

Occasionally, secondary recrystallization, or so-called **abnormal grain growth** can occur. This situation is characterized by the growth of just a few grains to very

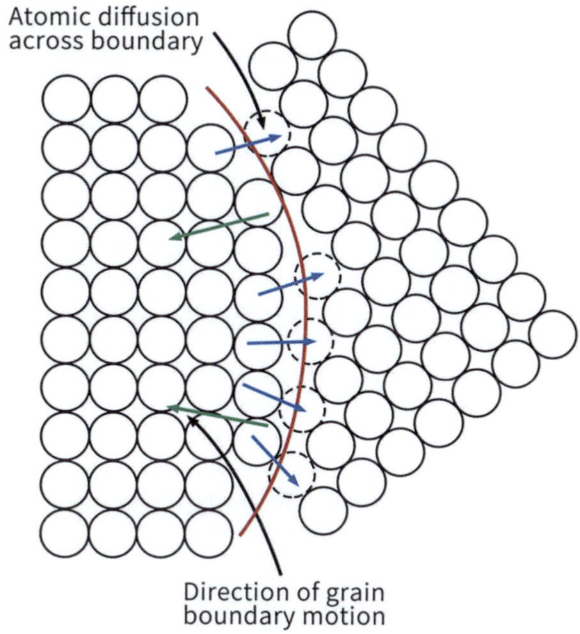

Fig. 12.32 Schematic representation of grain growth through atomic diffusion

large diameters. These grains then expand by consuming the surrounding grains until the fine grains are entirely replaced by the coarse ones. The abnormal grain growth generally occurs when the normal growth process of grains is strongly hindered by factors such as dispersed particles, texture or surface thermal erosion groove.

Grain growth depends upon several factors, such as the annealing temperature and time, the degree of previous cold working, the effect of insoluble inclusions on the metal, the addition of certain alloying elements, and the rate of heating and cooling. The grain size is considerably affected by the fine dispersion of second phase particles, which restricts grain boundary movement. There is a slight decrease in the hardness and strength values during grain growth. There is a further increase in the electrical conductivity and ductility of the metal. Schematics of grain growth and atomic motion are shown in Fig. 12.32.

12.4.5 Recrystallization Texture and Annealing Twins

① Recrystallization Texture

Recrystallization texture refers to the situation where the new grains of a metal with deformed texture, obtained after recrystallization, still have a preferred orientation. There are two main theories on the formation mechanism of recrystallization textures: the **directional growth theory** and the **directional nucleation theory**.

According to the directional growth theory, crystal nuclei of various orientations are formed during the recrystallization process, but only some of them have special orientations. These nuclei may quickly grow into the deformed matrix, leading to the formation of a recrystallization texture. In a deformed matrix, most of the grain orientations are similar, making it difficult for the grains to grow. However, some recrystallized nuclei have a special orientation relationship with the deformed texture, resulting in grain boundaries with very high migration speeds. As a result, preferential growth occurs, gradually consuming the surrounding deformed matrix and forming a recrystallization texture with a different orientation from the original deformed texture.

The directional nucleation theory, on the other hand, proposes that severe deformation texture in a metal structure causes recrystallized nucleation to have a preferred orientation. The nuclei then grow to form a recrystallized texture consistent with the original texture due to the similarity of each subgrain's orientation.

② Annealing Twins

When certain FCC metals and alloys, such as copper, nickel alloys, and austenitic stainless steels, undergo cold working and annealing, annealing twins can form within the crystal grains, as shown in Fig. 12.33a. The three typical annealing twin crystal morphologies, represented by *A*, *B*, and *C* in the figure, show an annealing twin at the corner of the grain boundary, a complete annealing twin that penetrates the grain, and a twin that terminates within the crystal at one end of an incomplete annealing twin, respectively. The twin boundaries on both sides of the twinning band are parallel and belong to the coherent boundaries, which are composed of (111) planes. The twin boundaries that terminate within the grains and the steps at the coherent twin boundaries are incoherent boundaries. These phenomena are observed in the aforementioned metals and alloys due to their specific crystal structures and the deformation mechanisms that occur during the cold working and annealing processes.

The formation of annealing twins in FCC crystals necessitates the presence of stacking faults in the stacking sequence of {111} planes. Specifically, the normal stacking sequence ABCA BCABC… is altered to ABCBACABC…, as shown in Fig. 12.33b. The interface and the crystals between them generate an annealing twinning band, with the twin boundaries on both sides of C representing coherent twins.

Annealing twins are generally believed to be formed during grain growth, as shown in Fig. 12.34. When crystal grains grow via grain boundary migration, the stacking sequence of the atomic layer on the (111) plane at the grain boundary corner can accidentally pile up to form coherent twin boundaries and subsequently annealing twins. These twins move and grow through high-angle grain boundary migrations. During this process, if the atoms pile up again on the surface of (111)

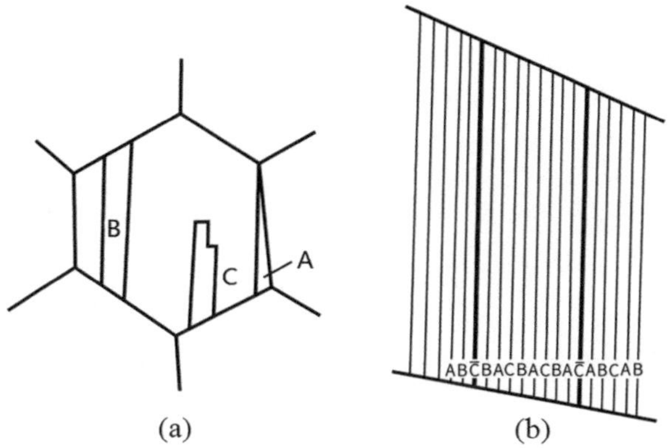

Fig. 12.33 **a** Schematic diagram of annealing twins and **b** the stacking sequence of (111) plane in a FCC crystal to form annealing twins

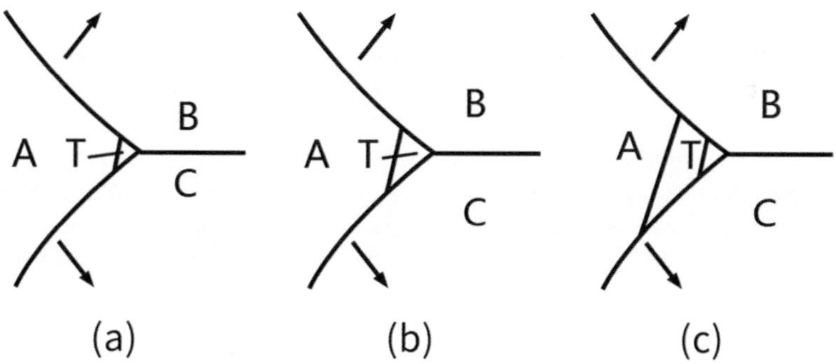

Fig. 12.34 Formation and growth of an annealing twin at the corner of a grain boundary during grain growth

and restore the original stacking sequence, the second coherent twin boundary forms, and a twinning band is obtained. The formation of annealing twins requires satisfying specific energy conditions, with crystals exhibiting lower stacking fault energy more likely to form annealing twins.

12.4.6 Thermal Deformation and Superplasticity

At high temperatures and low stresses, the deformation of fine-grained materials occurs through the mutually accommodating processes of grain-boundary sliding and matter transport. Depending on which mechanism provides the greater amount of strain, the deformation is referred to as diffusion creep or grain-boundary sliding, but the two mechanisms are not completely distinct. Grain-boundary sliding accommodated by diffusion creep is described by similar equations to those used to describe diffusion creep. In most materials, when the grain size is small and stable, there is a range of strain rates where the strain-rate sensitivity is higher than that of dislocation creep, and tensile deformation can occur in a stable manner up to very large strains. This range is known as the superplastic domain. Models for superplastic flow take into account the high strain-rate sensitivity and the potential for large strains, which are enabled by grain-boundary sliding during grain-shifting events, locally accommodated by diffusion creep or the climb and glide of grain-boundary dislocations in the mantle of the grains.

① Hot deformation

Hot working processes involve plastically deforming metals above their recrystallization temperature. This allows the material to recover and recrystallize during deformation through **dynamic recovery** and **dynamic recrystallization**. The importance of recrystallization lies in preventing strain hardening, which would ultimately result in low yield strength and hardness but high ductility.

Metals with high stacking fault energy, such as Al, α-Fe, Zr, Mo, and W, exhibit cross slip of screw dislocations and the climbing of edge dislocations, leading to the annihilation of dislocations with different signs. As a result, the dislocation density in the microstructure is low, and the residual strain energy is insufficient to cause dynamic recrystallization. Therefore, dynamic recovery plays a dominant role in the softening of metals during hot working. Metals with low stacking fault energy, such as Cu, Ni, γ-Fe, and stainless steel, exhibit a high tendency for dynamic recrystallization during hot working. Dynamic recrystallization occurs through nucleation and growth stages, and the corresponding mechanism may be similar to that described in previous sections.

② Superplasticity

Superplasticity is a phenomenon where materials undergo uniform plastic deformation with elongation of 500–2000% under certain conditions, without necking. To achieve superplasticity, materials must generally meet the following three conditions:

1. The material must have a very small and stable grain size of less than 10 μm, which does not significantly increase during superplastic deformation.
2. Superplastic deformation occurs only within the temperature range of 0.5 ~ 0.65 T_m, as it is a diffusion-controlled process.
3. The low strain rate must be within the range of 10^{-2}~10^{-4} s^{-1} to ensure smooth grain boundary diffusion.

Characteristics of Superplasticity

At high temperatures, the flow stress of a material is not only dependent on strain and temperature, but also sensitive to the strain rate ($\dot{\varepsilon}$), and follows the following relationship:

$$\sigma(\varepsilon, T) = K\dot{\varepsilon}^m \tag{12.39}$$

where K is a constant and m is called the strain rate sensitivity index. At room temperature, the m value for common metals typically ranges from 0.01 ~ 0.04. However, as the temperature increases and the grain size becomes finer, the m value may increase significantly. To achieve superplasticity, the value of m must be larger than 0.3, which highlights the importance of microstructure refinement. The m value is a critical parameter for assessing the potential superplasticity of materials, with higher values of m generally resulting in greater elongation of the material.

The Mechanism of Superplasticity

The widely accepted theory for the essence of superplasticity involves the rotation of grain boundaries and grains. Figure 12.35 illustrates how superplastic materials can maintain equiaxed crystal positions even after undergoing significant strains. If a group of hexagonal grains are subjected to tensile stress along the longitudinal direction, they experience pressure in the transverse direction. As a result of these stresses, the grains undergo grain boundary slip, migration, and directional diffusion, transitioning from initial state I to intermediate state II and eventually to final state III. While the initial and final grains have the same shape, their positions are altered, and the entire sample elongates longitudinally, causing the entire sample to deform accordingly.

Superplastic deformation induces changes in the microstructure of materials, which exhibit the following characteristics:
(a) There is no significant increase in dislocation density or occurrence of intragranular slip. (b) With prolonged superplastic deformation at high temperatures, grain growth may occur. (c) Despite undergoing significant deformation, the grain shape remains equiaxed. (d) In the case of refined-grain alloys prepared through cold deformation and recrystallization, and exhibiting texture, the texture will disappear following superplastic deformation.

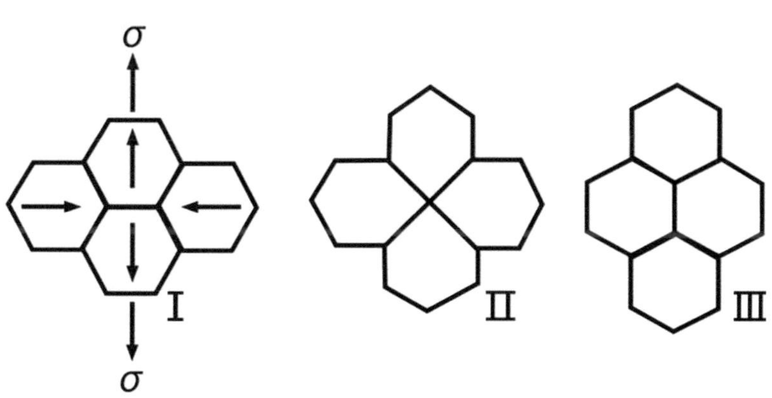

Fig. 12.35 Schematic sketch showing the mechanism of superplasticity

12.5 Deformation and Strength of Ceramics

12.5.1 Deformation and Modulus of Ceramic Materials

The classical view of ceramic materials is that they are brittle and that dislocations are not important because they do not move. It is also believed that ceramics fracture along grain boundaries if they are polycrystalline. However, the modern view of ceramics is quite different because they can be used as thin films where stresses can be very high. In addition, deformation at high temperatures can be significant, and in some special "new" ceramics, displacive transformations become important.

Deformation in ceramics is generally limited to elastic deformation at temperatures of 1000 °C or below. The strength of ceramics can be calculated using $\tau = F/S$ Eq. 12.3-point bending: $\sigma_r = \frac{3PL}{2BW^2}$ Eq. 12.6 for an elastic material. The elastic constants for some ceramics are given in Table 12.2.

The Young's modulus of composite materials containing ceramic substances and other components can be estimated using the rule of mixture given in $E{=}E_{\mathrm{a}}V_{\mathrm{a}} + E_{\mathrm{b}}V_{\mathrm{b}}$ Eq. 12.40.

$$E = E_a V_a + E_b V_b \tag{12.40}$$

wherein V_a and V_b are the respective volume fractions. Measurements of Young's modulus are often carried out at high temperatures. For alumina in the temperature range up to 600 °C, the following equation can be used to account for porosity (P) in the material, with E_0 being the Young's modulus of the dense material without porosity:

$$E = E_0(1 - 1.9P + 0.9P^2) \tag{12.41}$$

Young's modulus in ceramics is largely affected by porosity. For polycrystalline ceramics with porosity P, the elastic modulus decreases because the pores have zero stiffness. Empirical values for the constants a and b can be determined to establish the relation between Young's modulus and porosity, as shown in Table 12.3. It is worth noting that at temperatures of 1000 °C or below, deformation of ceramics, especially single crystals, is entirely elastic, with no plastic processing possible after sintering. However, creep can occur at temperatures beyond 1000°C. In the 1960s, Spriggs conducted a series of experiments to investigate the influence of porosity on the mechanical characteristics of ceramics.

12.5.2 Tensile and Compressive Strengths of Ceramic Materials

Ceramics are frequently required to support tensile loads in practical applications. For instance, steel reinforced concrete and safety glass use ceramics prestressed

Table 12.2 Elastic constants of selected polycrystalline ceramics (20 °C)

Material	Crystal type	*G*/GPa	*B*/GPa	*E*/GPa	*ν*
Carbides					
C	Cubic	468	416	1022	0.092
SiC	Cubic	170	210	402	0.181
TaC	Cubic	118	217	300	0.270
TiC	Cubic	182	242	437	0.199
ZrC	Cubic	170	223	407	0.196
Oxides					
Al_2O_3	Trigonal	163	251	402	0.233
$Al_2O_3 \cdot MgO$	Cubic	107	195	271	0.268
$BaO \cdot TiO_2$	Tetragonal	67	177	178	0.332
BeO	Tetragonal	165	224	397	0.204
CoO	Cubic	70	185	186	0.332
$FeO \cdot Fe_2O_3$	Cubic	91	162	230	0.263
Fe_2O_3	Trigonal	93	98	212	0.140
MgO	Cubic	128	154	300	0.175
$2MgO \cdot SiO_2$	Orthorhombic	81	128	201	0.239
MnO	Cubic	66	154	173	0.313
SrO	Cubic	59	82	143	0.210
$SrO \cdot TiO_2$	Cubic	266	183	538	0.010
TiO_2	Tetragonal	113	206	287	0.268
UO_2	Cubic	87	212	230	0.319
ZnO	Hexagonal	45	143	122	0.358
$ZrO_2 \cdot 12Y_2O_3$	Cubic	89	204	233	0.310
SiO_2	Trigonal	44	38	95	0.082
Chalcogenides					
CdS	Hexagonal	15	59	42	0.38
PbS	Cubic	33	62	84	0.27
ZnS	Cubic	33	78	87	0.31
PbTe	Cubic	22	41	56	0.27
Fluorides					
BaF_2	Cubic	25	57	65	0.31
CaF_2	Cubic	42	88	108	0.29
SrF_2	Cubic	35	70	90	0.29
LiF	Cubic	48	67	116	0.21
NaF	Cubic	31	49	77	0.24

(continued)

Table 12.2 (continued)

Material	Crystal type	*G*/GPa	*B*/GPa	*E*/GPa	*ν*
Other halides					
CsBr	Cubic	8.8	16	23	0.26
CsCl	Cubic	10	18	25	0.27
CsI	Cubic	7.1	13	18	0.27
KCl	Cubic	10	18	25	0.27
NaBr	Cubic	11	19	28	0.26
NaCl	Cubic	15	25	38	0.25
NaI	Cubic	8.5	15	20	0.27
RbCl	Cubic	7.5	16	21	0.29

* All values were calculated from single-crystal data

Table 12.3 Relations between porosity, *P*, and Young's modulus, *E*

$E = E_0(1-aP)$	Linear decrease in Young's modulus with porosity when P is small ($a \approx 4$)
$E = E_0(1-aP + bP^2)$	For a low concentration of spherical pores ($a \approx 1.9, b \approx 0.9$)
$E = E_0(1-aP)^b$	For solid foams with very high porosity $P > 0.7$ ($a = 1, b = 2$)
$E = E_0[(1-P)^2/1 + (a-1)P]$	"a" is a shape factor with values depending on porosity: $a = 2.5$ for interconnected porosity ($1/a = 0.4$) $a = 3.3 \sim 1.4$ for porosity that resembles ribbons ($1/a = 0.3 \sim 0.7$) $a = 0.6 \sim 1.0$ for isolated pores ($1/a = 0.6 \sim 1.0$) ($1/a$ is known as the Nielson shape factor)
$E = E_0 \exp(-aP)$	Empirical for oxides with porosity in the range 0~40% ($a \approx 4$)

in compression to increase their ability to bear tensile loads. However, the tensile test, which is commonly used to determine the tensile strength of metals, is unsuitable for ceramics because of their inherent brittleness. Tensile tests typically use "dog bone"-shaped samples, where the cross-sectional area is reduced in the gauge length. Connecting ceramic samples to the test machine using a screw thread can be difficult and may result in breakage. Moreover, ceramics fail after only approximately 0.1% strain, so precise specimen alignment is critical.

The compressive fracture strength of ceramics is typically 10~15 times greater than their tensile fracture strength. When the testing requirements cannot be met, the bending stresses (σ_f) should be used instead of the tensile strength (σ_t).

Ceramics display similar stress–strain curves in tension and compression, and their Young's modulus is identical because the curves have the same slope. Ceramics are generally stronger in compression and can withstand high compressive

Table 12.4 Ratio of compressive strength σ_c to bending strength σ_f

Ceramic	Grain size/μm	σ_c/σ_f
TiB_2	20–50	4–6
ZrB_2	20–50	4–6
B_4C	1	7
WC	1–6	4–6
Al_2O_3	1–100	4–30
$MgAl_2O_4$	1	7
ThO_2	4–60	13–17
UO_2	20–50	5–18

loads. Table 12.4 lists the ratios of compressive strength to tensile strength of a few ceramics.

However, reliable compressive strength data for ceramics are limited. One widely tested ceramic in compression is concrete, which is a ceramic-matrix composite made of aggregates (stone and sand) and a cementitious hydrated cement matrix. The aggregate provides the strength of hardened concrete, while the matrix controls the workability of fresh concrete. Due to the typical brittleness of concrete used in construction, it is designed to bear compressive loads.

When comparing the properties of metals and ceramics, significant differences can be observed between them. The primary reason for this gap is the difference in the way interatomic bonds form. In metallic bonds, electrons are not strongly bound by attraction to the nucleus, which allows plastic deformation based on their dislocations to readily occur. Consequently, metals are less likely to fracture due to stress concentration. In contrast, the ionic and covalent bonds in ceramics give the substance a structure in which atoms and ions cannot move easily. Thus, relaxation under stress is less likely to occur in ceramics, which can result in breakage from even comparatively small faults and the sample may shatter suddenly without warning. However, ceramics do have the advantage of stability: they do not change their properties even when exposed to chemicals, and their shape remains unchanged even when external forces are applied.

12.6 Molecular Motion and Transition of Polymers

Polymer chains exhibit flexibility and adopt random coil configurations due to the internal rotation of single σ-bonds in their main chains. This molecular motion plays a pivotal role in establishing a connection between the structure and properties of polymers, imparting them with distinct characteristics that set them apart from small molecular materials. For instance, polymethyl methacrylate is a rigid plastic at room temperature, but it becomes flexible when exposed to temperatures exceeding 100 °C, owing to the increased molecular motion at elevated temperatures. These alterations in properties, encompassing but not limited to mechanical and thermal aspects, are commonly referred to as transitions in polymers. Consequently, comprehending the mechanical behavior, thermal transformations, and relaxation phenomena of polymers at different temperatures and the various factors influencing the transition temperature becomes imperative for judicious material selection, determination of processing conditions, and material modification.

12.6.1 Characteristics of Molecular Motion of Polymers

① Multiplicity of Motion Units

The intricate nature of polymer structures gives rise to various motion units within them, and the movement of polymer molecules can be primarily categorized into five types based on the size of the motion units.

1. **Whole-chain motion**: This refers to the overall movement of the mass center of the polymer chain. Examples of whole-chain motion include the flow of a macroscopic melt and the crystallization process of polymer.
2. **Segment motion:** In segment motion, the mass center of the polymer remains stationary while a portion of the chain's segments moves in relation to one another through the internal rotation of single bonds. A segment represents the smallest independent moving unit within the polymer chain and typically spans several to several dozen structural units. Segment motion is a distinctive characteristic of polymers, differentiating them from small molecules. It results in the straightening or curling of macromolecules, leading to high elasticity (stretching and recovery) observed in rubber in macro mechanics. Additionally, the whole-chain motion of polymers is achieved through the coordinated movement of individual segments.
3. **Monomeric-unit motion**: This refers to the movement of monomeric units, which are smaller than a segment, while the segment itself remains immobile. It includes the crank motion of a main chain consisting of four or more monomeric units, as well as the motion of hybrid monomeric units within a heterochain polymer's main chain.
4. **Side-group or side-chain motion**: Side-group or side-chain motion involves the turning, internal rotation, and similar movements of side groups or side chains attached to the main chain at lower temperatures. These motions require less energy compared to segment motion and are also known as secondary relaxation.
5. **Crystalline-region motion**: This refers to the molecular motion occurring in processes such as crystal transformation, flaw extension, or recovery of the crystalline region within a crystalline polymer. These motions occur locally in response to external forces or environmental factors.

Usually, the whole polymer chain is referred to as a large-size motion unit, and the corresponding motion is known as Brownian motion. On the other hand, chain segments and motion units smaller than the chain segment are considered small-size motion units, and their corresponding motion is termed Micro-Brownian motion.

② Time-Dependence of Molecular Motion

Motion inherently requires time, and the duration for an object to transition from one equilibrium state to another through molecular motion under an external field is known as the relaxation time (τ). This process is called the relaxation process. In the case of small molecules, with relaxation times typically ranging from 10^{-10} to 10^{-8} s, molecular motion occurs swiftly, almost instantaneously. However, in the realm of polymers, due to their structure multiplicity, relaxation times extend from 10^{-8} s to days, weeks, and even months or years.

For instance, consider the relaxation of deformation: when a polymer is stretched by a displacement Δx under an external force, the polymer does not instantaneously revert to its original length upon removal of the force, i.e., Δx does not immediately become zero. As shown in Fig. 12.36, the polymer initially undergoes a relatively rapid contraction, which gradually decelerates over time. This behavior arises from the fact that when a polymer material is stretched, the chains of molecules transition from a coiled state to an extended state. After the external force is removed, various motion units within the molecules rely on thermal motion to return to the coiled state and complete the relaxation process. The relationship between the deformation Δx and the time t can be described as follows:

$$\Delta x(\mathrm{t}) = \Delta x_0 \cdot \mathrm{e}^{-\mathrm{t}/\tau} \tag{12.42}$$

where Δx_0 represents the initial increase in length of the polymer before the external force is removed, $\Delta x(t)$ represents the increase in length of the polymer at a

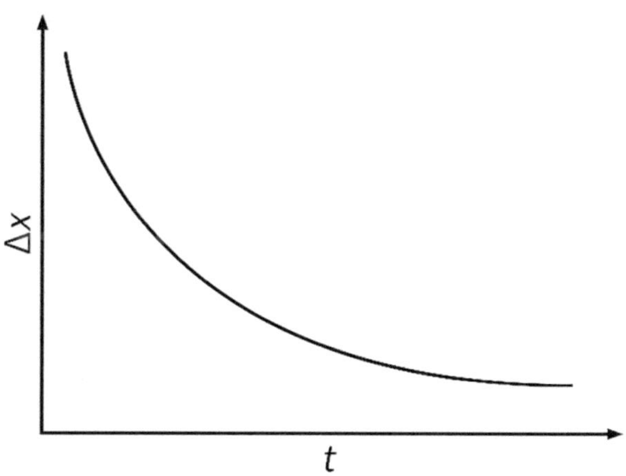

Fig. 12.36 Relaxation curve of a polymer

specific time point t after removing the external force, and τ represents the relaxation time. When $t = \tau$, the relationship $\Delta x(t) = \Delta x_0/\mathrm{e}$ is established. Therefore, from a macroscopic perspective, the relaxation time τ refers to the time required for $\Delta x(t)$ to decrease to 1/e of Δx_0. Consequently, the relaxation time τ serves as a physical quantity that characterizes the rate of the relaxation process.

In addition to deformation relaxation, there are other types of relaxation processes, such as stress relaxation, dielectric relaxation, volume relaxation, and more.

③ Temperature Dependence of Molecular Motion

The motion of molecules is inherently influenced by temperature. For small molecules, the temperature dependence of molecular motion follows the Arrhenius Equation, which can be expressed as:

$$\tau = \tau_0 \cdot \mathrm{e}^{\Delta \mathrm{E}/\mathrm{RT}} \tag{12.43}$$

However, in the case of polymers, the temperature dependence conforms to the Arrhenius Equation only at high temperature (corresponding to whole-chain motion) and low temperature (associated with the motion of monomeric units and smaller units). When the temperature reaches the glass transition temperature T_g, the motion units of polymers primarily consist of chain segments. The unique temperature dependence of molecular motion in this regime does not follow the Arrhenius Equation but instead conforms to the well-known Williams-Landel-Ferry (WLF) Equation:

$$\ln\left(\frac{\tau}{\tau_g}\right) = \frac{-C_1(T - T_g)}{C_2 + (T - T_g)} \tag{12.44}$$

In the equation, τ_g represents the relaxation time at the glass transition temperature T_g, while C_1 and C_2 are empirical constants (typically, $C_1 = 17.44$ and $C_2 = 51.6$).

This equation indicates that temperature rise would shorten the relaxation time.

12.6.2 Relationship Between Polymer Molecular Motion and Mechanical States

Molecular motion serves as a vital connection between the structures and properties of polymers, establishing a correspondence between molecular motion and the mechanical state of polymers. The mechanical state refers to the physical state of a polymer resulting from deformation under an external force. As temperature increases or decreases, polymers can undergo three distinct mechanical states and two transitions, each reflecting different molecular motions. Using amorphous linear polymer as an example, the relationship between deformation, elastic modulus, and temperature is illustrated in Fig. 12.37.

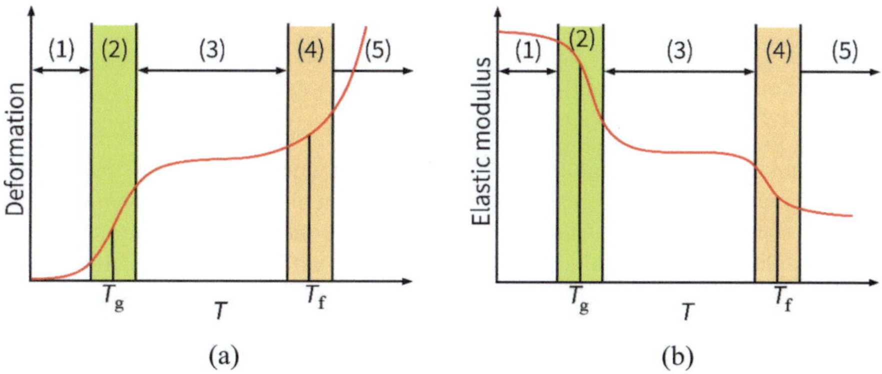

Fig. 12.37 **a** Deformation-temperature curve and **b** elastic modulus-temperature curve of an amorphous linear polymer

1. **Glassy state:** At temperatures below T_g, the motions of the whole polymer chain and chain segments are locked in a glassy state, with limited vibration occurring at fixed positions. In terms of mechanical properties, the polymer resembles glass and tends to be brittle. The modulus of a glassy state polymer is $E = 10^0 \sim 10^{0.5}$ GPa.
2. **Glass transition** (glass–rubber transition): The transition from the glassy state to the rubbery state is known as the glass transition of polymers, and the temperature at which this occurs is referred to as T_g. Around T_g, although the whole polymer chain remains immobile, the motion of segments is activated, resulting in a rapid decline in modulus, that is, $E = 10^{-3.3} \sim 10^0$ N/m^2.
3. **Rubbery state**: At higher temperatures in rubbery state, the motion of segments is completely free, but the motion of the whole polymer chain is not available yet. The polymer exhibits the characteristic high elasticity of polymers, and the modulus remains nearly constant within the range of $E = 10^{-3.6} \sim 10^{-3.3}$ GPa.
4. **Viscous flow transition**: At even higher temperatures, the polymer undergoes a transition from rubbery state to the viscous flow state, with the transition temperature denoted as T_f. At this temperature, intensified thermal motion allows the polymer chain to start moving as a whole, resulting in a sharp decrease in modulus to $E < 10^{-3.6}$ GPa.
5. **Viscous flow state**: At temperatures higher than T_f, the overall motion of the polymer chain becomes completely unrestricted, transforming the polymer into a melt-like flow, with a very low modulus.

12.6.3 Glass Transition of Polymers

① Definition of Glass Transition

From the perspective of molecular motion, the glass transition refers to the activation of molecular segment motion in amorphous polymers. In terms of mechanical state, the glass transition corresponds to the transition of amorphous polymers from a glassy state to a rubbery state as the temperature increases, or from a rubbery state to a glassy state as the temperature decreases. In the case of crystalline polymers, the glass transition refers to the transition occurring within the amorphous regions of the polymer. The temperature at which the glass transition occurs is referred to as the glass temperature, denoted as T_g.

Note that, in crystalline polymers, the crystalline region's presence significantly influences the amorphous region's molecular motion. As a result, determining the T_g for highly crystalline polymers, such as polyethylene, remains a subject of debate. Therefore, the discussion here primarily focuses on the glass transition of amorphous polymers.

T_g is a characteristic temperature of polymers. In theory, T_g is an indicator of polymer chain flexibility (see Sect. 4.6.2). Lower T_g values correspond to greater molecular chain flexibility. In practical terms, T_g marks the boundary between plastics and rubbers. When the temperature exceeds T_g, the polymer behaves like rubber and exhibits high elasticity. Conversely, when the temperature falls below T_g, the polymer becomes rigid and behaves like hard plastic. Therefore, T_g represents the upper limit of the service temperature for amorphous thermoplastic and the lower limit of the service temperature for rubbers. For commonly used polymers, we classify those with T_g below room temperature as rubbers and those with T_g above room temperature as plastics.

It is important to note that many physical and mechanical properties of polymers, such as specific volume, refractive index, elastic modulus, and dielectric loss, undergo significant changes at T_g.

② Theories for Glass Transition

The glass transition is a highly complex phenomenon, and to date, no single theory provides a complete and accurate explanation that aligns perfectly with all observed facts. There are four commonly recognized theories: the equal free volume theory, the equal viscosity theory, the thermodynamic theory, and the kinetic theory. This book briefly introduces the equal free volume theory, which is the most renowned among them.

According to the equal free volume theory, the volume of a solid or liquid substance consists of two parts, i.e., the volume occupied by the molecules themselves and the volume of voids, the latter of which is also known as free volume. It is the existence of free volume that enables the movement of molecular chains. As depicted in Fig. 12.38, when the temperature is higher than T_g, the free volume increases with temperature. However, the free volume remains constant when the temperature is below Tg. Therefore, T_g represents the temperature at which the

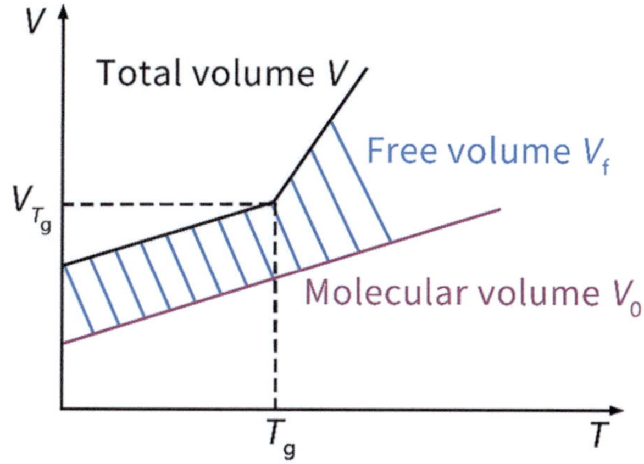

Fig. 12.38 Illustration of free volume in a polymer

free volume reaches a critical value, allowing for the movement of polymer segments. The free volume is the same for all polymers The equal free volume theory suggests that all polymers have the same free volume, giving rise to the concept of equal free volume.

According to the WLF Equation $\lg \alpha_T = \frac{-C_1(T-T_g)}{C_2+(T-T_g)}$, , for all polymers, the fraction of free volume (f_g) at T_g is approximately 0.025. In other words, during the glass transition, the free volume of the polymer occupies approximately 2.5% of the total volume.

③ Influencing Factors on Glass Transition

Flexibility of the Main Chain

The flexibility of the polymer's main chain plays a significant role in determining its glass transition behavior. The ease of internal rotation within the main chain directly affects the polymer's flexibility and, consequently, its glass transition temperature. Table 12.5 shows the internal rotation potential barriers (U_0) for the main chains of polydimethylsiloxane (PDMS), polyethylene (PE), and polytetrafluoroethylene (PTFE). Although PE has a highly flexible molecular chain, it

Table 12.5 Relationship between the molecular chain's internal rotation potential energy (U_0) and the glass transition temperature (T_g)

Polymer	PDMS	PE	PTFE
Structural Unit	$—Si(CH_3)_2—O—$	$—CH_2—CH_2—$	$—CF_2—CF_2—$
U_0 /kJ·mol^{-1}	1~2	13.8	19.7
T_g /°C	−123	−68*	−50

* There are varying opinions regarding the glass transition temperature of PE, each with its own rationale

exhibits plastic behavior at room temperature due to its ordered arrangement and crystallization propensity.

12.6.4 Side Group or Side Chain

The presence of side groups or side chains in monosubstituted polymers significantly impacts T_g. The introduction of side groups hinders the internal rotation of the molecular chain, reducing chain flexibility and increasing the T_g. As the size of the side group or side chain increases, the T_g also rises (Table 12.6). Furthermore, the polarity of the side group influences intermolecular interactions. More vital polarity leads to stronger intermolecular interactions, further hindering internal rotation and reducing chain flexibility, resulting in an elevated T_g (Table 12.7). Note that the T_g of polyacrylic acid is higher than that of polyvinyl chloride and polyacrylonitrile due to the presence of hydrogen bonding in addition to polar interactions with the carboxyl group, further reducing the molecular chain's flexibility.

For symmetrically disubstituted polymers, symmetry impacts the molecular chain's internal rotation potential barrier, leading to increased flexibility and a decrease in T_g, as shown in Table 12.8.

As shown in Table 12.9, when the side group molecular chain is longer and more flexible, the T_g of the polymer decreases. This can be attributed to the fact that the increase in flexibility resulting from the longer and more flexible side group outweighs the decrease in flexibility caused by the enlargement of the side group.

Interaction Between Polymer Chains

The interaction between polymer chains plays a significant role in influencing the glass transition. In general, strong intermolecular interactions hinder the motion of chain segments, reducing the flexibility of the molecular chains and increasing the T_g. An example of this is the high T_g of polyacrylic acid, as shown in Table 12.7, which is attributed to the presence of hydrogen bonding and polarity, leading to stronger intermolecular interactions.

Apart from the factors mentioned above, other factors such as molecular weight, external forces (such as tension or pressure), frequency of external force application, cooling rate, and various other factors can also affect T_g. These additional factors contribute to the complex nature of the glass transition phenomenon.

Table 12.6 Relationship between T_g and side group size

Polymere	Polyethylene	Polypropylene	Polystyrene	Poly(α-methylstyrene)	Polyvinylcarbazole
Side group	—H	$—CH_3$	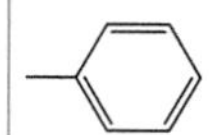	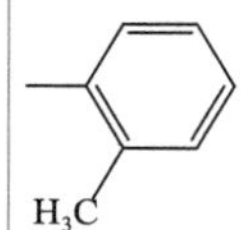	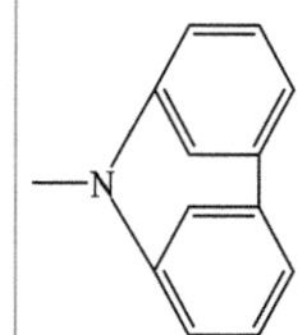
T_g /°C	−68	−10	97	115	208

Table 12.7 Relationship between the polarity of substituent and T_g

Polymers	Polyethylene	Polypropylene	Polyacrylic acid	Polyvinyl chloride	Polyacrylonitrile
Side group	—H	—CH_3	—COOH	—Cl	—CN
T_g /°C	−68	−10	106	87	104

Table 12.8 Relationship between T_g and substituents symmetry. T_g decreases due to symmetric substituents

Polymers	Polyvinyl chloride	polyvinylidene chloride	polypropylene	polyisobutylene
	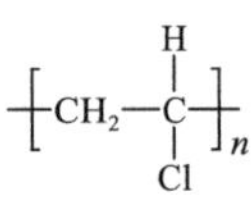	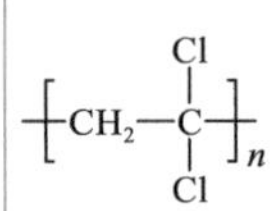	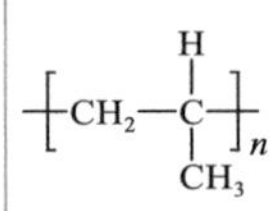	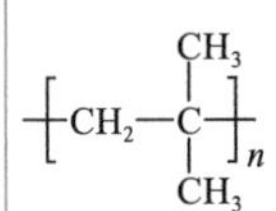
T_g /°C	87	–17	–10	–70

Table 12.9 Relationship between substituent length in polymethacrylate and T_g

Polymers	polymethyl methacrylate	polyethyl methacrylate	polypropyl methacrylate	polybutyl methacrylate
	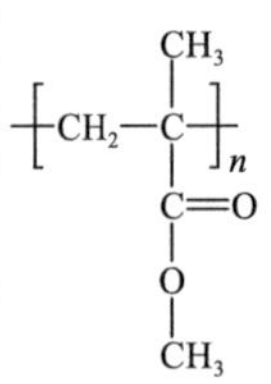	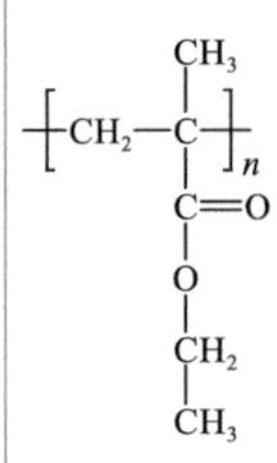	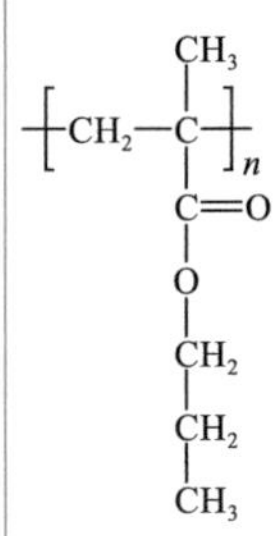	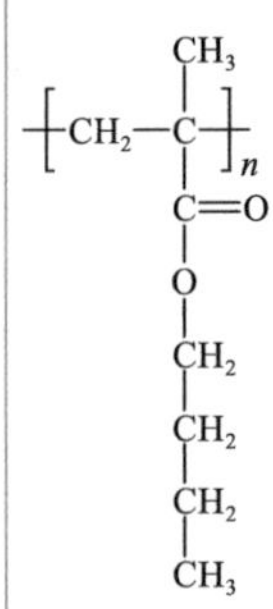
T_g /°C	100~120	65	35	20

12.7 Deformation and Strength of Polymers

12.7.1 High Elasticity

As discussed in Sect. 12.5, when the temperature is above T_g, the polymer is in a state of high elasticity and has unique physical and mechanical properties. For example, polymers with high elasticity maintain a stable dimension, and their response to small deformations (< 5%) follows Hooke's law, resembling that of solids. However, the coefficient of thermal expansion and isothermal compressibility of these polymers are similar in magnitude to those of liquid, indicating that the interaction between macromolecules in the high elasticity state resembles that of liquid. Moreover, in the high elasticity state, the stress required for deformation increases with temperature, similar to how gas pressure increases with temperature.

① Characteristics of High Elasticity

In terms of mechanical properties, polymers with high elasticity have the following characteristics when compared to metals and/or ceramics:

1. The reversible elastic deformation can reach up to 1000%, whereas it does not exceed 1% for most metals.
2. The elastic modulus (referred to as the high elastic modulus) is relatively small, ranging from approximately 10^{-4} to 10^{-3} GPa, while typical metals have elastic moduli ranging from 10 to 10^2 GPa;
3. For polymers, the elastic modulus at the high elastic stage increases with increasing temperature, whereas it decreases with temperature for metals.
4. During rapid stretching in a thermally insulated system, rubbers will heat up due to heat release, whereas metals cool down due to heat absorption.
5. High elasticity in polymers is closely associated with entropy change, whereas the instantaneous elasticity of metals and ceramics is linked to energy change.

② Thermodynamic analysis of high elasticity

Consider a slightly crosslinked rubber sample with an initial length of l_0, and slowly stretch it to $l_0 + dl$ under a constant temperature while applying tension f. Slow stretching indicates that the sample remains in a thermodynamic equilibrium conformation, and the deformation is in an equilibrium state without any relaxation process.

According to the first law of thermodynamics, the change in internal energy of the system, dU, during the stretching process is given by:

$$dU = dQ - dW \tag{12.45}$$

Here, dQ represents the heat absorbed by the system, which is related to the entropy change dS and the temperature T according to the second law of thermodynamics for a constant temperature reversible process:

$$dQ = TdS \tag{12.46}$$

where dS represents the entropy change of the system.

dW represents the external work done on the system, which includes the expansion work pdV due to volume change and the elongation work $-fdl$ due to stretch deformation (since external work is performed on the system during stretching, the elongation work is negative):

$$dW = pdV - fdl \tag{12.47}$$

Consequently, the following equation is obtained:

$$dU = TdS - pdV + f\,dl \tag{12.48}$$

Assuming constant volume during the stretching process ($pdV = 0$), the equation simplifies to:

$$dU = TdS + f\,dl \tag{12.49}$$

Further calculations with partial differentiation of l under the constant temperature and volume conditions lead to the following:

$$f = \left(\frac{\partial U}{\partial l}\right)_{T,V} - T\left(\frac{\partial S}{\partial l}\right)_{T,V} \tag{12.50}$$

This equation is called the thermodynamic equation of state for isothermally stretched rubber. It indicates that during stretch deformation, the equilibrium stress in the rubber comprises two components: the internal energy change ΔU and the entropy change ΔS of the rubber.

From the molecular motion perspective, polymers' high elastic deformation arises from the relative motion of chain segments, which in turn depends on the internal rotation of single bonds. Under ideal conditions, assuming complete freedom of internal rotation, where all conformations of the molecular chain have the same energy, and there are no intramolecular or intermolecular forces, when there is no change in the internal energy of the system ($dU = 0$), the following equation is obtained:

$$f = -T\left(\frac{\partial S}{\partial l}\right)_{T,V} \tag{12.51}$$

This equation implies that the elastic restoring force in ideal rubber during isothermal stretch primarily arises from the entropy change of the system. Under tension, the polymer chains change from their original curled state to an extended state, decreasing conformational entropy. However, due to thermal motion, the polymer chains tend to spontaneously recover their original curled state, resulting in an elastic restoring force. The recovery tendency of conformational entropy strengthens with increasing material temperature, thereby increasing the elastic stress. In addition, the conformational entropy decreases ($dS < 0$), and dQ is negative. In other words, rubber releases heat during stretching, causing its temperature to rise.

12.7.2 Viscoelasticity

Viscoelasticity is a unique mechanical behavior characterized by the simultaneous presence of solid elasticity and liquid viscosity. While all natural objects possess both elasticity and viscosity, their dominant behavior—elasticity or viscosity—varies depending on external conditions such as loading time and temperature. However, in polymers, elasticity, and viscosity are distinctly observed simultaneously.

As mentioned in Sect. 12.5, the molecular motion of polymers exhibits time dependence. The viscoelastic nature of polymers is primarily manifested through the dependence of their mechanical behavior on the duration of external forces, representing the macroscopic manifestation of molecular motion's time dependence. The changes in the mechanical behavior of polymers over time are collectively referred to as mechanical relaxation. Mechanical relaxation phenomena in polymers include creep, stress relaxation, hysteresis, and mechanical loss. In this section, we will focus on creep and stress relaxation, followed by a brief introduction to the theoretical models of viscoelasticity and the concept of time–temperature equivalence.

① Creep Behavior

Creep is a phenomenon where the deformation of an object gradually increases with time under a constant external force and reaches equilibrium at a specific temperature. When the external force is removed after a specific time, the deformation changes over time, known as creep recovery.

An ideal elastic body (Hooke elastic body) responds instantaneously under constant stress, resulting in instantaneous elastic deformation. Upon removal of the external force, the strain immediately disappears, and the object returns to its original state. An example of creep phenomenon is observed in a soft polyvinyl chloride wire elongating slowly under the influence of a weight and gradually retracting upon the weight's removal. Figure 12.39 illustrates the creep curve depicting this process, where t_1 is the time when the stress is applied, and t_2 is the time when the stress is removed. It is evident that the creep behavior of polymers differs significantly from that of metals.

From the perspective of molecular motion, the creep process encompasses three types of deformation:

Instantaneous elastic deformation: When a polymer is subjected to an external force, the bond length and bond angle within the molecular chain immediately change, resulting in slight deformation that completely recovers after removing the external force. This phenomenon is known as instantaneous elastic deformation (represented by ε_1), as shown in Fig. 12.40. Instantaneous elastic deformation follows Hooke's law: $\varepsilon_1 = \sigma/E_1$, where σ represents stress, and E_1 represents the modulus of instantaneous elastic deformation.

High elastic deformation: High elastic deformation occurs as the molecular chain gradually extends through the motion of chain segments, resulting in significantly

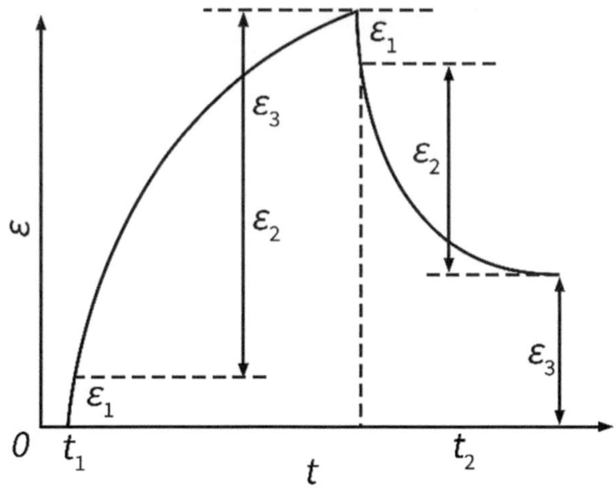

Fig. 12.39 Creep curve of a linear polymer

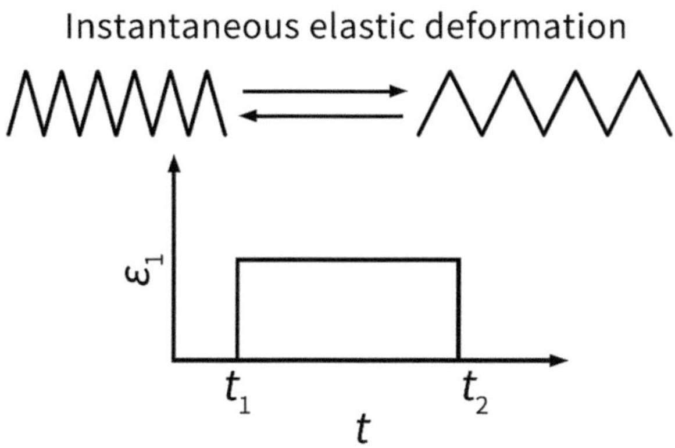

Fig. 12.40 Schematic diagram of instantaneous elastic deformation

more significant deformation compared to instantaneous elastic deformation. A relationship following an exponential decay is established between high elastic deformation (represented by ε_2) and time: $\varepsilon_2 = \sigma/E_2(1\text{-}e^{-t/\tau})$, where τ represents relaxation time related to the viscosity (η_2) of segment motion and the high elastic modulus E_2. Specifically, $\tau = \eta_2/E_2$. Upon removing the external force, the high elastic deformation recovers gradually. Figure 12.41 demonstrates the process of high elastic deformation.

Viscous flow: In linear polymers, where no chemical bonds exist between molecular chains, external forces induce relative slip between the chains, known as viscous flow (represented by ε_3). It can be expressed as $\varepsilon_3 = \sigma/\eta_3 t$, where η_3 represents bulk viscosity. Figure 12.42 provides a schematic diagram of viscous flow.

When a polymer is subjected to an external force, these three types of deformation occur simultaneously, resulting in a total deformation $\varepsilon = \varepsilon_1 + \varepsilon_2 + \varepsilon_3 = \sigma/$

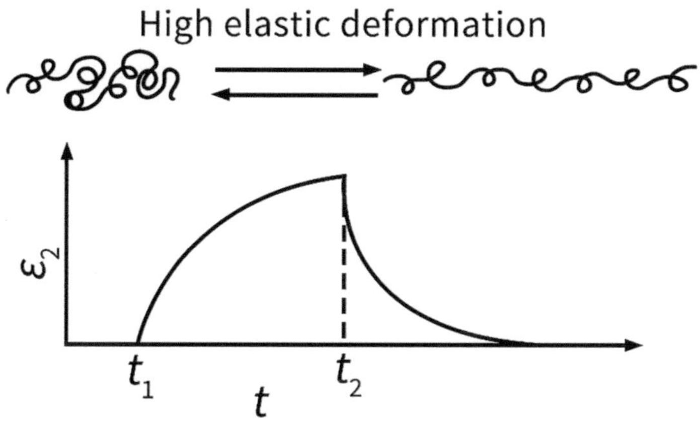

Fig. 12.41 Schematic diagram of high elastic deformation

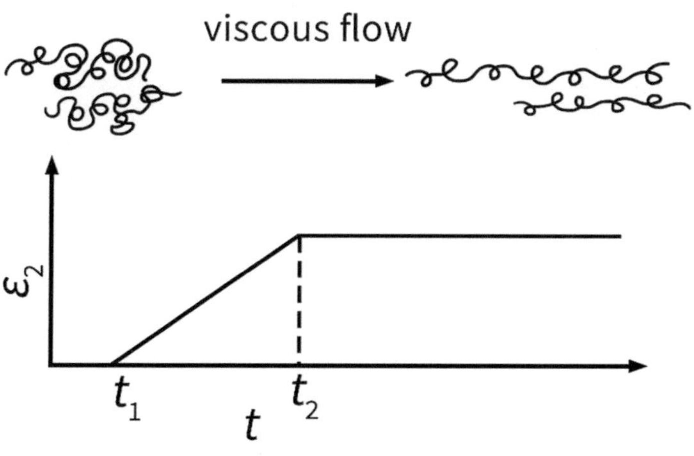

Fig. 12.42 Schematic diagram of viscous flow

$E_1 + \sigma/E_2(1-e^{-t/\tau}) + \sigma/\eta_3 t$. The proportion of these deformations varies depending on specific conditions. When the temperature is below T_g, segment motions are frozen, leading to a large relaxation time τ and a small ε_2. The internal friction resistance between molecules (large η_3) results in a small ε_3, leading to overall minimal deformation (ε). When the temperature is above T_g, chain segments begin to move, and τ decreases with the increase in temperature. As a result, ε_2 increases, and ε_3 is relatively smaller. At temperatures equal to or higher than the viscous flow temperature, τ becomes very small, the viscosity η_3 decreases, and both ε_2 and ε_3 become prominent. Since viscous flow cannot be fully recovered, when

Fig. 12.43 Relationship of creep with temperature and external force

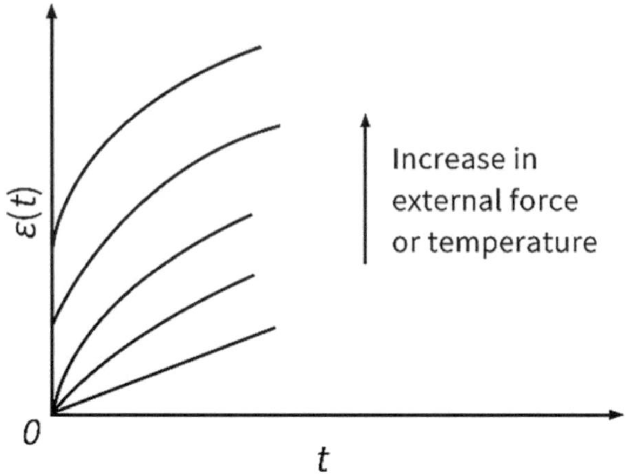

the external force is removed, there is always a portion of irrecoverable deformation in linear polymers upon removing the external force, known as permanent deformation.

As shown in Fig. 12.43, the creep phenomenon is influenced by temperature and external force. At very low temperatures or with very small external forces, creep is minimal and slow, making it difficult to observe within a short time frame. Conversely, at high temperatures or with large external forces, deformation develops rapidly, impeding the observation of creep. Typically, chain segments can slowly move under suitable external force at temperatures slightly above T_g due to reduced internal friction, allowing for the conspicuous observation of creep phenomena.

In addition, the creep phenomenon is influenced by the polymer's structure. Figure 12.44 compares the creep properties of several polymers at 23 °C. Polymers with rigid molecular chains have good creep resistance and can serve as replacements for metal materials in mechanical components. However, materials with significant creep require appropriate measures to mitigate the issue. For example, when using polyvinyl chloride as a pipeline material, additional support should be incorporated to prevent creep.

② Stress Relaxation

Stress relaxation refers to the phenomenon where the stress exerted on a polymer gradually decreases over time under constant temperature and deformation conditions. For example, as shown in Fig. 12.45, when a piece of crosslinked rubber is stretched to a certain length and held at that length, the resilience of the rubber gradually diminishes over time. Stress follows an exponential relationship with time: $\sigma = \sigma_0 e^{-t/\tau}$, where σ_0 represents initial stress and τ represents the relaxation time.

Stress relaxation and creep are two aspects of the same phenomenon, both reflecting the molecular movements within polymers. Initially, when a polymer is

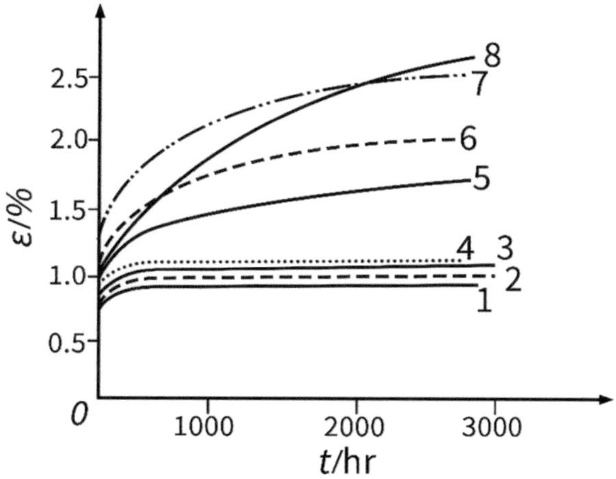

Fig. 12.44 Comparison of creep properties of several polymers at 23 °C: 1-polysulfone; 2-polyphenylene ether; 3-polycarbonate; 4-modified polyphenylene ether; 5-(heat resistant) ABS; 6-polyformaldehyde; 7-Nylon; 8-ABS

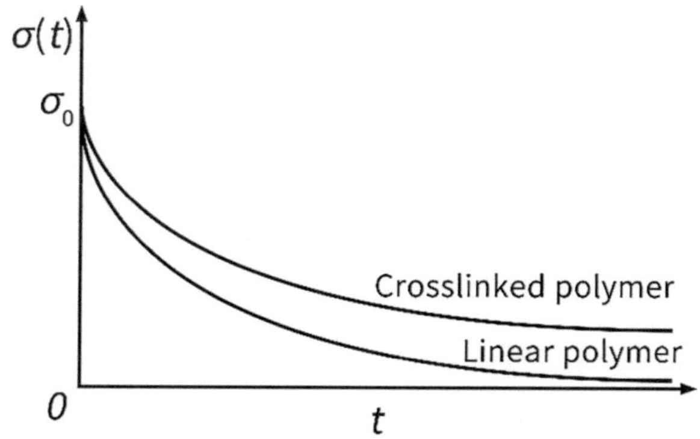

Fig. 12.45 Typical stress relaxation curves of polymers

subjected to stretching, the molecules are in an unbalanced conformation, gradually transitioning to an equilibrium conformation. In other words, the chain segments move in the direction of the external force to reduce or eliminate the stress. If the temperature is significantly higher than T_g, such as at room temperature for rubber, the internal friction during the chain segment movement is minimal. This results in rapid stress relaxation, sometimes occurring too quickly to be easily observed. Conversely, if the temperature is much lower than T_g, such as at room temperature for plastics, although the chain segments experience significant stress, their ability to move is limited due to high internal friction. As a result, stress relaxation is slow and not easily observable. The stress relaxation phenomenon becomes apparent only in the temperature range close to T_g (Fig. 12.46). An example of noticeable stress relaxation is when using polyvinyl chloride wire containing plasticizer to secure objects: initially, the wire is tightly bound, but it becomes loose over time. In the case of crosslinked polymers, where molecular slipping is restricted, stress relaxation occurs only to a certain value but not to zero. This is why rubber products are crosslinked.

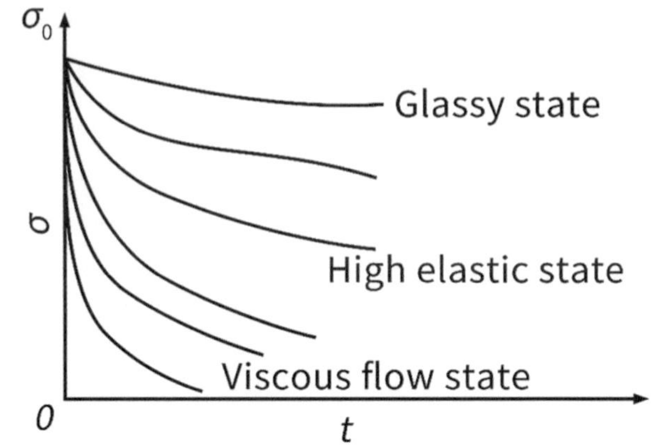

Fig. 12.46 Stress relaxation curves of polymers at different temperatures

Since both creep and stress relaxation are temperature-dependent and reflect molecular motion within polymers, studying the temperature dependence of these phenomena can provide insights into polymer molecular motion and transformations.

③ Mechanical Models of Viscoelasticity

The elasticity of an ideal elastomer follows Hooke's law ($\sigma = E\varepsilon$), and this mechanical behavior can be represented by a spring, as shown in Fig. 12.47a. On the other hand, the viscosity of an ideal liquid adheres to Newton's law ($\sigma = \eta \mathrm{d}\varepsilon/\mathrm{d}t$), which can be represented by a dashpot, as shown in Fig. 12.47b. However, the mechanical behavior of polymers under external forces often exhibits viscoelasticity, which lies between elasticity and viscosity and does not strictly adhere to either Hooke's law or Newton's law. In viscoelastic materials, stress depends on both strain and strain rate simultaneously. Therefore, different types of viscoelastic behavior can be described using combinations of springs and dashpots. When viscoelasticity is a combination of ideal elasticity and viscosity, it is called linear viscoelasticity. Otherwise, it is known as nonlinear viscoelasticity.

Maxwell Model

The Maxwell model consists of a Hooke spring (with elastic modulus E) and a dashpot filled with Newtonian liquid (with viscosity η) which are connected in series, as shown in Fig. 12.48. When subjected to an external force, the Hooke spring undergoes instantaneous deformation, and the elastic force is calculated according to Hooke's law: $\sigma_1 = E\varepsilon_1$. At the same time, the piston in the dashpot starts to move, and Newton's law of viscous fluid determines the viscous force: $\sigma_2 = \eta \mathrm{d}\varepsilon_2/\mathrm{d}t$. Since the spring and the dashpot are connected in series, the stress is the same ($\sigma = \sigma_1 = \sigma_2$), and the total strain is the sum of the two ($\varepsilon = \varepsilon_1 + \varepsilon_2$).

The Maxwell model can qualitatively describe the general characteristics of polymer viscoelasticity and define the main parameter of polymer viscoelasticity, which is the relaxation time τ. However, due to the model's equivalence of the

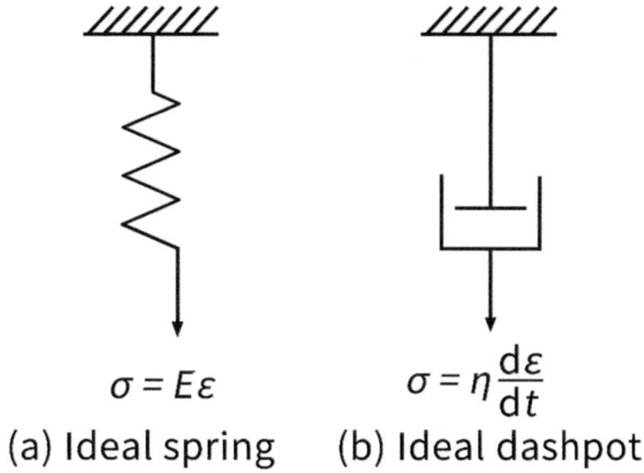

Fig. 12.47 Analog elements representing elasticity and viscosity: (a) ideal spring and (b) ideal dashpot

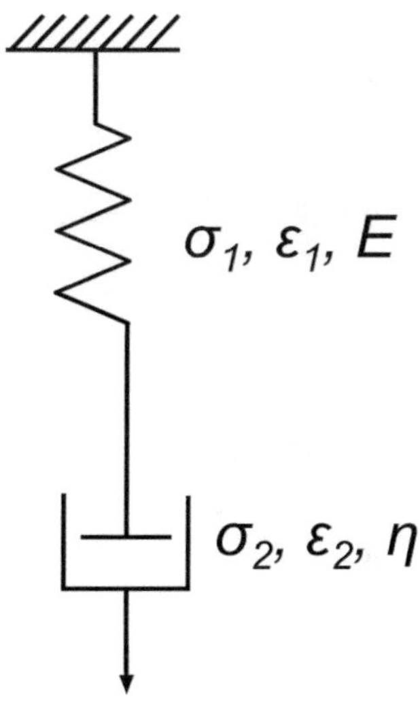

Fig. 12.48 Schematic diagram of the Maxwell model

creep process to the viscous flow of Newtonian fluids and the complexity of polymer creep, it is not suitable for describing creep processes. Similarly, the model cannot simulate the stress relaxation process of cross-linked polymers.

Vogit Model

The Voigt model consists of a Hooke spring (with elastic modulus E) and a Newtonian dashpot (with viscosity η) which are connected in parallel, as shown in Fig. 12.49. Because of the parallel connection of these elements, the stress applied to the model is shared by both elements ($\sigma = \sigma_1 + \sigma_2$), and the strain is the same across both elements ($\varepsilon = \varepsilon_1 = \varepsilon_2$). The Voigt model is suitable for simulating the creep process of crosslinked polymers. When an external force is applied, the presence of the dashpot restricts the immediate rctraction of the spring, causing the

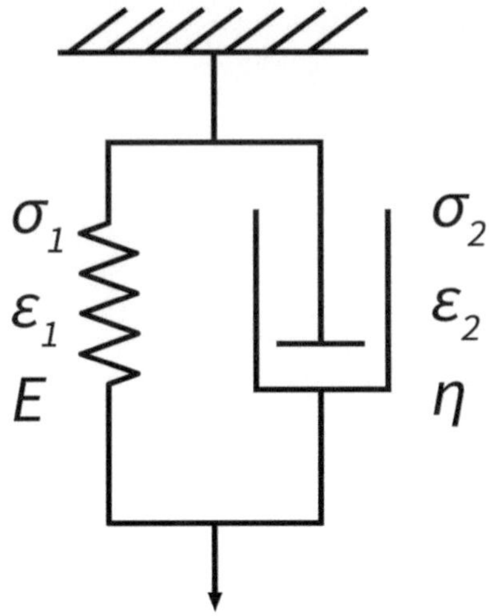

Fig. 12.49 Schematic diagram of the Voigt model

deformation to develop gradually. Upon removal of the external force, the spring's restoring force gradually leads to the recovery of the entire model's deformation, which aligns with the creep process of polymers.

Four-Element Model

As previously mentioned, the creep process involves three types of deformation. The first is the instantaneous elastic deformation caused by changes in bond length and bond angle in molecules, which can be simulated using a rigid spring (E_1) due to its instantaneous deformation. The second is the high elastic deformation resulting from extension and curling of chain segments, which changes over time and can be simulated with spring (E_2) and dashpot (η_2) connected in parallel, as described earlier. The third is the viscous flow caused by mutual slip of polymer chains, which develops linearly with time and can be simulated with a dashpot (η_3). Since the total deformation of the polymer is the sum of these three parts, the model connects them in series. The resulting four-element model can be regarded as a series connection of the Maxwell model and the Voigt model, as shown in Fig. 12.50. By combining these four elements, the coexistence of elasticity and viscosity in polymers can be explained from the viewpoint of polymer structure. The four-element model is particularly suitable for describing the creep process of linear polymers.

④ Relationship of Viscoelasticity with Time and Temperature (Time–Temperature Equivalence Principle)

As discussed in Sect. 5.1, the molecular motion of polymers depends on both time and temperature, and the macroscopic manifestation of molecular motion is considered viscoelasticity. Therefore, the influence of time and temperature on polymer viscoelasticity is equivalent. By altering the time scale, the material properties

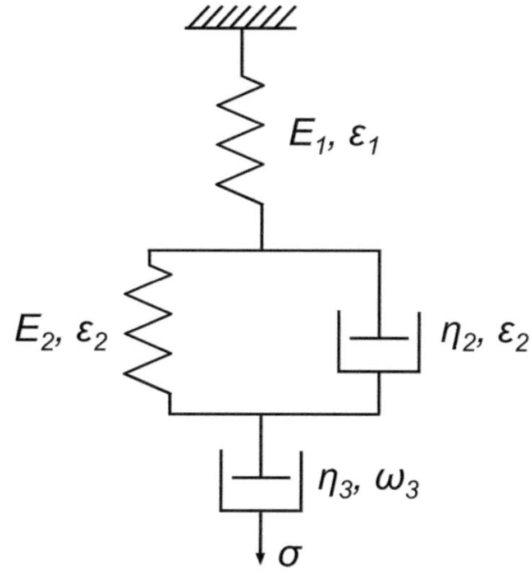

Fig. 12.50 Schematic diagram of the four-element model

at different temperatures can be made equivalent to one another. This principle, known as the time–temperature equivalence principle, implies that the viscoelastic properties of a polymer at a high temperature for a short time can be replicated at a low temperature for a longer time. This equivalence principle applies not only to viscoelastic properties but also to various other properties of polymer materials.

The time–temperature equivalence principle allows researchers to determine material properties over a broader temperature and time range based on measurements obtained within a limited time frame. The fundamental expression of this principle follows the WLF equation. In this equation, shift factor α_T is defined as the ratio of the actual time τ required for the mechanical properties (such as modulus, mechanical loss factor, etc.) to undergo a certain change at the measurement temperature (T) to the time τ_s required for the exact change at the reference temperature (T_s). The shift factor α_T is depicted in Fig. 12.51.

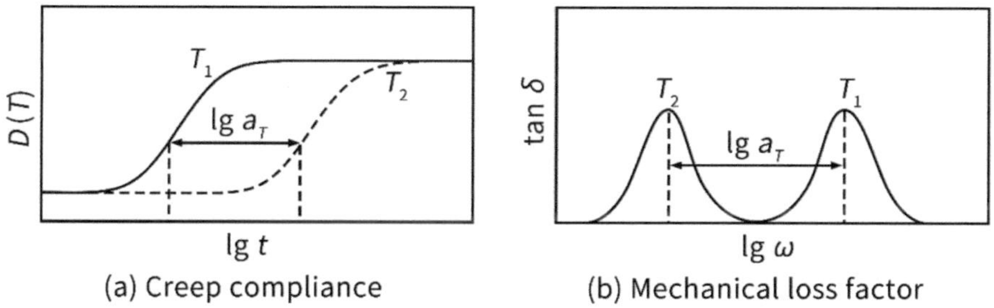

Fig. 12.51 Schematic diagram illustrating the graphical representation of shift factor α_T: **a** creep compliance and **b** mechanical loss factor

12.7.3 Yield and Stress–Strain Curve of Polymer

Under the action of a significant external load, a material undergoes plastic deformation, indicating that it has yielded. Yielding results in a noticeable change in the overall shape of the specimen. From an application perspective, plastic deformation renders the material unusable, which is particularly critical for polymers. Polymers are essentially ductile materials, and the ultimate limit of ductile materials is typically defined by their yield strength rather than their ultimate strength.

The yielding behavior of polymers is analyzed using the stress–strain curve. This curve provides valuable indices for evaluating material properties, such as Young's modulus, yield strength, tensile strength, and elongation at break. It also assists in assessing the strength, hardness, and brittleness of a polymer.

The typical tensile stress–strain curve of an amorphous polymer is shown in Fig. 12.52. It also assists in assessing the strength, hardness, and brittleness of a polymer. Prior to the yield point, the polymer exhibits elastic behavior (*OY* section), where the deformation can fully recover without any permanent deformation after unloading. Therefore, the yield point represents the critical point at which the polymer maintains complete elasticity after unloading, and the stress corresponding to the yield point *Y* is referred to as the yield strength or yield stress (σ_y). Beyond the yield point, the polymer enters the plastic region and displays typical plastic characteristics. Deformation in this region does not fully recover after unloading, resulting in permanent deformation or residual strain. The stress–strain relationship of the polymer in the plastic region is complex. Initially, there is a short duration of strain softening, causing a slight decrease in stress (*YA* section) with increasing strain. Subsequently, the sample experiences plastic instability, characterized by necking, where the strain continues to increase while the stress remains relatively constant (*AC* section). Following this, the polymer undergoes orientation hardening, leading to a sharp increase in stress (*CB* section) until it eventually fractures at point *B*. The stress corresponding to point *B* is called the tensile strength, while the corresponding plastic strain is referred to as the elongation at break.

The stress–strain curves of polymers exhibit diverse forms due to the wide range of polymer types. Based on the characteristics of the yield point, elongation, and fracture during the stretching process, polymers can be roughly categorized into five types: (a) hard and brittle, (b) hard and ductile, (c) hard and strong, (d) soft and ductile, and (e) soft and weak. The corresponding stress–strain curves for each of these polymer types are shown in Fig. 12.53.

Hard and brittle polymers, such as polystyrene, polymethylmethacrylate, and many phenolic resins, possess a high modulus and considerable tensile strength, but they exhibit minimal elongation before breaking. These polymers lack a distinct yield point, and their elongation at break is generally less than 2%.

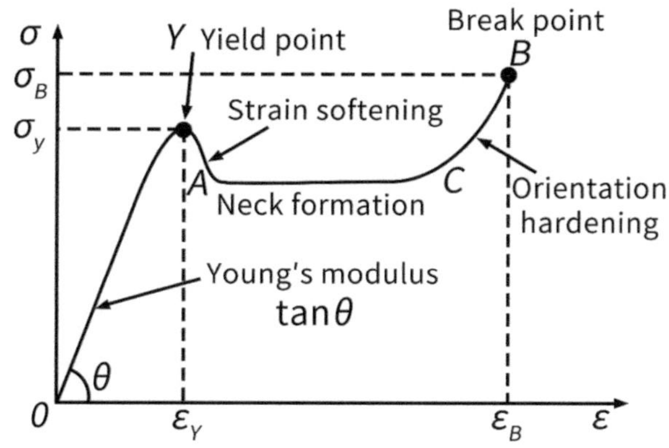

Fig. 12.52 Typical tensile stress–strain curve of an amorphous polymer

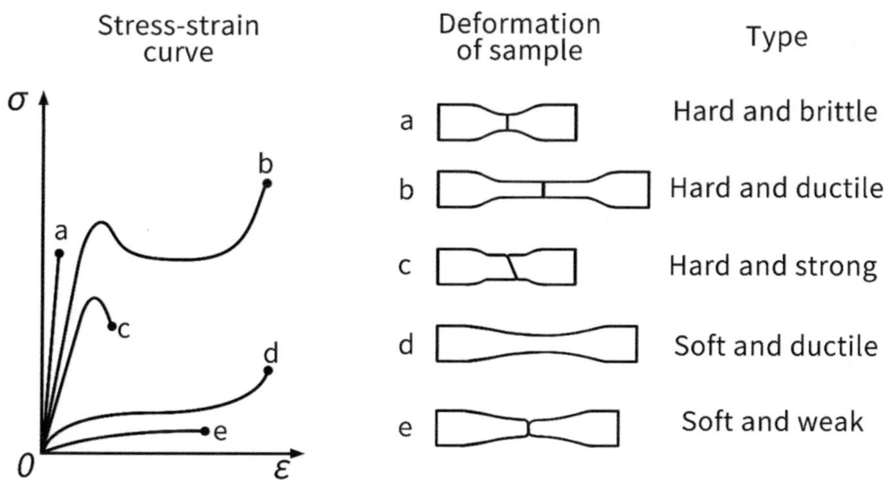

Fig. 12.53 Types and characteristics of stress–strain curves of polymers

Hard and ductile polymers, such as nylon and polycarbonate, exhibit a high modulus, high yield point, substantial tensile strength, and significant elongation at break. These polymers undergo thin necking during the stretching, which forms the basis for fiber and film stretching technology.

Hard and strong polymers have a high Young's modulus, high tensile strength, and an elongation before fracture of approximately 5%. This category includes certain blends of unplasticized polyvinyl chloride and polystyrene with varying compositions.

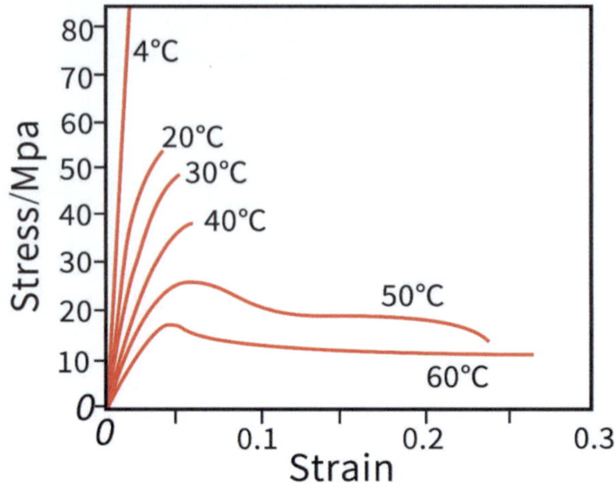

Fig. 12.54 Influence of temperature on the stress–strain curve of PMMA

Rubber and plasticized polyvinyl chloride belong to the catalog of soft and ductile polymers. They have a low modulus and lack a noticeable yield point, with only prominent bending regions observed on the stress–strain curve. Soft and ductile polymers exhibit a large elongation at break (20~1000%) and high breaking strength.

Soft and weak polymers encompass specific soft polymer gels that cannot withstand external loads, making them rarely utilized as materials.

The stretch process of polymers is significantly influenced by external conditions such as test temperature and strain rate due to their viscoelastic nature. Altering the test temperature or strain rate can lead to corresponding changes in the type of stress–strain curve exhibited by the polymer. This effect is illustrated in Fig. 12.54, demonstrating temperature's substantial influence on the stress–strain curve of polymethyl methacrylate. Consequently, it is essential to comprehend how the stress–strain curve of a polymer changes under various factors during the stretching process. This understanding enables the selection of appropriate materials for design and application, considering the requirements of the operating environment. Relying solely on stress–strain curves measured at a single temperature and strain rate is insufficient as a basis for design purposes.

12.7.4 Fracture and Strength of Polymers

As polymer science has advanced, the mechanical properties of polymers have been greatly improved, allowing them to be successfully employed as structural materials. Fracture phenomena in polymers encompass ductile fracture, brittle fracture, and fatigue fracture, all of which are associated with the initiation, combination, and propagation of cracks.

① Brittle Fracture and Ductile Fracture

Ductile materials have the ability to yield and absorb energy when subjected to stress or deformation. This ductility enables polymer materials to withstand significant deformation or impact without sustaining damage during practical applications. However, due to the viscoelastic nature and structural complexity of polymers, their ductility is conditional, and brittle fracture frequently occurs. Under different conditions, the same polymer material can exhibit either ductile fracture or brittle fracture. There is no strict demarcation between ductile fracture and brittle fracture, and one approach to differentiate between them is by examining the shape of the stress–strain curve, as illustrated in Figs. 12.52 and 12.53. Generally, brittle fracture is considered to occur prior to material yielding. In brittle fracture, the material undergoes only instantaneous elastic deformation, exhibiting a linear or nearly linear stress–strain relationship. The extent of deformation is minimal, with an elongation at break of less than 5%. Microcracks develop rapidly under tensile stress, ultimately resulting in brittle fractures. Conversely, in ductile fracture, the material yields first, followed by substantial deformation. The stress–strain relationship is nonlinear, with an elongation at break typically exceeding 10%, and ductile fracture is prompted by the development of a yield shear zone.

Another common method for distinguishing ductile fracture from brittle fracture is through examination of the fracture surface morphology. As depicted in Fig. 12.55, the fracture surface of brittle fracture appears as a flat and smooth surface perpendicular to the applied external force, with the cross-sectional area remaining almost unchanged. In contrast, the fracture surface of ductile fracture is irregular and rough, with a reduced cross-sectional area.

The occurrence of ductile fracture and brittle fracture in polymers is also significantly influenced by test conditions, primarily temperature and strain rate. The ductility of polymer materials increases with rising temperature and decreases with higher strain rate. The brittleness and ductility of polymers can interchange

Fig. 12.55 Three types of fracture surface

depending on the prevailing conditions. Moreover, the fracture behavior of polymers is influenced by many other factors, such as polymer structure, molecular weight, shape, plasticization, and more.

② Fracture and Theoretical Strength of Polymers

The resistance of polymers to external forces primarily relies on the chemical bond forces within molecules, as well as the van der Waals forces and hydrogen bonds between molecules, from a molecular structure perspective. The theoretical strength of polymers can be calculated on a microscopic level without considering other complex influencing factors. Comparing the theoretical results with the actual strength of polymers helps in understanding the disparity between them, which guides further research and exploration on improving the actual strength of polymers.

To simplify the problem, the microscopic process of polymer fracture can be classified into three types (Fig. 12.56). If the alignment of polymer bonds is parallel to the applied force, the fracture may result from the breakage of chemical bonds or slippage between molecules. If the alignment of polymer chains is perpendicular to the applied force, the fracture may occur due to the disruption of the van der Waals force or hydrogen bond.

The theoretical strength of a polymer can be estimated based on the strength of the chemical bonds within the polymer chains and the strength of the interaction force between the chains. One semi-empirical formula is $\sigma = 0.1E$, where σ represents the fracture strength, and E denotes Young's modulus. Generally,

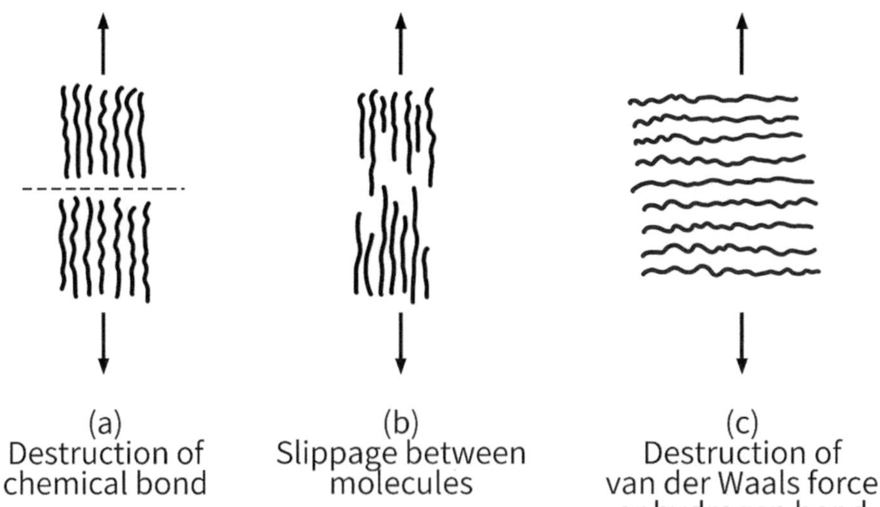

Fig. 12.56 Schematic diagram of three models for the microscopic process of polymer fracture: **a** breakage of chemical bonds, **b** slippage between molecules, and **c** destruction of van der Waals force or hydrogen bonds

the theoretical strength estimated using this formula is 100 to 1000 times higher than the actual strength of the polymer. Therefore, there is significant potential for enhancing the actual strength of polymers.

The notable difference between theoretical strength and actual strength arises from the fact that the molecular chains of polymers are not perfectly aligned as assumed. The force experienced by each molecular chain is non-uniform, and not all chains reach the theoretical strength simultaneously during fracture. In reality, material damage often originates from the weakest part, leading to stress concentration and further propagation of the damage——consequently, the material fractures well below its theoretical strength.

The weak spots in the material typically consist of complex particles (such as impurities and fillers) or internal defects (such as cracks, notches, voids, bubbles, etc.). Stress concentration occurs when the stress acting on a small volume of the material is much higher than the average stress of the material, resulting in macroscopic fracture. For instance, brittle fracture usually occurs in the form of cracks. During stretching, numerous tiny cracks develop rapidly on the material's surface, progressing from the outside to the inside in a direction perpendicular to the tensile stress, ultimately causing the material to break into two or more pieces.

Polymers inherently contain numerous defects and cracks, some of which are minute and distributed throughout the polymer. Transparent plastics like acrylic and polystyrene may exhibit glittering filamentous cracks (known as crazes), which typically reduce transparency and tensile strength. When solvent infiltrates and accelerates aging, these crazes may further propagate and evolve into cracks, resulting in material damage.

Surface cracks have the greatest impact on strength and are considered the most critical defects in a sample. The probability of encountering such critical defects and cracks in larger samples is higher compared to smaller samples, meaning that larger sample sizes yield lower measured strength. This explains the standard requirements regarding sample size during material strength determination.

12.7.5 Factors Affecting Polymer Strength

The factors influencing polymer strength can be categorized into internal and external factors. Internal factors primarily involve polymer molecular structure, crystallization, and orientation, while external factors include temperature, applied force speed, plasticizer, and more. The following brief overview focuses on the effects of molecular structure, crystallization, and orientation on polymer strength. The impacts of temperature and applied force speed on brittle fracture and ductile fracture of polymers are discussed in Sect. 6.4.1.

Effect of Molecular Structure

The chemical structure of polymers is a fundamental factor influencing polymer strength, with the following main effects:

1. Polarity and hydrogen bonds: Increasing polarity or introducing hydrogen bonds can enhance strength. Higher density of polar groups or hydrogen bonds corresponds to higher strength.
2. Polymers with aromatic heterocycles in the main chain: They exhibit higher strength and modulus compared to aliphatic polymers. Introducing aromatic side groups also improves strength and modulus.
3. Polymers with branched molecular chains: Branching increases the distance between molecules. The higher the degree of branching, the lower the tensile strength but the higher the impact strength of polymers.
4. Moderately crosslinked polymers: Moderate crosslinking can improve tensile strength.
5. Effect of molecular weight: Strength increases with increasing molecular weight. However, when the molecular weight is sufficiently large, tensile strength becomes independent of the molecular weight while the impact strength increases.
6. Effect of molecular weight distribution: The low molecular weight fraction reduces strength.

Effect of Crystallization and Orientation

Crystallization significantly influences the mechanical properties of polymer materials, with essential factors including crystallinity, grain size, and crystal structure. Molecular orientation refers to the arrangement of all or part of the molecular chains in one or two directions. In the orientation direction, material properties are enhanced, resulting in anisotropic behavior. The effects of crystallization and orientation are as follows:

1. Effect of crystallinity: Increased crystallinity benefits tensile strength, flexural strength, and elastic modulus. However, excessively high crystallinity will reduce impact strength and elongation at break.
2. Effect of spherulite structure: Larger spherulites generally decrease impact strength and elongation at break, whereas smaller spherulites improve tensile strength, elastic modulus, and elongation at break.
3. Effect of crystal morphology: Fibrous crystals composed of straight chains exhibit superior tensile properties compared to those composed of folded chains.
4. Effect of orientation: Orientation imparts anisotropy to polymers, resulting in increased strength in the direction of orientation. In the oriented direction, the tensile strength, yield stress, and modulus are 2 ~ 5 times that of non-orientated cases. However, in the perpendicular direction, the strength is approximately 1/3 ~ 1/2 of the strength in the non-oriented case.

12.8 Questions

Q12.1 Clarify the quantitative relationship among some classical parameters related to deformation or mechanical property of solid materials.

Q12.2 Compare the difference on deformation and mechanical behavior of some typical materials (e.g. metals, ceramics and polymers).

Q12.3 Compare the deformation mechanisms among slip, twinning and kinking in single crystals from the corresponding atomic mechanisms.

Q12.4 Prove that there are 12 potential slip systems in the face-centered cubic structure and 48 potential slip systems in the body-centered cubic structure.

Q12.5 Consider a single crystal of BCC iron oriented such that a tensile stress is applied along a [010] direction. (a) Compute the resolved shear stress along a (110) plane and in a direction [111] when a tensile stress of 52 MPa is applied. (b) If slip occurs on a (110) plane and in a direction $[11\bar{1}]$, and the critical resolved shear stress is 30 MPa, calculate the magnitude of the applied tensile stress necessary to initiate yielding.

Q12.6 Explain the difference between resolved shear stress and critical resolved shear stress.

Q12.7 The copper single crystal suffers a tensile deformation. The tensile axis is [001] and the stress is 10^4 Pa. Calculate the resolved shear stress on (111) plane along $[\bar{1}\,01]$ direction.

Q12.8 The critical resolved shear stress of aluminum is 2.4×10^5 Pa. If the tensile axis is [001]. Calculate the tensile stress to start yielding.

Q12.9 Analyze the rotation direction of single crystal during tension and explain the phenomena.

Q12.10 Schematically show the structural variation in FCC structure if every two layers of (111) plane slip a distance of Shockley partial dislocation.

Q12.11 Describe the different driving forces for recovery, recrystallization and grain growth stages.

Q12.12 Determine the steady-state creep rate and rupture time from the creep curve of the specified materials.

Q12.13 What are the characteristics of Molecular Motion of Polymers?

Q12.14 Draw the deformation-temperature curve and modulus-temperature curve of linear amorphous polymer, and briefly describe the relationship between mechanical behavior and molecular motion of each region and turning point?

Q12.15 The relaxation activation energy of natural rubber is 1.05 kJ mol^{-1}. If a piece of natural rubber is heated from 27 to 127 °C, how much is the relaxation time shortened?

Q12.16 Draw the stress–strain curve of a typical glassy amorphous polymer, and mark the tensile (breaking) strength, elongation at break, yield point, and modulus on the curve.

Bibliography

Callister WD Jr, Rethwisch DG (2010) Materials science and engineering: an introduction, 8th edn. John Wiley & Sons Inc., New York
He PS (2021a) Structure and properties of polymer, 2nd edn. Sci Press, Beijing
He MJ (2021b) Polymer physics, 3rd edn. Fudan Univ Press, Shanghai
Hu GX, Cai X, Rong YH (2010) Fundamentals of materials science, 3rd edn. Shanghai Jiao Tong Univ Press, Shanghai
Pan ZR (2011) Polymer chemistry, 5th edn. Chem Ind Press, Beijing
Pan JS, Tong JM, Tian MB (2011) Fundamentals of materials science. Tsinghua Univ Press, Re.ed. Beijing
Shi DK (2003) Fundamentals of materials science, 2nd edn. China Mach Press, Beijing
Tao J, Yao ZJ, Xue F (2018) Fundamentals of materials science, 2nd edn. Chem Ind Press, Beijing
Yu YN (2012) Fundamentals of materials science, 2nd edn. Higher Edu Press, Beijing

13 Computational Materials Science

Learning Objectives

1. Recognize reasonable approximations and simplifications required to solve Schrödinger's equation.
2. Transform from a multi-particle system into a multi-electron system, and from multi-electron system to single-electron system.
3. Distinguish between local density approximation (LDA) and generalized gradient approximation (GGA) exchange–correlation functional and know their advantages and disadvantages.
4. Master the classic mechanical laws of molecular dynamics and write basic equations.
5. List the four basic algorithms of molecular dynamics and know their advantages and disadvantages.
6. Cite different types of potential functions and know their corresponding applications.
7. Derive the three main calculation steps of the Monte Carlo method.
8. List and describe the three basic principles of the Monte Carlo method.

13.1 Introduction

Computational materials science is the intersection of materials science and computer science, involving multiple disciplines such as materials, chemistry, physics, and computing. The four main elements of materials science and engineering are the composition, structure, properties, and performance of materials. The purposeful synthesis of structures with desired properties is a key goal of materials

Z. Sun et al., *Fundamentals of Materials Science*,
https://doi.org/10.1007/978-981-96-3904-5_13

science. However, there is no one-to-one relationship between composition, structure, and properties of materials. Traditional materials research has limitations due to its reliance on laboratory-based experimentation, which can be time-consuming, resource-intensive, and often leads to blind alleys, and it is no longer possible to meet the requirements of modern materials research and development by relying solely on the laboratory. The emergence of computational materials science has brought new possibilities for on-demand materials preparation.

Computational materials science usually refers to the use of computers to study the inherent properties of materials through theory and computation. This field involves two main aspects: computational simulations and computer design of materials. In computational simulations, mathematical models are built using experimental data, and data calculations are used to simulate actual processes. This approach allows experimental results to be translated into general and quantitative theory. Computer design of materials involves predicting material properties directly from theoretical models and calculations, which is forward-looking and innovative, and can effectively improve the efficiency of research. Computational materials science thus serves as a bridge between theory and experiment in materials science, facilitating the development of new materials with desired properties.

In 2011, the US launched the Materials Genome Initiative, which identified data processing, computational tools, and experimental tools as three infrastructures for materials innovation. The aim of this initiative is to significantly reduce the research and development cycle and cost of new materials by promoting data sharing and using computational technologies. In 2012, the "Materials Science System Engineering Development Strategy Research—China's version of the Materials Genome Initiative" was launched. The development of machine learning and artificial intelligence has shortened the cycle of materials research and development, reduced investment and accelerated the process of the whole field.

13.2 Density Functional Theory

13.2.1 Fundamentals of Quantum Mechanics

At the beginning of the twentieth century, scientists found that classical mechanics for macro world was inapplicable to describe the laws of the micro world. Through years of scientists' exploration, quantum mechanics was established as the framework to describe the objective laws of the micro world. Schrödinger equation is the most basic equation of quantum mechanics. The motion of micro particles follows Schrödinger equation.

$$\hat{H}\Psi = \mathrm{i}\hbar\frac{\partial}{\partial t}\Psi \tag{13.1}$$

where $\widehat{H}$ is the Hamiltonian operator representing the total energy of the wave function, Ψ is the wave function, i is the imaginary unit, $\hbar$ is the reduced Planck constant, and t is time. In a system without time t, the eigen equation of $\widehat{H}$ is the stationary Schrödinger equation.

$$\widehat{H}\Psi = E\Psi \tag{13.2}$$

where E is the eigenvalue. As the solution of Schrödinger equation, the wave function of a system contains all the information of a system in a certain state. Obtaining the wave function of a system allows us to determine its properties. However, the solution of wave function is extremely difficult, and only a few systems without interaction terms, such as hydrogen-like systems (H, He^+, Li^{2+}, Be^{3+}, etc.) can be solved. The hydrogen atom consists of a single proton nucleus and an electron, exhibiting spherical symmetry. In spherical coordinates, the wave function of the system can be separated into radial and angular parts. The radial part depends only on the distance from the nucleus, while the angular part describes the electron's distribution on the spherical surface. Since the Schrödinger equation is a partial differential equation, using an appropriate coordinate system can simplify its form. In spherical coordinates, the radial part of the equation can be separated from the angular part, making it more convenient to solve. In practice, the material is an interacting multi particle system (including the system of atomic nucleus and electron), and its Hamiltonian includes the kinetic energy of electrons, the Coulomb interaction between electrons, the kinetic energy of nuclei, the interaction energy between nuclei and the interaction energy between electrons and nuclei, which cannot be solved like the hydrogen-like system. To solve the system with interaction, reasonable approximation and simplification are needed.

In an atom, the mass of electrons is significantly smaller compared to that of the nuclei, resulting in the electrons moving at much higher speeds than the nuclei. As a consequence, when the nuclei undergo slight movements, the fast-moving electrons can quickly adjust to the new positions, allowing for an approximation where the nuclei are considered fixed while the electrons move. Exploiting this notion, the motion of electrons and nuclei in the atom can be treated separately, leading to the simplification of the system from a multi-particle to a multi-electron system. This approximation is known as the Born–Oppenheimer approximation (adiabatic approximation).

Within the framework of the Born–Oppenheimer approximation, the wave function can be expressed as a product of the electron wave function and the nuclear wave function:

$$\Psi(r, R) = \psi(r, R)\nu(R) \tag{13.3}$$

where ψ and ν are the electron wave function and the nuclear wave function, respectively. Schrödinger equation is expressed as:

$$\widehat{H}\Psi = \left[\widehat{T} + \widehat{V} + \widehat{U}\right]\Psi \tag{13.4}$$

where $\widehat{T}$ is the kinetic energy term, $\widehat{V}$ is the potential energy of the external field (nucleus), and $\widehat{U}$ is the interaction term between electrons. For the N-electron system, each electron has three spatial coordinates, which means that the Schrödinger equation contains $3N$ variables. However, the interaction between electrons is still very complex and further simplification is necessary.

The emergence of Hartree–Fock approximation solves the above issue by taking into account the fact that the electron–electron interactions always exist. Each electron can be approximated to be moving in the average potential field of ion potential field and other electrons, a concept known as the average field approximation. This method simplifies the multi electron system into a single electron system, making the solution possible. The wave function of the multi electron system can be regarded as the product of the wave function of a single electron. At the same time, the electrons of fermions have exchange symmetry. The wave function of N electrons can be approximated as the product of the wave function of N single electrons by Slater determinant:

$$\boldsymbol{\Psi}_{HF} = \frac{1}{\sqrt{N!}} \begin{vmatrix} \psi_1(x_1) & \psi_1(x_2) & \cdots & \psi_1(x_N) \\ \psi_2(x_1) & \psi_2(x_1) & \cdots & \psi_2(x_1) \\ \vdots & \vdots & & \vdots \\ \psi_N(x_1) & \psi_N(x_1) & \cdots & \psi_N(x_1) \end{vmatrix} \tag{13.5}$$

where $\psi_i(x_i)$ represents the i-th single electron in x_i. The wave function at i is called the spin orbit; coordinate x_i contains the spatial position and spin state of the i-th electron. Hartree–Fock equation is:

$$\left[-\frac{\hbar^2}{2m_e}\nabla^2 + V(r) + \sum_j \int \frac{e^2}{|r - r'|} \psi_j^*(r')\psi_j(r')\mathrm{d}r' \right] \psi_i(r) - \sum_{j,\parallel} \psi_j(r) \int \frac{e^2}{|r - r'|} \psi_j^*(r')\psi_i(r')r' = \varepsilon_i \psi_i(r) \tag{13.6}$$

Hartree–Fock method transforms the high-dimensional unsolvable multi electron problem into a solvable single electron moving in the effective potential field. However, this method has some limitations. First, the electrons in the model do not interact directly and instantaneously, but indirectly through the mean field, ignoring the dynamic correlation of electrons. Second, the static correlation of electrons is ignored. Although Hartree–Fock approximation is suitable for molecular, or cluster systems dominated by exchange interaction, it is not accurate for periodic systems where the correlation effect between electrons cannot be ignored. Scientists were motivated to discover a new method that could solve the single electron problem without interaction, while simultaneously improving the rigor and accuracy of the Hatree-Fock method and reducing the computational complexity. Thus, density functional theory was born.

13.2.2 Development History of Density Functional Theory

What is density functional theory? Density refers to electron density, functional refers to energy as a function of electron density, and electron density is a function of spatial coordinates. Density functional theory takes electron density as the only basic observation to solve the ground state properties of multi electron systems. Unlike the Hartree–Fock method, density functional theory uses electron density instead of electron wave function to describe a multi electron system. This reduces the degree of freedom of N electron system from $3N$ to 3, significantly reducing the computational complexity. Density functional theory, renowned for its remarkable accuracy and computationally efficient nature, has revolutionized the calculation of properties for practical materials based on the first principle. In physics and chemistry, the term "first principles" often referred to as "ab initio" (a Latin term meaning "from the beginning"), signifies a theoretical approach that eschews assumed approximations, empirical data, or fitting parameters. Instead, it originates from fundamental physical laws and fundamental constants. The calculation of electronic structure based on density functional theory is widely used in the fields of condensed matter physics and computational chemistry. Walter Kohn won the Nobel Prize in chemistry in 1998 for his contributions to the development of density functional theory.

In 1927, Thomas and Fermi proposed using a simple single electron to solve the Schrodinger equation. They used statistical models to approximate the distribution of electrons in atoms, and established Thomas-Fermi model based on uniform electron gas, which served as a predecessor of density functional theory. In this model, they described a multi electron system using the uniform electron gas model and proposed to calculate the electronic structure with the electron density as the basic variable of the system, rather than the wave function. The total energy of an electron is the sum of its kinetic energy and potential energy:

$$E_{\mathrm{TF}}[\rho] = \frac{3}{10}\left(3\pi^2\right)^{\frac{2}{3}} \int \rho(r)^{\frac{5}{3}} \mathrm{d}r + \int \rho(r) V_{\mathrm{ext}} \mathrm{d}r + \frac{1}{2} \iint \frac{\rho(r_i)\rho(r_j)}{\left|r_i - r_j\right|} \mathrm{d}r_1 \mathrm{d}r_2 \tag{13.7}$$

where $\rho(r)$ represents the electron density. The first term of the formula denotes the kinetic energy term of the uniform electron gas, the second term denotes the energy of the electron in the external potential field, and the third term represents the Coulomb interaction.

The Thomas–Fermi model, although simple and not ideal for precise calculations of atomic systems, has its limitations. The main source of error lies is the approximation used for expressing the kinetic energy, followed by errors in the treatment of exchange energy and the complete neglect of electron exchange interaction. However, despite these limitations, the model offers valuable insights into solving the energy of a system solely based on its electron density. This concept has served as an inspiration for researchers to further explore and develop density functional theory. It was the groundbreaking work of Hohenberg and Kohn that brought renewed attention to the Thomas–Fermi model.

In 1964, based on Thomas–Fermi model, Hohenberg and Kohn strictly proved two theorems, which laid the foundation for density functional theory.

The first theorem: for a multiparticle system interacting in a system with an external potential of $V_{\text{ext}}(r)$, the ground-state properties are uniquely determined by the ground-state electron density, making it the only functional of the ground-state electron density.

The second theorem: the energy of the system is a functional of the electron density distribution, and the ground-state energy of the system is the global minimum of the functional when the electron density distribution is the ground-state electron density distribution.

Within the framework of Hobenberg-Kohn theorem, the energy functional takes the form:

$$E[\rho] = T[\rho] + E_{ee}[\rho] + \int V_{\text{ext}}(r)\rho(r)\mathrm{d}r = F[\rho] + \int V_{\text{ext}}(r)\rho(r)\mathrm{d}r \quad (13.8)$$

where $F[\rho]$ represents the sum of kinetic energy and interaction energy between electrons, independent of the external potential, and solely dependent on the ground-state electron density ρ. The first theorem guarantees that the ground-state properties of multi electron systems are functional of electron density, and the second theorem provides a variational method to calculate the total ground-state energy and electron density of the system. While the ground-state properties of the system can be obtained by the ground-state charge density in principle, the specific form of the functional remains unknown.

To solve this problem, Kohn and Sham proposed the Kohn–Sham equation in 1965. They assumed that any interacting multi electron system has an equivalent non-interacting electron system, whose ground-state charge density is strictly equal to that of the interacting electron system. Based on this assumption, the formula can be expressed as:

$$E[\rho] = T_s[\rho] + E_H[\rho] + \int V_{\text{ext}}(r)\rho_0(r)\mathrm{d}r + E_{xc}[\rho] \quad (13.9)$$

where ρ is expressed by the wave function of the non-interacting electron system, $T_s[\rho]$ and $E_H[\rho]$ represent the kinetic energy and classical Coulomb potential energy of the non-interacting electron system, respectively. $E_{xc}[\rho]$ is the exchange–correlation functional.

$$E_{xc}[\rho] = (T[\rho] - T_s[\rho]) + (E_{ee}[\rho] - E_H[\rho])E_H[\rho] \quad (13.10)$$

$E_{xc}[\rho]$ serves as a corrective term that accounts for unknown errors introduced by approximating the kinetic and interaction energy of the interacting electronic system. In the approximation, these energies are replaced with the kinetic energy and Coulomb potential of a non-interacting electronic system. Additionally, $E_{xc}[\rho]$ encompasses all multi-body effects beyond the Hartree–Fock approximation.

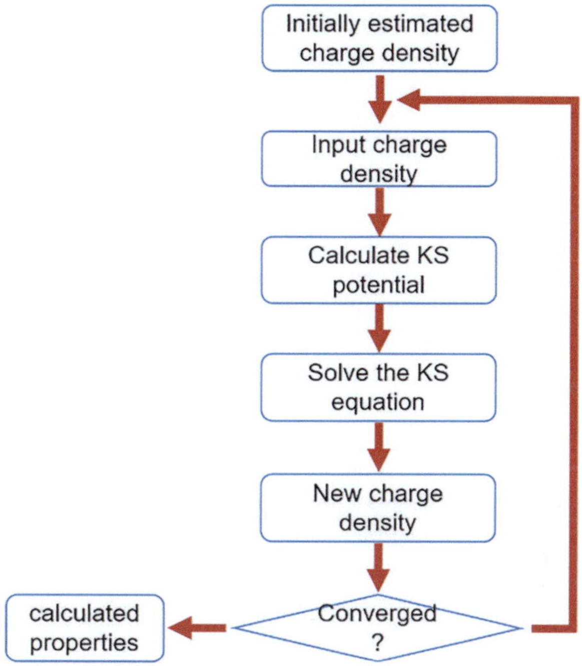

Fig. 13.1 Flow chart of the Kohn–Sham system of self-consistent equations solution process

The Kohn–Sham equation is defined as follows:

$$\left[-\frac{\hbar^2}{2m_e}\nabla^2 + V_{\text{ext}}(r) + V_H(r) + V_{xc}(r)\right]\psi_i(r) = \varepsilon_i\psi_i(r) \tag{13.11}$$

The flow chart for solving the Kohn–Sham equation is shown in Fig. 13.1. Hartree potential and exchange–correlation potential are:

$$V_H(r) = \int \frac{e^2}{|r - r\prime|}\rho(r\prime)\mathrm{d}r\prime \tag{13.12}$$

$$V_{xc}(r) = \frac{\delta E_{xc}[\rho]}{\delta\rho(r)} \tag{13.13}$$

The eigenvalues of Kohn–Sham equation, ε_i, represent a Lagrange multiplier and have no practical physical meaning. From the above development of density functional theory, we can get the ground-state properties of the system if we can get the exact exchange–correlation functional. However, the development of commutative correlation functional is not smooth, since the exact functional can only be approximated step by step. At present, the most widely used approximations are generalized gradient approximation and local density approximation.

13.2.3 Exchange–Correlation Functional

Theoretically, the Kohn–Sham equation provides a rigorous and accurate framework for describing multi electron system, but the exact form of exchange-correlation energy functional is still unknown. It may be extremely complex, and even it is difficult to find an accurate expression in the future, so researchers can only make some parametric approximation. Despite Kohn-sham's related theory being proposed in 1965, it was not until the 1980s that the actual form of exchange-correlation functional could be fitted by quantum Monte Carlo method, leading to gradual popularity of the that Kohn–Sham density functional. Therefore, one of the main directions of development for density functional theory is to find the appropriate form of exchange-correlation functional.

Local Density Approximation (LDA)

The LDA approximation is the result of using uniform electron gas, and the exchange-correlation functional is regarded as a weighted sum of the functional exchange-correlation energy density of the discrete local charge density.

$$E_{xc}[\rho(r)] = \int \varepsilon_{xc}[\rho(r)]\rho(r)\mathrm{d}r \tag{13.14}$$

$E_{xc}[\rho(r)]$ can be divided into exchange part and correlation part. The exchange part is derived from the uniform electron gas model, while the correlation part is accurately approximated by Monte Carlo simulation of uniform electron gas based on Ceperley and Alder. There are two commonly used forms of this approximation: Perdew-Zunger function and Vosko-Wilk-Nusair function. However, LDA uses the result of uniform electron gas, only considering the local charge density, and cannot deal with the sudden change in exchange-correlation term potential during electron transition. This approximation is reasonable for systems with uniform electron distribution, but it cannot give reasonable results for covalent systems. LDA tends to overestimate the binding energy and elastic constant, underestimates the band gap and lattice constant of semiconductors or insulators, and cannot effectively deal with van der Waals force and strong correlation system.

Generalized Gradient Approximation (GGA)

Compared to the LDA approximation, the GGA approximation includes the electron density gradient and higher-order terms into exchange-correlation term:

$$E_{xc}[\rho(r)] = \int \rho(r)\varepsilon_{xc}[\rho(r)]F_{xc}[\rho(r), \nabla\rho(r)]\mathrm{d}r \tag{13.15}$$

$F_{xc}[\rho(r), \nabla\rho(r)]$ is the enhancement factor, which is the correction for the non-local and non-uniform terms of the uniform electron gas model. However, there are no analytical expressions for the exchange and correlation energy in GGA. Currently, there are two main schools of construction methods: one led by Becke who advocates for introducing many parameters to fit a more accurate functional, and the other led by Perdew who believes that the form of functional should follow basic physical laws and focuses more on the scaling relationship and asymptotic behavior. The commonly used GGA functionals include Perdew-Burke-Ernzerhof (PBE), Perdew–Wang (PW91), and Becke-Lee–Yang–Parr (BLYP) functionals.

Hybrid Functional

LDA and GGA functionals tend to underestimate the band gap of semiconductors, while Hartree–Fock method tends to overestimate the band gap which only considers the exchange between electrons and ignores the screen effect. Therefore, a hybrid functional can be used by mixing the GGA functional and the exchange energy in Hartree–Fock to obtain a more accurate band gap. The simplest hybrid functional form is:

$$E_{xc}^{\text{hybrid}} = \alpha E_x^{\text{HF}} + (1-\alpha)E_x^{\text{GGA}} + E_c^{\text{GGA}} \tag{13.16}$$

PBE0 is a hybrid functional widely utilized for investigating the electronic structure and properties of molecules and materials due to its balanced combination of computational efficiency and accuracy. With an α value of 0.25, the PBE0 functional falls within the category of non-empirical hybrid functionals, eliminating the need for parameters fitting based on experimental data. This characteristic enhances its versatility and makes it applicable across various systems and scenarios. The commutative term in Hartree–Fock equation, that is, Fock term, is an unshielded long-range interaction, which significantly increases the computational cost and convergence difficulty. To solve these problems, Hedy et al. proposed the HSE06 functional, which divides the exchange action into long-range and short-range parts. Only the exchange energy of the short-range part is mixed, and the long-range part remains unchanged:

$$E_{xc}^{\text{hybrid}} = \alpha E_x^{\text{HF,SR}} + (1-\alpha)E_x^{\text{PBE,SR}} + E_x^{\text{PBE,LR}} + E_c^{\text{PBE}} \tag{13.17}$$

The HSE06 functional is widely recognized as one of the most extensively employed functionals in solid material calculations. It outperforms the PBE functional in its ability to accurately describe intricate systems, such as transition metal compounds, and yields band gap predictions that closely align with experimental values. Nonetheless, the drawback of the HSE06 functional lies in its significantly longer calculation time compared to PBE, rendering its application challenging for large-scale systems.

13.3 Molecular Dynamics

13.3.1 Basics of Molecular Dynamics

Molecular dynamics is a powerful and extensively used method in molecular mechanics. Since 1970, due to the rapid development of molecular mechanics, force fields have been developed and subsequently many force field systems have been established for biochemical molecular systems, polymers, metallic and non-metallic materials, which have greatly improved the capability and accuracy for calculating the structure and some thermodynamic and spectroscopic properties of complex systems. Molecular dynamics simulations are computational methods developed by applying these force fields and based on the principles of Newtonian mechanics. The classical molecular dynamics approach, first proposed by Alder and Wainwright in 1957 and 1959, was applied to the ideal "hard sphere" liquid model. The basic principle is to calculate the trajectories of individual atoms in a physical system by Newtonian classical mechanics, and then use certain statistical methods to calculate the mechanical, thermodynamic, and kinetic properties of the system.

In molecular dynamics, the system of N particles is first abstracted into N interacting masses with coordinates (usually in a Cartesian coordinate system), mass, charge and bonding mode, and the initial velocity of each mass is randomly specified according to the Boltzmann distribution at the target temperature. The interaction energy between the masses and the force on each mass are calculated according to the corresponding bonding and non-bonding energy expressions in the chosen force field. The acceleration and velocity of each mass point are calculated using Newtonian mechanics, and the new coordinates and velocity of each mass point are obtained after a specified integration step (time step, usually 1 fs), so that the mass point moves. After a certain number of integration steps, the mass point has a trajectory, which is saved by setting the time interval. Finally, the trajectories can be analyzed for various structural, energy, thermodynamic, kinetic, mechanical, etc., to obtain the calculation results of interest. Molecular dynamics has many advantages, including a correct physical basis, high accuracy, the ability to obtain the dynamic and thermodynamic statistics of the system simultaneously, and being applicable to explore various systems and their characteristics.

The computational techniques of molecular dynamics simulations have been improved in many ways. For example, in 1967, Verlet proposed the famous Verlet's algorithm, a stepwise algorithm for the displacement, velocity and acceleration of particle motion in molecular dynamics simulations. In 1981, Parrinello and Rahman introduced a constant-pressure molecular dynamics model, which extended the isobaric molecular dynamics to allow the shape of the cell change with the motion of particles within it, marking a milestone in the development of molecular dynamics. In recent years, with the rapid development of computer technology and computational mathematical algorithms, molecular dynamics has become an essential research tool in fields such as physics, chemistry, materials science, biology, and pharmaceutical research.

Classical Mechanical Law

Molecular dynamics simulation is a deterministic method that calculates the equilibrium and transfer properties of a many-body system. Classical means that the motion of the particles in system follows the laws of classical mechanics. In simple terms, the example of a system treated in molecular dynamics obeys Newton's equations, i.e.

$$\boldsymbol{F}_{\mathrm{i}}(t) = \mathrm{m}_{\mathrm{i}}\boldsymbol{a}_{\mathrm{i}}(t) \tag{13.18}$$

where $\boldsymbol{F}_{\mathrm{i}}(t)$ is the force on the particle, m_{i} is the mass of the particle, and $\boldsymbol{a}_{\mathrm{i}}(t)$ is the acceleration of atom i. The force $\boldsymbol{F}_{\mathrm{i}}(t)$ on atom i can be directly expressed as the first order derivative of the potential energy function with respect to the coordinate $\boldsymbol{r}_{\mathrm{i}}$, i.e., $\boldsymbol{F}_{\mathrm{i}}(t) = -\frac{\partial U}{\partial \boldsymbol{r}_{\mathrm{i}}}$, where U is the potential energy function.

Thus, for a system of N particles each particle has

$$\begin{cases} m_{\mathrm{i}}\frac{\partial v_{\mathrm{i}}}{\partial t} = F = -\frac{\partial U}{\partial r_{\mathrm{i}}} + \ldots \\ \dot{r}(t) = v(t) \end{cases} \tag{13.19}$$

In these equations, $\boldsymbol{v}$ is the velocity vector and m_{i} is the mass of the particle. The solution of these equations generally requires a numerical approach (analytical methods can only solve the simplest form of the potential function, which is not meaningful in practical simulations), and these numerical solutions yield a series of position and velocity pairs $\{x^n, v^n\}$, with n denoting some column of discrete times, $t = n\Delta t$, and Δt denoting the time interval (time step). To solve this system of equations, the initial coordinates and velocity of each particle in the system must be specified.

The classical equations of motion are deterministic, implying that given the initial coordinates and velocity of a unit, the coordinates and velocity at any subsequent moment can be determined. In molecular dynamics, the paths followed by the coordinates and velocities throughout the simulation are referred to as trajectories. The finite difference method is the conventional numerical technique employed to solve ordinary differential equations and is widely used in the numerical integration of the equation of motion in molecular dynamics simulations.

Working Flow Diagram of Molecular Dynamics

Given the coordinates and velocity at time t and other kinetic information, then the coordinates and velocity at time $t + \Delta t$ can be calculated. Δt is the time step, which is related to the integration method and the system. Figure 13.2 illustrates the flowchart of the calculation of the molecular dynamic method. The potential function governing the interatomic or intermolecular interaction forces must be selected at start-up. Once the potential function of the model is determined, the conservation quantities in the simulation can be derived based on the laws of physics. After the potential function is selected, some reasonable initial conditions can be given to accelerate the system's convergence to equilibrium and to obtain a good accuracy. After determining the initial and boundary conditions, the equations of motion can be solved and molecular dynamics simulations can be performed.

The main criteria of a good integration algorithm in molecular dynamics include:

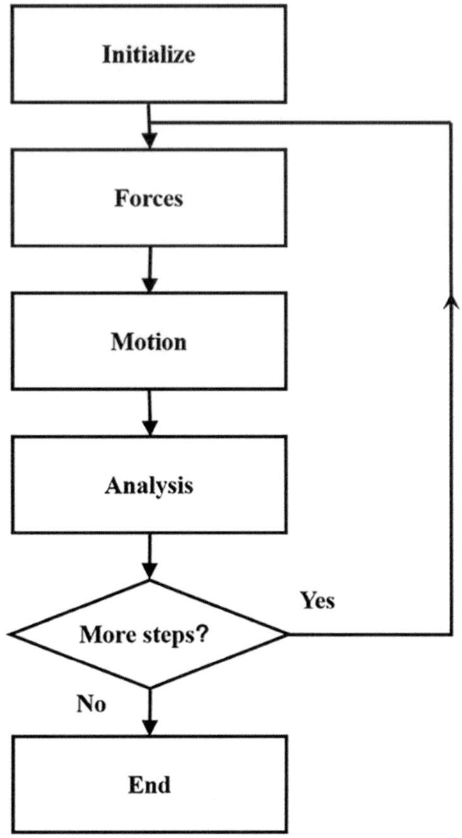

Fig. 13.2 Working flow diagram of molecular dynamics

1. Fast computational speed;
2. Requires less computer memory;
3. Allowing the use of long-time steps;
4. Good energy conservation.

13.3.2 Classification of the Heterogeneous System

The ensemble is a concept of statistical mechanics, which was created and completed by Gibbs in 1901. An ensemble is not an actual object; the system that makes up the tether is the actual object. In molecular dynamics simulations, the system under study is a multi-particle system, but due to the limitation of computer memory and computational speed, the number of particles that can be simulated is limited. Despite this limitation, the laws of statistical physics still apply, enabling the simulation of multi-particle system that can be described by the laws of statistical physics. When performing molecular dynamics simulation, it is essential to choose the appropriate ensemble for the system under investigation.

Common ensembles used in molecular dynamics simulations include the following: isothermal isobaric ensemble (NPT), isothermal isobaric enthalpy ensemble (NPH), microcanonical ensemble (NVE) and canonical ensemble (NVT).

1. Isothermal isobaric ensemble (NPT): As one of the most common ensembles, the atomic number (N), pressure (P) and temperature (T) of the system are kept constant within this ensemble. By constantly adjusting the velocity of the particles in the system, the temperature of the system can be kept constant. However, keeping the pressure constant is slightly more complicated. This is because the pressure P of the system and its volume V are co-constant quantities, and it is possible to maintain a fixed pressure value by changing the volume of the system.
2. Isothermal isobaric enthalpy ensemble (NPH): Compared with other ensemble, this ensemble is only applicable to periodic structure systems and can only guarantee constant atomic number (N), pressure (P) and enthalpy (H). The pressure and enthalpy can be stabilized by controlling the cell size and temperature within a specified range. However, this adjustment technique can be challenging to implement, which is why isothermal isobaric enthalpy ensemble is not commonly used in practice.
3. Microcanonical ensemble (NVE): The microcanonical ensemble as an isolated conservative system where the atomic number (N), volume (V) and energy (E) of the system are kept constant and the system energy is guaranteed to evolve according to the set orbit. In general, the energy under the initial conditions cannot be precisely determined. To adjust the system to the desired state, it is usually necessary to continuously increase or decrease the energy value until it meets the requirements. Sufficient computational time must be allocated to allow the microcanonical ensemble to reach the equilibrium state, or else unnecessary systematic errors will occur.
4. Canonical ensemble (NVT): Under this system, the total energy of the system is not constant, and constantly exchanges energy with the outside world. To maintain a constant temperature of the system, it is necessary to keep the external heat bath and the study system in a specific thermal equilibrium by means of temperature regulation. There are several methods to regulate the temperature, including the Nose-hoorer heat bath method, the Gaussian heat bath method, and the velocity temperature scale method.

13.3.3 Numerical Algorithms

Algorithms are involved in the process of molecular dynamics simulations. To adapt to the requirements of different system simulations, several algorithms have been developed to obtain more accurate results at an adapted rate.

1. Verlet Algorithm: The Taylor Expansion of the Position Coordinates of the Particles in Terms of the Position Coordinates of the Moments Respectively

Has:

$$\begin{cases} r(t+\Delta t) = r(t) + \Delta t v(t) + (\Delta t)^2 \frac{a(t)}{2} + L \\ r(t-\Delta t) = r(t) + \Delta t v(t) + (\Delta t)^2 \frac{a(t)}{2} + L \end{cases} \tag{13.20}$$

The position of the particle at the moment can be obtained from Eq. 13.21 as:

$$r(t+\Delta t) = 2r(t) - r(t-\Delta t) + (\Delta t)^2 \frac{F(t)}{m} \tag{13.21}$$

The corresponding speed is:

$$v(t) = \frac{r(t+\Delta t) - r(t-\Delta t)}{2\Delta t} \tag{13.22}$$

where m, $v(t)$, $a(t)$ and $F(t)$ are the mass, velocity, acceleration, and force on the atom, respectively.

2. Velocity-Verlet algorithm: This algorithm is proposed by Swope, which gives the position, velocity, and acceleration without sacrificing accuracy compared to the Verlet algorithm and gives the apparent velocity term. This algorithm is more convenient for the same amount of computation, so it is more widely used in simulation.
3. "Leap-frog" algorithm: This algorithm was proposed by Hockney. The "Leap-frog" algorithm evolved from Verlet's algorithm, which inherits Verlet's algorithm and designs the velocity of half-time interval, i.e.:

$$\begin{cases} r(t+\delta t) = r(t) + \delta t v + v(t+\delta t/2) \\ v(t+\delta t/2) = \delta t a(t) + v(t-\delta t/2) \end{cases} \tag{13.23}$$

The velocity at moment t is given by the following equation:

$$v(t) = \frac{[(t+\delta t/2) + v(1-\delta t/2)]}{2} \tag{13.24}$$

Therefore, the "Leap-frog" algorithm provides three advantages over the Verlet algorithm: it has an apparent velocity term, convergence is fast, and the computational effort is small. However, this algorithm also has the obvious drawback of unsynchronized position and velocity.

4. Gear's prediction-correction algorithm: This algorithm is one of the more complex ones and is usually performed in three steps. First, the new position, velocity, and acceleration of the atoms in the simulated system are predicted using Taylor expansion. Then, the acceleration is calculated based on the new potential function formula. In the third step, this acceleration is compared with the acceleration in the Taylor expansion. Finally, the error is corrected based on the difference between the two, thus correcting the position and velocity terms.

A major drawback of this algorithm is its complexity, which results in large memory consumption of the computer.

Besides these commonly used algorithms, there are also Beeman's algorithm and Rahman's algorithm, etc., not included herein.

13.3.4 Potential Function

The potential function, also known as the force field, is a function that represents the interactions between atoms (molecules). Interatomic interactions control the behavior of interatomic interactions and fundamentally determine all properties of materials, which are specifically described by the potential function. The study and development of potential functions are crucial in the development of molecular dynamics. In this subsection, some common interatomic potential functions are presented. Table 13.1 lists the commonly used atomic interaction potential models.

In 1903, G. Mie conducted a study on the interaction potential between two particles and proposed that the potential function should comprise two components: interatomic repulsion and interatomic attraction. In 1924, J. E. Lennard–Jones published the analytic form of the famous negative power exponential Lennard–Jones potential function. In 1929, P. M. Morse published the exponential. In 1931, M. Born and J. E. Mayer published the Born–Mayer potential function describing ionic crystals. With the development of computers, molecular dynamics began to be applied in scientific research in the late 1950s and early 1960s, where the selection of interatomic interactions was the key to molecular dynamics simulations. In this model, the hard sphere moves in a uniform linear motion, all collisions are perfectly elastic, and collisions occur when the distance between the centers of the two spheres is equal to the sphere diameter. There are various forms of potential functions, with a few described as follows.

Pair Potential

Pair potentials are interactions depending solely on the coordinates of two atoms, and these types of interactions can more adequately describe the interactions in all inorganic compounds in semiconductors and metals. During the early stages of molecular dynamics simulations, pair potentials were extensively used. This

Table 13.1 Atomic interaction potential model and common potential functions

Functional potential	Application object	Typical representative
Pair potential	Inorganic compound	• Lennard–Jones potential • Morse potential • Johnson potential
Pair functional potential	Metal	• EAM/MEAM potential
Combination potential	Organic compound	• SW potential
Combination functional potential	Semiconductor ceramics	• Abell-Tersoff potential • CheliKowsky-Phillips potential

potential considers the interaction between atoms as existing between two, independent of other particles, resulting in the force between being calculated without considering the effect of other particles. As this potential has been mentioned in the previous plot, it will not be explained here.

Morse potential and Johnson potential.

Morse potential: $\Phi_{ij}(r_{ij}) = A\left[e^{-2\alpha(r_{ij}-r_0)} - 2e^{-\alpha(r_{ij}-r_0)}\right]$;

Johnson potential: $\Phi_{ij}(r_{ij}) = -A_n(r_{ij} - B_n)^3 + C_n r_{ij} - D_n$

These two potential functions are often used to describe metallic solids, with the former mostly for copper and the latter mostly for ferrite. The potential well of the Morse potential is larger than that of the Johnson potential, resulting in the stronger forces being described by the former describes. Moreover, the Morse potential has a longer range of forces than the Johnson potential, leading to better ductility of Morse potential solids as compared to Johnson potential solids.

Many Body Potential-EAM, MEAM

Prior to 1982, the development and application of the "pair potential" was the primary focus of attention due to its simplicity and ease of simulation, and the results obtained were basically consistent with the macroscopic physical laws. However, the potential has an inherent drawback that it does not lead to a fundamental breakthrough. The potential leads to the Cauchy relation, i.e., $C_{12} = C_{44}$, which is not satisfied by most crystalline metals. Thus, the pair potential cannot accurately describe the elastic properties of the crystal, and its simulation results can only be qualitative. In practical research, the object of study is often a multi-particle system with strong interactions, where changes in the state of one particle will affect the changes of the others, not simply a two-by-two interaction, but a multi-body interaction. The inherent drawbacks of the potential have led to the search for new potential functions to overcome these drawbacks. The many-body potential emerged in the early 1980s, and the Embedded Atom Method (EAM) was first proposed by Daw and Baskes.

Embedded Atom Method

The basic idea of the EAM potential divides the total potential energy of the crystal into two parts: the interaction pair potential between the atoms located on the lattice dot matrix, and the embedding energy of the atoms embedded in the background of the electron cloud, which represents the many-body interactions. The functional forms of both the pair potential and the mosaic potential that constitute the EAM potential are chosen empirically. In the embedded atom method, the total potential energy of the crystal can be expressed as:

$$U = \sum_i F_i(\rho_i) + \frac{1}{2}\sum_{j \neq i} \Phi_{ij}(r_{ij}) \tag{13.25}$$

where, the first term F_i is the embedding energy; the second term is the pair potential term, which can take different forms as needed. ρ_i is the sum of the electron cloud densities produced at the i th atom by all but the i th atom's out-of-nuclear electrons, which can be expressed as:

$$\rho_i = \sum_{j \neq i} f_i(r_{ij}) \tag{13.26}$$

$f_i(r_{ij})$ is the charge density contributed by the out-of-nuclear electron of the j atom at the i th atom, and r_{ij} is the distance between the i atom and the j atom. For different metals, the embedding energy function and the pair potential function need to be determined by fitting the macroscopic parameters of the metal.

Modified Embedded Atom Method

The EAM potential framework assumes a spherically symmetric distribution of electron density, which deviates from the actual situation in certain cases, such as transition group (Fe, Co, Ni) elements with unsatisfied d-electron orbitals, semiconductor elements with diamond structure, and systems with orbital hybridization. As a result, the obtained results can be vastly different from the actual ones, and the embedding function cannot deal with Cauchy negative pressure metals and alloys such as Cr and Cs. To extend the EAM potential to covalently bonded, transition metal materials, it is necessary to consider the non-spherical symmetry of the electron cloud. To address this, Baskes et al. proposed the modified embedded atom method (MEAM). The improvement of the original EAM potential by Baskes and Johnson retains the framework of the original theory and improve it for the assumption of spherically symmetric charge density of the atoms, by engaging the angular dependence of the atomic electron density distribution. The summation of the substrate electron density for the s, p, and d states of the electron distribution densities is calculated separately, but the total electron density is still equal to a linear superposition of the various electron densities. In addition, Jacobsen et al. proposed another form of many-body potential function based on the equivalent medium principle, which has also been widely used due to its simplicity and effectiveness.

13.4 Monte Carlo Method

Monte Carlo method (MC method) simulates the motion process of objects by constantly generating random number sequence. Many processes in nature are random, such as particle decay process, particle transport process in the medium and other physical phenomena. Although the MC method dates back to the end of the nineteenth century, it did not get rapid development and application until the 1940s with the popularization of computers. In the World War II, the MC method was first applied to the development of atomic bombs by American scientists. At present, this method has been widely used in systems engineering, scientific management, biological genetics, social science, and other disciplines, fully demonstrates the originality and superiority of this method completely different from other research methods.

13.4.1 Calculation Framework of Monte Carlo Method

In the Monte Carlo method, firstly, a probabilistic model or stochastic process must be established, with parameters equal to the solution of the problem. Next, the statistical characteristics of the desired parameters are calculated by observing the model or process, or sampling experiment. Finally, an approximate value of the solution is obtained. Monte Carlo method can be summarized as the following three main steps:

Construct or Describe Random Processes
In problems with stochastic nature, such as particle transport problem, it is necessary to accurately describe and simulate the stochastic process. For deterministic problems, like calculating definite integrals or solving linear equations and boundary value problems of partial differential equations, an artificial stochastic process must be constructed in beforehand, with some of its parameters exactly representing the solutions of the desired problem. Random models can range from simple and intuitive to complex and abstract, depending on the problem being solved. They can be random models composed of discrete random variables or a continuous random variable. They can be a process composed of one random variable or a random model composed of multiple random variables. The process can also be a sequence of random variables or a more complex random model consisting of random functions. The complexity of stochastic model determines the complexity of Monte Carlo calculation.

Implement Sampling from a Known Distribution
Monte Carlo simulation experiments rely on generating random variables (or random vectors) with known probability distributions, as various probability models can be regarded as composed of these distributions. Thus, the Monte Carlo method is also known as random sampling technique. The simplest, most important and most basic probability distribution is the uniform distribution (or rectangular distribution) on (0, 1), and random number is a random variable with this uniform distribution. A sequence of independent random variables with this distribution is a random number sequence, which is a simple subsample of the population with this distribution. Generating random numbers is a sampling problem that obeys this distribution.

Set Up Various Estimators
Once a probabilistic model has been constructed and sampled from it, that is, after simulation experiments have been carried out, a random variable must be determined as the estimator of the solution of the desired problem. If the expected value of this random variable is exactly the solution to the problem, we call it

an unbiased estimate. The establishment of various estimators is equivalent to the investigation and registration of the results of simulation experiments, which lead to the solution of the problem. The most commonly used estimate in Monte Carlo calculations is the unbiased estimate. Of course, other types of estimation can be introduced, such as maximum likelihood estimation, asymptotically biased estimation and so on.

13.4.2 Basic Principles of Monte Carlo Method

Random Process

In a mechanical system, the set of characteristic variables that determine the motion state of the system at a certain moment is called dynamic variables, expressed by the vector $\boldsymbol{q}$. For example, in a system consisting of N gas molecules in irregularly reciprocating motion, the position coordinates $(r_1, r_2, \cdots, r_N)$ and molecular momentum $(p_1, p_2, \cdots, p_N)$ constitute the so-called dynamic variable $\boldsymbol{q}$ of the system. One of the most basic topics in the Monte Carlo method is to study how the dynamic variables of the system are distributed in the thermal equilibrium or ground state. We do not need to care about the specific trajectory that the $\boldsymbol{q}$ should give in accordance with the deterministic equation, but only need to study the value of the dynamic variable at a certain time under certain conditions. For this reason, the system can be divided into regions for observation, or observations can be made in the same region at different times, which is equivalent to sampling the whole set of all possible values of a variable in the system. Thus, the dynamic variable $\boldsymbol{q}$ has a different value for each sample, and even the same sample has a different value at different times. Through such sampling, the distribution of $\boldsymbol{q}$, sometimes called the random variable, can be studied. A particular value taken by a random variable is called a random number.

To study the dynamic behavior of dynamic variables, a time covariant sequence of dynamic variables is usually adopted, known as time series. This sequence of dynamic variables $\boldsymbol{q}_1, \boldsymbol{q}_2, \cdots, \boldsymbol{q}_L$ obtained through L times of observation of the system at time $t_1, t_2, \cdots, t_L$. When the selection of time and time series is probabilistic at the same time, such a time series is generally called random process or probabilistic process. For the dynamical behavior of multi-particle system, the Monte Carlo method is used to determine how the state of the system evolves with time in phase space through the study of stochastic process. As a random simulation method, the Monte Carlo method can use random numbers to update dynamic variables. At this point, a sample obtained from a simulation can be used as the above dynamic variable group, i.e., time series. In the actual Monte Carlo method simulation, in most cases M samples are obtained by repeating the above process several times (such as M times).

Markov Chain

Markov Process is a typical stochastic process. Let $x(t)$ be a random process, the state of which is known at time t_0. The state at time t $(t > t_0)$ has dependence on any state prior time t_0, and this property is called non-aftereffect. A random process with no aftereffect is called a Markov process. Time and states in Markov processes can be either continuous or discrete. The Markov process of time discrete and state discrete is called Markov Chain, in which the state transition at each moment is controlled by a probability matrix of state transition.

Markov processes or Markov chains are important concepts in stochastic processes, and all algorithms are based on Markov processes when using Monte Carlo methods to simulate real systems. For example, in Markov process, the value of dynamic variable at the subsequent time T has no relationship with the state of the system before time t_0, where time t_0 has dynamic variable $\boldsymbol{q}_0$. It is a random walk process, meaning the system's behavior at any stage of the walk is unrestricted by the history of the previous walk process. In other words, the points in the region can be visited multiple times (Fig. 13.3).

Sampling Method

Monte Carlo method takes the random process of probability theory as the starting point and research object of the problem. Therefore, except in special cases, it is insufficient to discuss only one sample. Usually, enough samples should be adopted, and their average should be taken. So, in Monte Carlo one must take multiple samples to simulate random processes. There are two sampling methods: simple sampling and important sampling.

Simple Sampling

Simple sampling refers to those integration methods that pick random numbers from a uniform distribution and treating each sample as independent. When considering the accuracy of Monte Carlo method, we cannot simply reduce the variance

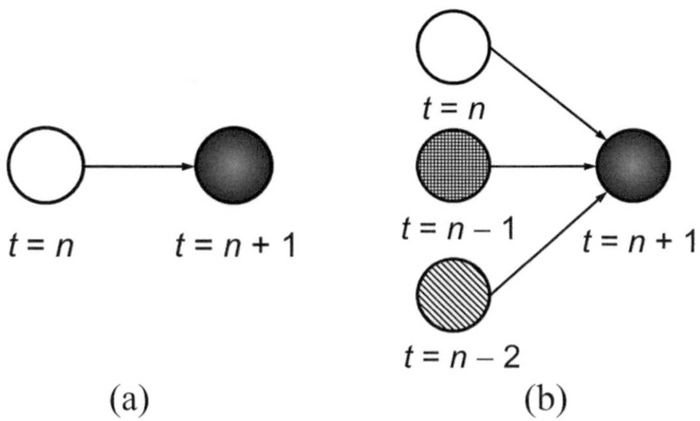

Fig. 13.3 **a** Markov processes, the new state is determined only by the present state; **b** non-Markov processes, the new state is also directly dependent on the past state

of a function or variable, but also consider the calculation cost, that is, the simulation times (random points). The product of variance and calculation cost is usually used as a standard to measure the merits of the calculation method. The concept of accuracy of the Monte Carlo method is not convergence to truth in the general sense, but convergence to truth in a certain confidence or probabilistic sense. In other words, the accuracy is random, and we can only know how likely it is to have a certain precision, but we can't be sure that there is a certain precision. For well-defined random process, a simulation algorithm using Monte Carlo method to directly sample is called simple sampling.

Monte Carlo method was first applied to the simulation of random diffusion process. This process can be decomposed into a series of unrelated processes, which is the problem of random walking of particles in the field of diffusion. No outfield, no infield can dictate the direction of diffusion, it is arbitrary. An example of simple sampling is the process of guessing a circle area integral. See Fig. 13.4. The first step is to inscribe a circle of radius 1 into a square ($-1 < x < 1$, $-1 < y < 1$). The second step is to determine the number of experimental points through random number selection. The third step is to determine whether the selected point falls inside or outside the circle. By calculating the number of shots n_0 and the total number of shots n, we can estimate the value of the integral of the circle area, that is, the circle area/square area = n_0 / n. Obviously, the accuracy of this approximation depends on the total number of experiments.

Importance Sampling

Simple sampling is a highly effective method for calculating the numerical integration of the mean value of the physical quantity A (q) when it is approximately uniformly distributed. However, when the distribution function becomes a localized Gaussian-like distribution with maximum value, it becomes challenging to calculate the thermal mean distribution of A. To overcome this difficulty, for statistics at finite temperature, samples that play an important role in temperature should

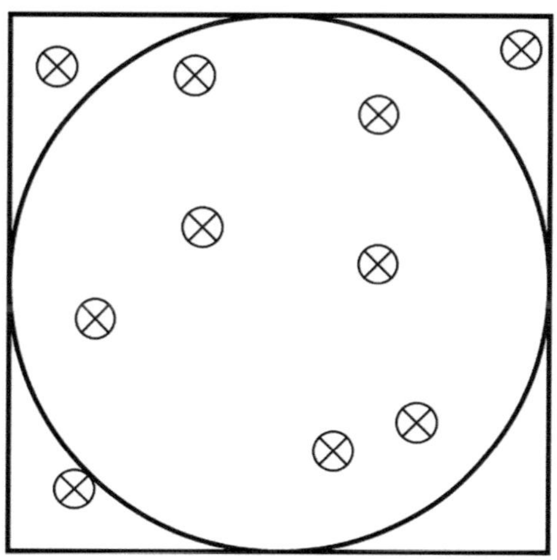

Fig. 13.4 Drop point determine the area of the circle

not be randomly generated, but preferentially sampled. In statistical physics, the mean value of $A(\{\underline{r}_N\})$ of a thermodynamic equilibrium system is often required:

$$< A >\equiv \frac{\int \mathrm{d}^{3N} r A(\{\underline{r}_N\}) \mathrm{e}^{-\beta U(\{\underline{r}_N\})}}{\int \mathrm{d}^{3N} r \mathrm{e}^{-\beta U(\{\underline{r}_N\})}} \tag{13.27}$$

This calculation is the average of the thermodynamic ensemble, where A depends on the positions of all the particles in the system. A regular thermodynamic ensemble is characterized by the Boltzmann factor $\mathrm{e}^{-U/k_B T}$, where U is the potential energy of the system, k_B is the Boltzmann constant, and T is the temperature of the system. The integration is performed over all particle positions (N particles, $3N$ coordinates), with the denominator in Eq. 13.27 serving as a normalization factor and being an important thermodynamic quantity, known as the partition function.

Imagine two ways to implement the above integral. The first method involves randomly sample N_c thermodynamic states, and summing the positions of all particles sampled from the regular distribution according to the results obtained by Eq. 13.27, that is

$$< A >= \frac{\sum_{j=1}^{N_\mathrm{c}} A(\{\underline{r}_N\}_j) \mathrm{e}^{-\beta U(\{\underline{r}_N\}_j)}}{\sum_{j=1}^{N_\mathrm{c}} \mathrm{e}^{-\beta U(\{\underline{r}_N\}_j)}} \tag{13.28}$$

In practice, this approach is inefficient, as it is easy to sample a high energy state $U \gg k_\mathrm{B} T$, in which the exponential factor $\mathrm{e}^{-U/k_B T}$ causes contribution for the result are negligible. That is, the expected value $< A >$ is determined by just a few sampled thermodynamic states. This outcome is not ideal. To solve this difficulty, another solution can be adopted, in which the N_c thermodynamic states are extracted according to the distribution $\mathrm{e}^{-\beta U}$, rather than being randomly selected. In this case, $< A >$ is determined by averaging the contributions of each state

$$\langle A \rangle = \frac{1}{N_\mathrm{c}} \sum_{j=1}^{N_\mathrm{c}} A\left(\left\{\underline{r}_N\right\}_j'\right) \tag{13.29}$$

Here, the $\{\underline{r}_N\}_j'$ are states sampled from the distribution $\mathrm{e}^{-\beta U(\{\underline{r}_N\}_j')}$. In this way, the calculation of the average is reduced to a simple algebraic average.

13.5 Questions

Q13.1 What are the advantages of computational materials science compared to traditional experiments?

Q13.2 Investigate three kinds of density functional theory software, and briefly describe their advantages and disadvantages.

Q313. Solve the time-independent Schrödinger equation for the hydrogen atom.

Q13.4 The Monte Carlo method is generally divided into three steps. Please illustrate the realization process of monte Carlo method simulation with examples.

Q13.5 Please list several mathematical methods of importance sampling and evaluates their efficiency and characteristics in the process of computer sampling.

Q13.6 Please write the flow chart and main program of dynamic Monte Carlo simulation algorithm for film growth according to Sect. 13.4.2.

Q13.7 List and briefly describe the general steps of the molecular dynamics.

Q13.8 Briefly explain the development and use of EAM and MEAM.

Bibliography

Fermi E (1927) Statistical method to determine some properties of atoms. Rend Accad Naz Lincei 6:5

Guzmán O, de Pablo JJ (2003) An effective-colloid pair potential for Lennard-Jones colloid–polymer mixtures. J Chem Phys 118:2392–2397

Hohenberg P, Kohn W (1964) Density functional theory (DFT). Phys Rev 136:B864

Kohn W, Sham LJ (1965) Self-Consistent equations including exchange and correlation effects. Phys Rev 140:A1133–A1138

Li ZS, Zhao YJ, Jia X (2008) Progress in computer simulation of molecular dynamics. Mech Manag Dev 23:174–176

Parrinello M, Rahman A (1981) Polymorphic transitions in single crystals: A new molecular dynamics method[J]. J Appl Phys 52:7182–7190

Perdew JP, Burke K, Ernzerhof M (1996) Generalized gradient approximation made simple. Phys Rev Lett 77:3865–3868

Rapaport DCR (2004) The art of molecular dynamics simulation[M]. Cambridge University Press, Cambridge

Verlet L (1967) Computer "Experiments" on Classical Fluids. I. Thermodyn Prop Lennard-Jones Mol. Phys Rev, 159: 98–103

Nobel Prizes and Materials Science 14

14.1 Introduction

The Nobel Prize was established upon the death of Swedish inventor and entrepreneur Alfred Nobel, who bequeathed the majority of his fortune to establish prizes in physics, chemistry, physiology or medicine, literature, and peace. The prizes are awarded annually to individuals or organizations who have made the greatest impact for the betterment of humankind in the previous year. In 1968, Sweden's central bank added the Sveriges Riksbank Prize in Economic Sciences in Memory of Alfred Nobel.

The first Nobel Prizes were awarded in 1901 and have been distributed annually ever since, with the exception of certain years during World Wars I and II. As of 2022, the Nobel Prizes and the Sveriges Riksbank Prize in Economic Sciences have been awarded 615 times to 989 recipients.

Although materials science is not a category for the Nobel Prize, many of the laureates have backgrounds in materials science and have made significant contributions to physics, chemistry, and other disciplines, leading to advancements in technology, such as the development of lithium-ion batteries and optical fibers.

In this chapter, we will examine recent Nobel Prizes in Physics and Chemistry that are closely related to materials science. These discoveries demonstrate the impact of semiconductors on the electronics industry, the potential of superconductors, the relationship between computational methods and quantum mechanics, the revolution of communication brought about by optical fibers, the challenge and reward of blue light-emitting diodes, and the transformative effect of lithium-ion batteries on the automobile and energy storage industries. These stories and the underlying science may inspire students to delve further into the exciting field of materials science.

Z. Sun et al., *Fundamentals of Materials Science*,
https://doi.org/10.1007/978-981-96-3904-5_14

14.2 Semiconductors and Transistor Effect (1956, The Nobel Prize in Physics)

14.2.1 The Prize

The Nobel Prize in Physics 1956 was awarded to William Bradford Shockley, John Bardeen, and Walter Houser Brattain for their pioneering work in the field of semiconductors and their discovery of the transistor effect.

Transistor is a semiconductor device, one of its important functions is to amplify electronic signals and power or switch electrical signals. Transistors, as one of the basic components of modern electronic products, are composed of semiconductor materials and usually contain at least three terminals connecting to external circuits. By adjusting the voltage or current applied to a pair of terminals of the transistor, the current of the other pair of terminals can be controlled, which has the effect of amplifying the signal. In 1947, **John Bardeen** and **Walter Brattain** in Bell Telephone Laboratories, attempted to recognize the properties of electrons at the contact between a metal and a semiconductor. By placing two-point contacts close enough to each other, they managed to make a three-terminal device—the first "point contact" transistor as illustrated in Fig. 14.1. A point-contact transistor consists of a piece of germanium, a semiconductor, and two gold contacts that are very closely spaced. Brattain put a small strip of gold foil on the tip of the plastic triangle-to build a point contact diode. He cut the gold foil at the tip of the triangle to create two very close but electrically isolated gold contacts. Two gold electrodes with a spacing of 50 μm are pressed on the germanium semiconductor, and the tiny electrical signal enters the germanium semiconductor (base) from one gold electrode (emitter) and is significantly amplified, and then is output through the other gold electrode (collector). This device soon leaded to the generation of audio amplifiers. the power of this new technology quickly ignited huge research interests in solid-state electronics.

14.2.2 Brief History of Transistors

This discovery marked the dawn of a new transistor-led era in the electronics industry. In 1952, the point-contact transistor became widely accessible in commercial use and was instrumental in manufacturing telephone systems. In an attempt to improve the transistor design of Bardeen and Brattain, William Shockley fabricated the junction transistor from germanium in 1951. While the point-contact and junction transistor relied on germanium, researchers soon thereafter noted that the component broke down at 180°F (about 82.2 °C)because too many free electrons were introduced in transistors. This drawback inspired Gordon Teal to invent the first-ever silicon transistor in 1954, which could withstand high temperatures. The development of the silicon transistor led to the invention of more silicon-based transistors, such as metal–oxide–semiconductor transistors (MOSFETs). The first MOSFET was fabricated by Bell Labs researcher John Atalla in 1960. The design

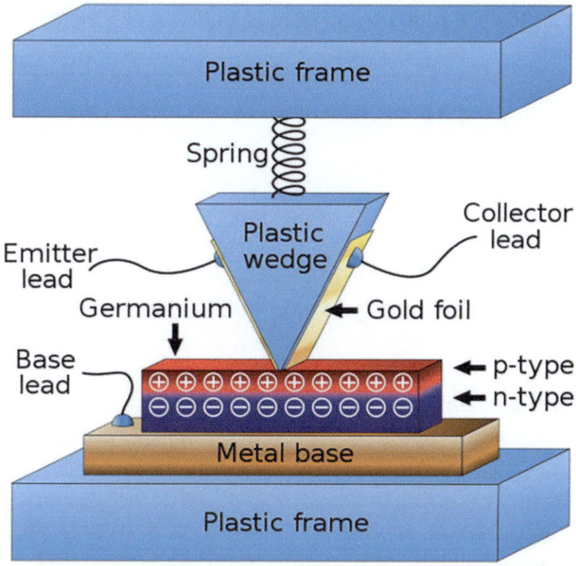

Fig. 14.1 The first "point contact" transistor

was based on Shockley's field-effect theories. Today, researchers are aiming to downsize transistors to the nanometer scale to enhance the number of transistors per unit area in an integrated circuit. However, there are several challenges. For instance, a nano-sized MOSFET might experience short-channel effects or high channel leakage currents that adversely affect the transistor's performance. Researchers are now looking into 2D ultrathin monolayer materials to fabricate transistors, such as molybdenum disulfide. Carbon nanotube and graphene are also promising materials to replace silicon in transistors.

14.2.3 The Future of Transistors

Today, transistors are everywhere where electronic devices can be found, including cell phones, computers, satellites, and spacecraft. The transistor is an important part of the development of electronic technology, and it marks the beginning of the digital age. The entire semiconductor-based industry is built based on transistors. In fact, telecommunications would not be possible without transistors. Transistors revolutionized the field of electronics, and paved the way for smaller and cheaper radios, calculators, and computers, among other things. In the future, it is theoretically possible that a transistor can be small enough that it can operate with only one electron. Such transistors are known, not surprisingly, as single electron transistors, and they'd be considered "on" or "off" depending on whether they were holding an electron. Transistors at this level would be solely used as switches for binary coding rather than an amplifier. This would open up the door of the next-generation computer.

14.3 Ceramic Superconductor (1987, The Nobel Prize in Physics)

14.3.1 The Prize

The Nobel Prize in Physics 1987 was awarded jointly to J. Georg Bednorz and K. Alex Müller "for their important break-through in the discovery of superconductivity in ceramic materials." Their discovery involved a superconducting ceramic with a record-breaking transition temperature of 35 K, which prompted numerous research groups to pursue the development of superconducting materials with even higher transition temperatures. The discovery was a significant milestone in the field of superconductivity and has since led to important technological advancements.

J. Georg Bednorz received his master's degree from the University of Münster in 1976 and his PhD from the Swiss Federal Institute of Technology, ETH Zurich, in 1982. In January 1982 he joined the IBM Zurich Research Laboratory in Rüschlikon, Switzerland, as a Research Staff Member. His research focused on solid state physics, particularly on the ferroelectric properties and structural phase transitions of perovskite-type oxides. Starting in 1983, he shifted his focus to the study of metallic oxides in the pursuit of developing superconductors with high transition temperatures.

K. Alex Müller received his doctorate from the Swiss Federal Institute of Technology in 1958 and beginning in 1963 he performed research in solid-state physics at the IBM Zürich Research Laboratory, heading the physics department there for several years and becoming an IBM fellow in 1982.

Since the discovery of superconductivity in 1911, great efforts have been devoted to searching for superconductors, especially at higher transition temperatures. In 1983, Karl Alex Müller and his associate, Johannes Georg Bednorz, started to test newly developed ceramic materials (known as metal oxides) to search materials with superconductivity. After three years of efforts, they succeeded in finding superconductivity in a Ba-La-Cu oxide at a temperature of 35 K, 12 K higher than the highest transition temperature had previously been achieved. Their discovery immediately inspired researchers worldwide to develop superconducting oxides with even higher transition temperatures.

14.3.2 Brief History

The history of superconductivity can be tracked back to 1911, when Dutch physicist Heike Kamerlingh Onnes discovered that when mercury was cooled below 4 K, its electrical resistance became zero, marking the first observation of superconductivity. For this work, Onnes was awarded the Nobel Prize in Physics in 1913. At that time, it was believed that superconductors had the same properties as normal materials, except for the lack of electrical resistance. It wasn't until 1933 that German physicists Walther Meissner and Robert Ochsenfeld

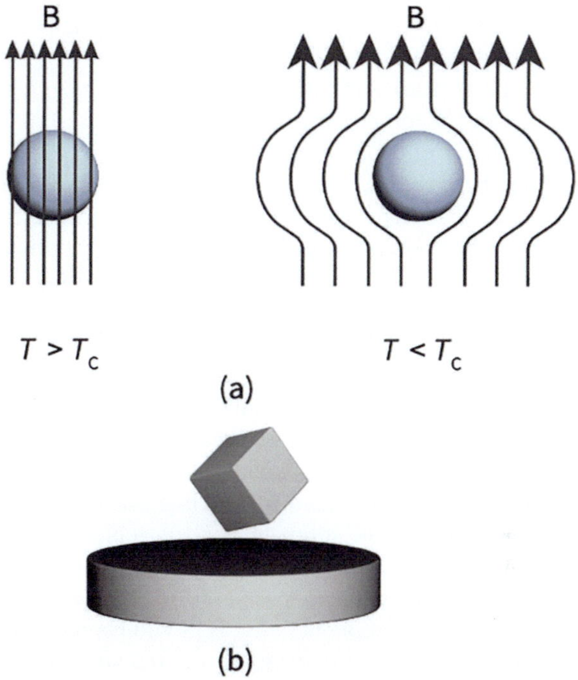

Fig. 14.2 Meissner effect **a** Diagram of the Meissner effect. When a superconductor is cooled below its critical temperature, it expels magnetic field lines, which are depicted as arrows; **b** Magnetic levitation

observed that superconductors exhibited diamagnetism, repelling magnetic fields. This phenomenon, known as the Meissner effect or Meissner-Ochsenfeld effect (Fig. 14.2), is fundamental to the magnetic suspension behavior applied in levitated superconducting magnet trains.

In 1950, theoretical physicists Lev Landau and Vitaly Ginzburg proposed the Landau-Ginzburg theory to explain how superconductors work. However, their theory focused on the large-scale behaviors of superconductors without examining their microscopic properties. In 1957, physicists John Bardeen, Leon N. Cooper and Robert Schrieffer proposed the BCS theory of superconductivity. According to this theory, According to this theory, electron pairs with opposite spin and momentum are attracted to each other through the exchange of lattice vibration phonons, forming Cooper pairs that can travel unimpeded through the lattice without electrical resistance. However, the Cooper pairs are easily broken up by thermal energy, which explains why superconductivity typically occurs at very low temperatures. In 1972, John Bardeen, Leon N. Cooper and J. Robert Schrieffer were awarded the Nobel Prize in Physics for their work on the BCS theory. This theory, however, cannot explain the phenomenon of "high-temperature" superconductors, which occurs at approximately 80 K (−193 °C) and above and requires other mechanisms to be invoked.

Although a large number of theoretical physicists attempted to describe the underlying mechanism of superconductivity, the experimental progress was extremely slow before the 1980s. For many years, its critical transition temperature was fixed at 23 K.

In 1986, Karl Alex Müller and his associate, Johannes Georg Bednorz, succeeded in observing superconductivity in a Ba-La-Cu–O (whose crystal structure is shown in Fig. 14.3) at a temperature of 35 K, 12 K higher than the highest transition temperature had previously been achieved. These superconductors are ceramic materials that have a great guiding role in the research of new superconductors, opening the prelude to the development of high-temperature superconductors. They were awarded the Nobel Prize in Physics in 1987 for their pioneering work. This discovery immediately inspired researchers worldwide to explore the possibility of achieving even higher transition temperatures. Within a year, superconducting oxides with transition temperatures approaching 100 K were reported. However, this oxide high-temperature superconductor cannot be explained by BCS theory, and no one currently knows its microscopic mechanism.

In the next century, researchers have discovered an increasing number of superconductors, and the record for the highest transition temperature has continued to progress towards the ultimate goal of achieving superconductivity at room temperature. In 2014, Ma Yanming's research team and Cui Tian's research team at Jilin University predicted, based on BCS theory, that hydrogen sulfide compounds could exhibit superconductivity: H_2S will become superconductor at 160 GPa, with a transition temperature of 80 K; the transition temperature for the H_3S structure composed of H_2S and H_2 is between 191 and 204 K at 200 GPa. In 2015, A. P. Drozdov, M. I. Eremets, and colleagues at the Max Planck Institute of Chemistry discovered that superconductivity at a temperature of 203 K could be achieved by compressing H_2S to a pressure close to 2 million times the atmospheric pressure

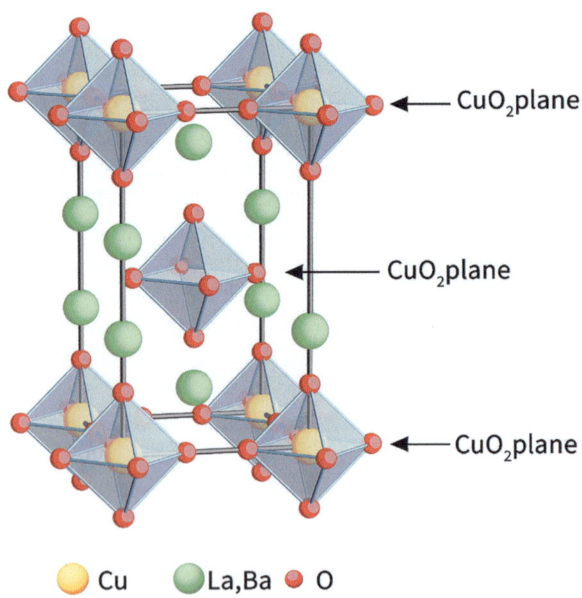

Fig. 14.3 Crystal structure of $La_{2-x}Ba_xO_4$

(155 GPa), and they argued that the H_3S phase was responsible for the superconductivity. Four years later, they made self-breakthroughs and found that LaH_{10} could become a superconductor at a temperature of 250 K under a pressure of 170 GPa.

14.3.3 The Future of Superconductivity

Superconductivity is an area of research with tremendous potential for the future. The development of new superconducting materials and technologies has the potential to revolutionize many fields, including energy storage, transportation, and computing. Here are some of the exciting developments and potential future applications of superconductivity:

[1] High-Temperature Superconductivity: One of the main challenges of superconductivity is achieving it at high temperatures, which would allow for more practical applications. Recent advances in materials science have led to the discovery of new classes of high-temperature superconductors that operate at temperatures as high as −70 °C. Continued research in this area could lead to the development of even higher-temperature superconductors that operate at or near room temperature, which would be a game-changer for many industries.
[2] Energy Storage: Superconductivity could be used for high-capacity energy storage systems. Superconducting magnetic energy storage systems use superconducting coils to store energy in a magnetic field. These systems have the potential to store large amounts of energy for long periods of time and release it quickly when needed.
[3] Magnetic Levitation: Superconductors can be used to create powerful magnetic fields that can levitate objects, including high-speed trains. This technology is already in use in Japan's high-speed Maglev trains, which can reach speeds of up to 600 km/h. Further advances in superconducting materials could lead to even faster and more efficient transportation systems.
[4] Quantum Computing: Superconducting materials are a promising candidate for the development of quantum computing. Quantum computers use the principles of quantum mechanics to perform calculations that are impossible for classical computers. Superconducting quantum bits have shown promising results in recent experiments, and continued research in this area could lead to the development of practical quantum computers.
[5] Medical Imaging: Superconductivity is already used in medical imaging systems such as MRI (magnetic resonance imaging). Advances in superconducting materials could lead to even more sensitive and precise imaging systems for medical diagnosis and treatment.

In conclusion, superconductivity has the potential to revolutionize many fields in the future, and continued research and development in this area will be crucial to unlocking its full potential.

14.4 Fullerene (1996, The Nobel Prize in Chemistry)

14.4.1 The Prize

The 1996 Nobel prize in Chemistry was awarded to Robert F. Curl, Jr., Sir Harold W. Kroto and Richard E. Smalley, for their discovery of fullerenes. This new form of element carbon, in which the atoms are arranged in closed shells, was discovered in 1985. Beforehand, six forms of crystallized carbon were discovered, known as two types of graphite, two types of diamonds, chaoit and carbon(IV).

At that time **Harold W. Kroto** was expert in microwave spectroscopy and especially interested in carbon-rich giant stars. During his research in atmospheres, he noticed a sort of long-chained molecule made of only carbon and nitrogen and wished to study the formation of such molecules. He then got in touch with **Richard E. Smalley**, who was a researcher in chemical physics. Smalley had designed and built a special apparatus to vaporize materials into a plasma of atoms, thus to study the design and distribution of the formed clusters. Together with **Robert F. Curl**, whose background was in spectroscopy, they performed this experiment by obtaining gaseous carbon through the special apparatus then cooling it down to analyze the carbon clusters with mass spectrometry. During the experiment, the clusters of 60 carbon atoms were the most abundant, in which they found the structure of high stability and great symmetry. According to speculation, the clusters of 60 atoms, known now as C_{60}, could be a "truncated icosahedron cage", a structure of polyhedron with 20 hexagonal surfaces and 12 pentagonal surfaces (Fig. 14.4).

Since the geodetic dome for the 1967 Montreal World Exhibition had exactly the same structure and the designer was R. Buckminister Fuller (Fig. 14.5), the newly-found C_{60} was named "*buckministerfullerene*" after him. The authors concluded that, "For such a unique and centrally important molecular structure, a more

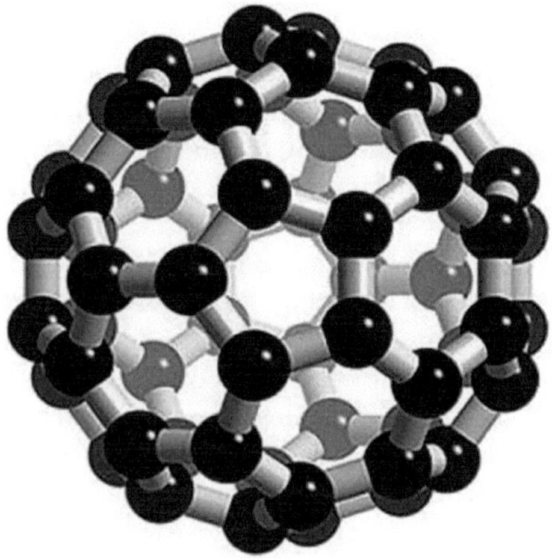

Fig. 14.4 Proposed structure of C_{60}

Fig. 14.5 The geodetic dome for the 1967 Montreal World Exhibition

concise name would be useful. A number of alternatives come to mind (for example, ballene, spherene, soccerene, carbosoccer), but we prefer to let this issue of nomenclature be settled by consensus".

14.4.2 History of Fullerene

In the year of 1966, David Jones first speculated about a structure for carbon hexagons of "a closed spherical shell, resilient like little footballs". Later, in 1969, Eiji Osawa theorized a structure of truncated icosahedron for 60 carbon atoms, which was expected to be extremely stable. After 15 years, Andrew Kaldor and his group watched a series of carbon clusters by mass spectrometry and actually identified peaks that corresponded to clusters containing the even number of 60 atoms. Unfortunately, they didn't realize the significance of these data at that time. Even after Curl, Kroto and Smalley published their findings on C_{60} in 1985, only one year later, Kaldor and his group still disputed this structure. They argued that the molecule of C_{60} should be planar, and further evidence for the icosahedral structure should leave to the development of new characterization techniques to confirm. Despite Kaldor' opinion, the paper received a mixed reception, both criticism and enthusiastic acceptance. As the paper provided no image of C_{60}, yet only a photo of real football, no researchers would believe that carbon atoms could be arranged in such symmetrical form.

In 1988, Curl, Kroto and Smalley responded with two papers, answered doubts towards reactivity, stability and other issues of C_{60} and further elaborated that the proposed structure was the key to these properties. However, no direct proof for C_{60} being spherical rather than planar was given. It was until 1990 that Walter Krätschmer and his group managed to take photos of C_{60} with scanning tunneling microscope, in which thin platelets, rods and stars of hexagonal symmetry were clearly seen (Fig. 14.6). Finally, the tale of C_{60} came to a happy ending.

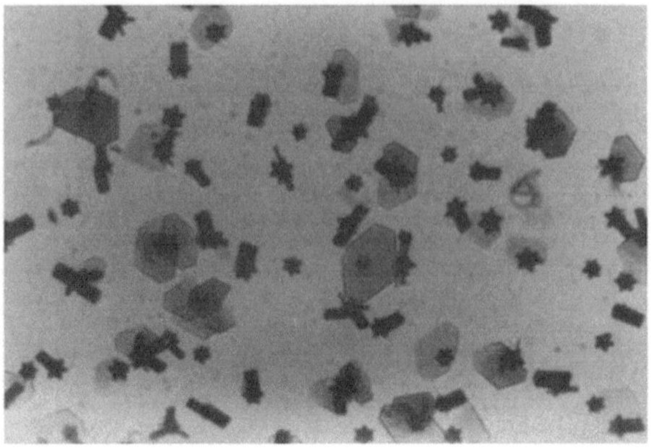

Fig. 14.6 Typical crystals of C_{60}

14.4.3 Future of Fullerene

Although fullerene didn't play an important role in the field of application, it did have a great impact in basic science. For example, fullerene showed wave-particle duality, and was "the most massive and complex object" in the observing of wave behaviour. This finding was regarded as a landmark in quantum mechanics. Single-molecule transistors involving a full zero-dimensional fullerene was realized and proved of superconductivity. Furthermore, fullerene was the first in the series of new carbon nanomaterials over the past two decades—carbon nanotubes became the focus of research in the 1990s with countless publications, and graphene is still one of the hottest issues nowadays. By contrast, fullerene may have been beaten in potential applications, but it still remains to surprise the world as it has been reported to be detected in space. As the early work of such molecule was motivated by Kroto's findings in astrophysical chemistry, the story of fullerene has now turned full circle.

14.5 Computational Methods from Quantum Mechanics (1998, The Nobel Prize in Chemistry)

14.5.1 The Prize

In 1998, the Nobel Prize in Chemistry was shared by Prof. Walter Kohn and Prof. John A. Pople. The former developed the density-functional theory (DFT) and the latter developed computational methods in quantum chemistry. Due to their pioneering work, computer-based methods can now be used for theoretical studies of molecules and materials with a relatively large size. Thanks to the fast development of computational power and algorithms, methods based on the fundamental laws of quantum mechanism have become a workhorse for the study and prediction of the structure, energetics, and functionalities of matter in detail. The predicted results are frequently used to supplement real experiments and provide deeper insights into a range of phenomena.

Mathematically, conventional calculations based on many electron wavefunctions are extremely hard. To tackle this challenge, **Walter Kohn** and his colleagues showed that one does not really need to describe the motion of individual electrons in the system. Instead, the electron density, as a fundamental variable, can be employed to treat the many-body problem smartly. In this case, the total energy of the system is a function of the electron density, and then the complex many-body problem is simplified by an equivalent single-particle problem. This is main concept and the beauty of the density-functional theory. The simplicity of the DFT method makes it possible to study large systems with hundreds and even thousands of atoms. Indeed, DFT is now an indispensable tool for the study of materials science, chemistry, physis and many other subjects.

John Pople was devoted to developing more sophisticated ab initio quantum chemistry methods. At early time, these methods are extremely time-consuming to perform since they are on the basis of quantum mechanics. One of the greatest contributions of Pople is that he designed a famous computer program, named Gaussian, that allows people to carry out quantum-mechanics calculations effectively and accurately. Using the Gaussian code, researchers can readily predict the properties of molecules, revisit the chemical reaction processes, and uncover the underlying the mechanism of reactions. The first version of the Gaussian program was released in 1970, and the latest series of programs provide state-of-the-art capabilities for electronic structure modeling.

14.5.2 Brief History of Computer-Based Calculations

The DFT method was developed on the basis of quantum mechanics, which began at the 1900s. In 1926, Erwin Schrödinger developed a wave equation (i.e., the famous Schrödinger equation) that is the fundamental equation of quantum mechanics. Based on this equation, one can describe the form of the wave function that govern the motion of small particles. Although in principle everything is known, the Schrödinger equation cannot be solved exactly for relatively complex systems, due to the complex quantum many-body problem. This significantly hinders the real application of the equation in materials science.

In 1927, Born and Oppenheimer simplified this problem by assuming that the motion of the electrons and the nuclear are decoupled, since they exhibit dramatically distinct dynamics. Correspondingly, the wavefunction of a molecule is divided into two aspects: the electronic and nuclear components. This excellent approximation allows us to compute the wavefunction in two relatively "simple" steps.

In 1960s, Kohn and Hohenberg proposed two Hohenberg–Kohn theorems (HK), which are the firm theoretical footing for DFT. By using functional of the electron density, one can reduce the many-body problem with $3N$ (N is the number of electrons) spatial coordinates to only 3 spatial coordinates. Kohn and Sham further developed the HK theorem and produced the well-known Kohn–Sham DFT (short for KS DFT). In this case, the intractable many-body problem of interacting electrons is replaced by an auxiliary independent-particle problem. Nowadays, the Kohn–Sham DFT is the most widely used method for electronic structure calculations.

14.5.3 The Application and Future of Computer Simulations

Thanks to the pioneering work of Walter Kohn and John Pople, computer simulations have widely accepted as the third paradigm of material research (the other two are experiment and theory). Along with the fast development of high-performance computer clusters, one can readily use DFT methods to "predict" new molecules or materials that have not been synthesize or fabricated in experiments. Besides that, DFT can be used to "understand" the properties and behaviors of materials at the atomic and electronic level. By establishing the relationship between measurement data with calculated results, computer simulations help people to explore the underlying mechanisms that are not easily, if not impossible, detected in real experiments.

The DFT is exact in principle. However, approximations must be made to treat the interactions between electrons, so-called exchange–correlation functionals. Despite recent improvements, there are still some situations where the DFT methods with "standard" exchange–correlation functionals do not work well. The most typical examples include the problems of van der Waals forces (dispersion), charge transfer excitations, and strongly correlated systems. Many efforts are performing in computer simulation field to overcome these challenges.

14.6 Conductive Polymers (2000, The Nobel Prize in Chemistry)

14.6.1 The Prize

When considering materials that conduct electricity, metals often come to mind rather than synthetic polymers, such as plastics. Despite the diverse physical properties exhibited by different polymers, prior to the 1970s, nearly all polymers discovered were electrical insulators. However, in the 1960s, it was demonstrated that certain classes of organic polymers, as well as their corresponding plastics, could be made to exhibit semiconducting behavior, with conductivity values rivaling those of metals by the late 1970s. This breakthrough earned three researchers - Alan Heeger from the University of California, Santa Barbara, Alan MacDiarmid from the University of Pennsylvania in Philadelphia, and Hideki Shirakawa from the University of Tsukuba in Japan - the Nobel Prize in Chemistry in 2000, for discovering that plastics can be made electrically conductive. Their discovery paved the way for groundbreaking applications, including full-color displays for cellular phones and plastic electronics for computerized merchandise.

Alan J. Heeger, (born January 22, 1936, Sioux City, Iowa, U.S.), American chemist. He received his Ph.D. in physics from the University of California at Berkeley in 1961 and subsequently taught and conducted research at the University of Pennsylvania until 1982. He then became a professor at the University of California at Santa Barbara and served as director of its Institute for Polymers and Organic Solids until 1999. In 1990, Heeger founded UNIAX Corporation to develop and manufacture light-emitting displays based on conducting polymers, which was acquired by DuPont in 2000. In 2001, he co-founded Konarka Technologies to produce thin, flexible solar cells made of plastic. However, the company filed for bankruptcy protection in 2012 and was eventually liquidated. Heeger's research in conducting polymers and their applications in optoelectronics and renewable energy has earned him numerous accolades, including the Nobel Prize in Chemistry in 2000, which he shared with two other scientists.

Shirakawa Hideki, (born August 20, 1936, Tokyo, Japan), Japanese chemist. He received his Ph.D. from the Tokyo Institute of Technology in 1966 and began his career as a faculty member in the Institute of Materials Science at the University of Tsukuba the same year. He was appointed as a professor of chemistry in 1982 and retired as a professor emeritus in 2000.

Alan G. MacDiarmid, (born April 14, 1927, Masterson, N.Z.—died Feb. 7, 2007, Drexel Hill, Pa., U.S.), New Zealand-born American chemist. He earned Ph.D.'s in chemistry at the University of Wisconsin at Madison (1953) and the University of Cambridge (1955). He then joined the faculty of the University of Pennsylvania, where he became a full professor in 1964 and the Blanchard Professor of Chemistry in 1988.

14.6.2 Brief History of Conductive Polymers

Polymeric sulfur nitride, or (SN)x, composed of monomers made up of sulfur and nitrogen, was reported in the 1960s to exhibit unusual electrical properties. In 1975, Alan MacDiarmid and Alan Heeger, both professors at the University of Pennsylvania in chemistry and physics, respectively, decided to collaborate and investigate this compound. That same year, while MacDiarmid was a visiting professor in Japan, he gave a talk on (SN)x at the Tokyo Institute of Technology. After the lecture, MacDiarmid and Hideki Shirakawa discussed their shared fascination with conductive materials. Shirakawa revealed a sample of a silvery polymer that he and his colleagues had serendipitously created. By accidentally adding too much Ziegler–Natta catalyst to acetylene that was to be polymerized, instead of the expected black powder, they produced ragged pieces of silvery polyacetylene film. As it is well-known in chemistry that a substance with a metallic sheen is likely to conduct electricity, MacDiarmid promptly invited Shirakawa to the University of Pennsylvania for a year to work with him and Heeger on conductive polymers.

Previously, MacDiarmid and Heeger discovered that the conductivity of (SN)x could be increased tenfold by adding bromine atoms through a doping procedure similar to that used in silicon transistors (as seen in Gordon Moore's work). Doping is a technique that introduces impurities into a substance to enhance its conductivity (in this case, bromine atoms were introduced). In the case of Shirakawa's polyacetylene, the increase in conductivity was even more remarkable, as the introduction of bromine through doping resulted in a ten-million-fold increase in conductivity compared to before doping.

Unlike (SN)x, polyacetylene is an organic polymer. The discovery of highly conductive polyacetylene by MacDiarmid, Heeger, and Shirakawa in 1976 was groundbreaking, as organic polymers are cheaper, easier to produce, lighter, and more flexible than their inorganic counterparts. This discovery opened the door to a surge of research into conducting and semiconducting organic polymers. These polymers have been applied in various fields, including flexible "plastic" transistors and electrodes, plastic batteries, electromagnetic interference shielding, static elimination, corrosion inhibition, and electroluminescent-polymer displays. The field of organic conducting polymers remains promising, and many researchers are actively exploring it.

14.6.3 The Future of Conductive Polymers

The ability to modulate conductivity based on the synthetic and doping processes is an interesting and useful property of conducting polymers. Specifically, the reversible switching between insulating and conducting forms is an intriguing feature. The functionalization of conducting polymers, which enables smart materials that can interact with and respond to their environment, presents an exciting opportunity. Additionally, the preparation of self-doped polymers is a promising way to overcome the inherent limitations of poor ionic charge transportation during redox switching. In the future, the integration of conductive polymers with other functional components is expected to be an important direction for research.

There are high expectations for the controlled manipulation of nanomaterial morphology, such as nanofibers, nanorods, and other nanostructures, through the supramolecular self-assembly of conducting polymers. This may involve adopting surfactants or sacrificing templates, or utilizing specific polymerization techniques. The manipulation of polymer microstructures has the potential to improve the electrical or mechanical properties of conductive polymers. This will require the efforts of chemists and electrochemists in the production and characterization of new materials, as well as theoreticians to explain and predict phenomena, and engineers to provide the final form of conductive polymer-based materials. In addition, conducting polymers are eco-friendly and economically available, and improvements in their properties could greatly enhance mass production, making them more widely applicable in many industries. For instance, the development of conductive ink, electrical circuits, solar cells, batteries, and flexible displays based on conducting polymers opens up new horizons for materials science and engineering.

14.7 Optical Fiber (2009, The Nobel Prize in Physics)

14.7.1 The Prize

The Nobel Prize in Physics 2009 was awarded in one half to Charles Kuen Kao "for groundbreaking achievements concerning the transmission of light in fibers for optical communication" and the other half jointly to Willard S. Boyle and George E. Smith "for the invention of an imaging semiconductor circuit—the CCD sensor."

In 1966, an invention proposed by Charles K. Kao led to a breakthrough in fiber optics. He found a technique to transfer light signals over long distances through optical glass fibers. With a fiber made of high-purity glass, the light signals could be sent over 100 km, which is significantly improved from the value of 20 m for the fibers fabricated in the 1960s. Kao's outstanding work stimulated other researchers to share his idea of the future opportunity of fiber optics. In 1970, just four years later, the high-purity fiber was successfully made first.

Today, optical fibers set up the circulatory system and sustain our communication society. These low-loss glass fibers allow global communication, such as the internet. light transports in thin threads of glass, and transfer almost the telephone and data transportation in all directions. Text, image, music and video can be instantly conveyed to the whole world. If all of the glass fibers that link the globe were unwrapped, we would get a single string over one billion kilometers long that can enclose the globe more than 25,000 times and is growing by thousands of kilometers every hour.

14.7.2 Early History of Glass Fibers

Glass has been drawn into fibers in Roman times. Until the 1790s, the French Chappe brothers invented the first "optical telegraph." It was a system containing lots of lights set up on towers where operators would transmit a message between towers. Great improvements were made in optical science in the next century.

Physicists Daniel Collodon and Jacques Babinet realized that light could be directd along water steam for fountain shows in the 1840s. In 1854, a British physicist John Tyndall proposed that light could pass through a bowed water jet, confirming that the light signal could be bent. To prove this finding, he arranged a water box with a pipe extending out from one side. When the water poured from the pipe, he shined light into the tank and found a bow of light went after the flowing water.

Alexander Graham Bell invented an optical telephone system in 1880, which is called the photophone. However, his earlier creation of the telephone was proven to be more applicable. William Wheeler created a system of light pipes lined with a highly reflective coating in the same year. These pipes lightened the near homes by utilizing a light source from an electric arc lamp in the basement and guiding the light through the pipes. Doctors Roth and Reuss of Vienna utilized bent glass rods to illuminate the cavities within bodies in 1888. French engineer Henry Saint-Rene devised a system of bent glass rods for guiding optical images in seven years later as an early effort at television. In 1898, American David Smith patented a dental illuminator using bent glass rods.

In the 1920s, John Logie Baird applied a patent of using arrays of semitransparent rods to transmit images for television, and Clarence W. Hansell invented the similar device for facsimile equipment. Heinrich Lamm, however, was the first inventor to transfer an image of the light bulb filament through many optical fibers in 1930. His initial purpose was to judge the inaccessible organs of the body, while he was forced to relocate to America and discarded his dream to be a professor in medicine due to the rise of the Nazis. His attempt to apply a patent was declined due to Hansell's British patent.

In 1951, Holger Moeller requested for a Danish patent on fiber optics for imaging transmission, where he proposed the cladding glass or plastic fibers with translucent materials but was also rejected due to Baird and Hansell's patents. In 1954, Abraham Van Heel and Harold H. Hopkins put forward imaging bundles in the journal Nature at exclusive times. Van Heel later devised a cladded fiber system that greatly reduced signal interference of fibers.

Additionally, the "maser", i.e. "microwave amplification by stimulated emission of radiation", was invented by Charles Townes and his colleagues in 1954 at Columbia University. The laser, namely "light amplification by stimulated emission of radiation", was proclaimed in 1958 as a high-efficiency source of light. The concept was proposed by Charles Townes and Arthur Schawlow to illustrate that the maser could be fabricated for optical and infrared applications. Basically, light is reflected repeatedly in an empowered medium to produce amplified light, exciting gas molecules to generate radio waves as in the maser.

In 1960, the continuously working helium–neon gas laser was first designed. In the same year, an operable laser was invented, which exploided a synthetic pink ruby crystal as the empowered medium to generate a pulse of light. In 1961, Elias Snitzer published a theoretical depiction of single-mode fibers whose tiny core could carry light with only one-wave-guided mode. Snitzer demonstrated that a laser guided through a thin glass fiber was sufficient for medical applications, but the light loss was too great for communication applications. Charles Kao and George Hockham of Standard Communications Laboratories in England issued a paper in 1964, demonstrating that light loss in glass fibers could be diminished dramatically by decreasing impurities.

In 1970, the aim of making single-mode fibers with attenuation less than 20 dB/km was achieved by scientists at Corning Glass Works. This was realized through doping silica glass with titanium. Additionally, Morton Panish and Izuo Hayashi of Bell Laboratories worked with a group from the Ioffe Physical Institute in Leningrad to demonstrate that a semiconductor diode laser is capable of emitting continuous waves at room temperature in 1970. In 1973, Bell Laboratories exploided a modified chemical vapor deposition process that heats chemical vapors and oxygen to make ultra-transparent glass that can be produced into low-loss fiber optics massively. This process persists the standard method for fiber-optic cable manufacturing. The first fiber-optic link was settled down by the Dorset police in UK in 1975. Two years later, the first live telephone traffic by fiber optics appeared in Long Beach, California. In the late 1970s and early 1980s, telephone companies initiated to apply fibers extensively to rebuild their communications infrastructure.

The company Sprint was founded on the first national, complete digital, fiber-optic network in the mid-1980s. An erbium-doped fiber amplifier was invented in 1986 by David Payne of the University of Southampton and Emmanuel Desurvire at Bell Labratories, which lowered the cost of long-distance fiber systems by removing the need for optical-electrical-optical repeaters. Based on Desurvire's optimization on laser amplification technology, the first transatlantic telephone cable began to run in 1988.

In 1991, Desurvire and Payne suggested optical amplifiers could be built by the fiber-optic cable itself. The all-optic system could bring 100 times more information than cables with electronic amplifiers. In addition, photonic crystal fiber was discovered in 1991. This fiber directs light by the diffraction from a periodic structure rather than the total internal reflection, which permits power to be carried more efficiently than traditional fibers.

The first all-optic fiber cable (TPC-5), using optical amplifiers, was placed across the Pacific Ocean in 1996. The Fiber Optic Link Around the Globe (FLAG) became the longest single-cable network around the world in the following years and offered infrastructure for the next generation of Internet applications.

14.7.3 Present and Future Applications

Today, a variety of industries, including the telecommunication, medical instruments, military, network and broadcasts, are able to apply and use the fiber optic technology in a variety of applications.

Telecommunication Uses

Fiber optical cables are excellent mediums for telecommunication networks due to their flexibility and durability, as shown in Fig. 14.7. Optics fiber is applied by many telecommunications companies to convey telephone signals, Internet communication, and television signals. The fiber is behaved for transmitting and receiving purposes. Unlike electrical cables, fiber optics experience relatively low loss of intensity for signals transmission, allowing to transport information far distances with few repeaters. Fiber-optic cables can contain a lot of different signals concurrently through a technique called wavelength division multiplexing, which increases the efficiency for transporting large quantities of independent signals. Their effectiveness is further expedited by their immunity to electrical interference.

Optical fibers are ideally proper for transferring digital information, which is especially helpful in computer and cellular networks. The higher transferring capacity, less signal loss, low power, light weight and flexibility make fiber optics ideal media for telecommunications.

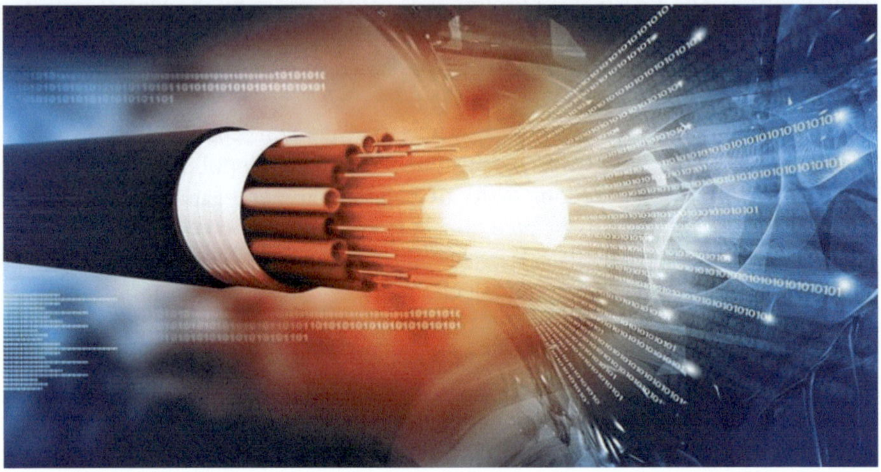

Fig. 14.7 Fiber optical cables as mediums for telecommunication networks

Biomedical Applications

In medical instruments, optical fibers enable physicians to examine the inside body through tiny incisions without surgery. They are fabricated for endoscope instruments for observing the interior of hollow organs in the body, as shown in Fig. 14.8. Most endoscopes have two sets of fibers: an outer ring of incoherent fibers for the light resource and an inner coherent fibers for the image transmission. For example, physicians handle arthroscopy to examine knees, shoulders, and other joints. In some models, a third set of fibers conveys a laser beam to prevent bleeding or to burn away diseased tissue. They can also be utilized for insertion into blood vessels to obtain a quick, accurate assessment of blood chemistry.

Fiber-optic cables are widely utilized in hospitals and doctor offices. They produce the backbone of advanced imaging techniques for digital diagnostics because

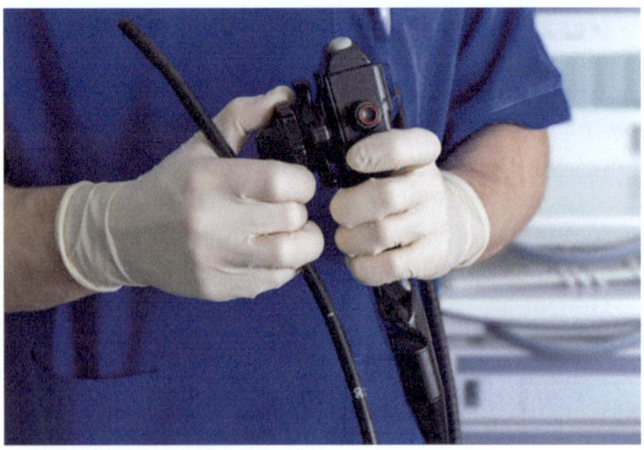

Fig. 14.8 Biomedical Spectroscopy made with fiber optics

they can efficiently transfer large quantities of sensitive data. Their inertness and flexibility makes fiber optic cables the natural choice for endoscopes utilized in minimally invasive surgical procedures. The growing application of endoscopy has replaced invasive surgery in diagnosing difficult medical cases.

Fiber Optic Cable Sensors

Fiber-optic cable sensors are manipulated to evaluate many kinds of physical properties, such as mechanical strain, temperature, and pressure. Some fiber optic sensors determine these properties directly by modulated light. Other sensors use fiber optics as carriers to transport light from harsh environments to subtle sensors. For example, the fiber optic gyroscope can detect mechanical rotation without moving parts. A major profit of extrinsic sensors is their ability to reach places that are unreachable. An example is the measurement of temperature inside aircraft jet engines by using a fiber to convey radiation into a pyrometer outside the engine. Extrinsic sensors can also measure the internal temperature of electrical transformers, where the extreme electromagnetic fields make other measurement techniques impossible difficult.

14.8 Graphene (2010, The Nobel Prize in Physics)

14.8.1 The Prize

The 2010 award of the Nobel Prize in Physics was awarded to Andre Geim and Konstantin Novoselov, two Russian born physicists working at Manchester University in the United Kingdom, "for groundbreaking experiments regarding the two-dimensional material graphene". Together with the 1996 Noble Prize in Chemistry for Fullerenes and the 2008 Kavli Prize in nanoscience for carbon nanotubes, this prize completes a trifecta of mega-awards to three different topologies of hexagonally-bonded carbon.

Andre Geim, well-known in the world of physics for his off-the-wall ideas, is more of an inventor than a conventional scientist. His style is to take 10% of his time to work on some so-called "Friday evening experiments", and the study on graphene was also one of them, which he didn't expect much. At the time of 2002, graphite of 10 or 100 layers seemed to be reasonable, and researchers believed it was impossible for single-layer graphene to be stable. Andre Geim and Konstantin Novoselov, who was a PhD student at that time, tried to thin the graphite with polisher at first, yet ultimately failed. After countless attempts, they extracted graphene from a piece of graphite such as is found in ordinary pencils, and by using regular adhesive tape they managed to obtain a flake of graphene with a thickness of only one atom (Fig. 14.9). This finding struck the whole world for making experiments possible that give new twists to the phenomena in quantum physics. Playfulness is one of the hallmarks for Andre Geim and Konstantin Novoselov, one always learns something in the process and even hit the jackpot

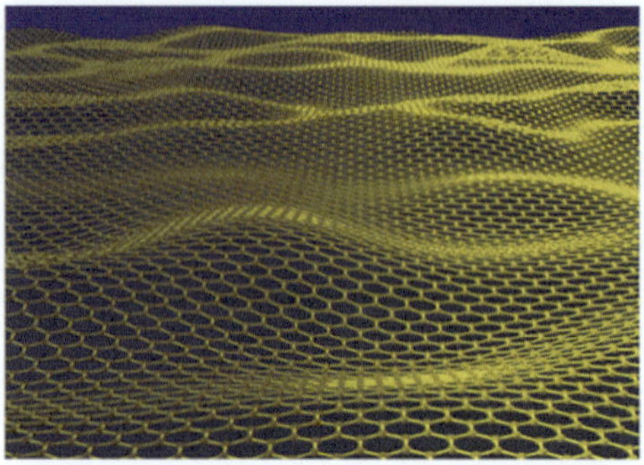

Fig. 14.9 Atomic structure of graphene

unexpectedly. Like now when they, with graphene, write themselves into the annals of science.

14.8.2 Brief History of Graphene

The lamellar structure of thermally reduced graphite was first discovered by Benjamine Collins Brodie in 1859. The structure of graphene was detected in 1916 and the theory was explored in 1924 (Fig. 14.10). In 1962, Boehm et al. reported the first detailed study on few-layer graphite with monolayer flakes of reduced graphene oxide (rGO). In 1990, ultra-thin films of graphite were synthesized by mechanical exfoliation, yet no graphite thinner than 50 layers was successfully synthesized before 2004.

In 2004, Andre Geim, Konstantin Novoselov and their group from the University of Manchester managed to pattern samples containing only a few layers of

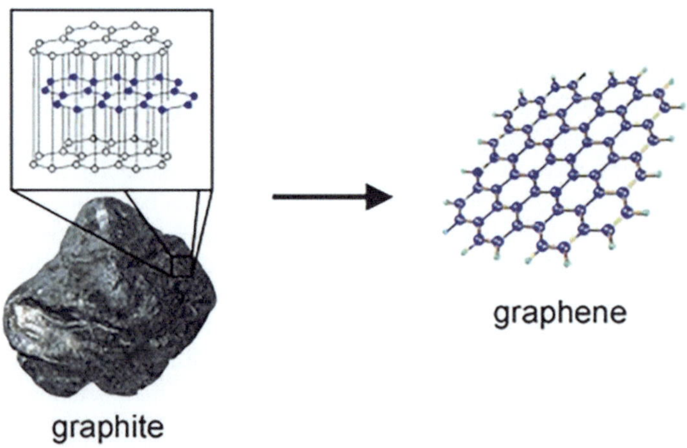

Fig. 14.10 Graphite and graphene

graphene into a hall bar. By connecting electrodes to it, they were able to measure both the (longitudinal) resistance and the Hall resistance. Once the technology to fabricate and attach electrodes to the graphene was established, researchers around the world quickly made a large number of new experiments on the synthesis of graphene. W.A. de Heer et al. reported a method by heating the SiC crystal to 1300 °C, burning silicon off the surface and leaving a thin layer of carbon behind. A P. Kim et al. attached a graphite crystal to the tip of an atomic force microscope and dragged along surface to make thin graphite to approximately 10 layers.

Since 2005, this research area has literally exploded, producing a growing number of papers concerning graphene and its properties. Studies of higher magnetic fields have been carried out to investigate the fractional quantum Hall effect in graphene. Further mechanical studies have shown that graphene is mechanically strong, even a hundred times stronger than steel. Double layers of graphene, which have different properties comparing to single-layer graphene, have also been studied thoroughly during this period. The development of this new material creates new and exciting possibilities. It is the first crystalline 2D-material, and its unique properties make it interesting for both fundamental science and future applications.

14.8.3 Future of Graphene

Graphene nanoribbons, with the width less than 10 nm, are a type of quasi 1D materials and can be used to produce field effect transistor with excellent switching speed and carrier mobility. Graphene membranes, one atom thick only, are the thinnest membrane available today and widely used for gas separation. Graphene can also be used as window electrode for solar cells. While the efficiency is close, its flexibility outperforms traditional ITO electrode.

By combining graphene with existing materials through the functionalization of the covalent and non-covalent bonds on it, graphene composite materials are created. At present, the research on graphene composite materials mainly focuses on graphene polymer composites, graphene inorganic nanocomposites and graphene metal composites, which open many unique opportunities in a wide range of applications. A new type of electrode material based on graphene and transition metal oxides can improve the the problems of low power density, long charging time and poor cycle stability of traditional Li-ion batteries. The large specific surface area, regular lamellar structure and ultra-high electron mobility of graphene help to accelerate the speed of ion (de)adsorption. When graphene is used as electrode for supercapacitors, it can obtain double-layer capacitors with higher power density, lower internal resistance, larger capacity and longer cycle life (Fig. 14.11). Apart from the field of energy, graphene composites are also applied in thermoacoustic, biomedicine, inks and paint and so on.

Fig. 14.11 Flexible graphene-based electrode

14.9 Blue Light-Emitting Diodes (2014, The Nobel Prize in Physics)

14.9.1 The Prize

The Nobel Prize in Physics 2014 was awarded jointly to Isamu Akasaki, Hiroshi Amano and Shuji Nakamura "for the invention of efficient blue light-emitting diodes which has enabled bright and energy-saving white light sources."

Light-emitting diodes (leds) are narrow-band light sources based on semiconductor devices with wavelengths ranging from infrared to ultraviolet. It requires the development of high-quality crystal growth technology and the ability to control the p-doping of high-bandgap semiconductors which is necessary to produce heterojunctions. Gallium nitride (GaN) with high crystal quality and good optical properties was only realized in the late 1980s, while its heterojunctions were realised in the early 1990s. To realize high-efficiency blue light, people need to prepare GaN-based alloys of different compositions and integrate them into multilayer structures such as heterojunctions and quantum wells. The generation of high-efficiency blue LEDs can greatly save energy which is of great benefit to mankind (Fig. 14.12).

When the PN junction is forward biased, the injected minority carriers recombine with the majority carriers near the interface and release energy. In ordinary diodes, energy is mostly released in the form of lattice vibration (thermal energy), while in light-emitting diodes, part of the energy released by carrier recombination is released in the form of light. Commonly used LED materials, such as gallium nitride (GaN), are direct band gap semiconductors. Their conduction band electrons can directly transition to the valence band to recombine with holes, and only electron–hole pairs are involved in the process and a photon is emitted. Therefore, their radiation efficiency is very high, and the internal quantum efficiency can be

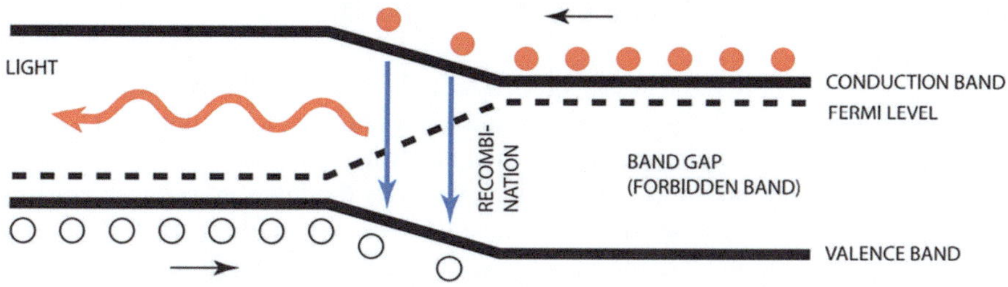

Fig. 14.12 The principle of light-emitting diodes (light emission from a p–n junction)

close to 100%, that is, a photon is generated by the recombination of an electron–hole pair. Indirect band gap semiconductors such as silicon and germanium have a large amount of non-radiative recombination during the recombination process, and the quantum efficiency is not high. For direct band gap semiconductors, the energy of the emitted photons is directly related to the forbidden band width of the semiconductor. Theoretically, the forbidden band width of GaN is 3.44 eV, and the energy of the emitted photon is also 3.44 eV, which is equivalent to near-ultraviolet light with a wavelength of about 360 nm. This is the theoretical basis for using GaN-based semiconductor in short-wave visible light emitting diodes by Nakamura et al.

14.9.2 Brief History of Blue Light-Emitting Diodes

In August 1989, Cree introduced the first commercially available blue LED based on the indirect bandgap semiconductor, silicon carbide (SiC). SiC LEDs had very low efficiency, no more than about 0.03%, but did emit in the blue portion of the visible light spectrum. The first blue-violet LED using magnesium-doped gallium nitride was made at Stanford University in 1972 by Herb Maruska and Wally Rhines. Today, magnesium-doping of gallium nitride remains the basis for all commercial blue LEDs and laser diodes. In the early 1970s, these devices were too dim for practical use, and research into gallium nitride devices slowed. In the late 1980s, key breakthroughs in GaN epitaxial growth and p-type doping ushered in the modern era of GaN-based optoelectronic devices. Then came a pivotal moment for the LED lighting industry. In 1994, Shuji Makamura, working for Nichia, invented the high-brightness blue LED that still serves as the foundation for nearly all commercial LED lighting products and for which he, along with his co-researchers, Isamu Akasaki and Hiroshi Amano, won a Nobel Prize in 20 This advance enabled the creation of white light simply by applying a phosphor coating to the lens covering the blue LED(s).

14.9.3 The Future of Blue Light-Emitting Diodes

Today, the main application areas of LEDs include traffic lights, lighting, full-color LED displays, car lights, background light sources and white LED solid-state lighting. Future applications may include water purification using AlGaN/GaN LEDs emitting ultraviolet light, which destroys the DNA of bacteria, viruses and micro-organisms. In countries with inadequate or no electricity grid, electricity from solar panels stored in batteries during the day powers white LEDs at night. Scientists may use nanostructured metals to improve LED efficiency and explored the potential of metal nanostructures to generate strong light-matter interactions to control the brightness, color, and directionality of light emitted by LEDs. Different metallic nanostructures they offer new possibilities for efficient light management of next-generation LED devices.

14.10 Li-Ion Battery (2019, The Nobel Prize in Chemistry)

14.10.1 The Prize

The Nobel Prize in Chemistry 2019 was awarded jointly to John B. Goodenough, M. Stanley Whittingham and Akira Yoshino for "the development of lithium-ion batteries." The three scientists pioneered in the chemistry of novel cathodes and anodes that led to the explosive growth of lithium-ion batteries (LIBs). This lightweight, rechargeable and powerful battery is now used in everything from mobile phones to laptops and electric vehicles. It can also store significant amounts of energy from renewable energies such as solar and wind power, making possible a fossil fuel-free society.

Storing electrical energy in batteries is a key factor in solving the world's energy supply. Lithium is useful in batteries since it willingly releases electrons. The foundation of the lithium-ion battery was laid during the oil crisis in the 1970s. Stanley Whittingham, with Exxon at that time, worked on developing methods that could lead to fossil fuel-free energy technologies. He started to research superconductors and discovered an energy-rich material, which he used to create an innovative cathode in a lithium battery. This was made from titanium disulfide (TiS_2), which has spaces capable of housing lithium ions, known as intercalation. The anode of the battery was made from metallic lithium with a strong drive to release electrons, resulting in a battery that literally had great potential, over two volts. The schematic of Whittingham's battery is illustrated in Fig. 14.13. However, metallic lithium is reactive, and the battery was too explosive to be viable, partially due to the formation of lithium whiskers during the charge–discharge cycles.

John Goodenough, a 97-year-old Nobel Laureate, the oldest thus far in history, with Oxford at that time, predicted that the cathode would have greater potential if it was made using a metal oxide instead of a metal sulfide. In 1980, he demonstrated that cobalt oxide with intercalated lithium ions can produce as much as four volts, a much higher voltage than early batteries. The schematic of Goodenough's

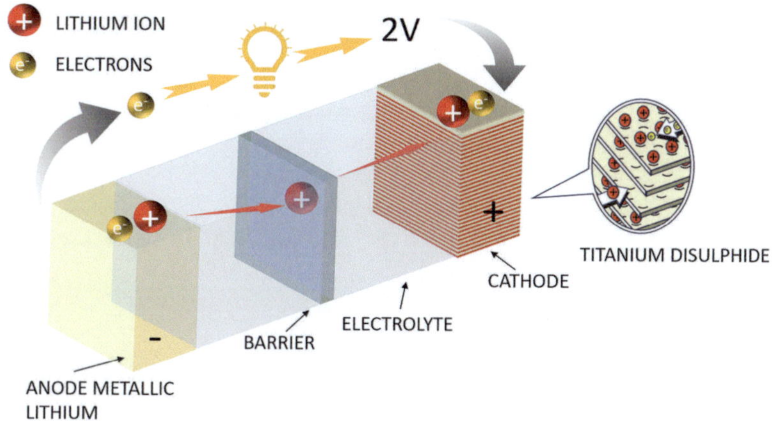

Fig. 14.13 Whittingham's battery

battery is illustrated in Fig. 14.14 This was an important breakthrough and would lead to much more powerful batteries.

Based on the cobalt oxide cathode developed by Goodenough, Akira Yoshino, at Asahi Kasei Corporation, created the first commercially viable lithium-ion battery in 1985. Rather than using reactive lithium in the anode, he used petroleum coke, a carbon material that can intercalate lithium ions, such as the cathode's cobalt oxide. The schematic of Yoshino's battery is illustrated in Fig. 14.15.

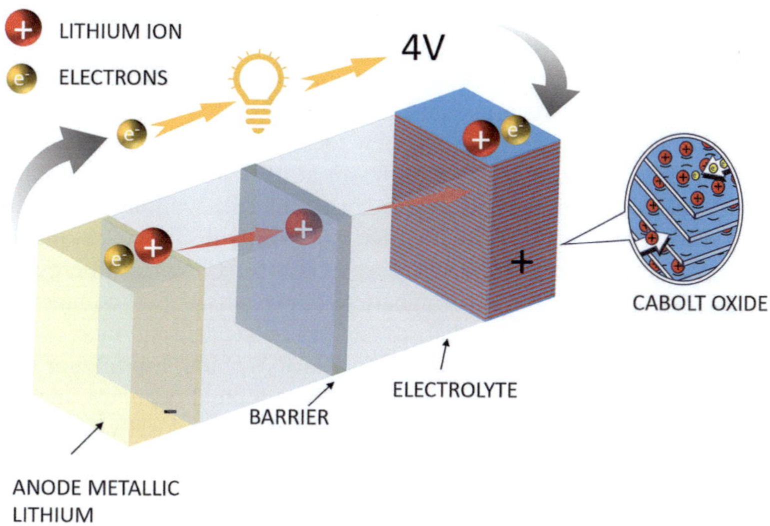

Fig. 14.14 Goodenough's battery

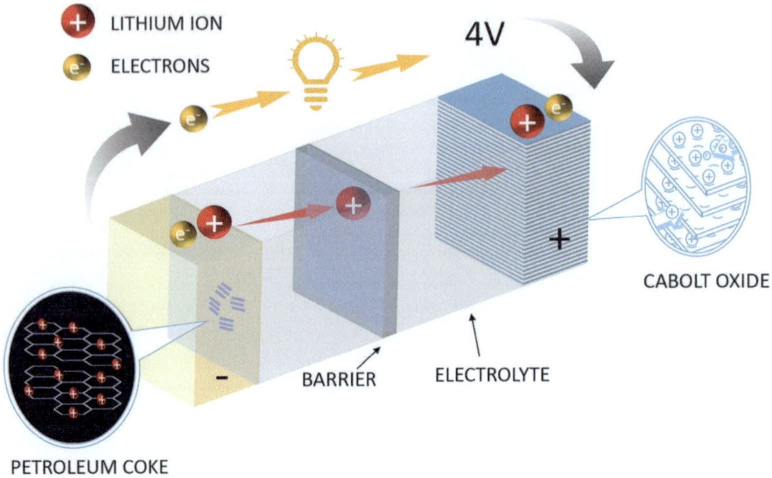

Fig. 14.15 Yoshino's battery

The result was a lightweight, hardwearing battery that could be charged hundreds of times before its performance deteriorated. The advantage of lithium-ion batteries is that they are not based upon chemical reactions that breakdown the electrodes but upon lithium ions flowing back and forth between the anode and cathode. Lithium-ion batteries first entered the market in 1991, which revolutionized our lives. They have laid the foundation of a wireless, fossil fuel-free society and are of the greatest benefit to humankind.

The success of commercial LIBs was a result of intensive research and contributions by many great scientists over the past decades. The three scientists receiving the Nobel Prize in Chemistry deserve recognition for their fundamental contribution in the areas of solid-state chemistry, electrochemistry, and engineering. The three laureates are still active, sparkling new ideas and enormous enthusiasm. The 97-year-old Goodenough, currently a professor at the University of Texas, Austin, USA, told that "I am in the lab every day. I am still working". Recently, he developed a prototype of a solid-state battery along with his collaborator. Stanley Whittingham, the Director of Northeast Centre for Chemical Energy Storage and Professor of Chemistry from Binghamton University, USA, commented that "I think it passes on the message that we can have a cleaner and nicer world, and perhaps start to solve the climate problem". Akira Yoshino, a professor at Meijo University, Japan, advised his students that "When a child becomes interested in something after being given the right stimulus, they will naturally find their own way of pursuing it."

14.10.2 Brief History of Batteries

In 1748, Benjamin Franklin adopted the term "battery," originally associated with military operations involving coordinated weaponry, to describe multiple Leyden jars in analogy to a battery of cannons. Subsequently, in 1786, Italian physician Galvani conducted an experiment involving the dissection of a frog using two dissimilar metal tools, iron and bronze. To his amazement, the frog's leg exhibited spontaneous movement, leading Galvani to proclaim his discovery of animal electricity. The year 1800 marked a significant milestone in the field of electrochemistry when Italian physicist Alessandro Volta constructed and documented the inaugural electrochemical battery, known as the voltaic pile (Fig. 14.16). This device comprised a stack of copper and zinc plates separated by paper disks soaked in brine, demonstrating the ability to generate a consistent electrical current for extended periods.

Over the ensuing centuries, primary batteries (nonrechargeable) underwent continual refinement, enhancement, and miniaturization. While early batteries served valuable purposes in experimental contexts, they suffered from voltage fluctuations and an inability to deliver sustained high current output. The Daniell cell, innovated by British chemist John Frederic Daniell in 1836, emerged as the first practical source of electricity. It quickly gained recognition as an industry standard and found widespread application in powering electrical telegraph networks.

Fig. 14.16 The very first battery, a voltaic pile

Towards the end of the nineteenth century, the first rechargeable zinc battery was developed, utilizing liquid zinc bromide as its basis. In 1901, Thomas Edison obtained the first patent for Zn-Ni batteries. The lead-acid battery, initially conceived in 1859 by French physicist Gaston Planté, remains in use today as the primary means to initiate the engines of most internal combustion automobiles.

These wet cells used liquid electrolytes, which were prone to leakage and spillage if poorly handled. This made wet cells unsuitable for portable appliances. Near the end of the nineteenth century, the invention of dry cell batteries, which replaced the liquid electrolyte with a paste, made portable electrical devices practical.

In the second half of the twentieth century, zinc rechargeable batteries were increasingly outdone by higher performance systems such as lead-acid (car batteries, stationary storage), nickel–cadmium (power tools, toys), nickel-metal hydride (cellphones, toys, hybrid electric vehicles) and, of course, ubiquitous lithium-ion batteries (portable electronics, electric vehicles, storage). Today, batteries come in a wide range of sizes—from large Megawatt capacities, which store power from solar farms or wind farms or substations to guarantee a stable power supply, to tiny batteries such as those used in watches.

14.10.3 The Future of Batteries

The flammable liquid electrolyte in an LIB is a major challenge in comparison with a fossil fuel source. With a focus on safety, solid-state batteries consisting exclusively of solid components are being developed. These solid-state batteries focus on low volume and higher energy density at a lighter weight, without the thermal issues presently associated with LIBs.

Electric vehicles have been promising cleaner and healthier alternatives to gasoline-powered cars since they first came to use more than one hundred years ago. The commercialization of LIBs in 1991 established a feasible platform for the growth of electric vehicles. Advances in battery performance have made it possible for electric cars to have the range required to make them viable and competitive with internal combustion engine vehicles. Notable cars are early-generation electric cars such as Tesla's Roadster and Nissan's Leaf. In China, many electric vehicle manufacturers have emerged in recent years, such as NIO (Fig. 14.17), XPENG, and ONE. A severe challenge electric vehicles still face is the lack of sustainable and economically viable means of recycling LIBs. There are also serious concerns about environmental hazards pertaining to the disposal of waste.

Cathodes with two or three 3D metals and polyanions are promising, and consequently, new electrolytes are required. Battery management systems to enhance the safety of large battery packs are needed to expand the market of LIB-powered vehicles. Nonflammable LIBs based on aqueous electrolytes or ceramic electrolytes and all-solid-state batteries are other possibilities. Next-generation LIBs may use high voltage (5 V) cathodes and high capacity anodes, such as Si- or Sn-based anodes. The impact of the present materials on the environment is still uncertain.

Fig. 14.17 A NIO-EC6 electric car

Quite a few issues remain before achieving next-generation LIBs. Safety and sustainability are still the major factors to go beyond LIBs in the future. The chemistry to replace the existing LIB technology is still challenging. To increase their energy density, electrode couples with high specific capacities and high operating cell voltages must be found, where Si- and Sn-based anodes could be potential candidates. $LiCoO_2$ is expensive and toxic. The increasingly popular $LiFePO_4$ has a low capacity. Although NiCoMn-based cathodes are highly attractive in capacity, their safety problem is still a concern. From an environmental perspective, it is desirable to avoid Co, Ni or other toxic elements.

To develop batteries with higher energy density, Li-air and Li-sulfur have attracted considerable attention. Although Zn-air and high-temperature Na-sulfur batteries have been known for some time, rechargeable Li-O_2 and Li–S batteries have yet to cross technical barriers. Sodium-ion batteries (SIBs) are being considered for large-scale stationary energy storage systems. Unlike lithium, graphite is not suitable for reversible incorporation of sodium. However, selected carbon nanomaterials show promising Na storage. Mg-ion batteries have the advantages of negligible dendrite formation and offer high-energy density at a significantly lower cost. Better safety features in Mg-ion batteries make them a step closer to commercialization. Al-ion batteries, with three-electron transfer, could achieve three times higher energy density than LIBs, but the use of expensive ionic liquids limits their application. Rechargeable Zn-air batteries could be more promising than Li-air batteries.

Bibliography

All About Circuits (2022) The history of the transistor marches on [EB/OL]. https://www.allaboutcircuits.com/news/history-of-transistor-marches-on

Arndt M et al (1999) Wave-particle duality of C-60 molecules. Nat 401(6754):680–682

Ashuri M, He Q, Shaw LL (2016) Silicon as a potential anode material for Li-ion batteries: Where size, geometry, and structure matter. Nanoscale 8(1):74–103
Bardeen J, Cooper LN, Schrieffer JR (1957) Microscopic theory of superconductivity. Phys Rev 106:162–164
Bednorz JG, Müller KA (1986) Possible high Tc superconductivity in the Ba-La-Cu-O system. Z Für Phys B: Condens Matter 64(2):189–193
Berger C et al (2004) Ultrathin epitaxial graphite: 2D electron gas properties and a route toward graphene-based nanoelectronics. J Phys Chem B 108(52):19912–19916
Boehm HP et al (1962) Das Adsorptionsverhalten sehr dünner Kohlenstoff-Folien. Z Anorg Allg Chem 316(3/4):119–127
Bolotin KI et al (2009) Observation of the fractional quantum Hall effect in graphene. Nat 462(7270):196–199
Born M, Oppenheimer R (1927) Zur quantentheorie der molekeln. Ann Phys 389(20):457–484
Braga MH et al (2017) Alternative strategy for a safe rechargeable battery. Energy Environ Sci 10(1):331–336
Braga MH et al (2018) Extraordinary dielectric properties at heterojunctions of amorphous ferroelectrics. J Am Chem Soc 140(51):17968–17976
Braga MH et al (2019) Low-temperature performance of a ferroelectric glass electrolyte rechargeable cell. ACS Appl Energy Mater 2(7):4943–4953
Cami J et al (2010) Detection of C-60 and C-70 in a young planetary nebula. Sci 329(5996):1180–1182
Cheng XB et al (2017) Toward safe lithium metal anode in rechargeable batteries: A review. Chem Rev 117(15):10403–10473
Cox DM et al (1986) C60LA—A deflated soccer ball. J Am Chem Soc 108(9):2457–2458
Curl RF, Smalley RE (1988) Probing C-60. Sci 242(4881):1017–1022
Dalvesco RR (2017) Buckminster fuller, the expo ’67 pavilion and the atoms for peace program. Leonardo 50(5):486–492
DiVincenzo DP, Mele EJ (1984) Self-consistent effective-mass theory for intralayer screening in graphite-intercalation compounds. Phys Rev B 29(4):1685–1694
Dong X, Wang Y, Xia Y (2014) Re-building Daniell cell with a Li-ion exchange film. Sci Rep 4:6503
Dreyer DR, Ruoff RS, Bielawski CW (2010) From conception to realization: An historical account of graphene and some perspectives for its future. Angew Chem Int Ed 49(49):9336–9344
Drozdov AP et al (2015) Conventional superconductivity at 203 kelvin at high pressures in the sulfur hydride system. Nature 525(7567):73–76
Drozdov AP et al (2019) Superconductivity at 250 K in lanthanum hydride under high pressures. Nature 569(7757):528–531
Du X et al (2009) Fractional quantum Hall effect and insulating phase of Dirac electrons in graphene. Nat 462(7270):192–195
Duan DF et al (2014) Pressure-induced metallization of dense $(H_2S)_2H_2$ with high-T_c superconductivity. Sci Rep 4:6968
Duffner F et al (2021) Post-lithium-ion battery cell production and its compatibility with lithium-ion cell production infrastructure. Nat Energy 6(2):123–134
Garche J (1990) On the historical development of the lead-acid battery, especially in Europe. J Power Sources 31(1/4):401–406
Ginzburg VL, Landau LD (1950) On the theory of superconductivity. J Exp Theor Phys 20:1064–1082
Goodenough JB (2018) How we made the Li-ion rechargeable battery. Nat Electron 1(3):204–204
Goodenough JB, Gao H (2019) A perspective on the Li-ion battery. Sci China Chem 62(12):1555–1556
Hohenberg P, Kohn W (1964) Inhomogeneous electron gas. Phys Rev B 864:136
Huang S et al (2018) Single-layer graphene membranes by crack-free transfer for gas mixture separation. Nat Commun 9:2005

Kohn W, Sham LJ (1965) Self-consistent equations including exchange and correlation effects. Phys Rev 140:566–569
Kratschmer W et al (1990) Solid C-60 - A new form of carbon. Nat 347(6291):354–358
Kroto H (1988) Space, stars, C-60, and soot. Sci 242(4882):1139–1145
Kroto HW et al (1985) C-60—Buckminsterfullerene. Nature 318(6042):162–163
Lee C et al (2008) Measurement of the elastic properties and intrinsic strength of monolayer graphene. Sci 321(5887):385–388
Li Y, Dai H (2014) Recent advances in zinc-air batteries. Chem Soc Rev 43(15):5257–5275
Li YW, Hao J, Liu H et al (2014) The metallization and superconductivity of dense hydrogen sulfide. J Chem Phys 140(17):174712
Lin YM et al (2010) 100-GHz transistors from wafer-scale epitaxial graphene. Sci 327(5966):662–662
Mauger A et al (2020) Tribute to Michel Armand: From rocking chair Li-ion to solid-state lithium batteries. J Electrochem Soc 167(7):070520
Meissner W, Ochsenfeld R (1933) Ein neuer Effekt bei Eintritt der Supraleitfähigkeit. Naturwissenschaften, (21):787–788
Nitta N et al (2015) Li-ion battery materials: Present and future. Mater Today 18(5):252–264
Nobel Prize org (2024) Nobel prize in physics 1987. [EB/OL]. https://www.nobelprize.org/prizes/physics/1987
Novoselov KS et al (2004) Electric field effect in atomically thin carbon films. Sci 306(5696):666–669
Novoselov KS et al (2006) Unconventional quantum Hall effect and Berry's phase of 2 pi in bilayer graphene. Nat Phys 2(3):177–180
Osawa E (1993) The evolution of the football structure for the C-60 molecule - A retrospective. Philos Trans R Soc London, Ser A 343(1667):1–8
Park H et al (2000) Nanomechanical oscillations in a single-C-60 transistor. Nat 407(6800):57–60
Poizot P et al (2000) Nano-sized transition-metal oxides as negative-electrode materials for lithium-ion batteries. Nat 407(6803):496–499
Ponrouch A et al (2019) Multivalent rechargeable batteries. Energy Storage Mater 20:253–262
Ramanan A (2019) Nobel prize in chemistry 2019. Reson 24(12):1381–1395
Rohlfing EA, Cox DM, Kaldor A (1984) Production and characterization of supersonic carbon cluster beams. J Chem Phys 81(7):3322–3330
Saslow WM (2021) Voltaic cells: The good (Faraday), the bad (Volta), and the ugly (Galvani). Phys Teach 59(1):22–26
Service R F (2009) Materials science: Carbon sheets an atom thick give rise to graphene dreams. Sci 324(5929):875–877
The rise and rise of grapheme (2010) Nature Nanotechnology, 5(11), 755
Wang X, Zhi LJ, Mullen K (2008) Transparent, conductive graphene electrodes for dye-sensitized solar cells. Nano Lett 8(1):323–327
Wang T et al (2018) Electrode materials for sodium-ion batteries: Considerations on crystal structures and sodium storage mechanisms. Electrochem Energy Rev 1(2):200–237
Whittingham MS (1978) Chemistry of intercalation compounds: Metal guests in chalcogenide hosts. Prog Solid State Chem 12(1):41–99
Whittingham MS (2012) History, evolution, and future status of energy storage. Proc IEEE 100:1518–1534
Whittingham MS (2020) Lithium batteries: 50 years of advances to address the next 20 years of climate issues. Nano Lett 20(12):8435–8437
Whittingham MS (2021) Lithium titanium disulfide cathodes. Nat Energy 6(2):214–214
Winkelmann CB et al (2009) Superconductivity in a single-C-60 transistor. Nat Phys 5(12):876–879
Winter M, Barnett B, Xu K (2018) Before Li-ion batteries. Chem Rev 118(23):11433–11456
Wu Q et al (2010) Supercapacitors based on flexible graphene/polyaniline nanofiber composite films. ACS Nano 4(4):1963–1970

Ying H, Han WQ (2017) Metallic Sn-based anode materials: Application in high-performance lithium-ion and sodium-ion batteries. Adv Sci 4(11):1700214
Yoshino A (2012) The birth of the lithium-ion battery. Angew Chem Int Ed 51(24):5798–5800
Zhang SS (2020) Problems and their origins of Ni-rich layered oxide cathode materials. Energy Storage Mater 24:247–254
Zhang YB et al (2005) Experimental observation of the quantum Hall effect and Berry's phase in graphene. Nat 438(7065):201–204
Zhu Y et al (2011) Carbon-based supercapacitors produced by activation of graphene. Sci 332(6037):1537–1541

MIX
Papier aus verantwortungsvollen Quellen
Paper from responsible sources
FSC® C105338
FSC
www.fsc.org

If you have any concerns about our products,
you can contact us on
ProductSafety@springernature.com

In case Publisher is established outside the EU,
the EU authorized representative is:
Springer Nature Customer Service Center GmbH
Europaplatz 3, 69115 Heidelberg, Germany

Printed by Libri Plureos GmbH
in Hamburg, Germany